Classical Electrodynamics

Lecture notes

Classical Electrodynamics

Lecture notes

Konstantin K Likharev

IOP Publishing, Bristol, UK

ISBN 978-0-7503-1404-6 (ebook)
ISBN 978-0-7503-1405-3 (print)
ISBN 978-0-7503-1406-0 (mobi)

DOI 10.1088/978-0-7503-1404-6

Version: 20180501

IOP Expanding Physics
ISSN 2053-2563 (online)
ISSN 2054-7315 (print)

British Library Cataloguing-in-Publication Data: A catalogue record for this book is available from the British Library.

Published by IOP Publishing, wholly owned by The Institute of Physics, London

IOP Publishing, Temple Circus, Temple Way, Bristol, BS1 6HG, UK

US Office: IOP Publishing, Inc., 190 North Independence Mall West, Suite 601, Philadelphia, PA 19106, USA

Contents

Preface to the EAP Series

Essential Advanced Physics

Essential Advanced Physics (EAP) is a series of lecture notes and problems with solutions, consisting of the following four parts[1]:

- *Part CM*: *Classical Mechanics* (a one-semester course),
- *Part EM*: *Classical Electrodynamics* (two semesters),
- *Part QM*: *Quantum Mechanics* (two semesters), and
- *Part SM*: *Statistical Mechanics* (one semester).

Each part includes two volumes: *Lecture Notes* and *Problems with Solutions*, and an additional file *Test Problems with Solutions*.

Distinguishing features of this series—in brief

- condensed lecture notes (~250 pp per semester)—much shorter than most textbooks
- emphasis on simple explanations of the main notions and phenomena of physics
- a focus on problem solution; extensive sets of problems with detailed model solutions
- additional files with test problems, freely available to qualified university instructors
- extensive cross-referencing between all parts of the series, which share style and notation

Level and precursors

The goal of this series is to bring the reader to a general physics knowledge level necessary for professional work in the field, regardless on whether the work is theoretical or experimental, fundamental or applied. From the formal point of view, this level (augmented by a few special topic courses in a particular field of concentration, and of course by an extensive thesis research experience) satisfies the typical PhD degree requirements. Selected parts of the series may be also valuable for graduate students and researchers of other disciplines, including astronomy, chemistry, mechanical engineering, electrical, computer and electronic engineering, and material science.

The entry level is a notch lower than that expected from a physics graduate from an average US college. In addition to physics, the series assumes the reader's familiarity with basic calculus and vector algebra, to such an extent that the meaning of the formulas listed in appendix A, 'Selected mathematical formulas' (reproduced at the end of each volume), is absolutely clear.

[1] Note that the (very ambiguous) term *mechanics* is used in these titles in its broadest sense. The acronym *EM* stems from another popular name for classical electrodynamics courses: *Electricity and Magnetism*.

Origins and motivation

The series is a by-product of the so-called 'core physics courses' I taught at Stony Brook University from 1991 to 2013. My main effort was to assist the development of students' problem-solving skills, rather than their idle memorization of formulas. (With a certain exaggeration, my lectures were not much more than introductions to problem solution.) The focus on this main objective, under the rigid time restrictions imposed by the SBU curriculum, had some negatives. First, the list of covered theoretical methods had to be limited to those necessary for the solution of the problems I had time to discuss. Second, I had no time to cover some core fields of physics—most painfully general relativity[2] and quantum field theory, beyond a few quantum electrodynamics elements at the end of *Part QM*.

The main motivation for putting my lecture notes and problems on paper, and their distribution to students, was my desperation to find textbooks and problem collections I could use, with a clear conscience, for my purposes. The available graduate textbooks, including the famous *Theoretical Physics* series by Landau and Lifshitz, did not match the minimalistic goal of my courses, mostly because they are far too long, and using them would mean hopping from one topic to another, picking up a chapter here and a section there, at a high risk of losing the necessary background material and logical connections between the course components—and the students' interest with them. In addition, many textbooks lack even brief discussions of several traditional and modern topics that I believe are necessary parts of every professional physicist's education[3].

On the problem side, most available collections are not based on particular textbooks, and the problem solutions in them either do not refer to any background material at all, or refer to the included short sets of formulas, which can hardly be used for systematic learning. Also, the solutions are frequently too short to be useful, and lack discussions of the results' physics.

Style

In an effort to comply with the Occam's Razor principle[4], and beat Malek's law[5], I have made every effort to make the discussion of each topic as clear as the time/space (and my ability :-) permitted, and as simple as the subject allowed. This effort has resulted in rather succinct lecture notes, which may be thoroughly read by a student during the semester. Despite this briefness, the introduction of every new

[2] For an introduction to this subject, I can recommend either a brief review by S Carroll, *Spacetime and Geometry* (2003, New York: Addison-Wesley) or a longer text by A Zee, *Einstein Gravity in a Nutshell* (2013, Princeton University Press).

[3] To list just a few: the statics and dynamics of elastic and fluid continua, the basics of physical kinetics, turbulence and deterministic chaos, the physics of computation, the energy relaxation and dephasing in open quantum systems, the reduced/RWA equations in classical and quantum mechanics, the physics of electrons and holes in semiconductors, optical fiber electrodynamics, macroscopic quantum effects in Bose–Einstein condensates, Bloch oscillations and Landau–Zener tunneling, cavity quantum electrodynamics, and density functional theory (DFT). All these topics are discussed, if only briefly, in my lecture notes.

[4] *Entia non sunt multiplicanda praeter necessitate*—Latin for 'Do not use more entities than necessary'.

[5] 'Any simple idea will be worded in the most complicated way'.

physical notion/effect and of every novel theoretical approach is always accompanied by an application example or two.

The additional exercises/problems listed at the end of each chapter were carefully selected[6], so that their solutions could better illustrate and enhance the lecture material. In formal classes, these problems may be used for homework, while individual learners are strongly encouraged to solve as many of them as practically possible. The few problems that require either longer calculations, or more creative approaches (or both), are marked by asterisks.

In contrast with the lecture notes, the model solutions of the problems (published in a separate volume for each part of the series) are more detailed than in most collections. In some instances they describe several alternative approaches to the problem, and frequently include discussions of the results' physics, thus augmenting the lecture notes. Additional files with sets of shorter problems (also with model solutions) more suitable for tests/exams, are available for qualified university instructors from the publisher, free of charge.

Disclaimer and encouragement

The prospective reader/instructor has to recognize the limited scope of this series (hence the qualifier *Essential* in its title), and in particular the lack of discussion of several techniques used in current theoretical physics research. On the other hand, I believe that the series gives a reasonable introduction to the *hard core* of physics—which many other sciences lack. With this hard core knowledge, today's student will always feel at home in physics, even in the often-unavoidable situations when research topics have to be changed at a career midpoint (when learning from scratch is terribly difficult—believe me :-). In addition, I have made every attempt to reveal the remarkable logic with which the basic notions and ideas of physics subfields merge into a wonderful single construct.

Most students I taught liked using my materials, so I fancy they may be useful to others as well—hence this publication, for which all texts have been carefully reviewed.

[6] Many of the problems are original, but it would be silly to avoid some old good problem ideas, with long-lost authorship, which wander from one textbook/collection to another one without references. The assignments and model solutions of all such problems have been re-worked carefully to fit my lecture material and style.

Preface to *Classical Electrodynamics: Lecture Notes*

The structure of this classical electrodynamics course is quite traditional[7]; in order to address the most important problems of the field—which involve not only charged point particles, but also conducting, dielectric, and magnetic media—the electromagnetic interactions of charges are discussed in parallel with the simplest models of the electric and magnetic properties of materials.

Also following tradition, I use this volume (mostly chapter 2) as a convenient platform for the discussion of various methods of the solution of partial differential equations, using:

- the polar, cylindrical, spherical, and other curvilinear orthogonal coordinates (sections 2.3 and 2.4), including conformal mapping (section 2.4);
- the variable separation method[8], also in various coordinates (sections 2.5–2.8), involving the Bessel functions (section 2.7), and the Legendre polynomials and the associated Legendre functions (section 2.8)[9];
- the spatial (section 2.10), temporal (section 7.2)[10], and spatial-temporal (chapter 6) Green's functions; and
- the finite-difference variety of numerical methods, with a very brief discussion of the finite-element approach (section 2.11).

I hope the discussion of these methods of mathematical physics, using the platform of specific problems of electrodynamics, makes them more clear and vivid.

One more traditional part of classical electrodynamics is an introduction to special relativity, because—although this field includes a substantial classical mechanics component—it is the electrodynamics which makes the relativistic approach obligatory. The narrative logic accepted in this course dictates the discussion of the special relativity effects in the very end of the course—chapters 9 and 10.

[7] A notable exception from tradition is the famous *Theoretical Physics* series by L Landau and E Lifshitz, with two different volumes (#2 and #8) devoted to, respectively, electromagnetic interactions in free space and the electrodynamics of continua. I believe that such approach may only be justified if the first topic is discussed (as it is in the Landau and Lifshitz series) together with general relativity, for which I unfortunately could not find time in this series.

[8] This method was already briefly discussed in *Part CM* section 6.5, and then used in *Part CM* sections 6.6 and 8.4. One more family of special functions, the Fresnel integrals, are discussed in section 8.6.

[9] Due to their key importance for quantum mechanics, these functions will be also discussed in *Part QM* section 3.5.

[10] Such functions were also discussed in *Part CM* section 5.1.

Acknowledgments

I am extremely grateful to my faculty colleagues and other readers of the preliminary (circa 2013) version of this series, who provided feedback on certain sections; here they are listed in alphabetical order[11]: A Abanov, P Allen, D Averin, S Berkovich, P-T de Boer, M Fernandez-Serra, R F Hernandez, A Korotkov, V Semenov, F Sheldon, and X Wang. (Obviously, these kind people are not responsible for any remaining deficiencies.)

A large part of my scientific background and experience, as reflected in these materials, came from my education, and then research, in the Department of Physics of Moscow State University from 1960 to 1990. The Department of Physics and Astronomy of Stony Brook University provided a comfortable and friendly environment for my work during the following 25+ years.

Last but not least, I would like to thank my wife Lioudmila for all her love, care, and patience—without these, this writing project would have been impossible.

I know very well that my materials are still far from perfection. In particular, my choice of covered topics (always very subjective) may certainly be questioned. Also, it is almost certain that despite all my efforts, not all typos have been weeded out. This is why all remarks (however candid) and suggestions from readers will be greatly appreciated. All significant contributions will be gratefully acknowledged in future editions.

Konstantin K Likharev
Stony Brook, NY

[11] I am very sorry for not keeping proper records from the beginning of my lectures at Stony Brook, so I cannot list all the numerous students and TAs who have kindly attracted my attention to typos in earlier versions of these notes. Needless to say, I am very grateful to all of them as well.

Notation

Abbreviations	Fonts	Symbols
c.c. complex conjugate	F, $\mathcal{F}$ scalar variables[12]	$\dot{}$ time differentiation operator (d/dt)
h.c. Hermitian conjugate	$\mathbf{F}$, $\boldsymbol{\mathcal{F}}$ vector variables	∇ spatial differentiation vector (*del*)
	$\hat{F}$, $\hat{\mathcal{F}}$ scalar operators	$\approx$ approximately equal to
	$\hat{\mathbf{F}}$, $\hat{\boldsymbol{\mathcal{F}}}$ vector operators	$\sim$ of the same order as
	F matrix	$\propto$ proportional to
	$F_{jj'}$ matrix element	$\equiv$ equal to by definition (or evidently)
		$\cdot$ scalar ('dot-') product
		$\times$ vector ('cross-') product
		$\overline{}$ time averaging
		$\langle\ \rangle$ statistical averaging
		[,] commutator
		{ , } anticommutator

Prime signs

The prime signs (′, ″, etc) are used to distinguish similar variables or indices (such as j and j' in the matrix element above), rather than to denote derivatives.

Parts of the series

Part CM: Classical Mechanics *Part EM: Classical Electrodynamics*
Part QM: Quantum Mechanics *Part SM: Statistical Mechanics*

Appendices

Appendix A: Selected mathematical formulas
Appendix B: Selected physical constants

Formulas

The abbreviation Eq. may mean any displayed formula: either the equality, or inequality, or equation, etc.

[12] The same letter, typeset in different fonts, typically denotes different variables.

IOP Publishing

Classical Electrodynamics
Lecture notes
Konstantin K Likharev

Chapter 1

Electric charge interaction

This brief chapter describes the basics of electrostatics, the study of interactions between static (or slowly moving) electric charges. Much of this material should be known to the reader from his or her undergraduate studies; because of that, the explanations will be very brief[1].

1.1 The Coulomb law

A serious discussion of any part of classical electrodynamics, starting from the electrostatics, requires a common agreement on the meaning of the following notions[2]:

- *electric charges* q_k, as revealed, most explicitly, by observation of the *electrostatic interaction* between the charged particles;
- the *electric charge conservation*, meaning that the algebraic sum of q_k of all particles inside any closed volume is conserved, unless the charged particles cross the volume's border; and
- a *point charge*, meaning an ultimately small ('point') charged particle, whose position in space is completely described (in a given reference frame) by its radius-vector **r**.

I will assume that these notions are well known to the reader—although my strong advice is to give some thought to their vital importance. Using these notions, the *Coulomb law*[3] for the interaction of two point, stationary charges may be formulated as follows:

---

[1] For remedial reading, virtually any undergraduate text on electricity and magnetism may be used; I can recommend, for example, [1].

[2] On top of the more general notions of the *classical Cartesian space*, *point particles*, and *forces*, which are used in classical mechanics—see, e.g. *Part CM* section 1.1.

[3] Discovered experimentally in the early 1780s, and formulated in 1785 by C-A de Coulomb.

doi:10.1088/978-0-7503-1404-6ch1

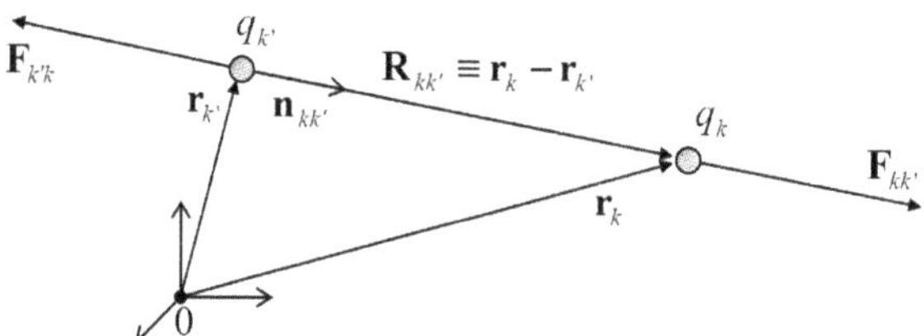

Figure 1.1. Coulomb force directions (for the case $q_k q_{k'} > 0$).

$$\mathbf{F}_{kk'} = \kappa q_k q_{k'} \frac{\mathbf{r}_k - \mathbf{r}_{k'}}{|\mathbf{r}_k - \mathbf{r}_{k'}|^3} \equiv \kappa \frac{q_k q_{k'}}{R_{kk'}^2} \mathbf{n}_{kk'}, \quad \text{with } \mathbf{R}_{kk'} \equiv \mathbf{r}_k - \mathbf{r}_{k'}, \tag{1.1}$$

where $\mathbf{F}_{kk'}$ denotes the electrostatic ('Coulomb') force exerted on charge number k by charge number k', and $\mathbf{n}_{kk'} \equiv \mathbf{R}_{kk'}/R_{kk'}$ is a unit vector directed along the vector $\mathbf{R}_{kk'}$ of the relative position of the charges—see figure 1.1.

I am confident that this law is very familiar to the reader, but a few comments may still be due:

(i) Flipping indices k and k', we see that Eq. (1.1) complies with Newton's third law: the reciprocal force is equal in magnitude but opposite in direction: $\mathbf{F}_{k'k} = -\mathbf{F}_{kk'}$.

(ii) Since the vector $\mathbf{R}_{kk'} \equiv \mathbf{r}_k - \mathbf{r}_{k'}$ is directed from point $\mathbf{r}_{k'}$ toward point $\mathbf{r}_k$ (figure 1.1), Eq. (1.1) implies that charges of the same sign (i.e. with $q_k q_{k'} > 0$) repulse, while those with opposite signs ($q_k q_{k'} < 0$) attract each other.

(iii) In some textbooks, the Coulomb law (1.1) is given with the qualifier 'in free space' or 'in a vacuum'. However, Eq. (1.1) actually remains valid even in the presence of other charges, for example the charges of quasi-continuous media that may surround the two charges (number k and k') under consideration. The confusion stems from the fact (to be discussed in detail in chapter 3) that in some cases it is convenient to *formally* represent the effect of other charges as an effective modification of the Coulomb law.

(iv) The constant κ in Eq. (1.1) depends on the system of units we use. In the *Gaussian* units, κ is set to 1, for the price of introducing a special unit of charge (the *statcoulomb*) that would fit the experimental data for (1.1), for forces $\mathbf{F}_{kk'}$ measured in the Gaussian units (*dynes*). On the other hand, in the *International System* (SI) of units, the charge unit is one *coulomb* (abbreviated C)[4], close to 3×10^9 statcoulombs, and κ is different from unity[5]:

[4] In the formal metrology, one coulomb is defined as the charge carried over by a constant current of one ampere (see chapter 5 for its definition) during 1 s.

[5] ε_0 is called either the *electric constant* or the *free-space permittivity*; from Eq. (1.2) with the free-space speed of light $c \approx 3 \times 10^8$ m c^{-1}, $\varepsilon_0 \approx 8.85 \times 10^{-12}$ SI units. For more accurate values of the constants, and their brief discussion, see appendix B.

$$\kappa|_{\text{SI}} = \frac{1}{4\pi\varepsilon_0} \equiv 10^{-7}c^2, \tag{1.2}$$

where c is the speed of light. Unfortunately, the continuing struggle between zealot proponents of these two systems resembles the not-so-nice features of a religious war, with a similarly slim chance of any side winning in the foreseeable future. In my humble view, each of these systems has its advantages and handicaps (to be noted below on several occasions), and every educated physicist should have no problem with using either of them. Following the insistent recommendations of international scientific unions, I am using SI units throughout this series. However, for the readers' convenience, in this course (where the difference between the Gaussian and SI systems is particularly significant) I will write the most important formulas with the constant (1.2) clearly displayed—for example, Eq. (1.1) as

$$\mathbf{F}_{kk'} = \frac{q_k q_{k'}}{4\pi\varepsilon_0} \frac{\mathbf{r}_k - \mathbf{r}_{k'}}{|\mathbf{r}_k - \mathbf{r}_{k'}|^3}, \tag{1.3}$$

so that the transfer to the Gaussian units may be performed just by the formal replacement $4\pi\varepsilon_0 \to 1$. Moreover, in the cases when the transfer is not obvious, I will duplicate such formulas in the Gaussian units.

In addition to Eq. (1.3), another key experimental law of electrostatics is the *linear superposition principle*: the electrostatic forces exerted on some point charge (say, q_k) by other charges do not affect each other and add up as vectors to form the net force:

$$\mathbf{F}_k = \sum_{k' \neq k} \mathbf{F}_{kk'}, \tag{1.4}$$

where the summation is extended over all charges but q_k, and the partial force $\mathbf{F}_{kk'}$ is described by Eq. (1.3). The fact that the sum is restricted to $k' \neq k$ means that a *point charge, in statics, does not interact with itself*. This fact may look trivial from Eq. (1.3), whose right-hand side diverges at $\mathbf{r}_k \to \mathbf{r}_{k'}$, but becomes less evident (though still true) in quantum mechanics where the charge of even an elementary particle is effectively spread around some volume, together with the particle's wavefunction[6].

Now we may combine Eqs. (1.3) and (1.4) to obtain the following expression for the net force $\mathbf{F}$ acting on some charge q located at point $\mathbf{r}$:

$$\mathbf{F}(\mathbf{r}) = q\frac{1}{4\pi\varepsilon_0} \sum_{\mathbf{r}_{k'} \neq \mathbf{r}} q_{k'} \frac{\mathbf{r} - \mathbf{r}_{k'}}{|\mathbf{r} - \mathbf{r}_{k'}|^3}. \tag{1.5}$$

This equality implies that it makes sense to introduce the notion of the *electric field* at point $\mathbf{r}$ (as an entity independent of the *probe charge* q), characterized by the following vector:

[6] Moreover, there are some widely used approximations, e.g. the Kohn–Sham equations in the density functional theory (DFT) of multi-particle systems, which essentially violate this law, thus limiting the accuracy and applicability of these approximations—see, e.g. *Part QM* section 8.4.

$$\mathbf{E}(\mathbf{r}) \equiv \frac{\mathbf{F}(\mathbf{r})}{q}, \tag{1.6}$$

formally called the *electric field strength*—but much more frequently, just the 'electric field'. In these terms, Eq. (1.5) becomes

$$\mathbf{E}(\mathbf{r}) = \frac{1}{4\pi\varepsilon_0} \sum_{\mathbf{r}_{k'} \neq \mathbf{r}} q_{k'} \frac{\mathbf{r} - \mathbf{r}_{k'}}{|\mathbf{r} - \mathbf{r}_{k'}|^3}. \tag{1.7}$$

The notion of field is just convenient is electrostatics, but becomes virtually unavoidable for description of time-dependent phenomena (such as electromagnetic waves), where the electromagnetic field shows up as a specific form of matter, with zero rest mass, and hence different from the usual 'material' particles.

Many real-world problems involve multiple point charges located so closely that it is possible to approximate them with a continuous charge distribution. Indeed, let us consider a group of many ($dN \gg 1$) close charges, located at points $\mathbf{r}_{k'}$, all within an elementary volume d^3r'. For relatively distant field observation points, with $|\mathbf{r} - \mathbf{r}_{k'}| \gg dr'$, the geometrical factor in the corresponding terms of Eq. (1.7) is essentially the same. As a result, these charges may be treated as a single elementary charge $dQ(\mathbf{r}')$. Since at $dN \gg 1$ this elementary charge is proportional to the elementary volume d^3r', we can define the local 3D *charge density* $\rho(\mathbf{r}')$ by the relation[7]

$$\rho(\mathbf{r}')d^3r' \equiv dQ(\mathbf{r}') \equiv \sum_{r_{k'} \in d^3r'} q_{k'}, \tag{1.8}$$

and rewrite Eq. (1.7) as an integral (over the volume containing all essential charges):

$$\mathbf{E}(\mathbf{r}) = \frac{1}{4\pi\varepsilon_0} \int \rho(\mathbf{r}') \frac{\mathbf{r} - \mathbf{r}'}{|\mathbf{r} - \mathbf{r}'|^3} d^3r'. \tag{1.9}$$

Note that for a continuous, smooth charge density $\rho(\mathbf{r}')$, the integral in Eq. (1.9) does not diverge at $\mathbf{R} \equiv \mathbf{r} - \mathbf{r}' \to 0$, because in this limit the fraction under the integral increases as R^{-2}, i.e. slower than the decrease of the elementary volume d^3r', proportional to R^3.

Let me emphasize the dual role of Eq. (1.9). In the case if $\rho(\mathbf{r})$ is a continuous function representing the average charge, defined by Eq. (1.8), Eq. (1.9) is not valid at distances $|\mathbf{r} - \mathbf{r}_{k'}|$ of the order of the distance between the adjacent point charges, i.e. does not describe rapid variations of the electric field at these distances. Such an *approximate*, smoothly changing field $\mathbf{E}(\mathbf{r})$ is called *macroscopic*; we will repeatedly return to this notion[8] in the following chapters. On the other hand, Eq. (1.9) may be

[7] The 2D (areal) charge density σ and the 1D (linear) density λ may be defined absolutely similarly: $dQ = \sigma d^2r$, $dQ = \lambda dr$. Note that a finite value of σ and λ means that the volumic density ρ is formally infinite at the charge location; for example for a plane $z = 0$, charged with a constant areal density σ, $\rho = \sigma\delta(z)$, where $\delta(z)$ is the Dirac delta-function.

[8] It was formally introduced by H Lorentz in the early 1900 s for the description of dielectrics—see section 3.3.

also used for the description of the *exact* ('microscopic') field of discrete point charges, employing the notion of Dirac's δ-function[9], which is the mathematical approximation for a very sharp function equal to zero everywhere but one point, and still having a finite integral (equal to 1). Indeed, in this formalism, a set of point charges $q_{k'}$ located in points $\mathbf{r}_{k'}$ may be represented by the pseudo-continuous density

$$\rho(\mathbf{r}') = \sum_{k'} q_{k'}\delta(\mathbf{r}' - \mathbf{r}_{k'}). \tag{1.10}$$

Plugging this expression into Eq. (1.9), we return to its exact, discrete version (1.7). In this sense, Eq. (1.9) is exact, and we may use it as the general expression for the electric field.

1.2 The Gauss law

Due to this extension of Eq. (1.9) to point ('discrete') charges, it may seem that we do not need anything else for solving any problem of electrostatics. In practice, however, this is not quite true, first of all because the direct use of Eq. (1.9) frequently leads to complex calculations. Indeed, let us try to solve a very simple problem: finding the electric field produced by a spherically symmetric charge distribution with density $\rho(r')$. We may immediately use the problem symmetry to argue that the electric field should also be spherically symmetric, with only one component in the spherical coordinates: $\mathbf{E}(\mathbf{r}) = E(r)\mathbf{n}_r$, where $\mathbf{n}_r \equiv \mathbf{r}/r$ is the unit vector in the direction of the field observation point $\mathbf{r}$ (figure 1.2).

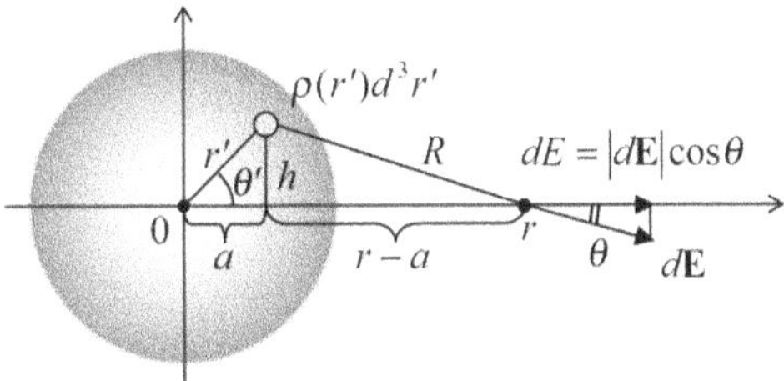

Figure 1.2. One of the simplest problems of electrostatics: an electric field produced by a spherically symmetric charge distribution.

Taking this direction as the polar axis of a spherical coordinate system, we can use the evident independence of the elementary radial field dE, created by the elementary charge $\rho(r')d^3r' = \rho(r')r'^2\sin\theta\, dr'd\theta'd\phi'$, of the azimuthal angle ϕ', and reduce the integral (1.9) to

$$E = \frac{1}{4\pi\varepsilon_0}2\pi\int_0^{\pi} \sin\theta' d\theta' \int_0^{\infty} r'^2 dr' \frac{\rho(r')}{R^2}\cos\theta, \tag{1.11}$$

where θ and R are the geometrical parameters marked in figure 1.2. Since they may all be readily expressed via r' and θ', using the auxiliary parameters a and h,

[9] See, e.g. appendix A, section A.14.

$$\cos\theta = \frac{r - a}{R}, \quad R^2 = h^2 + (r - r'\cos\theta)^2, \qquad \text{where } a \equiv r'\cos\theta', \quad h \equiv r'\sin\theta', \tag{1.12}$$

Eq. (1.11) may be eventually reduced to an explicit integral over r' and θ', and worked out analytically, but that would require some effort.

For other problems, the integral (1.9) may be much more complicated, defying analytical solution. One could argue that with the present-day abundance of computers and numerical algorithm libraries, one can always resort to numerical integration. This argument may be enhanced by the fact that numerical *integration* is based on the replacement of the integral by a discrete sum, and summation is much more robust to the (unavoidable) discretization and rounding errors than the finite-difference schemes typical in the numerical solution of *differential* equations. These arguments, however, are only partly justified, since in many cases the numerical approach runs into a problem sometimes called the *curse of dimensionality*—the exponential dependence of the number of needed calculations on the number of independent parameters of the problem[10]. Thus, despite the proliferation of numerical methods in physics, analytical results have an ever-lasting value, and we should try to get them whenever we can. For our current problem of finding the electric field generated by a fixed set of electric charges, large help may come from the so-called *Gauss law*.

Let us consider a single point charge q inside a smooth, closed surface S (figure 1.3), and calculate product $E_n d^2r$, where d^2r is an elementary area of the surface (which may be well approximated with a plane fragment of that area), and E_n is the component of the electric field in that point, normal to that plane.

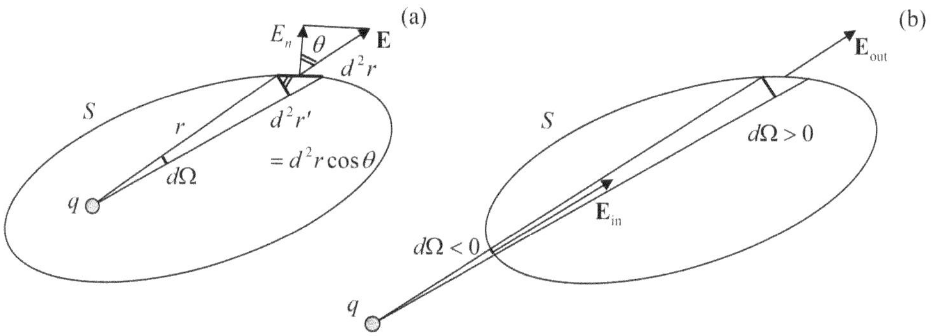

Figure 1.3. Deriving the Gauss law: a point charge q is (a) inside volume V and (b) outside of that volume.

This component may be calculated as $E\cos\theta$, where θ is the angle between the vector **E** and the unit vector **n** normal to the surface. (Equivalently, E_n may be

[10] For a more detailed discussion of this problem, see, e.g. *Part CM* section 5.8.

represented as the scalar product $\mathbf{E}\cdot\mathbf{n}$.) Now let us notice that the product $\cos\theta\, d^2r$ is nothing more than the area d^2r' of the projection of d^2r onto the plane perpendicular to vector $\mathbf{r}$ connecting charge q with this point of the surface (figure 1.3), because the angle between the planes d^2r' and d^2r is also equal to θ. Using the Coulomb law for $\mathbf{E}$, we obtain

$$E_n d^2r = E\cos\theta\, d^2r = \frac{1}{4\pi\varepsilon_0}\frac{q}{r^2}d^2r'. \tag{1.13}$$

But the ratio d^2r'/r^2 is nothing more than the elementary solid angle $d\Omega$ under which the areas d^2r' and d^2r are seen from the charge point, so that $E_n d^2r$ may be represented as just a product of $d\Omega$ by a constant $(q/4\pi\varepsilon_0)$. Summing these products over the whole surface, we obtain

$$\oint_S E_n d^2r = \frac{q}{4\pi\varepsilon_0}\oint_S d\Omega \equiv \frac{q}{\varepsilon_0}, \tag{1.14}$$

since the full solid angle equals 4π. (The integral in the left-hand side of this relation is called the *flux of electric field* through surface S.)

The relation (1.14) expresses the Gauss law for one point charge. However, it is only valid if the charge is located *inside* the volume limited by the surface. In order to find the flux created by a charge *outside* the volume, we still can use Eq. (1.13), but have to be careful with the signs of the elementary contributions $E_n dA$. Let us use the common convention to direct the unit vector $\mathbf{n}$ out of the closed volume we are considering (the so-called *outer normal*), so that the elementary product $E_n d^2r = (\mathbf{E}\cdot\mathbf{n})d^2r$ and hence $d\Omega = E_n d^2r'/r^2$ is positive if the vector $\mathbf{E}$ is pointing out of the volume (as in the example shown in figure 1.3a and the upper-right area in figure 1.3b), and negative in the opposite case (for example, in the lower-left area in figure 1.3b). As the latter figure shows, if the charge is located outside the volume, for each positive contribution $d\Omega$ there is always an equal and opposite contribution to the integral. As a result, at the integration over the solid angle the positive and negative contributions cancel exactly, so that

$$\oint_S E_n d^2r = 0. \tag{1.15}$$

The real power of the Gauss law is revealed by its generalization to the case of many charges within volume V. Since the calculation of flux is a linear operation, the linear superposition principle (1.4) means that the flux created by several charges is equal to the (algebraic) sum of individual fluxes from each charge, for which either Eq. (1.14) or Eq. (1.15) is valid, depending on the charge position (in or out of the volume). As the result, for the total flux we obtain:

$$\oint_S E_n d^2r = \frac{Q_V}{\varepsilon_0} \equiv \frac{1}{\varepsilon_0}\sum_{\mathbf{r}_j\in V} q_j \equiv \frac{1}{\varepsilon_0}\int_V \rho(\mathbf{r}')d^3r', \tag{1.16}$$

where Q_V is the net charge inside volume V. This is the full version of the Gauss law.

In order to appreciate the problem-solving power of the law, let us return to the problem illustrated by figure 1.2, i.e. a spherical charge distribution. Due to its symmetry, which has already been discussed above, if we apply Eq. (1.16) to a sphere of radius r, the electric field should be perpendicular to the sphere at each of its points (i.e. $E_n = E$), and its magnitude the same at all points: $E_n = E = E(r)$. As a result, the flux calculation is elementary:

$$\oint E_n d^2r = 4\pi r^2 E(r). \tag{1.17}$$

Now, applying the Gauss law (1.16), we obtain

$$4\pi r^2 E(r) = \frac{1}{\varepsilon_0}\int_{r'<r} \rho(r')d^3r' = \frac{4\pi}{\varepsilon_0}\int_0^r r'^2\rho(r')dr', \tag{1.18}$$

so that, finally,

$$E(r) = \frac{1}{r^2\varepsilon_0}\int_0^r r'^2\rho(r')dr' = \frac{1}{4\pi\varepsilon_0}\frac{Q(r)}{r^2}, \tag{1.19}$$

where $Q(r)$ is the full charge inside the sphere of radius r:

$$Q(r) \equiv 4\pi\int_0^r \rho(r')r'^2 dr'. \tag{1.20}$$

In particular, this formula shows that the field *outside* a sphere of a finite radius R is exactly the same as if all its charge $Q = Q(R)$ was concentrated in the sphere's center. (Note that this important result is only valid for a spherically symmetric charge distribution.) For the field *inside* the sphere, finding the electric field still requires an explicit integration (1.20), but this 1D integral is much simpler than the 2D integral (1.11), and in some important cases may be readily worked out analytically. For example, if the charge Q is uniformly distributed inside a sphere of radius R,

$$\rho(r') = \rho = \frac{Q}{V} = \frac{Q}{(4\pi/3)R^3}, \tag{1.21}$$

the integration is elementary:

$$E(r) = \frac{\rho}{r^2\varepsilon_0}\int_0^r r'^2 dr' = \frac{\rho r}{3\varepsilon_0} = \frac{1}{4\pi\varepsilon_0}\frac{Qr}{R^3}. \tag{1.22}$$

We see that in this case the field is growing linearly from the center to the sphere's surface, and only at $r > R$ starts to decrease in agreement with Eq. (1.19) with constant $Q(r) = Q$. Another important observation is that the results for $r \leqslant R$ and $r \geqslant R$ give the same value $(Q/4\pi\varepsilon_0 R^2)$ at the charged sphere's surface, $r = R$, so that the electric field is continuous.

In order to underline the importance of the last fact, let us consider one more elementary but very important example of the application of the Gauss law. Let a thin plane sheet (figure 1.4) be charged uniformly, with an areal density $\sigma = \text{const}$. In

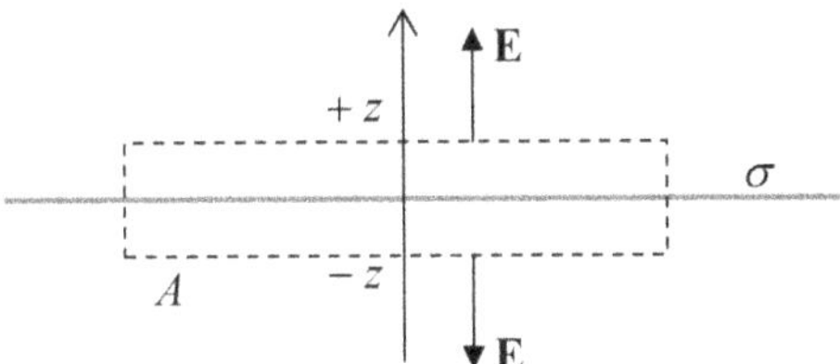

Figure 1.4. The electric field of a charged plane.

this case, it is fruitful to use the Gauss volume in the form of a planar 'pillbox' of thickness $2z$ (where z is the Cartesian coordinate perpendicular to the charged plane) and certain area A—see the dashed lines in figure 1.4. Due to the symmetry of the problem, it is evident that the electric field should be: (i) directed along axis z, (ii) constant on each of the upper and bottom sides of the pillbox, (iii) equal and opposite on these sides, and (iv) parallel to the side surfaces of the box. As a result, the full electric field flux through the pillbox's surface is just $2AE(z)$, so that the Gauss law (1.16) yields $2AE(z) = Q_A/\varepsilon_0 \equiv \sigma A/\varepsilon_0$, and we get a very simple but important formula

$$E(z) = \frac{\sigma}{2\varepsilon_0} = \text{const.} \tag{1.23}$$

Notice that, somewhat counter-intuitively, the field magnitude does not depend on the distance from the charged plane. (From the point of view of the Coulomb law (1.5), this result may be explained as follows: the farther the observation point from the plane, the weaker the effect of each elementary charge, $dQ = \sigma d^2r$, but the more such elementary charges give contributions to the vertical component of vector **E**.)

Note also that though the magnitude $E \equiv |\mathbf{E}|$ of the electric field is constant, its component E_n normal to the plane (for our coordinate choice, E_z) changes its sign at the plane, experiencing a *discontinuity* (jump) equal to

$$\Delta E_n = \frac{\sigma}{\varepsilon_0}. \tag{1.24}$$

This jump disappears if the surface is not charged ($\sigma = 0$). Returning for a minute to our charged sphere problem, very close to its surface it may be considered planar, so that the electric field should indeed be continuous, as it is.

Admittedly, the *integral form* (1.16) of the Gauss law is immediately useful only for highly symmetrical geometries, such as in the two problems discussed above. However, it may be recast into an alternative, *differential form* whose field of useful applications is much wider. This form may be obtained from Eq. (1.16) using the *divergence theorem* that, according to the vector algebra, is valid for any space-differentiable vector, in particular **E**, and for volume V limited by any closed surface S[11]:

$$\oint_S E_n d^2r = \int_V (\nabla \cdot \mathbf{E}) d^3r, \tag{1.25}$$

where ∇ is the *del* (or 'nabla') *operator* of spatial differentiation[12]. Combining Eq. (1.25) with the Gauss law (1.16), we obtain

$$\int_V \left(\nabla \cdot \mathbf{E} - \frac{\rho}{\varepsilon_0} \right) d^3r = 0. \tag{1.26}$$

For a given distribution of electric charge (and hence of the electric field), this equation should be valid for any choice of volume V. This can hold only if the function under the integral vanishes at each point, i.e. if[13]

$$\nabla \cdot \mathbf{E} = \frac{\rho}{\varepsilon_0}. \tag{1.27}$$

Note that in a sharp contrast to the integral form (1.16), Eq. (1.27) is *local*: it relates the electric field divergence to the charge density *at the same point*. This equation, being the differential form of the Gauss law, is frequently called one of the *Maxwell equations*.

Another, homogeneous Maxwell equation's 'embryo' (valid for the stationary case only!) may be obtained by noticing that the curl of the point charge's field, and hence that of *any* system of charges, equals zero[14]:

$$\nabla \times \mathbf{E} = 0. \tag{1.28}$$

(We will arrive at two other Maxwell equations, for the magnetic field, in chapter 5, and then generalize all the equations to their full, time-dependent form by the end of chapter 6. However, Eq. (1.27) would stay the same.)

Just to obtain a better gut feeling of Eq. (1.27), let us apply it to the same example of a uniformly charged sphere (figure 1.2). The vector algebra tells us that the divergence of a spherically symmetric vector function $\mathbf{E}(\mathbf{r}) = E(r)\mathbf{n}_r$ may be simply expressed in spherical coordinates[15]: $\nabla \cdot \mathbf{E} = [d(r^2E)/dr]/r^2$. As a result, Eq. (1.27) yields a linear, ordinary differential equation for the scalar function $E(r)$:

---

[11] See, e.g. Eq. (A.79). Note that the scalar product under the integral in Eq. (1.25) is nothing more that the divergence of vector **E**—see, e.g. Eq. (A.53).

[12] See, e.g. appendix A, sections A.8–A.10.

[13] In Gaussian units, just as in the initial Eq. (1.6), ε_0 has to be replaced with $1/4\pi$, so that the Maxwell equation (1.27) looks like $\nabla \cdot \mathbf{E} = 4\pi\rho$, while Eq. (1.28) stays the same.

[14] This follows, for example, from the direct application of Eq. (A.69) to any spherically symmetric vector function of the type $\mathbf{f}(\mathbf{r}) = f(r)\mathbf{n}_r$ (in particular, to the electric field of a point charge placed at the origin), giving $f_\theta = f_\phi = 0$ and $\partial f_r/\partial\theta = \partial f_r/\partial\phi = 0$, so that all components of the vector $\nabla \times \mathbf{f}$ vanish. Since nothing prevents us from placing the reference frame's origin at the point charge's location, this result remains valid for any position of the charge.

[15] See, e.g. Eq. (A.68) for this particular case (when $\partial/\partial\theta = \partial/\partial\phi = 0$).

$$\frac{1}{r^2}\frac{d}{dr}(r^2E) = \begin{cases} \rho/\varepsilon_0, & \text{for} \quad r \leqslant R, \\ 0, & \text{for} \quad r \geqslant R, \end{cases} \tag{1.29}$$

which may be readily integrated on each of the segments:

$$E(r) = \frac{1}{\varepsilon_0}\frac{1}{r^2} \times \begin{cases} \rho \int r^2 dr = \rho r^3/3 + c_1, & \text{for } r \leqslant R, \\ c_2, & \text{for } r \geqslant R. \end{cases} \tag{1.30}$$

In order to determine the integration constant c_1, we can use the following boundary condition: $E(0) = 0$. (It follows from the problem's spherical symmetry: in the center of the sphere, the electric field has to vanish, because otherwise, where would it be directed?) This requirement gives $c_1 = 0$. The second constant, c_2, may be found from the continuity condition $E(R - 0) = E(R + 0)$, which has already been discussed above, giving $c_2 = \rho R^3/3 \equiv Q/4\pi$. As a result, we arrive at our previous results (1.19) and (1.22).

We can see that in this particular, highly symmetric case, using the differential form of the Gauss law is more complex than its integral form. (For our second example, shown in figure 1.4, it would be even less natural.) However, Eq. (1.27) and its generalizations are more convenient for asymmetric charge distributions, and invaluable in the cases where the charge distribution $\rho(\mathbf{r})$ is not known *a priori* and has to be found in a self-consistent way. (We will start discussing such cases in the next chapter.)

1.3 Scalar potential and electric field energy

One more help for solving electrostatics (and more complex) problems may be obtained from the notion of the *electrostatic potential*, which is just the electrostatic potential energy U of a probe point charge placed into the field in question, normalized by its charge:

$$\phi \equiv \frac{U}{q}. \tag{1.31}$$

As we know from classical mechanics[16], the notion of U (and hence ϕ) makes most sense for the case of *potential forces*, for example those depending just on the particle's position. Eqs. (1.6) and (1.9) show that, in static situations, the electric field clearly falls into this category. For such a field, the potential energy may be defined as a scalar function $U(\mathbf{r})$ that allows the force to be calculated as its gradient (with the opposite sign):

$$\mathbf{F} = -\nabla U. \tag{1.32}$$

[16] See, e.g. *Part CM* section 1.4.

Dividing both sides of this equation by the probe charge, and using Eqs. (1.6) and (1.31), we obtain[17]

$$\mathbf{E} = -\nabla\phi. \tag{1.33}$$

In order to calculate the scalar potential, let us start from the simplest case of a single point charge q placed at the origin. For it, Eq. (1.7) takes a simple form

$$\mathbf{E} = \frac{1}{4\pi\varepsilon_0}q\frac{\mathbf{r}}{r^3} = \frac{1}{4\pi\varepsilon_0}q\frac{\mathbf{n}_r}{r^2}. \tag{1.34}$$

It is straightforward to verify that the last fraction in the right-hand side of Eq. (1.34) is equal to $-\nabla(1/r)$.[18] Hence, according to the definition (1.33), for this particular case

$$\phi = \frac{1}{4\pi\varepsilon_0}\frac{q}{r}. \tag{1.35}$$

(In the Gaussian units, this result is spectacularly simple: $\phi = q/r$.) Note that we could add an arbitrary constant to this potential (and indeed to *any* other distribution of ϕ discussed below) without changing the force, but it is convenient to define the potential energy to approach zero at infinity.

In order to justify the further exploration of U and ϕ, let me demonstrate (perhaps, unnecessarily) how useful the notions are, on a very simple example. Let two similar charges q be launched from afar, with the same initial speed $v_0 \ll c$ each, straight toward each other (i.e. with the zero impact parameter)—see figure 1.5. Since, according to the Coulomb law, the charges repel each other with increasing force, they will stop at some minimum distance $r_{\min}$ from each other, and then fly back. We could of course find $r_{\min}$ directly from the Coulomb law. However, for that we would need to write Newton's second law for each particle (actually, due to the problem symmetry, they would be similar), then integrate them over time to find the particle velocity v as a function of distance, and then recover $r_{\min}$ from the requirement $v = 0$.

The notion of potential allows this problem to be solved in one line. Indeed, in the field of potential forces the system's total energy $\mathscr{E} = T + U \equiv T + q\phi$ is conserved. In our non-relativistic case $v_0 \ll c$, the kinetic energy T is just $mv^2/2$. Hence, equating the total energy of two particles in the points $r = \infty$ and $r = r_{\min}$, and using Eq. (1.35) for ϕ, we obtain

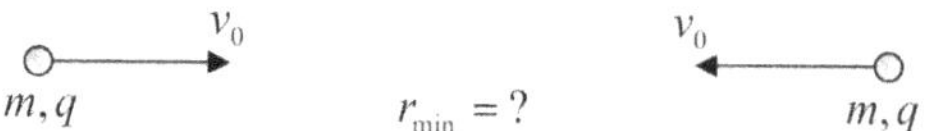

Figure 1.5. A simple problem of charged particle motion.

[17] Eq. (1.28) could also be derived from this relation, because according to vector algebra, any gradient field has vanishing curl—see, e.g. Eq. (A.71).

[18] This may be done either by Cartesian components or using the well-known expression $\nabla f = (df/dr)\mathbf{n}_r$ valid for any spherically symmetric scalar function $f(r)$—see, e.g. Eq. (A.66) for the particular case $\partial/\partial\theta = \partial/\partial\phi = 0$.

$$2\frac{mv_0^2}{2} + 0 = 0 + \frac{1}{4\pi\varepsilon_0}\frac{q^2}{r_{\min}}, \tag{1.36}$$

immediately giving us the final answer: $r_{\min} = q^2/4\pi\varepsilon_0 m{v_0}^2$. So, the notion of the scalar potential is indeed very useful.

With this motivation, let us calculate ϕ for an arbitrary configuration of charges. For a single charge in an arbitrary position (say, $\mathbf{r}_{k'}$), r in Eq. (1.35) should be evidently replaced for $|\mathbf{r} - \mathbf{r}_{k'}|$. Now, the linear superposition principle (1.3) allows for an easy generalization of this formula to the case of an arbitrary set of discrete charges,

$$\phi(\mathbf{r}) = \frac{1}{4\pi\varepsilon_0}\sum_{\mathbf{r}_{k'}\neq\mathbf{r}}\frac{q_{k'}}{|\mathbf{r} - \mathbf{r}_{k'}|}. \tag{1.37}$$

Finally, using the same arguments as in section 1.1, we can use this result to argue that in the case of an arbitrary continuous charge distribution

$$\phi(\mathbf{r}) = \frac{1}{4\pi\varepsilon_0}\int\frac{\rho(\mathbf{r}')}{|\mathbf{r} - \mathbf{r}'|}d^3r'. \tag{1.38}$$

Again, the Dirac's delta-function allows the use of the last equation to recover Eq. (1.37) for discrete charges as well, so that Eq. (1.38) may be considered the general expression for the electrostatic potential.

For most practical calculations, using this expression and then applying Eq. (1.33) to the result, is preferable to using Eq. (1.9), because ϕ is a scalar, while $\mathbf{E}$ is a 3D vector, mathematically equivalent to 3 scalars. Still, this approach still may lead to technical problems similar to those discussed in section 1.2. For example, applying it to the spherically symmetric distribution of charge (figure 1.2), we obtain the integral

$$\phi = \frac{1}{4\pi\varepsilon_0}2\pi\int_0^\pi \sin\theta' d\theta' \int_0^\infty r'^2 dr' \frac{\rho(r')}{R}\cos\theta, \tag{1.39}$$

which is not much simpler than Eq. (1.11).

The situation may be much improved by re-casting Eq. (1.38) into a differential form. For that, it is sufficient to plug the definition of ϕ, Eq. (1.33), into Eq. (1.27):

$$\nabla\cdot(-\nabla\phi) = \frac{\rho}{\varepsilon_0}. \tag{1.40}$$

The left-hand side of this equation is nothing more than the Laplace operator of ϕ (with the minus sign), so that we obtain the famous *Poisson equation*[19] for the electrostatic potential:

$$\nabla^2\phi = -\frac{\rho}{\varepsilon_0}. \tag{1.41}$$

(In Gaussian units, the Poisson equation is $\nabla^2\phi = -4\pi\rho$.)[20] This differential equation is so convenient for applications that even its particular case for $\rho = 0$,

$$\nabla^2\phi = 0, \tag{1.42}$$

has earned a special name—the *Laplace equation*[21].

In order to obtain a feeling of the Poisson equation's value as a problem-solving tool, let us return to the spherically symmetric charge distribution (figure 1.2) with a constant charge density ρ. Using the symmetry, we can represent the potential as $\phi(\mathbf{r}) = \phi(r)$, and hence use the following simple expression for its Laplace operator[22]:

$$\nabla^2\phi = \frac{1}{r^2}\frac{d}{dr}\left(r^2\frac{d\phi}{dr}\right), \tag{1.43}$$

so that for the points inside the charged sphere ($r \leqslant R$) the Poisson equation yields

$$\frac{1}{r^2}\frac{d}{dr}\left(r^2\frac{d\phi}{dr}\right) = -\frac{\rho}{\varepsilon_0}, \quad \text{i.e.} \quad \frac{d}{dr}\left(r^2\frac{d\phi}{dr}\right) = -\frac{\rho}{\varepsilon_0}r^2. \tag{1.44}$$

Integrating the last form of the equation over r once, with the natural boundary condition $d\phi/dr|_{r=0} = 0$ (because of the condition $E(0) = 0$, which has been discussed above), we obtain

$$\frac{d\phi}{dr}(r) = -\frac{\rho}{r^2\varepsilon_0}\int_0^r r'^2 dr' = -\frac{\rho\, r}{3\varepsilon_0} = -\frac{1}{4\pi\varepsilon_0}\frac{Qr}{R^3}. \tag{1.45}$$

Since this derivative is nothing more than $-E(r)$, in this formula we can readily recognize our previous result (1.22). Now we may like to carry out the second integration to calculate the potential itself:

$$\phi(r) = -\frac{Q}{4\pi\varepsilon_0 R^3}\int_0^r r'dr' + c_1 = -\frac{Qr^2}{8\pi\varepsilon_0 R^3} + c_1. \tag{1.46}$$

[19] Named after S D Poisson (1781–1840), also famous for the *Poisson distribution*—one of the central results of probability theory—see, e.g. *Part SM* section 5.2.

[20] As a sanity check, a point charge q at the origin, according to Eq. (1.10), $\rho(\mathbf{r}) = q\delta(\mathbf{r})$, and with the well-known (or readily verifiable) identity $\nabla^2(1/r) = -4\pi\delta(\mathbf{r})$, Eq. (1.41) immediately yields Eq. (1.35).

[21] After mathematician (and astronomer) P S de Laplace (1749–1827) who, together with A Clairault, is credited for the development of the very concept of potential.

[22] See, e.g. Eq. (A.67) for $\partial/\partial\theta = \partial/\partial\phi = 0$.

Before making any judgment on the integration constant c_1, let us solve the Poisson equation (in this case, just the Laplace equation) for the range outside the sphere ($r > R$):

$$\frac{1}{r^2}\frac{d}{dr}\left(r^2\frac{d\phi}{dr}\right) = 0. \tag{1.47}$$

Its first integral,

$$\frac{d\phi}{dr}(r) = \frac{c_2}{r^2}, \tag{1.48}$$

also gives the electric field (with the minus sign). Now using Eq. (1.45) and requiring the field to be continuous at $r = R$, we obtain

$$\frac{c_2}{R^2} = -\frac{Q}{4\pi\varepsilon_0 R^2}, \quad \text{i.e.} \ \frac{d\phi}{dr}(r) = -\frac{Q}{4\pi\varepsilon_0 r^2}, \tag{1.49}$$

in an evident agreement with Eq. (1.19). Integrating this result again,

$$\phi(r) = -\frac{Q}{4\pi\varepsilon_0}\int\frac{dr}{r^2} = \frac{Q}{4\pi\varepsilon_0 r} + c_3, \quad \text{for } r > R, \tag{1.50}$$

we can select $c_3 = 0$, so that $\phi(\infty) = 0$, in accordance with the usual (though not compulsory) convention. Now we can finally determine the constant c_1 in Eq. (1.46) by requiring that this equation and Eq. (1.50) give the same value of ϕ at the boundary $r = R$. (According to Eq. (1.33), if the potential had a jump, the electric field at that point would be infinite.) The final answer may be represented as

$$\phi(r) = \frac{Q}{4\pi\varepsilon_0 R}\left(\frac{R^2 - r^2}{2R^2} + 1\right), \quad \text{for } r \leqslant R. \tag{1.51}$$

We see that using the Poisson equation to find the electrostatic potential distribution for highly symmetric problems may be more cumbersome than directly finding the electric field—say, from the Gauss law. However, we will repeatedly see below that if the electric charge distribution is not fixed in advance, using Eq. (1.41) may be the only practicable way to proceed.

Returning now to the general theory of electrostatic phenomena, let us calculate the potential energy U of an arbitrary system of point electric charges q_k. Despite the apparently straightforward relation (1.31) between U and ϕ, the result is not that straightforward. Indeed, let us assume that the charge distribution has a finite spatial extent, so that at large distances from it (formally, at $\mathbf{r} = \infty$) the electric field tends to zero, so that the electrostatic potential tends to a constant. Selecting this constant, for convenience, to equal zero, we may calculate U as a sum of the energy increments ΔU_k

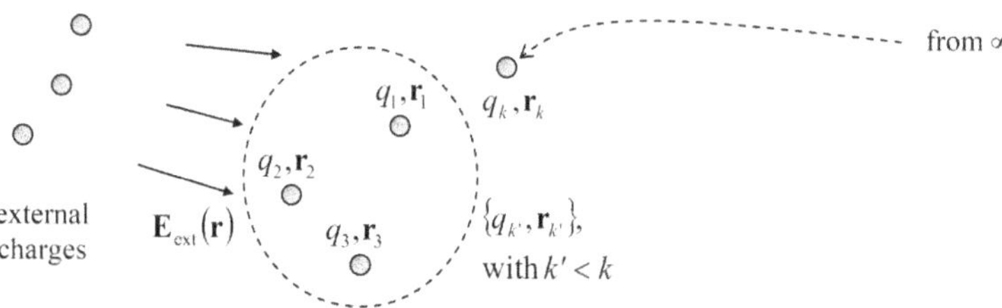

Figure 1.6. Deriving Eqs. (1.55) and (1.60) for the potential energies of a system of electric charges.

created by bringing the charges, one by one, from infinity to their final positions $\mathbf{r}_k$—see figure 1.6[23]. According to the integral form of Eq. (1.32), such a contribution is

$$\Delta U_k = -\int_{\infty}^{\mathbf{r}_k} \mathbf{F}(\mathbf{r}) \cdot d\mathbf{r} = -q_k \int_{\infty}^{\mathbf{r}_k} \mathbf{E}(\mathbf{r}) \cdot d\mathbf{r} \equiv q_k \phi(\mathbf{r}_k), \tag{1.52}$$

where $\mathbf{E}(\mathbf{r})$ is the total electric field, and $\phi(\mathbf{r})$ is the total electrostatic potential during this process, besides the field created by the very charge q_k that is being moved.

This expression shows that the increment ΔU_k, and hence the total potential energy U, depend on the source of the electric field $\mathbf{E}$. If the field is dominated by an *external field* $\mathbf{E}_{\text{ext}}$, induced by some *external* charges, not being a part of the charge configuration under our analysis (whose energy we are calculating, see figure 1.6), the spatial distribution $\phi(\mathbf{r})$ is determined by this field, i.e. does not depend on how many charges we have already brought in, so that Eq. (1.52) is reduced to

$$\Delta U_k = q_k \phi_{\text{ext}}(\mathbf{r}_k), \quad \text{where } \phi_{\text{ext}}(\mathbf{r}) \equiv -\int_{\infty}^{\mathbf{r}} E_{\text{ext}}(\mathbf{r}') \cdot d\mathbf{r}'. \tag{1.53}$$

Summing up these contributions, we get what is called the charge system's energy *in* the external field[24]:

$$U_{\text{ext}} \equiv \sum_k \Delta U_k = \sum_k q_k \phi_{\text{ext}}(\mathbf{r}_k). \tag{1.54}$$

Now repeating the argumentation that has led us to Eq. (1.9), we see that for a continuously distributed charge, this sum turns into an integral:

$$U_{\text{ext}} = \int \rho(\mathbf{r}) \phi_{\text{ext}}(\mathbf{r}) d^3 r. \tag{1.55}$$

(As was discussed above, using the delta-functional representation of point charges, we may always return from here to Eq. (1.54), so that Eq. (1.55) may be considered as a final, universal result.)

---

[23] Indeed, by the very definition of the potential energy of a system, it should not depend on the way we are arriving at the its final configuration.

[24] An alternative, perhaps more accurate term for U_{ext} is the energy of the system's *interaction with* the external field.

The result is different in the opposite limit, when the electric field $\mathbf{E}(\mathbf{r})$ is created only by the very charges whose energy we are calculating. In this case, $\phi_k(\mathbf{r}_k)$ in Eq. (1.52) is the potential created only by the charges with numbers $k' = 1, 2, ..., (k - 1)$ already in place when the kth charge is moved in (in figure 1.6, the charges inside the dashed boundary), and we may use the linear superposition principle to write

$$\Delta U_k = q_k \sum_{k'<k} \phi_{k'}(\mathbf{r}_k), \quad \text{so that } U = \sum_k U_k = \sum_{\substack{k,k' \\ (k'<k)}} q_k \phi_{k'}(\mathbf{r}_k). \tag{1.56}$$

This result is so important that it is worth rewriting it in several other forms. First, we may use Eq. (1.35) to represent Eq. (1.56) in a more symmetric form:

$$U = \frac{1}{4\pi\varepsilon_0} \sum_{\substack{k,k' \\ (k'<k)}} \frac{q_k q_{k'}}{|\mathbf{r}_k - \mathbf{r}_{k'}|}. \tag{1.57}$$

The expression under this sum is evidently symmetric with respect to the index swap, so that it may be extended into a different form,

$$U = \frac{1}{4\pi\varepsilon_0} \frac{1}{2} \sum_{\substack{k',k \\ (k'\neq k)}} \frac{q_k q_{k'}}{|\mathbf{r}_k - \mathbf{r}_{k'}|}, \tag{1.58}$$

where the interaction between each couple of charges is described by two, equal terms under the sum, and the front coefficient ½ is used to compensate this *double-counting*. The convenience of the last form is that it may be readily generalized to the continuous case:

$$U = \frac{1}{4\pi\varepsilon_0} \frac{1}{2} \int d^3r \int d^3r' \frac{\rho(\mathbf{r})\rho(\mathbf{r}')}{|\mathbf{r} - \mathbf{r}'|}. \tag{1.59}$$

(As before, in this case the restriction expressed in the discrete charge case as $k \neq k'$ is not important, because if the charge density is a continuous function, the integral (1.59) does not diverge at point $\mathbf{r} = \mathbf{r}'$.)

To represent this result in one more form, let us notice that according to Eq. (1.38), the inner integral over r' in (1.59), divided by $4\pi\varepsilon_0$, is just the full electrostatic potential at point $\mathbf{r}$, and hence

$$U = \frac{1}{2} \int \rho(\mathbf{r}) \phi(\mathbf{r}) d^3r. \tag{1.60}$$

For the discrete charge case, this result becomes

$$U = \frac{1}{2} \sum_k q_k \phi(\mathbf{r}_k), \tag{1.61}$$

but now it is important to remember that here the 'full' potential's value $\phi(\mathbf{r}_k)$ should exclude the (infinite) contribution of the point charge k itself. Comparing the last two formulas with Eqs. (1.54) and (1.55), we see that the electrostatic energy of charge interaction within the system, as expressed via the charge-by-potential product, is twice less than that of the energy of charge interaction with a fixed ('external') field. This is evidently the result of the fact that in the case of mutual interaction of the charges, the electric field **E** in the basic Eq. (1.52) is proportional to the charge magnitude, rather than constant[25].

Now we are ready to address an important conceptual question: can we locate this interaction energy in space? Expressions (1.58)–(1.61) seem to imply that contributions to U come only from the regions where the electric charges are located. However, one of the most beautiful features of physics is that sometimes completely different interpretations of the same mathematical result are possible. In order to obtain an alternative view of our current result, let us write Eq. (1.60) for a volume V so large that the electric field on the limiting surface S is negligible, and plug into it the charge density expressed from the Poisson equation (1.41):

$$U = -\frac{\varepsilon_0}{2}\int_V \phi\, \nabla^2\phi d^3r. \qquad (1.62)$$

This expression may be integrated by parts as[26]

$$U = -\frac{\varepsilon_0}{2}\left[\oint_S \phi\, (\nabla\phi)_n d^2r - \int_V (\nabla\phi)^2 d^3r\right]. \qquad (1.63)$$

According to our condition of negligible field $\mathbf{E} = -\nabla\phi$ on the surface, the first integral vanishes, and we obtain a very important formula

$$U = \frac{\varepsilon_0}{2}\int (\nabla\phi)^2 d^3r = \frac{\varepsilon_0}{2}\int E^2 d^3r. \qquad (1.64)$$

This result, represented in the following equivalent form[27]:

$$U = \int u(\mathbf{r})d^3r, \quad \text{with } u(\mathbf{r}) \equiv \frac{\varepsilon_0}{2}E^2(\mathbf{r}), \qquad (1.65)$$

certainly invites an interpretation very much different than Eq. (1.60): it is natural to interpret $u(\mathbf{r})$ as the *spatial density of the electric field energy*, which is continuously distributed over all the space where the field exists—rather than just its part where the charges are located.

Let us have a look how these two alternative pictures work for our testbed problem, a uniformly charged sphere. If we start from Eq. (1.60), we may limit the integration by the sphere volume ($0 \leqslant r \leqslant R$) where $\rho \neq 0$. Using Eq. (1.51), and the spherical symmetry of the problem ($d^3r = 4\pi r^2 dr$), we get

---

[25] The nature of this additional factor ½ is absolutely the same as in the well-known formula $U = (½)\kappa x^2$ for the potential energy of an elastic spring, providing the returning force $F = -\kappa x$ proportional to the deviation x from equilibrium.

[26] This transformation follows from the divergence theorem (A.79) applied to the vector function $\mathbf{f} = \phi\nabla\phi$, taking into account the differentiation rule (A.74*a*): $\nabla \cdot (\phi\, \nabla\phi) = (\nabla\phi) \cdot (\nabla\phi) + \phi\, \nabla \cdot (\nabla\phi) = (\nabla\phi)^2 + \phi\nabla^2\phi$.

[27] In Gaussian units, the standard replacement $\varepsilon_0 \to 1/4\pi$ turns the last of Eqs. (1.65) into $u(\mathbf{r}) = E^2/8\pi$.

$$U = \frac{1}{2}4\pi\int_0^R \rho\phi\, r^2dr = \frac{1}{2}4\pi\rho\frac{Q}{4\pi\varepsilon_0 R}\int_0^R\left(\frac{R^2 - r^2}{2R^2} + 1\right)r^2dr$$
$$= \frac{6}{5}\frac{1}{4\pi\varepsilon_0 R}\frac{Q^2}{2}. \tag{1.66}$$

On the other hand, if we use Eq. (1.65), we need to integrate the energy density everywhere, i.e. both inside and outside of the sphere:

$$U = \frac{\varepsilon_0}{2}4\pi\left(\int_0^R E^2r^2dr + \int_R^\infty E^2r^2dr\right). \tag{1.67}$$

Using Eqs. (1.19) and (1.22) for, respectively, the external and internal regions, we obtain

$$U = \frac{\varepsilon_0}{2}4\pi\left[\int_0^R\left(\frac{Qr}{4\pi\varepsilon_0}\right)^2 r^2dr + \int_R^\infty\left(\frac{Q}{4\pi\varepsilon_0 r^2}\right)^2 r^2dr\right] = \left(\frac{1}{5} + 1\right)\frac{1}{4\pi\varepsilon_0 R}\frac{Q^2}{2}. \tag{1.68}$$

This is (fortunately:-) the same answer as given by Eq. (1.66), but to some extent it is more informative, because it shows how exactly the electric field energy is distributed between the interior and exterior of the charged sphere[28].

We see that, as we could expect, within the realm of *electrostatics*, Eqs. (1.60) and (1.65) are equivalent. However, when we examine *electrodynamics* (in chapter 6 and beyond), we will see that the latter equation is more general, and that it is more adequate to associate the electric energy with the field itself rather than its sources—in our current case, the electric charges.

Finally, let us calculate the potential energy of a system of charges in the general case when both the internal interaction of the charges, and their interaction with an external field are important. One might fancy that such a calculation should be very difficult, since in both ultimate limits, when one of these interactions dominates, we have obtained different results. However, once again we get help from the almighty linear superposition principle: in the general case, for the total electric field we may write

$$\mathbf{E}(\mathbf{r}) = \mathbf{E}_{\text{int}}(\mathbf{r}) + \mathbf{E}_{\text{ext}}(\mathbf{r}), \qquad \phi(\mathbf{r}) = \phi_{\text{int}}(\mathbf{r}) + \phi_{\text{ext}}(\mathbf{r}), \tag{1.69}$$

where the index 'int' now marks the field induced by the charge system under analysis, i.e. the variables participating (without indices) in Eqs. (1.56)–(1.68). Now let us imagine that our system is being built up in the following way: first, the charges are brought together at $\mathbf{E}_{\text{ext}} = 0$, giving the potential energy U_{int} expressed by Eq. (1.60), and then $\mathbf{E}_{\text{ext}}$ is slowly increased. Evidently, the energy contribution from

[28] Note that $U \to \infty$ at $R \to 0$. Such divergence appears at application of Eq. (1.65) to any point charge. Since it does not affect the force acting on the charge, the divergence does not create any technical difficulty for analysis of charge statics or non-relativistic dynamics, but it points to a conceptual problem of classical electrodynamics as the whole. This issue will be discussed in the very end of the course (section 10.6).

the latter process cannot depend on the internal interaction of the charges, and hence may be expressed in the form (1.55). As the result, the total potential energy[29] is the sum of these two components:

$$U = U_{\text{int}} + U_{\text{ext}} = \frac{1}{2}\int \rho(\mathbf{r})\phi_{\text{int}}(\mathbf{r})d^3r + \int \rho(\mathbf{r})\phi_{\text{ext}}(\mathbf{r})d^3r. \tag{1.70}$$

Now making, in the first integral, the transition from the potentials to the fields, absolutely similar to that performed in Eqs. (1.62)–(1.65), we may rewrite this expression as

$$U = \int u(\mathbf{r})d^3r, \quad \text{with } u(\mathbf{r}) \equiv \frac{\varepsilon_0}{2}[E_{\text{int}}^2(\mathbf{r}) + 2\mathbf{E}_{\text{int}}(\mathbf{r}) \cdot \mathbf{E}_{\text{ext}}(\mathbf{r})]. \tag{1.71}$$

One might think that this result, more general than Eq. (1.65) and perhaps less familiar to the reader, is something entirely new; however, it is not. Indeed, let us add and subtract $E_{\text{ext}}^2(\mathbf{r})$ from the sum in the brackets, and use Eq. (1.69) for the total electric field $\mathbf{E}(\mathbf{r})$; then Eq. (1.71) takes the form

$$U = \frac{\varepsilon_0}{2}\int E^2(\mathbf{r})d^3r - \frac{\varepsilon_0}{2}\int E_{\text{ext}}^2(\mathbf{r})d^3r. \tag{1.72}$$

Hence, in the most important case when we are using the potential energy to analyze the statics and dynamics of a system of charges in a fixed external field, i.e. when the second term in Eq. (1.72) may be considered as a constant, we may still use for U an expression similar to the familiar Eq. (1.65), but with the field $\mathbf{E}(\mathbf{r})$ being the sum (1.69) of the internal and external fields.

Let us see how this works in a very simple problem. A uniform external electric field $\mathbf{E}_{\text{ext}}$ is applied normally to a very broad, planar layer that contains a very large and equal number of free electric charges of both signs—see figure 1.7. What is the equilibrium distribution of the charges over the layer?

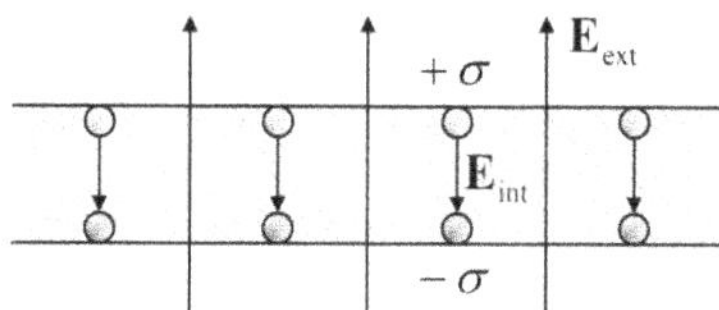

Figure 1.7. A simple model of the electric field screening in a conductor.

Since the equilibrium distribution should minimize the total potential energy of the system, Eq. (1.72) immediately gives the answer: the distribution should provide $\mathbf{E} \equiv \mathbf{E}_{\text{int}} + \mathbf{E}_{\text{ext}} = 0$ inside the whole layer—the effect called the electric field *screening*. The only way to ensure this equality is to have enough free charges of opposite signs residing on the layer's surfaces to induce a uniform field $\mathbf{E}_{\text{int}} = -\mathbf{E}_{\text{ext}}$, exactly compensating the external field at each point inside the layer—see figure 1.7.

[29] This total U (or rather its part dependent on our system of charges) is sometimes called the *Gibbs potential energy* of the system. (I will discuss this notion in detail in section 3.5.)

According to Eq. (1.24), the areal density of these surface charges should equal $\pm\sigma$, with $\sigma = E_{ext}/\varepsilon_0$. This is a rudimentary but reasonable model of the conductor's *polarization*—to be discussed in detail in the next chapter.

1.4 Problems

Problem 1.1. Calculate the electric field created by a thin, long, straight filament, electrically charged with a constant linear density λ, using two approaches:

(i) directly from the Coulomb law, and
(ii) using the Gauss law.

Problem 1.2. Two thin, straight parallel filaments, separated by distance ρ, carry equal and opposite uniformly distributed charges with linear density λ—see the figure below. Calculate the electrostatic force (per unit length) of the Coulomb interaction between the wires. Compare the result with the Coulomb law for the force between the point charges.

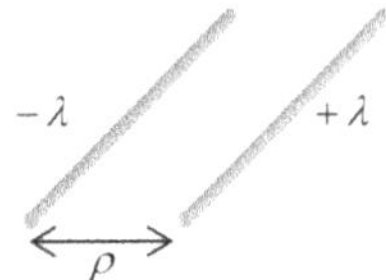

Problem 1.3. A sphere of radius R, whose volume has been charged with a constant density ρ, is split with a very narrow, planar gap passing through its center. Calculate the force of mutual repulsion of the resulting two hemispheres.

Problem 1.4. A thin spherical shell of radius R, which has been charged with a constant areal density σ, is split into two equal halves by a very narrow, planar cut passing through the sphere's center. Calculate the force of electrostatic repulsion between the resulting hemispheric shells, and compare the result with that of the previous problem.

Problem 1.5. Calculate the distribution of the electrostatic potential created by a straight, thin filament of finite length $2l$, charged with a constant linear density λ, and explore the result in the limits of very small and very large distances from the filament.

Problem 1.6. A thin plane sheet, perhaps of an irregular shape, carries electric charge with a constant areal density σ.

(i) Express the electric field's component normal to the plane, at a certain distance from it, via the solid angle Ω at which the sheet is visible from the observation point.
(ii) Use the result to calculate the field in the center of a cube, with one face charged with constant density σ.

Problem 1.7. Can one create, in a non-vanishing region of space, electrostatic fields with the Cartesian components proportional to the following products of Cartesian coordinates $\{x, y, z\}$,

(i) $\{yz, xz, xy\}$,
(ii) $\{xy, xy, yz\}$?

Problem 1.8. Distant sources have been used to create different electric fields on two sides of a wide and thin metallic membrane, with a round hole of radius R in it—see the figure below. Besides the local perturbation created by the hole, the fields are uniform:

$$\mathbf{E}(\mathbf{r})|_{r\gg R} = \mathbf{n}_z \times \begin{cases} E_1, & \text{at } z < 0, \\ E_2, & \text{at } z > 0. \end{cases}$$

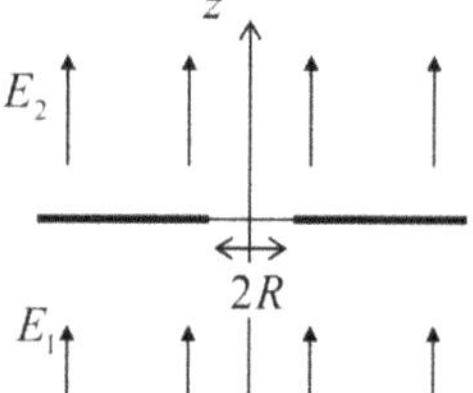

Prove that the system may serve as an electrostatic lens for charged particles flying along axis z, at distances $\rho \ll R$ from it, and calculate the focal distance f of the lens. Spell out the conditions of validity of your result.

Problem 1.9. Eight equal point charges q are located at the corners of a cube of side a. Calculate all Cartesian components E_j of the electric field, and their spatial derivatives $\partial E_j/\partial r_{j'}$, at the cube's center, where r_j are the Cartesian coordinates oriented along the cube's sides—see the figure below. Are all your results valid for the center of a planar square, with four equal charges in its corners?

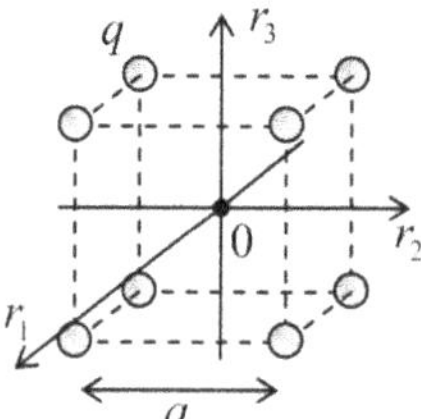

Problem 1.10. By a direct calculation, find the average electric potential of the spherical surface of radius R, created by a point charge q located at a distance $r > R$ from the sphere's center. Use the result to prove the following general *mean value theorem*: the electric potential at any point is always equal to its average value on any spherical surface with the center at that point, and containing no electric charges inside it.

Problem 1.11. Two similar thin, circular, coaxial disks of radius R, separated by distance $2d$, are uniformly charged with equal and opposite areal densities $\pm\sigma$—see the figure below. Calculate and sketch the distribution of the electrostatic potential and the electric field of the disks along their common axis.

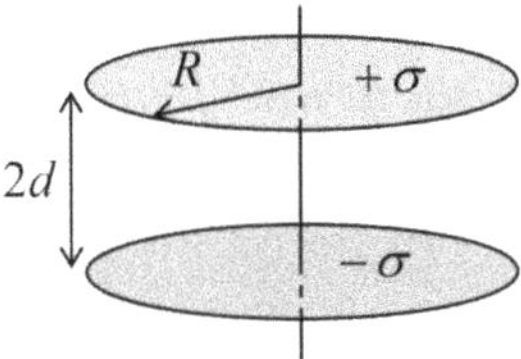

Problem 1.12. In a certain reference frame, the electrostatic potential created by some electric charge distribution, is

$$\phi(\mathbf{r}) = C\left(\frac{1}{r} + \frac{1}{2r_0}\right)\exp\left\{-\frac{r}{r_0}\right\},$$

where C and r_0 are constants, and $r \equiv |\mathbf{r}|$ is the distance from the origin. Calculate the charge distribution in space.

Problem 1.13. A thin flat sheet, cut in a form of a rectangle of size $a \times b$, is electrically charged with a constant areal density σ. Without an explicit calculation of the spatial distribution $\phi(\mathbf{r})$ of the electrostatic potential induced by this charge, find the ratio of its values at the center and at the corners of the rectangle.

Hint: Consider partitioning the rectangle into several similar parts and using the linear superposition principle.

Problem 1.14. Calculate the electrostatic energy per unit area of the system of two thin, parallel planes with equal and opposite charges of a constant areal density σ, separated by distance d.

Problem 1.15. The system analyzed in the previous problem (two thin, parallel, oppositely charged planes) is now placed into an external, uniform, normal electric field $E_{ext} = \sigma/\varepsilon_0$—see the figure below. Find the force (per unit area) acting on each plane, by two methods:

(i) directly from the electric field distribution, and
(ii) from the potential energy of the system.

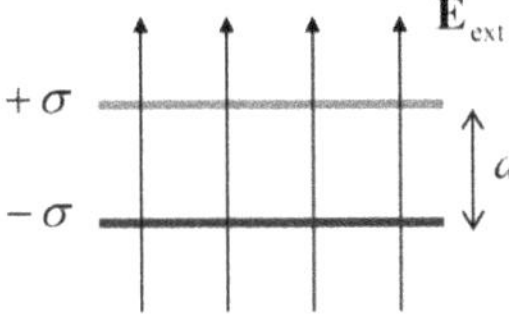

Problem 1.16. Explore the relation between the Laplace equation (1.42) and the condition of minimum of the electrostatic field energy (1.65).

Problem 1.17. Prove the following *reciprocity theorem of electrostatics*[30]: if two spatially confined charge distributions $\rho_1(\mathbf{r})$ and $\rho_2(\mathbf{r})$ create respective distributions $\phi_1(\mathbf{r})$ and $\phi_2(\mathbf{r})$ of the electrostatic potential, then

$$\int \rho_1(\mathbf{r})\, \phi_2(\mathbf{r}) d^3r = \int \rho_2(\mathbf{r})\, \phi_1(\mathbf{r}) d^3r.$$

Hint: Consider integral $\int \mathbf{E}_1 \cdot \mathbf{E}_2 d^3r$.

Problem 1.18. Calculate the energy of electrostatic interaction of two spheres, of radii R_1 and R_2, each with a spherically symmetric charge distribution, separated by distance $d > R_1 + R_2$.

Problem 1.19. Calculate the electrostatic energy U of a (generally, thick) spherical shell, with a charge Q uniformly distributed through its volume—see the figure below. Analyze and interpret the dependence of U on the inner cavity's radius R_1, at fixed Q and R_2.

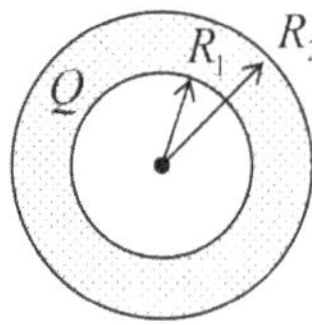

Reference

[1] Griffiths D 2015 *Introduction to Electrodynamics* 4th edn (Pearson)

[30] This is only the simplest of several reciprocity theorems in electromagnetism—see, e.g. section 6.8.

IOP Publishing

Classical Electrodynamics
Lecture notes
Konstantin K Likharev

Chapter 2

Charges and conductors

This chapter will start a discussion of the very common situations where the electric charge distribution in space is not known a priori, but rather should be calculated in a self-consistent way together with the electric field it creates. The simplest situations of this kind involve conductors, and lead to the so-called boundary problems, in which partial differential equations describing the field distribution have to be solved with appropriate boundary conditions. Such problems are also broadly used in other areas of electrodynamics (and indeed in other fields of physics as well), so that following tradition, I will use this chapter's material as a playground for the discussion of various methods of boundary problem solution, and the special functions most frequently encountered in the process.

2.1 Polarization and screening

The basic principles of electrostatics outlined in chapter 1 present the conceptually full solution for the problem of finding the electrostatic field (and hence Coulomb forces) induced by electric charges distributed over space with density $\rho(\mathbf{r})$. However, in most practical situations this function is not known but should be found self-consistently with the field. For example, if a volume of relatively dense material is placed into an external electric field, it is typically *polarized*, i.e. acquires some local charges of their own, which contribute to the total electric field $\mathbf{E}(\mathbf{r})$ inside, and even outside it—see figure 2.1a.

The full solution of such problems should satisfy not only the fundamental Eq. (1.7), but also the so-called *constitutive relations* between various macroscopic variables describing the body's material. Due to the atomic character of real materials, such relations may be very involved. In this part of my series, I will have time to address these relations, for various materials, only very superficially[1],

[1] More detailed discussions may be found, e.g. in section 13.5 of [1], or the section on electric field screening in chapter 17 of [2].

doi:10.1088/978-0-7503-1404-6ch2

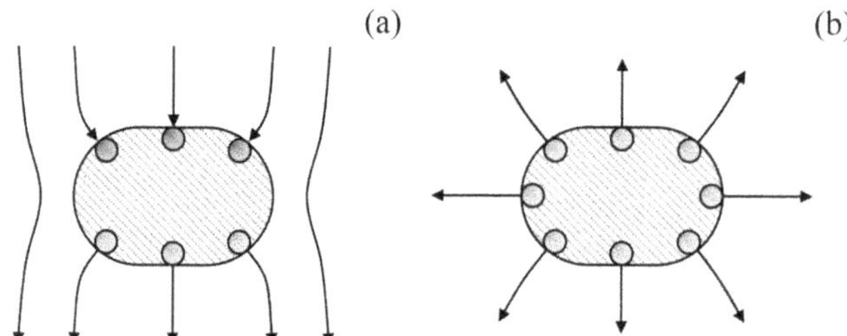

Figure 2.1. Schematic representations of two typical electrostatic situations involving conductors: (a) the polarization by an external field and (b) the conductor's own charge redistribution over its surface. Here and below, the red and blue points are used to show charges of opposite signs.

focusing on their simple approximations. Fortunately, in most practical cases such approximations work very well.

In particular, for the polarization of good conductors, a very reasonable approximation is given by the so-called *macroscopic model*, in which the free charges in the conductor are treated as a charged continuum, which is free to move under the effect of the average force $\mathbf{F} = q\mathbf{E}$ exerted by the averaged, *macroscopic* electric field $\mathbf{E}$—see the discussion of this notion at the end of section 1.1. In electrostatics (which excludes the case of dc currents, to be discussed in chapter 4), there should be no such motion, so that everywhere inside the conductor the macroscopic electric field should vanish:

$$\mathbf{E} = 0. \tag{2.1a}$$

This is the *electric field screening*[2] *effect*, meaning, in particular, that the conductor's polarization in an external electric field has the extreme form shown (schematically) in figure 2.1a, with the field of the induced surface charges completely compensating the external field in the conductor's bulk. Eq. (2.1*a*) may be rewritten in another, frequently more convenient form:

$$\phi = \text{const}, \tag{2.1b}$$

where ϕ is the *macroscopic* electrostatic potential, related to the macroscopic electric field by Eq. (1.33). Note, however, that if a problem includes several unconnected conductors, the constant in Eq. (2.1*b*) may be different for each of them.

Now let us examine what we can say about the electric field just *outside* a conductor, within the same macroscopic model. At a close proximity, any smooth surface (in our current case, that of a conductor) looks planar. Let us integrate Eq. (1.28) over a narrow ($d \ll l$) rectangular loop C encircling a part of such plane conductor's surface (see the dashed line in figure 2.2a), and apply to the electric field vector $\mathbf{E}$ the well-known vector algebra equality, the *Stokes theorem*[3]

[2] This term, used for the *electric* field, should not be confused with *shielding*—the word used for the description of the *magnetic* field reduction by magnetic materials—see chapter 5.

[3] See, e.g. Eq. (A.78).

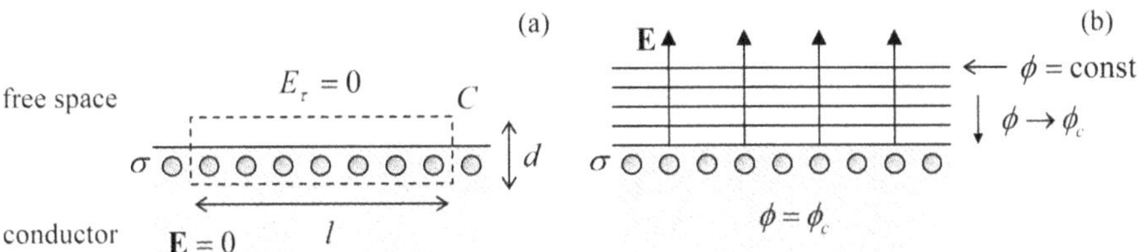

Figure 2.2. (a) The surface charge layer at a conductor's surface, and (b) the electric field lines and equipotential surfaces near it.

$$\oint_S (\nabla \times \mathbf{E})_n d^2 r = \oint_C \mathbf{E} \cdot d\mathbf{r}, \tag{2.2}$$

where S is any surface limited by the contour C. In our case the contour is dominated by two straight lines of length l, so that if l is much smaller that the characteristic scale of field change, but much larger that the inter-atomic distances, the right-hand side of Eq. (2.2) may be approximated as $[(E_\tau)_{\text{in}} - (E_\tau)_{\text{out}}]l$, where (E_τ) is the component of the corresponding macroscopic field, parallel to the surface. On the other hand, according to Eq. (1.28), the left-hand side of Eq. (2.2) equals zero. Hence, E_τ should be continuous at the surface, and in order to satisfy Eq. (2.1*a*) inside the conductor, the component has to vanish immediately outside it: $(E_\tau)_{\text{out}} = 0$.

This means that the electrostatic potential immediately outside a conducting surface does not change along it. In other words, the equipotential surfaces outside a conductor should 'lean' to the conductor's surface, with their potential values approaching the constant potential—see figure 2.2b.

So, the electrostatic field just outside the conductor has be normal to its surface. In order to find this normal field, we may use the universal relation (1.24). Since in our current case $E_n = 0$ inside the conductor, we obtain

$$\sigma = \varepsilon_0 (E_n)_{\text{out}} = -\varepsilon_0 (\nabla \phi)_n \equiv -\varepsilon_0 \frac{\partial \phi}{\partial n}, \tag{2.3}$$

where σ is the areal density of conductor's surface charge. Note that in deriving this universal relation between the normal component of the field and the surface charge density, we have not used any cause-versus-effect arguments, so that Eq. (2.3) is valid regardless of whether the surface charge is induced by an externally applied field (the case of polarization of the conductor, shown in figure 2.1a), or the electric field is induced by the electric charge placed on the conductor and then self-redistributed over its surface (figure 2.1b), or it is some mixture of both effects.

Before starting to use the macroscopic model for the solution of particular problems of electrostatics, let me use the balance of this section to briefly discuss its limitations. (The reader in a rush may skip this discussion and proceed to section 2.2; however, I believe that every educated physicist has to understand when the model works, and when it does not.)

Since the argumentation which has led to Eq. (1.24) and hence Eq. (2.3) is valid for any thickness d of the Gauss pillbox, within the macroscopic model, the whole surface charge is located within an infinitely thin surface layer. This is of course impossible physically: for one, this would require an infinite volume density ρ of the charge. In reality the charged layer (and hence the region of electric field's crossover from the finite value (2.3) to zero) has a non-vanishing thickness λ. At least three effects contribute to λ.

(i) *Atomic structure of matter.* Within each atom, and frequently between the adjacent atoms as well, the genuine ('microscopic') electric field is highly non-uniform. Thus Eq. (2.1) is valid only for the *macroscopic field* in a conductor (see the discussion following Eq. (1.9) in section 1.1), averaged over distances much larger than the atomic size scale $a_0 \sim 10^{-10}$ m,[4] and cannot be applied to the field changes on that scale. As a result, the surface layer of charges cannot be much thinner than a_0.

(ii) *Thermal excitation.* In the conductor's bulk, the number of protons of atomic nuclei (n) and electrons (n_e) and per unit volume are balanced, so that the net charge density, $\rho = e(n - n_e)$, vanishes[5]. However, if an external electric field penetrates a conductor, free electrons can shift in or out of its affected part, depending on the field addition to their potential energy, $\Delta U = q_e\phi = -e\phi$. (Here the arbitrary constant in ϕ is chosen to give $\phi = 0$ well inside the conductor.) In the classical statistics, this change is described by the Boltzmann distribution[6]:

$$n_e(\mathbf{r}) = n \exp\left\{-\frac{U(\mathbf{r})}{k_B T}\right\}, \qquad (2.4)$$

where $k_B \approx 1.38 \times 10^{-23}$ J K^{-1} is the Boltzmann constant, and T is the absolute temperature in SI units (kelvins). As a result, the net charge density is

$$\rho(\mathbf{r}) = en\left(1 - \exp\left\{\frac{e\phi(\mathbf{r})}{k_B T}\right\}\right) \qquad (2.5)$$

The penetrating electric field polarizes the atoms as well. As will be discussed in the next chapter, such polarization results in the reduction of the electric field by a material-specific dimensionless factor κ (larger, but typically not too much larger than 1), called the *dielectric constant.* As a result, the Poisson equation (1.41) takes the form[7],

---

[4] This scale originates from the quantum-mechanical effects of electron motion, characterized by the *Bohr radius* $r_B \equiv \hbar^2/m_e(e^2/4\pi\varepsilon_0) \approx 0.53 \times 10^{-10}$ m—see, e.g. *Part QM* Eq. (1.13). It also defines the scale $E_B = e/4\pi\varepsilon_0 r_B^2 \sim 10^{12}$ SI units (V m^{-1}) of the microscopic electric fields inside the atoms. (Please note how large these fields are.)

[5] In this series, e denotes the fundamental charge, $e \approx 1.6 \times 10^{-19}$ C > 0, so that the electron's charge equals $(-e)$.

[6] See, e.g. *Part SM* section 3.1.

[7] This equation and/or its straightforward generalization to the case of charged particles (ions) of several kinds is frequently (especially in the theories of electrolytes and plasmas) called the *Debye–Hückel equation.*

$$\frac{d^2\phi}{dz^2} = -\frac{\rho}{\kappa\varepsilon_0} = \frac{en}{\kappa\varepsilon_0}\left(\exp\left\{\frac{e\phi}{k_BT}\right\} - 1\right), \tag{2.6}$$

where we have taken advantage of the 1D geometry of the system to simplify the Laplace operator, with the axis z normal to the surface.

Even with this simplification, Eq. (2.6) is a nonlinear differential equation allowing an analytical but rather bulky solution. Since our current goal is just to estimate of the field penetration depth λ, let us simplify the equation further by considering the low-field limit: $e|\phi| \sim e|E|\lambda \ll k_BT$. In this limit we can extend the exponent into the Taylor series, and keep only two leading terms (of which the first one cancels with the unity). As a result, Eq. (2.6) becomes linear,

$$\frac{d^2\phi}{dz^2} = \frac{en}{\varepsilon\varepsilon_0}\frac{e\phi}{k_BT}, \quad \text{i.e.} \quad \frac{d^2\phi}{dz^2} = \frac{1}{\lambda^2}\phi, \tag{2.7}$$

where the constant λ in this case is equal to the so-called *Debye screening length* λ_D, defined by the relation

$$\lambda_D^2 \equiv \frac{\kappa\varepsilon_0 k_BT}{e^2n} \tag{2.8}$$

As the reader certainly knows, Eq. (2.7) describes an exponential decrease of the electric potential, with the characteristic length λ_D: $\phi \propto \exp\{-z/\lambda_D\}$, where axis z is directed inside the conductor. Plugging in the fundamental constants, we obtain the following estimate: $\lambda_D[\text{m}] \approx 70 \times (\kappa \times T[\text{K}]/n[\text{m}^{-3}])^{1/2}$. According to this formula, in semiconductors at room temperature, the Debye length may be rather substantial. For example, in silicon ($\kappa \approx 12$) doped to the free charge carrier concentration $n = 3 \times 10^{24}\ \text{m}^{-3}$ (the value typical for modern integrated circuits)[8], $\lambda_D \approx 2$ nm, still well above the atomic size scale a_0. However, for typical good metals ($n \sim 10^{29}\ \text{m}^{-3}$, $\kappa \sim 10$) the same formula gives an estimate $\lambda_D \sim 10^{-11}$ m, less than a_0. In this case Eq. (2.8) should not be taken literally, because it is based on the assumption of a continuous charge distribution.

(iii) *Quantum statistics*. Actually, the last estimate is not valid for good metals (and highly doped semiconductors) for one more reason: their free electrons obey the quantum (*Fermi–Dirac*) statistics rather that the Boltzmann distribution (2.4)[9]. As a result, at all realistic temperatures the electrons form a degenerate quantum gas,

[8] There is a good reason for making an estimate of λ_D for this case: the electric field created by the gate electrode of a field-effect transistor, penetrating into doped silicon by a depth $\sim\lambda_D$, controls the electric current in this most important electronic device—on whose back all the current information technology rides. Because of that, λ_D establishes the possible scale of semiconductor circuit shrinking, which is the basis of the well-known Moore's law. (Practically, the scale is determined by integrated circuit patterning techniques, and Eq. (2.8) may be used to find the proper charge carrier density n and hence the necessary level of silicon doping.)

[9] See, e.g. *Part SM* section 2.8. For a more detailed derivation of Eq. (2.10), see *Part SM* chapter 3.

occupying all available energy states below certain energy level $\mathscr{E}_F \gg k_B T$, called the *Fermi energy*. In these conditions, the screening of relatively low electric field may be described by replacing Eq. (2.5) with

$$\rho = e(n - n_e) = -eg(\mathscr{E}_F)(-U) = -e^2 g(\mathscr{E}_F)\phi, \tag{2.9}$$

where $g(\mathscr{E})$ is the density of quantum states (per unit volume) at electron's energy $\mathscr{E}$. At the Fermi surface, the density is of the order of $n/\mathscr{E}_F$.[10] As a result, we again obtain the second of Eqs. (2.7), but with a different characteristic scale λ, defined by the following relation:

$$\lambda_{TF}^2 \equiv \frac{\kappa\varepsilon_0}{e^2 g(\mathscr{E}_F)} \sim \frac{\kappa\varepsilon_0\mathscr{E}_F}{e^2 n}, \tag{2.10}$$

and called the *Thomas–Fermi screening length*. Since for most good metals, n is of the order of 10^{29} m^{-3}, and $\mathscr{E}_F$ is of the order of 10 eV, Eq. (2.10) typically gives λ_{TF} close to a few a_0, and makes the Thomas–Fermi screening theory valid at least semi-quantitatively.

To summarize, the electric field penetration into good conductors is limited to a depth λ ranging from fractions of a nanometer to a few nanometers, so that for problems with the characteristic size much larger than that scale, the macroscopic model (2.1) gives a very good accuracy, and we will use them in the rest of this chapter. However, the reader should remember that in many situations involving semiconductors, as well as at some nanoscale experiments with metals, the electric field penetration effect should be taken into account.

Another important condition of the macroscopic model's validity is imposed on the electric field's magnitude, which is especially significant for semiconductors. Indeed, as Eq. (2.6) shows, Eq. (2.7) is only valid if $e\phi \ll k_B T$, so that $E \sim \phi/\lambda_D$ should be much lower than $k_B T/e\lambda_D$. In the example given above ($\lambda_D \approx 2$ nm, $T = 300$ K), this means $E \ll E_t \sim 10^7$ V m^{-1} $\equiv 10^5$ V cm^{-1}—a value readily reachable in the lab. At larger fields, the field penetration becomes nonlinear, leading to the important effect of *carrier depletion*; it will be discussed in *Part SM* chapter 6. For typical metals, such linearity limit, $E_t \sim \mathscr{E}_F/e\lambda_{TF}$ is much higher, $\sim 10^{11}$ V m^{-1}, but the model may be violated at lower fields (also $\sim 10^7$ V m^{-1}) by other effects, such as the impact-ionization, leading to *electric breakdown*.

2.2 Capacitance

Let us start with the systems consisting of charged conductors only, with no stand-alone charges in the space outside them. Our goal here is calculating the distributions of electric field $\mathbf{E}$ and potential ϕ in space, and the distribution of the surface

[10] See, e.g. *Part SM* section 3.3.

charge density σ over the conductor surfaces. However, before doing that for particular situations, let us see if there are any integral measures of these distributions, that should be our primary focus.

The simplest case is of course a single conductor in the otherwise free space. According to Eq. (2.1*b*), all its volume should have a constant electrostatic potential ϕ, evidently providing one convenient global measure of the situation. Another integral measure is provided by the total charge

$$Q \equiv \int_V \rho \, d^3r \equiv \oint_S \sigma \, d^2r, \tag{2.11}$$

where the latter integral is extended over the whole surface S of the conductor. In the general case, what can we tell about the relation between Q and ϕ? At $Q = 0$, there is no electric field in the system, and it is natural (although not necessary) to select the arbitrary constant in the electrostatic potential to have $\phi = 0$. Then, if the conductor is charged with a finite Q, according to the Coulomb law, the electric field in any point of space is proportional to Q. Hence the electrostatic potential everywhere, including its value ϕ on the conductor, is also proportional to Q:

$$\phi = p Q. \tag{2.12}$$

The proportionality coefficient p, which depends on the conductor's size and shape, but not on ϕ or Q, is called the *reciprocal capacitance* (or, not too often, 'electric elastance'). Usually, Eq. (2.12) is rewritten in a different form,

$$Q = C\phi, \quad \text{with } C \equiv \frac{1}{p}, \tag{2.13}$$

where C is called *self-capacitance*. (Frequently, C is called just *capacitance*, but as we will see very soon, for more complex situations the latter term may be too ambiguous.)

Before calculating C for particular geometries, let us have a look at the electrostatic energy U of a single conductor. In order to calculate it, of the several relations discussed in chapter 1, Eq. (1.61) is the most convenient, because all elementary charges q_k are now parts of the conductor's surface charge, and hence reside at the same potential ϕ—see Eq. (2.1*b*) again. As a result, the equation becomes very simple:

$$U = \frac{1}{2}\phi \sum_k q_k = \frac{1}{2}\phi \, Q. \tag{2.14}$$

Moreover, using the linear relation (2.13), the same result may be re-written in two more forms:

$$U = \frac{Q^2}{2C} = \frac{C}{2}\phi^2. \tag{2.15}$$

We will discuss several ways to calculate C in the next sections, and right now will have a quick look at just the simplest example, for which we have calculated everything necessary in the previous chapter: a conducting sphere of radius R. Indeed, we already know the electric field distribution: according to Eq. (2.1), $E = 0$ inside the sphere, while Eq. (1.19), with $Q(r) = Q$, describes the field distribution outside it, because of the evident spherical symmetry of the surface charge distribution. Moreover, since the latter formula is exactly the same as for the point charge placed in the sphere's center, the potential distribution in space can be obtained from Eq. (1.35) by replacing q with the sphere's full charge Q. Hence, on the surface of the sphere (and, according to Eq. (2.1b), through its interior),

$$\phi = \frac{1}{4\pi\varepsilon_0}\frac{Q}{R}. \tag{2.16}$$

Comparing this result with the definition (2.13), for the self-capacitance we obtain a very simple formula

$$C = 4\pi\varepsilon_0 R. \tag{2.17}$$

This formula, which should be very familiar to the reader[11], is convenient to obtain some feeling of how large the SI unit of capacitance (1 *farad*, abbreviated as F) is: the self-capacitance of Earth ($R_E \approx 6.34 \times 10^6$ m) is below 1 mF! Another important note is that while Eq. (2.17) is not exactly valid for a conductor of arbitrary shape, it implies an important estimate

$$C \sim 2\pi\varepsilon_0 a \tag{2.18}$$

where a is the scale of the linear size of any conductor[12].

Now proceeding to a system of two arbitrary conductors, we immediately see why we should be careful with the capacitance definition: one constant C is insufficient to describe such system. Indeed, here we have two, generally different, conductor potentials, ϕ_1 and ϕ_2, that may depend on both conductor charges, Q_1 and Q_2. Using the same arguments as for the single-conductor case, we may conclude that the dependence is always linear:

$$\begin{aligned}\phi_1 &= p_{11}Q_1 + p_{12}Q_2,\\ \phi_2 &= p_{21}Q_1 + p_{22}Q_2,\end{aligned} \tag{2.19}$$

but now has to be described by more than one coefficient. Actually, it turns out that there are three rather than four different coefficients in these relation, because

[11] In Gaussian units, using the standard replacement $4\pi\varepsilon_0 \to 1$, this relation takes an even simpler form: $C = R$, easy to remember. Generally, in Gaussian units (but not in the SI system!) the capacitance has the dimensionality of length, i.e. is measured in centimeters. Note also that a fractional SI unit, 1 picofarad (10^{-12} F), is very close to the Gaussian unit: 1 pF = $(1 \times 10^{-12})/(4\pi\varepsilon_0 \times 10^{-2})$ cm $\approx$ 0.8998 cm. So, this unit is rather close to the capacitance of a metallic ball with a 1 cm radius, making it very convenient for human-scale systems.

[12] These arguments are somewhat insufficient to say which size should be used for a in the case of narrow, extended conductors, e.g. a thin, long wire. Very soon we will see that in such cases the electrostatic energy, and hence C, depends mostly on the *larger* size of the conductor.

$$p_{12} = p_{21}. \tag{2.20}$$

This relation may be proved in several ways, for example, using the general *reciprocity theorem of electrostatics* (whose proof was the subject of one of the problems of chapter 1):

$$\int \rho_1(\mathbf{r})\, \phi^{(2)}(\mathbf{r}) d^3r = \int \rho_2(\mathbf{r})\, \phi^{(1)}(\mathbf{r}) d^3r, \tag{2.21}$$

where $\phi^{(1)}(\mathbf{r})$ and $\phi^{(2)}(\mathbf{r})$ are the potential distributions induced, respectively, by two electric charge distributions, $\rho_1(\mathbf{r})$ and $\rho_2(\mathbf{r})$. In our current case, each of these integrals is limited to the volume (or, more exactly, the surface) of the corresponding conductor, where each potential is constant, and may be taken out of the integral. As a result, Eq. (2.21) is reduced to

$$Q_1 \phi^{(2)}(\mathbf{r}_1) = Q_1 \phi^{(1)}(\mathbf{r}_2). \tag{2.22}$$

In terms of Eq. (2.19), $\phi^{(2)}(\mathbf{r}_1)$ is just $p_{12}Q_2$, while $\phi^{(1)}(\mathbf{r}_2)$ equals $p_{21}Q_1$. Plugging these expressions into Eq. (2.22), and cancelling the products Q_1Q_2, we arrive at Eq. (2.20).

Hence the 2×2 matrix of coefficients $p_{jj'}$ (called the *reciprocal capacitance matrix*) is always symmetric, and using the natural notation $p_{11} \equiv p_1$, $p_{22} \equiv p_2$, $p_{12} = p_{21} \equiv p$, we may write it in a simple form

$$\begin{pmatrix} p_1 & p \\ p & p_2 \end{pmatrix}. \tag{2.23}$$

Plugging the relation (2.19), in this new notation, into Eq. (1.61), we see that the full electrostatic energy of the system may be expressed by a quadratic form:

$$U = \frac{p_1}{2} Q_1^2 + p Q_1 Q_2 + \frac{p_2}{2} Q_2^2. \tag{2.24}$$

It is evident that the middle term on the right-hand side of this equation describes the electrostatic coupling of the conductors. (Without it, the energy would be just a sum of two independent electrostatic energies of conductors 1 and 2.[13]) Still, even with this simplification, Eqs. (2.19) and (2.20) show that in the general case of arbitrary charges Q_1 and Q_2, the system of two conductors should be characterized by three coefficients, rather than just one coefficient ('the capacitance'). This is why we may attribute a certain single capacitance to the system only in some particular cases.

[13] This is why systems with $p \ll p_1, p_2$ are called *weakly coupled*, and may be analyzed using approximate methods—see, e.g. figure 2.4 and its discussion below.

For practice, the most important of them is when the system as the whole is electrically neutral: $Q_1 = -Q_2 \equiv Q$. In this case the most important function of Q is the difference of the conductors' potentials, called the *voltage*[14]:

$$V \equiv \phi_1 - \phi_2, \tag{2.25}$$

For that function, the subtraction of the two Eqs. (2.19) gives

$$V = \frac{Q}{C}, \quad \text{with } C \equiv \frac{1}{p_1 + p_2 - 2p}, \tag{2.26}$$

where the coefficient C is called the *mutual capacitance* between the conductors—or, again, just 'capacitance', if the term's meaning is absolutely clear from the context. The same coefficient describes the electrostatic energy of the system. Indeed, plugging Eqs. (2.19) and (2.20) into Eq. (2.24), we see that both forms of Eq. (2.15) are reproduced if ϕ is replaced with V, Q_1 with Q, and with C meaning the mutual capacitance:

$$U = \frac{Q^2}{2C} = \frac{C}{2}V^2. \tag{2.27}$$

The best known system for which the mutual capacitance C may be readily calculated is the *plane* (or 'parallel-plate') *capacitor*, a system of two conductors separated with a narrow plane gap of thickness d and area $A \sim a^2 \gg d^2$—see figure 2.3.

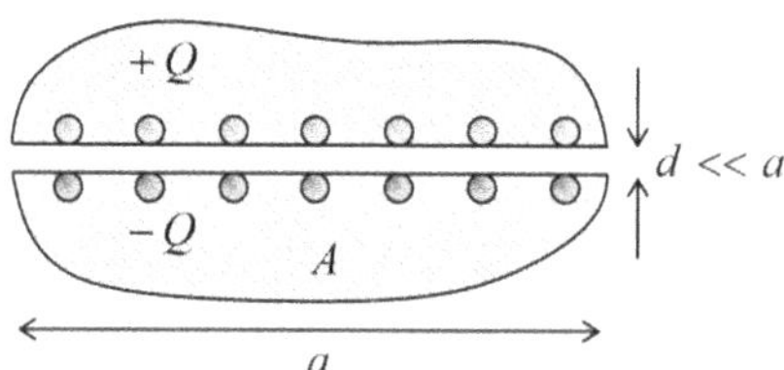

Figure 2.3. A schematic representation of the plane capacitor.

Since the surface charges, that contribute to the opposite charges $\pm Q$ of the conductors of this system, attract each other, in the limit $d \ll a$ they sit entirely on the opposite surfaces limiting the gap, so there is virtually no electric field outside of the gap, while (according to the discussion in section 2.1) inside the gap it is normal to the surfaces. According to Eq. (2.3), the magnitude of this field is $E = \sigma/\varepsilon_0$. Integrating this field across thickness d of the narrow gap, we obtain $V \equiv \phi_1 - \phi_2 = Ed = \sigma d/\varepsilon_0$, so that $\sigma = \varepsilon_0 V/d$. But due the constancy of the potential of each

[14] A word of caution: in condensed matter physics and electrical engineering, voltage is frequently defined as the difference of *electrochemical* rather than *electrostatic* potentials. These two notions coincide if the conductors have equal *workfunctions*—for example, if they are made of the same material. In this course this condition will be implied, and the difference between the two voltages ignored—to be discussed in detail in *Part SM* section 6.4.

electrode, V should not depend on the position in the gap area. As a result, σ should be also constant over all the gap area A, and hence $Q = \sigma A = \varepsilon_0 V/d$. Thus, we may write $V = Q/C$, with

$$C = \frac{\varepsilon_0}{d} A. \tag{2.28}$$

Let me offer a few comments on this well-known formula. First, it is valid even if the gap is not quite planar, for example if it gently curves on a scale much larger than d. Second, Eq. (2.28), valid if $A \sim a^2$ is much larger than d^2, ignores the electric field deviations from uniformity[15] at distances $\sim d$ near the gap edges. Finally, the same condition ($A \gg d^2$) assures that C is much larger than the self-capacitance $C_j \sim \varepsilon_0 a$ of each conductor—see Eq. (2.18). The opportunities open by this fact for electronic engineering and experimental physics practice are rather astonishing. For example, a very realistic 3 nm layer of high-quality aluminum oxide (which may provide a nearly perfect electric insulation between two thin conducting films) with an area of 0.1 m^2 (which is a typical area of silicon wafers used in semiconductor industry) provides $C \sim 1$ mF,[16] larger than the self-capacitance of the whole planet Earth!

In the case shown in figure 2.3, the electrostatic coupling of the two conductors is evidently strong. As an opposite example of a weakly coupled system, let us consider two conducting spheres of the same radius R, separated by a much larger distance d (figure 2.4).

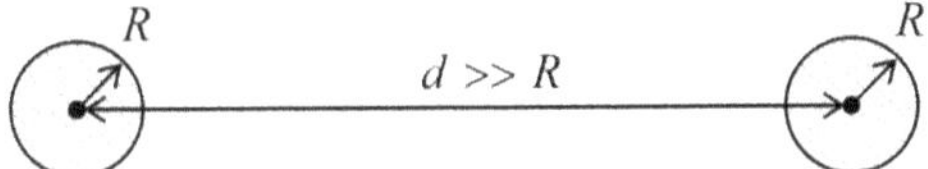

Figure 2.4. A system of two well separated, similar conducting spheres.

In this case the diagonal components of the matrix (2.23) may be approximately found from Eq. (2.16), i.e. by neglecting the coupling altogether:

$$p_1 = p_2 \approx \frac{1}{4\pi\varepsilon_0 R}. \tag{2.29}$$

Now, if we had just one sphere (say, number 1), the electric potential at distance d from its center would be given by Eq. (2.16): $\phi = Q_1/4\pi\varepsilon_0 d$. If we move into this point a small ($R \ll d$) sphere without its own charge, we may expect that its potential should not be too far from this result, so that $\phi_2 \approx Q_1/4\pi\varepsilon_0 d$. Comparing this expression with the second of Eq. (2.19) (taken for $Q_2 = 0$), we obtain

$$p \approx \frac{1}{4\pi\varepsilon_0 d} \ll p_{1,2}. \tag{2.30}$$

[15] Frequently referred to as 'fringe' fields resulting in an additional 'stray' capacitance $C' \sim \varepsilon_0 a \ll C \sim \varepsilon_0 a \times (a/d)$.

[16] Just as in section 2.1, in order for the estimate to be realistic, I took into account the additional factor κ (for aluminum oxide, close to 10) which should be included into the nominator of Eq. (2.28) to make it applicable to dielectrics—see chapter 3 below.

From here and Eq. (2.26) the mutual capacitance

$$C \approx \frac{1}{p_1 + p_2} \approx 2\pi\varepsilon_0 R. \tag{2.31}$$

We see that (somewhat counter-intuitively), in this case C does not depend substantially on the distance between the spheres, i.e. does *not* describe their electrostatic coupling. The off-diagonal coefficients of the reciprocal capacitance matrix (2.20) play this role much better—see Eq. (2.30).

Now let us consider the case when only one conductor of the two is charged, for example $Q_1 \equiv Q$, while $Q_2 = 0$. Then Eqs. (2.19) and (2.20) yield

$$\phi_1 = p_1 Q_1. \tag{2.32}$$

Now, we may follow Eq. (2.13) and define $C_1 \equiv 1/p_1$ (and $C_2 \equiv 1/p_2$), just to see that such *partial capacitances* of the conductors of the system differ from its mutual capacitance C—cf Eq. (2.26). For example, in the case shown in figure 2.4, $C_1 = C_2 \approx 4\pi\varepsilon_0 R \approx 2C$.

Finally, let us consider one more frequent case when one of the conductors carries a certain charge (say, $Q_1 = Q$) but the potential of its counterpart is a sustained constant, say $\phi_2 = 0$.[17] (This condition is particularly easy to implement if the second conductor is much larger that the first one. Indeed, as the estimate (2.18) shows, in this case it would take a much larger charge Q_2 to make the potential ϕ_2 comparable with ϕ_1.) In this case the second of Eq. (2.19), with an account of Eq. (2.20), yields $Q_2 = -(p/p_2)Q_1$. Plugging this relation into the first of those equations, we obtain

$$Q_1 = C_1^{\text{ef}}\phi_1, \quad \text{with } C_1^{\text{ef}} \equiv \left(p_1 - \frac{p^2}{p_2}\right)^{-1} \equiv \frac{p_2}{p_1 p_2 - p^2}. \tag{2.33}$$

Thus, this *effective capacitance* of the first conductor, is generally different both from both its partial capacitance C_1 and the mutual capacitance C of the system, emphasizing again how accurate one should be using this term.

Note also that none of these capacitances is equal to any element of the matrix reciprocal to the matrix (2.23):

$$\begin{pmatrix} p_1 & p \\ p & p_2 \end{pmatrix}^{-1} = \frac{1}{p^2 - p_1 p_2}\begin{pmatrix} -p_2 & p \\ p & -p_1 \end{pmatrix}. \tag{2.34}$$

For this reason, the last matrix (sometimes the called the *physical capacitance matrix*), which expresses the vector of conductor charges via the vector of their potentials, is less useful for applications than the reciprocal capacitance matrix (2.23). The same conclusion is valid for multi-conductor systems, which are most

[17] In electrical engineering, such a constant-potential conductor is called the *ground*. This term stems from the fact that in many cases the Earth surface may be considered a good electric ground, because its potential is unaffected by laboratory-scale electric charges.

conveniently characterized by an evident generalization of Eq. (2.19). Indeed, in this case even the mutual capacitance between two selected conductors may depend on the electrostatics conditions of other components of the system.

Logically, at this point I would need to discuss the particular, but practically very important, case when the regions where the electric field between each pair of conductors is most significant, do not overlap—such as in the example shown in figure 2.5a. In this case the system's properties may be discussed using the equivalent circuit language, representing each such region as a *lumped* (localized) *capacitor*, with a certain mutual capacitance C, and the whole system as some connection of these capacitors by conducting 'wires', whose length and geometry are not important—see figure 2.5b.

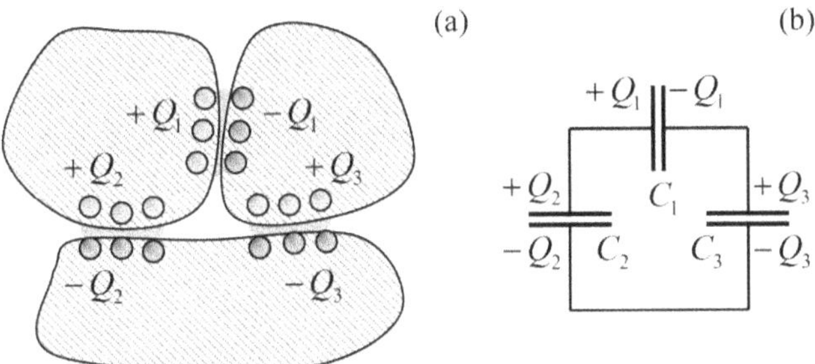

Figure 2.5. (a) A simple system of conductors, with three well-localized regions of the electric field and surface charge concentration, and (b) its representation with an equivalent circuit of three lumped capacitors.

Since the analysis such equivalent circuits is covered in typical introductory physics courses, I will save time by skipping their discussion. However, since such circuits are very frequently encountered in physical experiments and electrical engineering practice, I would urge the reader to self-test his or her understanding of this topic by solving a couple of problems given at the end of this chapter[18], and if their solution present any difficulty, review the corresponding section in an undergraduate textbook.

2.3 The simplest boundary problems

In the general case when the electric field distribution in the free space between the conductors cannot be easily found from the Gauss law or by any other special methods, the best approach is to try to solve the differential Laplace equation (1.42), with boundary conditions (2.1*b*):

$$\nabla^2\phi = 0, \quad \phi|_{S_k} = \phi_k, \tag{2.35}$$

where S_k is the surface of the kth conductor of the system. After such a *boundary problem* has been solved, i.e. the spatial distribution $\phi(\mathbf{r})$ has been found at all points

[18] These problems have been selected to emphasize the fact that not every circuit may be reduced to the simplest connections of the capacitors in parallel and/or in series.

outside the conductors, it is straightforward to use Eq. (2.3) to find the surface charge density, and finally the total charge

$$Q_k = \oint_{S_k} \sigma \, d^2r \tag{2.36}$$

of each conductor, and hence any component of the reciprocal capacitance matrix. As an illustration, let us implement this program for three very simple problems.

(i) *Plane capacitor* (figure 2.3). In this case, the easiest way to solve the Laplace equation is to use the linear (Cartesian) coordinates with one coordinate axis (say, z), normal to the conductor surfaces—see figure 2.6. In these coordinates, the Laplace operator is just the sum of three second derivatives[19]. It is evident that due to problem's translational symmetry in the $\{x, y\}$ plane, deep inside the gap (i.e. at the lateral distance from the edges much larger than d) the electrostatic potential may only depend on the coordinate perpendicular to the gap surfaces: $\phi(\mathbf{r}) = \phi(z)$. For such a function, the derivatives over x and y vanish, and the boundary problem (2.35) is reduced to a very simple ordinary differential equation

$$\frac{d^2\phi}{dz^2}(z) = 0, \tag{2.37}$$

with boundary conditions

$$\phi(0) = 0, \quad \phi(d) = V. \tag{2.38}$$

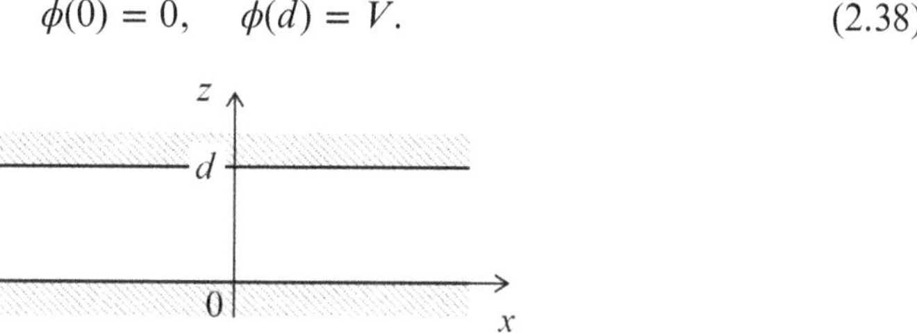

Figure 2.6. The plane capacitor as the system for the simplest illustration of the boundary problem (2.35) and its solution.

(For the sake of notation simplicity, I have used the discretion of adding a constant to the potential to make one of the potentials vanish, and also the definition (2.25) of the voltage V.) The general solution of Eq. (2.37) is a linear function: $\phi(z) = c_1 z + c_2$, whose constant coefficients $c_{1,2}$ may be readily found from the boundary conditions (2.38). The final solution is

$$\phi = V\frac{z}{d}. \tag{2.39}$$

[19] See, e.g. Eq. (A.56).

From here the only non-vanishing component of the electric field is

$$E_z = -\frac{d\phi}{dz} = -\frac{V}{d}, \tag{2.40}$$

and the surface charge of the capacitor plates

$$\sigma = \varepsilon_0 E_n = \mp\varepsilon_0 E_z = \pm\varepsilon_0 \frac{V}{d}, \tag{2.41}$$

where the upper and lower sign correspond to the upper and lower plate, respectively. Since σ does not depend on coordinates x and y, we can obtain the full charges $Q_1 = -Q_2 \equiv Q$ of the surfaces by its multiplication by the gap area A, giving us again the already known result (2.28) for the mutual capacitance $C \equiv Q/V$. I believe that this calculation, though very easy, may serve as a good introduction to the boundary problem solution philosophy.

(ii) *Coaxial-cable capacitor*. The *coaxial cable* is a system of two round cylindrical, coaxial conductors, with the cross-section as shown in figure 2.7.

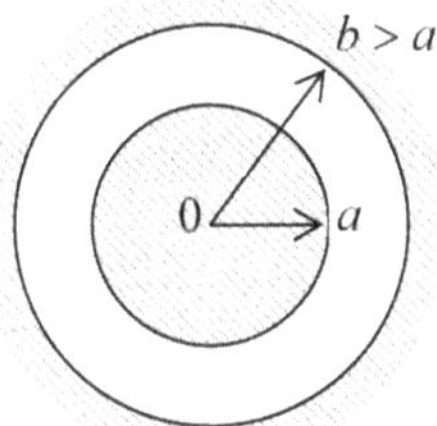

Figure 2.7. The cross-section of a coaxial capacitor.

Evidently, in this case the cylindrical coordinates $\{\rho, \varphi, z\}$, with the axis z coinciding with the common axis of the cylinders, are most convenient. Due to the axial symmetry of the problem, in these coordinates $\mathbf{E}(\mathbf{r}) = \mathbf{n}_\rho E(\rho)$, $\phi(\mathbf{r}) = \phi(\rho)$, so that in the general expression for the Laplace operator[20] we can take $\partial/\partial\varphi = \partial/\partial z = 0$. As a result, only the first (radial) term of the operator survives, and the boundary problem (2.35) takes the form

$$\frac{1}{\rho}\frac{d}{d\rho}\left(\rho\frac{d\phi}{d\rho}\right) = 0, \quad \phi(a) = V, \quad \phi(b) = 0. \tag{2.42}$$

The sequential integration of this ordinary differential equation is elementary (and similar to that of the Poisson equation in spherical coordinates, carried out in section 1.3), giving

$$\rho\frac{d\phi}{d\rho} = c_1, \quad \phi(\rho) = c_1 \int_a^\rho \frac{d\rho\rho}{\rho'} + c_2 = c_1 \ln\frac{\rho}{a} + c_2. \tag{2.43}$$

[20] See, e.g. Eq. (A.61).

The constants $c_{1,2}$ may be found using boundary conditions (2.42):

$$V = c_2, \qquad 0 = c_1 \ln \frac{b}{a} + c_2, \tag{2.44}$$

giving $c_1 = -V/\ln(b/a)$, so that solution (2.43) takes the following form:

$$\phi(\rho) = V\left(1 - \frac{\ln(\rho/a)}{\ln(b/a)}\right). \tag{2.45}$$

Next, for our axial symmetry the general expression for the gradient of a function is reduced to its radial derivative, so that

$$E(\rho) \equiv -\frac{d\phi(\rho)}{d\rho} = \frac{V}{\rho \ln(b/a)}. \tag{2.46}$$

This expression, plugged into Eq. (2.2), allows us to find the density of the conductors' surface charges. For example, for the inner electrode

$$\sigma_a = \varepsilon_0 E_a = \frac{\varepsilon_0 V}{a \ln(b/a)}, \tag{2.47}$$

so that its full charge (per unit length of the system) is

$$\frac{Q}{l} = 2\pi a \sigma_a = \frac{2\pi\varepsilon_0 V}{\ln(b/a)}. \tag{2.48}$$

(It is straightforward to check that the charge of the outer electrode is equal and opposite.) Hence, by the definition of the mutual capacitance, its value per unit length is

$$\frac{C}{l} \equiv \frac{Q}{lV} = \frac{2\pi\varepsilon_0}{\ln(b/a)}. \tag{2.49}$$

This expression shows that the total capacitance C is proportional to the systems length l (if $l \gg a,b$), while being only logarithmically dependent on the dimensions of its cross-section. Since the logarithm of a very large argument is an extremely slow function (sometimes called a *quasi-constant*), if the external conductor is made large ($b \gg a$) the capacitance diverges, but very weakly. Such a logarithmic divergence may be cut by any miniscule additional effect, for example by the finite length l of the system. This allows one to obtain a crude but very useful estimate of the self-capacitance of a *single* round wire of radius a:

$$C \approx \frac{2\pi\varepsilon_0 l}{\ln(l/a)}, \qquad \text{for } l \gg a. \tag{2.50}$$

On the other hand, if the gap between the conductors is very narrow, $b = a + d$ with $d \ll a$, then $\ln(b/a) = \ln(1 + d/a)$ may be approximated as d/a, and Eq. (2.49) is reduced to $C \approx 2\pi\varepsilon_0 aL/d$, i.e. to Eq. (2.28) for the plane capacitor, with the appropriate area $A = 2\pi al$.

(iii) *Spherical capacitor*. This is a system of two conductors, with the *central* cross-section similar to that of the coaxial cable (figure 2.7), but now with the spherical rather than axial symmetry. This symmetry implies that we would be better off using spherical coordinates, so that potential ϕ depends only on one of them, the distance r from the common center of the conductors: $\phi(\mathbf{r}) = \phi(r)$. As we already know from section 1.3, in this case the general expression for the Laplace operator is reduced to its first (radial) term, so that the Laplace equation takes the simple form (1.47). Moreover, we have already found the general solution to this equation—see Eq. (1.50):

$$\varphi(r) = \frac{c_1}{r} + c_2. \tag{2.51}$$

Now acting exactly as above, i.e. determining the constant c_1 from the boundary conditions $\phi(a) = V$, $\phi(b) = 0$, we obtain

$$V = c_1\left(\frac{1}{a} - \frac{1}{b}\right), \quad \text{so that } \phi(r) = \frac{V}{r}\left(\frac{1}{a} - \frac{1}{b}\right)^{-1} + c_2. \tag{2.52}$$

Next, we can use the spherical symmetry to find electric field, $\mathbf{E}(\mathbf{r}) = \mathbf{n}_r E(r)$, with

$$E(r) = -\frac{d\phi}{dr} = \frac{V}{r^2}\left(\frac{1}{a} - \frac{1}{b}\right)^{-1}, \tag{2.53}$$

and hence its values on conductors' surfaces, and then the surface charge density σ from Eq. (2.3). For example, for the inner conductor's surface,

$$\sigma_a = \varepsilon_0 E(a) = \varepsilon_0 \frac{V}{a^2}\left(\frac{1}{a} - \frac{1}{b}\right)^{-1}, \tag{2.54}$$

so that, finally, for the full charge of that conductor we obtain the following result:

$$Q = 4\pi a^2 \sigma = 4\pi\varepsilon_0\left(\frac{1}{a} - \frac{1}{b}\right)^{-1} V. \tag{2.55}$$

(Again, the charge of the outer conductor is equal and opposite.) Now we can use the definition (2.26) of the mutual capacitance to obtain the final result

$$C \equiv \frac{Q}{V} = 4\pi\varepsilon_0\left(\frac{1}{a} - \frac{1}{b}\right)^{-1} = 4\pi\varepsilon_0 \frac{ab}{b - a}. \tag{2.56}$$

For $b \gg a$, this result coincides with Eq. (2.17) for the self-capacitance of the inner conductor. On the other hand, if the gap between two conductors is narrow, $d \equiv b - a \ll a$,

$$C = 4\pi\varepsilon_0 \frac{a(a+d)}{d} \approx 4\pi\varepsilon_0 \frac{a^2}{d}, \tag{2.57}$$

i.e. the capacitance approaches that of the planar capacitor of area $A = 4\pi a^2$—as it should.

All this seems (and is) very straightforward, but let us contemplate what the reason was for such easy successes. In each of the cases (i)–(iii) we have managed to find such coordinates that both the Laplace equation and the boundary conditions involve only one of them. The necessary condition for the former fact is that the coordinates are *orthogonal*. This means that three vector components of the local differential $d\mathbf{r}$, due to small variations of the new coordinates (say, dr, $d\theta$, and $d\varphi$ for the spherical coordinates), are mutually perpendicular.

2.4 Using other orthogonal coordinates

Since the cylindrical and spherical coordinates used above are only the simplest examples of the *curvilinear orthogonal* (or just 'orthogonal') coordinates, this methodology may be extended to other coordinate systems of this type. As an example, let us calculate the self-capacitance of a thin, round conducting disk. The cylindrical or spherical coordinates would not give too much help here because, although they have the appropriate axial symmetry about axis z, they would make the boundary condition on the disk too complex—involving two coordinates, either ρ and z, or r and θ. The help comes from noting that the flat disk, i.e. the area $z = 0$, $r < R$, may be thought of as the limiting case of an *axially symmetric ellipsoid*—the result of rotation of the usual ellipse about one of its axes—in our case, the symmetry axis of the disk (in figure 2.8, axis z)[21].

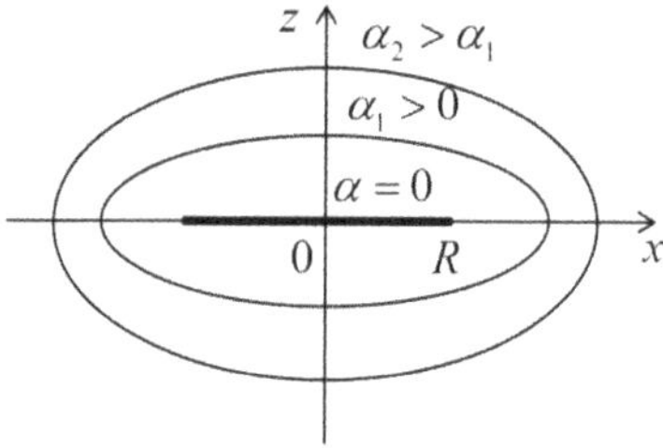

Figure 2.8. Solving the disk's capacitance problem. (The cross-section of the system by the vertical plane $y = 0$.)

Analytically, such an ellipsoid may be described by the following equation:

[21] Alternative names for this surface are the 'degenerate ellipsoid', 'ellipsoid of rotation', and 'spheroid'.

$$\frac{x^2 + y^2}{a^2} + \frac{z^2}{b^2} = 1, \tag{2.58}$$

where a and b are the so-called *major semi-axes* whose ratio determines the ellipse *eccentricity* (the degree of squeezing). For our problem, we will only need *oblate* ellipsoids with $a \geqslant b$; according to Eq. (2.58), they may be represented as surfaces of constant α in the *degenerate ellipsoidal* (or 'spheroidal') *coordinates* $\{\alpha, \beta, \varphi\}$, which are related to the Cartesian coordinates as follows:

$$\begin{aligned} x &= R \cosh \alpha \sin \beta \cos \varphi, \\ y &= R \cosh \alpha \sin \beta \sin \varphi, \\ z &= R \sinh \alpha \cos \beta. \end{aligned} \tag{2.59}$$

Such ellipsoidal coordinates are the evident generalization of the spherical coordinates, which correspond to the limit $\alpha \gg 1$ (i.e. $r \gg R$). In the opposite limit, the surface of constant $\alpha = 0$ describes our thin disk of radius R, with the coordinate β describing the distance $\rho \equiv (x^2 + y^2)^{1/2} = R \sin \beta$ of its point from the axis z. It is almost evident (and easy to prove) that the curvilinear coordinates (2.59) are also orthogonal, so that the Laplace operator may be expressed as a sum of three independent terms:

$$\begin{aligned} \nabla^2 = & \frac{1}{R^2(\cosh^2 \alpha - \sin^2 \beta)} \\ & \times \left[\frac{1}{\cosh \alpha} \frac{\partial}{\partial \alpha} \left(\cosh \alpha \frac{\partial}{\partial \alpha} \right) + \frac{1}{\sin \beta} \frac{\partial}{\partial \beta} \left(\sin \beta \frac{\partial}{\partial \beta} \right) + \left(\frac{1}{\sin^2 \beta} - \frac{1}{\cosh^2 \alpha} \right) \frac{\partial^2}{\partial \varphi^2} \right]. \end{aligned} \tag{2.60}$$

Although this expression may look a bit intimidating, let us notice that in our current problem, the boundary conditions depend only on coordinate α:[22]

$$\phi|_{\alpha=0} = V, \quad \phi|_{\alpha=\infty} = 0. \tag{2.61}$$

Hence there is every reason to believe that the electrostatic potential in all space is a function of α alone; in other words, that all ellipsoids $\alpha = \text{const}$ are equipotential surfaces. Indeed, acting on such a function $\phi(\alpha)$ by the Laplace operator (2.60), we see that the two last terms in the square brackets vanish, and the Laplace Eq. (2.35) is reduced to a simple ordinary differential equation

$$\frac{d}{d\alpha}\left[\cosh \alpha \frac{d\phi}{d\alpha} \right] = 0. \tag{2.62}$$

Integrating it twice, just as we did in the previous problems, we obtain

[22] I have called disk's potential V, to distinguish it from the potential ϕ at an arbitrary point of space.

$$\phi(\alpha) = c_1 \int \frac{d\alpha}{\cosh \alpha}. \tag{2.63}$$

This integral may be readily worked out, using the substitution $\xi \equiv \sinh \alpha$ (with $d\xi \equiv \cosh \alpha \, d\alpha$, $\cosh^2 \alpha = 1 + \sinh^2 \alpha = 1 + \xi^2$):

$$\phi(\alpha) = c_1 \int_0^{\sinh \alpha} \frac{d\xi}{1 + \xi^2} + c_2 = c_1 \tan^{-1}(\sinh \alpha) + c_2. \tag{2.64}$$

The integration constants $c_{1,2}$ are again simply found from boundary conditions, in this case Eqs. (2.61), and we arrive at the following final expression for the electrostatic potential:

$$\phi(\alpha) = V\left[1 - \frac{2}{\pi} \tan^{-1}(\sinh \alpha)\right]. \tag{2.65}$$

This solution satisfies both the Laplace equation and the boundary conditions. Mathematicians tell us that the solution of any boundary problem of the type (2.35) is *unique*, so we do not need to look any further.

Now we may use Eq. (2.3) to find the surface density of electric charge, but in the case of thin disk, it is more natural to add up such densities on its top and bottom surfaces at the same distance $\rho = (x^2 + y^2)^{1/2}$ from the disk's center (which are evidently equal, due to the problem symmetry about the plane $z = 0$): $\sigma = 2\varepsilon_0 E_n|_{z=+0}$. According to Eq. (2.65), the electric field on the upper surface is

$$\begin{aligned} E_{n|\alpha=+0} &= -\left.\frac{\partial \phi}{\partial z}\right|_{z=+0} = -\left.\frac{\partial \phi(\alpha)}{\partial (R \sinh \alpha \cos \beta)}\right|_{\alpha=+0} \\ &= \frac{2}{\pi} V \frac{1}{R \cos \beta} = \frac{2}{\pi} V \frac{1}{(R^2 - \rho^2)^{1/2}}, \end{aligned} \tag{2.66}$$

and we see that the charge is distributed along the disk very non-uniformly:

$$\sigma = \frac{4}{\pi} \varepsilon_0 V \frac{1}{(R^2 - \rho^2)^{1/2}}, \tag{2.67}$$

with a singularity at the disk edge. Below we will see that such singularities are very typical for sharp edges of conductors[23]. Fortunately, in our current case the divergence is integrable, giving a finite disk charge:

$$\begin{aligned} Q &= \int_{\substack{\text{disk} \\ \text{surface}}} \sigma \, d^2\rho = \int_0^R \sigma(\rho) 2\pi\rho \, d\rho = \frac{4}{\pi} \varepsilon_0 V 2\pi \int_0^R \frac{\rho \, d\rho}{(R^2 - \rho^2)^{1/2}} \\ &= 4\varepsilon_0 VR \int_0^1 \frac{d\xi}{(1 - \xi)^{1/2}} = 8\varepsilon_0 RV. \end{aligned} \tag{2.68}$$

[23] If you seriously worry about the formal infinity of the charge density at $\rho \to R$, please remember that this mathematical artifact disappears for any non-vanishing disk thickness.

Thus, for disk's self-capacitance we get a very simple result,

$$C = 8\varepsilon_0 R \equiv \frac{2}{\pi}4\pi\varepsilon_0 R, \tag{2.69}$$

a factor of $2/\pi \approx 0.64$ lower than that for the conducting sphere of the same equal radius, but still complying with the general estimate (2.18).

Can we always find such a 'good' system of orthogonal coordinates? Unfortunately, the answer is *no*, even for highly symmetric geometries. This is why the practical value of this approach is limited, and other, more general methods of boundary problem solution are clearly needed. Before proceeding to their discussion, however, let us note that in the case of 2D problems (i.e. cylindrical geometries[24]), the orthogonal coordinate method gets much help from the following *conformal mapping* approach.

Let us consider a pair of Cartesian coordinates $\{x, y\}$ in the cross-section plane as a complex variable $z = x + iy$,[25] where i is the imaginary unity ($i^2 = -1$), and let $w(z) = u + iv$ be an *analytic complex function* of z.[26] For our current purposes, the most important property of an analytic function is that its real and imaginary parts obey the following *Cauchy–Riemann relations*[27]:

$$\frac{\partial u}{\partial x} = \frac{\partial v}{\partial y}, \quad \frac{\partial v}{\partial x} = -\frac{\partial u}{\partial y}. \tag{2.70}$$

For example, for the function

$$w = z^2 \equiv (x + iy)^2 \equiv (x^2 - y^2) + 2ixy, \tag{2.71}$$

whose real and imaginary parts are

$$u \equiv \operatorname{Re} w = x^2 - y^2, \quad v \equiv \operatorname{Im} w = 2xy, \tag{2.72}$$

we immediately see that $\partial u/\partial x = 2x = \partial v/\partial y$, and $\partial v/\partial x = 2y = -\partial u/\partial y$, in accordance with (2.70).

Let us differentiate the first of Eqs. (2.70) over x again, then change the order of differentiation, and after that use the latter of those equations:

[24] Let me remind the reader that the term *cylindrical* describes any surface formed by a translation, along a straight line, of an *arbitrary* curve, and hence more general than the usual circular cylinder. (In this terminology, for example, a prism is also a particular form of cylinder, formed by a translation of a polygon.)

[25] The complex variable z should not be confused with the (real) third spatial coordinate z! We are considering 2D problems now, with the potential independent of z.

[26] The analytic (or 'holomorphic') function may be defined as the one that may be expanded into the complex Taylor series, i.e. is infinitely differentiable in the given point. (Almost all 'regular' functions, such as z^n, $z^{1/n}$, $\exp z$, $\ln z$, etc, and their combinations are analytic at all z, maybe apart from certain special points.) If the reader needs to brush up his or her background on this subject, I can recommend a popular (and very inexpensive:-) textbook [3]

[27] These relations may be used, in particular, to prove the Cauchy integral theorem—see, e.g. Eq. (A.91).

$$\frac{\partial^2 u}{\partial x^2} = \frac{\partial}{\partial x}\frac{\partial u}{\partial x} = \frac{\partial}{\partial x}\frac{\partial v}{\partial y} = \frac{\partial}{\partial y}\frac{\partial v}{\partial x} = -\frac{\partial}{\partial y}\frac{\partial u}{\partial y} = -\frac{\partial^2 u}{\partial y^2}, \tag{2.73}$$

and similarly for v. This means that the sum of the second-order partial derivatives of each of the real functions $u(x,y)$ and $v(x,y)$ is zero, i.e. that both functions obey the 2D Laplace equation. This mathematical fact opens a nice way of solving problems of electrostatics for (relatively simple) 2D geometries. Imagine that for a particular boundary problem we have found a function $w(z)$ for which either $u(x, y)$ or $v(x, y)$ is constant on all electrode surfaces. Then all lines of constant u (or v) represent equipotential surfaces, i.e. the problem of the potential distribution has been essentially solved.

As a simple example, consider a practically important problem: the *quadrupole electrostatic lens*—a system of four cylindrical electrodes with hyperbolic cross-sections, whose boundaries obey the following relations:

$$x^2 - y^2 = \begin{cases} +a^2, & \text{for the left and right electrodes,} \\ -a^2, & \text{for the top and bottom electrodes,} \end{cases} \tag{2.74}$$

voltage-biased as shown in figure 2.9a. Comparing these relations with Eq. (2.72), we see that each electrode surface corresponds to a constant value of the function u: $u = \pm a^2$. Moreover, the potentials of both surfaces with $u = +a^2$ are equal to $+V/2$, while those with $u = -a^2$ are equal to $-V/2$. Hence, we may conjecture that the electrostatic potential at each point is a function of u alone; moreover, a simple linear function,

$$\phi = c_1 u + c_2 = c_1(x^2 - y^2) + c_2, \tag{2.75}$$

is a valid (and hence the unique) solution of our boundary problem. Indeed, it does satisfy the Laplace equation, while its constants $c_{1,2}$ may be selected in a way to satisfy all the boundary conditions shown in figure 2.9a:

$$\phi = \frac{V}{2}\frac{x^2 - y^2}{a^2}. \tag{2.76}$$

so that the boundary problem has been solved.

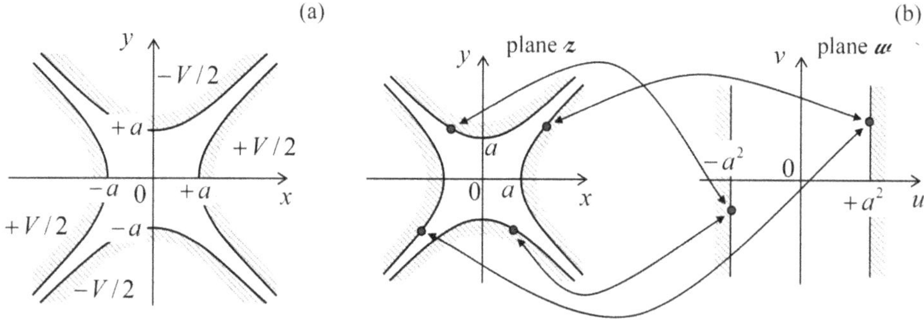

Figure 2.9. (a) The quadrupole electrostatic lens's cross-section and (b) its conformal mapping.

According to Eq. (2.76), all equipotential surfaces are hyperbolic cylinders, similar to those of the electrode surfaces. What remains is to find the electric field at an arbitrary point inside the system:

$$E_x = -\frac{\partial \phi}{\partial x} = -V\frac{x}{a^2}, \quad E_y = -\frac{\partial \phi}{\partial y} = V\frac{y}{a^2}. \tag{2.77}$$

These formulas show that if charged particles (e.g. electrons in an electron optics system) are launched to fly ballistically through the lens, along axis z, they experience a force pushing them toward the symmetry axis and proportional to particle's deviation from the axis (and thus equivalent in action to an optical lens with a positive refraction power) in one direction, and a force pushing them out (negative refractive power) in the perpendicular direction. One can show that letting charged particles fly through several such lenses, with alternating voltage polarities, in series enables beam focusing[28].

Hence, we have reduced the 2D Laplace boundary problem to that of finding the proper analytic function $w(z)$. This task may be also understood as that of finding a *conformal map*, i.e. a correspondence between components of any point pair, $\{x, y\}$ and $\{u, v\}$, residing, respectively, on the initial Cartesian plane z and the plane w of the new variables. For example, Eq. (2.71) maps the real electrode configuration onto a plane capacitor of an infinite area (figure 2.9b), and the simplicity of Eq. (2.75) is due to the fact that for the latter system the equipotential surfaces are just parallel planes.

For more complex geometries, the suitable analytic function $w(z)$ may be hard to find. However, for conductors with piece-linear cross-section boundaries, substantial help may be obtained from the following *Schwarz–Christoffel integral*

$$w(z) = \text{const} \times \int \frac{dz}{(z - x_1)^{k_1}(z - x_2)^{k_2} \ldots (z - x_{N-1})^{k_{N-1}}}, \tag{2.78}$$

that provides the conformal mapping of the interior of an arbitrary N-sided polygon on the plane $w = u + iv$, onto the upper-half ($y > 0$) of the plane $z = x + iy$. In Eq. (2.78), x_j ($j = 1, 2, N - 1$) are the points of axis $y = 0$ (i.e. of the boundary of the mapped region on plane z) to which the corresponding polygon vertices are mapped, while k_j are the exterior angles at the polygon vertices, measured in the units of π, with $-1 \leqslant k_j \leqslant +1$—see figure 2.10[29]. Of the points x_j, two may be selected arbitrarily (because their effects may be compensated by the multiplicative constant in Eq. (2.78), and the additive constant of integration), while all the others have to be adjusted to provide the correct mapping.

---

[28] See, e.g. the textbook [4] or the review collection [5], in particular the review by K-J Hanszen and R Lauer, pp 251–307.

[29] The integral (2.78) includes only (N – 1) rather than N poles, because a polygon's shape is completely determined by (N – 1) positions w_j of its vertices and (N – 1) angles πk_j. In particular, since the algebraic sum of all external angles of a polygon equals π, the last angle parameter $k_j = k_N$ is uniquely determined by the set of the previous ones.

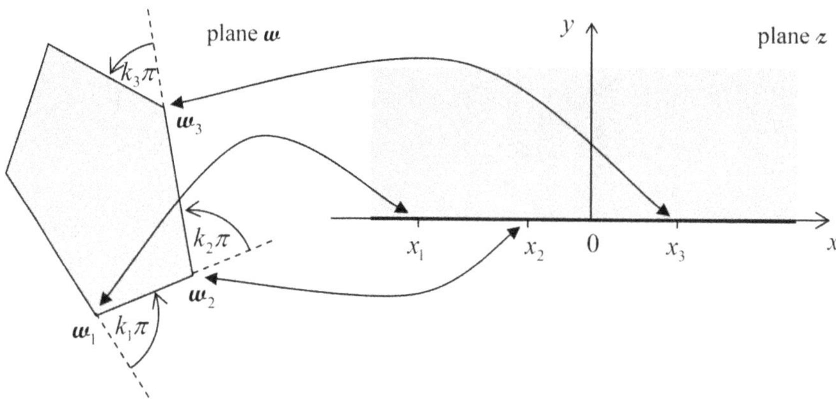

Figure 2.10. The Schwartz–Christoffel mapping of a polygon's interior onto the upper half-plane.

In the general case, the complex integral (2.78) may be difficult to tackle. However, in some important cases, in particular those with right angles ($k_j = \pm ½$) and/or with some points w_j at infinity, the integrals may be readily worked out, giving explicit analytical expressions for the mapping functions $w(z)$. For example, let us consider a semi-infinite strip, defined by restrictions $-1 \leqslant u \leqslant +1$ and $0 \leqslant v$, on the plane w—see the left-hand panel of figure 2.11.

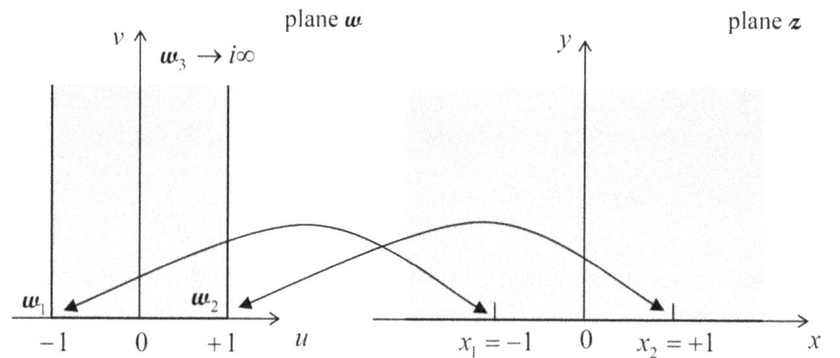

Figure 2.11. A semi-infinite strip mapped onto the upper half-plane.

The strip may be considered as a polygon, with one vertex at the infinitely distant vertical point $w_3 = 0 + i\infty$. Let us map the polygon on the upper half of plane z, shown in the right-hand panel of figure 2.11, with the vertex $w_1 = -1 + i0$ mapped onto the point $x_1 = -1$, $y_1 = 0$, and the vertex $w_2 = +1 + i0$ mapped onto the point $x_2 = +1$, $y_2 = 0$. Since both external angles are equal to $+\pi/2$, and hence $k_1 = k_2 = +½$, Eq. (2.78) is reduced to

$$w(z) = \text{const} \times \int \frac{dz}{(z+1)^{1/2}(z-1)^{1/2}} \equiv \text{const} \times \int \frac{dz}{(z^2-1)^{1/2}} \equiv \text{const} \times i \int \frac{dz}{(1-z^2)^{1/2}}. \tag{2.79}$$

This complex integral may be worked out, just as for real z, by the substitution $z = \sin\xi$, giving

$$w(z) = \text{const}' \times \int^{\sin^{-1} z} d\xi = c_1 \sin^{-1} z + c_2. \tag{2.80}$$

Determining the constants $c_{1,2}$ from the required mapping, i.e. from the equations $w(-1 + i0) = -1 + i0$ and $w(+1 + i0) = +1 + i0$ (see the arrows in figure 2.11), we finally obtain

$$w(z) = \frac{2}{\pi} \sin^{-1} z, \quad \text{i.e. } z = \sin \frac{\pi w}{2}. \tag{2.81a}$$

Using the well-known expression for the sine of a complex argument[30], we may rewrite this elegant result in either of the following two forms for the real and imaginary components of z and w:

$$\begin{aligned} u &= \frac{2}{\pi} \sin^{-1} \frac{2x}{[(x+1)^2 + y^2]^{1/2} + [(x-1)^2 + y^2]^{1/2}}, \\ v &= \frac{2}{\pi} \cosh^{-1} \frac{[(x+1)^2 + y^2]^{1/2} + [(x-1)^2 + y^2]^{1/2}}{2}, \\ x &= \sin \frac{\pi u}{2} \cosh \frac{\pi v}{2}, \quad y = \cos \frac{\pi u}{2} \sinh \frac{\pi v}{2}. \end{aligned} \tag{2.81b}$$

It is amazing how perfectly the last formula manages to keep $y \equiv 0$ at the different borders of our w-region (figure 2.11): at its side borders ($u = \pm 1$, $0 \leqslant v < \infty$), this is performed by the first multiplier, while at the bottom border ($-1 \leqslant u \leqslant +1$, $v = 0$), the equality is insured by the second multiplier.

This mapping may be used to solve several electrostatics problems with the geometry shown in figure 2.11; probably the most surprising of them is the following one. A straight gap of width $2t$ is cut in a very thin conducting plane, and voltage V is applied between the resulting half-planes—see the bold straight lines in figure 2.12.

Selecting a Cartesian coordinate system with axis z directed along the cut, axis y perpendicular to the plane, and the origin in the middle of the cut (figure 2.12), we can write the boundary conditions of this Laplace problem as

$$\phi = \begin{cases} +V/2, & \text{at } x > t,\ y = 0, \\ -V/2, & \text{at } x < -t,\ y = 0. \end{cases} \tag{2.82}$$

(Due to the problem's symmetry, we may expect that in the middle of the gap, i.e. at $-t < x < +t$ and $y = 0$, the electric field is parallel to the plane and hence $\partial\phi/\partial y = 0$.) The comparison of figures 2.11 and 2.12 shows that if we normalize our coordinates $\{x, y\}$ to t, Eqs. (2.81) provide the conformal mapping of our system on the plane z to the plane capacitor on plane w, with voltage V between two planes $u = \pm 1$. Since

[30] See, e.g. Eq. (A.20).

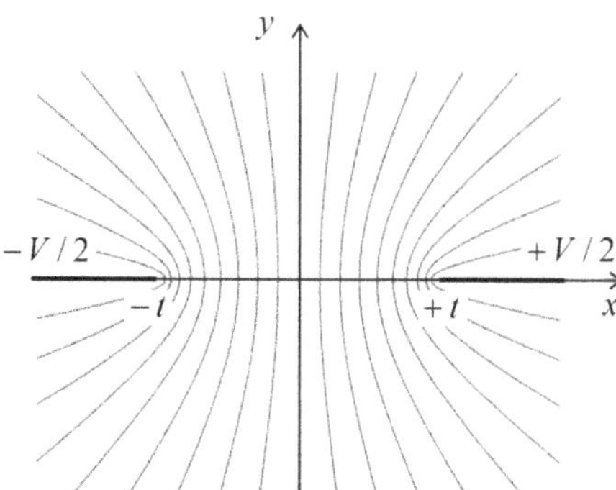

Figure 2.12. The equipotential surfaces of the electric field between two thin conducting semi-planes (or rather their cross-sections by the perpendicular plane z = const).

we already know that in that case $\phi = (V/2)u$, we may immediately use the first of Eq. (2.81*b*) to write the final solution of the problem[31]:

$$\phi = \frac{V}{2}u = \frac{V}{\pi}\sin^{-1}\frac{2x}{[(x+t)^2+y^2]^{1/2}+[(x-t)^2+y^2]^{1/2}}. \qquad (2.83)$$

The thin lines in figure 2.12 show the corresponding equipotential surfaces[32]; it is evident that the electric field concentrates at the gap edges, just as it did at the edge of the thin disk (figure 2.8). Let me leave the remaining calculation of the surface charge distribution and the mutual capacitance between the half-planes (per unit length) for the reader's exercise.

2.5 Variable separation—Cartesian coordinates

The general approach of the methods discussed in the last two sections was to satisfy the Laplace equation by a function of a single variable that also satisfies the boundary conditions. Unfortunately, in many cases this cannot be done (at least, using practicably simple functions). In this case, a very powerful method, called *variable separation*[33], may work, frequently producing 'semi-analytical' results in the form of series (infinite sums) of either elementary or well-studied special functions. Its main idea is to express the solution of the general boundary problem (2.35) as the sum of partial solutions,

[31] This result may be also obtained using the so-called *elliptical* (not ellipsoidal!) coordinates, and by the Green's function method, to be discussed in section 2.10 below.

[32] Another graphical representation of the electric field distribution, by *field lines*, is much less convenient. As a reminder, the field lines are defined as lines to whom the (in our current case, electrostatic) field vectors are tangential at each point. By this definition, the field lines are always normal to the equipotential surfaces, so that it is always straightforward to sketch them from the equipotential surface pattern—such as shown in figure 2.12.

[33] Again, this method was already discussed in *Part CM* section 6.5, and then used in *Part CM* sections 6.6 and 8.4. However, the method is so important that I need to repeat its discussion in this part of my series, for the benefit of the readers who have skipped my Classical Mechanics course for any reason.

$$\phi = \sum_k c_k \phi_k, \tag{2.84}$$

where each function ϕ_k satisfies the Laplace equation, and then select the set of coefficients c_k to satisfy the boundary conditions. More specifically, in the variable separation method the partial solutions ϕ_k are looked for in the form of a product of functions, each depending of just one spatial coordinate.

Let us discuss this approach on the classical example of a rectangular box with conducting walls (figure 2.13), with the same potential (that I will take for zero) at all the walls, but a different potential V fixed at the top lid. Moreover, in order to demonstrate the power of the variable separation method, let us carry out all the calculations for a more general case when the top lead potential is an arbitrary 2D function $V(x, y)$[34].

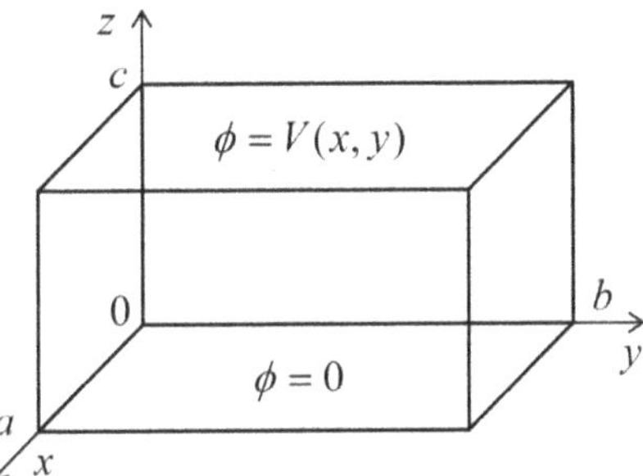

Figure 2.13. The standard playground for the variable separation discussion: a rectangular box with five conducting, grounded walls and a fixed potential distribution $V(x,y)$ on the top lid.

For this geometry, it is natural to use the Cartesian coordinates $\{x, y, z\}$ and hence represent each of the partial solutions in Eq. (2.84) as a product

$$\phi_k = X(x)Y(y)Z(z). \tag{2.85}$$

Plugging it into the Laplace equation expressed in the Cartesian coordinates,

$$\frac{\partial^2 \phi_k}{\partial x^2} + \frac{\partial^2 \phi_k}{\partial y^2} + \frac{\partial^2 \phi_k}{\partial z^2} = 0, \tag{2.86}$$

and dividing the result by the product XYZ, we obtain

$$\frac{1}{X}\frac{d^2X}{dx^2} + \frac{1}{Y}\frac{d^2Y}{dy^2} + \frac{1}{Z}\frac{d^2Z}{dz^2} = 0. \tag{2.87}$$

Here comes the punchline of the variable separation method: since the first term of this sum may depend only on x, the second one only on y, etc, Eq. (2.87) may be satisfied everywhere in the volume only if each of these terms equals a constant. Shortly we will see that for our current problem (figure 2.13), these constant x- and

[34] Such distributions may be implemented in practice using the so-called *mosaic electrodes* consisting of many electrically insulated and individually biased panels.

y-terms have to be negative; hence let us denote these *variable separation constants* as $(-\alpha^2)$ and $(-\beta^2)$, respectively. Now Eq. (2.87) shows that the constant z-term has to be positive; if we denote it as γ^2, we obtain the following relation:

$$\alpha^2 + \beta^2 = \gamma^2. \tag{2.88}$$

Now the variables are separated in the sense that for the functions $X(x)$, $Y(y)$, and $Z(z)$ we have obtained separate ordinary differential equations,

$$\frac{d^2X}{dx^2} + \alpha^2 X = 0, \qquad \frac{d^2Y}{dy^2} + \beta^2 Y = 0, \qquad \frac{d^2Z}{dz^2} - \gamma^2 Z = 0, \tag{2.89}$$

which are related only by Eq. (2.88) for their parameters. Let us start from the equation for function $X(x)$. Its general solution is the sum of functions $\sin\alpha x$ and $\cos\alpha x$, multiplied by arbitrary coefficients. Let us select these coefficients to satisfy our boundary conditions. First, since $\phi \propto X$ should vanish at the back vertical wall of the box (i.e. with the choice of coordinate origin shown in figure 2.13, at $x = 0$ for any y and z), the coefficient at $\cos\alpha x$ should be zero. The remaining coefficient (at $\sin\alpha x$) may be included in the general factor c_k in Eq. (2.84), so that we may take X in the form

$$X = \sin \alpha x. \tag{2.90}$$

This solution satisfies the boundary condition at the opposite wall ($x = a$) only if its argument αa is a multiple of π, i.e. if α is equal to any of the following numbers (commonly called *eigenvalues*)[35]:

$$\alpha_n = \frac{\pi}{a}n, \qquad n = 1, 2, \ldots \tag{2.91}$$

(Terms with negative values of n would not be linearly independent from those with positive n, and may be dropped from the sum (2.84). The value $n = 0$ is formally possible, but would give $X = 0$, i.e. $\phi_k = 0$, at any x, i.e. no contribution to sum (2.84), so it may be dropped as well.) Now we see that we indeed had to take α real, (i.e. α^2 positive); otherwise, instead of the oscillating function (2.90) we would have a sum of two exponential functions, which cannot equal zero in two independent points of axis x.

Since Eq. (2.89) for function $Y(y)$ is similar to that for $X(x)$, and the boundary conditions on the walls perpendicular to axis y ($y = 0$ and $y = b$) are similar to those for x-walls, the absolutely similar reasoning gives

$$Y = \sin \beta y, \quad \beta_m = \frac{\pi}{b}m, \qquad m = 1, 2, \ldots, \tag{2.92}$$

[35] Note that according to Eqs. (2.91) and (2.92), as the spatial dimensions a and b of the system are increased, the distances between the adjacent eigenvalues tend to zero. This fact implies that for spatially infinite, non-periodic systems, the eigenvalue spectra are continuous, so that the sums of the type (2.84) become integrals. A few problems of this type are provided in section 2.9 for the reader's exercise.

where the choice of the integer m is independent of that of n. Now we see that according to Eq. (2.88), the separation constant γ depends on two indices, n and m, so that the relation may be rewritten as

$$\gamma_{nm} = \left[\alpha_n^2 + \beta_m^2\right]^{1/2} = \pi\left[\left(\frac{n}{a}\right)^2 + \left(\frac{m}{b}\right)^2\right]^{1/2}. \tag{2.93}$$

The corresponding solution of the differential equation for Z may be represented as a sum of two exponents $\exp\{\pm\gamma_{nm}z\}$, or alternatively as a linear combination of two hyperbolic functions, $\sinh\gamma_{nm}z$ and $\cosh\gamma_{nm}z$, with arbitrary coefficients. At our choice of coordinate origin, the latter option is preferable, because $\cosh\gamma_{nm}z$ cannot satisfy the zero boundary condition at the bottom lid of the box ($z = 0$). Hence, we may take Z in the form

$$Z = \sinh\gamma_{nm}z, \tag{2.94}$$

which automatically satisfies that condition.

Now it is the right time to combine Eqs. (2.84) and (2.85) in a more explicit form, replacing the temporary index k for full set of possible eigenvalues, in our current case of two integer indices n and m:

$$\phi(x, y, z) = \sum_{n,m=1}^{\infty} c_{nm}\sin\frac{\pi nx}{a}\sin\frac{\pi my}{b}\sinh\gamma_{nm}z, \tag{2.95}$$

where γ_{nm} is given by Eq. (2.93). This solution satisfies our boundary conditions on all walls of the box, apart from the top, for arbitrary coefficients c_{nm}. The only job left is to choose these coefficients from the top-lid requirement:

$$\phi(x, y, c) \equiv \sum_{n,m=1}^{\infty} c_{nm}\sin\frac{\pi nx}{a}\sin\frac{\pi my}{b}\sinh\gamma_{nm}c = V(x, y). \tag{2.96}$$

It looks bad to have just one equation for the infinite set of coefficients c_{nm}. However, decisive help comes from the fact that the functions of x and y that participate in Eq. (2.96) form *full, orthogonal* sets of 1D functions. The last term means that the integrals of the products of the functions with different integer indices over the region of interest equal zero. Indeed, a direct integration gives

$$\int_0^a \sin\frac{\pi nx}{a}\sin\frac{\pi n'x}{a}dx = \frac{a}{2}\delta_{nn'}, \tag{2.97}$$

where $\delta_{nn'}$ is the Kronecker symbol[36], and similarly for y (with the evident replacements $a \to b$, $n \to m$). Hence, a fruitful way to proceed is to multiply both sides of Eq. (2.96) by the product of the basis functions, with arbitrary indices n' and m', and integrate the result over x and y:

[36] Let me hope that the reader knows what it is; if not, see Eq. (A.82).

$$\sum_{n,m=1}^{\infty} c_{nm} \sinh \gamma_{nm} c \int_0^a \sin \frac{\pi n x}{a} \sin \frac{\pi n' x}{a} dx \int_0^b \sin \frac{\pi m y}{b} \sin \frac{\pi m' y}{b} dy$$
$$= \int_0^a dx \int_0^b dy \; V(x, y) \sin \frac{\pi n' x}{a} \sin \frac{\pi m' y}{b}. \tag{2.98}$$

Due to Eq. (2.97), all terms on the left-hand side of the last equation, apart from those with $n = n'$ and $m = m'$, vanish, and (replacing n' with n, and m' with m) we finally obtain

$$c_{nm} = \frac{4}{ab \sinh \gamma_{nm} c} \int_0^a dx \int_0^b dy V(x, y) \sin \frac{\pi n x}{a} \sin \frac{\pi m y}{b}. \tag{2.99}$$

The relations (2.93), (2.95), and (2.99) give the complete solution of the posed boundary problem; we can see both good and bad news here. The first bit of bad news is that in the general case we still need to work out the integrals (2.99)—formally, an infinite number of them. In some cases, it is possible to achieve this analytically. For example, if the top lid in our problem is a single conductor, i.e. has a constant potential, we may take $V(x,y) = \text{const} \equiv V_0$, and both 1D integrations are elementary; for example

$$\int_0^a \sin \frac{\pi n x}{a} dx = \frac{2a}{\pi n} \times \begin{cases} 1, & \text{for } n \text{ odd}, \\ 0, & \text{for } n \text{ even}, \end{cases} \tag{2.100}$$

and similarly for the integral over y, so that

$$c_{nm} = \frac{16 V_0}{\pi^2 nm \sinh \gamma_{nm} c} \times \begin{cases} 1, & \text{if both } n \text{ and } m \text{ are odd}, \\ 0, & \text{otherwise}. \end{cases} \tag{2.101}$$

The second item of bad news is that even at such a happy occasion, we still have to sum up the series (2.95), so that our result may only be called analytical with some reservations, because in most cases we need a computer to obtain the final numbers or plots.

Now the first *good* news. Computers are very efficient for both operations (2.95) and (2.99), i.e. for the summation and integration. (As was discussed in section 1.2, random errors are averaged out at these operations.) As an example, figure 2.14 shows the plots of the electrostatic potential in a cubic box ($a = b = c$), with an equipotential top lid ($V = V_0 = \text{const}$), obtained by a numerical summation of the series (2.95), using the analytical expression (2.101). The remarkable feature of this calculation is a very fast convergence of the series; for the middle cross-section of the cubic box ($z/c = 0.5$), already the first term (with $n = m = 1$) gives an accuracy about 6%, while the sum of four leading terms (with $n, m = 1, 3$) reduces the error to just 0.2%. (For a longer box, $c > a, b$, the convergence is even faster—see the discussion below.) Only close to the corners between the top and the side walls, where the

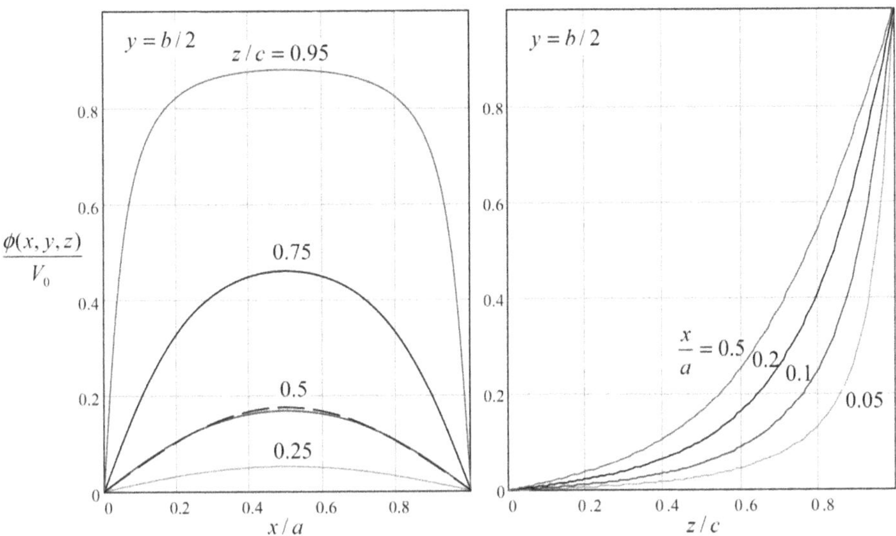

Figure 2.14. The electrostatic potential's distribution inside a cubic box ($a = b = c$) with a constant voltage V_0 on the top lid (figure 2.13), calculated numerically from Eqs. (2.93), (2.95), and (2.101). The dashed line on the left panel shows the contribution of the main term of the series (with $n = m = 1$) to the full result, for $z/c = 0.5$.

potential changes very rapidly, are several more terms necessary to obtain a reasonable accuracy.

The second bit of good news is that our 'semi-analytical' result allows its ultimate limits to be explored analytically. For example, Eq. (2.93) shows that for a very flat box ($c \ll a, b$), $\gamma_{n,m}z \leqslant \gamma_{n,m}c \ll 1$ at least for the lowest terms of series (2.95), with n, $m \ll c/a$, c/b. In these terms, sinh functions in Eqs. (2.96) and (2.99) may be well approximated with their arguments, and their ratio by z/c. So, if we limit the summation to these terms, Eq. (2.95) gives a very simple result

$$\phi(x, y) \approx \frac{z}{c} V(x, y), \tag{2.102}$$

which means that each segment of the flat box behaves just as a plane capacitor. Only near the side walls, the higher terms in the series (2.95) are important, producing some deviations from Eq. (2.102). (For the general problem with an arbitrary function $V(x,y)$, this is also true at all regions where this function changes sharply.)

In the opposite limit ($a, b \ll c$), Eq. (2.93) shows that, in contrast, $\gamma_{n,m}c \gg 1$ for all n and m. Moreover, the ratio sinh $\gamma_{n,m}z$/sinh $\gamma_{n,m}c$ drops sharply if either n or m is increased, if z is not too close to c. Hence in this case a very good approximation may be obtained by keeping just the leading term, with $n = m = 1$, in Eq. (2.95), so that the issue of summation disappears. (As was discussed above, this approximation works reasonably well even for a cubic box.) In particular, for the constant potential of the top lid, we can use Eq. (2.101) and the exponential asymptotic for both sinh functions, to obtain a very simple formula:

$$\phi = \frac{16}{\pi^2} \sin\frac{\pi x}{a} \sin\frac{\pi y}{b} \exp\left\{-\pi\frac{(a^2+b^2)^{1/2}}{ab}(c-z)\right\}. \tag{2.103}$$

These results may be readily generalized to some other problems. For example, if all walls of the box shown in figure 2.13 have an arbitrary potential distribution, one can use the linear superposition principle to argue that the electrostatic potential distribution inside the box is the sum of six partial solutions of the type of Eq. (2.95), each with one wall biased by the corresponding voltage and all others grounded ($\phi = 0$).

To summarize, the results given by the variable separation method in Cartesian coordinates are closer to what we could call a genuinely analytical solution than to purely numerical solutions. Now, let us explore the issues that arise when this method is applied in other orthogonal coordinate systems.

2.6 Variable separation—polar coordinates

If a system of conductors is cylindrical, the potential distribution is independent of the coordinate z along the cylinder axis: $\partial\phi/\partial z = 0$, and the Laplace equation becomes two-dimensional. If conductor's cross-section is rectangular, the variable separation method works best in Cartesian coordinates $\{x, y\}$, and is just a particular case of the 3D solution discussed above. However, if the cross-section is circular, much more compact results may be obtained by using the polar coordinates $\{\rho, \varphi\}$. As we already know from the last section, these 2D coordinates are orthogonal, so that the 2D Laplace operator is a sum of two separable terms[37]. Requiring, just as we have done above, each component of the sum (2.84) to satisfy the Laplace equation, we obtain

$$\frac{1}{\rho}\frac{\partial}{\partial\rho}\left(\rho\frac{\partial\phi_k}{\partial\rho}\right) + \frac{1}{\rho^2}\frac{\partial^2\phi_k}{\partial\varphi^2} = 0. \tag{2.104}$$

In a full analogy with Eq. (2.75), let us represent each particular solution as a product: $\phi_k = \mathcal{R}(\rho)\mathcal{F}(\varphi)$. Plugging this expression into Eq. (2.104) and then dividing all its parts by $\mathcal{R}\mathcal{F}/\rho^2$, we obtain

$$\frac{\rho}{\mathcal{R}}\frac{d}{d\rho}\left(\rho\frac{d\mathcal{R}}{d\rho}\right) + \frac{1}{\mathcal{F}}\frac{d^2\mathcal{F}}{d\varphi^2} = 0. \tag{2.105}$$

Following the same reasoning as for the Cartesian coordinates, we obtain two separated ordinary differential equations

$$\rho\frac{d}{d\rho}\left(\rho\frac{d\mathcal{R}}{d\rho}\right) = \nu^2\mathcal{R}, \tag{2.106}$$

[37] See, e.g. Eq. (A.61) with $\partial/\partial z = 0$.

$$\frac{d^2\mathcal{F}}{d\varphi^2} + \nu^2\mathcal{F} = 0, \tag{2.107}$$

where ν^2 is the variable separation constant.

Let us start their analysis from Eq. (2.106), plugging into it a probe solution $\mathcal{R} = c\rho^\alpha$, where c and α are some constants. The elementary differentiation shows that if $\alpha \neq 0$, the equation is indeed satisfied for any c, with just one requirement on constant α, namely $\alpha^2 = \nu^2$. This means that the following linear superposition

$$\mathcal{R} = a_\nu\rho^{+\nu} + b_\nu\rho^{-\nu}, \quad \text{for } \nu \neq 0, \tag{2.108}$$

with any constant coefficients a_ν and b_ν, is also a solution to Eq. (2.106). Moreover, the general theory of linear ordinary differential equations tells us that the solution of a second-order equation such as Eq. (2.106) may only depend on just two constant factors that scale two linearly independent functions. Hence, for all values $\nu^2 \neq 0$, Eq. (2.108) presents the *general* solution of that equation. The case when $\nu = 0$, in which functions $\rho^{+\nu}$ and $\rho^{-\nu}$ are just constants and hence are *not* linearly independent, is special, but in this case the integration of Eq. (2.106) is straightforward[38], giving

$$\mathcal{R} = a_0 + b_0 \ln \rho, \quad \text{for } \nu = 0. \tag{2.109}$$

In order to specify the separation constant, let us explore Eq. (2.107), whose general solution is

$$\mathcal{F} = \begin{cases} c_\nu \cos \nu\varphi + s_\nu \sin \nu\varphi, & \text{for } \nu \neq 0, \\ c_0 + s_0\varphi, & \text{for } \nu = 0. \end{cases} \tag{2.110}$$

There are two possible cases here. In many boundary problems solvable in cylindrical coordinates, the free space region, in which the Laplace equation is valid, extends continuously around the origin point $\rho = 0$. In this region, the potential has to be continuous and uniquely defined, so that $\mathcal{F}$ has to be a 2π-periodic function of angle φ. For that, one needs the product $\nu(\varphi + 2\pi)$ to be equal to $\nu\varphi + 2\pi n$, with n an integer, immediately giving us a discrete spectrum of possible values of the variable separation constant:

$$\nu = n = 0, \pm 1, \pm 2, \ldots \tag{2.111}$$

In this case both functions $\mathcal{R}$ and $\mathcal{F}$ may be labeled with the integer index n. Taking into account that the terms with negative values of n may be summed up with those with positive n, and that s_0 should equal zero (otherwise the 2π-periodicity of function $\mathcal{F}$ would be violated), we see that the general solution to the 2D Laplace equation may be represented as

[38] Actually, we have already done this in section 2.3—see Eq. (2.43).

$$\phi(\rho, \varphi) = a_0 + b_0 \ln \rho + \sum_{n=1}^{\infty}\left(a_n\rho^n + \frac{b_n}{\rho^n}\right)(c_n \cos n\varphi + s_n \sin n\varphi). \quad (2.112)$$

Let us see how all this machinery works on the famous problem of a round cylindrical conductor placed into an electric field that is uniform and perpendicular to the cylinder's axis at large distances (say, created by a large plane capacitor)—see figure 2.15a. First of all, let us explore the effect of system's symmetries on the coefficients in Eq. (2.112). Selecting the coordinate system as shown in figure 2.15a, and taking the cylinder's potential for zero, we immediately obtain $a_0 = 0$. Moreover, due to the mirror symmetry about the plane $[x, z]$, the solution has to be an even function of the angle φ, and hence all coefficients s_n should also equal zero. Also, at large distances ($\rho \gg R$) from the cylinder axis its effect on the electric field should vanish, and the potential should approach that of the uniform field $\mathbf{E} = E_0\mathbf{n}_x$:

$$\phi \to -E_0 x = -E_0\rho \cos \varphi, \quad \text{for } \rho \to \infty. \quad (2.113)$$

This is only possible if in Eq. (2.112), $b_0 = 0$, and also all coefficients a_n with $n \neq 1$ vanish, while the product a_1c_1 should be equal to $(-E_0)$. Thus the solution is reduced to the following form

$$\phi(\rho, \varphi) = -E_0\rho \cos \varphi + \sum_{n=1}^{\infty}\frac{B_n}{\rho^n}\cos n\varphi, \quad (2.114)$$

in which the coefficients $B_n \equiv b_nc_n$ should be found from the boundary condition on the cylinder's surface, i.e. at $\rho = R$:

$$\phi(R, \varphi) = 0. \quad (2.115)$$

This requirement yields the following equation,

$$\left(-E_0R + \frac{B_1}{R}\right)\cos \varphi + \sum_{n=2}^{\infty}\frac{B_n}{R^n}\cos n\varphi = 0, \quad (2.116)$$

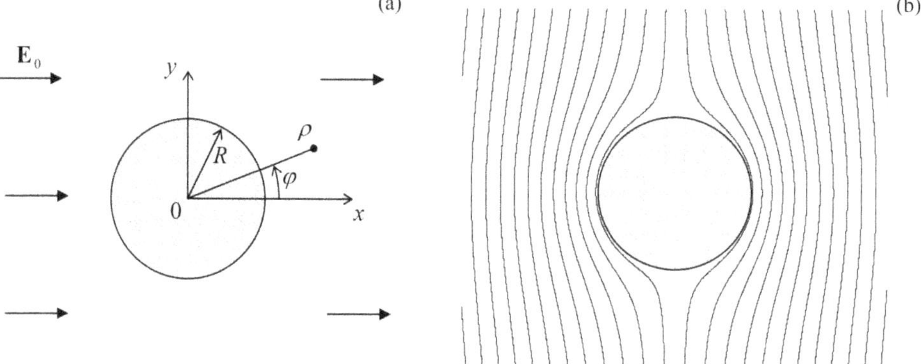

Figure 2.15. A conducting cylinder inserted into an initially uniform electric field perpendicular to is axis: (a) the problem's geometry and (b) the equipotential surfaces given by Eq. (2.117).

which should be satisfied for all φ. This equality, read backwards, may be considered as an expansion of a function identically equal to zero into a series over mutually orthogonal functions $\cos n\varphi$.[39] It is of course valid if all coefficients of the expansion, including $(-E_0R + B_1/R)$ and all B_n for $n \geqslant 2$, are equal to zero. As a result, our final answer (valid only outside of the cylinder, i.e. for $\rho \geqslant R$), is

$$\phi(\rho, \varphi) = -E_0\left(\rho - \frac{R^2}{\rho}\right)\cos\varphi \equiv -E_0\left(1 - \frac{R^2}{x^2 + y^2}\right)x. \tag{2.117}$$

This result (figure 2.15b) shows a smooth transition between the uniform field Eq. (2.113) far from the cylinder, to the equipotential surface of the cylinder (with $\phi = 0$). Such smoothening is very typical for Laplace equation solutions. Indeed, as we know from chapter 1, these solutions correspond to the lowest potential energy (1.65), i.e. the lowest integral of the potential gradient's square, possible at the given boundary conditions.

To complete the problem, let us calculate the distribution of the surface charge density over the cylinder's cross-section, using Eq. (2.3):

$$\sigma = \varepsilon_0 E_{n|\text{surface}} \equiv -\varepsilon_0 \frac{\partial \phi}{\partial \rho}\bigg|_{\rho=R} = \varepsilon_0 E_0 \cos\varphi \frac{\partial}{\partial \rho}\left(\rho - \frac{R^2}{\rho}\right)\bigg|_{\rho=R} = 2\varepsilon_0 E_0 \cos\varphi. \tag{2.118}$$

This very simple formula shows that with the field direction shown in figure 2.15a ($E_0 > 0$), the surface charge is positive on the right side of the cylinder and negative on its left side, thus creating a field directed from the right to the left, which compensates the external field inside the conductor, where the net field is zero. (Please take one more look at the schematic figure 2.1a.) Note also that the net electric charge of the cylinder is zero, in the correspondence with the problem symmetry. Another useful by-product of the calculation (2.118) is that the surface electric field equals $2E_0\cos\varphi$, and hence its largest magnitude is twice the field far from the cylinder. Such electric field concentration is very typical for all convex conducting surfaces.

The last observation receives additional confirmation for the second possible topology, when Eq. (2.110) is used to describe problems with no angular periodicity. A typical example of this situation is a cylindrical conductor with a cross-section that features a corner limited by straight lines (figure 2.16). Indeed, at we may argue that at $\rho < R$ (where R is the scale of radial extension of the straight sides of the corner, see figure 2.16), the Laplace equation may be satisfied by a sum of partial solutions $\mathcal{R}(\rho)\mathcal{F}(\varphi)$, if the angular components of the products satisfy the boundary conditions on the corner sides. Taking (just for the simplicity of notation) the conductor's potential to be zero, and one of the corner's sides as the axis x ($\varphi = 0$), these boundary conditions are

[39] Mathematics tells us that such expansions are unique, so this is the only possible solution of Eq. (2.116).

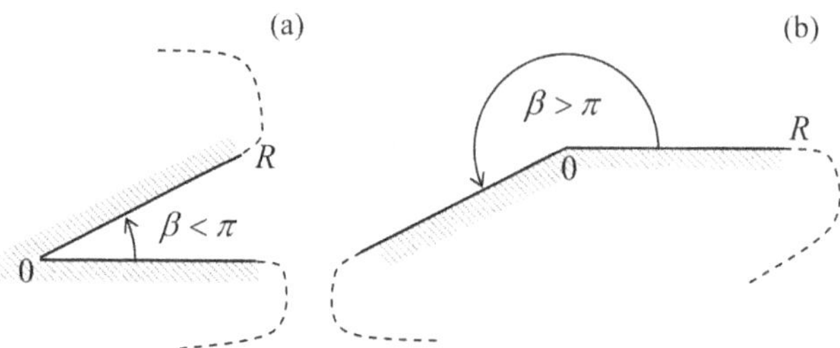

Figure 2.16. The cross-sections of cylindrical conductors with (a) a corner and (b) a wedge.

$$\mathcal{P}(0) = \mathcal{P}(\beta) = 0, \tag{2.119}$$

where the angle β may be anywhere between 0 and 2π—see figure 2.16.

Comparing this condition with Eq. (2.110), we see that it requires all c_ν to vanish, and ν to take one of the values of the following discrete spectrum:

$$\nu_m\beta = \pi m, \quad \text{with } m = 1, 2, \ldots. \tag{2.120}$$

Hence the full solution of the Laplace equation takes the form

$$\phi = \sum_{m=1}^{\infty} a_m \rho^{\pi m/\beta} \sin\frac{\pi m\varphi}{\beta}, \quad \text{for } \rho < R, \quad 0 \leqslant \varphi \leqslant \beta, \tag{2.121}$$

where constants s_ν have been incorporated into a_m. The set of constants a_m cannot be universally determined, because it depends on the exact shape of the conductor outside the corner and the externally applied electric field. However, whatever the set is, in the limit $\rho \to 0$, the solution (2.121) is almost[40] always dominated by the term with lowest ν (corresponding to $m = 1$),

$$\phi \to a_1\rho^{\pi/\beta} \sin\frac{\pi}{\beta}\varphi, \tag{2.122}$$

because the higher terms go to zero faster. This potential distribution corresponds to the surface charge density

$$\sigma = \varepsilon_0 E_n|_{\text{surface}} = -\varepsilon_0\frac{\partial\phi}{\partial(\rho\varphi)}\Big|_{\rho=\text{const},\, \varphi\to+0} = -\varepsilon_0\frac{\pi a_1}{\beta}\rho^{(\pi/\beta-1)}. \tag{2.123}$$

(It is similar on the opposite face of the angle.)

The result (2.123) shows that if we are dealing with a usual, concave corner ($\beta < \pi$, see figure 2.16a), the charge density (and the surface electric field) tends to zero. On the other case, at a 'convex corner' with $\beta > \pi$ (actually, a wedge—see figure 2.16b), both the charge and the field's strength concentrate, formally diverging at $\rho \to 0$. (So,

[40] Exceptions are possible only for highly symmetric configurations when the external field is specially crafted to make $a_1 = 0$. In this case the solution at $\rho \to 0$ is dominated by the first non-vanishing term of the series (2.121).

do not sit on a roof's ridge during a thunderstorm; rather hide in a ditch!) We have already seen qualitatively similar effects for the thin round disk and the split plane.

2.7 Variable separation—cylindrical coordinates

Now, let us discuss whether it is possible to generalize our approach to problems whose geometry is still axially symmetric, but with a substantial dependence of the potential on the axial coordinate ($\partial\phi/\partial z \neq 0$). The classical example of such a problem is shown in figure 2.17. Here the side wall and the bottom lid of a hollow round cylinder are kept at a fixed potential (say, $\phi = 0$), but the potential V fixed at the top lid is different. Evidently, this problem is qualitatively similar to the rectangular box problem solved above (figure 2.13), and we will also try to solve it first for the case of arbitrary voltage distribution over the top lid: $V = V(\rho, \varphi)$.

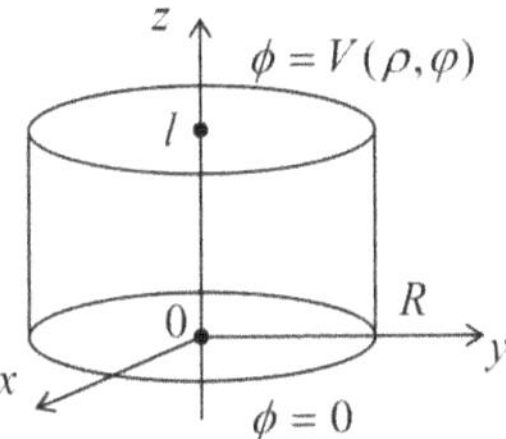

Figure 2.17. A cylindrical volume with conducting walls.

Following the main idea of the variable separation method, let us require that each partial function ϕ_k in Eq. (2.84) satisfies the Laplace equation, now in the full cylindrical coordinates $\{\rho, \varphi, z\}$:[41]

$$\frac{1}{\rho}\frac{\partial}{\partial\rho}\left(\rho\frac{\partial\phi_k}{\partial\rho}\right) + \frac{1}{\rho^2}\frac{\partial^2\phi_k}{\partial\varphi^2} + \frac{\partial^2\phi_k}{\partial z^2} = 0. \tag{2.124}$$

Plugging in ϕ_k in the form $\mathcal{R}(\rho)\mathcal{F}(\varphi)\mathcal{Z}(z)$ into Eq. (2.124) and dividing all terms by the product $\mathcal{R}\mathcal{F}\mathcal{Z}$, we obtain

$$\frac{1}{\rho\mathcal{R}}\frac{d}{d\rho}\left(\rho\frac{d\mathcal{R}}{d\rho}\right) + \frac{1}{\rho^2\mathcal{F}}\frac{d^2\mathcal{F}}{d\phi^2} + \frac{1}{\mathcal{Z}}\frac{d^2\mathcal{Z}}{dz^2} = 0. \tag{2.125}$$

Since the first two terms of Eq. (2.125) can only depend on the polar variables ρ and φ, while the third term only on z, at least that term should be a constant. Denoting it (just like in the rectangular box problem) by γ^2, we obtain, instead of Eq. (2.125), a set of two equations:

$$\frac{d^2\mathcal{Z}}{dz^2} = \gamma^2\mathcal{Z}, \tag{2.126}$$

[41] See, e.g. Eq. (A.61).

$$\frac{1}{\rho \mathcal{R}} \frac{d}{d\rho}\left(\rho \frac{d\mathcal{R}}{d\rho}\right) + \gamma^2 + \frac{1}{\rho^2 \mathcal{F}} \frac{d^2 \mathcal{F}}{d\varphi^2} = 0. \qquad (2.127)$$

Now, multiplying all terms of Eq. (2.127) by ρ^2, we see that the last term, $(d^2\mathcal{F}/d\varphi^2)/\mathcal{F}$, may depend only on φ, and thus should be constant. Calling that constant ν^2 (just as in section 2.6 above), we separate Eq. (2.127) into an angular equation,

$$\frac{d^2\mathcal{F}}{d\varphi^2} + \nu^2 \mathcal{F} = 0, \qquad (2.128)$$

and a radial equation:

$$\frac{d^2\mathcal{R}}{d\rho^2} + \frac{1}{\rho}\frac{d\mathcal{R}}{d\rho} + \left(\gamma^2 - \frac{\nu^2}{\rho^2}\right)\mathcal{R} = 0. \qquad (2.129)$$

We see that the ordinary differential equations for functions $\mathcal{Z}(z)$ and $\mathcal{F}(\varphi)$ (and hence their solutions) are identical to those discussed earlier. However, Eq. (2.129) for the radial function $\mathcal{R}(\rho)$ (called the *Bessel equation*) is more complex than in the 2D case, and depends on two independent constant parameters, γ and ν. The latter challenge may be readily overcome if we notice that any change of γ may be reduced to a re-scaling of the radial coordinate ρ. Indeed, introducing a dimensionless variable $\xi \equiv \gamma\rho$,[42] Eq. (2.129) may be reduced to an equation with just one parameter, ν:

$$\frac{d^2\mathcal{R}}{d\xi^2} + \frac{1}{\xi}\frac{d\mathcal{R}}{d\xi} + \left(1 - \frac{\nu^2}{\xi^2}\right)\mathcal{R} = 0. \qquad (2.130)$$

Moreover, we already know that for angle-periodic problems, the spectrum of eigenvalues of Eq. (2.128) is discrete: $\nu = n$, with integer n.

Unfortunately, even in this case, Eq. (2.130) cannot be satisfied by a single 'elementary' function, and is the canonical form of the Bessel equation. Its solutions that we need for our current problem are called the *Bessel function of the first kind, of order* ν, commonly denoted as $J_\nu(\xi)$. Let me review in brief the properties of these functions that are most relevant for our problem—and many other problems discussed in this series[43].

First of all, the Bessel function of a negative integer order is very simply related to that with the positive order:

[42] Note that this normalization is specific for each value of the variable separation parameter γ. Also, please notice that the normalization is meaningless for $\gamma = 0$, i.e. for the case $Z(z)$ = const. However, if we need the partial solutions for this particular value of γ, we can always use Eqs. (2.108) and (2.109).

[43] For a more complete discussion of these functions, see the literature listed in appendix A, section 2.16, for example, chapter 9 (written by F Olver) in [6].

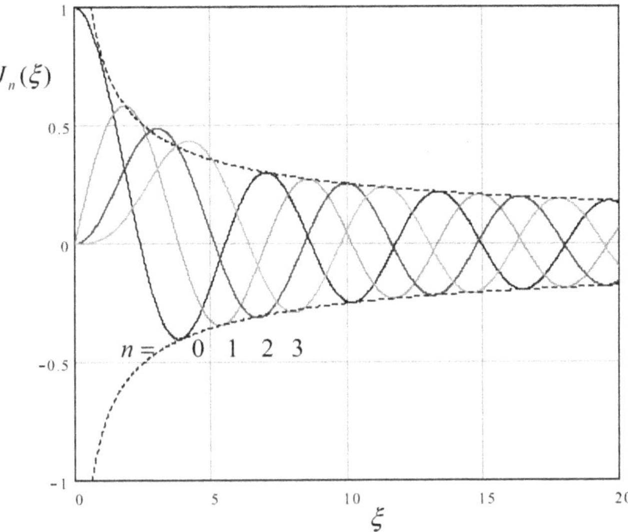

Figure 2.18. Several Bessel functions $J_n(\xi)$ of integer order. The dashed lines show the envelope of the asymptotes (2.135).

$$J_{-n}(\xi) = (-1)^n J_n(\xi), \tag{2.131}$$

enabling us to limit our discussion to the functions with $n \geqslant 0$. Figure 2.18 shows four of these functions with the lowest positive n.

As its argument is increased, each function is initially close to a power law: $J_0(\xi) \approx 1$, $J_1(\xi) \approx \xi/2 = \xi/2$, $J_2(\xi) \approx \xi^2/8$, etc. This behavior follows from the Taylor series

$$J_n(\xi) = \left(\frac{\xi}{2}\right)^n \sum_{k=0}^{\infty} \frac{(-1)^k}{k!(n+k)!} \left(\frac{\xi}{2}\right)^{2k}, \tag{2.132}$$

which is formally valid for any ξ, and may even serve as an alternative definition of the functions $J_n(\xi)$. However, this series is converging fast only at relatively small arguments, $\xi < n$, where its leading term is

$$J_n(\xi)\Big|_{\xi \to 0} \to \frac{1}{n!}\left(\frac{\xi}{2}\right)^n. \tag{2.133}$$

At $\xi \approx n + 1.86n^{1/3}$, the Bessel function reaches its maximum[44]

$$\max_\xi[J_n(\xi)] \approx \frac{0.675}{n^{1/3}}, \tag{2.134}$$

[44] These two formulas for the Bessel function peak are strictly valid for $n \gg 1$, but may be used for reasonable estimates starting already from $n = 1$; for example, $\max_\xi [J_1(\xi)]$ is close to 0.58 and is reached at $\xi \approx 2.4$, just about 30% away from the values given by the asymptotic formulas.

and then starts to oscillate with a period gradually approaching 2π, a phase shift that increases by $\pi/2$ with each unit increment of n, and an amplitude that decreases as $\xi^{-1/2}$. These features are described by the following asymptotic formula:

$$J_n(\xi) \to \left(\frac{2}{\pi\xi}\right)^{1/2} \cos\left(\xi - \frac{\pi}{4} - \frac{n\pi}{2}\right), \quad \text{for } \xi/n \to \infty, \tag{2.135}$$

which starts to give reasonable results very soon above the function peaks—see figure 2.18[45].

Now we are ready to return to our case study (figure 2.17). Let us select functions $\mathcal{Z}(z)$ to satisfy the Eq. (2.126) and the bottom-lid boundary condition $\mathcal{Z}(0) = 0$, i.e. proportional to $\sinh\gamma z$—cf Eq. (2.94). Then

$$\phi = \sum_{n=0}^{\infty}\sum_{\gamma} J_n(\gamma\rho)\left(c_{n\gamma}\cos n\varphi + s_{n\gamma}\sin n\varphi\right)\sinh\gamma z. \tag{2.136}$$

Next, we need to satisfy the zero boundary condition at the cylinder's side wall ($\rho = R$). This may be ensured by taking

$$J_n(\gamma R) = 0. \tag{2.137}$$

Since each function $J_n(x)$ has an infinite number of positive zeros (see figure 2.18), which may be numbered by an integer index $m = 1, 2, \ldots$, Eq. (2.137) may be satisfied with an infinite number of discrete values of the separation parameter γ:

$$\gamma_{nm} = \frac{\xi_{nm}}{R}, \tag{2.138}$$

where ξ_{nm} is the mth zero of the function $J_n(x)$—see the top numbers in the cells of table 2.1. (Very soon we will see what we need the bottom numbers for.)

Hence, Eq. (2.136) may be represented in a more explicit form:

$$\phi(\rho, \varphi, z) = \sum_{n=0}^{\infty}\sum_{m=1}^{\infty} J_n\left(\xi_{nm}\frac{\rho}{R}\right)\left(c_{nm}\cos n\varphi + s_{nm}\sin n\varphi\right)\sinh\left(\xi_{nm}\frac{z}{R}\right). \tag{2.139}$$

Here the coefficients c_{nm} and s_{nm} have to be selected to satisfy the only remaining boundary condition—that on the top lid:

$$\begin{aligned}\phi(\rho, \varphi, l) &\equiv \sum_{n=0}^{\infty}\sum_{m=1}^{\infty} J_n\left(\xi_{nm}\frac{\rho}{R}\right)\left(c_{nm}\cos n\varphi + s_{nm}\sin n\varphi\right)\sinh\left(\xi_{nm}\frac{L}{R}\right) \\ &= V(\rho, \varphi).\end{aligned} \tag{2.140}$$

[45] Eq. (2.135) and figure 2.18 clearly show the close analogy between the Bessel functions and the usual trigonometric functions, sine and cosine. In order to emphasize this similarity, and help the reader to develop a better gut feeling of the Bessel functions, let me mention one fact of elasticity theory: while sine functions describe, in particular, possible modes of standing waves on a guitar string, functions $J_n(\xi)$ describe, in particular, possible standing waves on an elastic round membrane, with $J_0(\xi)$ describing their lowest (fundamental) mode.

Table 2.1. Approximate values of a few first zeros, ξ_{nm}, of a few lowest-order Bessel functions $J_n(\xi)$ (the top number in each cell), and the values of $dJ_n(\xi)/d\xi$ at these points (the bottom number).

	$m = 1$	2	3	4	5	6
$n = 0$	2.40482	5.52008	8.65372	11.79215	14.93091	18.07106
	−0.51914	+0.34026	−0.27145	+0.23245	−0.20654	+0.18773
1	3.83171	7.01559	10.17347	13.32369	16.47063	19.61586
	−0.40276	+0.30012	−0.24970	+0.21836	−0.19647	+0.18006
2	5.13562	8.41724	11.61984	14.79595	17.95982	21.11700
	−0.33967	+0.27138	−0.23244	+0.20654	−0.18773	+0.17326
3	6.38016	9.76102	13.01520	16.22347	19.40942	22.58273
	−0.29827	+0.24942	−0.21828	+0.19644	−0.18005	+0.16718
4	7.58834	11.06471	14.37254	17.61597	20.82693	24.01902
	−0.26836	+0.23188	−0.20636	+0.18766	−0.17323	+0.16168
5	8.77148	12.33860	15.70017	18.98013	22.21780	25.43034
	−0.24543	+0.21743	−0.19615	+0.17993	−0.16712	+0.15669

To use it, let us multiply both parts of Eq. (2.140) by $J_n(\xi_{nm'}\rho/R)\cos n'\varphi$, integrate the result over the lid area, and use the following property of the Bessel functions:

$$\int_0^1 J_n(\xi_{nm}s)J_n(\xi_{nm'}s)\,s\,ds = \frac{1}{2}[J_{n+1}(\xi_{nm})]^2\delta_{mm'}. \qquad (2.141)$$

The relation (2.141) expresses a very specific ('2D') orthogonality of the Bessel functions with different indices m—do not confuse them with the function's order n, please[46]! Since it relates two Bessel functions with the same index n, it is natural to ask why its right-hand side contains the function with a different index $(n + 1)$. Some clue may come from one more very important property of the Bessel functions, the so-called *recurrence relations*[47]:

$$J_{n-1}(\xi) + J_{n+1}(\xi) = \frac{2nJ_n(\xi)}{\xi}, \qquad (2.142a)$$

[46] The Bessel functions of the *same argument* but of *different orders* are also orthogonal, but in a different way:

$$\int_0^\infty J_n(\xi)J_{n'}(\xi)\frac{d\xi}{\xi} = \frac{1}{n+n'}\delta_{nn'}.$$

[47] These relations provide, in particular, a convenient way for fast numerical computation of all $J_n(\xi)$ after $J_0(\xi)$ has been computed. (The latter is usually done with an algorithm using Eq. (2.132) for smaller ξ and an extension of Eq. (2.135) for larger ξ.) Note that most mathematical software packages, including all those listed in appendix A, section A.16(iv), include ready subroutines for calculation of the functions $J_n(\xi)$ and other special functions used in this lecture series. In this sense, the line separating these 'special functions' from 'elementary functions' is rather fine.

$$J_{n-1}(\xi) - J_{n+1}(\xi) = 2\frac{dJ_n(\xi)}{d\xi}, \qquad (2.142b)$$

which in particular yield the following relation (convenient for working out some Bessel function integrals):

$$\frac{d}{d\xi}[\xi^n J_n(\xi)] = \xi^n J_{n-1}(\xi). \qquad (2.143)$$

For our current purposes, let us apply the recurrence relations at the special points ξ_{nm}. At these points, J_n vanishes, and the system of two equations (2.142) may be readily solved to obtain, in particular,

$$J_{n+1}(\xi_{nm}) = -\frac{dJ_n}{d\xi}(\xi_{nm}), \qquad (2.144)$$

so that the square bracket on the right-hand side of Eq. (2.141) is just $(dJ_n/d\xi)^2$ at $\xi = \xi_{nm}$. Thus the values of the Bessel function derivatives at the zero points, given by the lower numbers in the cells of table 2.1, are as important for boundary problem solutions as the zeros themselves.

Since the angular functions cos $n\varphi$ are also orthogonal—both to each other,

$$\int_0^{2\pi} \cos(n\varphi)\cos(n'\varphi)\, d\varphi = \pi\delta_{nn'}, \qquad (2.145)$$

and to all functions sin $n\varphi$, the integration over the lid area kills all terms of both series in Eq. (2.140), apart from just one term proportional to $c_{n'm'}$, and hence gives an explicit expression for that coefficient. The counterpart coefficients $s_{n'm'}$ may be found by repeating the same procedure with the replacement of cos $n'\varphi$ by sin $n'\varphi$. This evaluation (left for the reader's exercise) completes the solution of our problem for an arbitrary lid potential $V(\rho,\varphi)$.

Still, before leaving the Bessel functions (for a while only:-), we need to address two important issues. First, we have seen that in our cylinder problem (figure 2.17), the set of functions $J_n(\xi_{nm}\rho/R)$ with different indices m (which characterize the degree of the Bessel function's stretch along axis ρ) play a role similar to that of functions $\sin(\pi nx/a)$ in the rectangular box problem shown in figure 2.13. In this context, what is the analog of functions $\cos(\pi nx/a)$ which may be important for some boundary problems? In a more formal language, are there any functions of the same argument $\xi \equiv \xi_{nm}\rho/R$ that would be linearly independent of the Bessel functions of the first kind, while satisfying the same Bessel Eq. (2.130)?

The answer is *yes*. For the definition of such functions, we first need to generalize our prior formulas for $J_n(\xi)$, and in particular Eq. (2.132), to the case of arbitrary, not necessarily real order ν. Mathematics says that the generalization may be performed in the following way:

$$J_\nu(\xi) = \left(\frac{\xi}{2}\right)^\nu \sum_{k=0}^{\infty} \frac{(-1)^k}{k!\Gamma(\nu + k + 1)}\left(\frac{\xi}{2}\right)^{2k}, \qquad (2.146)$$

where $\Gamma(s)$ is the so-called *gamma function* that may be defined as[48]

$$\Gamma(s) \equiv \int_0^{\infty} \xi^{s-1} e^{-\xi} \, d\xi. \tag{2.147}$$

The simplest, and the most important property of the gamma function is that for integer values of argument it gives the factorial of a number smaller by one:

$$\Gamma(n+1) = n! \equiv 1 \cdot 2 \cdot \ldots \cdot n, \tag{2.148}$$

so it is essentially a generalization of the notion of the factorial to all real numbers.

The Bessel functions defined by Eq. (2.146) satisfy, after the replacements $n \to \nu$ and $n! \to \Gamma(n+1)$, virtually all the relations discussed above, including the Bessel Eq. (2.130), the asymptotic formula (2.135), the orthogonality condition (2.141), and the recurrence relations (2.142). Moreover, it may be shown that $\nu \neq n$, functions $J_\nu(\xi)$ and $J_{-\nu}(\xi)$ are linearly independent of each other, and hence their linear combination may be used to represent a general solution of the Bessel equation. Unfortunately, as (2.131) shows, for $\nu = n$ this is not true, and a solution independent of $J_n(\xi)$ has to be formed in a different way.

The most common way of overcoming this difficulty is first to define, for all $\nu \neq n$, functions

$$Y_\nu(\xi) \equiv \frac{J_\nu(\xi)\cos \nu\pi - J_{-\nu}(\xi)}{\sin \nu\pi}, \tag{2.149}$$

called the *Bessel functions of the second kind*, or more often the *Weber functions*[49], and then to follow the limit $\nu \to n$. At this, both the nominator and denominator of the right-hand side of Eq. (2.149) tend to zero, but their ratio tends to a finite value called $Y_n(x)$. It may be shown that the resulting functions are still the solutions of the Bessel equation and are linearly independent of $J_n(x)$, though are related just as those functions if the sign of n changes:

$$Y_{-n}(\xi) = (-1)^n Y_n(\xi). \tag{2.150}$$

Figure 2.19 shows a few Weber functions of the lowest integer orders. The plots show that the asymptotic behavior is very much similar to that of $J_n(\xi)$,

$$Y_n(\xi) \to \left(\frac{2}{\pi\xi}\right)^{1/2} \sin\left(\xi - \frac{\pi}{4} - \frac{n\pi}{2}\right), \quad \text{for } \xi \to \infty, \tag{2.151}$$

but with the phase shift necessary to make these Bessel functions orthogonal to those of the first order—cf Eq. (2.135). However, for small values of argument ξ, the Bessel

48 See, e.g. Eq. (A.34). Note that $\Gamma(s) \to \infty$ at $s \to 0, -1, -2, \ldots$.

49 They are also sometimes called the *Neumann* functions, and denoted as $N_\nu(\xi)$.

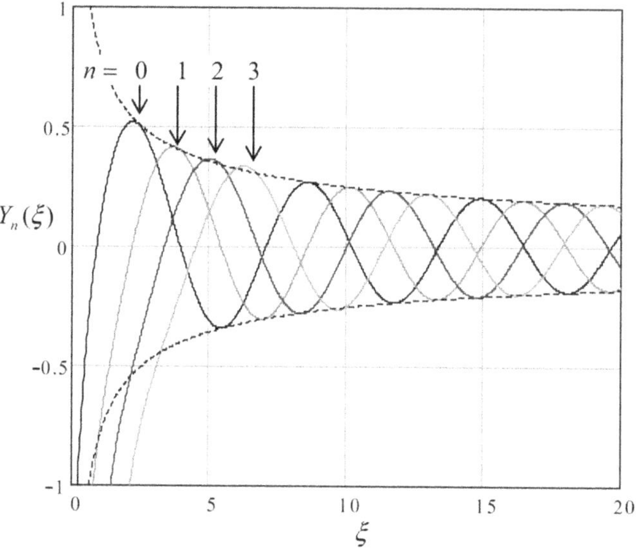

Figure 2.19. A few Bessel functions of the second kind (a.k.a. the Weber functions, a.k.a. the Neumann functions).

functions of the second kind behave completely differently from those of the first kind:

$$Y_n(\xi) \to \begin{cases} \dfrac{2}{\pi}\left(\ln \dfrac{\xi}{2} + \gamma\right), & \text{for } n = 0, \\ -\dfrac{(n-1)!}{\pi}\left(\dfrac{\xi}{2}\right)^{-n}, & \text{for } n \neq 0, \end{cases} \tag{2.152}$$

where γ is the so-called *Euler constant*, defined as follows:

$$\gamma \equiv \lim_{n\to\infty}\left(1 + \frac{1}{2} + \frac{1}{3} + \ldots + \frac{1}{n} - \ln n\right) \approx 0.577157\ldots \tag{2.153}$$

As Eq. (2.152) and figure 2.19 show, the functions $Y_n(\xi)$ diverge at $\xi \to 0$ and hence cannot describe the behavior of any physical variable, in particular the electrostatic potential. One may wonder: if this is true, when do we need these functions in physics? This does not happen too often, but still does.

Figure 2.20 shows an example of a simple boundary problem of electrostatics, whose solution by the variable separation method involves both functions $J_n(\xi)$ and $Y_n(\xi)$. Here two round, conducting coaxial cylindrical tubes are kept at the same (say, zero) potential, but at least one of two lids has a different potential. The problem is almost completely similar to that discussed above (figure 2.17), but now we need to find the potential distribution in the free space between the tubes, i.e. for $R_1 < \rho < R_2$. If we use the same variable separation as in the simpler counterpart problem, we need the radial functions $\mathcal{R}(\rho)$ to satisfy two zero boundary conditions: at $\rho = R_1$ and $\rho = R_2$. With the Bessel functions of just the first kind, $J_n(\gamma\rho)$, it is

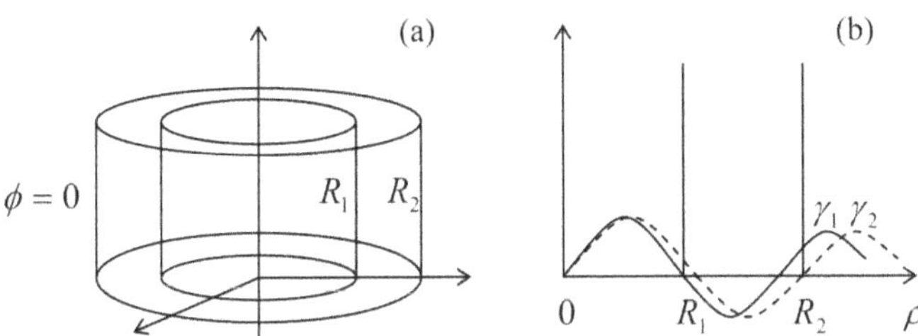

Figure 2.20. A simple boundary problem that cannot be solved using just one kind of Bessel functions.

impossible to do, because the two boundaries would impose two independent (and generally incompatible) conditions, $J_n(\gamma R_1) = 0$, and $J_n(\gamma R_2) = 0$, for one 'stetching parameter' γ. The existence of the Bessel functions of the second kind immediately saves the day, because if the radial function solution is represented as a linear combination,

$$\mathcal{R} = c_J J_n(\gamma\rho) + c_Y Y_n(\gamma\rho), \tag{2.154}$$

two zero boundary conditions give two equations for γ and ratio $c \equiv c_Y/c_J$.[50] (Due to the oscillating character of both Bessel functions, these conditions would be typically satisfied by an infinite set of discrete pairs $\{\gamma, c\}$.) Note, however, that generally none of these pairs would correspond to zeros of either J_n nor Y_n, so that having an analog of table 2.1 for the latter function would not help much. Hence, even simple problems of this kind (such as the one shown in figure 2.20) typically require numerical solutions of algebraic (transcendental) equations.

In order to complete the discussion of variable separation in the cylindrical coordinates, one more issue to address is the so-called *modified Bessel functions*: of the *first kind*, $I_\nu(\xi)$, and of the *second kind*, $K_\nu(\xi)$. They are two linearly independent solutions of the *modified Bessel equation*,

$$\frac{d^2\mathcal{R}}{d\xi^2} + \frac{1}{\xi}\frac{d\mathcal{R}}{d\xi} - \left(1 + \frac{\nu^2}{\xi^2}\right)\mathcal{R} = 0, \tag{2.155}$$

which differs from Eq. (2.130) 'only' by the sign of one of its terms. Figure 2.21 shows a simple problem that leads to this equation: a round conducting cylinder is sliced, perpendicular to its axis, to rings of equal height h, which are kept at equal but sign-alternating potentials.

[50] A pair of independent linear functions, used for representation of the general solution of the Bessel equation, may be also chosen in a different way, using the so-called *Hankel functions*

$$H_n^{(1,2)}(\xi) \equiv J_n(\xi) \pm iY_n(\xi).$$

For representing the general solution of Eq. (2.130), this alternative is completely similar, for example, to using the pair of complex functions $\exp\{\pm i\alpha x\} = \cos \alpha x \pm i\sin \alpha x$ instead of the pair of real functions $\{\cos \alpha x, \sin \alpha x\}$ for representing the general solution of Eq. (2.89) for $X(x)$.

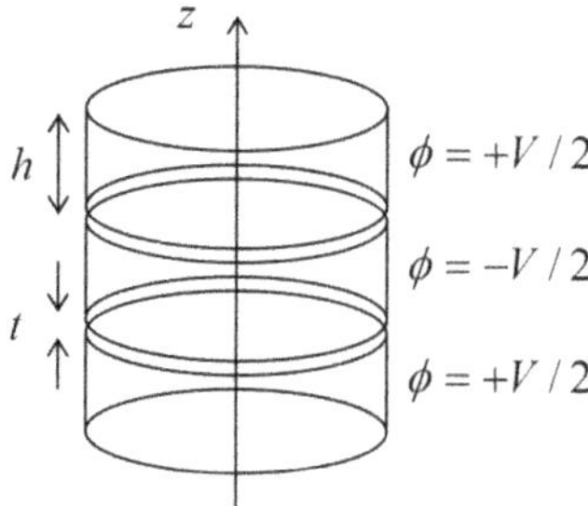

Figure 2.21. A typical boundary problem whose solution may be conveniently described in terms of the modified Bessel functions.

If the gaps between the sections are narrow, $t \ll h$, we may use the variable separation method for the solution to this problem, but now we evidently need periodic (rather than exponential) solutions along axis z, i.e. linear combinations of $\sin kz$ and $\cos kz$ with various real values of the constant k. Separating the variables, we arrive at a differential equation similar to Eq. (2.129), but with the negative sign before the separation constant:

$$\frac{d^2\mathcal{R}}{d\rho^2} + \frac{1}{\rho}\frac{d\mathcal{R}}{d\rho} - \left(k^2 + \frac{\nu^2}{\rho^2}\right)\mathcal{R} = 0. \tag{2.156}$$

The radial coordinate's normalization, $\xi \equiv k\rho$, immediately leads us to Eq. (2.155), and hence (for $\nu = n$) to the modified Bessel functions $I_n(\xi)$ and $K_n(\xi)$.

Figure 2.22 shows the behavior of such functions, of a few lowest orders. One can see that at $\xi \to 0$ the behavior is virtually similar to that of the 'usual' Bessel functions—cf Eqs. (2.132) and (2.152), with $K_n(\xi)$ multiplied (due to purely historical reasons) by an additional coefficient, $\pi/2$:

$$I_n(\xi) \to \frac{1}{n!}\left(\frac{\xi}{2}\right)^n, \quad K_n(\xi) \to \begin{cases} -\left[\ln\left(\frac{\xi}{2}\right) + \gamma\right], & \text{for } n = 0, \\ \frac{(n-1)!}{2}\left(\frac{\xi}{2}\right)^{-n}, & \text{for } n \neq 0, \end{cases} \tag{2.157}$$

However, the asymptotic behavior of the modified functions is very different, with $I_n(x)$ exponentially growing, and $K_n(\xi)$ exponentially dropping at $\xi \to \infty$:

$$I_n(\xi) \to \left(\frac{1}{2\pi\xi}\right)^{1/2} e^{\xi}, \quad K_n(\xi) \to \left(\frac{\pi}{2\xi}\right)^{1/2} e^{-\xi}. \tag{2.158}$$

To complete our brief survey of the Bessel functions, let me note that all them discussed so far may be considered as particular cases of *Bessel functions of the complex argument*, say $J_n(z)$ and $Y_n(z)$, or, alternatively, $H_n^{(1,2)}(z) = J_n(z) \pm iY_n(z)$.[51] The 'usual' Bessel functions $J_n(\xi)$ and $Y_n(\xi)$ may be considered as the

[51] These complex functions still obey the general relations (2.143) and (2.146), with ξ replaced with z.

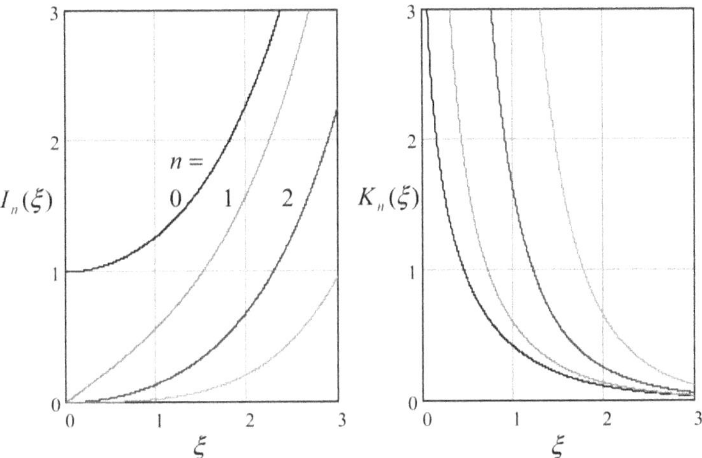

Figure 2.22. The modified Bessel functions of the first kind (left panel) and the second kind (right panel).

sets of values of these generalized functions on the real axis ($z = \xi$), while the modified functions as their particular case at $z = i\xi$, also with real ξ:

$$I_\nu(\xi) = i^{-\nu} J_\nu(i\xi), \quad K_\nu(\xi) = \frac{\pi}{2} i^{\nu+1} H_\nu^{(1)}(i\xi). \tag{2.159}$$

Moreover, this generalization of the Bessel functions to the whole complex plane z enables the use of their values along other directions on that plane, for example under angles $\pi/4 \pm \pi/2$. As a result, one arrives at the so-called *Kelvin functions*:

$$\begin{aligned} \text{ber}_\nu \xi + i \, \text{bei}_\nu \xi &\equiv J_\nu(\xi e^{-i\pi/4}), \\ \text{ker}_\nu \xi + i \, \text{kei}_\nu \xi &\equiv i \frac{\pi}{2} H_\nu^{(1)}(\xi e^{-i3\pi/4}), \end{aligned} \tag{2.160}$$

which are also useful for some important problems of physics and engineering. Unfortunately, I do not have time to discuss these problems in this course[52].

2.8 Variable separation—spherical coordinates

The spherical coordinates are very important in physics, because of the (approximate) spherical symmetry of many physical objects—from nuclei and atoms, to water drops in clouds, to planets and stars. Let us again require each component ϕ_k of Eq. (2.84) to satisfy the Laplace equation. Using the full expression for the Laplace operator in spherical coordinates[53], we obtain

[52] Later in the course we will also run into the so-called *spherical Bessel functions* $j_n(\xi)$ and $y_n(\xi)$, which may be expressed via the Bessel functions of a semi-integer order. Surprisingly enough, the spherical Bessel functions turn out to be much simpler than $J_n(\xi)$ and $Y_n(\xi)$.

[53] See, e.g. Eq. (A.67).

$$\frac{1}{r^2}\frac{\partial}{\partial r}\left(r^2\frac{\partial \phi_k}{\partial r}\right) + \frac{1}{r^2 \sin\theta}\frac{\partial}{\partial \theta}\left(\sin\theta\frac{\partial \phi_k}{\partial \theta}\right) + \frac{1}{r^2 \sin^2\theta}\frac{\partial^2 \phi_k}{\partial \varphi^2} = 0. \tag{2.161}$$

Let us look for a solution of this equation in the following variable-separated form:

$$\phi_k = \frac{\mathcal{R}(r)}{r}\mathcal{P}(\cos\theta)\mathcal{F}(\varphi), \tag{2.162}$$

Separating the variables one by one, just as has been done in cylindrical coordinates, we obtain the following equations for the functions participating in this solution:

$$\frac{d^2\mathcal{R}}{dr^2} - \frac{l(l+1)}{r^2}\mathcal{R} = 0, \tag{2.163}$$

$$\frac{d}{d\xi}\left[(1-\xi^2)\frac{d\mathcal{P}}{d\xi}\right] + \left[l(l+1) - \frac{\nu^2}{1-\xi^2}\right]\mathcal{P} = 0, \tag{2.164}$$

$$\frac{d^2\mathcal{F}}{d\varphi^2} + \nu^2\mathcal{F} = 0, \tag{2.165}$$

where $\xi \equiv \cos\theta$ is a new variable in lieu of θ (so that $-1 \leqslant \xi \leqslant +1$), while ν^2 and $l(l+1)$ are the separation constants. (The reason for selection of the latter one in this form will be clear in a minute.)

One can see that Eq. (2.165) is very simple, and is absolutely similar to the Eq. (2.107) we obtained for the cylindrical coordinates. Moreover, the equation for the radial functions is simpler than in the cylindrical coordinates. Indeed, let us look for its solution in the form cr^α—just as we have done with Eq. (2.106). Plugging this solution into Eq. (2.163), we immediately obtain the following condition on the parameter α:

$$\alpha(\alpha - 1) = l(l+1). \tag{2.166}$$

This quadratic equation has two roots, $\alpha = l + 1$ and $\alpha = -l$, so that the general solution to Eq. (2.163) is

$$\mathcal{R} = a_l r^{l+1} + \frac{b_l}{r^l}. \tag{2.167}$$

However, the general solution of Eq. (2.164) (called either the *general* or *associated Legendre equation*) cannot be expressed via what is usually called elementary functions—although there is no generally accepted line between 'elementary' and 'special' functions.

Let us start its discussion from the axially symmetric case, when $\partial\phi/\partial\varphi = 0$. This means $\mathcal{F}(\varphi) = \text{const}$, and thus $\nu = 0$, so that Eq. (2.164) is reduced to the so-called *Legendre differential equation*:

$$\frac{d}{d\xi}\left[(1-\xi^2)\frac{d\mathcal{P}}{d\xi}\right]+l(l+1)\mathcal{P}=0. \quad (2.168)$$

One can readily check that the solutions of this equation for integer values of l are specific (*Legendre*) polynomials[54] that may be described by the following *Rodrigues formula*:

$$\mathcal{P}_l(\xi)=\frac{1}{2^l l!}\frac{d^l}{d\xi^l}(\xi^2-1)^l, \quad l=0, 1, 2, \dots. \quad (2.169)$$

As this formula shows, the first few Legendre polynomials are pretty simple:

$$\begin{aligned}
\mathcal{P}_0(\xi) &= 1,\\
\mathcal{P}_1(\xi) &= \xi,\\
\mathcal{P}_2(\xi) &= \frac{1}{2}(3\xi^2-1),\\
\mathcal{P}_3(\xi) &= \frac{1}{2}(5\xi^3-3\xi),\\
\mathcal{P}_4(\xi) &= \frac{1}{8}(35\xi^4-30\xi^2+3), \dots,
\end{aligned} \quad (2.170)$$

though such explicit expressions become more and more bulky as l is increased. As figure 2.23 shows, all these polynomials, which are defined on the [−1, +1] segment, start at the same point, $\mathcal{P}_l(+1)=+1$, and end up either at the same point or at the

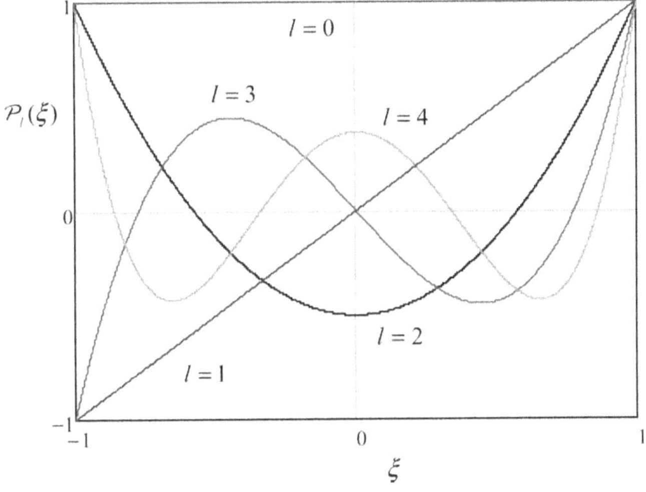

Figure 2.23. A few lowest Legendre polynomials $\mathcal{P}_l(\xi)$.

[54] For the reader's reference: if l is not integer, the general solution of Eq. (2.168) may be represented as a linear combination of the so-called *Legendre functions* (not polynomials!) *of the first and second kind*, $\mathcal{P}_l(\xi)$ and $\mathcal{Q}_l(\xi)$.

opposite point: $\mathcal{P}_l(-1) = (-1)^l$. Between these two end points, the lth polynomial has l zeros. It is straightforward to use Eq. (2.169) to prove that these polynomials form a full, orthogonal set of functions, with the following normalization rule:

$$\int_{-1}^{+1} \mathcal{P}_l(\xi)\mathcal{P}_{l'}(\xi)d\xi = \frac{2}{2l+1}\delta_{ll'}, \tag{2.171}$$

so that any function $f(\xi)$, defined on the segment $[-1, +1]$, may be represented as a unique series over the polynomials[55].

Thus, taking into account the additional division by r in Eq. (2.162), the general solution of any axially symmetric Laplace problem may be represented as

$$\phi(r, \theta) = \sum_{l=0}^{\infty}\left(a_l r^l + \frac{b_l}{r^{l+1}}\right)\mathcal{P}_l(\cos\theta). \tag{2.172}$$

Please note a strong similarity between this solution and Eq. (2.112) for the 2D Laplace problem in the polar coordinates. However, apart from the difference in angular functions, there is also a difference (by one) in the power of the second radial function, and this difference immediately shows up in problem solutions.

Indeed, let us solve a problem similar to that shown in figure 2.15: find the electric field around a conducting sphere of radius R, placed into an initially uniform external field $\mathbf{E}_0$ (whose direction I will take for axis z)—see figure 2.24a. If we select the arbitrary constant in the electrostatic potential so that $\phi|_{z=0} = 0$, then in Eq. (2.172) we should take $a_0 = b_0 = 0$. Now, just as has been argued for the cylindrical case, at $r \gg R$ the potential should approach that of the uniform field:

$$\phi \to -E_0 z = -E_0 r\cos\theta, \tag{2.173}$$

so that in Eq. (2.172), only one of the coefficients a_l survives: $a_l = -E_0\delta_{l1}$. As a result, from the boundary condition on the surface, $\phi(R, \theta) = 0$, we obtain:

$$0 = \left(-E_0 R + \frac{b_1}{R^2}\right)\cos\theta + \sum_{l\geqslant 2}\frac{b_l}{R^{l+1}}\mathcal{P}_l(\cos\theta). \tag{2.174}$$

Now repeating the argumentation that led to Eq. (2.117), we may conclude that Eq. (2.174) is satisfied if

$$b_l = E_0 R^3\delta_{l,1}, \tag{2.175}$$

so that, finally, Eq. (2.172) is reduced to

$$\phi = -E_0\left(r - \frac{R^3}{r^2}\right)\cos\theta. \tag{2.176}$$

[55] This is the reason why, at least for the purposes of this course, there is no reason in pursuing (more complex) solutions to Eq. (2.168) for non-integer values of l, which were mentioned in the previous footnote.

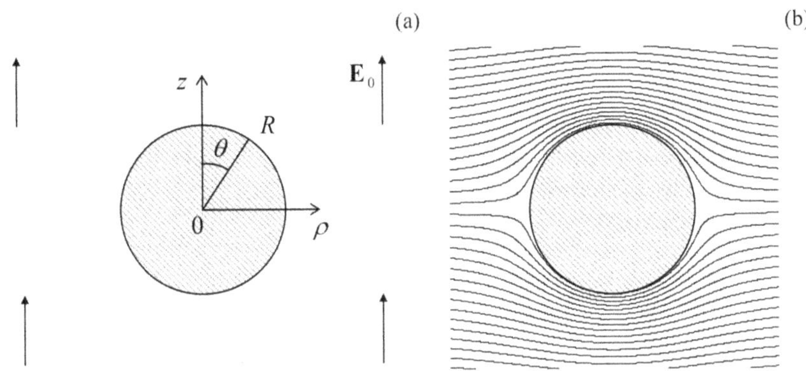

Figure 2.24. Conducting sphere in a uniform electric field: (a) the problem's geometry, and (b) the equipotential surface pattern given by Eq. (2.176). The pattern is qualitatively similar but quantitatively different from that for the conducting cylinder in a perpendicular field—cf figure 2.15.

This distribution, shown in figure 2.24b, is very much similar to Eq. (2.117) for the cylindrical case (cf figure 2.15b, with the account for a different plot orientation), but with a different power of the radius in the second term. This difference leads to a quantitatively different distribution of the surface electric field:

$$E_n = -\left.\frac{\partial \phi}{\partial r}\right|_{r=R} = 3E_0 \cos \theta, \tag{2.177}$$

so that its maximal value is a factor of 3 (rather than 2) larger than the external field.

Now let me briefly (mostly just for the reader's reference) mention the Laplace equation solutions in the general case (with no axial symmetry). If the free space surrounds the origin from all sides, the solutions to Eq. (2.165) have to be 2π-periodic, and hence $\nu = n = 0, \pm 1, \pm 2, \ldots$ Mathematics says that Eq. (2.164) with integer $\nu = n$ and a fixed integer l has a solution only for a limited range of n:[56]

$$-l \leqslant n \leqslant +l. \tag{2.178}$$

These solutions are called the *associated Legendre functions* (generally, they are not polynomials!) For $n \geqslant 0$, these functions may be defined via the Legendre polynomials, using the following formula[57]:

$$\mathcal{P}_l^n(\xi) = (-1)^n (1 - \xi^2)^{n/2} \frac{d^n}{d\xi^n} \mathcal{P}_l(\xi). \tag{2.179}$$

On the segment $\xi \in [-1, +1]$, each set of the associated Legendre functions with a fixed index n and non-negative l form a full, orthogonal set, with the normalization relation,

[56] In quantum mechanics, the letter n is typically reserved to be used for the 'principal quantum number', while the azimuthal functions are numbered by index m. However, here I will keep using n as their index, because for this course's purposes, this seems more logical in view of the similarity of the spherical and cylindrical functions.

[57] Note that some texts use different choices for the front factor (called the *Condon–Shortley phase*) in the functions $\mathcal{P}_l^m$, which do not affect the final results for the spherical harmonics Y_l^m—see below.

$$\int_{-1}^{+1} \mathcal{P}_l^n(\xi)\mathcal{P}_{l'}^n(\xi)d\xi = \frac{2}{2l+1}\frac{(l+n)!}{(l-n)!}\delta_{ll'}, \tag{2.180}$$

that is evidently a generalization of Eq. (2.171).

Since these relations may seem a bit intimidating, let me write down explicit expressions for a few $\mathcal{P}_l^n$ (cosθ) with the lowest values of l and $n \geqslant 0$, which are most important for applications.

$$l = 0: \quad \mathcal{P}_0^0(\cos\theta) = 1; \tag{2.181}$$

$$l = 1: \quad \begin{cases} \mathcal{P}_1^0(\cos\theta) = \cos\theta, \\ \mathcal{P}_1^1(\cos\theta) = -\sin\theta; \end{cases} \tag{2.182}$$

$$l = 2: \begin{cases} \mathcal{P}_2^0(\cos\theta) = (3\cos^2\theta - 1)/2, \\ \mathcal{P}_2^1(\cos\theta) = -2\sin\theta\cos\theta, \\ \mathcal{P}_2^2(\cos\theta) = -3\cos^2\theta\ . \end{cases} \tag{2.183}$$

The reader should agree there is not much to fear is these functions—they are just the sums of products of functions $\cos\theta \equiv \xi$ and $\sin\theta \equiv (1-\xi^2)^{1/2}$. Figure 2.25 shows the plots of a few lowest functions $\mathcal{P}_l^n$ (ξ).

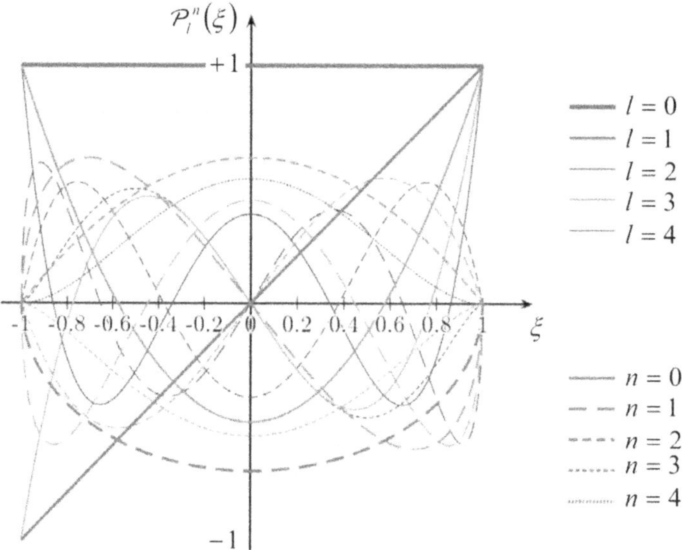

Figure 2.25. A few lowest associated Legendre functions. (Adapted from an original by *Geek3*, available at https://en.wikipedia.org/wiki/Associated_Legendre_polynomials, under the GNU Free Documentation License.)

Using the associated Legendre functions, the general solution (2.162) to the Laplace equation in the spherical coordinates may be expressed as

$$\phi(r, \theta, \varphi) = \sum_{l=0}^{\infty}\left(a_l r^l + \frac{b_l}{r^{l+1}}\right)\sum_{n=0}^{l} \mathcal{P}_l^n(\cos\theta)\mathcal{F}_n(\varphi),$$
$$\mathcal{F}_n(\varphi) = c_n \cos n\varphi + s_n \sin n\varphi. \tag{2.184}$$

Since the difference between angles θ and φ is somewhat artificial, physicists prefer to think not about functions $\mathcal{P}$ and $\mathcal{F}$ in separation, but directly about their products that participate in this solution[58].

As a rare exception for my courses, in order to save time, I will skip giving an example of using the associated Legendre functions here, because quite a few such examples of that will be given in the quantum mechanics part of these series.

2.9 Charge images

So far, we have discussed various methods of solution of the *Laplace* boundary problem (2.35). Let us now move on to the discussion of its generalization, the *Poisson* equation (1.41), that we need when besides the conductors, we also have 'stand-alone' charges with a known spatial distribution $\rho(\mathbf{r})$. (This will also allow us, better equipped, to revisit the Laplace problem again in the next section.)

Let us start with a somewhat limited, but very useful *charge image* (or 'image charge') *method.* Consider a very simple problem: a single point charge near a conducting half-space—see figure 2.26.

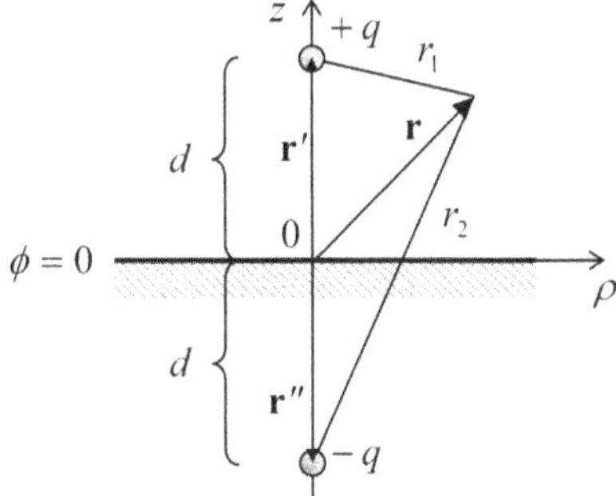

Figure 2.26. The simplest problem readily solvable by the charge image method. Point colors are used, as before, to denote the charges of the original (red) and opposite (blue) sign.

[58] In quantum mechanics, it is more convenient to use a slightly different, alternative set of basic functions, namely the following complex functions, called the *spherical harmonics*,

$$Y_l^n(\theta, \varphi) \equiv \left[\frac{2l+1}{4\pi}\frac{(l-n)!}{(l+n)!}\right]^{1/2} \mathcal{P}_l^n(\cos\theta)e^{in\varphi},$$

which are defined for both positive and negative n (within the limits $-l \leqslant n \leqslant +l$)—see, e.g. *Part QM* sections 3.6 and 5.6. Note again that in that field, the index n is traditionally denoted as m, and called the *magnetic quantum number*.

Let us prove that its solution, above the conductor's surface ($z \geqslant 0$), may be represented as,

$$\phi(\mathbf{r}) = \frac{1}{4\pi\varepsilon_0}\left(\frac{q}{r_1} - \frac{q}{r_2}\right) = \frac{q}{4\pi\varepsilon_0}\left(\frac{1}{|\mathbf{r} - \mathbf{r}'|} - \frac{1}{|\mathbf{r} - \mathbf{r}''|}\right), \tag{2.185}$$

or in a more explicit form, using the cylindrical coordinates shown in figure 2.26:

$$\phi(\mathbf{r}) = \frac{q}{4\pi\varepsilon_0}\left(\frac{1}{[\rho^2 + (z - d)^2]^{1/2}} - \frac{1}{[\rho^2 + (z + d)^2]^{1/2}}\right), \tag{2.186}$$

where ρ is the distance of the observation point from the 'vertical' line on which the charge is located. Indeed, this solution evidently satisfies both the boundary condition $\phi = 0$ at the surface of the conductor ($z = 0$), and the Poisson equation (1.41), with the single δ-functional source at point $\mathbf{r}' = \{0, 0, d\}$ on its right-hand side, because the second singularity of the solution, at point $\mathbf{r}'' = \{0, 0, -d\}$, is outside the region of the solution's validity ($z \geqslant 0$).

Physically, this solution may be interpreted as the sum of the fields of the actual charge ($+q$) at point $\mathbf{r}'$, and an equal but opposite charge ($-q$) at the 'mirror image' point $\mathbf{r}''$ (figure 2.26). This is the basic idea of the charge image method. Before moving on to more complex problems, let us discuss the situation shown in figure 2.26 in a little bit more detail, due to its fundamental importance. First, we can use Eqs. (2.3) and (2.186) to calculate the surface charge density:

$$\begin{aligned}\sigma &= -\varepsilon_0\frac{\partial\phi}{\partial z}\bigg|_{z=0} = -\frac{q}{4\pi}\frac{\partial}{\partial z}\left(\frac{1}{[\rho^2 + (z - d)^2]^{1/2}} - \frac{1}{[\rho^2 + (z + d)^2]^{1/2}}\right)_{z=0} \\ &= -\frac{q}{4\pi}\frac{2d}{(\rho^2 + d^2)^{3/2}}.\end{aligned} \tag{2.187}$$

The total surface charge is

$$Q = \int_A \sigma\, d^2r = 2\pi\int_0^\infty \sigma(\rho)\rho\, d\rho = -\frac{q}{2}\int_0^\infty \frac{d}{(\rho^2 + d^2)^{3/2}}2\rho\, d\rho. \tag{2.188}$$

This integral may be easily worked out using the substitution $\xi \equiv \rho^2/d^2$ (giving $d\xi = 2\rho\, d\rho/d^2$):

$$Q = -\frac{q}{2}\int_0^\infty \frac{d\xi}{(\xi + 1)^{3/2}} = -q. \tag{2.189}$$

This result is very natural, because the conductor 'wants' to bring as much surface charge from its interior to the surface as necessary to fully compensate the initial charge ($+q$) and hence to kill the electric field at large distances as efficiently as possible, hence reducing the total electrostatic energy (1.65) to the lowest possible value.

For a better feeling of this *polarization charge* of the surface, let us take our calculations to the extreme, to q equal to one elementary change e, and place a particle with this charge (for example, a proton) at a macroscopic distance—say 1 m—from the conductor's surface. Then, according to Eq. (2.189), the total polarization charge of the surface equals that of an electron, and according to Eq. (2.187), its spatial extent is of the order of $d^2 = 1\ \text{m}^2$. This means that if we consider a much smaller part of the surface, $\Delta A \ll d^2$, its polarization charge magnitude $\Delta Q = \sigma \Delta A$ is much *less than one electron*! For example, Eq. (2.187) shows that the polarization charge of quite a macroscopic area $\Delta A = 1\ \text{cm}^2$ right under the initial charge ($\rho = 0$) is $e\Delta A/2\pi d^2 \approx 1.6 \times 10^{-5}\ e$. Can this be true, or our theory is somehow limited to the charges q much larger than e? (After all, the theory is substantially based on the approximate macroscopic model (2.1); maybe this is the culprit?)

Surprisingly enough, the answer to this question has become clear (at least to some physicists:-) only as late as in the mid-1980s when several experiments demonstrated, and theorists accepted, some rather grudgingly, that the usual polarization charge formulas are valid for elementary charges as well, i.e. the polarization charge ΔQ of a macroscopic surface area can indeed be less than e. The underlying reason for this paradox is the nature of the polarization charge of a conductor's surface: as was discussed in section 2.1, it is due not to new charged particles brought into the conductor (such charge would be in fact quantized in the units of e), but to a small *shift* of the free charges of a conductor by a very small distance from their equilibrium positions that they had in the absence of the external field induced by charge q. This shift is not quantized, at least on the scale relevant for our problem, and neither is ΔQ.

This understanding has opened a way toward the invention and experimental demonstration of several new devices including so-called *single-electron transistors*[59], which are used, in particular, for ultrasensitive measurement of polarization charges as small as $\sim 10^{-6}\ e$. Another important class of single-electron devices is the current standards based on the fundamental relation $I = -ef$, where I is the dc current carried by electrons transferred with the frequency f. The experimentally achieved[60] relative accuracy of such standards is of the order of 10^{-7}, and is not too far from that provided by the competing approach based on the combination of the Josephson effect and the quantum Hall effect[61].

Second, let us find the potential energy U of the charge-to-surface interaction. For that we may use the value of the electrostatic potential (2.185) at the point of the real charge ($\mathbf{r} = \mathbf{r}'$), of course ignoring the infinite potential created by the charge itself, so that the remaining potential is that of the image charge

---

[59] Actually, this term (for which the author of these notes should be blamed:-) is misleading: the operation of the 'single-electron transistor' is based on the interplay of discrete charges (multiples of e) transferred between conductors, and *sub*-single-electron polarization charges—see, e.g. [7].

[60] See, e.g. [8, 9].

[61] See [10].

$$\phi_{\text{image}}(\mathbf{r}') = -\frac{1}{4\pi\varepsilon_0}\frac{q}{2d}. \qquad (2.190)$$

Looking at the definition of the electrostatic potential, given by Eq. (1.31), it may be tempting to immediately write $U = q\phi_{\text{image}} = -(1/4\pi\varepsilon_0)(q^2/2d)$ (WRONG!), but this would be incorrect. The reason is that potential ϕ_{image} is not independent of q, but is actually induced by this charge. This is why the correct approach is to calculate U from Eq. (1.61), with just one term:

$$U = \frac{1}{2}q\phi_{\text{image}} = -\frac{1}{4\pi\varepsilon_0}\frac{q^2}{4d}, \qquad (2.191)$$

giving a twice lower energy than in the wrong result cited above. In order to double-check Eq. (2.191), and also obtain a better feeling of the factor ½ that distinguishes it from the wrong guess, we can recalculate U as the integral of the force exerted on the charge by the conductor (i.e. in our formalism, by the image charge):

$$U = -\int_{\infty}^{d} F(z)dz = \frac{1}{4\pi\varepsilon_0}\int_{\infty}^{d}\frac{q^2}{(2z)^2}dz = -\frac{1}{4\pi\varepsilon_0}\frac{q^2}{4d}. \qquad (2.192)$$

This calculation clearly accounts for the gradual build-up of the force F, as the real charge is brought from afar (where we have opted for $U = 0$) toward the surface.

This result has several important applications. For example, let us plot the electrostatic energy U for an electron, i.e. particle with charge $q = -e$, near a metallic surface, as a function of d. For that, we may use Eq. (2.191) until our macroscopic model (2.1) becomes invalid, and U transitions to some negative constant value ($-\psi$) inside the conductor—see figure 2.27a. Since our calculation was for an electron with zero potential energy at infinity, at relatively low temperatures, $k_BT \ll \psi$, electrons in a metal may occupy only the states with energies below $-\psi$ (the so-called Fermi level[62]). The positive constant ψ is called the *workfunction*, because it describes the smallest work necessary to remove the electron from a metal. As was discussed in section 2.1, in good metals the electric field screening takes place at interatomic distances $a_0 \approx 10^{-10}$ m. Plugging $d = a_0$ and $q = -e$ into Eq. (2.191), we

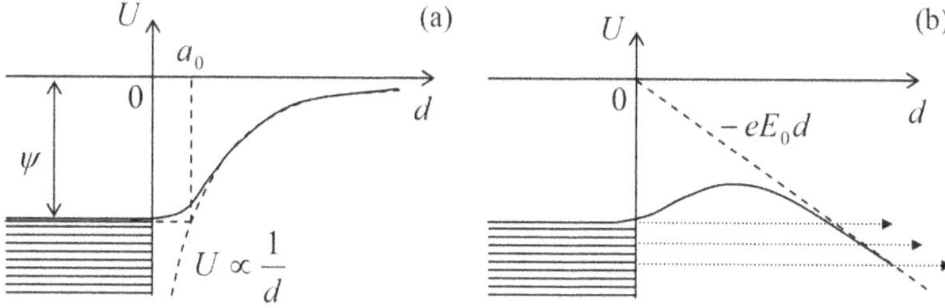

Figure 2.27. (a) Schematic representations of the origin of the workfunction and (b) the field emission of electrons.

[62] More discussion of these states may be found in *Part SM* sections 3.3 and 6.4.

obtain $\psi \approx 6 \times 10^{-19}$ J $\approx$ 3.5 eV. This crude estimate is in a surprisingly good agreement with the experimental values of the workfunction, ranging between 4 and 5 eV for most metals[63].

Next, let us consider the effect of an additional uniform external electric field $\mathbf{E}_0$ applied normally to a metallic surface, and toward it, on this potential profile. We can add the potential energy the field gives to the electron at distance d from the surface, $U_{\text{ext}} = -eE_0d$, to that created by the image charge. (As we know from Eq. (1.53), since field $\mathbf{E}_0$ is independent of the electron's position, its recalculation into the potential energy does not require the coefficient ½.) As the result, the potential energy of an electron near the surface becomes

$$U(d) = -eE_0d - \frac{1}{4\pi\varepsilon_0}\frac{e^2}{4d}, \qquad \text{for } d \gtrsim a_0, \tag{2.193}$$

with a similar crossover to $U = -\psi$ inside the conductor—see figure 2.27b. One can see that at the appropriate sign, and a sufficient magnitude of the applied field, it lowers the potential barrier that prevented the electron from leaving the conductor. At $E_0 \sim \psi/a_0$ (for metals, $\sim 10^{10}$ V m^{-1}), this suppression becomes so strong that electrons with energies at, and just below the Fermi level start quantum-mechanical tunneling through the remaining thin barrier. This is the *field emission* effect, which is used in vacuum electronics to provide efficient cathodes that do not require heating to high temperatures[64].

Returning to the basic electrostatics, let us consider some other geometries where the method of charge images may be effectively applied. First, let us consider a right corner (figure 2.28a). Reflecting the initial charge in the vertical plane we obtain the image shown in the top left corner of the panel. This image makes the boundary condition ϕ = const satisfied on the vertical surface of the corner. However, in order for the same to be true on the horizontal surface, we have to reflect *both* the initial charge *and* the image charge in the horizontal plane, flipping their signs. The final configuration of four charges, shown in figure 2.28a, evidently satisfies all the boundary conditions. The resulting potential distribution may be readily written as the generalization of Eq. (2.185). From it, the electric field and electric charge distributions, and the potential energy and forces acting on the charge may be calculated exactly as above—an easy exercise, left for the reader.

Next, consider a corner with angle $\pi/4$ (figure 2.28b). Here we need to repeat the reflection operation not two but four times before we arrive at the final pattern of eight positive and negative charges. (Any attempt to continue this process would lead to an overlap with the already existing charges.) This reasoning may be readily

[63] More discussion of the workfunction, and its effect on the electrons' kinetics, is given in *Part SM* section 6.4.

[64] The practical use of such 'cold' cathodes is affected by the fact that, as follows from our discussion in section 2.4, any nanoscale surface irregularity (a protrusion, an atomic cluster, or even a single 'adatom' stuck to the surface) may cause a strong increase of the local field well above the applied field E_0, making the emission reproducibility and stability in time significant challenges. In addition, the impact-ionization effects may lead to avalanche-type *electric breakdown* already at fields as low as $\sim 3 \times 10^6$ V m^{-1}.

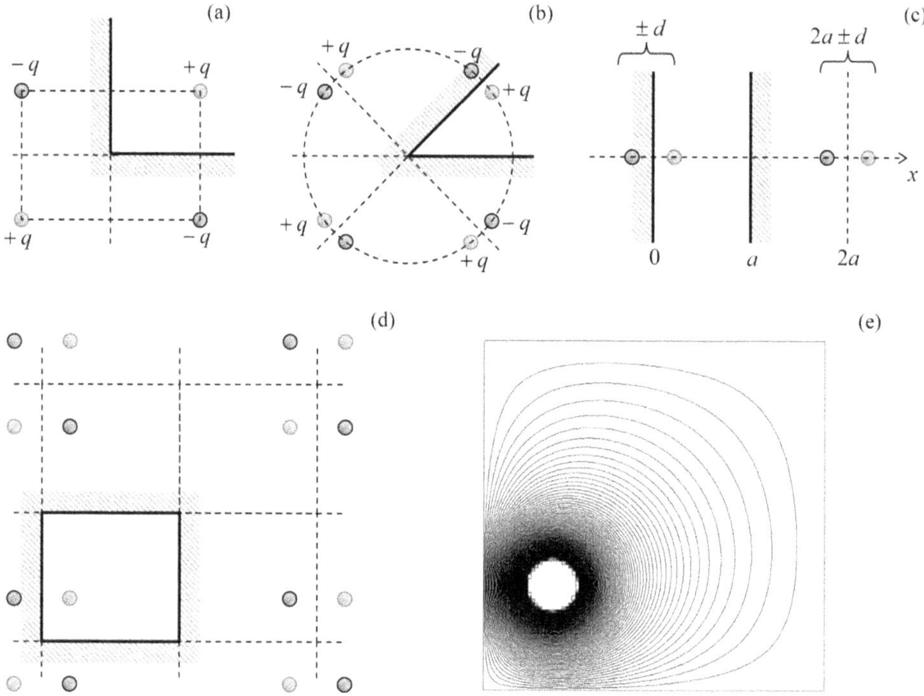

Figure 2.28. Charge images for (a, b) the corners with angles π and $\pi/2$, (c) a plane capacitor, and (d) a rectangular box. (e) The typical equipotential surfaces for the last system.

extended to a corner of angle $\beta = \pi/n$, with any integer n, which requires $2n$ charges (including the initial one) to satisfy all the boundary conditions.

Some configurations require an infinite number of images, but are still tractable. The most important of them is a system of two parallel conducting surfaces, i.e. a plane capacitor of an infinite area (figure 2.28c). Here the repeated reflection leads to an infinite system of charges $\pm q$ at points

$$x_j^{\pm} = 2aj \pm d, \tag{2.194}$$

where d (with $0 < d < a$) is the position of the initial charge, and j an arbitrary integer. The resulting infinite sum for the potential of the real charge q, created by the field of its images,

$$\begin{aligned}\phi(d) &= \frac{1}{4\pi\varepsilon_0}\left[-\frac{q}{2d} + \sum_{j\neq 0}\sum_{\pm}\frac{\pm q}{\left|d - x_j^{\pm}\right|}\right]\\ &\equiv -\frac{q}{4\pi\varepsilon_0}\left\{\frac{1}{2d} + \frac{d^2}{a^3}\sum_{j=1}^{\infty}\frac{1}{j[j^2 - (d/a)^2]}\right\},\end{aligned} \tag{2.195}$$

is converging (in its last form) very fast. For example, the exact value, $\phi(D/2) = -2\ln 2(q/4\pi\varepsilon_0 D)$, differs by less than 5% from the approximation using just the first term of the sum.

The same method may be applied to 2D (cylindrical) and 3D rectangular conducting boxes that require, respectively, a 2D or 3D infinite rectangular lattices of images; for example in a 3D box with sides a, b, and c, charges $\pm q$ are located at points (figure 2.28d)

$$\mathbf{r}^{\pm}_{jkl} = 2ja + 2kb + 2lc \pm \mathbf{r}', \tag{2.196}$$

where $\mathbf{r}'$ is the location of the initial (real) charge, and j, k, and l are arbitrary integers. Figure 2.28e shows the results of summation of the potentials of such charge set, including the real one, in a 2D box (within the plane of the real charge). One can see that the equipotential surfaces, concentric near the charge, are naturally leaning along the conducting walls of the box, which should be equipotential.

Even more surprisingly, the image charge method works very efficiently not only for rectilinear geometries, but also for spherical ones. Indeed, let us consider a point charge q at distance d from the center of a conducting, grounded sphere of radius R (figure 2.29a), and try to satisfy the boundary condition $\phi = 0$ for the electrostatic potential on sphere's surface using an imaginary charge q' located at some point located beyond the surface, i.e. inside the sphere.

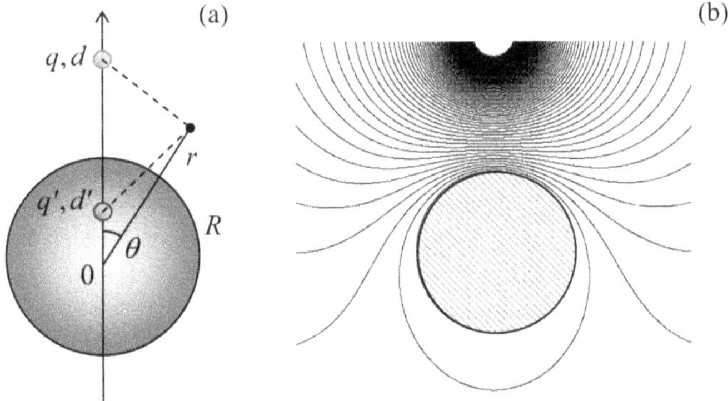

Figure 2.29. Method of charge images for a conducting sphere: (a) the concept and (b) the resulting potential distribution in the central plane containing the charge, for the particular case $d = 2R$.

From the problem's symmetry, it is clear that the point should be at the line passing through the real charge and the sphere's center, at some distance d' from the center. Then the total potential created by the two charges at an arbitrary point with $r \geqslant R$ (figure 2.29a) is

$$\phi(r, \theta) = \frac{1}{4\pi\varepsilon_0}\left[\frac{q}{(r^2 + d^2 - 2rd\cos\theta)^{1/2}} + \frac{q'}{(r^2 + d'^2 - 2rd'\cos\theta)^{1/2}}\right]. \tag{2.197}$$

This expression shows that we can make the two fractions to be equal and opposite at all points on the sphere's surface (i.e. for any θ at $r = R$), if we take[65]

[65] In geometry, such points, with $dd' = R^2$, are referred to as the result of mutual *inversion* in a sphere of radius R.

$$d' = \frac{R^2}{d}, \quad q' = -\frac{R}{d}q. \tag{2.198}$$

Since the solution to any Poisson boundary problem is unique, Eqs. (2.197) and (2.198) give us such a solution for this problem. Figure 2.29b shows a typical equipotential pattern calculated using Eqs. (2.197) and (2.198). It is surprising how formulas that simple may describe such a non-trivial field distribution.

Now let us calculate the total charge Q induced by charge q on the grounded sphere's surface. We could do this, as we have done for the conducting plane, by the brute-force integration of the surface charge density $\sigma = -\varepsilon_0 \partial\phi/\partial r|_r = R$. It is more elegant, however, to use the following Gauss law argument. The expression (2.197) is valid (at $r \geqslant R$) regardless of whether we are dealing with our real problem (charge q and the conducting sphere) or with the equivalent charge configuration (with the point charges q and q', but no sphere at all). Hence, according to Eq. (1.16), the Gaussian integral over a surface with radius $r = R + 0$, and the total charge inside the sphere should be also the same. Hence we immediately obtain

$$Q = q' = -\frac{R}{d}q. \tag{2.199}$$

The similar argumentation may be used to find the charge-to-sphere interaction force:

$$\begin{aligned} F = qE_{\text{image}}(d) &= q\frac{q'}{4\pi\varepsilon_0(d - d')^2} = -\frac{q^2}{4\pi\varepsilon_0}\frac{R}{d}\frac{1}{(d - R^2/d)^2} \\ &= -\frac{q^2}{4\pi\varepsilon_0}\frac{Rd}{(d^2 - R^2)^2}. \end{aligned} \tag{2.200}$$

(Note that this expression is legitimate only at $d > R$.) At large distances, $d \gg R$, this attractive force decreases as $1/d^3$. This unusual dependence arises because, as Eq. (2.198) specifies, the induced charge of the sphere, responsible for the force, is not constant but decreases as $1/d$. In the next chapter we will see that such force is also typical for the interaction between a point charge and a point dipole.

All the previous formulas were for a sphere that is grounded to keep its potential equal to zero. But what if we keep the sphere galvanically insulated, so that its net charge is fixed, for example, equals zero? Instead of solving this problem from scratch, let us use (again!) the linear superposition principle. For that, we may add to the previous problem an additional charge, equal to $-Q = -q'$, to the sphere, and argue that this addition gives, at all points, an additional, spherically symmetric potential that does not depend of the potential induced by charge q, and was calculated in section 1.2—see Eq. (1.19). For the interaction force, such addition yields

$$F = \frac{qq'}{4\pi\varepsilon_0(d - d')^2} + \frac{qq'}{4\pi\varepsilon_0 d^2} = -\frac{q^2}{4\pi\varepsilon_0}\left[\frac{Rd}{(d^2 - R^2)^2} - \frac{R}{d^3}\right]. \quad (2.201)$$

At large distances, the two terms proportional to $1/d^3$ cancel each other, giving $F \propto 1/d^5$. Such a rapid force decay is due to the fact that the field of the uncharged sphere is equivalent to that of two (equal and opposite) induced charges $+q'$ and $-q'$, and the distance between them ($d' = R^2/d$) tends to zero at $d \to \infty$. The potential energy of such an interaction behaves as $U \propto 1/d^6$ at $d \to \infty$; in the next chapter we will see that this is the general law of the induced dipole interaction.

2.10 Green's functions

I have spent so much time/space discussing the potential distributions created by a single point charge in various conductor geometries, because for any of the geometries, the generalization of these results to the arbitrary distribution $\rho(\mathbf{r})$ of free charges is straightforward. Namely, if a single charge q, located at point $\mathbf{r}'$, creates the electrostatic potential

$$\phi(\mathbf{r}) = \frac{1}{4\pi\varepsilon_0} qG(\mathbf{r}, \mathbf{r}'), \quad (2.202)$$

then, due to the linear superposition principle, an arbitrary charge distribution (either discrete or continuous) creates the potential

$$\phi(\mathbf{r}) = \frac{1}{4\pi\varepsilon_0}\sum_j q_j G(\mathbf{r}, \mathbf{r}_j) = \frac{1}{4\pi\varepsilon_0}\int \rho(\mathbf{r}')G(\mathbf{r}, \mathbf{r}')d^3r'. \quad (2.203)$$

The function $G(\mathbf{r}, \mathbf{r}')$ is called the (spatial) *Green's function*—the notion is very fruitful and hence popular in all fields of physics[66]. Evidently, as Eq. (1.35) shows, in the unlimited free space

$$G(\mathbf{r}, \mathbf{r}') = \frac{1}{|\mathbf{r} - \mathbf{r}'|}, \quad (2.204)$$

i.e. the Green's function depends only on one scalar argument—the distance between the field observation point $\mathbf{r}$ and the field-source (charge) point $\mathbf{r}'$. However, as soon as there are conductors around, the situation changes. In this course I will only discuss the Green's functions that vanish as soon as the radius-vector $\mathbf{r}$ points to the surface of any conductor[67]:

$$G(\mathbf{r}, \mathbf{r}')|_{\mathbf{r}\in A} = 0. \quad (2.205)$$

With this definition, it is straightforward to deduce the Green's functions for the solutions of the last section's problems in which conductors were grounded ($\phi = 0$).

[66] See, e.g. *Part CM* section 5.1, *Part QM* sections 2.2 and 7.4, and *Part SM* section 5.5.

[67] G thus defined is sometimes called the *Dirichlet function*.

For example, for a semi-space $z \geqslant 0$ limited by a conducting plane (figure 2.26), Eq. (2.185) yields

$$G = \frac{1}{|\mathbf{r} - \mathbf{r}'|} - \frac{1}{|\mathbf{r} - \mathbf{r}''|}, \quad \text{with } \rho'' = \rho' \text{ and } z'' = -z'. \qquad (2.206)$$

We see that in the presence of conductors (and, as we will see later, any other polarizable media), the Green's function may depend not only on the difference $\mathbf{r} - \mathbf{r}'$, but on each of these two arguments in a specific way.

So far, this looks just like re-naming our old results. The really non-trivial result of this formalism for electrostatics is that, somewhat counter-intuitively, the knowledge of the Green's function for a system with *grounded* conductors (figure 2.30a) enables the calculation of the field created by *voltage-biased* conductors (figure 2.30b), with the same geometry.

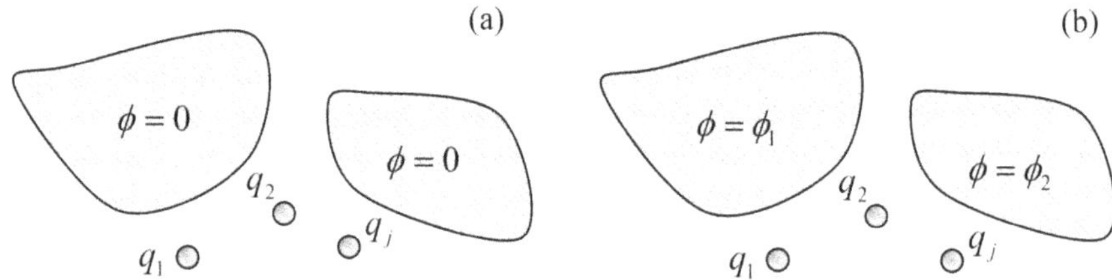

Figure 2.30. The Green's function method allows the solution of a simpler boundary problem (a) to be used to find the solution of a more complex problem (b), for the same conductor geometry.

In order to show this, let us use the so-called *Green's theorem* of the vector calculus[68]. The theorem states that for any two scalar, differentiable functions $f(\mathbf{r})$ and $g(\mathbf{r})$, and any volume V,

$$\int_V (f\nabla^2 g - g\nabla^2 f)d^3r = \oint_S (f\nabla g - g\nabla f)_n d^2r, \qquad (2.207)$$

where S is the surface limiting the volume. Applying the theorem to the electrostatic potential $\phi(\mathbf{r})$ and the Green's function G (also considered as a function of $\mathbf{r}$), let us use the Poisson equation (1.41) to replace $\nabla^2\phi$ with $(-\rho/\varepsilon_0)$, and notice that G, considered as a function of $\mathbf{r}$, obeys the Poisson equation with the δ-functional source:

$$\nabla^2 G(\mathbf{r}, \mathbf{r}') = -4\pi\delta(\mathbf{r} - \mathbf{r}'). \qquad (2.208)$$

(Indeed, according to its definition (2.202), this function may be formally considered as the field of a point charge $q = 4\pi\varepsilon_0$.) Now swapping the notation

[68] See, e.g. Eq. (A.80). Actually, this theorem is a ready corollary of the better-known divergence ('Gauss') theorem, Eq. (A.79).

of the radius-vectors, $\mathbf{r} \leftrightarrow \mathbf{r}'$, and using the Green's function symmetry, $G(\mathbf{r}, \mathbf{r}') = G(\mathbf{r}', \mathbf{r})$[69], we obtain

$$\begin{aligned}&-4\pi\phi(\mathbf{r}) - \int_V \left(-\frac{\rho(\mathbf{r}')}{\varepsilon_0}\right) G(\mathbf{r}, \mathbf{r}')d^3r' \\ &= \oint_S \left[\phi(\mathbf{r}')\frac{\partial G(\mathbf{r}, \mathbf{r}')}{\partial n'} - G(\mathbf{r}, \mathbf{r}')\frac{\partial \phi(\mathbf{r}')}{\partial n'}\right] d^2r'.\end{aligned} \tag{2.209}$$

Let us apply this relation to the volume V of *free space* between the conductors, and the boundary S drawn immediately outside of their surfaces. In this case, by its definition, the Green's function $G(\mathbf{r}, \mathbf{r}')$ vanishes at the conductor surface ($\mathbf{r} \in S$)—see Eq. (2.205). Now changing the sign of $\partial n'$ (so that it would be the outer normal for *conductors*, rather than free space volume V), dividing all terms by 4π, and partitioning the total surface S into the parts (numbered by index j) corresponding to different conductors (possibly, kept at different potentials ϕ_k), we finally arrive at the famous result[70]:

$$\phi(\mathbf{r}) = \frac{1}{4\pi\varepsilon_0}\int_V \rho(\mathbf{r}')G(\mathbf{r}, \mathbf{r}')d^3r' + \frac{1}{4\pi}\sum_k \phi_k \oint_{S_k} \frac{\partial G(\mathbf{r}, \mathbf{r}')}{\partial n'}d^2r'. \tag{2.210}$$

While the first term on the right-hand side of this relation is a direct and evident expression of the superposition principle, given by Eq. (2.203), the second term is highly non-trivial: it describes the effect of conductors with *non-zero* potentials ϕ_k (figure 2.30b), using the Green's function calculated for a similar system with *grounded* conductors, i.e. with all $\phi_k = 0$ (figure 2.29a). Let me emphasize that since our volume V excludes conductors, the first term on the right-hand side of Eq. (2.210) includes only the stand-alone charges in the system (in figure 2.30, marked q_1, q_2, etc), but not the surface charges of the conductors—which are taken into account, implicitly, by the second term.

In order to illustrate what a powerful tool Eq. (2.210) is, let us use to calculate the electrostatic field in two systems. In the first of them, a planar, circular, conducting disk, separated with a very thin cut from the remaining conducting plane, is biased with potential $\phi = V$, while the rest of the plane is grounded—see figure 2.31. If the width of the gap between the circle and rest of the plane is negligible, we may apply Eq. (2.210) without stand-alone charges, $\rho(\mathbf{r}') = 0$, and the Green's function for the uncut plane—see Eq. (2.206)[71]. In the cylindrical coordinates, with the origin at the disk's center (figure 2.31), the function is

[69] This symmetry, evident for the particular cases (2.204) and (2.206), may be readily proved for the general case, by applying Eq. (2.207) to functions $f(\mathbf{r}) \equiv G(\mathbf{r}, \mathbf{r}')$ and $g(\mathbf{r}) \equiv G(\mathbf{r}, \mathbf{r}'')$. With this substitution, the left-hand side becomes equal to $-4\pi\,[G(\mathbf{r}'', \mathbf{r}') - G(\mathbf{r}', \mathbf{r}'')]$, while the right-hand side is zero, due to Eq. (2.205).

[70] In some textbooks, the sign before the surface integral is negative, because their authors use the outer normal to the *free-space* region V rather than that *occupied by conductors*—as I do.

[71] Indeed, if all parts of the cut plane are grounded, a narrow cut does not change the field distribution, and hence the Green's function, significantly.

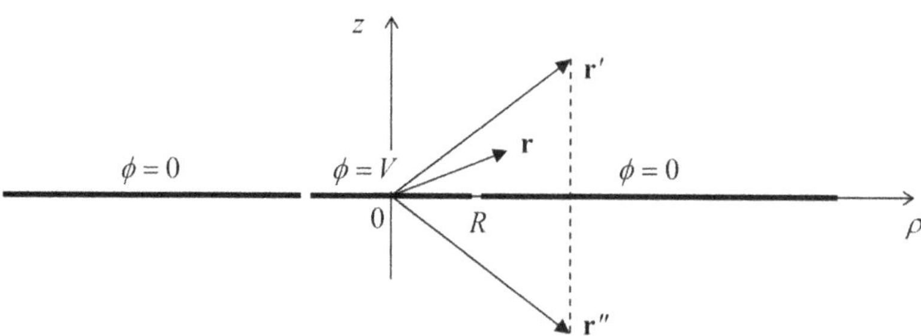

Figure 2.31. A voltage-biased conducting circle separated from the rest of a conducting plane.

$$G(\mathbf{r}, \mathbf{r}') = \frac{1}{[\rho^2 + \rho'^2 - 2\rho\rho' \cos(\varphi - \varphi') + (z - z')^2]^{1/2}} - \frac{1}{[\rho^2 + \rho'^2 - 2\rho\rho' \cos(\varphi - \varphi') + (z + z')^2]^{1/2}}. \qquad (2.211)$$

(The sum of the first three terms under the square roots in Eq. (2.211) is just the squared distance between the horizontal projections $\boldsymbol{\rho}$ and $\boldsymbol{\rho}'$ of vectors $\mathbf{r}$ and $\mathbf{r}'$ (or $\mathbf{r}''$), correspondingly, while the last terms are the squares of their vertical spacings.)

Now we can readily calculate the derivative participating in Eq. (2.210):

$$\left.\frac{\partial G}{\partial n'}\right|_S = \left.\frac{\partial G}{\partial z'}\right|_{z'=+0} = \frac{2z}{[\rho^2 + \rho'^2 - 2\rho\rho' \cos(\varphi - \varphi') + z^2]^{3/2}}. \qquad (2.212)$$

Due to the axial symmetry of the system, we can take φ for zero. With this, Eqs. (2.210) and (2.212) yield

$$\begin{aligned}\phi &= \frac{V}{4\pi} \oint_S \frac{\partial G(\mathbf{r}, \mathbf{r}')}{\partial n'} d^2r' \\ &= \frac{Vz}{2\pi} \int_0^{2\pi} d\varphi' \int_0^R \frac{\rho' d\rho'}{(\rho^2 + \rho'^2 - 2\rho\rho' \cos\varphi' + z^2)^{3/2}}.\end{aligned} \qquad (2.213)$$

This integral is not too pleasing, but may be readily worked out for points on the symmetry axis ($\rho = 0$):

$$\phi = Vz \int_0^R \frac{\rho' d\rho'}{(\rho'^2 + z^2)^{3/2}} = \frac{V}{2} \int_0^{R^2/z^2} \frac{d\xi}{(\xi + 1)^{3/2}} = V\left[1 - \frac{z}{(R^2 + z^2)^{1/2}}\right]. \qquad (2.214)$$

This expression shows that if $z \to 0$, the potential tends to V (as it should), while at $z \gg R$,

$$\phi \to V\frac{R^2}{2z^2}. \qquad (2.215)$$

Now, let us use the same Eq. (2.210) to solve the (in:-)famous problem of the cut sphere (figure 2.32). Again, if the gap between the two conducting semi-spheres is

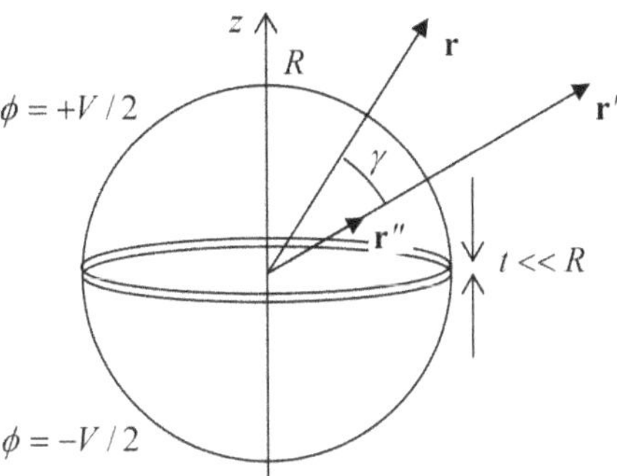

Figure 2.32. A system of two separated, oppositely biased semi-spheres.

very thin ($t \ll R$), we may use the Green's function for the grounded (and uncut) sphere. For a particular case $\mathbf{r}' = d\mathbf{n}_z$, this function is given by Eqs. (2.197) and (2.198); generalizing the former relation for an arbitrary direction of the vector $\mathbf{r}'$, we obtain

$$G = \frac{1}{[r^2 + r'^2 - 2rr' \cos \gamma]^{1/2}} - \frac{R/r'}{[r^2 + (R^2/r')^2 - 2r(R^2/r')\cos \gamma]^{1/2}}, \quad \text{for } r, r' \geqslant R, \tag{2.216}$$

where γ is the angle between the vectors $\mathbf{r}$ and $\mathbf{r}'$, and hence $\mathbf{r}''$—see figure 2.32.

Now, calculating the Green's function's derivative,

$$\left.\frac{\partial G}{\partial r'}\right|_{r'=R+0} = -\frac{(r^2 - R^2)}{R[r^2 + R^2 - 2Rr \cos \gamma]^{3/2}}, \tag{2.217}$$

and plugging it into Eq. (2.210), we see that the integration is again easy only for the field on the symmetry axis ($\mathbf{r} = r\mathbf{n}_z$, $\gamma = \theta$), giving:

$$\phi = \frac{V}{2}\left[1 - \frac{z^2 - R^2}{z(z^2 + R^2)^{1/2}}\right]. \tag{2.218}$$

For $z \to R$, $\phi \to V/2$ (just checking), while for $z \gg R$,

$$\phi \to V\frac{3R^2}{4z^2}. \tag{2.219}$$

As will be discussed in the next chapter, such a field is typical for an electric dipole.

2.11 Numerical methods

Despite the richness of analytical methods, for many boundary problems (in particular in geometries without a high degree of symmetry), numerical methods are the only way to the solution. Despite the current abundance of software codes and packages offering their automatic numerical solution[72], it is important for every educated physicist to understand 'what is under their hood', at least because most universal programs exhibit mediocre performance in comparison with custom codes written for particular problems, and sometimes do not converge at all, in particular for fast-changing (say, exponential) functions. The very brief discussion presented here[73] is a (hopefully, useful) fast glance under the hood, though it is certainly insufficient for professional numerical research work.

The simplest of the numerical approaches to the solution of partial differential equations, such as the Poisson or the Laplace equations (1.41) and (1.42), is the *finite-difference* method[74], in which the sought function of N scalar arguments $f(r_1, r_2, \ldots r_N)$ is represented by its values in discrete points of a rectangular grid (frequently called the *mesh*) of the corresponding dimensionality—see figure 2.33. Each partial second derivative of the function is approximated by the formula that readily follows from the linear approximations of the function f and then its partial derivatives—see figure 2.33a:

$$\begin{aligned}\frac{\partial^2 f}{\partial r_j^2} = \frac{\partial}{\partial r_j}\left(\frac{\partial f}{\partial r_j}\right) \approx \frac{1}{h}\left(\left.\frac{\partial f}{\partial r_j}\right|_{r_j+h/2} - \left.\frac{\partial f}{\partial r_j}\right|_{r_j-h/2}\right) \approx \frac{1}{h}\left[\frac{f_{\rightarrow} - f}{h} - \frac{f - f_{\leftarrow}}{h}\right] \\ = \frac{f_{\rightarrow} + f_{\leftarrow} - 2f}{h^2},\end{aligned} \tag{2.220}$$

where $f_{\rightarrow} \equiv f(r_j + h)$ and where $f_{\leftarrow} \equiv f(r_j - h)$. (The relative error of this approximation is of the order of $h^4\partial^4 f/\partial r_j^4$.) As a result, the action of a 2D Laplace operator on the function f may be approximated as

$$\frac{\partial^2 f}{\partial x^2} + \frac{\partial^2 f}{\partial y^2} = \frac{f_{\rightarrow} + f_{\leftarrow} - 2f}{h^2} + \frac{f_{\uparrow} + f_{\downarrow} - 2f}{h^2} = \frac{f_{\rightarrow} + f_{\leftarrow} + f_{\uparrow} + f_{\downarrow} - 4f}{h^2}, \tag{2.221}$$

and of the 3D operator, as

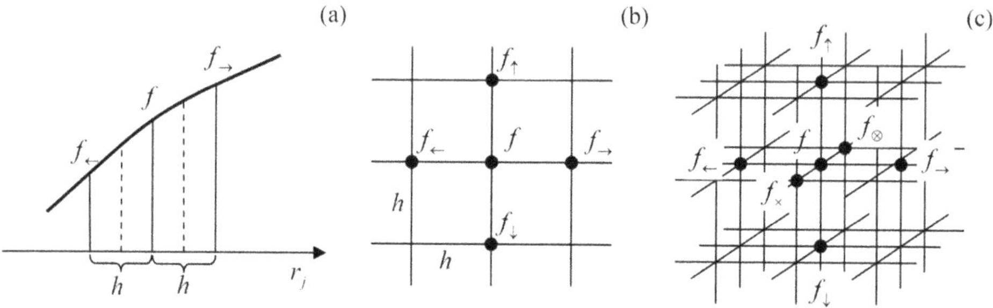

Figure 2.33. The general idea of the finite-difference method in (a) one, (b) two, and (c) three dimensions.

[72] See, for example, appendix A, sections A.16(iii) and (iv).

[73] It is almost similar to that given in *Part CM* section 8.5, and is reproduced here for the reader's convenience, being illustrated with examples from this (*Part EM*) course.

[74] For more details see, e.g. [11].

$$\frac{\partial^2 f}{\partial x^2} + \frac{\partial^2 f}{\partial y^2} + \frac{\partial^2 f}{\partial z^2} = \frac{f_{\rightarrow} + f_{\leftarrow} + f_{\uparrow} + f_{\downarrow} + f_{\otimes} + f_{\times} - 6f}{h^2}. \tag{2.222}$$

(The notation used in Eqs. (2.221) and (2.222) should be clear from figures 2.33b and c, respectively.)

Let us apply this scheme to find the electrostatic potential distribution inside a cylindrical box with conducting walls and square cross-section $a \times a$, using an extremely coarse mesh with step $h = a/2$ (figure 2.34). In this case our function, the electrostatic potential $\phi(x, y)$, equals zero at the side and bottom walls, and V_0 at the top lid, so that, according to Eq. (2.221), the 2D Laplace equation may be approximated as

$$\frac{0 + 0 + V_0 + 0 - 4\phi}{(a/2)^2} = 0. \tag{2.223}$$

The resulting value for the potential in the center of the box is $\phi = V_0/4$.

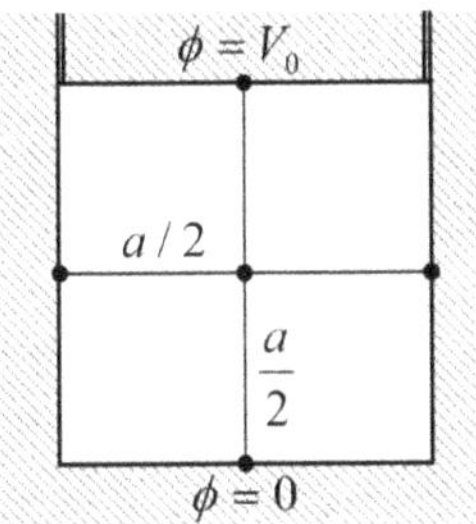

Figure 2.34. Numerically solving an internal 2D boundary problem for a conducting, cylindrical box with a square cross-section, using a very coarse mesh (with $h = a/2$).

Surprisingly, this is the *exact* value! This may be proved by solving this problem by the variable separation method, just as was done for the similar 3D problem in section 2.5. The result is

$$\phi(x, y) = \sum_{n=1}^{\infty} c_n \sin\frac{\pi n x}{a} \sinh\frac{\pi n y}{a}, \quad c_n = \frac{4V_0}{\pi n \sinh(\pi n)} \times \begin{cases} 1, & \text{if } n \text{ is odd,} \\ 0, & \text{otherwise.} \end{cases} \tag{2.224}$$

so that at the central point ($x = y = a/2$),

$$\begin{aligned} \phi &= \frac{4V_0}{\pi} \sum_{j=0}^{\infty} \frac{\sin[\pi(2j+1)/2] \sinh[\pi(2j+1)/2]}{(2j+1)\sinh[\pi(2j+1)]} \\ &\equiv \frac{2V_0}{\pi} \sum_{j=0}^{\infty} \frac{(-1)^j}{(2j+1)\cosh[\pi(2j+1)/2]}. \end{aligned} \tag{2.225}$$

The last series equals exactly to $\pi/8$, so that $\phi = V_0/4$.

For a similar 3D problem (a cubic box), with a similar 3D mesh, Eq. (2.222) yields

$$\frac{0 + 0 + V_0 + 0 + 0 + 0 - 6\phi}{(a/2)^2} = 0, \tag{2.226}$$

so that $\phi = V_0/6$. Unbelievably enough, this result is also exact! (This follows from our variable separation result expressed by Eqs. (2.95) and (2.99).)

Though such exact results should be considered as a happy coincidence rather than the general law, they still show that numerical methods, even with relatively crude meshes, may be more computationally efficient than the 'analytical' approaches, such as the variable separation method with its infinite-sum results that, in most cases, require computers anyway—at least for the result's comprehension and analysis.

A more powerful (but also much more complex) approach is the *finite-element* method in which the discrete point mesh, typically with triangular cells, is (automatically) generated in accordance with the system geometry[75]. Such mesh generators provide higher point concentration near sharp convex parts of conductor surfaces, where the field concentrates and hence the potential changes faster, and thus ensure a better accuracy-to-speed trade off than the finite-difference methods on a uniform grid. The price to pay for this improvement is the algorithm's complexity, which makes manual adjustments much harder. Unfortunately, I do not have time to go into the details of that method, and have to refer the reader to special literature on this subject[76].

2.12 Problems

Problem 2.1. Calculate the force (per unit area) exerted on a conducting surface by an external electric field, normal to it. Compare the result with the definition of the electric field, given by Eq. (1.6), and comment.

Problem 2.2. Certain electric charges Q_A and Q_B have been placed on two metallic, concentric spherical shells—see the figure below. What is the full charge of each of the surfaces S_1–S_4?

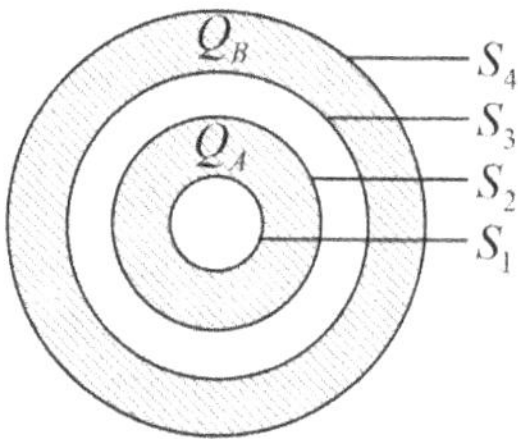

Problem 2.3. Calculate the mutual capacitance between the terminals of the lumped capacitor circuit shown in the figure below. Analyze and interpret the result for major particular cases.

[75] See, e.g. *Part CM* figure 8.14.

[76] See, e.g. [12] or [13].

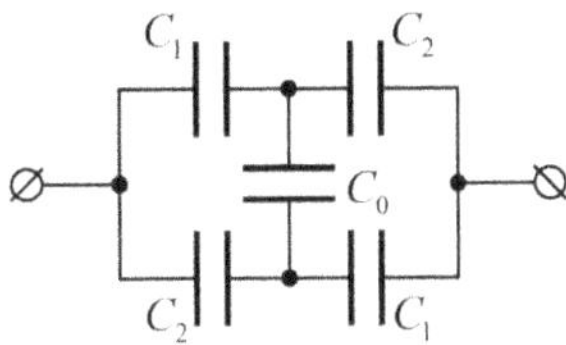

Problem 2.4. Calculate the mutual capacitance between the terminals of the semi-infinite lumped-capacitor circuit shown in the figure below, and the law of decay of the applied voltage along the system. Analyze and interpret the result.

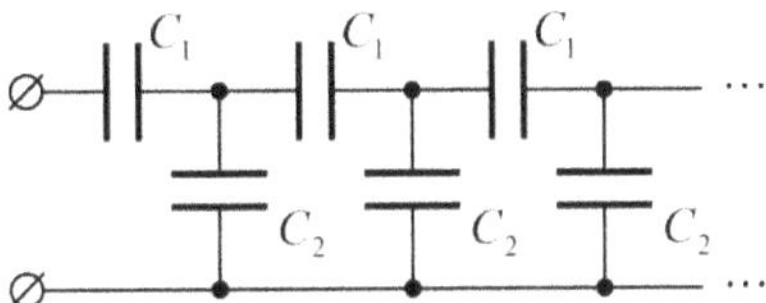

Problem 2.5. A system of two thin conducting plates is located over a ground plane as shown in the figure below, where A_1 and A_2 are the areas of the indicated plane parts, while d' and d'' are the distances between them. Neglecting the fringe effects, calculate:

(i) the effective capacitance of each plate, and
(ii) their mutual capacitance.

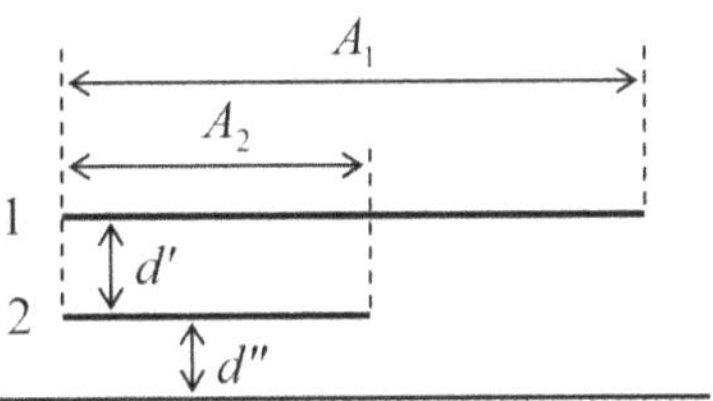

Problem 2.6. A wide, thin plane film, carrying a uniform electric charge density σ, is placed inside a plane capacitor whose plates are connected with a wire (see the figure below), and were initially electroneutral. Neglecting the edge effects, calculate the surface charges of the plates, and the net force acting on the film (per unit area).

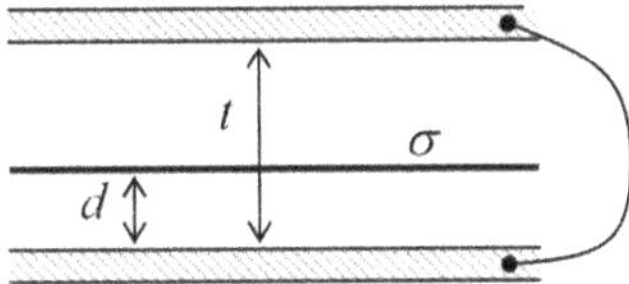

Problem 2.7. Following up the discussion of two weakly coupled spheres in section 2.2, find an approximate expression for the mutual capacitance (per unit length) between two very thin, parallel wires, both with round cross-sections, but each with its own radius. Compare the result with that for two small spheres, and interpret the difference.

Problem 2.8. Use the Gauss law to calculate the mutual capacitance of the following two-electrode systems, with the cross-section shown in figure 2.7 (reproduced below):

(i) a conducting sphere inside a concentric spherical cavity in another conductor, and
(ii) a conducting cylinder inside a coaxial cavity in another conductor. (In this case, we speak about the capacitance per unit length).

Compare the results with those obtained in section 2.2 using the Laplace equation solution.

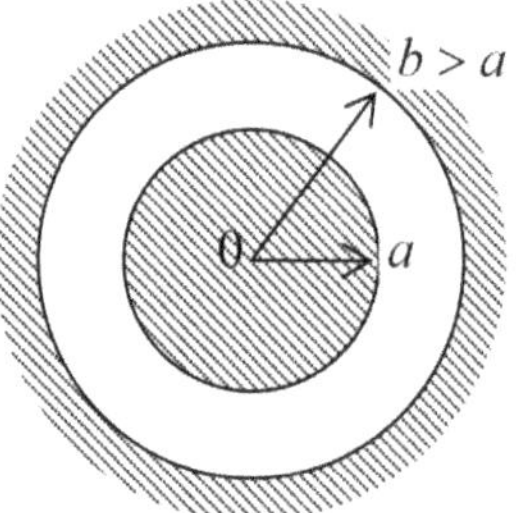

Problem 2.9. Calculate the electrostatic potential distribution around two barely separated conductors in the form of coaxial, round cones (see the figure below), with voltage V between them. Compare the result with that of a similar 2D problem, with the cones replaced by plane-face wedges. Can you calculate the mutual capacitances between the conductors in these systems? If not, can you estimate them?

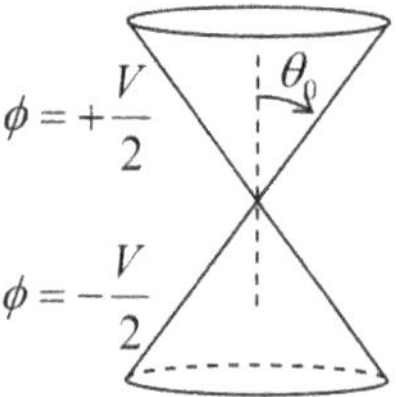

Problem 2.10. Calculate the mutual capacitance between two rectangular, planar electrodes of area $A = a \times l$, with a small angle $\varphi_0 \ll a/\rho_0$ between them—see the figure below.

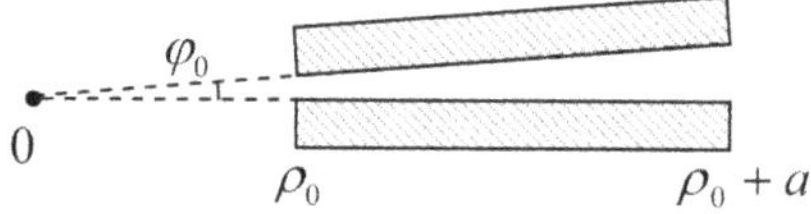

Problem 2.11. Using the results for a single thin round disk, obtained in section 2.4, consider a system of two such disks at a small distance $d \ll R$ from each other—see the figure below. In particular, calculate:

(i) the reciprocal capacitance matrix of the system,
(ii) the mutual capacitance between the disks,
(iii) the partial capacitance, and
(iv) the effective capacitance of one disk,

all in the first non-vanishing approximation in $d/R \ll 1$. Compare the results (ii)–(iv) and interpret their similarities and differences.

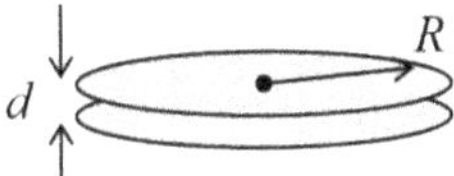

Problem 2.12.* Calculate the mutual capacitance (per unit length) between two cylindrical conductors forming a system with the cross-section shown in the figure below, in the limit $t \ll w \ll R$.

Hint: You may like to use *elliptical* (not 'ellipsoidal'!) *coordinates* $\{\alpha, \beta\}$ defined by the following equation:

$$x + iy = c\cosh(\alpha + i\beta), \tag{2.227}$$

with the appropriate choice of the constant c. In these orthogonal 2D coordinates, the Laplace operator is very simple[77]:

$$\nabla^2 = \frac{1}{c^2(\cosh^2\alpha - \cos^2\beta)}\left(\frac{\partial^2}{\partial\alpha^2} + \frac{\partial^2}{\partial\beta^2}\right).$$

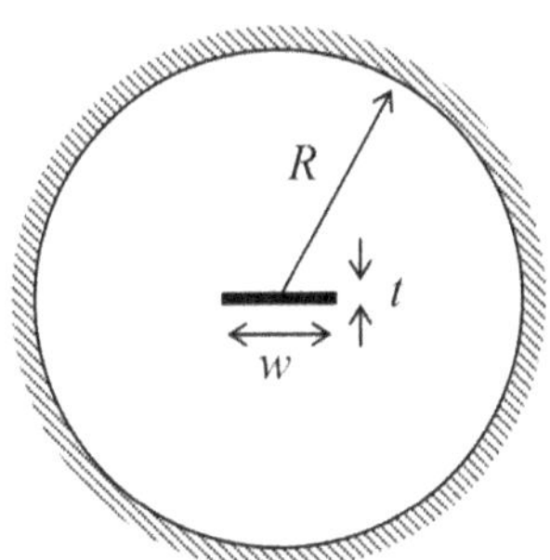

[77] This fact should not be surprising, because Eq. (2.227) is essentially the conformal map $z = c\cosh w$, where $z = x + iy$, and $w = \alpha + i\beta$—see the discussion in section 2.4.

Problem 2.13. Formulate 2D electrostatic problems that can be solved using each of the following analytic functions of the complex variable $z \equiv x + iy$:

(i) $w = \ln z$,
(ii) $w = z^{1/2}$,

and solve these problems.

Problem 2.14. On each side of a cylindrical volume with a rectangular cross-section $a \times b$, with no electric charges inside it, the electric field is uniform, normal to the side's plane, and opposite to that on the opposite side—see the figure below. Calculate the distribution of the electric potential inside the volume, provided that the field magnitude on the vertical sides equals E. Suggest a practicable method to implement such a potential distribution.

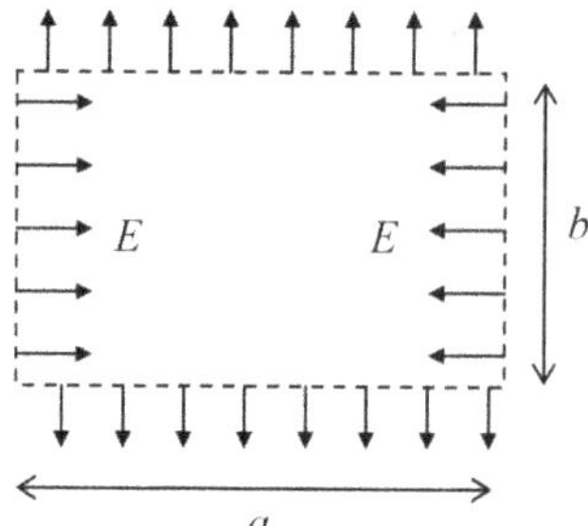

Problem 2.15. Complete the solution of the problem shown in figure 2.12, by calculating the distribution of the surface charge of the semi-planes. Can you calculate the mutual capacitance between the plates (per unit length)? If not, can you estimate it?

Problem 2.16.* A straight, long, thin, round-cylindrical metallic pipe has been cut, along its axis, into two equal parts—see the figure below.

(i) Use the conformal mapping method to calculate the distribution of the electrostatic potential, created by voltage V applied between the two parts, both outside and inside the pipe, and of the surface charge.
(ii) Calculate the mutual capacitance between pipe's halves (per unit length), taking into account a small width $2t \ll R$ of the cut.

Hints: In task (i) you may like to use the following complex function:

$$w = \ln\left(\frac{R + z}{R - z}\right),$$

while in task (ii) it is advisable to use the solution of the previous problem.

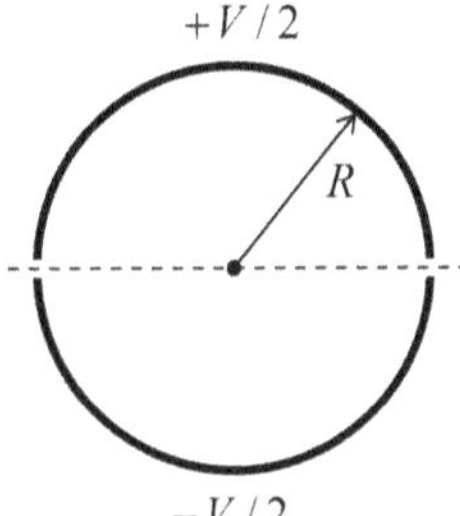

Problem 2.17. Solve task (i) of the previous problem using the variable separation method, and compare the results.

Problem 2.18. Use the variable separation method to calculate the potential distribution above the plane surface of a conductor, with a strip of width w separated by very thin cuts, and biased with voltage V—see the figure below.

Problem 2.19. The previous problem is now slightly modified: the cut-out and voltage-biased part of the conducting plane is now not a strip, but a square with side w. Calculate the potential distribution above the conductor's surface.

Problem 2.20. Each electrode of a large plane capacitor is cut into long strips of equal width w, with very narrow gaps between them. These strips are kept at the alternating potentials, as shown in the figure below. Use the variable separation method to calculate the electrostatic potential distribution, and explore the limit $w \ll d$.

Problem 2.21. Complete the cylinder problem started in section 2.7 (see figure 2.17), for the cases when the top lid's voltage is fixed as follows:

(i) $V = V_0 J_1(\xi_{11}\rho/R)\sin\varphi$, where $\xi_{11} \approx 3.832$ is the first root of the Bessel function $J_1(\xi)$;
(ii) $V = V_0 = \text{const}$.

For both cases, calculate the electric field at the centers of the lower and upper lids. (For task (ii), an answer including series and/or integrals is satisfactory.)

Problem 2.22. Solve the problem shown in figure 2.21. In particular:

(i) calculate and sketch the distribution of the electrostatic potential inside the system for various values of the ratio R/h, and
(ii) simplify the results for the limit $R/h \to 0$.

Problem 2.23. Use the variable separation method to find the potential distribution inside and outside a thin spherical shell of radius R, with a fixed potential distribution: $\phi(R, \theta, \varphi) = V_0 \sin\theta \cos\varphi$.

Problem 2.24. A thin spherical shell carries charge with areal density $\sigma = \sigma_0 \cos\theta$. Calculate the spatial distribution of the electrostatic potential and the electric field, both inside and outside the shell.

Problem 2.25. Use the variable separation method to calculate the potential distribution both inside and outside a thin spherical shell of radius R, separated with a very thin cut, along plane $z = 0$, into two halves, with voltage V applied between them—see the figure below. Analyze the solution; in particular, compare the field at the axis z, for $z > R$, with Eq. (2.218).

Hint: You may like to use the following integral of a Legendre polynomial with odd index $l = 1, 3, 5\ldots = 2n - 1$:[78]

$$I_n \equiv \int_0^1 P_{2n-1}(\xi)d\xi = \frac{1}{n!}\cdot\left(\frac{1}{2}\right)\cdot\left(-\frac{3}{2}\right)\cdot\left(-\frac{5}{2}\right)\ldots\left(\frac{3}{2}-n\right) \equiv (-1)^{n-1}\frac{(2n-3)!!}{2n(2n-2)!!}.$$

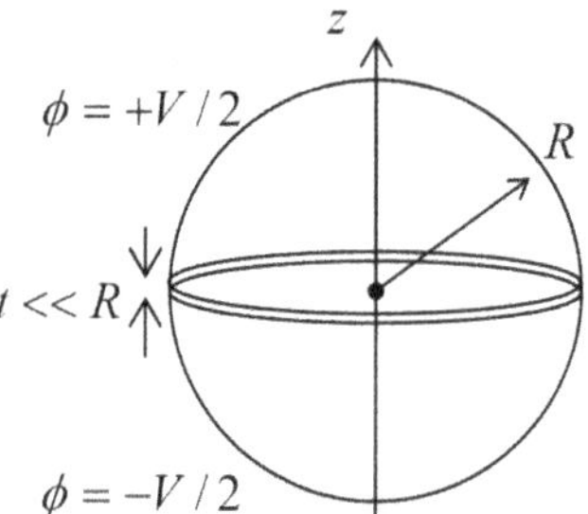

Problem 2.26. Calculate, up to the terms $O(1/r^2)$, the long-range electric field induced by a cut and voltage-biased conducting sphere—similar to that discussed in section 2.7 (see figure 2.32) and in the previous problem, but with the cut's plane at an arbitrary distance $d < R$ from the center—see the figure below.

---

[78] As a reminder, the *double factorial* (also called 'semi-factorial') operator (!!) is similar to the usual factorial operator (!), but with the product limited to numbers of the same parity as its argument (in our particular case, of the odd numbers in the nominator, and even numbers in the denominator).

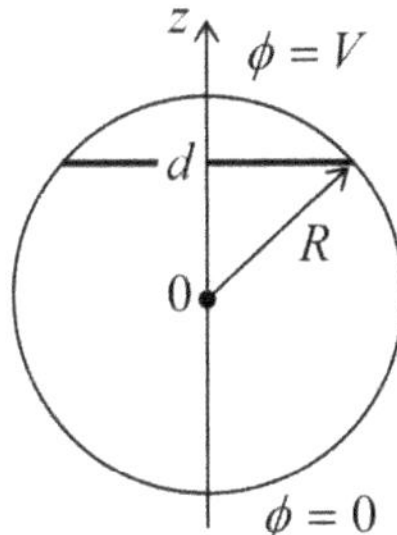

Problem 2.27. A small conductor (in this context, called the *single-electron island*) is placed between two conducting electrodes, with voltage V applied between them. The gap between the island and one of the electrodes is so narrow that electrons may tunnel quantum-mechanically through this 'junction'—see the figure below. Neglecting thermal excitations, calculate the equilibrium charge of the island as a function of V.

Hint: To solve this problem, you do not need to know much about the quantum-mechanical tunneling between conductors, except that such tunneling of an electron, followed by energy relaxation of the resulting excitations, may be considered as a single inelastic (energy-dissipating) event. At negligible thermal excitations, such an event takes place only when it decreases the total potential energy of the system[79].

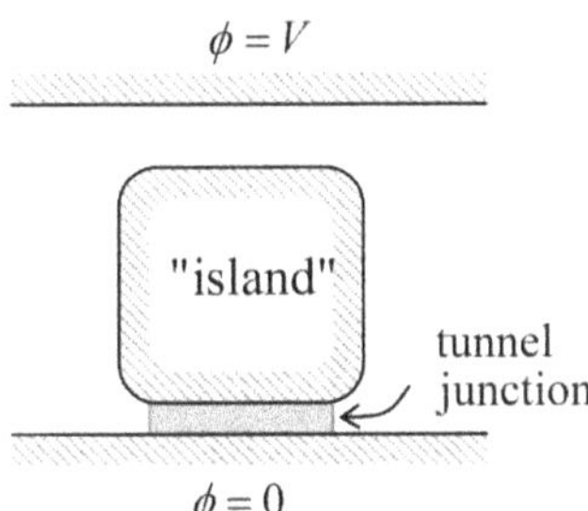

Problem 2.28. The system discussed in the previous problem is now generalized as shown in the figure to right. If the voltage V' applied between two bottom electrodes is sufficiently large, electrons can successively tunnel through two junctions of this system (called the *single-electron transistor*), carrying dc current between these electrodes. Neglecting thermal excitations, calculate the region of voltages V and V' where such a current is fully suppressed (*Coulomb-blocked*).

[79] Strictly speaking, this model, implying negligible quantum-mechanical coherence of the tunneling events, is correct only if the junction transparency is sufficiently low, so that its effective electric resistance is much higher than the quantum unit of resistance (see, e.g. *Part QM* section 3.2), $R_Q \equiv \pi\hbar/2e^2 \approx 6.5$ kΩ—which is usually the case.

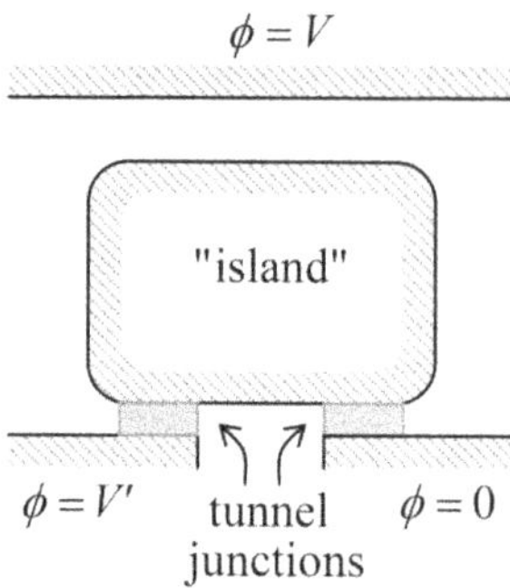

Problem 2.29. Use the image charge method to calculate the surface charges induced in the plates of a very broad plane capacitor of thickness D by a point charge q separated from one of the electrodes by distance d.

Problem 2.30. Prove the statement, made in section 2.9, that the 2D boundary problem, shown in the figure below, can be solved using a finite number of image charges if the angle β equals π/n, where $n = 1, 2,\ldots$

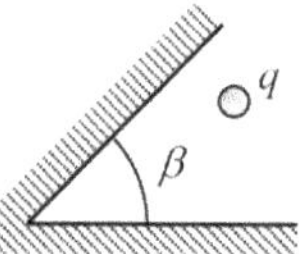

Problem 2.31. Use the image charge method to calculate the potential energy of the electrostatic interaction between a point charge placed in the center of a spherical cavity that was carved inside a grounded conductor, and the cavity's walls. Looking at the result, could it be obtained in a simpler way (or ways)?

Problem 2.32. Use the method of images to find the Green's function of the system shown in the figure below, where the bulge on the conducting plane has the shape of a semi-sphere of radius R.

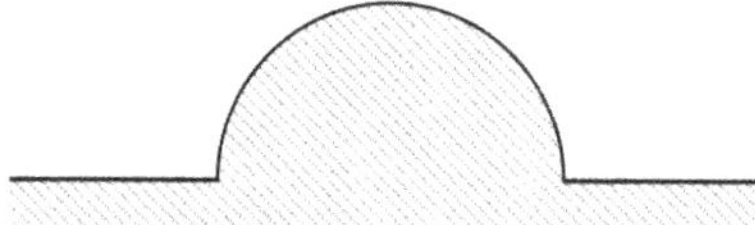

Problem 2.33.* Use the spherical inversion, expressed by Eq. (2.198), to develop an iterative method for more and more precise calculation of the mutual capacitance between two similar metallic spheres of radius R, with centers separated by distance $d > 2R$.

Problem 2.34.* A metallic sphere of radius R_1, carrying electric charge Q, is placed inside a spherical cavity of radius $R_2 > R_1$, cut inside another metal. Calculate the electric force exerted on the sphere if its center is displaced by a small distance $\delta \ll R_1, R_2 - R_1$ from that of the cavity—see the figure below.

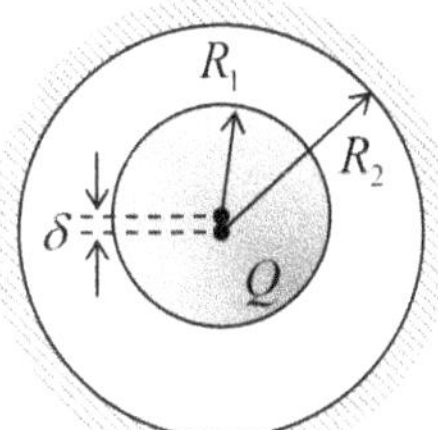

Problem 2.35. Within the simple models of the electric field screening in conductors, discussed in section 2.1, analyze the partial screening of the electric field of a point charge q by a plane, uniform conducting film of thickness $t \ll \lambda$, where λ is (depending on the charge carrier statistics) either the Debye or the Thomas–Fermi screening length—see, respectively, Eq. (2.8) or Eq. (2.10). Assume that the distance d between the charge and the film is much larger than t.

Problem 2.36. Suggest a convenient definition of the Green's function for 2D electrostatic problems and calculate it for:

(i) the unlimited free space, and
(ii) the free space above a conducting plane.

Use the latter result to re-solve problem 2.18.

Problem 2.37. Calculate the 2D Green's function for the free spaces:

(i) outside a round conducting cylinder, and
(ii) inside a round cylindrical hole in a conductor.

Problem 2.38. Solve task (i) of problem 2.16 (see also problem 2.17), using the Green's function method.

Problem 2.39. Solve the 2D boundary problem that was discussed in section 2.11 (figure 2.34) using:

(i) the finite difference method, with a finer square mesh, $h = a/3$, and
(ii) the variable separation method.

Compare the results at the mesh points, and comment.

References

[1] Hook J and Hall H 1991 *Solid State Physics* 2nd edn (Wiley)
[2] Ashcroft N and Mermin N 1976 *Solid State Physics* (Brooks Cole)

[3] Spiegel M *et al* 2009 *Complex Variables* 2nd edn (New York: McGraw-Hill)
[4] Grivet P 1972 *Electron Optics* 2nd edn (Pergamon)
[5] Septier A (ed) 1967 *Focusing Charged Particles* vol 1 (New York: Academic)
[6] Abramowitz M and Stegun I (eds) 1965 *Handbook of Mathematical Formulas* (New York: Dover)
[7] Likharev K 1999 *Proc. IEEE* **87** 606
[8] Keller M *et al* 1996 *Appl. Phys. Lett.* **69** 1804
[9] Stein F *et al* 2017 *Metrologia* **54** S1
[10] Brun-Pickard J *et al* 2016 *Phys. Rev.* X **6** 041051
[11] Leveque R 2007 *Finite Difference Methods for Ordinary and Partial Differential Equations* (SIAM)
[12] Johnson C 2009 *Numerical Solution of Partial Differential Equations by the Finite Element Method* (New York: Dover)
[13] Hughes T J R 2000 *The Finite Element Method* (New York: Dover)

IOP Publishing

Classical Electrodynamics
Lecture notes
Konstantin K Likharev

Chapter 3

Dipoles and dielectrics

In contrast to conductors, in dielectrics the charge motion is limited to the interior of an atom or a molecule, so that the electric polarization of these materials by an external field takes a different form. This issue is the main subject of this chapter, but in preparation for its analysis, we have to start with a general discussion of the electric field induced by a spatially restricted system of charges.

3.1 Electric dipole

Let us consider a localized system of charges, of a linear size scale a, and derive a simple but approximate expression for the electrostatic field induced by the system at a distant point $\mathbf{r}$. For that, let us select a reference frame with the origin either somewhere inside the system, or at a distance of the order of a from it (figure 3.1).

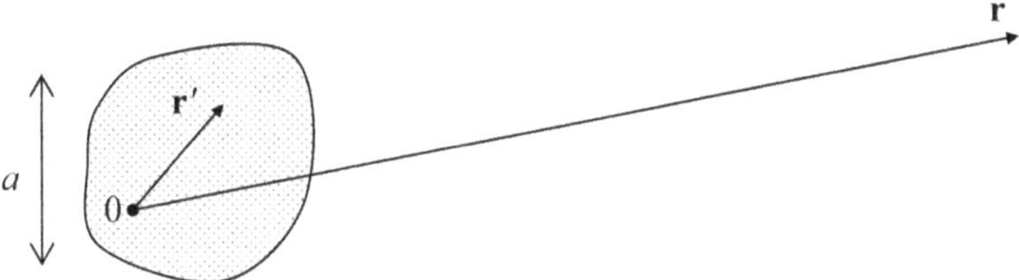

Figure 3.1. Deriving the approximate expression for the electrostatic field of a localized system of charges at a distant point ($r \gg r' \sim a$).

Then positions of all charges of the system satisfy the following condition:

$$r' \ll r. \tag{3.1}$$

Using this condition, we can expand the general expression (1.38) for the electrostatic potential $\phi(\mathbf{r})$ of the system into the Taylor series in small

doi:10.1088/978-0-7503-1404-6ch3

parameter $\mathbf{r}'$. For any spatial function of the type $f(\mathbf{r} - \mathbf{r}')$, the expansion may be represented as[1]

$$f(\mathbf{r} - \mathbf{r}') = f(\mathbf{r}) - \sum_{j=1}^{3} r_j' \frac{\partial f}{\partial r_j}(\mathbf{r}) + \frac{1}{2!} \sum_{j,j'=1}^{3} r_j' r_{j'}' \frac{\partial^2 f}{\partial r_j \partial r_{j'}}(\mathbf{r}) - \ldots. \tag{3.2}$$

Applying this formula to the free-space Green's function $1/|\mathbf{r} - \mathbf{r}'|$ in Eq. (1.38), we obtain the so-called *multipole expansion* of the electrostatic potential:

$$\phi(\mathbf{r}) = \frac{1}{4\pi\varepsilon_0} \left(\frac{1}{r} Q + \frac{1}{r^3} \sum_{j=1}^{3} r_j p_j + \frac{1}{2r^5} \sum_{j,j'=1}^{3} r_j r_{j'} \mathcal{Q}_{jj'} + \ldots \right), \tag{3.3}$$

whose $\mathbf{r}$-independent parameters are defined as follows:

$$Q \equiv \int \rho(\mathbf{r}') d^3r', \qquad p_j \equiv \int \rho(\mathbf{r}') r_j' d^3r', \qquad \mathcal{Q}_{jj'} \equiv \int \rho(\mathbf{r}') \left(3r_j' r_{j'}' - r'^2 \delta_{jj'} \right) d^3r'. \tag{3.4}$$

(Indeed, the two leading terms of the expansion (3.2) may be rewritten in the vector form $f(\mathbf{r}) - \mathbf{r}' \cdot \nabla f(\mathbf{r})$, and the gradient of such a spherically symmetric function $f(r) = 1/r$ is just $\mathbf{n}_r\, df/dr$, so that

$$\frac{1}{|\mathbf{r} - \mathbf{r}'|} \approx \frac{1}{r} - \mathbf{r}' \cdot \mathbf{n}_r \frac{d}{dr}\left(\frac{1}{r} \right) = \frac{1}{r} + \mathbf{r}' \cdot \frac{\mathbf{r}}{r^3}, \tag{3.5}$$

immediately giving the two first terms of Eq. (3.3). The proof of the third, quadrupole term in Eq. (3.3) is similar but a bit longer, and is left for the reader's exercise.)

Evidently, the scalar parameter Q in Eqs. (3.3) and (3.4) is just the total charge of the system. The constants p_j are the Cartesian components of a vector

$$\mathbf{p} \equiv \int \rho(\mathbf{r}') \mathbf{r}' d^3r', \tag{3.6}$$

called the system's *electric dipole moment*, and $\mathcal{Q}_{jj'}$ are the Cartesian components of a tensor—the system's *electric quadrupole moment*. If $Q \neq 0$, all higher terms on the right-hand side of Eq. (3.3), at large distances (3.1), are just small corrections to the first one, and in many cases may be ignored. However, the net charge of many systems is exactly zero, the most important examples being neutral atoms and molecules. For such neural systems, the second (dipole-moment) term, ϕ_d, in Eq. (3.3) is, most frequently, the leading one. (Such systems are called *electric dipoles*.) Due to its importance, let us rewrite the expression for this term in three other, equivalent forms:

$$\phi_d \equiv \frac{1}{4\pi\varepsilon_0} \frac{\mathbf{r} \cdot \mathbf{p}}{r^3} \equiv \frac{1}{4\pi\varepsilon_0} \frac{p \cos\theta}{r^2} \equiv \frac{1}{4\pi\varepsilon_0} \frac{pz}{[x^2 + y^2 + z^2]^{3/2}}, \tag{3.7}$$

[1] See, e.g. Eq. (A.14).

that are more convenient for some applications. Here θ is the angle between the vectors $\mathbf{p}$ and $\mathbf{r}$, and in the last (Cartesian) representation, the axis z is directed along the vector $\mathbf{p}$. Figure 3.2a shows equipotential surfaces of the dipole field (or rather their cross-sections by any plane in which the vector $\mathbf{p}$ resides).

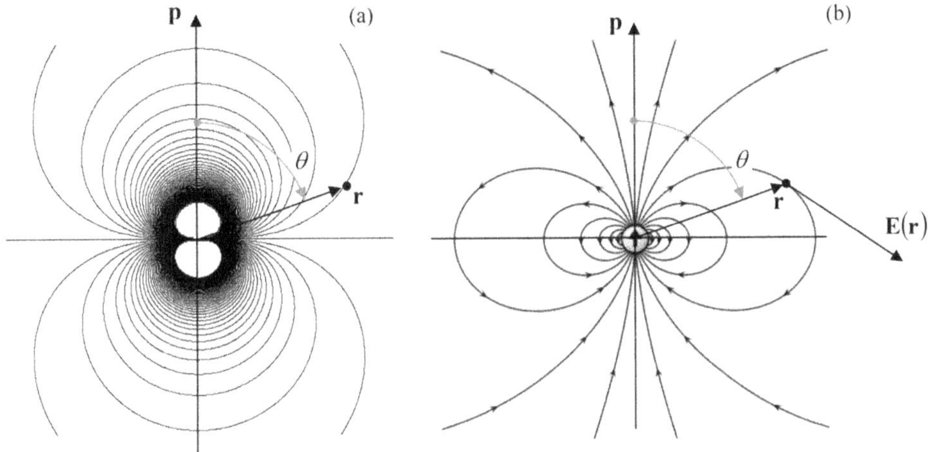

Figure 3.2. (a) The equipotential surfaces and (b) the electric field lines of a dipole. (Panel (b) adapted from http://en.wikipedia.org/wiki/Dipole under the GNU Free Documentation License.)

The simplest example of a system whose field, at large distances, approaches the dipole field (3.7) (and which gave the dipole the name), is a system of two equal but opposite point charges ('poles'), $+q$ and $-q$, with the radius-vectors, respectively, $\mathbf{r}_+$ and $\mathbf{r}_-$:

$$\rho(\mathbf{r}) = (+q)\delta(\mathbf{r} - \mathbf{r}_+) + (-q)\delta(\mathbf{r} - \mathbf{r}_-). \tag{3.8}$$

For this system, Eq. (3.5) yields

$$\mathbf{p} = (+q)\mathbf{r}_+ + (-q)\mathbf{r}_- = q(\mathbf{r}_+ - \mathbf{r}_-) = q\mathbf{a}, \tag{3.9}$$

where $\mathbf{a}$ is the vector connecting points $\mathbf{r}_-$ and $\mathbf{r}_+$. Note that in this case (and indeed for all systems with $Q = 0$), the dipole moment does not depend on the choice of the reference frame's origin.

A less trivial example of a dipole is a conducting sphere of radius R in a uniform external electric field $\mathbf{E}_0$. As a reminder, this problem was solved in section 2.8, and its result is expressed by Eq. (2.176). The first term in the parentheses of that relation describes just the external field (2.173), so that the field of the sphere itself (i.e. that of the surface charge induced by $\mathbf{E}_0$) is given by the second term:

$$\phi_s = \frac{E_0 R^3}{r^2}\cos\theta. \tag{3.10}$$

Comparing this expression with the second form of Eq. (3.7), we see that the sphere has an *induced* dipole moment

$$\mathbf{p} = 4\pi\varepsilon_0 \mathbf{E}_0 R^3. \tag{3.11}$$

This is an interesting example of an *almost* purely dipole field: at all points outside the sphere ($r > R$), the field has neither a quadrupole moment, nor any higher moments.

Other examples of dipole fields are given by two more systems discussed in chapter 2—see Eqs. (2.215) and (2.219). Those systems, however, do have higher-order multipole moments, so that for them Eq. (3.7) gives only the long-distance approximation.

Returning to the general properties of the dipole field (3.7), let us calculate its characteristics. First of all, we may use Eq. (3.7) to calculate the electric field of a dipole:

$$\mathbf{E}_\text{d} = -\nabla\phi_\text{d} = -\frac{1}{4\pi\varepsilon_0}\nabla\left(\frac{\mathbf{r}\cdot\mathbf{p}}{r^3}\right) = -\frac{1}{4\pi\varepsilon_0}\nabla\left(\frac{p\cos\theta}{r^2}\right). \tag{3.12}$$

The differentiation is easiest in the spherical coordinates, using the well-known expression for the gradient of a scalar function in these coordinates[2] and taking axis z parallel to the dipole moment $\mathbf{p}$. From the last form of Eq. (3.12) we immediately obtain

$$\mathbf{E}_\text{d} = \frac{p}{4\pi\varepsilon_0 r^3}(2\mathbf{n}_r\cos\theta + \mathbf{n}_\theta\sin\theta) \equiv \frac{1}{4\pi\varepsilon_0}\frac{3\mathbf{r}(\mathbf{r}\cdot\mathbf{p}) - \mathbf{p}r^2}{r^5}. \tag{3.13}$$

Figure 3.2b shows the electric field lines given by Eq. (3.13). The most important features of this result are a faster drop of the field's magnitude ($E_d \propto 1/r^3$, rather than $E \propto 1/r^2$ for a point charge), and the change of the signs of all field components as functions of the polar angle θ.

Next, let us use Eq. (1.55) to calculate the potential energy of interaction between a dipole and an external electric field. Assuming that the external field does not change much at distances of the order of a (figure 3.1), we may expand the external potential $\phi_\text{ext}(\mathbf{r})$ into the Taylor series, and keep only its two leading terms:

$$\begin{aligned} U &= \int\rho(\mathbf{r})\phi_\text{ext}(\mathbf{r})d^3r \approx \int\rho(\mathbf{r})[\phi_\text{ext}(0) + \mathbf{r}\cdot\nabla\phi_\text{ext}(0)]\,d^3r \\ &\equiv Q\phi_\text{ext}(0) - \mathbf{p}\cdot\mathbf{E}_\text{ext}. \end{aligned} \tag{3.14}$$

The first term is the potential energy the system would have if it were a point charge. If the net charge Q is zero that term disappears and the leading contribution is due to the dipole moment:

$$U = -\mathbf{p}\cdot\mathbf{E}_\text{ext}, \qquad \text{for } \mathbf{p} = \text{const}. \tag{3.15a}$$

Note that this result is only valid for a *fixed* dipole, with $\mathbf{p}$ independent of $\mathbf{E}_\text{ext}$. In the opposite limit, when the dipole is *induced* by the field, i.e. $\mathbf{p} \propto \mathbf{E}_\text{ext}$ (see again

[2] See, e.g. Eq. (A.66) with $\partial/\partial\varphi = 0$.

Eq. (3.11) as an example of such a proportionality), we need to start with Eq. (1.60) rather than Eq. (1.55), and obtain

$$U = -\frac{1}{2}\mathbf{p} \cdot \mathbf{E}_{\text{ext}}, \qquad \text{for } \mathbf{p} \propto \mathbf{E}_{\text{ext}}. \tag{3.15b}$$

In particular, combining Eqs. (3.13) and (3.15*a*), we may obtain the following important formula for the interaction of two independent dipoles

$$U_{\text{int}} = \frac{1}{4\pi\varepsilon_0}\frac{\mathbf{p}_1 \cdot \mathbf{p}_2 r^2 - 3(\mathbf{r} \cdot \mathbf{p}_1)(\mathbf{r} \cdot \mathbf{p}_2)}{r^5} = \frac{1}{4\pi\varepsilon_0}\frac{p_{1x}p_{2x} + p_{1y}p_{2y} - 2p_{1z}p_{2z}}{r^3}, \tag{3.16}$$

where $\mathbf{r}$ is the vector connecting the dipoles, and axis z is directed along this vector. It is easy to prove (this exercise is left for the reader) that if the magnitude of each dipole moment is fixed (an approximation valid, in particular, for weak interaction of so-called *polar molecules*), this potential energy reaches its minimum at, and hence favors the parallel orientation of the dipoles along, the line connecting them. Note also that in this case U_{int} is proportional to $1/r^3$. On the other hand, if each moment $\mathbf{p}$ has a random value plus a component due to its polarization by the electric field of its counterpart, $\Delta\mathbf{p}_{1,2} \propto \mathbf{E}_{2,1} \propto 1/r^3$, their average interaction energy (which may be calculated from Eq. (3.16) with the additional factor ½) is always negative and proportional to $1/r^6$. Such negative potential describes, in particular, the long-range, attractive part (the so-called *London dispersion force*) of the interaction between electrically neutral atoms and molecules[3].

According to Eqs. (3.15), in order to reach the minimum of U, the electric field 'tries' to align the dipole direction along its own. The quantitative expression of this effect is the torque $\boldsymbol{\tau}$ exerted by the field. The simplest way to calculate it is to sum up all the elementary torques $d\boldsymbol{\tau} = \mathbf{r} \times d\mathbf{F}_{\text{ext}} = \mathbf{r} \times \mathbf{E}_{\text{ext}}(\mathbf{r})\rho(\mathbf{r})d^3r$ exerted on all elementary charges of the system:

$$\boldsymbol{\tau} = \int \mathbf{r} \times \mathbf{E}_{\text{ext}}(\mathbf{r})\rho(\mathbf{r})d^3r \approx \mathbf{p} \times \mathbf{E}_{\text{ext}}(0), \tag{3.17}$$

where to make the last step, the spatial dependence of the external field was again neglected. The spatial dependence of $\mathbf{E}_{\text{ext}}$ cannot, however, be ignored at the calculation of the *total force* exerted by the field on the dipole (with $Q = 0$). Indeed, Eqs. (3.15) shows that if the field is constant, the dipole energy is the same at all spatial points, and hence the net force is zero. However, if the field has a non-zero gradient, a total force does appear; for a field-independent dipole,

$$\mathbf{F} = -\nabla U = \nabla(\mathbf{p} \cdot \mathbf{E}_{\text{ext}}), \tag{3.18}$$

[3] This force is calculated, using several models, in *Part QM* of this series.

where the derivative has to be taken at the dipole's position (in our notation, at $\mathbf{r} = 0$). If the dipole that is being moved in a field retains its magnitude and orientation, then the last formula is equivalent to[4]

$$\mathbf{F} = (\mathbf{p} \cdot \nabla)\mathbf{E}_{\text{ext}}. \tag{3.19}$$

Alternatively, the last expression may be obtained similarly to Eq. (3.14):

$$\begin{aligned}\mathbf{F} &= \int \rho(\mathbf{r})\mathbf{E}_{\text{ext}}(\mathbf{r})d^3r \approx \int \rho(\mathbf{r})[\mathbf{E}_{\text{ext}}(0) + (\mathbf{r} \cdot \nabla)\mathbf{E}_{\text{ext}}]d^3r \\ &= Q\mathbf{E}_{\text{ext}}(0) + (\mathbf{p} \cdot \nabla)\mathbf{E}_{\text{ext}}.\end{aligned} \tag{3.20}$$

Finally, let me add a note on the so-called *coarse-grain model* of the dipole. The dipole approximation explored above is asymptotically correct only *at large distances*, $r \gg a$. However, for some applications (including the forthcoming discussion of the *molecular field* effects in section 3.3) it is important to have an expression that would be approximately valid *everywhere* in space, though maybe without exact details at $r \sim a$, and also give the correct result for the space average of the electric field,

$$\bar{\mathbf{E}} \equiv \frac{1}{V}\int_V \mathbf{E}d^3r, \tag{3.21}$$

where V is a regularly shaped volume much larger than a^3, for example a sphere of a radius $R \gg a$, with the dipole at its center. For the field $\mathbf{E}_{\text{d}}$ given by Eq. (3.13), such an average is zero. Indeed, let us consider the Cartesian components of that vector in a reference frame with the axis z directed along vector $\mathbf{p}$. Due to the axial symmetry of the field, the averages of the components E_x and E_y evidently vanish. Let us use Eq. (3.13) to spell out the 'vertical' component of the field (parallel to the dipole moment vector):

$$\begin{aligned}E_z \equiv \mathbf{E}_{\text{d}} \cdot \frac{\mathbf{p}}{p} &= \frac{1}{4\pi\varepsilon_0 r^3}(2\mathbf{n}_r \cdot \mathbf{p}\cos\theta - \mathbf{n}_\theta \cdot \mathbf{p}\sin\theta) \\ &= \frac{p}{4\pi\varepsilon_0 r^3}(2\cos^2\theta - \sin^2\theta).\end{aligned} \tag{3.22}$$

Integrating this expression over the whole solid angle $\Omega = 4\pi$, at fixed r, using a convenient variable substitution $\cos\theta \equiv \xi$, we obtain

$$\begin{aligned}\oint_{4\pi} E_z\, d\Omega &= 2\pi\int_0^\pi E_z \sin\theta\, d\theta = \frac{p}{2\varepsilon_0 r^3}\int_0^\pi (2\cos^2\theta - \sin^2\theta)\sin\theta\, d\theta \\ &= \frac{p}{2\varepsilon_0 r^3}\int_{-1}^{+1}(3\xi^3 - \xi)\, d\xi = 0.\end{aligned} \tag{3.23}$$

[4] The equivalence may be proved, for example, by using Eq. (A.76) with $\mathbf{f} = \mathbf{p} = \text{const}$ and $\mathbf{g} = \mathbf{E}_{\text{ext}}$, taking into account that according to the general Eq. (1.28), $\nabla \times \mathbf{E}_{\text{ext}} = 0$.

On the other hand, the *exact* electric field of an *arbitrary* charge distribution, with the total dipole moment $\mathbf{p}$, obeys the following equality:

$$\int_V \mathbf{E}(\mathbf{r})d^3r = -\frac{\mathbf{p}}{3\varepsilon_0} \equiv -\frac{1}{4\pi\varepsilon_0}\frac{4\pi}{3}\mathbf{p}, \tag{3.24}$$

where the integration is over *any* sphere containing all the charges. A proof of this formula for the general case requires a straightforward, but somewhat tedious integration[5]. The origin of Eq. (3.24) is illustrated in figure 3.3 on the example of the dipole created by two equal but opposite charges—see Eqs. (3.8) and (3.9). The zero average (3.23) of the dipole field (3.13) does not take into account the contribution of the region between the charges (where Eq. (3.13) is not valid), which is directed mostly against the dipole vector (3.9).

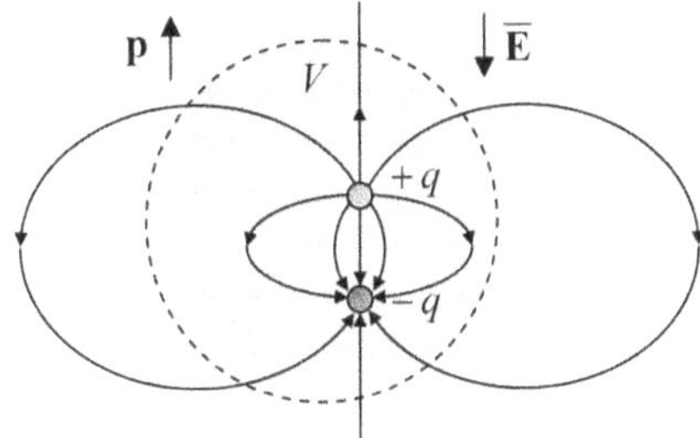

Figure 3.3. A sketch illustrating the origin of Eq. (3.24).

In order to be used as a reasonable coarse-grain model, Eq. (3.13) may be modified as follows:

$$\mathbf{E}_{\text{cg}} = \frac{1}{4\pi\varepsilon_0}\left[\frac{3\mathbf{r}(\mathbf{r}\cdot\mathbf{p}) - \mathbf{p}r^2}{r^5} - \frac{4\pi}{3}\mathbf{p}\delta(\mathbf{r})\right], \tag{3.25}$$

so that its integral satisfies Eq. (3.24). Evidently, such a modification does not change the field at large distances $r \gg a$, i.e. in the region where the expansion (3.3), and hence Eq. (3.13), are valid.

3.2 Dipole media

Let us generalize Eq. (3.7) to the case of several (possibly, many) dipoles $\mathbf{p}_j$ located at arbitrary points $\mathbf{r}_j$. Using the linear superposition principle, we obtain

$$\phi_{\text{d}}(\mathbf{r}) = \frac{1}{4\pi\varepsilon_0}\sum_j \mathbf{p}_j \cdot \frac{\mathbf{r} - \mathbf{r}_j}{\left|\mathbf{r} - \mathbf{r}_j\right|^3}. \tag{3.26}$$

If our system (medium) contains many similar dipoles, distributed in space with density $n(\mathbf{r})$, we may approximate the last sum with a *macroscopic potential*, which is the average of the 'microscopic' potential (3.26) over a local volume much larger than the distance between the dipoles, and as a result is given by the integral

[5] See, e.g. the end of section 4.1 in the textbook [1].

$$\phi_d(\mathbf{r}) = \frac{1}{4\pi\varepsilon_0}\int \mathbf{P}(\mathbf{r}')\cdot\frac{\mathbf{r}-\mathbf{r}'}{|\mathbf{r}-\mathbf{r}'|^3}d^3r', \qquad \text{with } \mathbf{P}(\mathbf{r}) \equiv n(\mathbf{r})\mathbf{p}, \tag{3.27}$$

where vector $\mathbf{P}(\mathbf{r})$, called the *electric polarization*, has the physical meaning of the net dipole moment per unit volume. (Note that by its definition, $\mathbf{P}(\mathbf{r})$ is also a 'macroscopic' field.)

Now comes a very impressive trick, which is the basis of all the theory of the 'macroscopic' electrostatics (and eventually, electrodynamics). Just as was done at the derivation of Eq. (3.5), Eq. (3.27) may be rewritten in the equivalent form

$$\phi_d(\mathbf{r}) = \frac{1}{4\pi\varepsilon_0}\int \mathbf{P}(\mathbf{r}')\cdot\nabla'\frac{1}{|\mathbf{r}-\mathbf{r}'|}d^3r', \tag{3.28}$$

where ∇' means the del operator (in this particular case, the gradient) acting in the 'source space' of vectors $\mathbf{r}'$. The right-hand side of Eq. (3.28), applied to any volume V limited by surface S, may be integrated by parts to give[6]

$$\phi_d(\mathbf{r}) = \frac{1}{4\pi\varepsilon_0}\oint_S \frac{P_n(\mathbf{r}')}{|\mathbf{r}-\mathbf{r}'|}d^2r' - \frac{1}{4\pi\varepsilon_0}\int_V \frac{\nabla'\cdot\mathbf{P}(\mathbf{r}')}{|\mathbf{r}-\mathbf{r}'|}d^3r'. \tag{3.29}$$

If the surface does not carry an infinitely dense (δ-functional) sheet of additional dipoles, or it is just very distant, the first term on the right-hand side is negligible[7]. Now comparing the second term with the basic equation (1.38) for the electric potential, we see that this term may be interpreted as the field of certain *effective* electric charges with density

$$\rho_{ef} = -\nabla\cdot\mathbf{P}. \tag{3.30}$$

Figure 3.4 illustrates the physics of this key relation for a cartoon model of a simple multi-dipole system: a layer of uniformly distributed two-point-charge units oriented perpendicular to the layer surface. (In this case $\nabla\cdot\mathbf{P} = dP/dx$.) One can see that the ρ_{ef} defined by Eq. (3.30) may be interpreted as the density of the uncompensated surface charges of polarized elementary dipoles.

Next, from section 1.2, we already know that Eq. (1.38) is equivalent to the inhomogeneous Maxwell equation (1.27) for the electric field, so that the *macroscopic* electric field of the dipoles (defined as $\mathbf{E}_d = -\nabla\phi_d$, where ϕ_d is given by Eq. (3.27)) obeys a similar equation, with the effective charge density (3.30).

Now let us consider a more general case when a system, in addition to the compensated charges of the dipoles, also has certain *stand-alone charges* (not parts of the dipoles already taken into account in the polarization $\mathbf{P}$)[8]. As was discussed in section 1.1, if we average this charge over the inter-charge distances,

[6] To prove this (almost evident) formula strictly, it is sufficient to apply the divergence theorem (A.79) to the vector function $\mathbf{f} = \mathbf{P}(\mathbf{r}')/|\mathbf{r}-\mathbf{r}'|$, in the 'source space' of radius-vectors $\mathbf{r}'$.

[7] Just as in the case of Eq. (1.9), if we want to describe such a dipole sheet, we may always do that using the second term in Eq. (3.29), by including a delta-functional part into the polarization distribution $\mathbf{P}(\mathbf{r}')$.

[8] In some texts, these charges are called 'free'. This term is somewhat misleading, because the stand-alone may well be bound, i.e. unable to move freely.

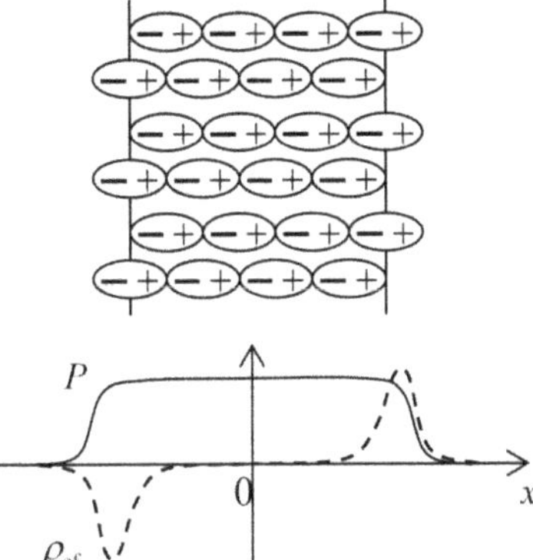

Figure 3.4. The spatial distributions of the polarization and effective charges in a layer of similar elementary dipoles (schematically).

i.e. approximate it with a continuous 'macroscopic' density $\rho(\mathbf{r})$, then its macroscopic electric field also obeys Eq. (1.27), but with the stand-alone charge density. Due to the linear superposition principle, for the *total macroscopic field* **E** of the charges and dipoles we may write

$$\nabla \cdot \mathbf{E} = \frac{1}{\varepsilon_0}(\rho + \rho_{ef}) = \frac{1}{\varepsilon_0}(\rho - \nabla \cdot \mathbf{P}). \tag{3.31}$$

This is already the main result of the 'macroscopic' electrostatics. However, it is evidently tempting (and very convenient for applications) to rewrite Eq. (3.31) in a different form by carrying the dipole-related term of this equation over to the left-hand side. The resulting equality is called the *macroscopic Maxwell equation for* **D**:

$$\nabla \cdot \mathbf{D} = \rho, \tag{3.32}$$

where $\mathbf{D}(\mathbf{r})$ is a new 'macroscopic' field, called the *electric displacement*, defined as[9]

$$\mathbf{D} \equiv \varepsilon_0 \mathbf{E} + \mathbf{P}. \tag{3.33}$$

The comparison of Eqs. (3.32) and (1.27) shows that **D** (or more strictly, the fraction $\mathbf{D}/\varepsilon_0$) may be interpreted as the 'would-be' macroscopic electric field that *would be* created by stand-alone charges in the absence of the dipole medium polarization. In contrast, the **E** participating in Eqs. (3.31) and (3.33) is the genuine macroscopic electric field, exact at distances much larger than that between the adjacent elementary point stand-alone charges and dipoles.

[9] Note that according to its definition (3.33), the dimensionality of **D** in the SI units is different from that of **E**. In contrast, in the Gaussian units the electric displacement is defined as $\mathbf{D} = \mathbf{E} + 4\pi\mathbf{P}$, so that $\nabla \cdot \mathbf{D} = 4\pi\rho$ (the relation $\rho_{ef} = -\nabla \cdot \mathbf{P}$ remains the same as in SI units), and the dimensionalities of **D** and **E** coincide. Philosophically, this coincidence is a certain perceptional handicap, because it is frequently convenient to consider the scalar components of **E** as generalized forces, and those of **D** as generalized coordinates (see section 3.5 below), and it is somewhat comforting to have their dimensionalities different.

In order to have a better look into the physics of the fields **E** and **D**, let us first rewrite the macroscopic Maxwell equation (3.32) in the integral form. Applying the divergence theorem to an arbitrary volume V limited by surface S, we obtain the following *macroscopic Gauss law*:

$$\oint_S D_n d^2r = \int_V \rho \ d^3r \equiv Q, \tag{3.34}$$

where Q is the total *stand-alone* charge inside volume V.

This general result may be used to find the boundary conditions for **D** at a sharp interface between two different dielectrics. (The analysis is clearly applicable to a dielectric/free-space boundary as well.) For that, let us apply Eq. (3.34) to a pillbox formed at the interface (see the solid rectangle in figure 3.5), which is sufficiently small on the spatial scales of the dielectrics' non-uniformity and the interface's curvature, but still containing many elementary dipoles. Assuming that the interface does not have stand-alone surface charges, we immediately obtain

$$(D_n)_1 = (D_n)_2, \tag{3.35}$$

i.e. the normal component of the electric displacement has to be continuous. Note that a similar statement for the macroscopic electric field **E** is generally not valid, because the polarization vector **P** may have, and typically does have a leap at a sharp interface (say, due to the different polarizability of the two different dielectrics), providing a surface layer of the effective charges (3.30)—see again the example in figure 3.4.

However, we still can make an important statement about the behavior of **E** at the interface. Indeed, the macroscopic field electric fields, defined by Eqs. (3.29) and (3.31), are evidently still potential ones, and hence obey another macroscopic Maxwell equation, similar to Eq. (1.28):

$$\nabla \times \mathbf{E} = 0. \tag{3.36}$$

Integrating this equation along to a narrow contour stretched along the interface (see the dashed rectangle in figure 3.5), we obtain

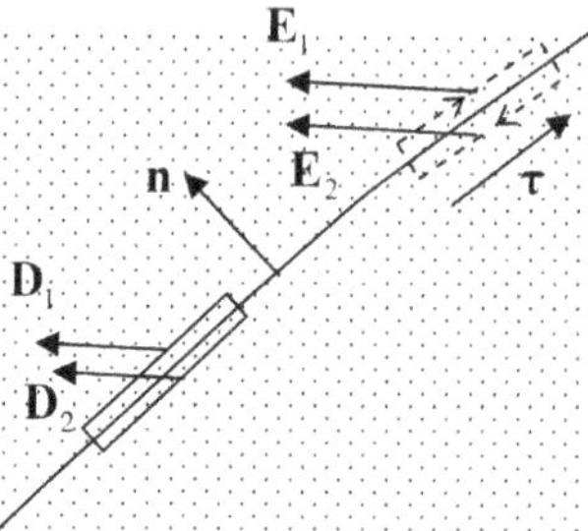

Figure 3.5. Deriving the boundary conditions at an interface between two dielectrics, using a Gauss pillbox (shown as a solid rectangle) and a circulation contour (dashed-line rectangle). **n** and **τ** are the unit vectors which are, respectively, normal and tangential to the interface.

$$(E_\tau)_1 = (E_\tau)_2. \tag{3.37}$$

Note that this condition is compatible with (and may be derived from) the continuity of the macroscopic electrostatic potential ϕ (related to the macroscopic field $\mathbf{E}$ by the relation similar to Eq. (1.33), $\mathbf{E} = -\nabla\phi$), at each point of the interface: $\phi_1 = \phi_2$.

In order to see how do these boundary conditions work, let us consider the simple problem shown in figure 3.6. A very broad plane capacitor, with zero voltage between its conducting plates (as may be fixed, e.g. by their connection with an external wire), is partly filled with a material with a constant polarization $\mathbf{P}_0$,[10] oriented normal to the plates. Let us calculate the spatial distribution of the fields $\mathbf{E}$ and $\mathbf{D}$, and also the surface charge density of each conducting plate.

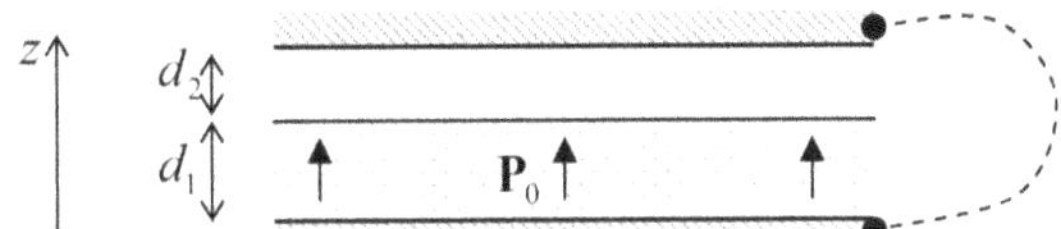

Figure 3.6. A simple system whose analysis requires Eq. (3.35).

Due to the symmetry of the system, vectors $\mathbf{E}$ and $\mathbf{D}$ are all evidently normal to the plates, and do not depend on the position in the capacitor's plane, so that we can limit the field analysis to the calculation of their z-components $E(z)$ and $D(z)$. In this case, Eq. (3.32) is reduced to $dD/dz = 0$ inside each layer (but not at their border!), so that within each of them D is constant—say, D_1 in the layer with $\mathbf{P} = \mathbf{P}_0$, and D_2 in the free-space layer (where $\mathbf{P} = 0$). As a result, according to Eq. (3.33), the (macroscopic) electric field inside each layer is also constant:

$$D_1 = \varepsilon_0 E_1 + P_0, \qquad D_2 = \varepsilon_0 E_2. \tag{3.38}$$

Since the voltage between the plates is zero, we may also require the integral of E, taken along a path connecting the plates, to vanish. This gives us one more relation:

$$E_1 d_1 + E_2 d_2 = 0. \tag{3.39}$$

Still, the three Eqs. (3.38) and (3.39) are insufficient to calculate the four fields in the system ($E_{1,2}$ and $D_{1,2}$). The decisive help comes from the boundary condition (3.35):

$$D_1 = D_2. \tag{3.40}$$

(Note that Eq. (3.35) is valid because the layer interface does not carry *stand-alone* electric charges, even though it has a *polarization surface charge*, whose areal density may be calculated by integrating Eq. (3.30) across the interface: $\sigma_{\text{ef}} = P_0$. Note also that in our simple system, Eq. (3.37) is identically satisfied due to the system's symmetry, and does not give any additional information.)

Now solving the resulting system of four equations, we readily obtain

[10] As will be discussed in the next section, this is a good approximation for the so-called *electrets*, and also *hard ferroelectrics*, in not very high electric fields.

$$E_1 = -\frac{P_0}{\varepsilon_0}\frac{d_2}{d_1 + d_2}, \quad E_2 = \frac{P_0}{\varepsilon_0}\frac{d_1}{d_1 + d_2}, \quad D_1 = D_2 = D = P_0\frac{d_1}{d_1 + d_2}. \tag{3.41}$$

The areal densities of the electrode surface charges may be readily calculated now by the integration of Eq. (3.32) across each surface:

$$\sigma_1 = -\sigma_2 = D = P_0\frac{d_1}{d_1 + d_2}. \tag{3.42}$$

Note that due to the spontaneous polarization of the lower layer's material, the capacitor plates are charged even in the absence of voltage between them, and that this charge is a function of the second electrode position (d_2)[11]. Also notice a substantial similarity between this system (figure 3.6) and the system whose analysis was the subject of problem 2.6.

3.3 Polarization of dielectrics

The general relations derived in the previous section may be used to describe the electrostatics of *dielectrics*—materials with bound electric charges (and hence with negligible dc electric conduction). However, in order to form the full system of equations necessary to solve electrostatics problems, they have to be complemented by certain relations between the vectors **P** and **E**.[12]

In most materials, in the absence of external electric field, the elementary dipoles **p** either equal zero or have a random orientation in space, so that the net dipole moment of each macroscopic volume (still containing many such dipoles) equals zero: $\mathbf{P} = 0$ at $\mathbf{E} = 0$. Moreover, if the field changes are sufficiently slow, most materials may be characterized by a unique dependence of **P** on **E**. Then using the Taylor expansion of function **P**(**E**), we may argue that in relatively low electric fields the function should be well approximated by a linear dependence between these two vectors. Such dielectrics are called *linear*. In an isotropic media, the coefficient of proportionality should be just a scalar. In the SI units, this scalar is defined by the following relation:

$$\mathbf{P} = \chi_e \varepsilon_0 \mathbf{E}, \tag{3.43}$$

with the dimensionless constant χ_e called the *electric susceptibility*. However, it is much more common to use, instead of χ_e, another dimensionless parameter[13],

[11] This effect is used in most modern microphones. In such a device, the sensed sound wave's pressure bends a thin conducting membrane playing the role of one of capacitor's plates, and thus modulates the thickness (in figure 3.6, d_2) of the air gap adjacent to an electret layer. This modulation produces a proportional electric charge variation that is picked up by device's electronics.

[12] This is one more example of *constitutive relations* (already mentioned in section 2.1). In the problem solved in the end of the previous section, the role of this relation was played by the equality $\mathbf{P}_0 = \text{const}$.

[13] In older physics literature, the dielectric constant is often denoted by letter ε_r, while in the electrical engineering literature, its notation is frequently K.

$$\kappa \equiv 1 + \chi_e, \tag{3.44}$$

which is sometimes called the 'relative electric permittivity', but much more often, the *dielectric constant.* This parameter is very convenient, because combining Eqs. (3.43) and (3.44),

$$\mathbf{P} = (\kappa - 1)\varepsilon_0\mathbf{E}, \tag{3.45}$$

and then plugging the resulting relation into the general Eq. (3.33), we obtain simply

$$\mathbf{D} = \kappa\varepsilon_0\mathbf{E}, \qquad \text{or} \quad \mathbf{D} = \varepsilon\mathbf{E}, \tag{3.46}$$

where another popular parameter[14],

$$\varepsilon \equiv \kappa\varepsilon_0 \equiv (1 + \chi_e)\varepsilon_0. \tag{3.47}$$

ε is called the *electric permittivity* of the material[15]. Table 3.1 gives approximate values of the dielectric constant for several representative materials.

In order to understand the range of these values, let me discuss (rather superficially) the two simplest mechanisms of electric polarization. The first of them is typical for liquids and gases of *polar* atoms/molecules, which have their own, spontaneous dipole moments **p**.[16] In the absence of the external electric field, the

Table 3.1. Dielectric constants of a few representative (and/or practically important) dielectrics.

Material	κ
Air (at ambient conditions)	1.00054
Teflon (polytetrafluoroethylene, C_nF_{2n})	2.1
Silicon dioxide (amorphous)	3.9
Glasses (of various compositions)	3.7–10
Castor oil	4.5
Silicon[a]	11.7
Water (at 100 °C)	55.3
Water (at 20 °C)	80.1
Barium titanate ($BaTiO_3$, at 20 °C)	~1,600

[a] Anisotropic materials, such as silicon crystals, require a *susceptibility tensor* to give an exact description of the linear relation of vectors **P** and **E**. However, most important crystals (including Si) are only weakly anisotropic, so that they may be reasonably well characterized with a scalar (angle-average) susceptibility.

---

[14] The reader may be perplexed by the use of three different but uniquely related parameters (χ_e, $\kappa \equiv 1 + \chi_e$, and $\varepsilon \equiv \kappa\varepsilon_0$) for the description of just one scalar property. Unfortunately, such redundancy is typical for physics, whose different sub-field communities have different, well-entrenched traditions.

[15] In the Gaussian units, χ_e is defined by the following relation: $\mathbf{P} = \chi_e\mathbf{E}$, while ε is defined just as in the SI units, $\mathbf{D} = \varepsilon\mathbf{E}$. Because of that, in the Gaussian units ε is dimensionless and equals $(1 + 4\pi\chi_e)$. As a result, $\varepsilon_{\text{Gaussian}} = (\varepsilon/\varepsilon_0)_{\text{SI}} \equiv \kappa$, so that $(\chi_e)_{\text{Gaussian}} = (\chi_e)_{\text{SI}}/4\pi$, sometimes creating a confusion with the numerical values of the latter parameter (dimensionless in both systems).

[16] A typical example is the water molecule H_2O, with the negative oxygen ion offset from the line connecting two positive hydrogen ions, thus producing a spontaneous dipole moment $p = ea$, with $a \approx 0.38 \times 10^{-10}$ m $\sim r_B$.

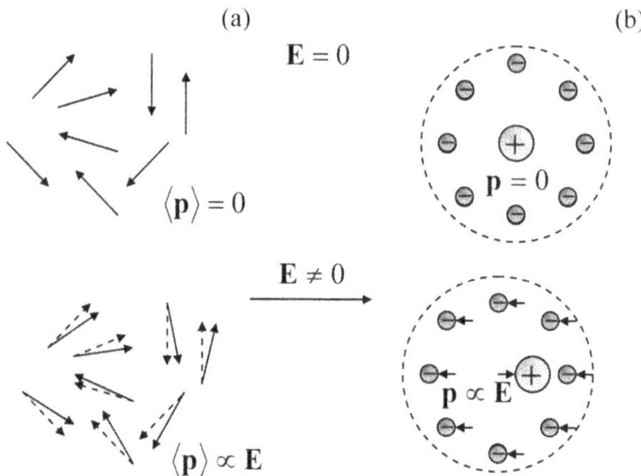

Figure 3.7. Crude cartoons of two mechanisms of the induced electrical polarization: (a) a partial ordering of spontaneous elementary dipoles and (b) an elementary dipole induction. The upper two panels correspond to $\mathbf{E} = 0$, and the lower two panels, to $\mathbf{E} \neq 0$.

orientation of such dipoles may be random, with the average polarization $\mathbf{P} = n\langle\mathbf{p}\rangle$ equal to zero—see the top panel of figure 3.7a.

A relatively weak external field does not change the magnitude of the dipole moments significantly but, according to Eqs. (3.15*a*) and (3.17), tries to orient them along the field and thus creates a non-zero vector average $\langle\mathbf{p}\rangle$ directed along the vector $\langle\mathbf{E}_m\rangle$, where $\mathbf{E}_m$ is the microscopic field at the point of the dipole's location—cf the two panels of figure 3.7a. If the field is not two high ($p\langle E_m\rangle \ll k_B T$), the induced average polarization $\langle\mathbf{p}\rangle$ is proportional to $\langle\mathbf{E}_m\rangle$. If we write this proportionality relation in the following traditional form,

$$\langle\mathbf{p}\rangle = \alpha\mathbf{E}_m, \tag{3.48}$$

where α is called the *atomic* (or, sometimes, 'molecular') *polarizability*, this means that α is positive. If the concentration n of such elementary dipoles is low, the contribution of their own fields into the microscopic field acting on each dipole is negligible, and we may identify $\langle\mathbf{E}_m\rangle$ with the macroscopic field $\mathbf{E}$. As a result, the second of Eq. (3.27) yields

$$\mathbf{P} \equiv n\langle\mathbf{p}\rangle = \alpha n\mathbf{E}. \tag{3.49}$$

Comparing this relation with Eq. (3.45), we obtain

$$\kappa = 1 + \frac{\alpha}{\varepsilon_0}n, \tag{3.50}$$

so that $\kappa > 1$ (i.e. $\chi_e > 0$). Note that at this particular polarization mechanism (illustrated on the lower panel of figure 3.7a), the thermal motion 'tries' to randomize the dipole orientation, i.e. reduce its ordering by the field, so that we may expect α, and hence $\chi_e \equiv \kappa - 1$ to increase as temperature T is decreased—the so-called *paraelectricity*. Indeed, the elementary statistical mechanics[17] shows that in this case the electric susceptibility follows the law $\chi_e \propto 1/T$.

The materials of the second, much more common class consist of *non-polar* atoms without intrinsic spontaneous polarization. A crude classical image of such an atom is an isotropic cloud of negatively charged electrons surrounding a positively charged nucleus—see the top panel of figure 3.7b. The external electric field shifts the positive charge in the direction of the vector **E**, and the negative charges in the opposite direction, thus creating a similarly directed average dipole moment $\langle \mathbf{p} \rangle$.[18] At relatively low fields, this average moment is proportional to **E**, so that we again arrive at Eq. (3.48), with $\alpha > 0$, and if the dipole concentration n is sufficiently low, also at Eq. (3.50), with $\kappa - 1 > 0$. So, the dielectric constant is larger than 1 for both polarization mechanisms—please have one more look at table 2.1.

In order to make a crude but physically transparent estimate of the magnitude of the difference $\kappa - 1$, let us consider the following toy model of a non-polar dielectric: a set of similar conducting spheres of radius R, distributed in space with a low density $n \ll 1/R^3$. At such density, the electrostatic interaction of the spheres is negligible, and we can use Eq. (3.11) for the induced dipole moment of a single sphere. Then the polarizability definition (3.48) yields $\alpha = 4\pi\varepsilon_0 R^3$, so that Eq. (3.50) gives

$$\kappa = 1 + 4\pi R^3 n. \tag{3.51}$$

Let us use this result for a crude estimate of the dielectric constant of air at the so-called *ambient conditions*, meaning the normal atmospheric pressure $\mathcal{P} = 1.013 \times 10^5$ Pa and temperature $T = 300$ K. At these conditions the molecular density n may be, with a few-percent accuracy, found from the equation of state of an ideal gas[19]: $n \approx \mathcal{P}/k_B T \approx (1.013 \times 10^5)/(1.38 \times 10^{-23} \times 300) \approx 2.45 \times 10^{25}\text{m}^{-3}$. The molecule of the air's main component, N_2, has a van der Waals radius[20] of 1.55×10^{-10} m. Taking this radius for the R of our crude model, we obtain $\chi_e \equiv \kappa - 1 \approx 1.15 \times 10^{-3}$. Comparing this number with the first line of table 3.1, we see that the model gives a surprisingly reasonable result: in order to obtain the experimental value, it is sufficient to decrease the effective R of the sphere by just ~30%, to $\sim 1.2 \times 10^{-10}$ m.[21]

This result may encourage us to try using Eq. (3.51) for a larger density n. For example, as a crude model for a non-polar crystal, let us assume that the conducting spheres form a simple cubic lattice with the period $a = 2R$ (i.e. the neighboring spheres virtually touch). With this $n = 1/a^3 = 1/8R^3$, and Eq. (3.44) yields $\kappa = 1 +$

---

[17] See, e.g. *Part SM* chapter 2.

[18] Realistically, these effects are governed by quantum mechanics, so that the average here should be understood not only in the statistical-mechanical, but also (and mostly) in the quantum-mechanical sense. Because of that, for non-polar atoms, α is typically a very weak function of temperature, at least on the usual scale $T \sim 300$ K.

[19] See, e.g. *Part SM* sections 1.4 and 3.1.

[20] Such radius is defined by the requirement that the volume of the corresponding sphere, if used in the van der Waals equation (see, e.g. *Part SM* section 4.1), gives the best fit to the experimental equation of state $n = n(P, T)$.

[21] As discussed in *Part QM* chapter 6, for a hydrogen atom in its ground state, the low-field polarizability may be calculated analytically, $\alpha = (9/2) \times 4\pi\varepsilon_0 r_B^3$, corresponding to our metallic-ball model with a close value of the effective radius: $R = (9/2)^{1/3} r_B \approx 1.65\ r_B \approx 0.87 \times 10^{-10}$ m.

$4\pi/8 \approx 2.5$. This estimate provides a reasonable semi-qualitative explanation for the values of κ listed in first few lines of table 3.1. However, at such small distances the electrostatic dipole–dipole interaction should already be essential, so that such a simple model cannot even approximately describe values of κ much larger than 1, listed in the last rows of the table.

Such high values may be explained by the *molecular field* effect: each elementary dipole is polarized not only by the external field (as Eq. (3.49) assumes), but by the field of neighboring dipoles as well. In 1850, O-F Mossotti and (perhaps, independently, but almost 30 years later) R Clausius suggested what is now known, rather unfairly, as the *Clausius–Mossotti formula*, which works remarkably well for a broad class of non-polar materials[22]. In our notation it reads[23]

$$\frac{\kappa - 1}{\kappa + 2} = \frac{\alpha n}{3\varepsilon_0}, \qquad \text{so that } \kappa = \frac{1 + 2\alpha n/3\varepsilon_0}{1 - \alpha n/3\varepsilon_0}. \tag{3.52}$$

If the dipole density is low in the sense $n \ll 3\varepsilon_0/\alpha$, this relation is reduced to Eq. (3.50) corresponding to independent dipoles. However, at higher dipole density, both κ and χ_e increase much faster, and tend to infinity as the density-polarizability product approaches a critical value, in the simple Clausius–Mossotti model equal to $3\varepsilon_0$.[24] This means that the zero-polarization state becomes unstable even in the absence of an external electric field.

This instability is a linear-theory (i.e. low-field) manifestation of the substantially nonlinear effect—the formation of a spontaneous polarization even in the absence of external electric field. Such materials are called *ferroelectrics* and may be experimentally recognized by the hysteretic behavior of their polarization as a function of the applied (external) electric field—see figure 3.8. As the plots show, the polarization of a ferroelectric depends on the applied field's history. For example, the direction of its spontaneous *remnant polarization* P_R may be switched by applying and then removing a sufficiently high field (larger than the so-called *coercive field* E_C—see figure 3.8). The physics of this switching is rather involved; the polarization vector **P** of a ferroelectric material is typically constant only within each of the spontaneously formed spatial regions (called *domains*), with a typical size of a few tenths of a micron, and different (frequently, opposite) directions of the vector **P** in adjacent domains. The change of the applied electric field results not in the switching of the direction of **P** inside each domain, but rather in a shift of the domain walls, resulting in the change of the average polarization of the sample.

[22] Applied to the high-frequency electric field, with κ replaced by the square of the refraction coefficient n at the field's frequency (see chapter 7), this formula is known as the *Lorenz–Lorentz relation*.

[23] I am leaving the proof of Eq. (3.52), with help from a formula that will be derived in the next section, for the reader's exercise.

[24] The Clausius–Mossotti model does not give qualitatively correct results for most ferroelectric materials. For a review of modern approaches to the theory of their polarization, see, e.g. the paper by Resta and Vanderbilt in the recent review collection [2].

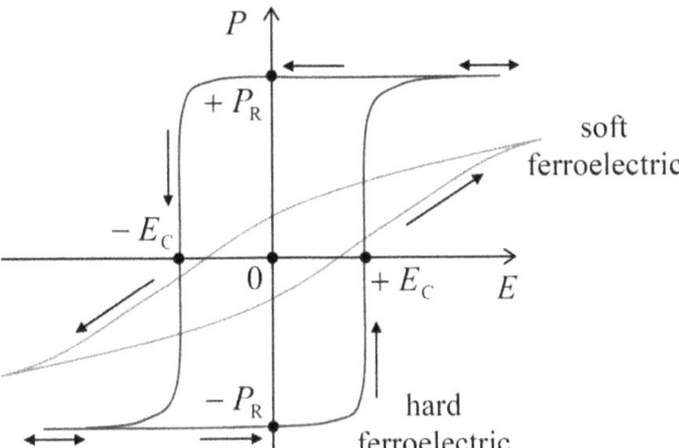

Figure 3.8. A schematic representation of the average polarization of soft and hard ferroelectrics as functions of the applied electric field.

Depending on the ferroelectric's material, temperature, and the sample's geometry (a solid crystal, a ceramic material, or a thin film), the hysteretic loops may be rather different, ranging from a rather smooth form in the so-called *soft ferroelectrics* (with include most ferroelectric thin films) to an almost rectangular form in *hard ferroelectrics*—see figure 3.8. In low fields, soft ferroelectrics behave essentially as linear paraelectrics, but with a very high average dielectric constant—see the bottom line of table 3.1 for such a classical material as $BaTiO_3$ (which is a soft ferroelectric at temperatures below $T_c \approx 120$ °C, and a paraelectric above this critical temperature). On the other hand, the polarization of a hard ferroelectric in the fields below its coercive field remains virtually constant, and the analysis of their electrostatics may be based on the condition $\mathbf{P} = \mathbf{P}_R$ = const—already used in the problem discussed at the end of the previous section[25]. This condition is even more applicable to the so-called *electrets*—synthetic polymers with a spontaneous polarization that remains constant even in very high electric fields.

Some materials exhibit even more complex polarization effects, for example *antiferroelectricity*, *helielectricity*, and (practically very valuable) *piezoelectricity*. Unfortunately, we do not have time for a discussion of these exotic phenomena in this course[26]; the main reason I am mentioning them is to emphasize again that the constitutive relation $\mathbf{P} = \mathbf{P}(\mathbf{E})$ is material-specific rather than fundamental. However, most insulators, in practicable fields, behave as linear dielectrics, so that the next section will be committed to the discussion of their electrostatics.

[25] Due to this property, hard ferroelectrics, such as the lead zirconate titanate (PZT) and strontium bismuth tantalite (SBT), with high remnant polarization P_R (up to ~1 C m^{-2}), may be used in non-volatile random-access memories (dubbed either FRAM or FeRAM)—see, e.g. [3]. In a cell of such a memory, binary information is stored in the form of one of two possible directions of spontaneous polarization at $\mathbf{E} = 0$ (see figure 3.8). Unfortunately, the time of spontaneous depolarization of ferroelectric thin films is typically well below 10 years—the industrial standard for data retention in non-volatile memories, and this time may be decreased even more by 'fatigue' from the repeated polarization recycling at information recording. Due to these reasons, the industrial production of FRAM is currently just a tiny fraction of the non-volatile memory market, which is dominated by floating-gate memories—see, e.g. section 4.2.

[26] For a detailed coverage of ferroelectrics, I can recommend the encyclopedic monograph [4] and the recent review collection [2].

3.4 Electrostatics of linear dielectrics

First, let us discuss the simplest problem: how is the electrostatic field of a set of *stand-alone* charges of density $\rho(\mathbf{r})$ modified by a uniform linear dielectric medium, which obeys Eq. (3.46) with a space-independent dielectric constant κ. For this case, we may combine Eqs. (3.32) and (3.46) to write

$$\nabla \cdot \mathbf{E} = \frac{\rho}{\varepsilon}. \tag{3.53}$$

As a reminder, in the free space we had the similar equation (1.27), but with a different constant, $\varepsilon_0 = \varepsilon/\kappa$. Hence all the results discussed in chapter 1 are valid inside a uniform linear dielectric, for the macroscopic field $\mathbf{E}$ (and the corresponding macroscopic electrostatic potential ϕ), if they are reduced by the factor of $\kappa > 1$. Thus, the most straightforward result of the induced polarization of a dielectric medium is the electric field reduction. This is a very important effect, in particular taking into account the very high values of κ in such dielectrics as water—see table 3.1. Indeed, this is the reduction of the attraction between positive and negative ions (called, respectively, *cations* and *anions*) in water that enables their substantial dissociation and hence almost all biochemical reactions, which are the basis of the biological cell functions—and hence of the life itself.

Let us apply this general result to the important particular case of the plane capacitor (figure 2.3) filled with a linear, uniform dielectric. Applying the macroscopic Gauss law (3.34) to a pillbox-shaped volume on the conductor surface, we obtain the following relation,

$$\sigma = D_n = \varepsilon E_n = -\varepsilon \frac{\partial \phi}{\partial n}, \tag{3.54}$$

which differs from Eq. (2.3) only by the replacement $\varepsilon_0 \to \varepsilon \equiv \kappa\varepsilon_0$. Hence the charge density, calculated for the free-space case, should be increased by the factor of κ—that is it. In particular, this means that all the mutual capacitance (2.28) has to be increased by this factor:

$$C = \frac{\kappa\varepsilon_0 A}{d} \equiv \frac{\varepsilon A}{d}. \tag{3.55}$$

(As a reminder, this increase of C by κ has already been incorporated, without derivation, into some estimates made in sections 2.1 and 2.2.)

If a linear dielectric is non-uniform, the situation is more complex. For example, let us consider the case of a sharp interface between two otherwise uniform dielectrics, free of stand-alone charges. In this case, we still may use (3.37) for the tangential component of the macroscopic electric field, and also Eq. (3.36), with $D_n = \varepsilon E_n$, which yields

$$(\varepsilon E_n)_1 = (\varepsilon E_n)_2, \quad \text{i.e.} \quad \varepsilon_1 \frac{\partial \phi_1}{\partial n} = \varepsilon_2 \frac{\partial \phi_2}{\partial n}. \tag{3.56}$$

Let us apply these boundary conditions, first of all, to the very illuminating case of two very thin ($t \ll d$) slits cut in a uniform dielectric with an initially uniform[27] electric field $\mathbf{E}_0$ (figure 3.9). In both cases, a slit with $t \to 0$ cannot modify the field distribution outside it substantially.

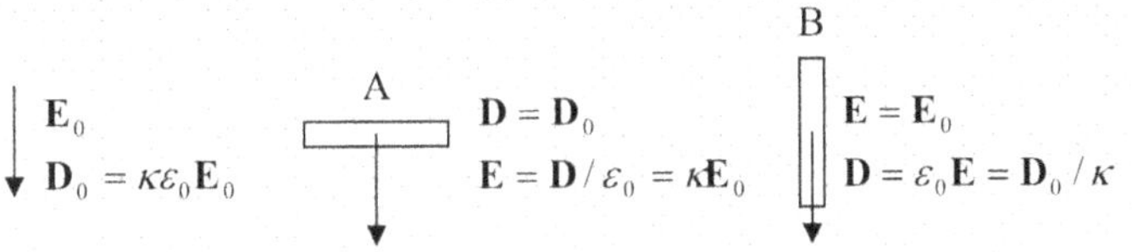

Figure 3.9. Fields inside two narrow slits cut in a linear dielectric with a uniform field $\mathbf{E}_0$.

For the slit A, with the plane normal to the applied field, we may apply Eq. (3.56) to the 'major' (broad) interfaces, shown horizontal in figure 3.9, to see that the vector $\mathbf{D}$ should be continuous. But according to Eq. (3.46), this means that in the free space inside the gap the electric field equals $\mathbf{D}/\varepsilon_0$, and hence is κ times higher than the applied field $\mathbf{E}_0 = \mathbf{D}/\kappa\varepsilon_0$. This field, and hence $\mathbf{D}$, may be measured by a sensor placed inside the gap, showing that the electric displacement is by no means a purely mathematical construct[28]. In contrast, for the slit B parallel to the applied field, we may apply Eq. (3.37) to the major (now, vertical) interfaces of the slit, to see that now the electric field $\mathbf{E}$ is continuous, while the electric displacement $\mathbf{D} = \varepsilon_0\mathbf{E}$ inside the gap is a factor of κ lower than its value in the dielectric. (Similarly to the case A, any perturbations of the field uniformity, caused by the compliance with Eq. (3.56) at the minor interfaces, are settled at distances $\sim t$ from them.)

For other problems with piecewise-constant ε and with more complex geometries we may need to apply the methods studied in chapter 2. As in the free space, in the simplest cases we can select such a set of orthogonal coordinates that the electrostatic potential depends on just one of them. Consider, for example, two types of plane capacitor filling with two different dielectrics—see figure 3.10.

In case (a), the voltage V between the electrodes is the same for each part of the capacitor and, at least far from the dielectric interface, the electric field is vertical, uniform, and similar ($E = V/d$). Hence the boundary condition (3.37) is satisfied even if such a distribution is valid near the surface as well, i.e. at any point of the system. The only effect of different values of ε in the two parts is that the electric displacement $D = \varepsilon E$ and hence the electrodes' surface charge densities $\sigma = D$ are different in the two parts. Thus we can calculate the electrode charges $Q_{1,2}$ of the two

[27] Actually, the following arguments and results are valid for *any* external field distribution, provided that the slits are much smaller than the characteristic scale of the field's change in space.

[28] Superficially, this result violates the boundary condition (3.37) at the vertical ('minor') interfaces. This apparent contradiction may be resolved, taking into account that the slit deforms the field outside it, near its edges. These fringe effects extend from the edges only by the horizontal distances $\sim t$, so that the above relations for $\mathbf{E}$ and $\mathbf{D}$ are valid at most of the slit area.

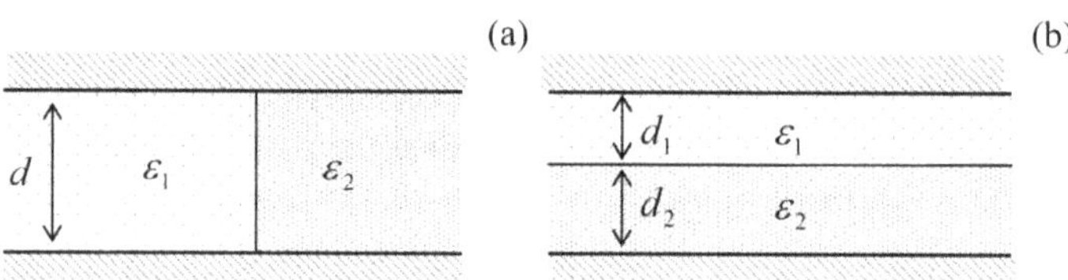

Figure 3.10. Plane capacitors filled with two different dielectrics.

parts independently, and then add up the results to obtain the total mutual capacitance

$$C = \frac{Q_1 + Q_2}{V} = \frac{1}{d}(\varepsilon_1 A_1 + \varepsilon_2 A_2). \tag{3.57}$$

Note that this formula may be interpreted as the total capacitance of two separate lumped capacitors connected (by wires) *in parallel*. This is natural, because we may cut the system along the dielectric interface, without any effect on the fields in either part, and then connect the corresponding electrodes by external wires, again without any effect on the system—besides very close vicinities of capacitor's edges.

Case (b) may be analyzed just as in the problem shown in figure 3.6 by applying Eq. (3.34) to a Gaussian pillbox with one lid inside (for example) the bottom electrode, and the other lid in any of the layers. From this we see that D anywhere inside the system should be equal to the surface charge density σ of the electrode, i.e. constant. Hence, according to Eq. (3.46), the electric field inside each dielectric layer is also constant: in the top layer $E_1 = D_1/\varepsilon_1 = \sigma/\varepsilon_1$, while in the bottom layer, $E_2 = D_2/\varepsilon_2 = \sigma/\varepsilon_2$. Integrating the field E across the whole capacitor, we obtain

$$V = \int_0^{d_1+d_2} E(z)dz = E_1 d_1 + E_2 d_2 = \left(\frac{d_1}{\varepsilon_1} + \frac{d_2}{\varepsilon_2}\right)\sigma, \tag{3.58}$$

so that the mutual capacitance per unit area

$$\frac{C}{A} \equiv \frac{\sigma}{V} = \left[\frac{d_1}{\varepsilon_1} + \frac{d_2}{\varepsilon_2}\right]^{-1}. \tag{3.59}$$

Note that this result is similar to the total capacitance of an *in-series* connection of two plane capacitors based on each of the layers. This is also natural, because we could insert an uncharged, thin conducting sheet (rather than a cut as in the previous case) at the layer interface, which is an equipotential surface, without changing the field distribution in any part of the system. Then we could thicken the conducting sheet as much as we like (turning it into a wire), also without changing the fields and hence the capacitance.

Proceeding to problems with more complex geometry, let us consider the system shown in figure 3.11a: a dielectric sphere placed into an initially uniform external electric field $\mathbf{E}_0$. According to Eq. (3.53) for the macroscopic electric field, and the definition of the macroscopic electrostatic potential, $\mathbf{E} = -\nabla\phi$, the potential satisfies

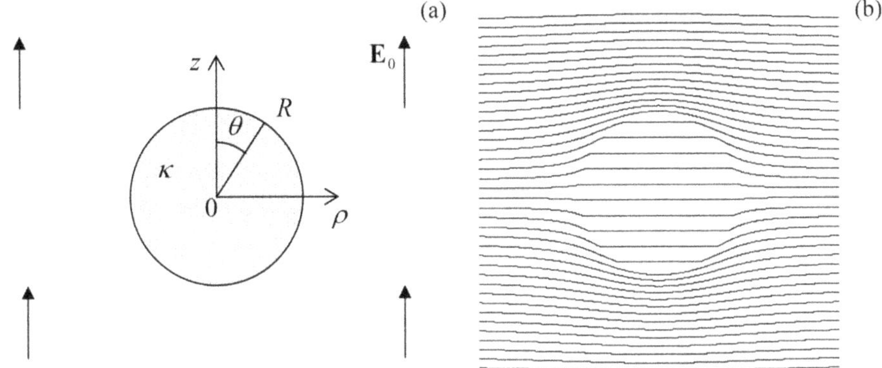

Figure 3.11. Dielectric sphere in an initially uniform electric field: (a) the problem and (b) the equipotential surfaces, as given by Eq. (3.62), for $\kappa = 3$.

the Laplace equation both inside and outside the sphere. Due to the spherical symmetry of the dielectric sample, this problem invites the variable separation method in spherical coordinates, which was discussed in section 2.8. From that discussion, we already know, in particular, the general solution (2.172) of the Laplace equation outside of the sphere. In order to avoid the divergence, and satisfy the uniform-field condition at $r \to \infty$, we have to reduce this solution to

$$\phi_{r\geqslant R} = -E_0 r \cos\theta + \sum_{l=1}^{\infty} \frac{b_l}{r^{l+1}} \mathcal{P}_l(\cos\theta). \tag{3.60}$$

Inside the sphere we can also use Eq. (2.172), but keeping only the radial functions finite at $r \to 0$:

$$\phi_{r\leqslant R} = \sum_{l=1}^{\infty} a_l r^l \mathcal{P}_l(\cos\theta). \tag{3.61}$$

Now, spelling out the boundary conditions (3.37) and (3.56) at $r = R$, we see that for all coefficients a_l and b_l with $l \geqslant 2$ we obtain (just as for the conducting sphere, discussed in section 2.8) homogeneous equations that have only trivial solutions. Hence, all these terms may be dropped, while for the only surviving angular harmonic, proportional to $\mathcal{P}_1(\cos\theta) \equiv \cos\theta$, we obtain two equations:

$$-E_0 - \frac{2b_1}{R^3} = \kappa a_1, \qquad -E_0 R + \frac{b_1}{R^2} = a_1 R. \tag{3.62}$$

Solving this simple system of linear equations for a_1 and b_1, and plugging the result into Eqs. (3.60) and (3.61), we obtain the final solution of the problem:

$$\phi_{r\geqslant R} = E_0\left(-r + \frac{\kappa - 1}{\kappa + 2}\frac{R^3}{r^2}\right)\cos\theta, \qquad \phi_{r\leqslant R} = -E_0 \frac{3}{\kappa + 2} r \cos\theta. \tag{3.63}$$

Figure 3.11b shows the equipotential surfaces given by this solution, for a particular value of the dielectric constant κ. Note that according to Eq. (3.62), at $r \geqslant R$ the dielectric sphere, just as the conducting sphere in a similar problem, produces (on the top of the uniform external field) a purely dipole field with the dipole moment

$$\mathbf{p} = 4\pi R^3 \frac{\kappa - 1}{\kappa + 2}\varepsilon_0 \mathbf{E}_0 \equiv 3V\frac{\kappa - 1}{\kappa + 2}\varepsilon_0 \mathbf{E}_0, \qquad \text{where } V = \frac{4\pi}{3}R^3, \tag{3.64}$$

an evident generalization of Eq. (3.11), to which Eq. (3.64) tends at $\kappa \to \infty$. By the way, this property is common: from the point of view of their electrostatic (but not transport!) properties, conductors may be adequately described as dielectrics with $\kappa \to \infty$.

Another remarkable feature of Eq. (3.63) is that the electric field and polarization inside the sphere are uniform, with R-independent values[29]

$$\mathbf{E} = \frac{3}{\kappa + 2}\mathbf{E}_0, \quad \mathbf{D} \equiv \kappa\varepsilon_0 \mathbf{E} = \varepsilon_0 \frac{3\kappa}{\kappa + 2}\mathbf{E}_0, \quad \mathbf{P} \equiv \mathbf{D} - \varepsilon_0 \mathbf{E} = 3\varepsilon_0 \frac{\kappa - 1}{\kappa + 2}\mathbf{E}_0. \tag{3.65}$$

In the limit $\kappa \to 1$ (the 'sphere made of free space', i.e. no sphere at all), the electric field inside the sphere naturally tends to the external one, and its polarization disappears. In the opposite limit $\kappa \to \infty$, the electric field inside the sphere vanishes. Curiously enough, in this limit the electric displacement inside the sphere remains finite: $\mathbf{D} \to 3\varepsilon_0 \mathbf{E}_0$.

More complex problems with piecewise-uniform dielectrics also may be addressed by the methods discussed in chapter 2, and I leave a few of them for the reader's exercise. Let me discuss just one of such problems, because it exhibits the new features of the charge image method which was discussed in section 2.9 (and is the basis of the Green's function approach—see section 2.10). Consider the system shown in figure 3.12, a point charge near a dielectric half-space, which evidently parallels the simplest problem discussed in section 2.9—see figure 2.26.

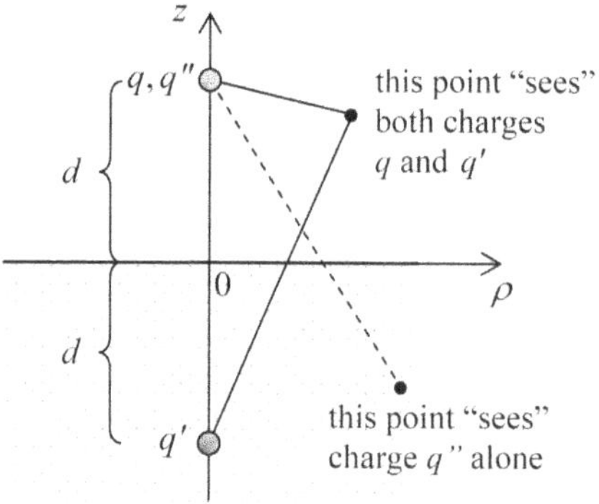

Figure 3.12. Charge images for a dielectric half-space.

[29] The first of these relations is used in the standard derivation of the Clausius–Mossotti formula (3.52).

As for the case of a conducting half-space, the Laplace equation for the electrostatic potential in the upper half-space $z > 0$ (besides the charge point $\rho = 0$, $z = d$) may be satisfied using a single image charge q' at point $\rho = 0$, $z = -d$, but now q' may differ from $(-q)$. In addition, in contrast to the case analyzed in section 2.9, we should also calculate the field inside the dielectric (at $z \leqslant 0$). This field cannot be contributed by the image charge q', because it would provide a potential divergence at its location. Thus, in that half-space we should try to use the real point source only, but maybe with a re-normalized charge q'' rather than the genuine charge q—see figure 3.12. As a result, we may look for the potential distribution in the form

$$\phi(\rho, z) = \frac{1}{4\pi\varepsilon_0} \times \begin{cases} \left[\dfrac{q}{[\rho^2 + (z-d)^2]^{1/2}} + \dfrac{q'}{[\rho^2 + (z+d)^2]^{1/2}}\right], & \text{for } z \geqslant 0, \\ \dfrac{q''}{[\rho^2 + (z-d)^2]^{1/2}}, & \text{for } z \leqslant 0, \end{cases} \tag{3.66}$$

at this stage with unknown q' and q''. Plugging this solution into the boundary conditions (3.37) and (3.56) at $z = 0$ (with $\partial/\partial n = \partial/\partial z$), we see that they are indeed satisfied (so that Eq. (3.66) does express the unique solution of the boundary problem), provided that the effective charges q' and q'' obey the following relations:

$$q - q' = \kappa q'', \quad q + q' = q''. \tag{3.67}$$

Solving this simple system of linear equations, we obtain

$$q' = -\frac{\kappa - 1}{\kappa + 1}q, \quad q'' = \frac{2}{\kappa + 1}q. \tag{3.68}$$

If $\kappa \to 1$, then $q' \to 0$, and $q'' \to q$—both facts are very natural, because in this limit (no polarization at all!) we have to recover the unperturbed field of the initial point charge in both semi-spaces. In the opposite limit $\kappa \to \infty$ (which, according to our discussion of the last problem, should correspond to a conducting half-space), $q' \to -q$ (repeating the result we have discussed in much detail in section 2.9) and $q'' \to 0$. The last result means that in this limit, the electric field **E** in the dielectric tends to zero—as it should.

3.5 Energy of electric field in a dielectric

In chapter 1, we have obtained two key results for the electrostatic energy: Eq. (1.55) for a charge interaction with an independent ('external') field, and a similarly structured formula (1.60), but with an additional factor ½, for the field produced by the charges under consideration. These relations are of course always valid for dielectrics as well, provided that the charge density includes *all* charges (including those bound into the elementary dipoles), but it is convenient to recast them into a form depending on the density $\rho(\mathbf{r})$ of only stand-alone charges.

If a field is created only by stand-alone charges under consideration, and is proportional to $\rho(\mathbf{r})$ (requiring that we deal with *linear* dielectrics only), we can repeat all the argumentation of the beginning of section 1.3, and again arrive at Eq. (1.60), provided that ϕ is now the macroscopic field's potential. Now we can recast this result in the terms of fields—essentially as was done in Eqs. (1.62)–(1.64), but now making a clear difference between the macroscopic electric field $\mathbf{E} = -\nabla\phi$ and the electric displacement field $\mathbf{D}$ that obeys the macroscopic Maxwell equation (3.32). Plugging $\rho(\mathbf{r})$, expressed from that equation, into Eq. (1.60), we obtain

$$U = \frac{1}{2}\int(\nabla \cdot \mathbf{D})\ \phi\ d^3r. \tag{3.69}$$

Using the fact[30] that for any differentiable functions ϕ and $\mathbf{D}$,

$$(\nabla \cdot \mathbf{D})\ \phi = \nabla \cdot (\phi\mathbf{D}) - (\nabla\phi) \cdot \mathbf{D}, \tag{3.70}$$

we may rewrite Eq. (3.69) as

$$U = \frac{1}{2}\int\nabla \cdot\ (\phi\ \mathbf{D})\ d^3r - \frac{1}{2}\int(\nabla\phi) \cdot \mathbf{D}\ d^3r. \tag{3.71}$$

The divergence theorem, applied to the first term on the right-hand side, reduces it to a surface integral of ϕD_n. (As a reminder, in Eq. (1.63) the integral was of $\phi(\nabla\phi)_n \propto \phi E_n$.) If the surface of the volume we are considering is sufficiently far, this surface integral vanishes. On the other hand, the gradient in the second term of Eq. (3.71) is just (minus) field $\mathbf{E}$, so that it gives

$$U = \frac{1}{2}\int\mathbf{E} \cdot \mathbf{D} d^3\ r = \frac{1}{2}\int E(\mathbf{r}) \cdot \varepsilon(\mathbf{r})E(\mathbf{r})\,d^3r = \frac{\varepsilon_0}{2}\int\kappa(\mathbf{r})E^2(\mathbf{r})\,d^3r. \tag{3.72}$$

This expression is a natural generalization of Eq. (1.65), and shows that we can, as we did in the free space, represent the electrostatic energy in a local form[31]

$$U = \int u(\mathbf{r})d^3r, \qquad \text{with } u = \frac{1}{2}\mathbf{E} \cdot \mathbf{D} = \frac{\varepsilon}{2}E^2 = \frac{D^2}{2\varepsilon}. \tag{3.73}$$

As a sanity check, in the trivial case $\varepsilon = \varepsilon_0$ (i.e. $\kappa = 1$), this result is reduced to Eq. (1.65).

Again, Eq. (3.73) is not valid for *nonlinear* dielectrics, because our starting point, Eq. (1.62), is only valid if ϕ is proportional to ρ. In order to make our calculation more general, we should intercept the calculations of section 1.3 at an earlier stage at which we have not yet used this proportionality. For example, the first of Eqs. (1.56) may be rewritten, in the continuous limit, as

[30] See, e.g. Eq. (A.74*a*).

[31] In the Gaussian units the last expression should be divided by 4π.

$$\delta U = \int \phi(\mathbf{r})\delta\rho(\mathbf{r})d^3r, \tag{3.74}$$

where symbol δ means a small variation of the function, e.g. its change in time, sufficiently slow to ignore the relativistic and magnetic-field effects. Applying such variation to Eq. (3.32) and plugging the resulting relation $\delta\rho = \nabla \cdot \delta\mathbf{D}$ into Eq. (3.74), we obtain

$$\delta U = \int (\nabla \cdot \delta\mathbf{D})\phi d^3r. \tag{3.75}$$

(Note that in contrast to Eq. (3.69), this expression does not have the front factor ½.) Now repeating the same calculations as in the linear case, for the energy density *variation* we get a remarkably simple (and general!) expression,

$$\delta u = \mathbf{E} \cdot \delta\mathbf{D} \equiv \sum_{j=1}^{3} E_j \delta D_j, \tag{3.76}$$

where the last expression uses the Cartesian components of the vectors **E** and **D**. This is as far as we can go for the general dependence **D**(**E**). If the dependence is linear and isotropic, as in Eq. (3.46), then $\delta\mathbf{D} = \varepsilon\delta\mathbf{E}$ and

$$\delta u = \varepsilon\mathbf{E} \cdot \delta\mathbf{E} \equiv \varepsilon\delta\left(\frac{E^2}{2}\right). \tag{3.77}$$

Integration of this expression over the variation, from the field equal to zero to a certain final distribution **E**(**r**), brings us back to Eq. (3.73).

An important role of Eq. (3.76), in its last form, is to indicate that the Cartesian coordinates of **E** may be interpreted as generalized forces, and those of **D** as generalized coordinates of the field effect on a unit dielectric's volume[32]. This allows one, in particular, to form the proper *Gibbs potential energy*[33] of a system with an electric field **E**(**r**) fixed, at every point, by some external source:

$$U_G = \int_V u_G(\mathbf{r})d^3r, \quad u_G(\mathbf{r}) = u(\mathbf{r}) - \mathbf{E}(\mathbf{r}) \cdot \mathbf{D}(\mathbf{r}). \tag{3.78}$$

As an analytical mechanics reminder, if a generalized external force (in our case, **E**) is fixed, the stable equilibrium of the system corresponds to the minimum of U_G, rather than of the potential energy U as such—in our case, that of the field in our system. As the simplest illustration, let us consider a very long cylinder (with an

---

[32] This is one of the cases where the SI units, prescribing different dimensionalities to the fields **E** and **D**, are more revealing than the Gaussian units.

[33] See, e.g. *Part CM* section 1.4, in particular Eq. (1.41), and *Part CM* section 2.1. Note that as Eq. (3.78) clearly illustrates, once again, that the difference between the potential energies U_G and U, usually discussed in courses of statistical physics and/or thermodynamics as the difference between the Gibbs and Helmholtz free energies (see, e.g. *Part SM* section 1.6), is more general than the effects of random thermal motion, addressed by these disciplines.

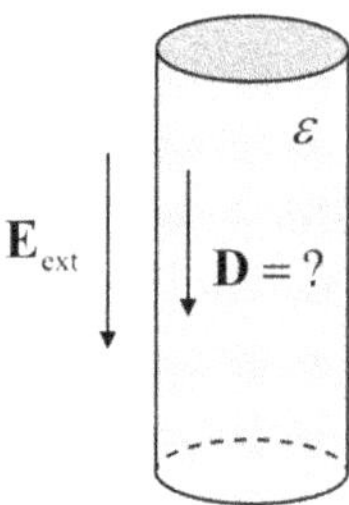

Figuree 3.13. A cylindrical dielectric sample in a longitudinal external electric field.

arbitrary cross-section's shape), made of a uniform linear dielectric, placed into a uniform external electric field, parallel to the cylinder's axis—see figure 3.13.

The equilibrium value of $\mathbf{D}$ inside the cylinder may, of course, be readily found without any appeal to energies. Indeed, the solution of the Laplace equation inside the cylinder, with the boundary condition (3.37), is evident: $\mathbf{E}(\mathbf{r}) = \mathbf{E}_{\text{ext}}$, and so that Eq. (3.46) immediately yields $\mathbf{D}(\mathbf{r}) = \varepsilon\mathbf{E}_{\text{ext}}$. However, one may wonder why the minimum of the potential energy U, given by Eq. (3.73) in its last form,

$$\frac{U}{V} = \frac{D^2}{2\varepsilon}, \tag{3.79}$$

corresponds to a different (zero) value of $\mathbf{D}$. The Gibbs potential energy (3.78) immediately removes the contradiction. For our uniform case, this energy per unit volume of the cylinder is

$$\frac{U_{\text{G}}}{V} = \frac{U}{V} - \mathbf{E}\cdot\mathbf{D} = \frac{D^2}{2\varepsilon} - \mathbf{E}\cdot\mathbf{D} \equiv \sum_{j=1}^{3}\left(\frac{D_j^2}{2\varepsilon} - E_jD_j\right), \tag{3.80}$$

and its minimum as a function of every Cartesian component of $\mathbf{D}$ corresponds to the correct value of the displacement: $D_j = \varepsilon E_j$, i.e. $\mathbf{D} = \varepsilon\mathbf{E} = \varepsilon\mathbf{E}_{\text{ext}}$. So, the minimum of the Gibbs potential energy indeed corresponds to the systems' equilibrium, and it may be very useful for analyses of the polarization dynamics. Note also that Eq. (3.80) at this equilibrium point may be rewritten as

$$\frac{U_{\text{G}}}{V} = \frac{U}{V} - \mathbf{E}\cdot\mathbf{D} = \frac{D^2}{2\varepsilon} - \frac{\mathbf{D}}{\varepsilon}\cdot\mathbf{D} \equiv -\frac{D^2}{2\varepsilon}, \tag{3.81}$$

i.e. formally coincides with Eq. (3.79), besides the opposite sign. Another useful general relation (*not* limited to linear dielectrics) may be obtained by taking the variation of the u_{G} expressed by Eq. (3.78), and then using Eq. (3.76):

$$\delta u_{\text{G}} = \delta u - \delta(\mathbf{E}\cdot\mathbf{D}) = \mathbf{E}\cdot\delta\mathbf{D} - (\delta\mathbf{E}\cdot\mathbf{D} + \mathbf{E}\cdot\delta\mathbf{D}) \equiv -\mathbf{D}\cdot\delta\mathbf{E}. \tag{3.82}$$

In order to see how do these expressions (with their perhaps counter-intuitive negative signs[34]) work, let us plug **D** from Eq. (3.33):

$$\delta u_{\mathrm{G}} = -(\varepsilon_0 \mathbf{E} + \mathbf{P}) \cdot \delta \mathbf{E} \equiv -\delta\left(\frac{\varepsilon_0 E^2}{2}\right) - \mathbf{P} \cdot \delta \mathbf{E}. \tag{3.83}$$

So far, this relation is general. In the particular case when the polarization **P** is field-independent, we may integrate Eq. (3.83) over the electric field, from 0 to some finite value **E**, obtaining

$$u_{\mathrm{G}} = -\frac{\varepsilon_0 E^2}{2} - \mathbf{P} \cdot \mathbf{E}. \tag{3.84}$$

Again, the Gibbs energy is relevant only if **E** is dominated by an external field $\mathbf{E}_{\text{ext}}$, independent of the orientation of the polarization **P**. If, in addition, $\mathbf{P}(\mathbf{r}) \neq 0$ only in some finite volume V, we may integrate Eq. (3.84) over the volume, obtaining

$$U_{\mathrm{G}} = -\mathbf{p} \cdot \mathbf{E}_{\text{ext}} + \text{const}, \qquad \text{with } \mathbf{p} \equiv \int_V \mathbf{P}(\mathbf{r}) d^3 r, \tag{3.85}$$

where 'const' means the terms independent of **p**. In this expression, we may readily recognize Eq. (3.15*a*) for an electric dipole **p**, which was obtained in section 3.1 in a different way.

This comparison shows again that U_{G} is nothing extraordinary; it is just the relevant part of the potential energy of the system in a fixed external field, including the energy of its interaction with the field. Still, I would strongly recommend the reader obtain a better gut feeling of the relation between the two potential energies, U and U_{G}—for example, by using them to solve a very simple problem: calculate the force of attraction between the plates of a plane capacitor.

3.6 Problems

Problem 3.1. Prove Eqs. (3.3) and (3.4), starting from Eqs. (1.38) and (3.2).

Problem 3.2. A planar thin ring of radius R is charged with a constant linear density λ. Calculate the exact electrostatic potential distribution along the symmetry axis of the ring, and prove that at large distances, $r \gg R$, the three leading terms of its multipole expansion are indeed correctly described by Eqs. (3.3)–(3.4).

Problem 3.3 In suitable reference frames, calculate the dipole and quadrupole moments of the following systems (see the figures below):

(i) four point charges of the same magnitude, but alternating signs, placed in the corners of a square;
(ii) a similar system, but with a pair charge sign alternation; and

[34] Some psychological relief may be provided by the fact that you may add to U_{G} (and U) any constant—positive if you like.

(iii) a point charge in the center of a thin ring carrying a similar but opposite charge, uniformly distributed along its circumference.

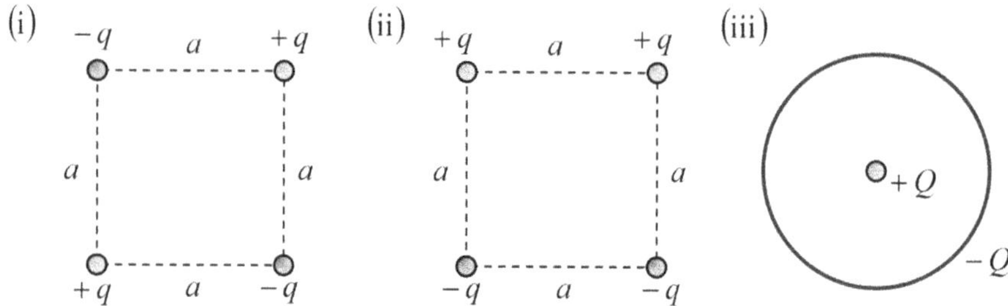

Problem 3.4. Without carrying out an exact calculation, can you predict the spatial dependence of the interaction between various electric multipoles, including point charges (in this context, frequently called *monopoles*), dipoles, and quadrupoles? Based on these predictions, what is the functional dependence of the interaction between dumbbell-shaped diatomic molecules such as H_2, N_2, O_2, etc, on the distance between them, if the distance is much larger than the molecular size?

Problem 3.5. Two similar electric dipoles of fixed magnitude p, located at a fixed distance r from each other, are free to rotate, changing their directions. What stable equilibrium position(s) may they take as a result of their electrostatic interaction?

Problem 3.6. An electric dipole is located above an infinite, grounded conducting plane (see the figure below). Calculate:

(i) the distribution of the induced charge in the conductor,
(ii) the dipole-to-plane interaction energy, and
(iii) the force and the torque acting on the dipole.

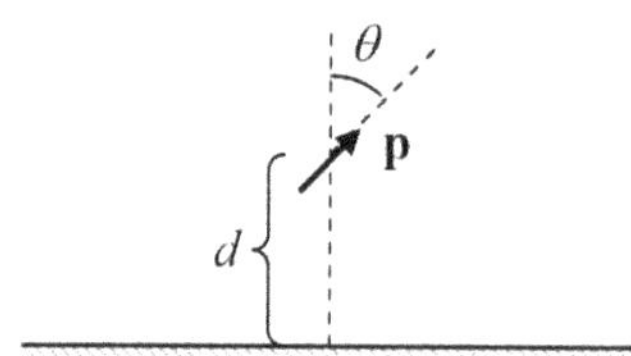

Problem 3.7. Calculate the net charge Q induced in a grounded conducting sphere of radius R by a dipole $\mathbf{p}$ located at point $\mathbf{r}$ outside the sphere—see the figure below.

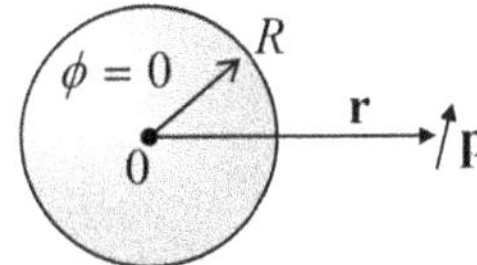

Problem 3.8. Use two different approaches to calculate the energy of interaction between a grounded conductor and an electric dipole $\mathbf{p}$, placed in the center of a spherical cavity of radius R, carved in the conductor.

Problem 3.9. A plane separating two parts of otherwise free space is densely and uniformly (with a constant areal density n) filled with dipoles, with similar dipole moments $\mathbf{p}$ oriented in a direction normal to the plane.

(i) Calculate the boundary conditions for the electrostatic potential on both sides of the plane.
(ii) Use the result of task (i) to calculate the potential distribution created in space by a spherical surface, with radius R, densely and uniformly filled with radially oriented dipoles.
(iii) What condition should be imposed on the dipole density n for your results to be qualitatively valid?

Problem 3.10. Prove the Clausius–Mossotti relation (3.52) for the case of a cubic lattice of similar dipoles obeying (3.48): $\mathbf{p} = \alpha\mathbf{E}_{\text{m}}$, where $\mathbf{E}_{\text{m}}$ is the microscopic electric field at the dipole's location point.

Hint: Use Eq. (3.65) to account for the difference between the external field and the macroscopic field.

Problem 3.11. A sphere of radius R is made of a material with a uniform, fixed polarization $\mathbf{P}_0$.

(i) Calculate the electric field everywhere in space—both inside and outside the sphere.
(ii) Compare the result for the internal field with Eq. (3.24).

Problem 3.12. Calculate the electric field at the center of a cube with side a, made of material with the uniform spontaneous polarization vector $\mathbf{P}_0$ parallel to one of cube's sides.

Problem 3.13. A stand-alone charge Q is distributed, in some way, in the volume of a body made of a uniform linear dielectric with a dielectric constant κ. Calculate the polarization charge Q_{ef} residing on the surface of the body, provided that it is surrounded by free space.

Problem 3.14. In two separate experiments, a thin, planar sheet of a linear dielectric with $\kappa = \text{const}$ is placed into a uniform external electric field $\mathbf{E}_0$:

(i) with sheet's surface parallel to the electric field, and
(ii) the surface normal to the field.

For each case, find the electric field **E**, the electric displacement **D**, and the polarization **P** inside the dielectric (far from sheet's edges).

Problem 3.15. A point charge q is located at a distance $r \gg R$ from the center of a uniform sphere of radius R, made of a uniform linear dielectric. In the first non-vanishing approximation in small parameter R/r, calculate the interaction force, and the energy of interaction between the sphere and the charge.

Problem 3.16. A fixed dipole $\mathbf{p}$ is placed in the center of a spherical cavity of radius R, cut inside a uniform, linear dielectric. Calculate the electric field distribution everywhere in the system (both at $r < R$ and at $r > R$).

Hint: You may start with the assumption that the field at $r > R$ has a distribution typical for a dipole (but be ready for surprises :-).

Problem 3.17. A spherical capacitor (see the figure below) is filled with a linear dielectric whose permittivity ε depends on spherical angles θ and φ, but not on the distance r from the system's center. Derive an explicit expression for its capacitance C.

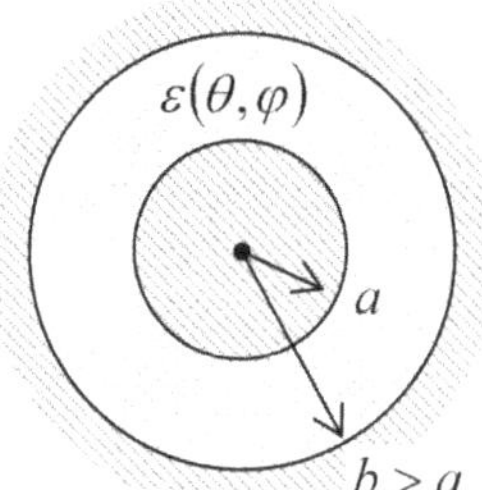

Problem 3.18. For each of the two capacitors shown in figure 3.10, calculate the electric forces (per unit area) exerted on the interface between the dielectrics, in terms of fields in the system.

Problem 3.19. A uniform electric field $\mathbf{E}_0$ has been created (by external sources) inside a uniform linear dielectric. Find the change of the electric field, created by cutting out a cavity in the shape of a round cylinder of radius R, with the axis perpendicular to the external field—see the figure below.

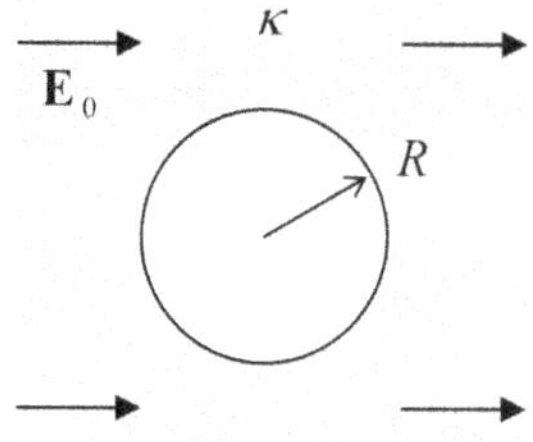

Problem 3.20. Similar small spherical particles, made of a linear dielectric, are dispersed in free space with a low concentration $n \ll 1/R^3$, where R is the particle's radius. Calculate the average dielectric constant of such a medium. Compare the result with the apparent, but wrong, answer

$$\bar{\kappa} - 1 = (\kappa - 1)\,nV \qquad \text{(WRONG!)}$$

(where κ is the dielectric constant of the particle's material and $V = (4\pi/3)R^3$ is its volume), and explain the origin of the difference.

Problem 3.21.* Calculate the spatial distribution of the electrostatic potential induced by a point charge q placed at distance d from a very wide parallel plate, of thickness D, made of a uniform linear dielectric—see the figure below.

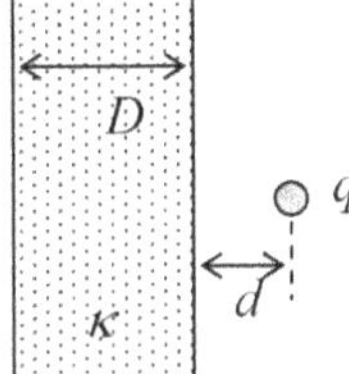

Problem 3.22. Discuss the physical nature of Eq. (3.76). Apply your conclusions to a material with a fixed (field-independent) polarization $\mathbf{P}_0$, and calculate the electric field's energy of the uniformly polarized sphere (see problem 3.11).

Problem 3.23. Use Eqs. (3.73) and (3.81) to calculate the force of attraction of the plane capacitor's plates (per unit area), for two cases:

(i) the capacitor is charged to a voltage V, and then disconnected from the battery, and

(ii) the capacitor remains connected to the battery.

References

[1] Jackson J 1999 *Classical Electrodynamics* 3rd edn (Wiley)

[2] Rabe K, Ahn C and Triscone J-M (eds) 2010 *Physics of Ferroelectrics: A Modern Perspective* (Berlin: Springer)

[3] Scott J 2000 *Ferroelectric Memories* (Berlin: Springer)

[4] Lines M and Glass A 2001 *Principles and Applications of Ferroelectrics and Related Materials* (Oxford: Oxford University Press)

IOP Publishing

Classical Electrodynamics
Lecture notes
Konstantin K Likharev

Chapter 4

DC currents

The goal of this chapter is to discuss the laws governing the distribution of stationary ('dc') currents inside conducing media. In the most important case of linear ('Ohmic') conductivity, the partial differential equation governing the distribution is reduced to the same Laplace and Poisson equations whose solution methods were discussed in detail in chapter 2—although sometimes with different boundary conditions. Because of this, the chapter is rather brief.

4.1 Continuity equation and the Kirchhoff laws

Until this point, our discussion of conductors has been limited to the cases when they are separated by *insulators* (meaning either the free space or some dielectric media), preventing any continuous motion of charges from one conductor to another, even if there is a non-zero voltage (and hence electric field) between them—see figure 4.1a.

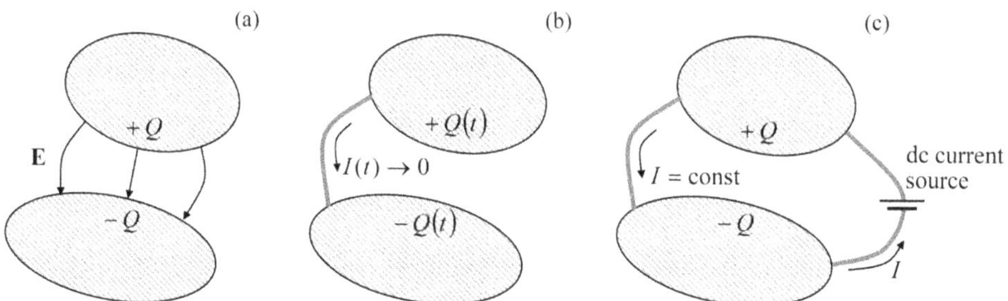

Figure 4.1. Two oppositely charged conductors: (a) in the electrostatic situation, (b) at the charge relaxation through an additional narrow conductor ('wire'), and (c) in a system sustaining a dc current I.

Now let us connect the two conductors with a *wire*—a thin, elongated conductor (figure 4.1b). Then the electric field causes the motion of charges in the wire—from the conductor with a higher electrostatic potential toward that with a lower

doi:10.1088/978-0-7503-1404-6ch4

potential, until the potentials equilibrate. Such process is called *charge relaxation.* The main equation governing this process may be obtained from the fundamental experimental fact (already mentioned in section 1.1) that electric charges cannot appear or disappear—though opposite charges may recombine with the conservation of the net charge. As a result, the charge Q in the top conductor may change only due to the *electric current* I through the wire:

$$\frac{dQ}{dt} = -I(t), \tag{4.1}$$

the relation that may be understood as the definition of the current[1].

Let us express Eq. (4.1) in a differential form, introducing the notion of the *current density* vector $\mathbf{j}(\mathbf{r})$. This vector may be defined via the following relation for the elementary current dI crossing an elementary area dA (figure 4.2):

$$dI = jdA\cos\theta = (j\cos\theta)dA = j_n dA, \tag{4.2}$$

where θ is the angle between the direction normal to the surface and the carrier motion direction (which is taken for the direction of vector $\mathbf{j}$).

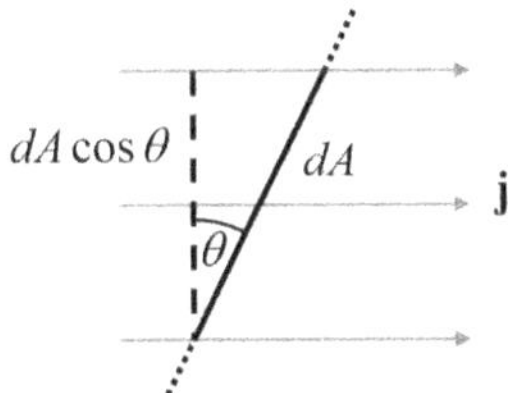

Figure 4.2. The current density vector $\mathbf{j}$.

With that definition, Eq. (4.1) may be re-written as

$$\frac{d}{dt}\int_V \rho\, d^3r = -\oint_S j_n\, d^2r, \tag{4.3}$$

where V is an arbitrary stationary volume limited by the closed surface S. Applying to this volume the same divergence theorem as was repeatedly used in previous chapters, we obtain

$$\int_V \left[\frac{\partial\rho}{\partial t} + \nabla\cdot\mathbf{j}\right] d^3r = 0. \tag{4.4}$$

Since volume V if arbitrary, this equation may be true only if

$$\frac{\partial\rho}{\partial t} + \nabla\cdot\mathbf{j} = 0. \tag{4.5}$$

---

[1] Just as a (hopefully, unnecessary) reminder, in the SI units the current is measured in amperes (A). In the legal metrology, the ampere (rather than the coulomb, which is defined as 1 C = 1 A × 1 s) is a primary unit. (Its formal definition will be discussed in the next chapter.) In the Gaussian units, Eq. (4.1) remains the same, so that the current's unit is the statcoulomb per second—the so-called *statampere*.

This is the fundamental *continuity equation*—which is true even for the time-dependent phenomena[2].

The charge relaxation, such as illustrated by figure 4.1b, is of course a dynamic, time-dependent process. However, electric currents may also exist in stationary situations, when a certain *current source*, for example a battery, drives the current against the electric field, and thus replenishes the conductor charges and sustains currents at a certain time-independent level—see figure 4.1c. (This process requires a persistent replenishment of the electrostatic energy of the system from either a source or a large storage of energy of a different kind—say, the chemical energy of the battery.) Let us discuss the laws governing the distribution of such *dc currents*. In this case ($\partial/\partial t = 0$), Eq. (4.5) reduces to a very simple equation

$$\nabla \cdot \mathbf{j} = 0. \tag{4.6}$$

This relation acquires an even simpler form in the particular but important case of *dc electric circuits* (figure 4.3), the systems that may be represented as connections of components of two types:

(i) small-size (*lumped*) *circuit elements* (also called 'two-terminal devices'), meaning a passive resistor, a current source, etc—generally, any 'black box' with two or more terminals, and

(ii) *perfectly conducting wires*, with negligible drop of the electrostatic potential along them, that are galvanically connected at certain points, called *nodes* (or 'junctions').

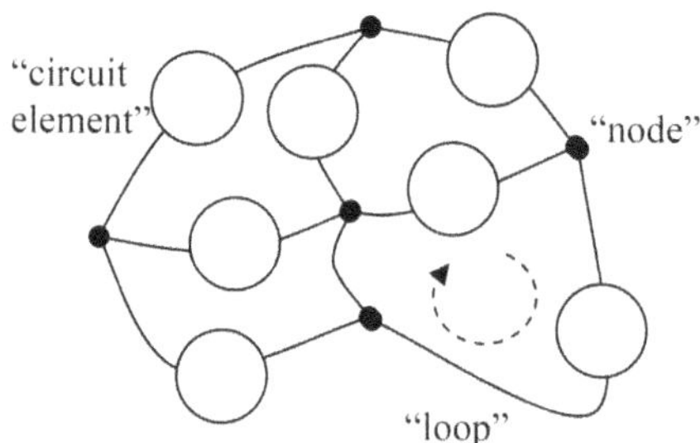

Figure 4.3. A typical system obeying the Kirchhoff laws.

In the standard circuit theory, the electric charges of the nodes are considered negligible[3], and we may integrate Eq. (4.6) over the closed surface drawn around any node to get a simple equality

[2] Similar differential relations are valid for the density of any conserved quantity, for example for mass in the classical fluid dynamics (see, e.g. *Part CM* section 8.3), and for the probability in the statistical physics (*Part SM* section 5.6) and quantum mechanics (*Part QM* section 1.4).

[3] In many cases, the charge accumulation may be described without an explicit violation of Eq. (4.7*a*), but just by adding other circuit elements, *lumped capacitors* (see figure 2.5 and its discussion), to the circuit under analysis. The resulting circuit may be used to describe not only the transient processes of charge accumulation/relaxation, but also ac currents. However, it will be more convenient for me to postpone the discussion of such *ac circuits* until chapter 6, where one more circuit element type, *lumped inductances*, will be introduced.

$$\sum_j I_j = 0, \tag{4.7a}$$

where the summation is over all the wires (numbered with index j) connected in the node. On the other hand, according to its definition (2.25), the voltage V_k across each circuit element may be represented as the difference of the electrostatic potentials of the adjacent nodes, $V_k = \phi_k - \phi_{k-1}$. Summing such differences around any closed loop of the circuit (figure 4.3), we obtain all terms cancelled, so that

$$\sum_k V_k = 0. \tag{4.7b}$$

These relations are called, respectively, the *first* and *second Kirchhoff laws*—or sometimes the *node rule* (4.7*a*) and the *loop rule* (4.7*b*). They may seem elementary, and their genuine power is in the mathematical fact that a set of Eqs. (4.7), covering every node and every circuit element of the system, gives a system of equations sufficient for the calculation of all currents and voltages in it—provided that the relation between current and voltage is known for each circuit element.

It is almost evident that in the absence of current sources, the system of equations (4.7) has only a trivial solution: $I_j = 0$, $V_k = 0$—with the exotic exception of superconductivity, to be discussed in section 6.3. The current sources, that allow non-vanishing current flows, may be described by their *electromotive forces* (e.m.f.) $\mathscr{V}_k$, having the dimensionality of voltage, which have to be taken into account in the corresponding terms V_k of the sum (4.7*b*). Let me hope that the reader has some experience of using Eqs. (4.7) for analyses of simple circuits—say consisting of several resistors and batteries, so that I can save time by skipping their discussion. Still, due to their practical importance, I would recommend the reader to carry out a self-test by solving the two problems given at the beginning of section 4.6.

4.2 The Ohm law

As was mentioned above, the relations spelled out in section 4.1 are sufficient for forming a closed system of equations for finding currents and an electric field in a system only if they are complemented with some constitutive relations between the scalars I and V in each lumped circuit element, i.e. between the vectors $\mathbf{j}$ and $\mathbf{E}$ in each point of the material of such an element. The simplest, and most frequently met relation of this kind is the famous *Ohm law* whose differential (or 'local') form is

$$\mathbf{j} = \sigma \mathbf{E}, \tag{4.8}$$

where σ is a constant called the *Ohmic conductivity* (or just the 'conductivity' for short)[4]. Although this is not a fundamental relation, and is approximate for *any* conducting medium, we can argue that if:

[4] In SI units, the conductivity is measured in S m^{-1}, where one siemens (S) is the reciprocal of one ohm: 1 S ≡ (1 Ω)$^{-1}$ ≡ 1 A/1 V. The constant reciprocal to conductivity, $1/\sigma$, is called *resistivity*, and is commonly denoted by letter ρ. I will, however, try to avoid using this notion, because I am already overusing this letter in these notes.

Table 4.1. Ohmic conductivities for some representative (or practically important) materials at 20 °C.

Material	σ (S/m)
Teflon ($[C_2F_4]_n$)	10^{-22}–10^{-24}
Silicon dioxide	10^{-16}–10^{-19}
Various glasses	10^{-10}–10^{-14}
Deionized water	$\sim 10^{-6}$
Sea water	5
Silicon n-doped to 10^{16} cm^{-3}	2.5×10^2
Silicon n-doped to 10^{19} cm^{-3}	1.6×10^4
Silicon p-doped to 10^{19} cm^{-3}	1.1×10^4
Nichrome (alloy 80% Ni + 20% Cr)	0.9×10^6
Aluminum	3.8×10^7
Copper	6.0×10^7
Zinc crystal along a-axis	1.65×10^7
Zinc crystal along c-axis	1.72×10^7

(i) there is no current at $\mathbf{E} = 0$ (mind superconductors!),
(ii) the medium is isotropic or almost isotropic (a notable exception: some organic conductors),
(iii) the mean free path l of current carriers is much smaller than the characteristic scale a of the spatial variations of $\mathbf{j}$ and $\mathbf{E}$,

then the Ohm law may be viewed as a result of the Taylor expansion of the local relation $\mathbf{j}(\mathbf{E})$ for relatively small fields, and thus is very common.

Table 4.1 gives the experimental values of the dc conductivity for some practically important (or just representative) materials. The reader can see that the range of its values is very broad, covering more than 30 orders of magnitude, even without going to such extremes as very pure metallic crystals at very low temperatures, where σ may reach $\sim 10^{12}$ s m^{-1}.

In order to obtain some feeling as to what these values mean, let us consider a very simple system (figure 4.4): a plane capacitor of area $A \gg d^2$, filled with a material that has not only a dielectric constant κ, but also some Ohmic conductivity σ, with much more conductive electrodes.

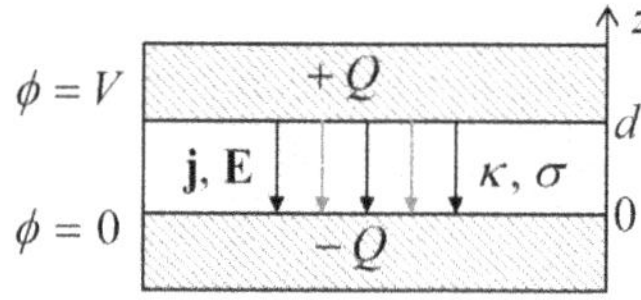

Figure 4.4. A 'leaky' plane capacitor.

Assuming that these properties are compatible with each other[5], we may assume that the distribution of the electric potential (not too close to the capacitor's edges) still obeys Eq. (2.39), so that the electric field is normal to the plates and uniform, with $E = V/d$. Then, according to Eq. (4.6), the current density is also uniform, $j = \sigma E = \sigma V/d$. From here, the total current between the plates is

$$I = jA = \sigma EA = \sigma\frac{V}{d}A. \tag{4.9}$$

On the other hand, from Eqs. (2.26) and (3.45), the instant value of the plate charge is $Q = CV = (\kappa\varepsilon_0 A/d)V$. Plugging these relations into Eq. (4.1), we see that the speed of charge (and voltage) relaxation does not depend on the geometric parameters A and d of the capacitor:

$$\frac{dV}{dt} = -\frac{V}{\tau_r}, \qquad \text{with } \tau_r \equiv \frac{\varepsilon_0\kappa}{\sigma} \equiv \frac{\varepsilon}{\sigma}, \tag{4.10}$$

so that where the *relaxation time constant* τ_r may be used to characterize the gap filling material as such.

As we already know (see table 3.1), for most practical materials the dielectric constant κ is within one order of magnitude from 10, so that the nominator of Eq. (4.10) is of the order of 10^{-10} (SI units). As a result, according to table 4.1, the charge relaxation time ranges from $\sim 10^{14}$ s (more than a million years!) for the best insulators such as teflon, to $\sim 10^{-18}$ s for the least resistive metals. What is the physics behind such a huge range of σ, and why, for some materials, does table 4.1 give them such a large uncertainty? As in chapters 2 and 3, I have time only for a brief, and admittedly superficial, discussion of these issues[6].

If the charge carriers move as classical particles (e.g. in plasmas or non-degenerate semiconductors), a very reasonable description of the conductivity is given by the famous *Drude formula*[7]. In his picture, due to a weak electric field, the charge carriers are accelerated in its direction (possibly on top of their random motion in all directions, i.e. with a vanishing average velocity vector),

$$\frac{d\mathbf{v}}{dt} = \frac{q}{m}\mathbf{E}, \tag{4.11}$$

and as a result their velocity acquires the average value

$$\langle\mathbf{v}\rangle = \frac{d\mathbf{v}}{dt}\tau = \frac{q}{m}\mathbf{E}\tau, \tag{4.12}$$

where the phenomenological parameter $\tau = l/v$ (not to be confused with τ_r!) may be understood as the effective average time between carrier scattering events. From here, the current density[8]:

---

[5] As will be discussed in chapter 6, such simple analysis is only valid if σ is not too high.

[6] A more detailed discussion may be found in *Part SM* chapter 6.

[7] It was suggested by P Drude in 1900.

[8] Note that $\mathbf{j}$ is usually defined as a macroscopic variable, by taking the area dA in Eq. (4.2) much larger than the square of inter-particle distances, so that no additional average sign is necessary in Eq. (4.13*a*).

$$\mathbf{j} = qn\langle \mathbf{v} \rangle = \frac{q^2 n\tau}{m}\mathbf{E}, \quad \text{i.e. } \sigma = \frac{q^2 n\tau}{m}. \tag{4.13a}$$

(Notice the independence of σ of the carrier charge sign.) Another form of the same result, more popular in the physics of semiconductors, is

$$\sigma = q^2 n\mu, \quad \text{with } \mu = \frac{\tau}{m}, \tag{4.13b}$$

where the parameter μ, defined by relation $\langle \mathbf{v} \rangle \equiv \mu\mathbf{E}$, is called the charge carrier *mobility*.

Most good conductors (e.g. metals) are essentially degenerate Fermi gases (or liquids), in which the average thermal energy of a particle, $k_B T$ is much lower that the Fermi energy ε_F. In this case, a quantum theory is needed for the calculation of σ. Such a theory was developed by the godfather of quantum physics A Sommerfeld in 1927 (and is sometimes called the *Drude–Sommerfeld model*). I have no time to discuss it in this course[9], and here will only note that for an ideal, isotropic Fermi gas the result is reduced to Eq. (4.13), with a certain effective value of τ, so it may be used for estimates of σ, with due respect to the quantum theory of scattering. In a typical metal, n is very high ($\sim 10^{23}$ cm^{-3}) and is fixed by the atomic structure, so that the sample quality may only affect σ via the scattering time τ.

At room temperature, the scattering of electrons by thermally excited lattice vibrations (*phonons*) dominates, so that τ and σ are high but finite, and do not change much from one sample to another. (Hence, the more accurate values given for metals in table 4.1.) On the other hand, at $T \rightarrow 0$, a perfect crystal should not exhibit scattering at all, and conductivity should be infinite. In practice, this is never true (for example, due to electron scattering from imperfect boundaries of finite-size samples), and the effective conductivity σ is infinite (or practically infinite, at least above the measurable value $\sim 10^{20}$ s m^{-1}) only in superconductors[10].

On the other hand, the conductivity of quasi-insulators (including the deionized water) and semiconductors depends mostly on the carrier density n, which is much lower than in metals. From the point of view of quantum mechanics, this happens because the ground-state eigenenergies of charge carriers are localized within an atom (or molecule) and separated from excited states, with space-extended wave-functions, by a large energy gap (called the *bandgap*). For example, in SiO_2 the bandgap approaches 9 eV, equivalent to ~4000 K. This is why, even at room temperature, the density of thermally excited free charge carriers in good insulators is negligible. In these materials, n is determined by impurities and vacancies, and may depend on a particular chemical synthesis or other fabrication technology, rather than on fundamental properties of the material. (In contrast, the carrier mobility μ in these materials is almost technology-independent.)

[9] For such a discussion see, e.g. *Part SM* section 6.3.

[10] The electrodynamic properties of superconductors are so interesting (and fundamentally important) that I will discuss them in more detail in chapter 6.

The practical importance of the technology may be illustrated by the following example. In the cells of the so-called *floating-gate memories*, in particular the *flash memories* which currently dominate non-volatile digital memory technology, data bits are stored as small electric charges ($Q \sim 10^{-16}$ C) of highly doped silicon islands (so-called *floating gates*) separated from the rest of the integrated circuit with a ~10 nm thick layer of the silicon dioxide, SiO_2. Such layers are fabricated by high-temperature oxidation of virtually perfect silicon crystals. The conductivity of the resulting high-quality (though amorphous) material is so low, $\sigma \sim 10^{-19}$ s m^{-1}, that the relaxation time τ_r, defined by Eq. (4.10), is well above 10 years—the industrial standard for data retention in non-volatile memories. In order to appreciate how good this technology is, the cited value should be compared with the typical conductivity $\sigma \sim 10^{-16}$ s m^{-1} of the usual, bulk SiO_2 ceramics[11].

4.3 Boundary problems

For an Ohmic conducting medium, we may combine equations (4.6) and (4.8) to obtain the following differential equation

$$\nabla \cdot (\sigma \nabla \phi) = 0. \tag{4.14}$$

For a uniform conductor ($\sigma = \text{const}$), Eq. (4.14) is reduced to the Laplace equation for the electrostatic potential ϕ. As we already know from chapters 2 and 3, its solution depends on the boundary conditions. These conditions depend on the interface type.

(i) *Conductor–conductor interface.* Applying the continuity equation (4.6) to a Gauss-type pillbox at the interface of two different conductors (figure 4.5), we obtain

$$(j_n)_1 = (j_n)_2, \tag{4.15}$$

so that if the Ohm law (4.8) is valid inside each medium, then

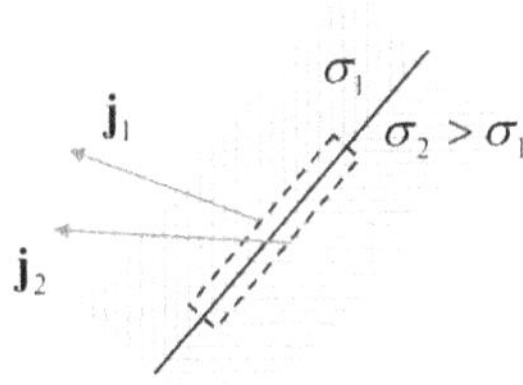

Figure 4.5. DC current's 'refraction' at the interface between two different conductors.

[11] This course is not an appropriate platform to discuss the details of floating-gate memory technology. However, I think that every educated physicist should know its basics, because such memories are currently the drivers of all semiconductor integrated circuit technology development, and hence of the entire progress of information technology. Perhaps the best available book is [1].

$$\sigma_1 \frac{\partial \phi_1}{\partial n} = \sigma_2 \frac{\partial \phi_2}{\partial n}. \tag{4.16}$$

Also, since the electric field should be finite, its potential ϕ has to be continuous across the interface—the condition that may also be written as

$$\frac{\partial \phi_1}{\partial \tau} = \frac{\partial \phi_2}{\partial \tau}. \tag{4.17}$$

Both these conditions (and hence the solutions of the boundary problems using them) are similar to those for the interface between two dielectrics—cf Eqs. (3.46) and (3.47).

Note that using the Ohm law, Eq. (4.17) may be rewritten as

$$\frac{1}{\sigma_1}(j_\tau)_1 = \frac{1}{\sigma_2}(j_\tau)_2. \tag{4.18}$$

Comparing it with Eq. (4.15) we see that, generally, the current density's magnitude changes at the interface: $j_1 \neq j_2$. It is also curious that if $\sigma_1 \neq \sigma_2$, the current line slope changes at the interface (figure 4.5), qualitatively to the refraction of light rays in optics—see chapter 7.

(ii) *Conductor–electrode interface.* An *electrode* is defined as a body made of a 'perfect conductor', i.e. of a medium with $\sigma \to \infty$. Then, at a fixed current density at the interface, the electric field in the electrode tends to zero, and hence it may be described by equation

$$\phi = \phi_j = \text{const}, \tag{4.19}$$

where constants ϕ_j may be different for different electrodes (numbered with index j). Note that with such boundary conditions, the Laplace boundary problem becomes exactly the same as in electrostatics—see Eq. (2.35)—and hence we can use all the methods (and some solutions) of chapter 2 for finding the dc current distribution.

(iii) *Conductor–insulator interface.* For the description of an insulator, we can use $\sigma = 0$, so that Eq. (4.16) yields the following boundary condition,

$$\frac{\partial \phi}{\partial n} = 0, \tag{4.20}$$

for the potential derivative *inside the conductor*. From the Ohm law (4.8) in the form $\mathbf{j} = -\sigma\nabla\phi$, we see that this is just the very natural requirement for the dc current not to flow into an insulator.

Now, note that this condition makes the Laplace problem inside the conductor completely well-defined, and independent of the potential distribution in the adjacent insulator. In contrast, due to the continuity of the electrostatic potential at the border, its distribution in the insulator has to follow that inside the conductor. Let us discuss this conceptual issue on the following (apparently, trivial) example: dc current in a uniform wire of length l and a cross-section of area A. The reader certainly knows the answer:

$$I = \frac{V}{R}, \quad \text{where } R \equiv \frac{V}{I} = \frac{l}{\sigma A}, \tag{4.21}$$

where the constant R is called the *resistance*[12]. However, let us derive this result formally from our theoretical framework. For the simple geometry shown in figure 4.6a, this is easy to do. Here the potential evidently has a linear 1D distribution

$$\phi = \text{const} - \frac{x}{l}V, \tag{4.22}$$

both in the conductor and the surrounding free space, with both boundary conditions (4.16) and (4.17) satisfied at the conductor–insulator interfaces, and the condition (4.20) satisfied at the conductor–electrode interface. As a result, the electric field is constant and has only one component $E_x = V/l$, so that inside the conductor

$$j_x = \sigma E_x, \quad I = j_x A, \tag{4.23}$$

giving us the well-known Eq. (4.21).

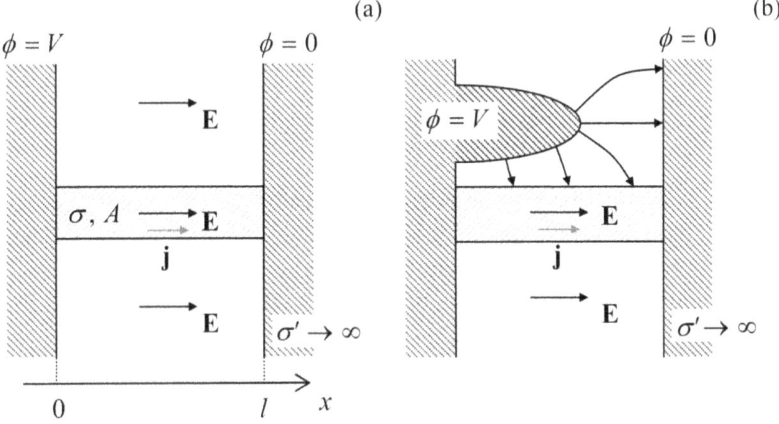

Figure 4.6. (a) An elementary problem and (b) a (slightly) less obvious problem of the field distribution at dc current flow (schematically).

However, what about the geometry shown in figure 4.6b? In this case the field distribution in the free space around the conductor is dramatically different, but according to boundary problem defined by equations (4.14) and (4.20), inside the conductor the solution is exactly the same as it was in the former case. Now, the Laplace equation in the surrounding insulator has to be solved with the boundary values of the electrostatic potential, 'dictated' by the distribution of the current (and hence potential) in the conductor. Note that as the result, the electric field lines are

[12] The first of Eqs. (4.21) is essentially the integral form of the Ohm law (4.8), and is valid not only for a uniform wire, but also for any Ohmic conductor with a geometry in which I and V may be clearly defined.

generally not normal to the conductor' surface, because the surface is not equipotential.

Let us solve a problem in which this *conduction hierarchy* may be followed analytically to the very end. Consider an empty spherical cavity cut in a conductor with an initially uniform current flow with a constant density $\mathbf{j}_0 = \mathbf{n}_z j_0$ (figure 4.7a). Following the conduction hierarchy, we have to solve the boundary problem in the conducting part of the system, i.e. outside the sphere ($r \geqslant R$), first. Since the problem is evidently axially symmetric, we already know the general solution of the Laplace equation—see Eq. (2.172). Moreover, we know that in order to match the uniform field distribution at $r \to \infty$, all coefficients a_l but one ($a_1 = -E_0 = -j_0/\sigma$) have to be zero, and that the boundary conditions at $r = R$ will give zero solutions for all coefficients b_l but one (b_1), so that

$$\phi = -\frac{j_0}{\sigma} r \cos\theta + \frac{b_1}{r^2}\cos\theta, \qquad \text{for } r \geqslant R. \tag{4.24}$$

In order to find the coefficient b_1, we have to use the boundary condition (4.20) at $r = R$:

$$\left.\frac{\partial \phi}{\partial r}\right|_{r=R} = \left(-\frac{j_0}{\sigma} - \frac{2b_1}{R^3}\right)\cos\theta = 0. \tag{4.25}$$

This gives $b_1 = -j_0 R^3/2\sigma$, so that, finally,

$$\phi(r, \theta) = -\frac{j_0}{\sigma}\left(r + \frac{R^3}{2r^2}\right)\cos\theta, \qquad \text{for } r \geqslant R. \tag{4.26}$$

(Note that this potential distribution corresponds to the dipole moment $\mathbf{p} = -\mathbf{E}_0 R^3/2$. It is straightforward to check that if the spherical cavity was cut in a dielectric, the potential distribution outside it would be similar, with $\mathbf{p} = -\mathbf{E}_0 R^3(\kappa - 1)/(\kappa + 2)$. In the limit $\kappa \to \infty$, these two results coincide, despite the rather different type of problem: in the dielectric case, there is no current at all.)

Now, as the second step in the conductivity hierarchy, we may find the electrostatic potential distribution $\phi(r,\theta)$ in the insulator, in this particular case inside the

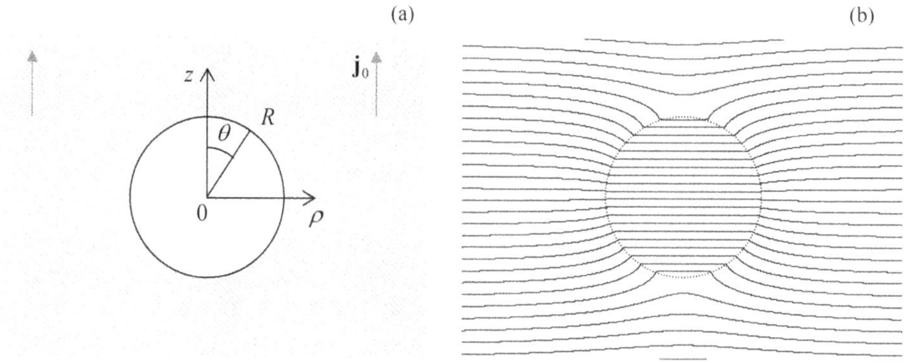

Figure 4.7. A spherical cavity cut in a uniform conductor: (a) the problem's geometry and (b) the equipotential surfaces as given by equations (4.26) and (4.28).

empty cavity ($r \leqslant R$). It should also satisfy the Laplace equation with the boundary conditions at $r = R$, 'dictated' by the distribution (4.26):

$$\phi(R, \theta) = -\frac{3}{2}\frac{j_0}{\sigma}R\cos\theta. \tag{4.27}$$

We could again solve this problem by the formal variable separation (keeping in the general solution (2.172) only the term proportional to a_1, which does not diverge at $r \to 0$), but if we notice that the boundary condition (4.27) depends on just one Cartesian coordinate, $z = R\cos\theta$, the solution may be just guessed:

$$\phi(r, \theta) = -\frac{3}{2}\frac{j_0}{\sigma}z = -\frac{3}{2}\frac{j_0}{\sigma}r\cos\theta, \quad \text{at } r \leqslant R. \tag{4.28}$$

It evidently satisfies the Laplace equation and the boundary condition (4.27), and corresponds to a constant electric field parallel to the vector $\mathbf{j}_0$, and equal to $3j_0/2\sigma$—see figure 4.7b. Again, the cavity surface is not equipotential, and the electric field lines at $r \leqslant R$ are not normal to it at almost all points.

The conductivity hierarchy says that static electrical fields and charges outside conductors (e.g. electric wires) do not affect currents flowing in the wires, and it is physically very clear why. For example, if a charge in the free space is slowly moved close to a wire, it (in accordance with the linear superposition principle) will only induce an additional surface charge (see chapter 2) that screens the external charge's field, without participating in the current flow inside the conductor.

Apart from this conceptual issue, the two examples given above may be considered as the further demonstrations of the first two methods discussed in chapter 2 (the orthogonal coordinates (figure 4.6) and the variable separation (figure 4.7)) to dc current distribution problems. If we have a glance at other methods discussed in that chapter, we may notice that there is also an analog of the method of charge images. Indeed, let us consider the spherically symmetric potential distribution of the electrostatic potential, similar to that given by the basic Eq. (1.35):

$$\phi = \frac{c}{r}. \tag{4.29}$$

As we know from chapter 1, this is a particular solution of the 3D Laplace equation at all points but $r = 0$. In the free space, this distribution would correspond to a point charge $q = 4\pi\varepsilon_0 c$; but what about the conductor? Calculating the corresponding electric field and current density,

$$\mathbf{E} = -\nabla\phi = \frac{c}{r^3}\mathbf{r}, \qquad \mathbf{j} = \sigma\mathbf{E} = \sigma\frac{c}{r^3}\mathbf{r}, \tag{4.30}$$

we see that the total current flowing from the origin through a sphere of an arbitrary radius r does not depend on the radius:

$$I = Aj = 4\pi r^2 j = 4\pi\sigma\, c. \tag{4.31}$$

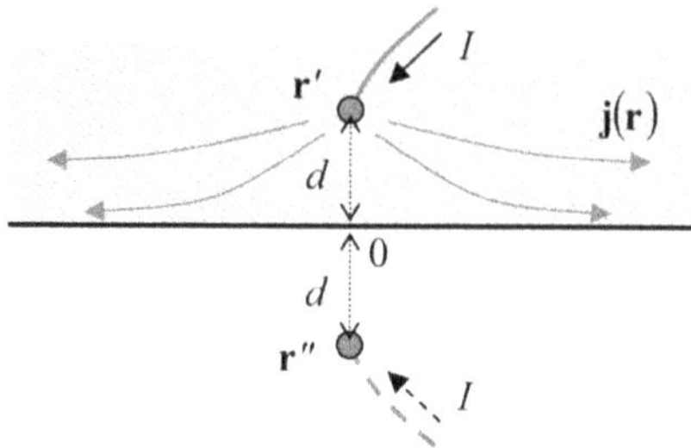

Figure 4.8. Applying the method of images for the current injection analysis.

Plugging the resulting coefficient c into Eq. (4.29), we obtain

$$\phi = \frac{I}{4\pi\sigma r}. \tag{4.32}$$

Hence the Coulomb-type distribution of the electric potential in a conductor is possible (at least at some distance from the singular point $r = 0$), and describes the dc current I flowing out of a small-size electrode—or *into* such a point, if the coefficient c is negative. Such *current injection* may be readily implemented experimentally; think for example about an insulated wire with a small bare end, inserted into a poorly conducting soil—an important method in geophysical research[13].

Now let the current injection point $\mathbf{r}'$ be close to a plane interface between the conductor and an insulator (figure 4.8). In this case, besides the Laplace equation, we should satisfy the boundary condition,

$$j_n = \sigma E_n = -\sigma\frac{\partial\phi}{\partial n} = 0, \tag{4.33}$$

at the interface. It is clear that this can be done by replacing the insulator with an imaginary similar conductor with an additional current injection point, at the mirror image point $\mathbf{r}''$. Note, however, that in contrast to the charge images, the sign of the imaginary current has to be *similar*, not opposite, to the initial one, so that the total electrostatic potential inside the conducting semi-space is

$$\phi(\mathbf{r}) = \frac{I}{4\pi\sigma}\left(\frac{1}{|r - r'|} + \frac{1}{|r - r''|}\right). \tag{4.34}$$

(The image current's sign would be opposite at the interface between a conductor with a moderate conductivity and a perfect conductor ('electrode'), whose potential should be virtually constant.)

This result may be readily used, for example, to calculate the current density at a plane surface of a uniform conductor, as a function of distance ρ from point 0 (the surface's point closest to the current injection site)—see figure 4.8. At such surface, Eq. (4.34) yields

[13] Such injection is even simpler in 2D situations—think about a wire soldered, in a small spot, to a thin conducting foil. (Note that here the current density distribution law is different, $j \propto 1/r$ rather than $1/r^2$.)

$$\phi = \frac{I}{2\pi\sigma} \frac{1}{(\rho^2 + d^2)^{1/2}}, \tag{4.35}$$

so that the current density is:

$$j_\rho = \sigma E_\rho = -\sigma \frac{\partial \phi}{\partial \rho} = \frac{I}{2\pi} \frac{\rho}{(\rho^2 + d^2)^{3/2}}. \tag{4.36}$$

Deviations from equations (4.35) and (4.36) may be used to find and characterize conductance inhomogeneities, say those due to mineral deposits in the Earth's crust[14].

4.4 Energy dissipation

Let me conclude this brief chapter with an ultra-short discussion of energy dissipation in conductors. In contrast to the electrostatics situations in insulators (vacuum or dielectrics), at dc conduction the electrostatic energy U is 'dissipated' (i.e. transferred to heat) at a certain rate $\mathscr{P} \equiv -dU/dt$, with the dimensionality of power[15]. This so-called *dissipation power* may be evaluated by calculating the power of the electric field's work on a single moving charge:

$$\mathscr{P}_1 = \mathbf{F} \cdot \mathbf{v} = q\mathbf{E} \cdot \mathbf{v}. \tag{4.37}$$

After the summation over all charges, Eq. (4.37) gives us the dissipation power. If the charge density n is uniform, multiplying by it the both parts of this relation, and taking into account that $qn\mathbf{v} = \mathbf{j}$, for the energy dissipation in a unit volume we obtain the *Joule law*

$$\mathscr{p} \equiv \frac{\mathscr{P}}{V} = \frac{\mathscr{P}_1 N}{V} = \mathscr{P}_1 n = q\mathbf{E} \cdot \mathbf{v}n = \mathbf{E} \cdot \mathbf{j}. \tag{4.38}$$

In the case of the Ohmic conductivity, this expression may also be rewritten in two other forms:

$$\mathscr{p} = \sigma E^2 = \frac{j^2}{\sigma}. \tag{4.39}$$

At the dc conduction, the electrostatic energy has to be permanently replenished by an equal flow of power from the current source(s). With our electrostatics background, it is also straightforward (and hence left for the reader's exercise) to prove that the dc current distribution in a uniform Ohmic conductor, at a fixed voltage applied at its borders, corresponds to the minimum of the total dissipation power

$$\mathscr{P} = \sigma \int_V E^2 d^3r. \tag{4.40}$$

[14] The current injection may be produced, due to electrochemical reactions, by an ore mass itself, so that one need only measure (and correctly interpret) the resulting potential distribution—the so-called *self-potential method*—see, e.g. section 6.1 in [2].

[15] Since the electric field and hence the electrostatic energy are time-independent, this means that the energy is replenished at the same rate from the current source(s).

4.5 Problems

Problem 4.1. A dc voltage V_0 is applied to the end of a semi-infinite chain of lumped Ohmic resistors, shown in the figure below. Calculate the voltage across the *j*th link of the chain.

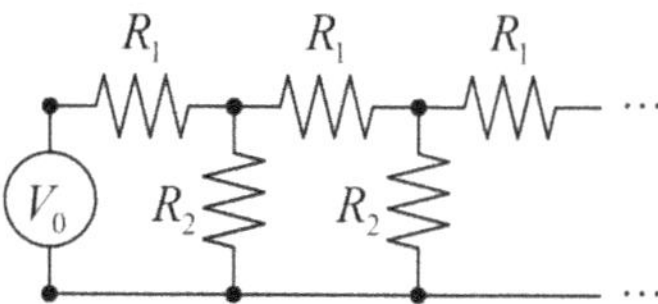

Problem 4.2. It is well known that properties of many dc current sources (e.g. batteries) may be reasonably well represented as a connection in series of a *perfect voltage source* and an Ohmic *internal resistance*. Discuss the option, and possible advantages, of using a different equivalent circuit that would include a *perfect current source*.

Problem 4.3. Calculate the resistance between two large, uniform Ohmic conductors separated with a very thin, planar, insulating partition, with a circular hole of radius R in it—see the figure below.

Hint: You may like to use the degenerate ellipsoidal coordinates, which had been used in section 2.4.

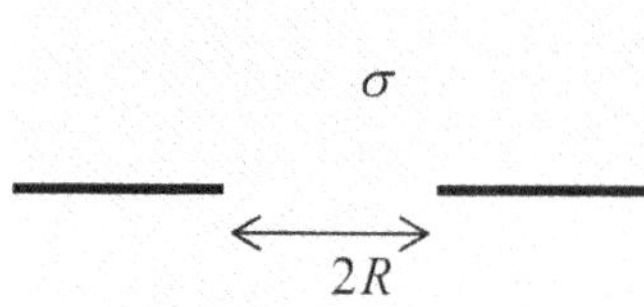

Problem 4.4. Calculate the effective (average) conductivity σ_{ef} of a medium with many empty spherical cavities of radius R, carved at random positions in a uniform Ohmic conductor (see the figure below), in the limit of a low density $n \ll R^{-3}$ of the spheres.

Hint: Try to use the analogy with a dipole medium—see, e.g. section 3.2.

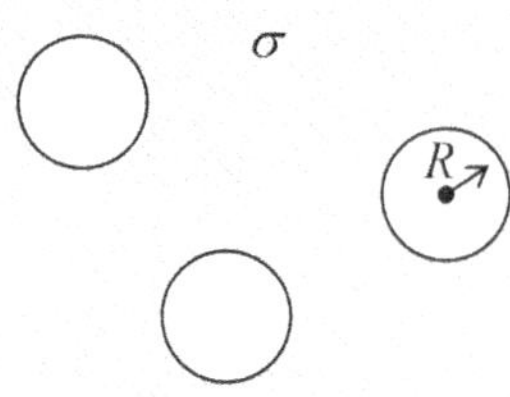

Problem 4.5. In two separate experiments, a narrow gap, possibly of an irregular width, between two close metallic electrodes is filled with some material—in the first case, with a uniform linear insulator with an electric permittivity ε, and in the second case, with a uniform conducting material with an Ohmic conductivity σ. Neglecting the fringe effects, calculate the relation between the mutual capacitance C between the electrodes (in the first case) and the dc resistance R between them (in the second case).

Problem 4.6. Calculate the voltage V across a uniform, wide resistive slab of thickness t, at distance l from the points of injection/pickup of the dc current I passed across the slab—see the figure below.

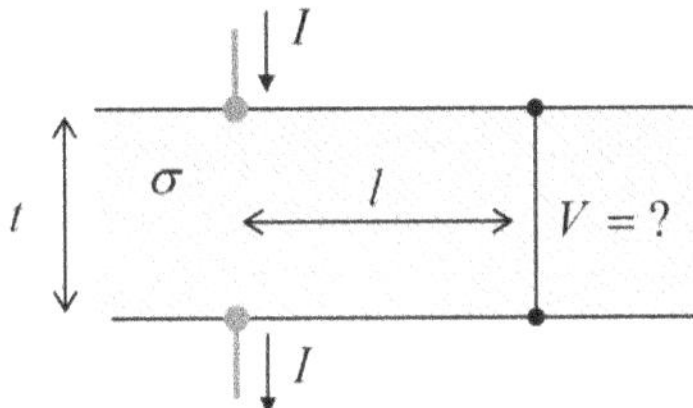

Problem 4.7. Calculate the voltage V between two corners of a square cut from a uniform, resistive sheet of a very small thickness t, induced by dc current I that is passed between its two other corners—see the figure below.

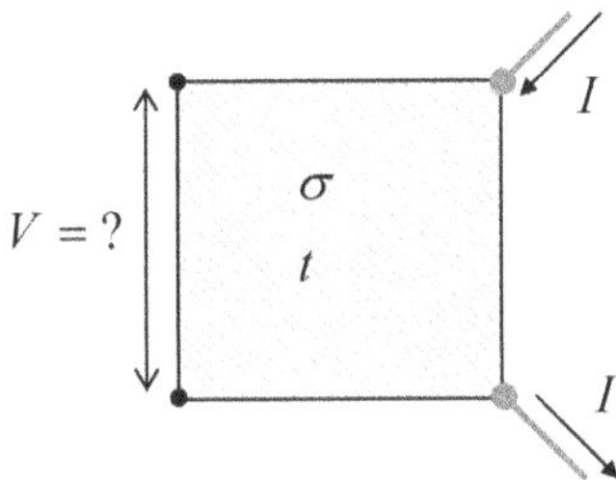

Problem 4.8. Calculate the distribution of the dc current's density in a thin, round, uniform resistive disk, if the current is inserted into a point at its rim, and picked up at the center.

Problem 4.9.* The simplest model of a vacuum diode consists of two planar, parallel metallic electrodes of area A, separated by a gap of thickness $d \ll A^{1/2}$: a 'cathode' which emits electrons to vacuum, and an 'anode' which absorbs the electrons arriving at its surface. Calculate the dc I–V curve of the diode, i.e. the stationary relation between the current I flowing between the electrodes and the voltage V applied between them, using the following simplifying assumptions:

(i) due to the effect of the negative space charge of the emitted electrons, the current I is much smaller than the emission ability of the cathode,
(ii) the initial velocity of the emitted electrons is negligible, and

(iii) the direct Coulomb interaction of electrons (besides the space charge effect) is negligible.

Problem 4.10.* Calculate the space-charge-limited current in a system with the same geometry, and using the same assumptions as in the previous problem, apart from assuming that now the emitted charge carriers move not ballistically, but drift in accordance with the Ohm law, with the conductivity given by (4.13): $\sigma = q^2\mu n$, with a constant mobility μ.

Hint: In order to obtain a realistic result, assume that the medium in which the charge carriers move[16] has a certain dielectric constant κ, unrelated to the carriers.

Problem 4.11. Prove that the distribution of dc currents in a uniform Ohmic conductor, at a fixed voltage applied at its boundaries, corresponds to the minimum of the total power dissipation ('Joule heat').

References

[1] Brewer J and Gill M (eds) 2008 *Nonvolatile Memory Technologies with Emphasis on Flash* (Piscataway, NJ: IEEE)

[2] Telford W *et al* 1990 *Applied Geophysics* 2nd edn (Cambridge University Press)

[16] As was mentioned in section 4.2 of the lecture notes, the assumption of a constant (charge-density-independent) mobility is most suitable for semiconductors.

IOP Publishing

Classical Electrodynamics
Lecture notes
Konstantin K Likharev

Chapter 5

Magnetism

Despite the fact that this chapter addresses a completely new type of electric charge interaction, its discussion (for the stationary case) will take not too much time/space, because it recycles many ideas and methods of electrostatics, although with a twist or two.

5.1 Magnetic interaction of currents

DC currents in conductors usually leave them *electroneutral*, $\rho(\mathbf{r}) = 0$, with very good precision, because even a minor disbalance of positive and negative charge density results in extremely strong Coulomb forces that restore their balance by an additional shift of free charge carriers[1]. This is why we start the discussion of magnetic interactions from the simplest case of two spatially separated, current-carrying, electroneutral conductors (figure 5.1).

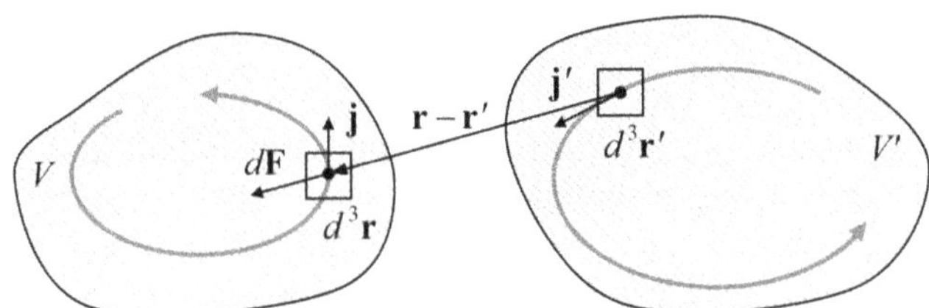

Figure 5.1. The magnetic interaction of two currents.

According to the Coulomb law, there should be no force between them. However, several experiments carried out in the early 1820s[2] proved that such non-Coulomb

[1] The most important case when the electroneutrality does not hold is the motion of electrons in vacuum. In this case, the magnetic forces we are going to discuss coexist with (typically, stronger) electrostatic forces—see Eq. (5.3) below and its discussion. In some semiconductor devices, local violations of electroneutrality also play an important role—see, e.g. *Part SM* chapter 6.

[2] Most notably, by H C Ørsted, J-B Biot and F Savart, and A-M Ampère.

forces do exist, and are the manifestation of a different, *magnetic* interaction between the currents. In the contemporary notation, their results may be summarized with just one formula, in SI units expressed as[3]:

$$\mathbf{F} = -\frac{\mu_0}{4\pi}\int_V d^3r \int_{V'} d^3r' [\mathbf{j}(\mathbf{r}) \cdot \mathbf{j}'(\mathbf{r}')] \frac{\mathbf{r} - \mathbf{r}'}{|\mathbf{r} - \mathbf{r}'|^3}. \tag{5.1}$$

Here the coefficient $\mu_0/4\pi$ (where μ_0 is called either the *magnetic constant* or the *free space permeability*), by definition, equals *exactly* 10^{-7} SI units, thus relating the electric current unit definition to that of force—see below.

Note an *almost* complete similarity of this expression to the Coulomb law (1.1), written for a continuous charge distribution, with the account of the linear superposition principle:

$$\mathbf{F} = \frac{1}{4\pi\varepsilon_0}\int_V d^3r \int_{V'} d^3r' \rho(\mathbf{r})\rho'(\mathbf{r}') \frac{\mathbf{r} - \mathbf{r}'}{|\mathbf{r} - \mathbf{r}'|^3}. \tag{5.2}$$

Apart from the different coefficient and a different sign, the 'only' difference between Eq. (5.1) and Eq. (5.2) is the scalar product of current densities, evidently necessary because of the vector character of the current density. We will see that this difference brings certain complications in applying the approaches discussed in the previous chapters to magnetostatics.

Before moving on to their discussion, let us have one more glance at the coefficients in Eqs. (5.1) and (5.2). To compare them, let us consider two objects with uncompensated charge distributions $\rho(\mathbf{r})$ and $\rho'(\mathbf{r})$, each moving parallel to each other as a whole, with certain velocities $\mathbf{v}$ and $\mathbf{v}'$, as measured in an inertial ('laboratory') frame. In this case, $\mathbf{j}(\mathbf{r}) = \rho(\mathbf{r})\mathbf{v}$, so that $\mathbf{j}(\mathbf{r}) \cdot \mathbf{j}'(\mathbf{r}) = \rho(\mathbf{r})\rho'(\mathbf{r})vv'$, and the integrals in Eqs. (5.1) and (5.2) become functionally similar, and differ only by the factor

$$\frac{F_{\text{magnetic}}}{F_{\text{electric}}} = -\frac{\mu_0 vv'}{4\pi}\Big/\frac{1}{4\pi\varepsilon_0} = -\frac{vv'}{c^2}. \tag{5.3}$$

(The last expression holds in any consistent system of units.) We immediately see that the magnetism is an essentially relativistic phenomenon, very weak in comparison with the electrostatic interaction at the human scale velocities, $v \ll c$, and may dominate only if the latter interaction vanishes—as it does in electroneutral systems[4].

[3] In the Gaussian units, the coefficient $\mu_0/4\pi$ is replaced with $1/c^2$ (i.e. implicitly with $\mu_0\varepsilon_0$) where c is the speed of light, in modern metrology considered *exactly* known—see, e.g. appendix B.

[4] The discovery and initial studies of such a subtle, relativistic phenomenon as magnetism in the early nineteenth century was much facilitated by the relative abundance of natural *ferromagnets*, materials with a spontaneous magnetic polarization, whose strong magnetic field may be traced back to relativistic effects (such as spin) in the constitute atoms. Ferromagnetism will be (briefly) discussed in section 5.5 below, and then in *Part SM* chapter 4.

Also, Eq. (5.3) points at an interesting paradox. Consider two electron beams moving parallel to each other, with the same velocity v with respect to a lab reference frame. Then, according to Eq. (5.3), the net force of their total (electric plus magnetic) interaction is proportional to $(1 - v^2/c^2)$, and tends to zero in the limit $v \to c$. However, in the reference frame moving together with the electrons, they are not moving at all, i.e. $v = 0$. Hence, from the point of view of such a moving observer, the electron beams should interact only electrostatically, with a repulsive force independent of the velocity v. Historically, this had been one of several paradoxes that led to the development of special relativity; its resolution will be discussed in chapter 9, which is devoted to this theory.

Returning to Eq. (5.1), in some simple cases the double integration in it may be carried out analytically. First of all, let us simplify this expression for the case of two thin, long conductors (wires) separated by a distance much larger than their thickness. In this case we may integrate the products $\mathbf{j}d^3r$ and $\mathbf{j}'d^3r'$ over the wires' cross-sections first, neglecting the corresponding change of $(\mathbf{r} - \mathbf{r}')$. Since the integrals of the current density over the cross-sections of the wires are just the currents I and I' flowing in the wires, and cannot change along their lengths (say, l and l', respectively), they may be taken out of the remaining integrals, reducing Eq. (5.1) to

$$\mathbf{F} = -\frac{\mu_0 II'}{4\pi} \oint_l \oint_{l'} (d\mathbf{r} \cdot d\mathbf{r}') \frac{\mathbf{r} - \mathbf{r}'}{|\mathbf{r} - \mathbf{r}'|^3}. \tag{5.4}$$

As the simplest example, consider two straight, parallel wires (figure 5.2), separated by distance d, with length $l \gg d$. In this case, due to symmetry, the vector of the magnetic interaction force has to:

(i) lie in the same plane as the currents, and
(ii) be perpendicular to the wires—see figure 5.2.

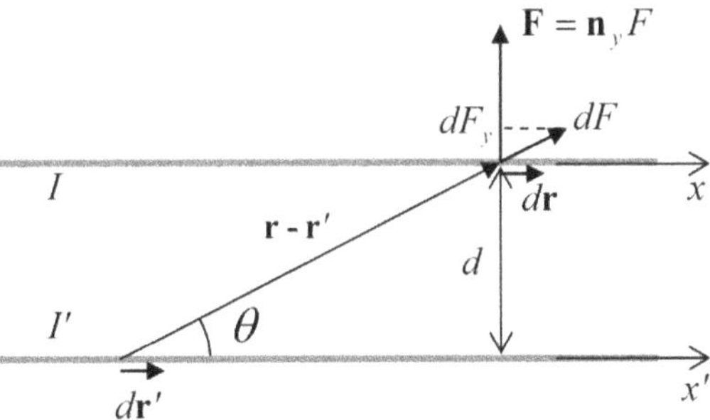

Figure 5.2. Magnetic force between two straight parallel currents.

Hence we can limit our calculations to just one component of the force, normal to the wires. Using the fact that with the coordinate choice shown in figure 5.2, the scalar product $d\mathbf{r} \cdot d\mathbf{r}'$ is just $dxdx'$, we obtain

$$\begin{aligned} F &= -\frac{\mu_0 II'}{4\pi}\int_{-\infty}^{+\infty} dx \int_{-\infty}^{+\infty} dx' \frac{\sin\theta}{d^2 + (x - x')^2} \\ &= -\frac{\mu_0 II'}{4\pi}\int_{-\infty}^{+\infty} dx \int_{-\infty}^{+\infty} dx' \frac{d}{[d^2 + (x - x')^2]^{3/2}}. \end{aligned} \tag{5.5}$$

Introducing, instead of x', a new, dimensionless variable $\xi \equiv (x - x')/d$, we may reduce the internal integral to a table integral which we have already met in this course:

$$F = -\frac{\mu_0 II'}{4\pi d}\int_{-\infty}^{+\infty} dx \int_{-\infty}^{+\infty} \frac{d\xi}{\left(1 + \xi^2\right)^{3/2}} = -\frac{\mu_0 II'}{2\pi d}\int_{-\infty}^{+\infty} dx. \tag{5.6}$$

The integral over x formally diverges, but this means merely that the interaction force *per unit length* of the wires is constant:

$$\frac{F}{l} = -\frac{\mu_0 II'}{2\pi d}. \tag{5.7}$$

Note that the force drops rather slowly (only as $1/d$) as the distance d between the wires is increased, and is *attractive* (rather than repulsive as in the Coulomb law) if the currents are of the same sign.

This is an important result[5], but again, the problems so simply solvable are few and far between, and it is intuitively clear that we would strongly benefit from the same approach as in electrostatics, i.e. from breaking Eq. (5.1) into a product of two factors via the introduction of a suitable *field*. Such decomposition may be done as follows:

$$\mathbf{F} = \int_V \mathbf{j}(\mathbf{r}) \times \mathbf{B}(\mathbf{r}) d^3r, \tag{5.8}$$

where the vector $\mathbf{B}$ is called the *magnetic field* (in our particular case, induced by current $\mathbf{j}'$)[6]:

$$\mathbf{B}(\mathbf{r}) \equiv \frac{\mu_0}{4\pi}\int_{V'} \mathbf{j}'(\mathbf{r}') \times \frac{\mathbf{r} - \mathbf{r}'}{|\mathbf{r} - \mathbf{r}'|^3} d^3r'. \tag{5.9}$$

[5] In particular, Eq. (5.7) is used for the legal definition of the SI unit of current, one ampere (A), via the SI unit of force (the newton, N), with the coefficient μ_0 considered exactly fixed as listed above.

[6] The SI unit of the magnetic field is called *tesla* (T) after N Tesla, an electrical engineering pioneer. In Gaussian units, the already discussed constant $1/c^2$ in Eq. (5.1) is equally divided between Eqs. (5.8) and (5.9), so that in them both, the constant before the integral is $1/c$. The resulting Gaussian unit of field $\mathbf{B}$ is called the *gauss* (G); taking into account the difference of units of electric charge and length, and hence the current density, 1 G equals exactly 10^{-4} T. Note also that in some textbooks, in particular old ones, $\mathbf{B}$ is called either the *magnetic induction*, or the *magnetic flux density*, while the term 'magnetic field' is reserved for the macroscopic vector $\mathbf{H}$, which will be introduced in section 5.5 below.

The last relation is called the *Biot–Savart law*, while the force **F** expressed by Eq. (5.8) is sometimes called the *Lorentz force*[7]. However, more frequently the later term is reserved for the full force,

$$\mathbf{F} = q(\mathbf{E} + \mathbf{v} \times \mathbf{B}), \tag{5.10}$$

exerted by electric and magnetic fields field on a point charge q, moving with velocity $\mathbf{v}$.[8]

Now we have to prove that the new formulation, given by Eqs. (5.8) and (5.9), is equivalent to Eq. (5.1). At first glance this seems unlikely. Indeed, first of all, Eqs. (5.8) and (5.9) involve vector products, while (5.1) is based on a scalar product. More profoundly, in contrast to Eq. (5.1), Eqs. (5.8) and (5.9) do *not* satisfy Newton's third law, applied to elementary current components $\mathbf{j}d^3r$ and $\mathbf{j}'d^3r'$, if these vectors are not parallel to each other. Indeed, consider the situation shown in figure 5.3.

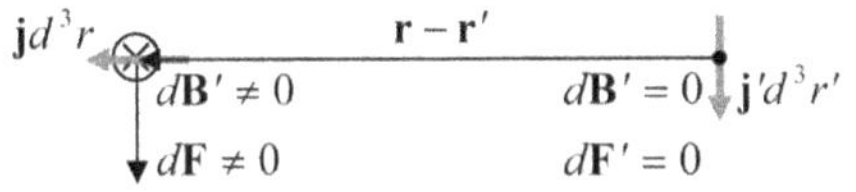

Figure 5.3. The apparent violation of Newton's third law in magnetism.

Here the vector $\mathbf{j}'$ is perpendicular to the vector $(\mathbf{r} - \mathbf{r}')$, and hence, according to Eq. (5.9), produces a non-zero contribution $d\mathbf{B}'$ to the magnetic field, directed (in figure 5.3) normal to the plane of drawing, i.e. is perpendicular to the vector $\mathbf{j}$. Hence, according to Eq. (5.8), this field provides a non-zero contribution to **F**. On the other hand, if we calculate the reciprocal force $\mathbf{F}'$ by swapping the indices in Eqs. (5.8) and (5.9), the latter equation immediately shows that $d\mathbf{B}(\mathbf{r}') \propto \mathbf{j} \times (\mathbf{r} - \mathbf{r}') = 0$, because the two operand vectors are parallel (figure 5.3). Hence, the current component $\mathbf{j}'d^3\mathbf{r}'$ does exert a force on its counterpart, while $\mathbf{j}d^3\mathbf{r}$ does not.

Despite this apparent problem, let us still go ahead and plug Eq. (5.9) into Eq. (5.8):

$$\mathbf{F} = \frac{\mu_0}{4\pi} \int_V d^3r \int_{V'} d^3r' \mathbf{j}(\mathbf{r}) \times \left(\mathbf{j}'(\mathbf{r}') \times \frac{\mathbf{r} - \mathbf{r}'}{|\mathbf{r} - \mathbf{r}'|^3} \right). \tag{5.11}$$

[7] Named after H Lorentz, already mentioned in section 3.3, but famous mostly for his numerous contributions to the development of special relativity—see chapter 9. To be fair, the magnetic part of the Lorentz force was first correctly calculated by O Heaviside.

[8] From the magnetic part of Eq. (5.10), Eq. (5.8) may be derived by the elementary summation of all forces acting on $n \gg 1$ particles in a unit volume, with $\mathbf{j} = qn\langle \mathbf{v} \rangle$—see the footnote for Eq. (4.13*a*). On the other hand, the reciprocal derivation of Eq. (5.10) from Eq. (5.8) with $\mathbf{j} = q\mathbf{v}\delta(\mathbf{r} - \mathbf{r}_0)$, where $\mathbf{r}_0$ is the current particle's position (so that $d\mathbf{r}_0/dt = \mathbf{v}$), requires care, and will be performed in chapter 9.

This double vector product may be transformed into two scalar products, using the vector algebraic identity called the *bac minus cab rule*, $\mathbf{a} \times (\mathbf{b} \times \mathbf{c}) = \mathbf{b}(\mathbf{a} \cdot \mathbf{c}) - \mathbf{c}(\mathbf{a} \cdot \mathbf{b})$.[9] Applying this relation, with $\mathbf{a} = \mathbf{j}$, $\mathbf{b} = \mathbf{j}'$, and $\mathbf{c} = \mathbf{R} \equiv \mathbf{r} - \mathbf{r}'$, to Eq. (5.11), we obtain

$$\mathbf{F} = \frac{\mu_0}{4\pi} \int_{V'} d^3r' \mathbf{j}'(\mathbf{r}') \left(\int_V d^3r \frac{\mathbf{j}(\mathbf{r}) \cdot \mathbf{R}}{R^3} \right) - \frac{\mu_0}{4\pi} \int_V d^3r \int_{V'} d^3r' \mathbf{j}(\mathbf{r}) \cdot \mathbf{j}'(\mathbf{r}') \frac{\mathbf{R}}{R^3}. \tag{5.12}$$

The second term on the right-hand side of this relation coincides with the right-hand side of Eq. (5.1), while the first term equals zero because its internal integral vanishes. Indeed, we may break the volumes V and V' into narrow *current tubes*—the stretched elementary volumes whose walls are not crossed by current lines (so that on their walls, $j_n = 0$). As a result, the elementary current in each tube, $dI = jdA = jd^2r$, is the same along its length and, just as in a thin wire, $\mathbf{j}d^2r$ may be replaced with $dId\mathbf{r}$, with the vector $d\mathbf{r}$ directed along $\mathbf{j}$. Because of this, each tube's contribution into the internal integral in the first term of Eq. (5.12) may be represented as

$$dI \oint_l d\mathbf{r} \cdot \frac{\mathbf{R}}{R^3} = -dI \oint_l d\mathbf{r} \cdot \nabla \frac{1}{R} = -dI \oint_l dr \frac{\partial}{\partial r} \frac{1}{R}, \tag{5.13}$$

where the operator ∇ acts in the $\mathbf{r}$ space, and the integral is taken along the tube's length l. Due to the current continuity, each loop should follow a closed contour, and an integral of a full differential of some scalar function (in our case, $1/r_{12}$) along such a contour equals zero.

So we have recovered Eq. (5.1). Returning for a minute to the paradox illustrated with figure 5.3, we may conclude that the apparent violation of Newton's third law was the artifact of our interpretation of Eqs. (5.8) and (5.9) as the sums of independent elementary components. In reality, due to the dc current continuity expressed by Eq. (4.6), these components are *not* independent. For the whole currents, Eqs. (5.8) and (5.9) do obey the third law—as follows from their already proved equivalence to Eq. (5.1).

Thus we have been able to break the magnetic interaction into two effects: the induction of the *magnetic field* $\mathbf{B}$ by one current (in our notation, $\mathbf{j}'$), and the effect of this field on the other current ($\mathbf{j}$). Now comes an additional experimental fact: other elementary components $\mathbf{j}d^3r'$ of the current $\mathbf{j}(\mathbf{r})$ also contribute to the magnetic field (5.9) acting on the component $\mathbf{j}d^3r$.[10] This fact allows us to drop the prime sign after $\mathbf{j}$ in Eq. (5.9), and rewrite Eqs. (5.8) and (5.9) as

$$\mathbf{B}(\mathbf{r}) = \frac{\mu_0}{4\pi} \int_{V'} \mathbf{j}(\mathbf{r}') \times \frac{\mathbf{r} - \mathbf{r}'}{|\mathbf{r} - \mathbf{r}'|^3} d^3r', \tag{5.14}$$

---

[9] See, e.g. Eq. (A.47).

[10] Just in electrostatics, one needs to exercise due caution at a transfer from these expressions to the limit of discrete classical particles, and extended wavefunctions in quantum mechanics, in order to avoid the (non-existent) magnetic interaction of a charged particle upon itself.

$$\mathbf{F} = \int_V \mathbf{j}(\mathbf{r}) \times \mathbf{B}(\mathbf{r}) d^3r, \tag{5.15}$$

Again, the field *observation* point $\mathbf{r}$ and the field *source* point $\mathbf{r}'$ have to be clearly distinguished. We immediately see that these expressions are similar to, but still different from the corresponding relations of the electrostatics, namely Eq. (1.9),

$$\mathbf{E}(\mathbf{r}) = \frac{1}{4\pi\varepsilon_0} \oint_{V'} \rho(\mathbf{r}') \frac{\mathbf{r} - \mathbf{r}'}{|\mathbf{r} - \mathbf{r}'|^3} d^3r', \tag{5.16}$$

and the distributed-charge version of Eq. (1.5):

$$\mathbf{F} = \oint_V \rho(\mathbf{r})\, \mathbf{E}(\mathbf{r}) d^3r. \tag{5.17}$$

(Note that the sign difference has disappeared, at the cost of the replacement of scalar-by-vector multiplications in electrostatics with cross-products of vectors in magnetostatics.)

For the frequent case of a field of a thin wire of length l', Eq. (5.14) may be re-written as

$$\mathbf{B}(\mathbf{r}) = \frac{\mu_0 I}{4\pi} \oint_{l'} d\mathbf{r}' \times \frac{\mathbf{r} - \mathbf{r}'}{|\mathbf{r} - \mathbf{r}'|^3}. \tag{5.18}$$

Let us see how this formula works for the simplest case of a straight wire (figure 5.4a). The magnetic field contribution $d\mathbf{B}$ due to any small fragment $d\mathbf{r}'$ of the wire's length is directed along the same line (perpendicular to both the wire and the perpendicular d dropped from the observation point to the wire line), and its magnitude is

$$dB = \frac{\mu_0 I}{4\pi} \frac{dx'}{|\mathbf{r} - \mathbf{r}'|^2} \sin\theta = \frac{\mu_0 I}{4\pi} \frac{dx'}{(d^2 + x^2)} \frac{d}{(d^2 + x^2)^{1/2}}. \tag{5.19}$$

Summing up all such elementary contributions, we obtain

$$B = \frac{\mu_0 I \rho}{4\pi} \int_{-\infty}^{\infty} \frac{dx}{(x^2 + d^2)^{3/2}} = \frac{\mu_0 I}{2\pi d}. \tag{5.20}$$

This is a simple but very important result. (Note that it is only valid for very long ($l \gg d$), straight wires.) It is particularly crucial to note the 'vortex' character of the

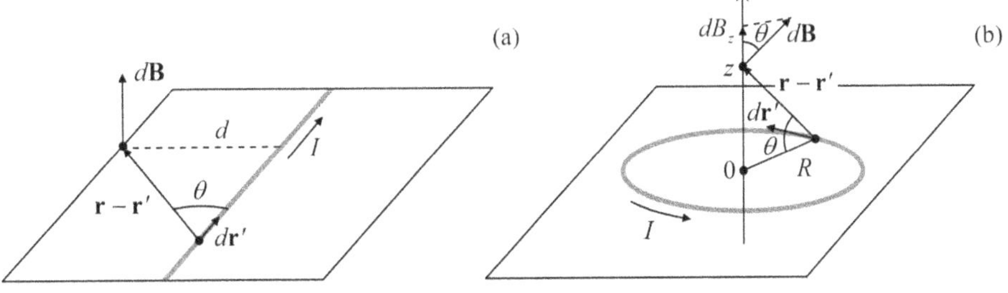

Figure 5.4. Magnetic fields of (a) a straight current and (b) a current loop.

field: its lines go around the wire, forming rings with their centers on the current line. This is in the sharp contrast to the electrostatic field lines that can only begin and end on electric charges and never form closed loops (otherwise the Coulomb force $q\mathbf{E}$ would not be conservative). In the magnetic case, the vortex *field* may be reconciled with the potential character of the magnetic *forces*, which is evident from Eq. (5.1), due to the vector products in Eqs. (5.14) and (5.15).

Now we may use Eq. (5.15), or rather its thin-wire version

$$\mathbf{F} = I\oint_l d\mathbf{r} \times \mathbf{B}(\mathbf{r}), \tag{5.21}$$

to apply Eq. (5.20) to the two-wire problem (figure 5.2). Since for the second wire vectors $d\mathbf{r}$ and $\mathbf{B}$ are perpendicular to each other, we immediately arrive at our previous result (5.7), which was obtained directly from Eq. (5.1).

The next important example of application of the Biot–Savart law (5.14) is the magnetic field at the axis of a circular current loop (figure 5.4b). Due to the problem's symmetry, the net field $\mathbf{B}$ has to be directed along the axis, but each of its components $d\mathbf{B}$ is tilted by angle $\theta = \tan^{-1}(z/R)$ to this axis, so that its axial component

$$dB_z = dB\cos\theta = \frac{\mu_0 I}{4\pi}\frac{dr'}{R^2 + z^2}\frac{R}{(R^2 + z^2)^{1/2}}. \tag{5.22}$$

Since the denominator of this expression remains the same for all wire components dr', the integration over $\mathbf{r}'$ is trivial ($\int dr' = 2\pi R$), giving finally

$$B = \frac{\mu_0 I}{2}\frac{R^2}{(R^2 + z^2)^{3/2}}. \tag{5.23}$$

Note that the magnetic field in the loop's center (i.e. for $z = 0$),

$$B = \frac{\mu_0 I}{2R}, \tag{5.24}$$

is π times higher than that due to a similar current in a straight wire, at distance $d = R$ from it. This difference is readily understandable, since all elementary components of the loop are at the same distance R from the observation point, while in the case of a straight wire, all its points but one are separated from the observation point by a distance larger than d.

Another notable fact is that at large distances ($z^2 \gg R^2$), the field (5.23) is proportional to z^{-3}:

$$B \approx \frac{\mu_0 I}{2}\frac{R^2}{|z|^3} = \frac{\mu_0}{4\pi}\frac{2m}{|z|^3}, \qquad \text{with } m \equiv IA, \tag{5.25}$$

where $A = \pi R^2$ is the loop area. Comparing this expression with Eq. (3.13), for the particular case $\theta = 0$, we see that such field is similar to that of an electric dipole (at least along its direction), with the replacement of the electric dipole moment magnitude p with m (besides the front coefficient). Indeed, such a plane current loop is the simplest example of a system whose field, at a large distance, is that of a *magnetic dipole*, with the *dipole moment* **m**—the notions that will be discussed in much more detail in section 5.4.

5.2 Vector-potential and the Ampère law

The reader can see that the calculations of the magnetic field using Eqs. (5.14) or (5.18) are still cumbersome even for the very simple systems we have examined. As we saw in chapter 1, similar calculations in electrostatics, at least for several important systems of high symmetry, could be substantially simplified using the Gauss law (1.16). A similar relation exists in magnetostatics as well, but has a different form, due to the vortex character of the magnetic field.

To derive it, let us notice that in an analogy with the scalar case, the vector product under integral (5.14) may be transformed as

$$\frac{j(\mathbf{r}') \times (\mathbf{r} - \mathbf{r}')}{|\mathbf{r} - \mathbf{r}'|^3} = \nabla \times \frac{\mathbf{j}(\mathbf{r}')}{|\mathbf{r} - \mathbf{r}'|}, \tag{5.26}$$

where the operator ∇ acts in the $\mathbf{r}$ space. (This equality may be really verified by its Cartesian components, noticing that the current density is a function of $\mathbf{r}'$ and hence its components are independent of $\mathbf{r}$.) Plugging Eq. (5.26) into Eq. (5.14), and moving the operator ∇ out of the integral over $\mathbf{r}'$, we see that the magnetic field may be represented as the curl of another vector field[11]:

$$\mathbf{B}(\mathbf{r}) = \nabla \times \mathbf{A}(\mathbf{r}), \tag{5.27}$$

namely of the so-called *vector-potential*, defined as

$$\mathbf{A}(\mathbf{r}) \equiv \frac{\mu_0}{4\pi} \int_{V'} \frac{\mathbf{j}(\mathbf{r}')}{|\mathbf{r} - \mathbf{r}'|} d^3r'. \tag{5.28}$$

Please note a beautiful analogy between Eqs. (5.27) and (5.28) and, respectively, Eqs. (1.33) and (1.38). This analogy implies that the vector-potential **A** plays, for the magnetic field, essentially the same role as the scalar potential ϕ plays for the electric field (hence the name 'potential'), with a due respect to the vortex character of **B**. This notion will be discussed in more detail below.

Now let us see what equations we may get for the spatial derivatives of the magnetic field. First, vector algebra says that the divergence of any curl is zero[12]. In application to Eq. (5.27), this means that

$$\nabla \cdot \mathbf{B} = 0. \tag{5.29}$$

Comparing this equation with Eq. (1.27), we see that Eq. (5.29) may be interpreted as the absence of a magnetic analog of an electric charge on which magnetic field lines could originate or end. Numerous searches for such hypothetical magnetic

[11] In the Gaussian units, Eq. (5.27) remains the same, and hence in Eq. (5.28), $\mu_0/4\pi$ is replaced with $1/c$.

[12] See, e.g. Eq. (A.72).

charges, called *magnetic monopoles*, using very sensitive and sophisticated experimental set-ups, have not given convincing evidence of their existence in Nature.

Proceeding to the alternative, vector derivative of the magnetic field (i.e. its curl) and using Eq. (5.28), we obtain

$$\nabla \times \mathbf{B}(\mathbf{r}) = \frac{\mu_0}{4\pi}\nabla \times \left(\nabla \times \int_{V'} \frac{\mathbf{j}(\mathbf{r}')}{|\mathbf{r} - \mathbf{r}'|} d^3r'\right). \tag{5.30}$$

This expression may be simplified by using the following general vector identity[13]:

$$\nabla \times (\nabla \times \mathbf{c}) = \nabla(\nabla \cdot \mathbf{c}) - \nabla^2\mathbf{c}, \tag{5.31}$$

applied to vector $\mathbf{c}(\mathbf{r}) = \mathbf{j}(\mathbf{r}')/|\mathbf{r} - \mathbf{r}'|$:

$$\nabla \times \mathbf{B} = \frac{\mu_0}{4\pi}\nabla \int_{V'} \mathbf{j}(\mathbf{r}') \cdot \nabla \frac{1}{|\mathbf{r} - \mathbf{r}'|} d^3r' - \frac{\mu_0}{4\pi}\int_{V'} \mathbf{j}(\mathbf{r}')\nabla^2 \frac{1}{|\mathbf{r} - \mathbf{r}'|} d^3r'. \tag{5.32}$$

As was already discussed during our study of electrostatics in section 3.1,

$$\nabla^2 \frac{1}{|\mathbf{r} - \mathbf{r}'|} = -4\pi\delta(\mathbf{r} - \mathbf{r}'), \tag{5.33}$$

so that the last term of Eq. (5.32) is just $\mu_0\mathbf{j}(\mathbf{r})$. On the other hand, inside the first integral we can replace ∇ with $(-\nabla')$, where prime means differentiation in the space of radius-vector $\mathbf{r}'$. Integrating that term by parts, we get

$$\nabla \times \mathbf{B} = -\frac{\mu_0}{4\pi}\nabla \oint_{S'} j_n(\mathbf{r}') \frac{1}{|\mathbf{r} - \mathbf{r}'|} d^2r' + \nabla \int_{V'} \frac{\nabla' \cdot \mathbf{j}(\mathbf{r}')}{|\mathbf{r} - \mathbf{r}'|} d^3r' + \mu_0\mathbf{j}(\mathbf{r}). \tag{5.34}$$

Applying this equation to the volume V' limited by a surface S' either sufficiently distant from the field concentration, or with no current crossing it, we may neglect the first term on the right-hand side of Eq. (5.34), while the second term always equals zero in statics, due to the dc charge continuity—see Eq. (4.6). As a result, we arrive at a very simple differential equation[14]

$$\nabla \times \mathbf{B} = \mu_0\mathbf{j}. \tag{5.35}$$

This is (the dc form of) the inhomogeneous Maxwell equation, which in magnetostatics plays a role similar to Eq. (1.27) in electrostatics. Let me display, for the first time in this course, this fundamental system of equations (at this stage, for statics only), and give the reader a minute to stare at their beautiful symmetry—which inspired so much of later physics development:

---

[13] See, e.g. Eq. (A.73).

[14] As in all earlier formulas for the magnetic field, in the Gaussian units the coefficient μ_0 in this relation is replaced with $4\pi/c$.

$$\nabla \times \mathbf{E} = 0, \qquad \nabla \times \mathbf{B} = \mu_0 \mathbf{j},$$
$$\nabla \cdot \mathbf{E} = \frac{\rho}{\varepsilon_0}, \qquad \nabla \cdot \mathbf{B} = 0. \tag{5.36}$$

Their only asymmetry, the two zeros on the right-hand sides (for the magnetic field's divergence and electric field's curl), is due to the absence in Nature of magnetic monopoles and their currents. I will discuss these equations in more detail in section 6.7, after the equations for the field curls have been generalized to their full (time-dependent) versions.

Returning now to our current, more mundane but important task of calculating the magnetic field induced by simple current configurations, we can benefit from an integral form of Eq. (5.35). For that, let us integrate this equation over an arbitrary surface S limited by a closed contour C, and apply to the result the Stokes theorem[15]. The resulting expression,

$$\oint_C \mathbf{B} \cdot d\mathbf{r} = \mu_0 \oint_S j_n d^2 r \equiv \mu_0 I, \tag{5.37}$$

where I is the net electric current crossing surface S, is called the *Ampère law*.

As the first example of its application, let us return to the current in a straight wire (figure 5.4a). With the Ampère law in our arsenal, we can readily pursue an even more ambitious goal then was achieved in the previous section—calculate the magnetic field both outside and inside a wire of an arbitrary radius R, with an arbitrary (albeit axially symmetric) current distribution $j(\rho)$—see figure 5.5.

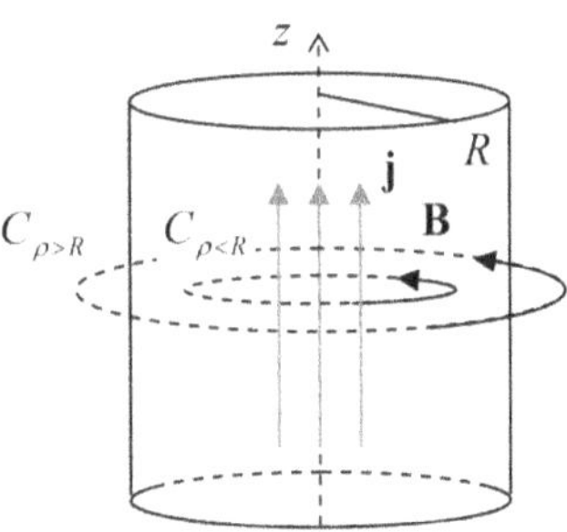

Figure 5.5. The simplest application of the Ampère law: the magnetic field of a straight current.

Selecting two contours C in the form of rings of some radius ρ in the plane perpendicular to the wire axis z, we have $\mathbf{B} \cdot d\mathbf{r} = B\rho d\varphi$, where φ is the azimuthal angle, so that the Ampère law (5.37) yields:

$$2\pi\rho B(\rho) = \mu_0 \times \begin{cases} 2\pi \int_0^{\rho} j(\rho')\rho' d\rho', & \text{for } \rho \leqslant R, \\ 2\pi \int_0^{R} j(\rho')\rho' \, d\rho' \equiv I, & \text{for } \rho \geqslant R. \end{cases} \tag{5.38}$$

[15] See, e.g. Eq. (A.78) with $\mathbf{f} = \mathbf{B}$.

Thus we have not only recovered our previous result (5.20), with the notation replacement $d \to \rho$, in a much simpler way, but could also find the magnetic field distribution inside the wire. (In the most common case when the wire's conductivity σ is constant, and hence the current is uniformly distributed along its cross-section, $j(\rho) = \text{const}$, the first of Eq. (5.38) immediately yields $B \propto \rho$ for $\rho \leqslant R$).

Another important system is a straight, long *solenoid* (figure 5.6a), with dense winding: $n^2 A \gg 1$, where n is the number of wire turns per unit length, and A is the area of the solenoid's cross-section.

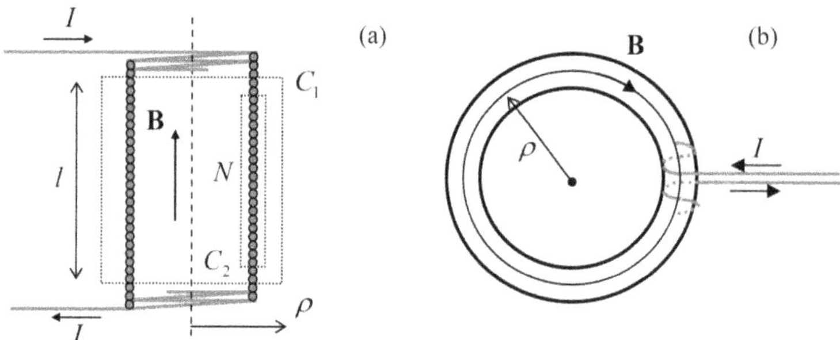

Figure 5.6. Magnetic fields of the (a) straight and (b) toroidal solenoids.

From the symmetry of this problem, the longitudinal (in figure 5.6a, vertical) component B_z of the magnetic field may only depend on the horizontal position ρ of the observation point. First taking a plane Ampère contour C_1, with both long sides outside the solenoid, we obtain $B_z(\rho_2) - B_z(\rho_1) = 0$, because the total current piercing the contour equals zero. This is only possible if $B_z = 0$ at any ρ outside of the (infinitely long!) solenoid[16]. With this result on hand, from the Ampère law applied to the contour C_2 we obtain the following relation for the only (z-) component of the internal field:

$$Bl = \mu_0 NI, \tag{5.39}$$

where N is the number of wire turns passing through the contour of length l. This means that regardless of the exact position on the internal side of the contour, the result is the same:

$$B = \mu_0 \frac{N}{l} I = \mu_0 nI. \tag{5.40}$$

Thus, the field inside an infinitely long solenoid (with an arbitrary shape of its cross-section) is uniform; in this sense, a long solenoid is a magnetic analog of a wide plane capacitor.

As should be clear from its derivation, the obtained results, in particular that the field outside the solenoid equals zero, are conditional on the solenoid length being

[16] Applying the Ampère law to a circular contour of radius ρ, coaxial with the solenoid, we see that the field outside (but not inside!) it has an azimuthal component B_φ, similar to that of the straight wire (see Eq. (5.38) above) and hence (at $N \gg 1$) much weaker than the longitudinal field inside the solenoid—see Eq. (5.40).

very large in comparison to its lateral size. (From Eq. (5.25), we may predict that for a solenoid of a finite length l, the close-range external field is only a factor of $\sim A/l^2$ lower than the internal one.) A much better suppression of this external ('fringe') field may be obtained using the *toroidal solenoid* (figure 5.6b). The application of Ampère law to this geometry shows that, in the limit of dense winding ($N \gg 1$), there is no fringe field at all (for any relation between two radii of the torus), while inside the solenoid, and distance ρ from the center,

$$B = \frac{\mu_0 NI}{2\pi\rho}. \tag{5.41}$$

We see that a possible drawback of this system for practical applications is that the internal field does depend on ρ, i.e. is not quite uniform; however, if the torus is thin, this problem is minor.

How should we solve the problems of magnetostatics for systems whose low symmetry does not allow obtaining easy results from the Ampère law? (The examples are of course too numerous to list; for example, we cannot use this approach even to reproduce Eq. (5.23) for a round current loop.) From the deep analogy with electrostatics, we may expect that in this case we could recover the field from the solution of a certain boundary problem for the field's potential, in this case the vector-potential $\mathbf{A}$ defined by Eq. (5.28). However, despite the similarity of this formula and Eq. (1.38) for ϕ, which was emphasized above, there are two additional issues we should tackle in the magnetic case.

First, calculating the vector-potential distribution means determining three scalar functions (say, A_x, A_y, and A_z), rather than one (ϕ). To reveal the second, deeper issue, let us plug Eq. (5.27) into Eq. (5.35),

$$\nabla \times (\nabla \times \mathbf{A}) = \mu_0 \mathbf{j}, \tag{5.42}$$

and then apply to the left-hand side of this equation the now-familiar identity (5.31). The result is

$$\nabla(\nabla \cdot \mathbf{A}) - \nabla^2 \mathbf{A} = \mu_0 \mathbf{j}. \tag{5.43}$$

On the other hand, as we know from electrostatics (please compare Eqs. (1.38) and (1.41)), the vector-potential $\mathbf{A}(\mathbf{r})$ defined by Eq. (5.28) has to satisfy a simpler ('vector-Poisson') equation

$$\nabla^2 \mathbf{A} = -\mu_0 \mathbf{j}, \tag{5.44}$$

which is just a set of three usual Poisson equations for each Cartesian component of $\mathbf{A}$.

In order to resolve this paradox, let us note that Eq. (5.43) is reduced to Eq. (5.44) if $\nabla \cdot \mathbf{A} = 0$. In this context, let us discuss what discretion we have in the choice of a potential. In electrostatics, we might add to the scalar function ϕ' that satisfied

Eq. (1.33) for the given field **E**, not only an arbitrary constant, but also an arbitrary function of time:

$$-\nabla[\varphi' + f(t)] = -\nabla\varphi' = \mathbf{E}. \qquad (5.45)$$

Similarly, using the fact that the curl of the gradient of any scalar function equals zero[17], we may add to any vector function $\mathbf{A}'$ that satisfies Eq. (5.27) for the given field **B**, not only any constant, but even a gradient of an arbitrary scalar function χ $(\mathbf{r}, t)$, because

$$\nabla \times (\mathbf{A}' + \nabla\chi) = \nabla \times \mathbf{A}' + \nabla \times (\nabla\chi) = \nabla \times \mathbf{A}' = \mathbf{B}. \qquad (5.46)$$

Such additions, which keep the fields intact, are called *gauge transformations*[18]. Let us see what such a transformation does to $\nabla \cdot \mathbf{A}'$:

$$\nabla \cdot (\mathbf{A}' + \nabla\chi) = \nabla \cdot \mathbf{A}' + \nabla^2\chi. \qquad (5.47)$$

For any choice of such a function $\mathbf{A}'$, we can always choose the function χ in such a way that it satisfies the Poisson equation $\nabla^2\chi = -\nabla \cdot \mathbf{A}'$, and hence makes the divergence of the transformed vector-potential, $\mathbf{A} = \mathbf{A}' + \nabla\chi$, to be equal to zero everywhere,

$$\nabla \cdot \mathbf{A} = 0, \qquad (5.48)$$

thus reducing Eq. (5.43) to Eq. (5.44).

To summarize, the set of distributions $\mathbf{A}'(\mathbf{r})$ that satisfy Eq. (5.27) for a given field $\mathbf{B}(\mathbf{r})$, is not limited to the vector-potential $\mathbf{A}(\mathbf{r})$ given by Eq. (5.44), but is reduced to it upon the additional *Coulomb gauge condition* (5.48). However, as we will see in a minute, even this condition still leaves some degrees of freedom in the choice of the vector-potential. In order to illustrate this fact, and also to obtain a better feeling of the vector-potential's distribution in space, let us calculate $\mathbf{A}(\mathbf{r})$ for two very basic cases.

First, let us revisit the straight wire problem shown in figure 5.5. As Eq. (5.28) shows, in this case vector **A** has just one component (along the axis z). Moreover, due to the problem's axial symmetry, its magnitude may only depend on the distance from the axis: $\mathbf{A} = \mathbf{n}_z A(\rho)$. Hence, the gradient of **A** is directed across the axis z, so that Eq. (5.48) is satisfied. For our symmetry ($\partial/\partial\varphi = \partial/\partial z = 0$), the Laplace operator, written in cylindrical coordinates, has just one term[19], reducing (5.44) to

$$\frac{1}{\rho}\frac{d}{d\rho}\left(\rho\frac{dA}{d\rho}\right) = -\mu_0 j(\rho). \qquad (5.49)$$

Multiplying both parts of this equation by ρ and integrating them over the coordinate once, we obtain

[17] See, e.g. Eq. (A.71).

[18] The use of the term 'gauge' (originally meaning 'a measure' or 'a scale') in this context is purely historic, so the reader should not try to find too much hidden sense in it.

[19] See, e.g. Eq. (A.61).

$$\rho \frac{dA}{d\rho} = -\mu_0 \int_0^{\rho} j(\rho')\rho' d\rho' + \text{const.} \tag{5.50}$$

Since in the cylindrical coordinates, for our symmetry[20], $B = -dA/d\rho$, Eq. (5.50) is nothing other than our old result (5.38) for the magnetic field[21]. However, let us continue the integration, at least for the region outside the wire, where the function $A(\rho)$ depends only on the full current I rather than on the current distribution. Dividing both parts of Eq. (5.50) by ρ, and integrating them over it again, we obtain

$$A(\rho) = -\frac{\mu_0 I}{2\pi} \ln \rho + \text{const}, \quad \text{where } I = 2\pi \int_0^R j(\rho)\rho d\rho, \quad \text{for } \rho \geqslant R. \tag{5.51}$$

As a reminder, we had similar logarithmic behavior for the electrostatic potential outside a uniformly charged straight line. This is natural, because the Poisson equations for both cases are similar.

Now let us find the vector-potential for the long solenoid (figure 5.6a), with its uniform magnetic field. Since Eq. (5.28) prescribes the vector $\mathbf{A}$ to follow the direction of the inducing current, we may start with looking for it in the form $\mathbf{A} = \mathbf{n}_\varphi A(\rho)$. (This is particularly natural if the solenoid's cross-section is circular.) With this orientation of $\mathbf{A}$, the same general expression for the curl operator in cylindrical coordinates yields $\nabla \times \mathbf{A} = \mathbf{n}_z(1/\rho)d(\rho A)/d\rho$. According to Eq. (5.27), this expression should be equal to $\mathbf{B}$, in our current case equal to $\mathbf{n}_z B$, with a constant B—see Eq. (5.40). Integrating this equality, and selecting such integration constant so that $A(0)$ is finite, we obtain

$$A(\rho) = \frac{B\rho}{2}, \qquad \text{i.e. } \mathbf{A} = \frac{B\rho}{2}\mathbf{n}_\varphi. \tag{5.52}$$

Plugging this result into the general expression for the Laplace operator in the cylindrical coordinates[22], we see that the Poisson equation (5.44) with $\mathbf{j} = 0$ (i.e. the Laplace equation), is satisfied again—which is natural since for this distribution, the Coulomb gauge condition (5.48) is satisfied: $\nabla \cdot \mathbf{A} = 0$.

However, Eq. (5.52) is not the unique (or even the simplest) vector-potential that gives the same uniform field $\mathbf{B} = \mathbf{n}_z B$. Indeed, using the well-known expression for the curl operator in Cartesian coordinates[23], it is straightforward to check that each of the vector functions $\mathbf{A}' = \mathbf{n}_y Bx$ and $\mathbf{A}'' = -\mathbf{n}_x By$ also has the same curl, and also satisfies the Coulomb gauge condition (5.48)[24]. If such solutions do not look very natural because of their anisotropy in the $[x,y]$ plane, please consider the fact that they represent the uniform magnetic field regardless of its source—for example, regardless of the shape of the long solenoid's cross-section. Such choices of the

---

[20] See, e.g. Eq. (A.63) with $\partial/\partial\varphi = \partial/\partial z = 0$.

[21] Since the magnetic field at the wire axis has to be zero (otherwise, being normal to the axis, where would it be directed?), the integration constant in Eq. (5.50) has to equal zero.

[22] See, e.g. Eq. (A.64).

[23] See, e.g. Eq. (A.54).

[24] The axially symmetric vector-potential (5.52) is just a weighed sum of these two functions: $\mathbf{A} = (\mathbf{A}' + \mathbf{A}'')/2$.

vector-potential may be very convenient for some problems, for example for the quantum-mechanical analysis of the 2D motion of a charged particle in the perpendicular magnetic field, giving the famous Landau energy levels[25].

5.3 Magnetic energy, flux, and inductance

Considering the currents flowing in a system as generalized coordinates, the magnetic forces (5.1) between them are their unique functions, and in this sense the energy U of their magnetic interaction may be considered a potential energy of the system. The apparent (but somewhat deceptive) way to derive an expression for this energy is to use the analogy between Eq. (5.1) and its electrostatic analog, Eq. (5.2). Indeed, Eq. (5.2) may be transformed into Eq. (5.1) with just three replacements:

(i) $\rho(\mathbf{r})\rho'(\mathbf{r}')$ should be replaced with $[\mathbf{j}(\mathbf{r}) \cdot \mathbf{j}'(\mathbf{r}')]$,
(ii) ε_0 should be replaced with $1/\mu_0$, and
(iii) the sign before the double integral has to be replaced with the opposite one.

Hence we may avoid repeating the calculation made in chapter 1, by making these replacements in Eq. (1.59), which gives the electrostatic potential energy of the system with $\rho(\mathbf{r})$ and $\rho'(\mathbf{r}')$ describing the same charge distribution, i.e. with $\rho'(\mathbf{r}) = \rho(\mathbf{r})$, to obtain the following expression for the magnetic potential energy in the system with, similarly, $\mathbf{j}'(\mathbf{r}) = \mathbf{j}(\mathbf{r})$[26]:

$$U_j = -\frac{\mu_0}{4\pi}\frac{1}{2}\int d^3r \int d^3r' \frac{\mathbf{j}(\mathbf{r}) \cdot \mathbf{j}(\mathbf{r}')}{|\mathbf{r} - \mathbf{r}'|}. \tag{5.53}$$

However, this is not the unique answer, and not even the most convenient answer. Actually, Eq. (5.53) describes the energy that is adequate (i.e. whose minimum corresponds to the stable equilibrium of the system) only in the case when the interacting currents are fixed—just as Eq. (1.59) is adequate when the interacting charges are fixed. Here comes a substantial difference between electrostatics and magnetostatics: due to the fundamental fact of charge conservation (already discussed in sections 1.1 and 4.1), keeping electric charges fixed does not require external work, while the maintenance of currents generally does. As a result, Eq. (5.53) describes the energy of the magnetic interaction *plus* of the system keeping the currents constant—or rather of its part depending on the system under our consideration[27].

Now in order to exclude from U_j the contribution due to the interaction with the current-supporting system(s), i.e. calculate the potential energy U of our system as

---

[25] See, e.g. *Part QM* section 3.2.

[26] As was repeatedly discussed above, for the interaction of two *independent* charge distributions $\rho(\mathbf{r})$ and $\rho'(\mathbf{r}')$, the front factor ½ has to be dropped; the same is true for the interaction of two independent current distributions $\mathbf{j}(\mathbf{r})$ and $\mathbf{j}'(\mathbf{r}')$.

[27] In the terminology already used in section 3.5 (see also a general discussion in *Part CM* section 1.4.), U_j is essentially the Gibbs potential energy of our magnetic system.

such, we need to know this contribution. The simplest way to do this is to use the *Faraday induction law*, which describes this interaction. This is why let me postpone the derivation until the beginning of the next chapter, and for now ask the reader to believe me that its account leads to an addition to U_j of a term of twice larger magnitude, so that the result is given by an expression similar to equation (5.53), but with the opposite sign:

$$U = \frac{\mu_0}{4\pi}\frac{1}{2}\int d^3r \int d^3r' \frac{\mathbf{j}(\mathbf{r}) \cdot \mathbf{j}(\mathbf{r}')}{|\mathbf{r} - \mathbf{r}'|}, \tag{5.54}$$

I promise to prove this fact in section 6.2 below.

Due to the importance of Eq. (5.54), let us rewrite it in several other forms, convenient for different applications. First of all, just as in electrostatics, it may be recast into a potential-based form. Indeed, using the definition (5.28) of the vector-potential $\mathbf{A}(\mathbf{r})$, Eq. (5.54) becomes[28]

$$U = \frac{1}{2}\int \mathbf{j}(\mathbf{r}) \cdot \mathbf{A}(\mathbf{r}) d^3r. \tag{5.55}$$

This formula, which is a clear magnetic analog of Eq. (1.60) of electrostatics, is very popular among field theorists, because it is very handy for their manipulations. However, for many calculations it is more convenient to have a direct expression of energy via the magnetic field. Again, this may be done very similarly to what was done for electrostatics in section 1.3, i.e. by plugging, into Eq. (5.55), the current density expressed from Eq. (5.35), and then transforming it as[29]

$$\begin{aligned} U &= \frac{1}{2}\int \mathbf{j} \cdot \mathbf{A} d^3r = \frac{1}{2\mu_0}\int \mathbf{A} \cdot (\nabla \times \mathbf{B})\, d^3r \\ &= \frac{1}{2\mu_0}\int \mathbf{B} \cdot (\nabla \times \mathbf{A}) d^3r - \frac{1}{2\mu_0}\int \nabla \cdot (\mathbf{A} \times \mathbf{B}) d^3r \end{aligned}. \tag{5.56}$$

Now using the divergence theorem, the second integral may be transformed into a surface integral of $(\mathbf{A} \times \mathbf{B})_n$. According to Eqs. (5.27) and (5.28) if the current distribution $\mathbf{j}(\mathbf{r})$ is localized, this vector product drops, at large distances, faster than $1/r^2$, so that if the integration volume is large enough, the surface integral is negligible. In the remaining first integral in Eq. (5.56) we may use Eq. (5.27) to recast $\nabla \times \mathbf{A}$ into the magnetic field. As a result, we obtain a very simple and fundamental formula.

$$U = \frac{1}{2\mu_0}\int B^2 d^3r. \tag{5.57a}$$

[28] This relation remains the same in the Gaussian units, because in those units, both Eqs. (5.28) and (5.54) should be stripped of their $\mu_0/4\pi$ coefficients.

[29] For that, we may use Eq. (A.77) with $\mathbf{f} = \mathbf{A}$ and $\mathbf{g} = \mathbf{B}$, giving $\mathbf{A} \cdot (\nabla \times \mathbf{B}) = \mathbf{B} \cdot (\nabla \times \mathbf{A}) - \nabla \cdot (\mathbf{A} \times \mathbf{B})$.

Just as with the electric field, this expression may be interpreted as a volume integral of the *magnetic energy density u*:

$$U = \int u(\mathbf{r})d^3r, \quad \text{with } u(\mathbf{r}) \equiv \frac{1}{2\mu_0}\mathbf{B}^2(\mathbf{r}), \tag{5.57b}$$

clearly similar to Eq. (1.65)[30]. Again, the conceptual choice between the spatial localization of magnetic energy—either at the location of electric currents only, as implied by Eqs. (5.54) and (5.55), or in all regions where the magnetic field exists, as apparent from Eq. (5.57*b*)—cannot be done within the framework of magnetostatics, and only the electrodynamics gives the decisive preference for the latter choice.

For the practically important case of currents flowing in several thin wires, Eq. (5.54) may be first integrated over the cross-section of each wire, just as was done at the derivation of Eq. (5.4). Again, since the integral of the current density over the kth wire's cross-section is just the current I_k in the wire, and cannot change along its length, it may be taken from the remaining integrals, giving

$$U = \frac{\mu_0}{4\pi}\frac{1}{2}\sum_{k,k'} I_k I_{k'} \oint_{l_k}\oint_{l_{k'}} \frac{d\mathbf{r}_k \cdot d\mathbf{r}_{k'}}{|\mathbf{r}_k - \mathbf{r}_{k'}|}, \tag{5.58}$$

where l is the full length of the wire loop. Note that Eq. (5.58) is valid if the currents I_k are independent of each other, because the double sum counts each current pair twice, compensating the coefficient ½ in front of the sum. It is useful to decompose this relation as

$$U = \frac{1}{2}\sum_{k,k'} I_k I_{k'} L_{kk'}, \tag{5.59}$$

where the coefficients $L_{kk'}$ are independent of the currents:

$$L_{kk'} \equiv \frac{\mu_0}{4\pi}\oint_{l_k}\oint_{l_{k'}} \frac{d\mathbf{r}_k \cdot d\mathbf{r}_{k'}}{|\mathbf{r}_k - \mathbf{r}_{k'}|}, \tag{5.60}$$

The coefficient $L_{kk'}$ with $k \neq k'$, is called the *mutual inductance* between current the loops with numbers k and k', while the diagonal coefficient $L_k \equiv L_{kk}$ is called the *self-inductance* (or just *inductance*) of the kth loop[31]. From the symmetry of Eq. (5.60)

[30] The transfer to the Gaussian units in Eqs. (5.77) and (5.78) may be accomplished by the usual replacement $\mu_0 \to 4\pi$, thus giving, in particular, $u = B^2/8\pi$.

[31] As is evident from Eq. (5.60), these coefficients depend only on the geometry of the system. Moreover, in the Gaussian units, in which Eq. (5.60) is valid without the factor $\mu_0/4\pi$, the inductance coefficients have the dimension of length (centimeters). The SI unit of inductance is called the *henry*, abbreviated as H, after J Henry, 1797–1878, who, in particular, discovered the effect of electromagnetic induction (see section 6.1) independently of M Faraday.

with respect to the index swap, $k \leftrightarrow k'$, it is evident that the matrix of coefficients $L_{kk'}$ is symmetric[32]:

$$L_{kk'} = L_{k'k}, \tag{5.61}$$

so that for the practically important case of two interacting currents I_1 and I_2, Eq. (5.59) reads

$$U = \frac{1}{2}L_1 I_1^2 + M I_1 I_2 + \frac{1}{2}L_2 I_2^2, \tag{5.62}$$

where $M \equiv L_{12} = L_{21}$ is the mutual inductance coefficient.

These formulas clearly show the importance of the self- and mutual inductances, so I will demonstrate their calculation for at least a few basic geometries. Before doing that, however, let me recast Eq. (5.58) into one more form that may facilitate such calculations. Namely, let us notice that for the magnetic field induced by current I_k in a thin wire, Eq. (5.28) is reduced to

$$\mathbf{A}_k(\mathbf{r}) = \frac{\mu_0}{4\pi} I_k \int_{l'} \frac{d\mathbf{r}_k}{|\mathbf{r} - \mathbf{r}_k|}, \tag{5.63}$$

so that Eq. (5.58) may be rewritten as

$$U = \frac{1}{2}\sum_{k,k'} I_k \oint_{l_k} \mathbf{A}_{k'}(\mathbf{r}_k) \cdot d\mathbf{r}_{k'}. \tag{5.64}$$

But according to the same Stokes theorem that was used earlier in this chapter to derive the Ampère law and Eq. (5.27), the integral in this expression is nothing more than the *magnetic field's flux* (more frequently called just the *magnetic flux*) through a surface S limited by the contour l[33]:

$$\oint_l \mathbf{A}(\mathbf{r}) \cdot d\mathbf{r} = \int_S (\nabla \times A)_n d^2r = \int_S B_n d^2r \equiv \Phi. \tag{5.65}$$

As a result, Eq. (5.64) may be rewritten as

$$U = \frac{1}{2}\sum_{k,k'} I_k \Phi_{kk'}, \tag{5.66}$$

where $\Phi_{kk'}$ is the flux of the field induced by the k'th current through the loop of the kth current. Comparing this expression with Eq. (5.59), we see that

[32] Note that the matrix of the mutual inductances $L_{jj'}$ is very similar to the matrix of *reciprocal* capacitance coefficients $p_{kk'}$—for example, compare Eq. (5.62) with Eq. (2.21).

[33] The SI unit of magnetic flux is called the *weber*, abbreviated as Wb, after W Weber, who, in particular, co-invented (with C Gauss) the electromagnetic telegraph, and in 1856 was the first, together with R Kohlrausch, to notice that the value of (in modern terms) $1/(\varepsilon_0\mu_0)^{1/2}$, derived from electrostatic and magnetostatic measurements, coincides with the independently measured speed of light c, giving an important motivation for Maxwell's theory.

$$\Phi_{kk'} \equiv \int_{S_k} (\mathbf{B}_{k'})_n d^2r = L_{kk'} I_{k'}, \tag{5.67}$$

This expression not only gives us one more means for calculating coefficients $L_{kk'}$, but also shows their physical sense: the mutual inductance characterizes what part of the magnetic field (colloquially, 'how many field lines'), induced by the current $I_{k'}$, pierces the k_{th} loop—see figure 5.7.

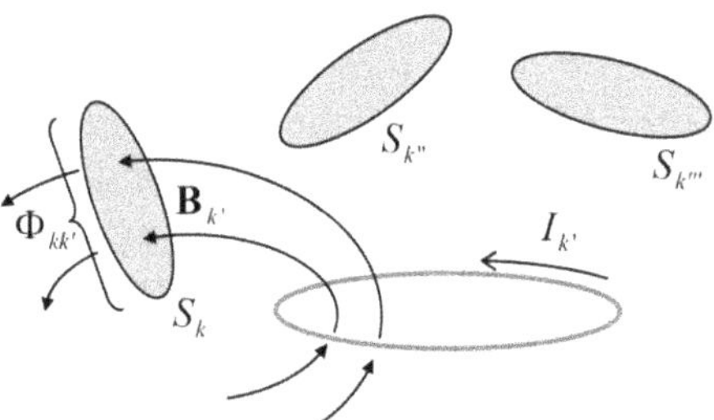

Figure 5.7. A schematic representation of the physical sense of the mutual inductance coefficient $L_{kk'} \equiv \Phi_{kk'}/I_{k'}$.

Due to the linear superposition principle, the total flux piercing kth loop may be represented as

$$\Phi_k \equiv \sum_{k'} \Phi_{kk'} = \sum_{k'} L_{kk'} I_{k'} \tag{5.68}$$

For example, for the system of two currents this expression is reduced to a clear analog of equations Eqs. (2.19):

$$\begin{aligned} \Phi_1 &= L_1 I_1 + M I_2, \\ \Phi_2 &= M I_1 + L_2 I_2. \end{aligned} \tag{5.69}$$

For the even simpler case of a single current,

$$\Phi = LI, \tag{5.70}$$

so that the magnetic energy of the current may be represented in several equivalent forms:

$$U = \frac{L}{2} I^2 = \frac{1}{2} I\Phi = \frac{1}{2L}\Phi^2. \tag{5.71}$$

These relations, similar to Eqs. (2.14) and (2.15) of electrostatics, show that the self-inductance L of a current loop may be considered as a measure of the system's magnetic energy, but, as we will see in section 6.1, this measure is adequate only if the flux Φ, rather than the current I, is fixed.

Now we are well equipped for the calculation of inductance coefficients, having three options. The first is to use Eq. (5.60) directly[34]. The second is to calculate the magnetic field energy from Eq. (5.57) as the function of all currents I_k in the system, and then use Eq. (5.59) to find all coefficients $L_{kk'}$. For example, for a system with just one current, Eq. (5.71) yields

$$L = \frac{U}{I^2/2}. \tag{5.72}$$

Finally, if the system consists of thin wires, so that the loop areas S_k and hence fluxes $\Phi_{kk'}$ are well defined, we may calculate them from Eq. (5.65) and then use Eq. (5.67) to find the inductances.

Actually, the first two options may have technical advantages over the third one even for some system of thin wires, in which the notion of magnetic flux is not quite apparent. As an important example, let us find the self-inductance of a long solenoid—see figure 5.6a again. We have already calculated the magnetic field inside it—see Eq. (5.40)—so that, due to the field uniformity, the magnetic flux piercing each wire turn is just

$$\Phi_1 = BA = \mu_0 nIA, \tag{5.73}$$

where A is the area of solenoid's cross-section—for example πR^2 for a round solenoid, although Eq. (5.40), and hence Eq. (5.73), are valid for any cross-section. Comparing Eq. (5.73) with Eq. (5.70), one might wrongly conclude that $L = \Phi_1/I = \mu_0 nA$ (WRONG!), i.e. that the solenoid's inductance is independent of its length. Actually, the magnetic flux Φ_1 pierces *each* wire turn, so that the total flux through the *whole* current loop, consisting of N turns, is

$$\Phi = N\Phi_1 = \mu_0 n^2 lAI, \tag{5.74}$$

and the correct expression for the long solenoid's self-inductance is

$$L = \frac{\Phi}{I} = \mu_0 n^2 lA, \tag{5.75}$$

i.e. the inductance per unit length is constant: $L/l = \mu_0 n^2 A$.

Since this reasoning may seem not quite apparent, it is prudent to verify it by using Eq. (5.72), with the full magnetic energy inside the solenoid (neglecting minor fringe and external field contributions) given by Eq. (5.57) with $\mathbf{B} = \text{const}$ within the internal volume $V = lA$, and zero outside it:

$$U = \frac{1}{2\mu_0} B^2 Al = \frac{1}{2\mu_0} (\mu_0 nI)^2 Al = \mu_0 n^2 lA \frac{I^2}{2}. \tag{5.76}$$

Plugging this relation into Eq. (5.72) immediately confirms the earlier result (5.75).

[34] Numerous applications of this *Neumann formula* to electrical engineering problems may be found, for example, in the classical text [1].

This approach becomes virtually inevitable for continuously distributed currents. As an example, let us calculate the self-inductance L of a long coaxial cable with the cross-section shown in figure 5.8[35], with the full current in the outer conductor equal and opposite to that (I) in the inner conductor.

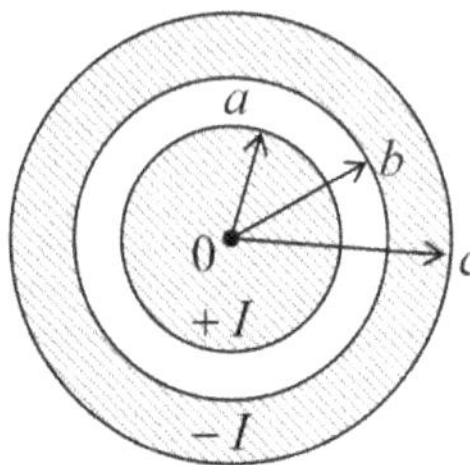

Figure 5.8. The cross-section of a coaxial cable.

Let us assume that the current is uniformly distributed over the cross-sections of both conductors. (As we know from the previous chapter, this is indeed the case if both the internal and external conductors are made of a uniform resistive material.) First, we should calculate the radial distribution of the magnetic field (that of course has only one, azimuthal component, because of the axial symmetry of the problem). This distribution may be immediately found from the application of the Ampère law to the circular contours of radii ρ within four different ranges:

$$2\pi\rho B = \mu_0 I|_{\text{piercing the contour}} = \mu_0 I \times \begin{cases} \rho^2/a^2, & \text{for } \rho < a, \\ 1, & \text{for } a < \rho < b, \\ (c^2 - \rho^2)/(c^2 - b^2), & \text{for } b < \rho < c, \\ 0, & \text{for } c < \rho. \end{cases} \tag{5.77}$$

Now, an elementary integration yields the magnetic energy per unit length of the cable:

$$\begin{aligned} \frac{U}{l} &= \frac{1}{2\mu_0}\int B^2 d^2r = \frac{\pi}{\mu_0}\int_0^\infty B^2 \rho d\rho \\ &= \frac{\mu_0 I^2}{4\pi}\left[\int_0^a \left(\frac{\rho}{a^2}\right)^2 \rho d\rho + \int_a^b \left(\frac{1}{\rho}\right)^2 \rho d\rho + \int_b^c \left(\frac{c^2 - \rho^2}{\rho(c^2 - b^2)}\right)^2 \rho d\rho\right] \\ &= \frac{\mu_0}{2\pi}\left[\ln\frac{b}{a} + \frac{c^2}{c^2 - b^2}\left(\frac{c^2}{c^2 - b^2}\ln\frac{c}{b} - \frac{1}{2}\right)\right]\frac{I^2}{2}. \end{aligned} \tag{5.78}$$

From here, and Eq. (5.72), we obtain the final answer:

$$\frac{L}{l} = \frac{\mu_0}{2\pi}\left[\ln\frac{b}{a} + \frac{c^2}{c^2 - b^2}\left(\frac{c^2}{c^2 - b^2}\ln\frac{c}{b} - \frac{1}{2}\right)\right]. \tag{5.79}$$

[35] As a reminder, the mutual capacitance C between the conductors of such a system was calculated in section 2.3.

Note that for the particular case of a thin outer conductor, $c - b \ll b$, this expression reduces to

$$\frac{L}{l} \approx \frac{\mu_0}{2\pi}\left(\ln\frac{b}{a} + \frac{1}{4}\right), \tag{5.80}$$

where the first term in the parentheses may be traced back to the contribution of the magnetic field energy in the free space between the conductors. This distinction is important for some applications, because in superconductor cables, as well as the normal-metal cables at high frequencies (to be discussed in the next chapter), the field does not penetrate the conductor's bulk, so that Eq. (5.80) is valid without the last term, 1/4, in the parentheses, which is due to the magnetic field energy inside the wire.

As the last example, let us calculate the *mutual* inductance between a long straight wire and a round wire loop adjacent to it (figure 5.9), neglecting the thickness of both wires. Here there is no problem with using the last of the approaches discussed above, based on the direct magnetic flux calculation. Indeed, as was discussed in section 5.1, the field $\mathbf{B}_1$ induced by the current I_1 at any point of the round loop is normal to its plane—e.g. to the plane of drawing of figure 5.9. In the Cartesian coordinates shown in that figure, Eq. (5.20) reads $B_1 = \mu_0 I_1/2\pi y$, giving the following magnetic flux through the loop:

$$\begin{aligned}\Phi_{21} &= \frac{\mu_0 I_1}{2\pi}\int_{-R}^{+R} dx \int_{R-(R^2-x^2)^{1/2}}^{R+(R^2-x^2)^{1/2}} dy\frac{1}{y} \\ &= \frac{\mu_0 I_1}{\pi}\int_0^R \ln\frac{R+(R^2-x^2)^{1/2}}{R-(R^2-x^2)^{1/2}}dx \\ &= \frac{\mu_0 I_1 R}{\pi}\int_0^1 \ln\frac{1+(1-\xi^2)^{1/2}}{1+(1-\xi^2)^{1/2}}d\xi.\end{aligned} \tag{5.81}$$

This is a table integral equal to π,[36] so that $\Phi_{21} = \mu_0 I_1 R$, and the final answer for the mutual inductance $M \equiv L_{12} = L_{21} = \Phi_{21}/I_1$ is finite (and very simple):

$$M = \mu_0 R, \tag{5.82}$$

despite the magnetic field's divergence at the lowest point of the loop ($y = 0$).

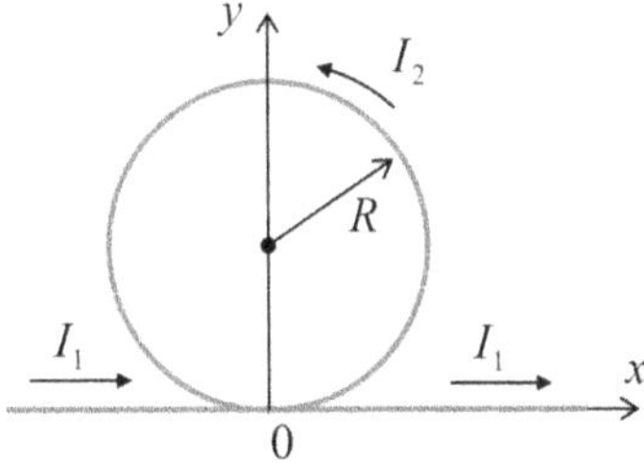

Figure 5.9. An example of the mutual inductance calculation.

[36] See, e.g. Eq. (A.40), with $a = 1$.

Note that in contrast with the finite *mutual* inductance of this system, the *self*-inductances of both wires are formally infinite in the thin-wire limit—see, e.g. Eq. (5.80), which in the limit $b/a \gg 1$ describes a thin straight wire. However, since this divergence is very weak (logarithmic), it is quenched by any deviation from this perfectly axial geometry. For example, a good estimate of the inductance of a wire of a large but finite length l may be obtained from Eq. (5.80) via the replacement of b with l:

$$L \sim \frac{\mu_0}{2\pi} l \ln \frac{l}{a}. \tag{5.83}$$

(Note, however, that the exact result depends on where from/to the current flows beyond that segment.) A close estimate, with l replaced with $2\pi R$, and b replaced with R, is valid for the self-inductance of the round loop. A more exact calculation of this inductance, which would be asymptotically correct in the limit $a \ll R$, is a very useful exercise, highly recommended to the reader[37].

5.4 Magnetic dipole moment, and magnetic dipole media

The most natural way in which the magnetic media description parallels that described in chapter 3 for dielectric media, is based on properties of *magnetic dipoles*—the notion close (but not identical!) to that of the electric dipoles discussed in section 3.1. To introduce this notion quantitatively, let us consider, just as in section 3.1, a spatially localized system with a current distribution $\mathbf{j}(\mathbf{r}')$, whose magnetic field is measured at relatively large distances $r \gg r'$ (figure 5.10).

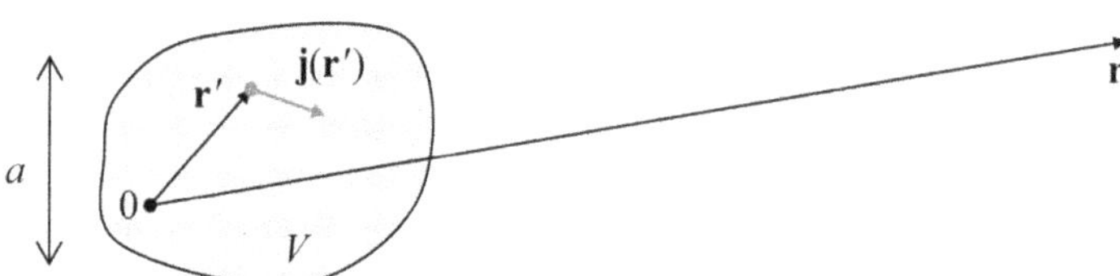

Figure 5.10. Calculating the magnetic field of localized currents, as observed from a distant point ($r \gg a$).

Applying the truncated Taylor expansion (3.5) of the fraction $1/|\mathbf{r} - \mathbf{r}'|$ to the vector potential given by Eq. (5.28), we obtain

$$\mathbf{A}(\mathbf{r}) \approx \frac{\mu_0}{4\pi}\left[\frac{1}{r}\int_V \mathbf{j}(\mathbf{r}')d^3r' + \frac{1}{r^3}\int_V (\mathbf{r}\cdot\mathbf{r}')\mathbf{j}(\mathbf{r}')d^3r'\right]. \tag{5.84}$$

Now, due to the vector character of this potential, we have to depart slightly from the approach of section 3.1 and use the following vector algebra identity[38]:

$$\int_V [f(\mathbf{j}\cdot\nabla g) + g(\mathbf{j}\cdot\nabla f)]\ d^3r = 0, \tag{5.85}$$

[37] Its solution may be found, for example, in section 34 of [2].

[38] See, e.g. Eq. (A.81) with the additional condition $j_n|_S = 0$, pertinent for space-restricted currents.

which is valid for any pair of smooth (differentiable) scalar functions $f(\mathbf{r})$ and $g(\mathbf{r})$, and any vector function $\mathbf{j}(\mathbf{r})$ that, as the dc current density, satisfies the continuity condition $\nabla \cdot \mathbf{j} = 0$ and whose normal component vanishes on the surface of the volume V. First, let us use Eq. (5.85) with $f = 1$ and g equal to any Cartesian component of the radius-vector $\mathbf{r}$: $g = r_l$ ($l = 1, 2, 3$). Then it yields

$$\int_V (\mathbf{j} \cdot \mathbf{n}_l) d^3r = \int_V j_l d^3r = 0, \tag{5.86}$$

so that for the vector as a whole

$$\int_V \mathbf{j}(\mathbf{r}) d^3r = 0, \tag{5.87}$$

showing that the first term on the right-hand side of Eq. (5.84) equals zero. Next, let us use Eq. (5.85) again, now with $f = r_l$, $g = r_{l'}$ ($l, l' = 1, 2, 3$); then it yields

$$\int_V \left(r_l j_{l'} + r_{l'} j_l\right) d^3r = 0, \tag{5.88}$$

so that the lth Cartesian component of the second integral in Eq. (5.84) may be transformed as

$$\begin{aligned}
\int_V (\mathbf{r} \cdot \mathbf{r}') j_l d^3r' &= \int_V \sum_{l'=1}^{3} r_{l'} r'_{l'} j_l d^3r' = \frac{1}{2} \sum_{l'=1}^{3} r_{l'} \int_V \left(r'_{l'} j_l + r'_{l'} j_l\right) d^3r' \\
&= \frac{1}{2} \sum_{l'=1}^{3} r_{l'} \int_V \left(r'_{l'} j_l - r'_{l} j_{l'}\right) d^3r' \\
&= -\frac{1}{2} \left[\mathbf{r} \times \int_V (\mathbf{r}' \times \mathbf{j}) d^3r'\right]_l .
\end{aligned} \tag{5.89}$$

As a result, Eq. (5.84) may be rewritten as

$$\mathbf{A}(\mathbf{r}) = \frac{\mu_0}{4\pi} \frac{\mathbf{m} \times \mathbf{r}}{r^3}, \tag{5.90}$$

where the vector $\mathbf{m}$, defined as[39]

$$\mathbf{m} \equiv \frac{1}{2} \int_V \mathbf{r} \times \mathbf{j}(\mathbf{r})\, d^3r, \tag{5.91}$$

is called the *magnetic dipole moment* of our system—that itself, within the long-range approximation (5.90), is called the *magnetic dipole*.

Note a close analogy between the $\mathbf{m}$ defined by Eq. (5.91) and the orbital[40] angular momentum of a non-relativistic particle with mass m_k:

[39] In the Gaussian units, the definition (5.91) is kept valid, so that Eq. (5.90) is stripped of the factor $\mu_0/4\pi$.

[40] This adjective is used, particularly in quantum mechanics, to distinguish the motion of a particle as a whole (not necessarily along a closed orbit) from its intrinsic angular momentum, the spin—see, e.g. *Part QM* chapter 4.

$$\mathbf{L}_k \equiv \mathbf{r}_k \times \mathbf{p}_k = \mathbf{r}_k \times m_k \mathbf{v}_k, \tag{5.92}$$

where $\mathbf{p}_k = m_k \mathbf{v}_k$ is its mechanical momentum. Indeed, for a continuum of such particles with the same electric charge q, with the spatial density n, $\mathbf{j} = qn\mathbf{v}$, and Eq. (5.91) yields

$$\mathbf{m} = \int_V \frac{1}{2} \mathbf{r} \times \mathbf{j}\, d^3r = \int_V \frac{nq}{2} \mathbf{r} \times \mathbf{v}\, d^3r, \tag{5.93}$$

while the total angular momentum of such a system of particles of the same mass ($m_k = m_0$) is

$$\mathbf{L} = \int_V n m_0 \mathbf{r} \times \mathbf{v} d^3r, \tag{5.94}$$

so that we obtain a very straightforward relation

$$\mathbf{m} = \frac{q}{2m_0} \mathbf{L}. \tag{5.95}$$

For the orbital motion, this classical relation survives in quantum mechanics for linear operators, and hence for the eigenvalues of the observables. Since the orbital angular momentum is quantized in units of the Plank's constant $\hbar$, for an electron the orbital magnetic moment is always a multiple of the so-called *Bohr magneton*

$$\mu_B \equiv \frac{e\hbar}{2m_e}, \tag{5.96}$$

where m_e is the free electron mass[41]. However, for particles with spin, such a universal relation between the vectors **m** and **L** is no longer valid. For example, the electron's spin $s = ½$ gives a contribution of $\hbar/2$ to the mechanical angular momentum, but its contribution to the magnetic moment it still very close to μ_B.

The next important example of a magnetic dipole is a *planar* thin-wire loop, limiting area A (of an arbitrary shape), and carrying current I, for which **m** has a surprisingly simple form,

$$\mathbf{m} = I\mathbf{A}, \tag{5.97}$$

where the modulus of vector **A** equals loop's area A, and its direction is normal to the loop's plane. This formula may be readily proved by noticing that if we select the coordinate origin on the plane of the loop (figure 5.11), then the elementary component of the magnitude of the integral (5.91),

$$m = \frac{1}{2}\left| \oint_C \mathbf{r} \times I d\mathbf{r} \right| = I \oint_C \left| \frac{1}{2} \mathbf{r} \times d\mathbf{r} \right| = I \oint_C \frac{1}{2} r^2 d\varphi, \tag{5.98}$$

is just the elementary area $dA = (1/2)rdh = (1/2)rd(r\sin\varphi) = r^2 d\varphi/2$.

[41] In SI units, $m_e \approx 0.91 \times 10^{-30}$ kg, so that $\mu_B \approx 0.93 \times 10^{-23}$ J/T.

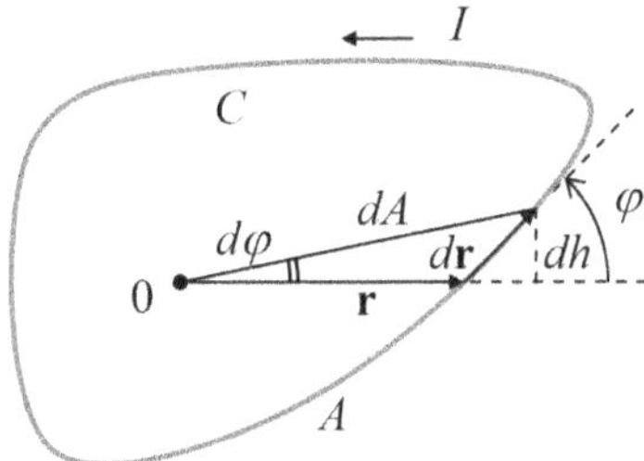

Figure 5.11. Calculating the magnetic dipole moment of a planar current loop.

The combination of Eqs. (5.96) and (5.97) allows a useful estimate of the scale of atomic currents, by finding what current I should flow in a circular loop of atomic size scale (the Bohr radius) $r_B \approx 0.5 \times 10^{-10}$ m, i.e. of area $A \approx 10^{-20}$ m^2, to produce a magnetic moment equal to μ_B.[42] The result is surprisingly macroscopic: $I \sim 1$ mA (quite comparable to the currents driving your earbuds). Although this estimate should not be taken too literally, due to the quantum-mechanical spread of the electron's wavefunctions, it is very useful for obtaining a feeling of how significant atomic magnetism is and hence why ferromagnets may provide such a strong field.

After these illustrations, let us return to equation (5.90). Plugging it into the general formula (5.27), we may calculate the magnetic field of a magnetic dipole[43]:

$$\mathbf{B}(\mathbf{r}) = \frac{\mu_0}{4\pi}\frac{3\mathbf{r}(\mathbf{r}\cdot\mathbf{m}) - \mathbf{m}r^2}{r^5}. \tag{5.99}$$

The structure of this formula *exactly* replicates that of Eq. (3.13) for the electric dipole field (including the sign). Because of this similarity, the energy of a dipole of a fixed magnitude m in an external field, and hence the torque and force exerted on it

[42] Another way to arrive at the same estimate is to take $I \sim ef = e\omega/2\pi$ with $\omega \sim 10^{16}$ s^{-1} being the typical frequency of radiation due to atomic interlevel quantum transitions.

[43] Similarly to the situation with the electric dipoles (see Eq. (3.24) and its discussion), it may be shown that the magnetic field of any closed current loop (or any system of such loops) satisfies the following equality:

$$\int_V \mathbf{B}(\mathbf{r})d^3r = (2/3)\mu_0\mathbf{m},$$

where the integral is over any sphere confining all the currents. On the other hand, as we know from section 3.1, for a field with the structure (5.99), derived from the long-range approximation (5.90), such integral vanishes. As a result, in order to obtain a course-grain description of the magnetic field of a small system, located at $r = 0$, which would give the correct average value of the magnetic field, Eq. (5.99) should be modified as follows:

$$\mathbf{B}_{cg}(\mathbf{r}) = \frac{\mu_0}{4\pi}\left(\frac{3\mathbf{r}(\mathbf{r}\cdot\mathbf{m}) - \mathbf{m}r^2}{r^5} + \frac{8\pi}{3}\mathbf{m}\delta(\mathbf{r})\right),$$

in a conceptual (but not quantitative) similarity to Eq. (3.25).

by a fixed external field, are also absolutely similar to the expressions for an electric dipole—see Eqs. (3.15)–(3.19)[44]:

$$U = -\mathbf{m} \cdot \mathbf{B}_{\text{ext}}, \tag{5.100}$$

and, as a result,

$$\boldsymbol{\tau} = \mathbf{m} \times \mathbf{B}_{\text{ext}}, \tag{5.101}$$

$$\mathbf{F} = \nabla(\mathbf{m} \cdot \mathbf{B}_{\text{ext}}). \tag{5.102}$$

Now let us consider a system of many magnetic dipoles (e.g. atoms or molecules), distributed in space with a macroscopic (i.e. average) density *n*. Then we can use Eq. (5.90) (generalized in the evident way for an arbitrary position, $\mathbf{r}'$, of a dipole), and the linear superposition principle, to calculate the macroscopic vector-potential $\mathbf{A}$:

$$\mathbf{A}(\mathbf{r}) = \frac{\mu_0}{4\pi}\int \frac{\mathbf{M}(\mathbf{r}') \times (\mathbf{r} - \mathbf{r}')}{|\mathbf{r} - \mathbf{r}'|^3} d^3r', \tag{5.103}$$

where $\mathbf{M} \equiv n\mathbf{m}$ is the *magnetization*, i.e. the average magnetic moment per unit volume. Transforming this integral absolutely similarly to how Eq. (3.27) had been transformed into Eq. (3.29), we obtain:

$$\mathbf{A}(\mathbf{r}) = \frac{\mu_0}{4\pi}\int \frac{\nabla' \times \mathbf{M}(\mathbf{r}')}{|\mathbf{r} - \mathbf{r}'|} d^3r'. \tag{5.104}$$

Comparing this result with Eq. (5.28), we see that $\nabla \times \mathbf{M}$ is equivalent, in its effect, to the density $\mathbf{j}_{\text{ef}}$ of a certain effective 'magnetization current'. Just as the electric-polarization charge ρ_{ef} discussed in section 3.2 (see figure 3.4), the vector $\mathbf{j}_{\text{ef}} = \nabla \times \mathbf{M}$ may be interpreted as the uncompensated part of the loop currents representing single magnetic dipoles $\mathbf{m}$ (figure 5.12). Note, however, that since the atomic dipoles may be due to the particles' spins, rather than the actual electric currents due to the orbital motion, the magnetization current's nature is not as direct than that of the polarization charge.

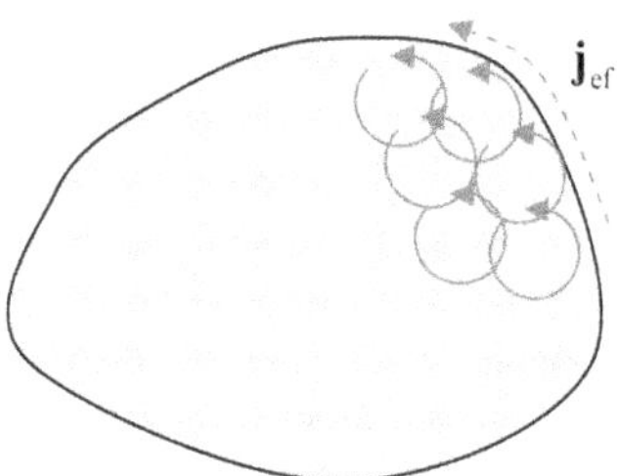

Figure 5.12. A cartoon illustrating the physical nature of the effective magnetization current $\mathbf{j}_{\text{ef}} = \nabla \times \mathbf{M}$.

[44] Note that the fixation of *m* and $\mathbf{B}_{\text{ext}}$ effectively means that the currents producing them are fixed—please have one more look at Eq. (5.35) and Eq. (5.97). As a result, Eq. (5.100) is a particular case of Eq. (5.53) rather than Eq. (5.54)—hence the minus sign.

Now, using Eq. (5.28) to add the possible contribution from 'stand-alone' currents **j**, not included into the currents of microscopic magnetic dipoles, we obtain the general equation for the vector-potential of the macroscopic field:

$$\mathbf{A}(\mathbf{r}) = \frac{\mu_0}{4\pi}\int \frac{[\mathbf{j}(\mathbf{r}') + \nabla' \times \mathbf{M}(\mathbf{r}')]}{|\mathbf{r} - \mathbf{r}'|} d^3r'. \tag{5.105}$$

Repeating the calculations that have led us from Eq. (5.28) to the Maxwell equation (5.35), with the account of the magnetization current term, for the macroscopic magnetic field **B** we obtain

$$\nabla \times \mathbf{B} = \mu_0(\mathbf{j} + \nabla \times \mathbf{M}). \tag{5.106}$$

Following the same philosophy as in section 3.2, we may recast this equation as

$$\nabla \times \mathbf{H} = \mathbf{j}, \tag{5.107}$$

where the field **H**, defined as

$$\mathbf{H} \equiv \frac{\mathbf{B}}{\mu_0} - \mathbf{M}, \tag{5.108}$$

for historic reasons (and very unfortunately) is also called the *magnetic field*[45]. It is crucial to remember that the physical sense of field **H** is very much different from field **B**.

In order to understand the difference better, let us use Eq. (5.107) to bring Eqs. (3.32), (3.36), (5.29), and (5.107) together, writing them as the system of the *macroscopic Maxwell equations* (again, so far for the stationary case $\partial/\partial t = 0$)[46]:

$$\begin{aligned} \nabla \times \mathbf{E} &= 0, \quad & \nabla \times \mathbf{H} &= \mathbf{j}, \\ \nabla \cdot \mathbf{D} &= \rho, \quad & \nabla \cdot \mathbf{B} &= 0. \end{aligned} \tag{5.109}$$

They clearly show that the roles of vector fields **D** and **H** are very similar: they both may be called the 'would-be fields'—which *would be* induced by the charges ρ and currents **j**, if the medium had not modified them by its dielectric and/or magnetic polarization.

Despite this similarity, let me note an important difference of signs in the relation (3.33) between **E, D**, and **P**, on one hand, and the relation (5.108) between **B**, **H**, and **M**, on the other hand. This is *not* just a matter of definition. Indeed, due to the similarity of equations (3.15) and (5.100), including similar signs, the electric and magnetic fields both try to orient the corresponding dipole moments along the field.

---

[45] This confusion is exacerbated by the fact that in Gaussian units, Eq. (5.108) has the form $\mathbf{H} = \mathbf{B} - 4\pi\mathbf{M}$, and hence the fields **B** and **H** have one dimensionality (and are equal in free space!)—although the unit of **H** has a different name (*oersted*, abbreviated as Oe). Mercifully, in the SI units, the dimensionality of **B** and **H** is different, with the unit of **H** being called the *ampere per meter*.

[46] Let me remind the reader once again that in contrast with the system (5.36) of the Maxwell equations for the genuine (microscopic) fields, the right-hand sides of Eq. (5.109) represent only the stand-alone charges and currents, not included in the microscopic electric and magnetic dipoles.

Hence, in the media that allow such an orientation (and as we will see momentarily, for magnetic media this is not always the case), the induced polarizations **P** and **M** are directed along, respectively, the vectors **E** and **B** of the genuine (although macroscopic) fields. According to Eq. (3.33), if the would-be field **D** is fixed—say, by a fixed stand-alone charge distribution $\rho(\mathbf{r})$—such a polarization *reduces* the electric field $\mathbf{E} = (\mathbf{D} - \mathbf{P})/\varepsilon_0$. On the other hand, Eq. (5.108) shows that in a magnetic medium with a fixed would-be field **H**, the magnetic polarization with **M** parallel to **B** *enhances* the magnetic field $\mathbf{B} = (\mathbf{H} + \mathbf{M})/\mu_0$. This difference may be traced back to the sign difference in the basic relations (5.1) and (5.2), i.e. to the fundamental fact that the electric charges of the same sign repulse, while the currents of the same direction attract each other.

5.5 Magnetic materials

In order to form a complete system of differential equations, the macroscopic Maxwell equations (5.109) have to be complemented with the constitutive relations describing the medium: $\mathbf{D} \leftrightarrow \mathbf{E}$, $\mathbf{j} \leftrightarrow \mathbf{E}$, and $\mathbf{B} \leftrightarrow \mathbf{H}$. In the previous two chapters we already discussed, in brief, the first two of them; let us proceed to the last one.

A major difference between the dielectric and magnetic constitutive relations **D**(**E**) and **B**(**H**) is that while a dielectric medium always *reduces* the external field, magnetic media may *either reduce or enhance* it. In order to quantify this fact, let us consider the most widespread materials—*linear magnetics* in which **M** (and hence **H**) are proportional to **B**. For isotropic materials, this proportionality is characterized by a scalar—either the *magnetic permeability* μ, defined by the following relation:

$$\mathbf{B} \equiv \mu \mathbf{H}, \tag{5.110}$$

or the *magnetic susceptibility*[47] defined as

$$\mathbf{M} = \chi_m \mathbf{H}. \tag{5.111}$$

Plugging these relations into Eq. (5.108), we see that these two parameters are not independent, but are related as

$$\mu = (1 + \chi_m)\mu_0. \tag{5.112}$$

Note that despite the superficial similarity between Eqs. (5.110)–(5.112) and relations (3.43)–(3.47) for linear dielectrics,

[47] According to Eqs. (5.110) and (5.112) (i.e. in SI units), χ_m is dimensionless, while μ has the same the same dimensionality as μ_0. In the Gaussian units, μ is dimensionless: $(\mu)_{\text{Gaussian}} = (\mu)_{\text{SI}}/\mu_0$, and χ_m is also introduced differently, as $\mu = 1 + 4\pi\chi_m$, Hence, just as for the electric susceptibilities, these dimensionless coefficients are different in the two systems: $(\chi_m)_{\text{SI}} = 4\pi(\chi_m)_{\text{Gaussian}}$. Note also that χ_m is formally called the volumic magnetic susceptibility, in order to distinguish it from the *atomic* (or 'molecular') susceptibility χ defined by a similar relation, $\langle \mathbf{m} \rangle \equiv \chi \mathbf{H}$, where **m** is the induced magnetic moment of a single dipole—e.g. an atom. Evidently, in a dilute medium, i.e. in the absence of a substantial dipole–dipole interaction, $\chi_m = n\chi$, where n is the dipole density. (Note that χ is an analog of the electric atomic polarizability α—see Eq. (3.48) and its discussion.)

$$\mathbf{D} = \varepsilon\mathbf{E}, \quad \mathbf{P} = \chi_e\varepsilon_0\mathbf{E}, \quad \varepsilon = (1 + \chi_e)\varepsilon_0, \tag{5.113}$$

there is an important conceptual difference between them. Namely, while the vector **E** on the right-hand sides of Eqs. (5.113) is the actual (although macroscopic) electric field, the vector **H** on the right-hand side of Eqs. (5.110) and (5.111) represents a 'would-be' magnetic field, in all aspects similar to **D** rather than **E**—see, for example, Eq. (5.109).

This historic difference in the traditional way to write the constitutive relations for the electric and magnetic fields is not without its physical reasons. Most key experiments with electric and magnetic materials have been performed by placing their samples into nearly uniform electric and magnetic fields, and the simplest systems for their implementation are, respectively, the plane capacitor (figure 2.3) and the solenoid (figure 5.6). The field in the former system may be most conveniently controlled by measuring the voltage V between its plates, which is proportional to the electric field **E**. On the other hand, the field provided by the solenoid may be controlled by the current I in it. According to Eq. (5.107), the field proportional to this stand-alone current is **H**, rather than **B**.[48]

Table 5.1 lists the values of magnetic susceptibility for several materials. It shows that in contrast to linear dielectrics whose susceptibility χ_e is always positive, i.e. the dielectric constant $\kappa = \chi_e + 1$ is always larger than 1 (see table 3.1), linear magnetics may be either *paramagnets* (with $\chi_m > 0$, i.e. $\mu > \mu_0$) or *diamagnets* (with $\chi_m < 0$, $\mu < \mu_0$).

Table 5.1. Magnetic susceptibility $(\chi_m)_{SI}$ of a few representative (and/or important) materials.[a]

'Mu-metal' (75% Ni + 15% Fe + a few %% of Cu and Mo)	~20 000[b]
Permalloy (80% Ni + 20% Fe)	~8000[b]
'Electrical' (or 'transformer') steel (Fe + a few %% of Si)	~4000[b]
Nickel	~100
Aluminum	$+2 \times 10^{-5}$
Oxygen (at ambient conditions)	$+0.2 \times 10^{-5}$
Water	-9×10^{-6}
Diamond	-2×10^{-5}
Copper	-7×10^{-5}
Bismuth (the strongest non-superconducting diamagnet)	-1.7×10^{-4}

[a] The table does not include bulk superconductors, which may be described, in a crude ('coarse-grain') approximation, as perfect diamagnets (with $\mathbf{B} = 0$, i.e. $\chi_m = -1$ and $\mu = 0$), although the actual physics of this phenomenon is more intricate—see section 6.3.

[b] The exact values of $\chi_m \gg 1$ for soft ferromagnetic materials (see, e.g., the upper three rows of the table) depend not only on their exact composition, but also on their thermal processing ("annealing"). Moreover, due to unintentional vibrations, the extremely high χ_m of such materials may somewhat decay with time, though may be restored to the original value by new annealing. The reason for that behavior is discussed below.

[48] This fact also explains the misleading term 'magnetic field' for **H**.

The reason for this difference is that in dielectrics, two different polarization mechanisms (schematically illustrated by figure 3.7) lead to the same sign of the average polarization—see the discussion in section 3.3. One of these mechanisms, illustrated by figure 3.7b, i.e. the ordering of spontaneous dipoles by the applied field, is also possible for magnetization—for the atoms and molecules with spontaneous internal magnetic dipoles of magnitude $m_0 \sim \mu_B$, due to their net spins. Again, in the absence of an external magnetic field the spins, and hence the dipole moments $\mathbf{m}_0$ may be disordered, but according to Eq. (5.100), the external magnetic field tends to align the dipoles along its direction. As a result, the average direction of the spontaneous elementary moments $\mathbf{m}_0$, and hence the direction of the arising magnetization $\mathbf{M}$, is the same as that of the microscopic field $\mathbf{B}$ at the points of the dipole location (i.e. for a diluted media, of $\mathbf{H} \approx \mathbf{B}/\mu_0$), resulting in a positive susceptibility χ_m, i.e. in the paramagnetism—such as that of oxygen and aluminum (see table 5.1).

However, in contrast to the electric polarization of atoms/molecules with no spontaneous electric dipoles, which gives the same sign of $\chi_e \equiv \kappa - 1$ (see figure 3.7a and its discussion), the magnetic materials with no spontaneous atomic magnetic dipole moments have $\chi_m < 0$—the effect called *orbital* (or 'Larmor'[49]) *diamagnetism*. As the simplest model of this effect, let us consider the orbital motion of an atomic electron about an atomic nucleus as that of a classical particle of mass m_0, with an electric charge q about an immobile attracting center. As classical mechanics tells us, the central attractive force does not change the particle's angular momentum $\mathbf{L} \equiv m_0\mathbf{r} \times \mathbf{v}$, but the applied magnetic field $\mathbf{B}$ (that may be taken as uniform on the atomic scale) does, due to the torque (5.101) it exerts on the magnetic moment (5.95):

$$\frac{d\mathbf{L}}{dt} = \boldsymbol{\tau} = \mathbf{m} \times \mathbf{B} = \frac{q}{2m_0}\mathbf{L} \times \mathbf{B}. \tag{5.114}$$

The vector diagram in figure 5.13 shows that in the limit of a relatively weak field, when the magnitude of the angular momentum $\mathbf{L}$ may be considered constant, this equation describes rotation (called the *torque-induced precession*[50]) of the vector $\mathbf{L}$ about the direction of the vector $\mathbf{B}$, with the angular frequency $\boldsymbol{\Omega} = -q\mathbf{B}/2m_0$, independent of the angle θ. According to Eqs. (5.91) and (5.114), the resulting additional (field-induced) magnetic moment $\Delta\mathbf{m} \propto q\boldsymbol{\Omega} \propto -q^2\mathbf{B}/m_0$ has, irrespectively of the sign of q, a direction *opposite* to the field. Hence, according to Eq. (5.111) with $\mathbf{H} \approx \mathbf{B}/\mu_0$, $\chi_m \propto \chi \equiv \Delta\mathbf{m}/\mathbf{H}$ is indeed negative. (Let me leave its quantitative estimate within this model for the reader's exercise.) The quantum-mechanical treatment confirms this qualitative picture of Larmor diamagnetism, while giving quantitative corrections to the classical result for χ_m.[51]

[49] After J Larmor (1857–1947) who was the first to describe the underlying mechanical effect, the torque-induced precession, mathematically.

[50] For a detailed discussion of the effect see, e.g. *Part CM* section 4.5.

[51] See, e.g. *Part QM* section 6.4. The quantum mechanics also shows why in the most important *s*-states, the contribution (5.95) of the basic angular momentum $\mathbf{L}$ to the average $\mathbf{m}$ vanishes—see, e.g. *Part QM* section 3.6.

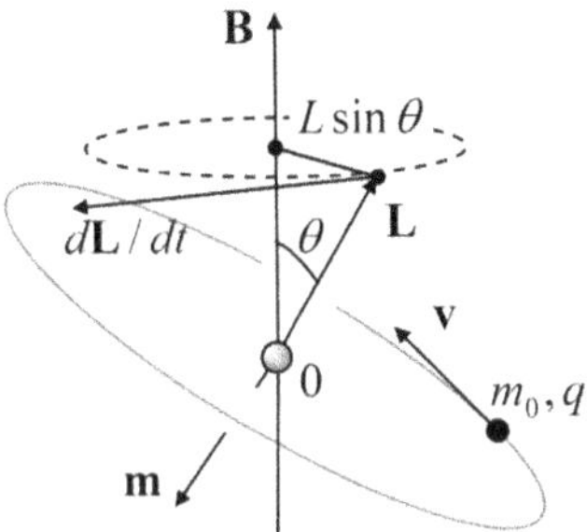

Figure 5.13. The torque-induced precession of a classical charged particle in a magnetic field.

A simple estimate (also left for the reader's exercise) shows that in atoms with uncompensated spins, the magnetic dipole orientation mechanism prevails over the orbital diamagnetism, so that the materials incorporating such atoms usually exhibit net paramagnetism—see table 5.1. Due to possible strong quantum interaction between the spin dipole moments, the magnetism of such materials is rather complex, with numerous interesting phenomena and elaborate theories. Unfortunately, all this physics is well outside the framework of this course, and I have to refer the interested reader to special literature[52], but still need to mention some key notions we will need.

Most importantly, a sufficiently strong dipole–dipole interaction may lead to their spontaneous ordering, even in the absence of an applied field. This ordering may correspond to either parallel alignment of the magnetic dipoles (*ferromagnetism*) or anti-parallel alignment of the adjacent dipoles (*antiferromagnetism*). Evidently, the external effects of ferromagnetism are stronger, because such a phase corresponds to a substantial spontaneous magnetization **M** even in the absence of an external magnetic field. (The corresponding magnitude of $\mathbf{B} = \mu_0\mathbf{M}$ is called the *remanence field*, B_R). The direction of the vector $\mathbf{B}_R$ may be switched by the application of an external magnetic field, with a magnitude above a certain value H_C called *coercivity*, leading to the well-known hysteretic loops on the $[B, H]$ plane (see figure 5.14 for a typical example), similar to those in ferroelectrics, already discussed in section 3.3.

Similarly to ferroelectrics, ferromagnets may also be *hard* or *soft*—in the magnetic sense. In hard ferromagnets (also called *permanent magnets*), the dipole interaction is so strong that B stays close to B_R in all applied fields below H_C, so that the hysteretic loops are virtually rectangular. Correspondingly, the magnetization **M** of a permanent magnet may be considered constant, with the magnitude B_R/μ_0. Such hard ferromagnetic materials, with high remanence fields (typically, from 1 to 2 T) have numerous practical applications. Let me give just two, perhaps the most important, examples.

First, permanent magnets are core components of most *electric motors*. By the way, this venerable (~150-year-old) technology is currently experiencing a quiet revolution, driven mostly by the development of electric cars. In the most advanced type of motors, called *permanent-magnet synchronous machines* (PMSM), the remnant magnetic field B_R of a permanent-magnet central part (called the *rotor*)

[52] See, e.g. [3] or [4].

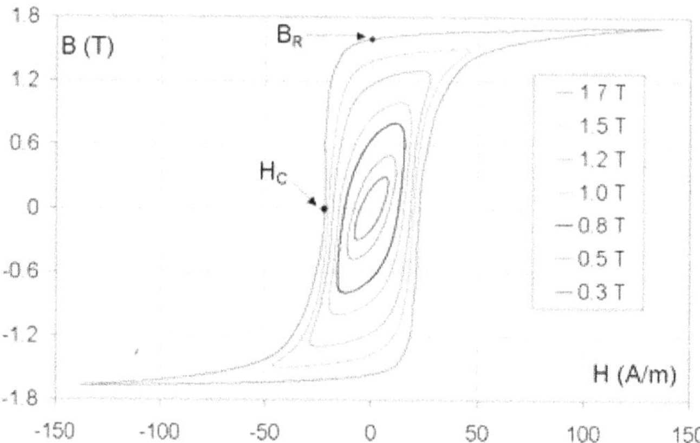

Figure 5.14. Experimental magnetization curves of specially processed (cold-rolled) electrical steel—a solid solution of a few Si in Fe. (Reproduced from www.thefullwiki.org/Hysteresis under the Creative Commons BY-SA 3.0 license.)

interacts with ac currents passed through wire windings in the external, static part of the motor (called the *stator*). The resulting torque drives the rotor to extremely high speeds, exceeding 10 000 rotations per minute, enabling the motor to deliver several kilowatts of mechanical power from each kilogram of its mass.

As the second important example, despite the decades of the exponential (*Moore's law*) progress of semiconductor electronics, most computer data storage systems are still based on *hard disk drives* whose active medium is a submicron-thin layer of a hard ferromagnet, with the data bits stored in the form of the direction of the permanent magnetization of small film spots. This technology has reached a fantastic sophistication[54], with a recorded data density of the order of 10^{12} bits per square inch. Only recently has it started to be seriously challenged by the so-called *solid state drives* based on the floating-gate semiconductor memories already mentioned in chapter 3.[55]

In contrast, in soft ferromagnets, with their lower magnetic dipole interactions, the magnetization is constant only within spontaneously formed magnetic domains, while the volume and shape of the domains are affected by the applied magnetic field. As a result, the hysteresis loop's shape of soft ferromagnets is dependent on the

[54] 'A magnetic head slider [the read/write head—KKL] flying over a disk surface with a flying height of 25 nm with a relative speed of 20 m s^{-1} [all realistic parameters—KKL] is equivalent to an aircraft flying at a physical spacing of 0.2 μm at 900 km h^{-1}.' B Bhushan, as quoted in the (generally good) book by G Hadjipanayis [5].

[55] High-frequency properties of hard ferromagnets are also very non-trivial. For example, according to Eq. (5.101), an external magnetic field $\mathbf{B}_{ext}$ exerts torque $\boldsymbol{\tau} = \mathbf{M} \times \mathbf{B}_{ext}$ on the spontaneous magnetic moment $\mathbf{M}$ of a ferromagnetic sample. In some nearly-isotropic, mechanically fixed ferromagnetic samples, this torque causes the precession around the direction of $\mathbf{B}_{ext}$ (very similar to that illustrated in figure 5.13) of not the sample as such, but of the magnetization $\mathbf{M}$ inside it, with a certain frequency ω_r. If the frequency ω of an additional RF field becomes very close to ω_r, its absorption sharply increases—the so-called *ferromagnetic resonance*. Moreover, if ω is somewhat higher than ω_r, the effective RF magnetic permeability $\mu(\omega)$ of the material for the RF field may become *negative*, enabling a series of interesting effects and practical applications. Most unfortunately, I do not have time for their discussion, and have to refer the interested reader to special literature, for example to the monograph [6].

cycled field's amplitude and cycling history—see figure 5.14. At high fields, their **B** (and hence **M**) are driven into saturation, with $B \approx B_R$, but at low cycled fields they behave essentially as linear magnetics with very high values of χ_m and hence μ—see the top rows of table 5.1. (The magnetic domain interaction and hence the low-field susceptibility of such soft ferromagnets are highly dependent on the material's fabrication technology and its post-fabrication thermal and mechanical treatment.) Due to these high values of μ, the soft ferromagnets, in particular iron and its alloys (e.g. various special steels), are extensively used in electrical engineering—for example in the cores of transformers—see the next section.

Due to the relative weakness of the magnetic dipole interaction in some materials, their ferromagnetic ordering may be destroyed by thermal fluctuations, if the temperature is increased above some value called the *Curie temperature* T_C. The transition between the ferromagnetic and paramagnetic phase at $T = T_C$ is the classical example of a *continuous phase transition*, with the average polarization **M** playing the role of the so-called *order parameter* that (in the absence of external fields) becomes different from zero only at $T < T_C$, increasing gradually with the further temperature reduction.[56]

5.6 Systems with magnetics

Similarly to the electrostatics of linear dielectrics, the magnetostatics of linear magnetics is very simple in the particular case when the stand-alone currents are embedded into a medium with a constant permeability μ. Indeed, let us assume that we know the solution $\mathbf{B}_0(\mathbf{r})$ of the magnetic pair of the genuine ('microscopic') Maxwell equations (5.36) in free space, i.e. when the genuine current density **j** coincides with that of stand-alone currents. Then the macroscopic Maxwell equations (5.109) and the linear constitutive equation (5.110) are satisfied with the pair of functions

$$\mathbf{H}(\mathbf{r}) = \frac{\mathbf{B}_0(\mathbf{r})}{\mu_0}, \qquad \mathbf{B}(\mathbf{r}) = \mu\mathbf{H}(\mathbf{r}) = \frac{\mu}{\mu_0}\mathbf{B}_0(\mathbf{r}). \tag{5.115}$$

Hence the only effect of the complete filling of a system of fixed currents with a uniform, linear magnetic is the change of the magnetic field **B** at all points by the same constant factor $\mu/\mu_0 \equiv 1 + \chi_m$, which may be either larger or smaller than 1. (As a reminder, a similar filling of a system of fixed stand-alone charges with a uniform, linear dielectric always leads to a reduction of the electric field **E** by a factor of $\varepsilon/\varepsilon_0 \equiv 1 + \chi_e$—the difference whose physics was already discussed at the end of section 5.4.)

However, this simple result is generally invalid in the case of non-uniform (or piece-wise uniform) magnetic samples. To analyze them, let us first integrate the macroscopic Maxwell equation (5.107) along a closed contour C limiting a smooth

[56] A quantitative discussion of such transitions may be found, in particular, in *Part SM* chapter 4.

surface S. Now using the Stokes theorem, we obtain the macroscopic version of the Ampère law (5.37):

$$\oint_C \mathbf{H} \cdot d\mathbf{r} = I. \tag{5.116}$$

Let us apply this relation to a sharp boundary between two regions with different magnetics, with no stand-alone currents on the interface, absolutely similarly to how this was done for field $\mathbf{E}$ in section 3.4—see figure 3.5. The result is similar as well:

$$H_\tau = \text{const}. \tag{5.117}$$

On the other hand, the integration of the Maxwell equation (5.29) over a Gaussian pillbox enclosing a border fragment (again just as shown in figure 3.5 for the field $\mathbf{D}$) yields the result similar to Eq. (3.35):

$$B_n = \text{const}. \tag{5.118}$$

For linear magnetics, with $\mathbf{B} = \mu\mathbf{H}$, the latter boundary condition is reduced to

$$\mu H_n = \text{const}. \tag{5.119}$$

Let us use these boundary conditions, first of all, to see what happens with a long cylindrical sample of a uniform magnetic material, placed parallel to a uniform external magnetic field $\mathbf{B}_0$—see figure 5.15. Such a sample cannot noticeably disturb the field in the free space outside it, at most of its length: $\mathbf{B}_{\text{ext}} = \mathbf{B}_0$, $\mathbf{H}_{\text{ext}} = \mu_0\mathbf{B}_{\text{ext}} = \mu_0\mathbf{B}_0$. Now applying Eq. (5.117) to the dominating side surfaces of the sample, we obtain $\mathbf{H}_{\text{int}} = \mathbf{H}_0$.[57] For a linear magnetic, these relations yield $\mathbf{B}_{\text{int}} = \mu\mathbf{H}_{\text{int}} = (\mu/\mu_0)\mathbf{B}_0$.[58] For the high-$\mu$, soft ferromagnetic materials, this means that $B_{\text{int}} \gg B_0$. This effect may be vividly represented as the concentration of the magnetic field lines in high-μ samples—see figure 5.15. (The concentration affects the external field distribution only at distances of the order of $(\mu/\mu_0)t \ll l$ near the sample's ends.)

Such concentration is widely used in such practically important devices as transformers, in which two multi-turn coils are wound on a ring-shaped (e.g. toroidal, see figure 5.6b) core made of a soft ferromagnetic material (such as the

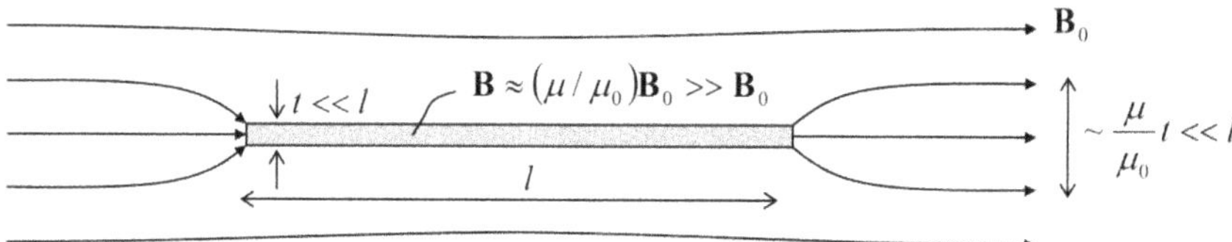

Figure 5.15. Schematic representation of the magnetic field concentration in long, high-μ magnetic samples.

[57] The fact of constancy of $\mathbf{H}$ in this geometry explains why this field's magnitude is used as the argument in the plots such as figure 5.14: such measurements are typically carried out by placing an elongated sample of the material under study into a long solenoid with a controllable current I, so that according to Eq. (5.116), $H_0 = nI$, regardless of the properties of the material under study.

[58] The reader is highly encouraged to carry out a similar analysis of the fields inside narrow gaps cut in a linear magnetic, similar to that carried in section 3.3 out for linear dielectrics—see figure 3.6 and its discussion.

transformer steel, see table 5.1) with $\mu \gg \mu_0$. This minimizes the number of 'stray' field lines, and makes the magnetic flux Φ piercing each wire turn (of either coil) virtually the same—the equality important for the secondary voltage induction—see the next chapter.

Samples of other geometries may create strong perturbations of the external field, extended to distances of the order of the sample's dimensions. In order to analyze such problems, we may benefit from a simple partial differential equation for a scalar function, e.g. the Laplace equation, because in chapter 2 we have learned how to solve it for many simple geometries. In magnetostatics, the introduction of a scalar potential is generally impossible due to the vortex-like magnetic field lines. However, if there are no stand-alone currents within the region we are interested in, then the macroscopic Maxwell equation (5.107) for the field $\mathbf{H}$ is reduced to $\nabla \times \mathbf{H} = 0$, similar to Eq. (1.28) for the electric field, showing that we may introduce the scalar potential of the magnetic field, ϕ_m, using the relation similar to Eq. (1.33):

$$\mathbf{H} = -\nabla\phi_m. \tag{5.120}$$

Combining it with the homogenous Maxwell equation (5.29) for the magnetic field, $\nabla \cdot \mathbf{B} = 0$, and Eq. (5.110) for a linear magnetic, we arrive at a single differential equation, $\nabla \cdot (\mu\nabla\phi_m) = 0$. For a uniform medium ($\mu(\mathbf{r}) = \text{const}$), it is reduced to our beloved Laplace equation:

$$\nabla^2\phi_m = 0. \tag{5.121}$$

Moreover, Eqs. (5.117) and (5.119) give us very familiar boundary conditions: the first of them

$$\frac{\partial\phi_m}{\partial\tau} = \text{const}, \tag{5.122a}$$

being equivalent to

$$\phi_m = \text{const}, \tag{5.122b}$$

with the second one giving

$$\mu\frac{\partial\phi_m}{\partial n} = \text{const}. \tag{5.123}$$

Indeed, these boundary conditions are absolutely similar for (3.37) and (3.56) of electrostatics, with the replacement $\varepsilon \to \mu$.[59]

Let us analyze the geometric effects on magnetization, first using the (too?) familiar structure: a sphere, made of a linear magnetic material, placed into a uniform external field $\mathbf{H}_0 \equiv \mathbf{B}_0/\mu_0$. Since the differential equation and the boundary conditions are

[59] This similarity may seem strange, because earlier we have seen that the parameter μ is physically more similar to $1/\varepsilon$. The reason for this paradox is that in magnetostatics, the introduced potential ϕ_m is traditionally used to describe the 'would-be field' $\mathbf{H}$, while in electrostatics, the potential ϕ describes the actual electric field $\mathbf{E}$. (This tradition persists from the days when $\mathbf{H}$ was perceived as a genuine magnetic field.)

similar to those of the corresponding electrostatics problem (see figure 3.11 and its discussion), we can use the above analogy to reuse the solution we already have—see Eqs. (3.63). Just as in the electric case, the field outside the sphere, with

$$\left(\phi_{\mathrm{m}}\right)_{r>R} = H_0\left(-r + \frac{\mu - \mu_0}{\mu + 2\mu_0}\frac{R^3}{r^2}\right)\cos\theta, \tag{5.124}$$

is a sum of the uniform external field $\mathbf{H}_0$, with the potential $-H_0 r\cos\theta \equiv -H_0 z$, and the dipole field (5.99) with the following induced magnetic dipole moment of the sphere[60]:

$$\mathbf{m} = 4\pi\frac{\mu - \mu_0}{\mu + 2\mu_0}R^3\mathbf{H}_0. \tag{5.125}$$

In contrast, the internal field is perfectly uniform, and directed along the external one:

$$\begin{aligned}\left(\phi_{\mathrm{m}}\right)_{r<R} &= -H_0\frac{3\mu_0}{\mu + 2\mu_0}r\cos\theta,\\ \text{so that}\quad \frac{H_{\mathrm{int}}}{H_0} &= \frac{3\mu_0}{\mu + 2\mu_0},\quad \frac{B_{\mathrm{int}}}{B_0} = \frac{\mu H_{\mathrm{int}}}{\mu_0 H_0} = \frac{3\mu}{\mu + 2\mu_0}.\end{aligned} \tag{5.126}$$

Note that the field $\mathbf{H}_{\mathrm{int}}$ inside the sphere is *not* equal to the applied external field $\mathbf{H}_0$. This example shows that the interpretation of $\mathbf{H}$ as the 'would-be' magnetic field generated by external currents $\mathbf{j}$ should not be exaggerated into saying that its distribution is independent of the magnetic bodies in the system. In the limit $\mu \gg \mu_0$, Eq. (5.126) yield $H_{\mathrm{int}}/H_0 \ll 1$, $B_{\mathrm{int}}/H_0 = 3\mu_0$, the factor 3 being specific for the particular geometry of the sphere. If a sample is strongly stretched along the applied field, with its length l much larger than the thickness scale t, this geometric effect is gradually decreased and B_{int} tends to its value $\mu H_0 \gg B_0$, as was discussed above—see figure 5.15.

Now let us calculate the field distribution in a similar, but slightly more complex (and practically important) system: a round cylindrical shell, made of a linear magnet, placed in a uniform external field $\mathbf{H}_0$ normal to its axis—see figure 5.16.

Since there are no stand-alone currents in the region of our interest, we can again represent the field $\mathbf{H}(\mathbf{r})$ by the gradient of the magnetic potential ϕ_{m}—see Eq. (5.120). Inside each of three constant-μ regions, i.e. at $\rho < b$, $a < \rho < b$, and $b < \rho$ (where ρ is the distance from the cylinder's axis), the potential obeys the Laplace equation (5.121). In the convenient, polar coordinates (see figure 5.16), we may, guided by the general solution (2.112) of the Laplace equation and our experience in its application to axially symmetric geometries, look for ϕ_{m} in the following form:

[60] To derive Eq. (5.125), we may either calculate the gradient of the ϕ_{m} given by Eq. (5.124), or use the similarity of Eqs. (3.13) and (5.99), to derive from Eq. (3.7) a similar expression for the magnetic dipole's potential:

$$\phi_{\mathrm{m}} = \frac{1}{4\pi}\frac{m\cos\theta}{r^2}.$$

Now comparing this formula with the second term of Eq. (5.124), we immediately obtain Eq. (5.125).

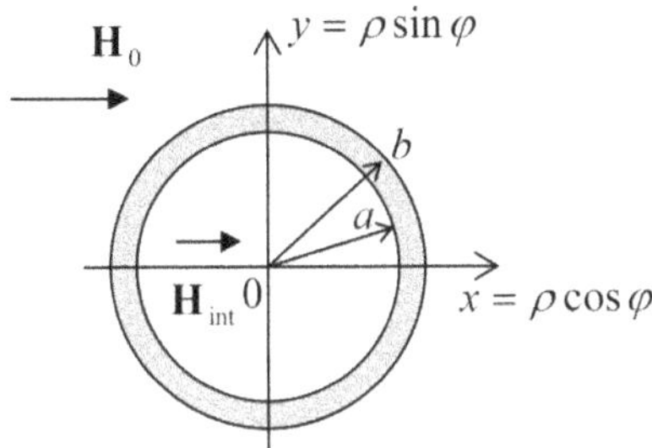

Figure 5.16. A cylindrical magnetic shield.

$$\phi_m = \begin{cases} (-H_0\rho + b_1'/\rho)\cos\varphi, & \text{for } b \leqslant \rho, \\ (a_1\rho + b_1/\rho)\cos\varphi, & \text{for } a \leqslant \rho \leqslant b, \\ -H_{\text{int}}\rho\cos\varphi, & \text{for } \rho \leqslant a. \end{cases} \tag{5.127}$$

Plugging this solution into the boundary conditions (5.122) and (5.123) at both interfaces ($\rho = b$ and $\rho = a$), we obtain the following system of four equations:

$$\begin{aligned} &-H_0 b + b_1'/b = a_1 b + b_1/b, && (a_1 a + b_1/a) = -H_{\text{int}}a, \\ &\mu_0\left(-H_0 - b_1'/b^2\right)H_0 = \mu(a_1 - b_1/b^2), && \mu(a_1 - b_1/a^2) = -\mu_0 H_{\text{int}}, \end{aligned} \tag{5.128}$$

for four unknown coefficients a_1, b_1, b_1', and H_{int}. Solving the system, we obtain, in particular:

$$\frac{H_{\text{int}}}{H_0} = \frac{\alpha_c - 1}{\alpha_c - (a/b)^2}, \qquad \text{with } \alpha_c \equiv \left(\frac{\mu + \mu_0}{\mu - \mu_0}\right)^2. \tag{5.129}$$

According to these formulas, at $\mu > \mu_0$, the field in the free space inside the cylinder is lower that the external field. This fact allows using such structures, made of high-μ materials such as permalloy (see table 5.1), for the passive shielding[61] from unintentional magnetic fields (e.g. the Earth's field)—a task very important for the design of many physical experiments. As Eq. (5.129) shows, the larger μ is, the closer α_c is to 1, and the smaller is the ratio H_{int}/H_0, i.e. the better is the shielding (for the same a/b ratio). On the other hand, for a given magnetic material, i.e. for a fixed parameter α_c, the shielding is improved by making the ratio $a/b < 1$ smaller, i.e. making the shield thicker. On the other hand, a smaller a leaves less space for the shielded equipment/samples, calling for a compromise.

Now let us discuss a curious (and practically important) approach to systems with relatively thin, closed magnetic cores made of sections of (possibly, different) high-μ magnetics, with the cross-section areas A_k much smaller than the squared lengths l_k of the sections—see figure 5.17. If all $\mu_k \gg \mu_0$, virtually all field lines are confined to the interior of the core. Then, applying the macroscopic Ampère law (5.116) to a contour C that follows a magnetic field line inside the core (see, for example, the

[61] A complementary approach to the undesirable magnetic fields' reduction is the 'active shielding'—the field compensation with the counter-field induced by controlled currents in specially designed wire coils.

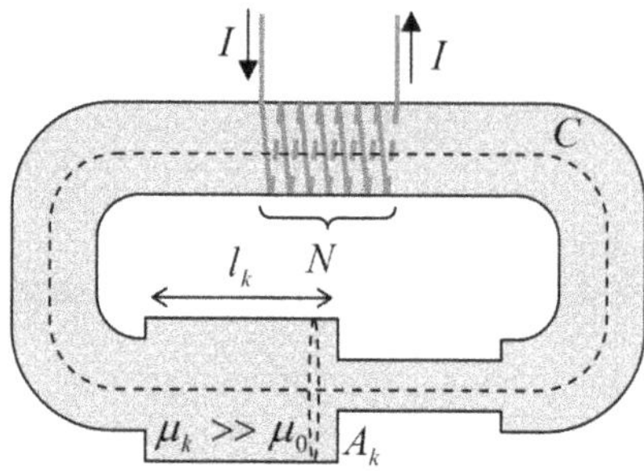

Figure 5.17. Deriving the 'magnetic Ohm law' (5.131).

dashed line in figure 5.17), we obtain the following approximate expression (exactly valid only in the limit μ_k/μ_0, $l_k^2/A_k \to \infty$):

$$\oint_C H_l dl \approx \sum_k l_k H_k \equiv \sum_k l_k \frac{B_k}{\mu_k} = NI. \tag{5.130}$$

However, since the magnetic field lines stay in the core, the magnetic flux $\Phi_k \approx B_k A_k$ should be the same ($\equiv \Phi$) for each section, so that $B_k = \Phi/A_k$. Plugging this condition into Eq. (5.130), we obtain

$$\Phi = \frac{NI}{\sum_k \mathscr{R}_k}, \qquad \text{where } \mathscr{R}_k \equiv \frac{l_k}{\mu_k A_k}. \tag{5.131}$$

Note a close analogy of the first of these equations with the usual Ohm law for several resistors connected in series, with the magnetic flux playing the role of electric current, with the product NI, of the voltage applied to the resistor chain. This analogy is fortified by the fact that the second of Eqs. (5.131) is similar to the expression for resistance $R = l/\sigma A$ of a long, uniform conductor, with the magnetic permeability μ playing the role of the electric conductivity σ. (In order to sound similar, but still different from the resistance R, the parameter $\mathscr{R}$ is called *reluctance*.) This is why Eq. (5.131) is called the *magnetic Ohm law*; it is very useful for approximate analyses of systems such as ac transformers, magnetic energy storage systems, etc.

Now let me proceed to a brief discussion of systems with permanent magnets. First of all, using the definition (5.108) of the field **H**, we may rewrite the Maxwell equation (5.29) for the field **B** as

$$\nabla \cdot \mathbf{B} \equiv \mu_0 \nabla \cdot (\mathbf{H} + \mathbf{M}) = 0, \qquad \text{i.e. as } \nabla \cdot \mathbf{H} = -\nabla \cdot \mathbf{M}, \tag{5.132}$$

While this relation is general, it is particularly convenient in permanent magnets, where **M** may be considered field-independent. In this case, Eq. (5.132) for **H** is an exact analog of Eq. (1.27) for **E**, with the fixed term $-\nabla \cdot \mathbf{M}$ playing the role of the fixed charge density (more exactly, of ρ/ε_0). For the scalar potential ϕ_m, defined by Eq. (5.120), this gives the Poisson equation

$$\nabla^2 \phi_m = \nabla \cdot \mathbf{M}, \tag{5.133}$$

similar to those solved, for many electrostatic situations, in previous chapters.

In the particular case when the magnetization vector **M** is not only field-independent, but also constant inside a permanent magnet's volume, then the right-hand sides of Eqs. (5.132) and (5.133) vanish both inside the volume and in the surrounding free space, and give a non-zero effective charge only on the magnet's surface. Integrating Eq. (5.132) along a short path normal to the surface and crossing it, we obtain the following boundary conditions:

$$\Delta H_n \equiv (H_n)_{\text{in free space}} - (H_n)_{\text{in magnet}} = M_n \equiv M \cos\theta, \tag{5.134}$$

where θ is the angle between the magnetization vector and the outer normal to the magnet's surface. This relation is an exact analog of Eq. (1.24) for the normal component of the field **E**, with the effective surface charge density (or rather σ/ε_0) equal to $M\cos\theta$.

This analogy between the magnetic field induced by a fixed, constant magnetization and the electric field induced by surface electric charges enables one to reuse quite a few problems considered in chapters 1–3. Leaving a few such problems for the reader's exercise (see section 5.7), let me demonstrate the power of this analogy with two examples specific to magnetic systems. First, let us calculate the force necessary to detach the flat ends of two long, uniform rod magnets, of length l and cross-section area $A \ll l^2$, with the saturated magnetization $\mathbf{M}_0$ directed along their length—see figure 5.18.

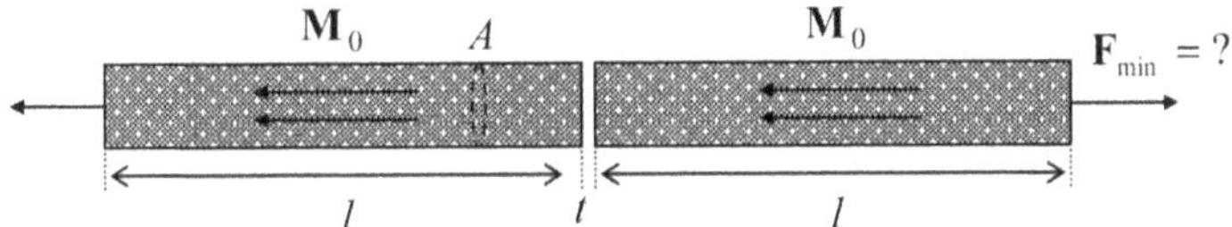

Figure 5.18. Detaching two magnets.

Let us assume we have succeeded in detaching the magnets by an infinitesimal distance $t \ll A^{1/2}, l$. Then, according to Eqs. (5.133) and (5.134), the distribution of the magnetic field near this small gap should be similar to that of the electric field in a system of two equal but opposite surface charges with the surface density σ proportional to M_0. From chapters 1 and 2, we know the properties of such a system very well: within the gap, the electric field is virtually constant, uniform, proportional to σ, and independent of t, while outside the gap it is negligible. (Due to the condition $A \ll l^2$, the effect of the similar effective charges at the 'outer' ends of the rods on the field near the gap t is negligible.) Hence the magnetic field H inside the gap is proportional to M_0, and independent of A and t. Specifically, for its magnitude, Eq. (5.134) gives simply $H = M_0$, and hence $B = \mu_0 M_0$.

Now we could calculate F_{min} as the force exerted by this field on the effective surface 'charges'. However, it is easier to find it from the following energy argument. Since the magnetic field energy localized inside the magnets and near their outer ends cannot depend on t, this small detachment may only alter the energy inside the gap. For this part of energy, Eq. (5.57) yields:

$$\Delta U = \frac{B^2}{2\mu_0}V = \frac{(\mu_0 M_0)^2}{2\mu_0}At \equiv \frac{\mu_0 M_0^2 A}{2}t. \tag{5.135}$$

The gradient of this potential energy is equal to the attraction force $\mathbf{F} = -\nabla(\Delta U)$, trying to reduce ΔU by decreasing the gap, with the magnitude

$$|F| = \frac{\partial(\Delta U)}{\partial t} = \frac{\mu_0 M_0^2 A}{2}. \tag{5.136}$$

The magnet detachment requires an equal and opposite force.

Now let us consider the situation where similar long permanent magnets (such as the *magnetic needles* used in magnetic compasses) are separated, in otherwise free space, by a larger distance $d \gg A^{1/2}$—see figure 5.19.

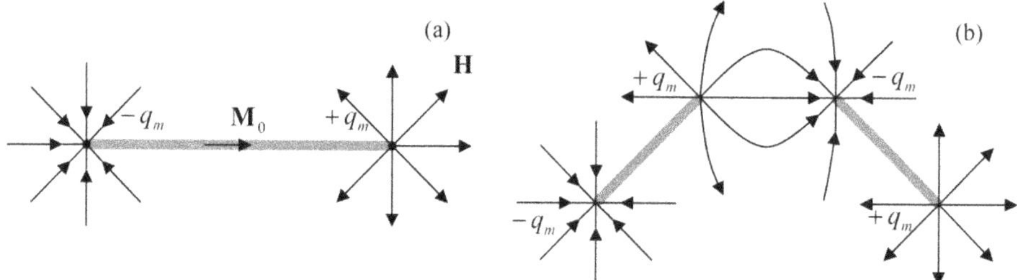

Figure 5.19. Schematic representations of (a) 'magnetic charges' at the ends of a thin permanent-magnet needle and (b) the result of its breaking into two parts.

For each needle (figure 5.19a) of length $l \gg A^{1/2}$, the right-hand side of Eq. (5.133) is substantially different from zero only in two relatively small areas at the needle's ends. Integrating the equation over each end, we see that at distances $r \gg A^{1/2}$ from each end, we may reduce Eq. (5.132) to

$$\nabla \cdot \mathbf{H} = q_\text{m}\delta(\mathbf{r} - \mathbf{r}_+) - q_\text{m}\delta(\mathbf{r} - \mathbf{r}_-), \tag{5.137}$$

where $\mathbf{r}_\pm$ are the ends' positions, and $q_\text{m} \equiv M_0 A$, with A being the needle's cross-section area. This equation is completely similar to Eq. (3.32) for the displacement $\mathbf{D}$, for the particular case of two equal and opposite point charges, i.e. with $\rho = q\delta(\mathbf{r} - \mathbf{r}_+) - q\delta(\mathbf{r} - \mathbf{r}_+)$, with the only replacement $q \to q_\text{m}$. Since we know the resulting electric field all too well (see, e.g. equation (1.7) for $\mathbf{E} \equiv \mathbf{D}/\varepsilon_0$), we may immediately write a similar expression for the field $\mathbf{H}$:

$$\mathbf{H}(\mathbf{r}) = \frac{1}{4\pi} q_\text{m} \left(\frac{\mathbf{r} - \mathbf{r}_+}{|\mathbf{r} - \mathbf{r}_+|^3} - \frac{\mathbf{r} - \mathbf{r}_-}{|\mathbf{r} - \mathbf{r}_-|^3} \right). \tag{5.138}$$

The resulting magnetic field $\mathbf{B}(\mathbf{r}) = \mu_0\mathbf{H}$ exerts on another 'magnetic charge' q'_m, located at point $\mathbf{r}'$, force $\mathbf{F} = q'_\text{m}\mathbf{B}(\mathbf{r}')$.[62] Hence if two ends of different needles are separated by an intermediate distance R ($A^{1/2} \ll R \ll l$, see figure 5.19b), we may neglect one term in Eq. (5.138), and obtain the following 'magnetic Coulomb law' for the interaction of the nearest ends:

[62] The simplest way to verify this (perhaps, obvious) expression is to check that for a system of two 'charges' $\pm q'_\text{m}$, separated by vector $\mathbf{a}$, placed into a uniform external magnetic field $\mathbf{B}_\text{ext}$, it yields the potential energy (5.100) with the correct magnetic moment $\mathbf{m} = q_\text{m}\mathbf{a}$—cf Eq. (3.9) for an electric dipole.

$$\mathbf{F} = \pm\frac{\mu_0}{4\pi}q_{\mathrm{m}}q'_{\mathrm{m}}\frac{\mathbf{R}}{R^3}. \tag{5.139}$$

The 'only' (but conceptually, crucial!) difference of this interaction from that of the electric point charges is that the 'magnetic charges' (quasi-monopoles) of one needle cannot be fully separated. For example, if we break a magnetic needle in the middle in an attempt to bring its two ends further apart, two new 'charges' appear—see figure 5.19b.

There are several solid state systems where more flexible structures, similar in their magnetostatics to the needles, may be implemented. First of all, certain ('type-II') superconductors may sustain so-called *Abrikosov vortices*—colloquially, flexible tubes with field-suppressed superconductivity inside, each carrying one magnetic flux quantum $\Phi_0 = \hbar/\pi e \approx 2 \times 10^{-15}$ Wb. Ending on the superconductor's surfaces, these tubes let their magnetic field lines spread into the surrounding free space, essentially forming magnetic monopole analogs—of course, with equal and opposite 'magnetic charges' q_{m} on each end of the tube. Such flux tubes are not only flexible but readily stretchable, resulting in several peculiar effects—see section 6.4 for more detail. Another, recently found, example of such paired quasi-monopoles includes *spin chains* in the so-called *spin ices*—crystals with paramagnetic ions arranged into a specific (pyrochlore) lattice—such as dysprosium titanate $Dy_2Ti_2O_7$.[63] Let me emphasize again that any reference to magnetic monopoles in such systems should not be taken literally.

In order to complete this section (and this chapter), let me briefly discuss the magnetic field energy U, for the simplest case of systems with *linear* magnetics. In this case we still may use Eq. (5.55), but if we want to operate with the macroscopic fields, and hence the stand-alone currents, we should repeat the manipulations that have led us to Eq. (5.57), replacing $\mathbf{j}$ not from Eq. (5.35), but from Eq. (5.107). As a result, instead of Eq. (5.57) we obtain

$$U = \int_V u(\mathbf{r})d^3r, \quad \text{with } u = \frac{\mathbf{B}\cdot\mathbf{H}}{2} = \frac{B^2}{2\mu} = \frac{\mu H^2}{2}. \tag{5.140}$$

This result is evidently similar to Eq. (3.73) of electrostatics.

As a simple but important example of its application, let us again consider a long solenoid (figure 5.6a), but now filled with a linear magnetic material with permeability μ. Using the macroscopic Ampère law (5.116), just as we used Eq. (5.37) for the derivation of Eq. (5.40), we obtain

$$H = In, \quad \text{and hence } B = \mu In, \tag{5.141}$$

where $n \equiv N/l$, just as in Eq. (5.40), is the winding density, i.e. the number of wire turns per unit length. (At $\mu = \mu_0$, we immediately return to that old result.) Now we may plug Eq. (5.141) into Eq. (5.140) to calculate the magnetic energy stored in the solenoid,

$$U = uV = \frac{\mu H^2}{2}lA = \frac{\mu(nI)^2lA}{2}, \tag{5.142}$$

[63] See, e.g. [7] and references therein.

and then use Eq. (5.72) to calculate its self-inductance[64]:

$$L = \frac{U}{I^2/2} = \mu n^2 l A. \qquad (5.143)$$

We see that $L \propto \mu V$, so that filling a solenoid with a high-μ material may allow one to make it more compact by preserving the same value of inductance. In addition, as the discussion of figure 5.15 has shown, such filling reduces the fringe fields near the solenoid's ends, which may be detrimental for some applications, particularly in precise physical experiments.

Still, we need to explore the issue of the energy in magnetics beyond Eq. (5.140), not only to obtain a general expression for it in materials with an arbitrary dependence **B**(**H**), but also to finally prove Eq. (5.54) and explore its relation with Eq. (5.53). I will do this at the beginning of the next chapter.

5.7 Problems

Problem 5.1. A circular wire loop, carrying a fixed dc current, has been placed inside a similar but larger loop, carrying a fixed current in the same direction—see the figure below. Use semi-quantitative arguments to analyze the mechanical stability of the coaxial, coplanar position of the inner loop with respect to its possible angular, axial, and lateral displacements relative to the outer loop.

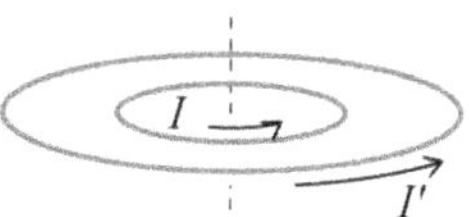

Problem 5.2. Two straight, planar, parallel, long, thin conducting strips of width w, separated by distance d, carry equal but oppositely directed currents I—see the figure below. Calculate the magnetic field in the plane located at the middle between the strips, assuming that the flowing currents are uniformly distributed across the strip widths.

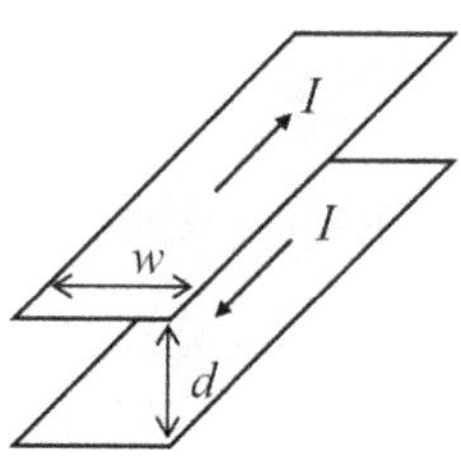

[64] Admittedly, we could obtain the same result just by arguing that since the magnetic material fills the whole volume of a substantial magnetic field in this system, the filling simply increases the vector **B** at all points and hence its flux Φ, and hence $L \equiv \Phi/I$ by a factor of μ/μ_0 in comparison with the free-space value (5.75).

Problem 5.3. For the system studied in the previous problem, but now only in the limit $d \ll w$, calculate:

(i) the distribution of the magnetic field (in the simplest possible way),
(ii) the vector-potential of the field,
(iii) the magnetic force (per unit length) exerted on each strip, and
(iv) the magnetic energy and self-inductance of the loop formed by the strips (per unit length).

Problem 5.4. Calculate the magnetic field distribution near the center of the system of two similar, planar, round, coaxial wire coils, fed by equal but oppositely directed currents—see the figure below.

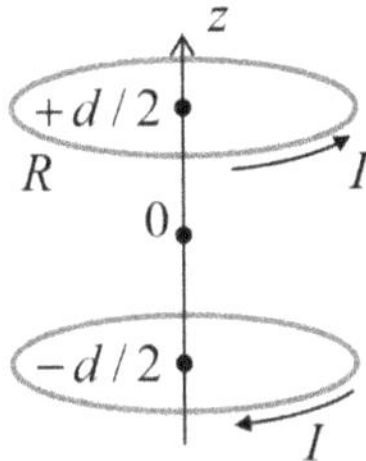

Problem 5.5. The two-coil-system, similar to that considered in the previous problem, now carries equal and similarly directed currents—see the figure below. Calculate what should be the ratio d/R for the second derivative $\partial^2 B_z/\partial z^2$ at $z = 0$ to vanish[65].

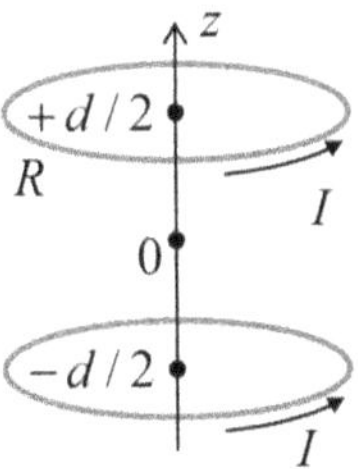

Problem 5.6. Calculate the magnetic field's distribution along the axis of a straight solenoid (see figure 5.6a, partly reproduced below) with a finite length l, and a round cross-section of radius R. Assume that the solenoid has many ($N \gg 1$, l/R) wire turns, uniformly distributed along its length.

[65] Such a system, producing a highly uniform field near its center, is called the *Helmholtz coils*, and is broadly used in physics experiment.

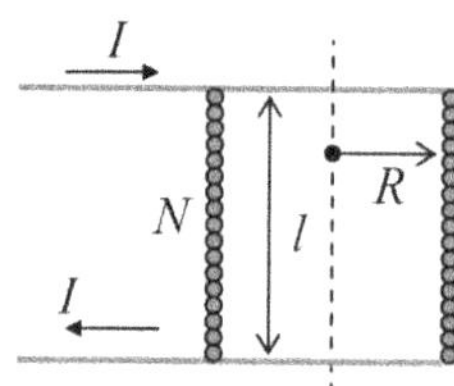

Problem 5.7. A thin round disk of radius R, carrying electric charge of a constant areal density σ, is being rotated around its axis with a constant angular velocity ω. Calculate:

(i) the induced magnetic field on the disk's axis,
(ii) the magnetic moment of the disk,

and relate these results.

Problem 5.8. A thin spherical shell of radius R, with charge Q uniformly distributed over its surface, rotates about its axis with angular velocity ω. Calculate the distribution of the magnetic field everywhere in space.

Problem 5.9. A sphere of radius R, made of an insulating material with a uniform electric charge density ρ, rotates about its diameter with angular velocity ω. Calculate the magnetic field distribution inside the sphere and outside it.

Problem 5.10. The reader is (hopefully:-) familiar with the classical Hall effect when it takes place in the usual rectangular *Hall bar* geometry—see the left panel of the figure below. However, the effect takes a different form in the so-called *Corbino disk*—see the right panel below. (Dark shading shows electrodes, with no appreciable resistance.) Analyze the effect in both geometries, assuming that in both cases the conductors are thin, planar, have a constant Ohmic conductivity σ and charge carrier density n, and that the applied magnetic field **B** is uniform and normal to the conductors' planes.

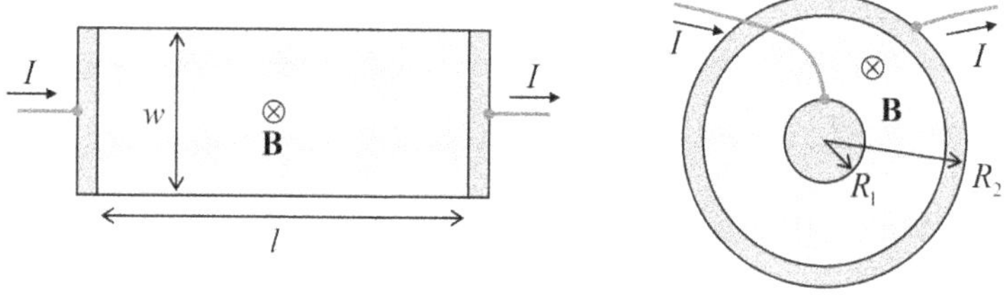

Problem 5.11.* The simplest model of the famous *homopolar motor*[66] is a thin, round conducting disk, placed in a uniform magnetic field normal to its plane, and fed by a dc current flowing from the disk's center to a sliding electrode ('brush') on its rim—see the figure below.

(i) Express the torque, rotating the disk, via its radius R, the magnetic field $\mathbf{B}$, and the current I.
(ii) If the disk is allowed to rotate about its axis, and the motor is driven by a battery with e.m.f. $\mathscr{V}$, calculate its stationary angular velocity ω, neglecting friction and the electric circuit's resistance.
(iii) Now assuming that the current circuit (battery + wires + contacts + disk itself) has a non-zero resistance $\mathscr{R}$, derive and solve the equation for the time evolution of ω, and analyze the solution.

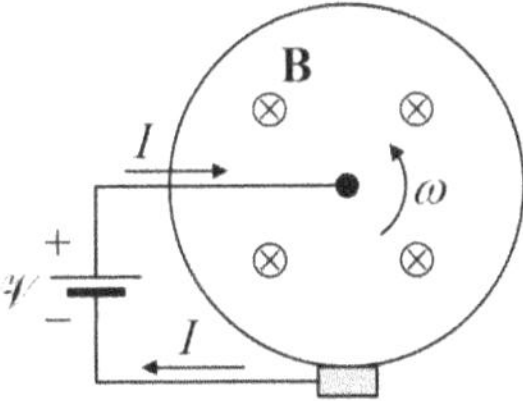

Problem 5.12. Current I flows in a thin wire bent into a planar, round loop of radius R. Calculate the net magnetic flux through the plane in which the loop is located.

Problem 5.13. Prove that:

(i) the self-inductance L of a current loop cannot be negative, and
(ii) any inductance coefficient $L_{kk'}$, defined by Eq. (5.60), cannot be larger than $(L_{kk}L_{k'k'})^{1/2}$.

Problem 5.14.* Estimate the values of magnetic susceptibility due to

(i) orbital diamagnetism, and
(ii) spin paramagnetism

for a medium with negligible interaction between the induced molecular dipoles. Compare the results.

Hints: For task (i), you may use the classical model described by Eq. (5.114)—see figure 5.13. For task (ii), assume the mechanism of ordering of spontaneous magnetic dipoles $\mathbf{m}_0$, with a magnitude m_0 of the order of the Bohr magneton μ_B, similar to the one sketched for electric dipoles in figure 3.7a.

[66] It was invented by M Faraday in 1821, i.e. well before his celebrated work on electromagnetic induction. The adjective 'homopolar' refers to the constant 'polarity' (sign) of the current; the alternative term is 'unipolar'.

Problem 5.15.* Use the classical picture of the orbital ('Larmor') diamagnetism, discussed in section 5.5, to calculate its (small) correction $\Delta\mathbf{B}(0)$ to the magnetic field $\mathbf{B}$ felt by an atomic nucleus, modeling the electrons of the atom by a spherically symmetric cloud with an electric charge density $\rho(r)$. Express the result via the value $\phi(0)$ of the electrostatic potential of the electron cloud, and use this expression for a crude numerical estimate of the relative correction, $\Delta B(0)/B$, for the hydrogen atom.

Problem 5.16. Calculate the (self-)inductance of a toroidal solenoid (see figure 5.6b) with the round cross-section of radius $r \sim R$ (see the figure below), with many ($N \gg 1$, R/r) wire turns uniformly distributed along the perimeter, and filled with a linear magnetic material of a magnetic permeability μ. Check your results by analyzing the limit $r \ll R$.

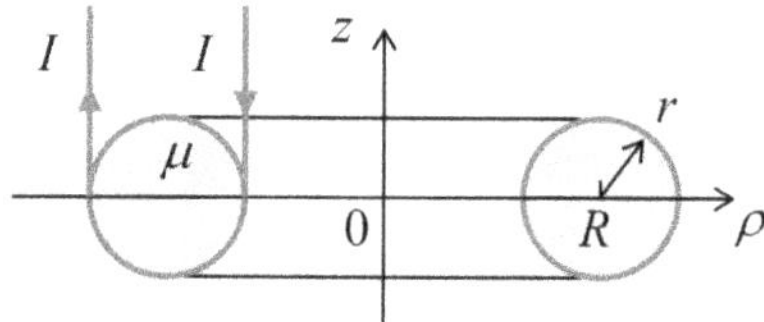

Problem 5.17. A long straight, thin wire, carrying current I, passes parallel to the plane boundary between two uniform, linear magnetics—see the figure below. Calculate the magnetic field everywhere in the system, and the force (per unit length) exerted on the wire.

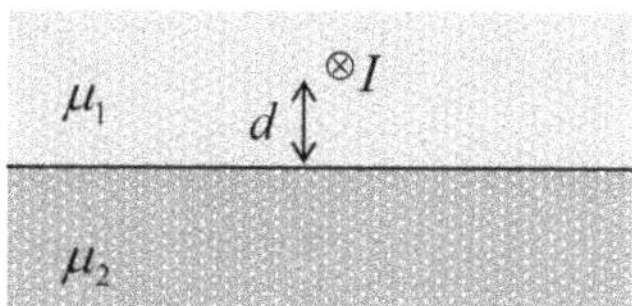

Problem 5.18. Solve the magnetic shielding problem similar to that discussed in section 5.6, but for a *spherical*, rather than cylindrical shell, with the central cross-section shown in figure 5.16. Compare the efficiency of these two shields, for the same permeability μ of the shell, and the same b/a ratio.

Problem 5.19. Calculate the magnetic field distribution around a spherical permanent magnet with a uniform magnetization $\mathbf{M}_0 = \text{const}$.

Problem 5.20. A limited volume V is filled with a magnetic material with a fixed (field-independent) magnetization $\mathbf{M}(\mathbf{r})$. Write explicit expressions for the magnetic field induced by the magnetization, and its potential, and recast these expressions into forms more convenient when $\mathbf{M}(\mathbf{r}) = \mathbf{M}_0 = \text{const}$ inside the volume V.

Problem 5.21. Use the results of the previous problem to calculate the distribution of the magnetic field $\mathbf{H}$ along the axis of a straight permanent magnet of length $2l$, with a round cross-section of radius R, and a uniform magnetization $\mathbf{M}_0$ parallel to the axis—see the figure below.

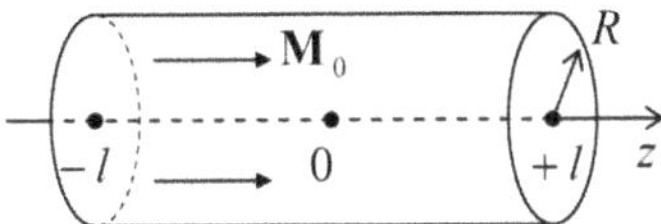

Problem 5.22. A very broad film of thickness $2t$ is permanently magnetized normally to its plane, with a periodic checkerboard pattern, with the square of area $a \times a$:

$$\mathbf{M}|_{|z|<t} = \mathbf{n}_z M(x, y), \quad \text{with } M(x, y) = M_0 \times \text{sgn}\left(\cos\frac{\pi x}{a} \quad \cos\frac{\pi y}{a}\right).$$

Calculate the magnetic field's distribution in space.[67]

Problem 5.23.* Based on the discussion of the quadrupole electrostatic lens in section 2.4, suggest permanent-magnet systems that may similarly focus particles moving close to the system's axis, and carrying:

(i) an electric charge,
(ii) no net electric charge, but a spontaneous magnetic dipole moment **m**.

References

[1] Grover F 1946 *Inductance Calculations* (New York: Dover)
[2] Landau L *et al* 1984 *Electrodynamics of Continuous Media* 2nd edn (Butterworth: Heinemann)
[3] Jiles D J 1998 *Introduction to Magnetism and Magnetic Materials* 2nd edn (Boca Raton FL: CRC Press)
[4] O'Handley R 1999 *Modern Magnetic Materials* (Wiley)
[5] Hadjipanayis G 2001 *Magnetic Storage Systems Beyond 2000* (Springer)
[6] Gurevich A and Melkov G 1996 *Magnetization Oscillations and Waves* (Boca Raton FL: CRC Press)
[7] Jaubert L and Holdworth P 2011 *J. Phys.: Condens. Matter* **23** 164222

[67] This problem is of an evident relevance for the *perpendicular magnetic recording* (PMR) technology, which currently dominates high-density digital magnetic recording.

IOP Publishing

Classical Electrodynamics
Lecture notes
Konstantin K Likharev

Chapter 6

Electromagnetism

This chapter discusses two major new effects that arise when the electric and magnetic fields are changing in time: the 'electromagnetic induction' of an additional electric field by the changing magnetic field, and the reciprocal effect of 'displacement currents'—actually, the induction of an additional magnetic field by the changing electric field. These two phenomena, which make the time-dependent electric and magnetic fields inseparable (hence the term 'electromagnetism'), are reflected in the full system of Maxwell equations, valid for an arbitrary electromagnetic process. On the way toward this system, I will make a pause for a brief review of the electrodynamics of superconductivity, which (in addition to its own significance), provides a perfect platform for the discussion of the general issue of gauge invariance.

6.1 Electromagnetic induction

As Eqs. (5.36) show, in static situations ($\partial/\partial t = 0$) the Maxwell equations describing the electric and magnetic fields are independent, and are coupled only implicitly, via the continuity equation (4.5) relating their right-hand sides ρ and $\mathbf{j}$. (In statics, this relation imposes a restriction only on the vector $\mathbf{j}$.) In dynamics, when the fields change in time, the situation is different.

Historically, the first discovered explicit coupling between the electric and magnetic fields was the effect of electromagnetic induction[1]. The summary of Faraday's numerous experiments has turned out to be very simple: if the magnetic flux, defined by Eq. (5.65),

$$\Phi \equiv \int_S B_n \, d^2r, \tag{6.1}$$

[1] The induction e.m.f. was discovered independently by J Henry, but is was the brilliant series of experiments by M Faraday, carried out mostly in 1831, which resulted in the general formulation of the induction law.

doi:10.1088/978-0-7503-1404-6ch6

through a surface S limited by contour C, changes in time for whatever reason (e.g. either due to a change of the magnetic field $\mathbf{B}$ (as in figure 6.1), or the contour's motion, or its deformation, or any combination of the above), it induces an additional, vortex-like electric field $\mathbf{E}_{\text{ind}}$—see figure 6.1.

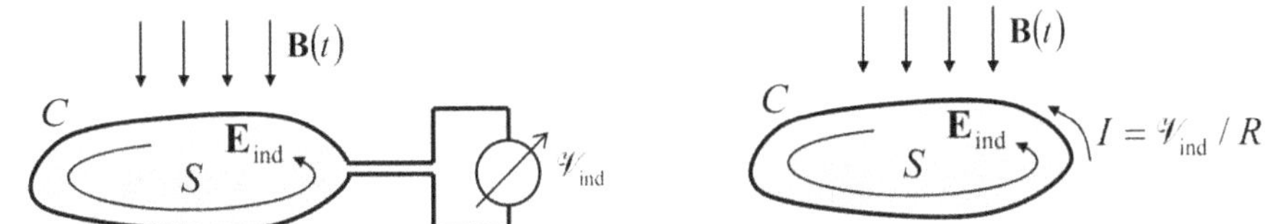

Figure 6.1. The two simplest ways to observe the Faraday electromagnetic induction.

The exact distribution of $\mathbf{E}_{\text{ind}}$ in space depends on system geometry details, but its integral along the contour C, called the *inductive electromotive force* (e.m.f.), obeys a very simple *Faraday induction law*[2]:

$$\mathscr{V}_{\text{ind}} \equiv \oint_C \mathbf{E}_{\text{ind}} \cdot d\mathbf{r} = -\frac{d\Phi}{dt}. \tag{6.2}$$

It is straightforward (and hence left for the reader's exercise) to show that this e.m.f. may be measured, for example, either by inserting a voltmeter into a conducting loop following contour C, or by measuring the small current $I = \mathscr{V}_{\text{ind}}/R$ it induces in a thin wire with a sufficiently large Ohmic resistance R,[3] whose shape follows the contour—see figure 6.1. (Actually, these methods are not entirely different, because some practical voltmeters measure voltage by the small Ohmic current it drives through a known high resistance.) In the context of the latter approach, it is easy to formulate the so-called *Lenz rule* used for the description of the minus sign in Eq. (6.2): the additional magnetic field of the induced current I provides a partial compensation of the *change* of the original flux $\Phi(t)$ with time[4].

In order to recast Eq. (6.2) in a differential form, let us apply to the contour integral in it, the same Stokes theorem that was repeatedly used in chapter 5. The result is

$$\mathscr{V}_{\text{ind}} = \int_S (\nabla \times \mathbf{E}_{\text{ind}})_n \, d^2r. \tag{6.3}$$

Now combining Eqs. (6.1)–(6.3), for a contour C whose shape does not change in time (so that the integration along it is interchangeable with the time derivative), we obtain

$$\int_S \left(\nabla \times \mathbf{E}_{\text{ind}} + \frac{\partial \mathbf{B}}{\partial t} \right)_n d^2r = 0. \tag{6.4}$$

[2] In Gaussian units, the right-hand side of this formula has the additional coefficient $1/c$.

[3] Such induced current is sometimes called the *eddy current*, although most often this term is saved for the distributed currents induced by a changing magnetic field in a bulk conductor—see section 6.3.

[4] Let me hope that the reader is also familiar with the paradox arising in attempts to measure $\mathscr{V}_{ind}$ with a voltmeter, without opening the wire loop; if not, I would highly recommend solving problem 6.2.

Since the induced electric field is additional to the gradient field (1.33) created by electric charges, for the net field we may write $\mathbf{E} = \mathbf{E}_{\text{ind}} - \nabla\phi$. However, since the curl of any gradient field is zero[5], $\nabla \times (\nabla\phi) = 0$, Eq. (6.4) remains valid for the net field $\mathbf{E}$. Since this equation should be correct for *any* closed area S, we may conclude that

$$\nabla \times \mathbf{E} + \frac{\partial \mathbf{B}}{\partial t} = 0 \tag{6.5}$$

at any point. This is the final (time-dependent) form of this Maxwell equation. Superficially, it may look as if Eq. (6.5) is less general than Eq. (6.2); for example it does not describe any electric field—and hence any e.m.f. in a moving loop—if the field $\mathbf{B}$ is constant in time, even if the magnetic flux (6.1) through the loop does change in time. However, this is not true; in chapter 9 we will see that in the reference frame moving with the loop such an e.m.f. does appear[6].

Now let us re-formulate Eq. (6.5) in terms of the vector-potential $\mathbf{A}$. Since the induction effect does not alter the fundamental relation $\nabla \cdot \mathbf{B} = 0$, we still may represent the magnetic field as prescribed by Eq. (5.27), i.e. as $\mathbf{B} = \nabla \times \mathbf{A}$. Plugging this expression into Eq. (6.5), and changing the order of the temporal and spatial differentiation, we obtain

$$\nabla \times \left(\mathbf{E} + \frac{\partial \mathbf{A}}{\partial t}\right) = 0. \tag{6.6}$$

Hence we can use the same argumentation as in section 1.3 (there applied to the vector $\mathbf{E}$ alone) to represent the expression in the parentheses as $-\nabla\phi$, so that we obtain

$$\mathbf{E} = -\frac{\partial \mathbf{A}}{\partial t} - \nabla\phi, \qquad \mathbf{B} = \nabla \times \mathbf{A}. \tag{6.7}$$

It is very tempting to interpret the first term on the right-hand side of the expression for $\mathbf{E}$ as describing the electromagnetic induction alone, and the second term representing a purely electrostatic field induced by electric charges. However, the separation of these two terms is, to a certain extent, conditional. Indeed, let us consider the gauge transformation already mentioned in section 5.2,

$$\mathbf{A} \to \mathbf{A} + \nabla\chi, \tag{6.8}$$

[5] See, e.g. Eq. (A.71).

[6] I have to admit that from the beginning of the course, I was carefully sweeping under the rug a very important question: in exactly which reference frame(s) are all the equations of electrodynamics valid? I promise to discuss this issue in detail later in the course (in chapter 9), and for now would like to get away with a very short answer: all the formulas discussed so far are valid in *any inertial* reference frame, as defined in classical mechanics—see, e.g. *Part CM* section 1.3. It is crucial, however, to have fields $\mathbf{E}$ and $\mathbf{B}$ measured *in the same* reference frame.

which, as we already know, does not change the magnetic field. According to Eq. (6.8), in order to keep the full electric field intact (*gauge-invariant*) as well, the scalar electric potential has to be transformed simultaneously as

$$\phi \to \phi - \frac{\partial \chi}{\partial t}, \tag{6.9}$$

leaving the choice of a time-independent addition to ϕ restricted only by the Laplace equation—since the full ϕ should satisfy the Poisson equation (1.41) with a gauge-invariant right-hand side. We will return to the discussion of gauge invariance in section 6.4.

6.2 Magnetic energy revisited

Now we are sufficiently equipped to return to the issue of magnetic energy, in particular to finally prove Eqs. (5.57) and (5.140) and discuss the dichotomy of the signs in Eqs. (5.53) and (5.54)[7]. For that, let us consider a sufficiently slow, small magnetic field variation $\delta\mathbf{B}$. If we want to neglect the kinetic energy of the system of electric currents under consideration, as well as the wave radiation effects, we need to prevent its acceleration by the arising induction field $\mathbf{E}_{\text{ind}}$. Let us suppose that we do this by the virtual balancing this field by an external electric field $\mathbf{E}_{\text{ext}} = -\mathbf{E}_{\text{ind}}$. According to Eq. (4.38), the work of that field[8] on the stand-alone currents of the system during time interval δt, and hence the change of the potential energy of the system, is

$$\delta U = \delta t \int_V \mathbf{j} \cdot \mathbf{E}_{\text{ext}}\, d^3r = -\delta t \int_V \mathbf{j} \cdot \mathbf{E}_{\text{ind}}\, d^3r, \tag{6.10}$$

where the integral is over the volume of the system. Now expressing the current density $\mathbf{j}$ from the macroscopic Maxwell equation (5.107), and then applying the vector algebra identity[9]

$$(\nabla \times \mathbf{H}) \cdot \mathbf{E}_{\text{ind}} \equiv \mathbf{H} \cdot (\nabla \times \mathbf{E}_{\text{ind}}) - \nabla \cdot (\mathbf{E}_{\text{ind}} \times \mathbf{H}), \tag{6.11}$$

we obtain

$$\delta U = -\delta t \int_V \mathbf{H} \cdot (\nabla \times \mathbf{E}) d^3r + \delta t \int_V \nabla \cdot (\mathbf{E} \times \mathbf{H}) d^3r. \tag{6.12}$$

According to the divergence theorem, the second integral on the right-hand side is equal to the flux of the vector $\mathbf{S} \equiv \mathbf{E} \times \mathbf{H}$ through the surface limiting the considered volume V. Later in the course we will see that this flux represents, in particular, the power of electromagnetic radiation through the surface. If such radiation is negligible (as it always is if the field variation is sufficiently slow) the surface may

[7] Actually, this dichotomy should not be too puzzling to the reader who has understood a similar sign duality in electrostatics—cf, e.g. Eqs. (3.73) and (3.81).

[8] As a reminder, the magnetic component of the Lorentz force (5.10), $\mathbf{v} \times \mathbf{B}$, is always perpendicular to the particle's velocity, so that the magnetic field $\mathbf{B}$ itself cannot perform any work on moving charges, i.e. on currents.

[9] See, e.g. Eq. (A.77) with $\mathbf{f} = \mathbf{E}_{\text{ind}}$ and $\mathbf{g} = \mathbf{H}$.

be selected sufficiently far such as that the flux of **S** is negligible. In this case, we may express $\nabla \times \mathbf{E}$ from the Faraday induction law (6.5) to obtain

$$\delta U = -\delta t \int_V \left(-\frac{\partial B}{\partial t}\right) \cdot \mathbf{H}\, d^3r = \int_V \mathbf{H} \cdot \delta \mathbf{B}\, d^3r. \tag{6.13}$$

Just as in electrostatics (see Eqs. (1.65) and (3.73), and their discussion), this relation may be interpreted as the variation of the magnetic field energy U of the system, and represented in the form

$$\delta U = \int_V \delta u(\mathbf{r}) d^3r, \quad \text{with} \quad \delta u \equiv \mathbf{H} \cdot \delta \mathbf{B}. \tag{6.14}$$

This is a keystone result; let us discuss it in some detail.

First of all, for a system filled with a linear magnetic material, we may use Eq. (6.14) together with Eq. (5.110): $\mathbf{B} = \mu \mathbf{H}$. Integrating the result over the variation of **B** from 0 to a certain final value, we obtain Eq. (5.140)—so important that it is worth rewriting again:

$$U = \int_V u(\mathbf{r}) d^3r, \quad \text{with} \quad u = \frac{B^2}{2\mu}. \tag{6.15}$$

In the simplest case of free space (no magnetics at all, so that **j** above is the complete current density), we may take $\mu = \mu_0$, and reduce Eq. (6.15) to Eq. (5.57). Now performing backwards the transformations that took us, in section 5.3, to derive that relation from Eq. (5.54), we finally have the latter formula proved—as was promised.

It is very important, however, to understand the limitations of Eq. (6.15). For example, let us try to apply it to a very simple problem, which was already analyzed in section 5.6 (see figure 5.15): a very long cylindrical sample of a linear magnetic material placed into a fixed external field $\mathbf{H}_{\text{ext}}$, parallel to the sample's axis. It is evident that in this simple geometry, the field **H** and hence the field $\mathbf{B} = \mu \mathbf{H}$ have to be uniform inside the sample, apart from negligible regions near its ends, so that Eq. (6.15) is reduced to

$$U = \frac{B^2}{2\mu} V, \tag{6.16}$$

where $V = Al$ is the cylinder's volume. Now if we try to calculate the static (equilibrium) value of the field from the minimum of this potential energy, we obtain evident nonsense: $\mathbf{B} = 0$ (WRONG!)[10].

The situation may be readily rectified by using the notion of the Gibbs potential energy, just as was done for the electric field in section 3.5 (and implicitly in the end of section 1.3). According to Eq. (6.14), in magnetostatics the Cartesian components of the field $\mathbf{H}(\mathbf{r})$ play the role of generalized forces, while those of the field $\mathbf{B}(\mathbf{r})$ play

[10] Note that this erroneous result cannot be corrected by just adding the energy (6.15) of the field outside the cylinder, because in the limit $A \to 0$, this field is not affected by the internal field **B**.

the role of generalized coordinates (per unit volume)[11]. As a result, the Gibbs potential energy, whose minimum corresponds to the stable equilibrium of the system under the effect of a fixed generalized force (in our current case, in a fixed external field $\mathbf{H}_{\text{ext}}$), is

$$U_{\text{G}} = \int_V u_{\text{G}}(\mathbf{r})d^3r, \quad \text{with} \quad u_{\text{G}}(\mathbf{r}) \equiv u(\mathbf{r}) - \mathbf{H}_{\text{ext}}(\mathbf{r}) \cdot \mathbf{B}(\mathbf{r}), \tag{6.17}$$

an expression to be compared with Eq. (3.78). For a system with linear magnetics, we may use Eq. (6.15) for u, obtaining the following Gibbs energy's density:

$$u_{\text{G}}(\mathbf{r}) = \frac{1}{2\mu}\mathbf{B} \cdot \mathbf{B} - \mathbf{H}_{\text{ext}} \cdot \mathbf{B} \equiv \frac{1}{2\mu}(\mathbf{B} - \mu\mathbf{H}_{\text{ext}})^2 + \text{const}, \tag{6.18}$$

where 'const' means a term independent of the field $\mathbf{B}$. For our simple cylindrical system, with its uniform fields, Eq. (6.18) gives the following full Gibbs energy of the sample:

$$U_{\text{G}} = \frac{(\mathbf{B}_{\text{int}} - \mu\mathbf{H}_{\text{ext}})^2}{2\mu}V + \text{const}, \tag{6.19}$$

whose minimum immediately gives the correct stationary value $\mathbf{B}_{\text{int}} = \mu\mathbf{H}_{\text{ext}}$, i.e. $\mathbf{H}_{\text{int}} \equiv \mathbf{B}_{\text{int}}/\mu = \mathbf{H}_{\text{ext}}$—which was already obtained in section 5.6 in a different way, from the boundary condition (5.117).

Now notice that with this result on hand, Eq. (6.18) may be rewritten in a different form,

$$u_{\text{G}}(\mathbf{r}) = \frac{1}{2\mu}\mathbf{B} \cdot \mathbf{B} - \frac{\mathbf{B}}{\mu} \cdot \mathbf{B} \equiv -\frac{B^2}{2\mu}, \tag{6.20}$$

similar to Eq. (6.15) for $u(\mathbf{r})$, but with an opposite sign. This sign dichotomy explains that in Eqs. (5.53) and Eq. (5.54); indeed, as was already noted in section 5.3, the former of these expressions gives the potential energy whose minimum corresponds to the equilibrium of a system with fixed currents. According to Eq. (5.107), $\nabla \times \mathbf{H} = \mathbf{j}$, this condition is fulfilled if the field $\mathbf{H}$ is fixed[12]. So, the energy U_j given by Eq. (5.53) is essentially the Gibbs energy U_{G} defined by Eqs. (6.17) and (for the case of linear magnetics, or no magnetic media at all) by Eq. (6.20), while Eq. (5.54) is just another form of Eq. (6.15)—as was explicitly shown in section 5.3[13].

Let me complete this section by stating that the difference between the energies U and U_{G} is not properly emphasized (or is even left obscure) in some textbooks, so that the reader is advised to seek additional clarity—for example, by spelling them

---

[11] Note that in this respect, the analogy with electrostatics is not quite complete. Indeed, according to Eq. (3.76), in electrostatics the role of a generalized coordinate is played by 'would-be' field $\mathbf{D}$, and that of the generalized force, by the actual electric field $\mathbf{E}$. This difference may be traced back to the fact that electric field $\mathbf{E}$ may perform work on a moving charged particle, while the magnetic field cannot. However, this difference does not affect the full analogy of expressions (3.73) and (6.15) for the field energy density in *linear* media.

[12] The opposite is not always true—see, e.g. the example discussed in section 5.6.

[13] As was already noted in section 5.4, one more example of the energy U_j is given by Eq. (5.100).

out for a simple case of a long straight solenoid (figure 5.6a), and then using these formulas to calculate the pressure exerted by the magnetic field on the solenoid's walls (windings) and the longitudinal forces exerted on its ends.

6.3 Quasi-static approximation and skin effect

Perhaps the most surprising experimental fact concerning time-dependent electromagnetic phenomena is that unless they are so fast that one more new effect, *displacement currents* (to be discussed in section 6.7 below), becomes noticeable, all formulas of electrostatics and magnetostatics remain valid, with only the exception: the generalization of Eq. (3.36) to Eq. (6.5), describing the Faraday induction. As a result, the system of macroscopic Maxwell equations (5.109) is generalized to

$$\begin{aligned} \nabla \times \mathbf{E} + \frac{\partial \mathbf{B}}{\partial t} = 0\,, \qquad & \nabla \times \mathbf{H} = \mathbf{j}\,, \\ \nabla \cdot \mathbf{D} = \rho\,, \qquad & \nabla \cdot \mathbf{B} = 0\,. \end{aligned} \tag{6.21}$$

(As follows from the discussions in chapters 3 and 5, the corresponding system of microscopic Maxwell equations for genuine fields $\mathbf{E}$ and $\mathbf{B}$ may be obtained from Eq. (6.16) by the formal substitutions $\mathbf{D} = \varepsilon_0\mathbf{E}$ and $\mathbf{H} = \mathbf{B}/\mu_0$, and the replacement of the stand-alone charge and current densities ρ and $\mathbf{j}$ with their full densities[14].) These equations, whose range of validity will be quantified in section 6.7, define the so-called *quasi-static approximation* of electromagnetism, and are sufficient for adequate description of a broad range of physical effects.

Let us use them first of all for an analysis of the so-called *skin effect*, the phenomenon of self-shielding of the alternating (ac) magnetic field by the eddy currents induced by the field in an Ohmic conductor. In order to form a complete system of equations, Eq. (6.21) should be augmented by constituent equations describing the medium. Let us take them, for a conductor, in the simplest (and simultaneously, most common) linear and isotropic forms already discussed in chapters 4 and 5:

$$\mathbf{j} = \sigma\mathbf{E}, \quad \mathbf{B} = \mu\mathbf{H}. \tag{6.22}$$

If the conductor is uniform, i.e. the coefficients σ and μ are constant inside it, the whole system of Eqs. (6.21) and (6.22) may be reduced to just one equation. Indeed, a sequential substitution of these equations into each other, using a well-known vector-algebra identity[15] in the middle, yields:

[14] Obviously, in free space the last replacement is unnecessary, because all charges and currents may be treated as 'stand-alone' ones.

[15] See, e.g. Eq. (A.73).

$$\begin{aligned}\frac{\partial \mathbf{B}}{\partial t} &= -\nabla \times \mathbf{E} = -\frac{1}{\sigma}\nabla \times \mathbf{j} = -\frac{1}{\sigma}\nabla \times (\nabla \times \mathbf{H}) \\ &= -\frac{1}{\sigma\mu}\nabla \times (\nabla \times \mathbf{B}) \equiv -\frac{1}{\sigma\mu}[\nabla(\nabla \cdot \mathbf{B}) - \nabla^2\mathbf{B}] \\ &= \frac{1}{\sigma\mu}\nabla^2\mathbf{B}.\end{aligned} \tag{6.23}$$

Thus we have arrived, without any further assumptions, at a rather simple partial differential equation. Let us use it to analyze the skin effect in the simplest geometry (figure 6.2a) when an external source (which, at this point, does not need to be specified) has produced, near a plane surface of a bulk conductor, a spatially uniform ac magnetic field $\mathbf{H}^{(0)}(t)$ parallel to the surface[16].

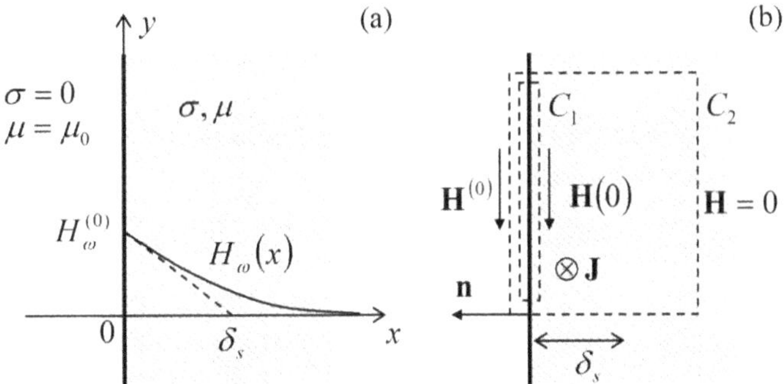

Figure 6.2. (a) The skin effect in the simplest, planar geometry and (b) two Ampère contours, C_1 and C_2, for deriving the 'microscopic' (C_1) and the 'macroscopic' (C_2) boundary conditions for **H**.

Selecting the coordinate system as shown in figure 6.2a, we may express this condition as

$$\mathbf{H}|_{x=-0} = H^{(0)}(t)\mathbf{n}_y. \tag{6.24}$$

The translational symmetry of our simple problem within the surface plane [y, z] implies that inside the conductor $\partial/\partial y = \partial/\partial z = 0$ as well, and $\mathbf{H} = H(x, t)\mathbf{n}_y$ even at $x \geqslant 0$, so that Eq. (6.23) for the conductor's interior is reduced to a differential equation for just one scalar function $H(x, t) = B(x, t)/\mu$:

$$\frac{\partial H}{\partial t} = \frac{1}{\sigma\mu}\frac{\partial^2 H}{\partial x^2}, \quad \text{for } x \geqslant 0. \tag{6.25}$$

[16] Due to the simple linear relation $\mathbf{B} = \mu\mathbf{H}$ between the fields **B** and **H**, it does not matter too much which of them is used for the solution of this problem, with a slight preference for **H**, due to the simplicity of Eq. (5.117)—the only boundary condition relevant for this simple geometry.

This equation may be further simplified by noticing that due to its linearity, we may use the linear superposition principle for the time dependence of the field[17], via expanding it, as well as the external field (6.24), into the Fourier series:

$$H(x,t)=\sum_{\omega}H_{\omega}(x)e^{-i\omega t},\quad \text{for}\quad x\geqslant 0,$$
$$H^{(0)}(t)=\sum_{\omega}H_{\omega}^{(0)}e^{-i\omega t},\quad \text{for}\quad x=0,\tag{6.26}$$

and arguing that if we know the solution for each frequency component of the series, the whole field may be found through the elementary summation (6.26) of these solutions. For each a single-frequency component, Eq. (6.25) is immediately reduced to an ordinary differential equation for the complex amplitude $H_\omega(x)$:[18]

$$-i\omega H_{\omega}=\frac{1}{\sigma\mu}\frac{d^2}{dx^2}H_{\omega}.\tag{6.27}$$

From the theory of linear ordinary differential equations we know that Eq. (6.27) has the following general solution,

$$H_{\omega}(x)=H_{+}e^{\kappa_{+}x}+H_{-}e^{\kappa_{-}x},\tag{6.28}$$

where constants $\kappa_\pm$ are roots of the characteristic equation that may be obtained by the substitution of any of these two exponents into the initial differential equation. For our particular case, the characteristic equation, following from Eq. (6.27), is

$$-i\omega=\frac{\kappa^2}{\sigma\mu}\tag{6.29}$$

and its roots are complex constants

$$\kappa_{\pm}=(-i\mu\omega\sigma)^{1/2}\equiv\pm\frac{1-i}{\sqrt{2}}(\mu\omega\sigma)^{1/2}.\tag{6.30}$$

For our problem, the field cannot grow exponentially at $x\to+\infty$, so that only one of the coefficients, namely H_- corresponding to the decaying exponent with $\operatorname{Re}\kappa<0$ (i.e. $\kappa=\kappa_-$), may be non-vanishing, so that $H_\omega(x)=H_\omega(0)\exp\{\kappa_-x\}$. In order to find the constant factor $H_\omega(0)$, we can integrate the macroscopic Maxwell equation $\nabla\times\mathbf{H}=\mathbf{j}$ along a pre-surface contour—say, the contour C_1 shown in figure 6.2b. The right-hand side's integral is negligible, because the stand-alone current density $\mathbf{j}$ does

[17] Another important way to exploit the linearity of Eq. (6.25) is to use the *spatial-temporal Green's function* approach to explore the dependence of its solutions on various initial conditions. Unfortunately, due to a lack of time, I have to leave an analysis of this opportunity to a reader's exercise.

[18] Let me hope that the reader is not intimidated by the (very convenient) use of such complex variables; their imaginary parts always disappear at the final summation (6.26). For example, if the external field is purely sinusoidal, with the actual (positive) frequency ω, each sum in Eq. (6.26) has just two terms, with complex amplitudes H_ω and $H_{-\omega}=H_\omega{}^*$, so that their sum is always real. (For a more detailed discussion of this issue, see, e.g. *Part CM* section 5.1.)

not include the 'genuinely surface' currents responsible for the magnetic properties of the material (see figure 5.12). As a result, we obtain the 'microscopic'[19] boundary condition similar to Eq. (5.117) for the stationary magnetic field, $H_\tau = \text{const}$ at $x = 0$, i.e.

$$H(0, t) = H^{(0)}(t), \qquad \text{i.e.} \quad H_\omega(0) = H_\omega^{(0)}, \tag{6.31}$$

so that the final solution of our problem may be represented as

$$H_\omega(x) = H_\omega^{(0)} \exp\left\{-\frac{x}{\delta_s}\right\} \exp\left\{-i\left(\omega t - \frac{x}{\delta_s}\right)\right\}, \tag{6.32}$$

where the constant δ_s, with the dimension of length, is called the *skin depth*:

$$\delta_s \equiv -\frac{1}{\operatorname{Re}\kappa_-} = \left(\frac{2}{\mu\sigma\omega}\right)^{1/2}. \tag{6.33}$$

This solution describes the *skin effect*: the penetration of the ac magnetic field of frequency ω into a conductor only to a finite depth of the order of δ_s.[20] Let me give a few numerical examples of the skin depth. For copper at room temperature, $\delta_s \approx 1$ cm at the standard ac power distribution frequency of 60 Hz, and is of the order of just 1 μm at a few GHz, i.e. at typical frequencies of cell phone signals and kitchen microwave magnetrons. For sea water, δ_s is close to 250 m at just 1 Hz (with big implications for radio communications with submarines), and is of the order of 1 cm at a few GHz (explaining the non-uniform heating of a soup bowl in a microwave oven).

In order to complete the skin effect discussion, let us consider what happens with the induced eddy currents[21] and the electric field at this effect. When deriving our basic equation (6.23), we have used, in particular, relations $\mathbf{j} = \nabla \times \mathbf{H} = \mu^{-1}\nabla \times \mathbf{B}$, and $\mathbf{E} = \mathbf{j}/\sigma$. Since a spatial differentiation of an exponent yields a similar exponent, the electric field and current density have the same spatial dependence as the magnetic field, i.e. penetrate inside the conductor only by distances of the order of $\delta_s(\omega)$, but their vectors are directed perpendicularly to $\mathbf{B}$, while still being parallel to the conductor surface:

$$\mathbf{j}_\omega(x) = \kappa_- H_\omega(x)\, \mathbf{n}_z, \qquad \mathbf{E}_\omega(x) = \frac{\kappa_-}{\sigma} H_\omega(x)\, \mathbf{n}_z. \tag{6.34}$$

We may use these expressions to calculate the time-averaged power density (4.39) of the energy dissipation, for the important case of a sinusoidal ('monochromatic')

[19] This common name is awkward, because Eq. (6.31) results from the macroscopic Maxwell equations (6.21), but is partly justified as the counterpart to the 'macroscopic' (better called 'coarse-grain') boundary conditions (6.38) and (6.59)—to be discussed in a minute.

[20] Let me hope that the physical intuition of the reader makes it evident that the ac field penetrates into a sample of any *shape* only by a distance of the order of δ_s.

[21] The loop (vortex) character of the induced current lines, responsible for the term 'eddy', is not very apparent in the 1D geometry explored above, with the near-surface currents (figure 6.2b) looping only implicitly, at $z \to \pm\infty$.

field $H(x, t) = |H_\omega(x)| \cos(\omega t + \varphi)$, and hence sinusoidal eddy currents: $j(x, t) = |j_\omega(x)| \cos(\omega t + \varphi')$:

$$\begin{aligned}\overline{p}(x) = \frac{\overline{j^2(x, t)}}{\sigma} = \frac{|j_\omega(x)|^2\, \overline{\cos^2(\omega t + \varphi')}}{\sigma} = \frac{|j_\omega(x)|^2}{2\sigma} \\ = \frac{|\kappa_-|^2 |H_\omega(x)|^2}{2\sigma} = \frac{|H_\omega(x)|^2}{\delta_s^2 \sigma} \equiv \frac{H_\omega(x) H_\omega^*(x)}{\delta_s^2 \sigma}.\end{aligned} \tag{6.35}$$

Now the (elementary) integration of this expression along the normal to the surface, i.e. along axis x (through all the skin depth), using the exponential law (6.32), gives us the following average power of energy loss per unit area:

$$\frac{d\overline{\mathscr{P}}}{dA} \equiv \int_0^\infty \overline{p}(x)dx = \frac{1}{2\delta_s \sigma} |H_\omega^{(0)}|^2 \equiv \frac{\mu\omega\delta_s}{4} |H_\omega^{(0)}|^2. \tag{6.36}$$

We will extensively use this expression in the next chapter to calculate the energy losses in waveguides and resonators with conducting (practically, metallic) walls, but for now just note that according to Eqs. (6.33) and (6.36) at a fixed applied field the losses grow with frequency as $\omega^{1/2}$.

One more important remark concerning Eq. (6.34): integrating the first of them over x, with the help of Eq. (6.32), we may readily prove that the *linear density* $\mathbf{J}$ of the surface currents (measured in A m^{-1}) is simply and fundamentally related to the applied magnetic field:

$$\mathbf{J}_\omega \equiv \int_0^\infty \mathbf{j}_\omega(x)dx = H_\omega^{(0)} \mathbf{n}_z. \tag{6.37}$$

Since this relation does not have any frequency-dependent factors, we may sum it up for all frequency components, and obtain a universal relation

$$\mathbf{J}(t) = H^{(0)}(t)\mathbf{n}_z \equiv H^{(0)}(t)(-\mathbf{n}_y \times \mathbf{n}_x) = \mathbf{H}^{(0)}(t) \times (-\mathbf{n}_x) = \mathbf{H}^{(0)}(t) \times \mathbf{n}, \tag{6.38}$$

where $\mathbf{n} = -\mathbf{n}_x$ is the outer normal to the surface—see figure 6.2b. This simple relation (whose last form is independent of the choice of coordinate axes) is independent of the used constituent relations (6.22) and is by no means occasional. Indeed, it may be readily obtained from the macroscopic Ampère law Eq. (5.116) applied to a contour drawn around a fragment of the surface, extending under it much deeper than the skin depth—see the contour C_2 in figure 6.2b—regardless of the exact law of the field penetration. The 'coarse-grain' relation (6.38) is sometimes called the 'macroscopic' boundary condition for the magnetic field near a conductor's surface, to distinguish it from the 'microscopic' boundary condition (6.31).

For the skin effect, this fundamental relation between the linear current density and the external magnetic field implies that the skin effect's implementation does not necessarily require a dedicated ac magnetic field source. For example, the effect takes place in any wire that carries an ac current, leading to a current concentration in a surface sheet of thickness $\sim\delta_s$. (Of course, the quantitative analysis of this problem in a wire with an arbitrary cross-section may be technically complicated, because it requires solving Eq. (6.23) for a 2D geometry; even for a round cross-section, the solution involves the Bessel functions.) In this case, the ac magnetic field outside the conductor, which still obeys Eq. (6.38), is better interpreted as an effect of, rather than the reason for, the ac current flow.

Finally, the reader should mind the validity limits of all these results—apart from the universal Eq. (6.38). First, in order for the quasi-static approximation to be valid, the field frequency ω should not be too high, so that the displacement current effects are negligible. (Again, this condition will be quantified in section 6.7; it will show that for usual metals, the condition is violated only at extremely high frequencies above $\sim 10^{18}$ s^{-1}.) A more practical upper limit on ω is that the skin depth δ_s should stay much larger than the mean free path l of charge carriers[22]. Beyond this point, the relation between the vectors $\mathbf{j}(\mathbf{r})$ and $\mathbf{E}(\mathbf{r})$ becomes essentially *non-local*. Both theory and experiment show that at $\delta_s < l$ the skin effect still persists, but acquires a frequency dependence slightly different from Eq. (6.33), $\delta_s \propto \omega^{-1/3}$ rather than $\omega^{-1/2}$. This so-called *anomalous skin effect* has useful applications, for example, for experimental measurements of the Fermi surface of metals[23].

6.4 Electrodynamics of superconductivity and gauge invariance

The effect of superconductivity[24] takes place (in certain materials only, mostly metals) when temperature T is reduced below a certain *critical temperature* T_c, specific for each material. For most metallic superconductors, T_c is typically of the order of a few kelvins, although several compounds (the so-called *high-temperature superconductors*) with T_c above 100 K have been found since 1987. The most notable property of superconductors is the absence at $T < T_c$ of measurable resistance to (not very high) dc currents. However, the electromagnetic properties of superconductors cannot be described by just taking $\sigma = \infty$ in our previous results. Indeed, for this case, Eq. (6.33) would give $\delta_s = 0$, i.e. no ac magnetic field penetration at all, while for the dc field we would have the uncertainty $\sigma\omega \rightarrow$? Experiments show something substantially different: weak magnetic fields do penetrate into superconductors by a material-specific distance $\delta_L \sim 10^{-7}$–10^{-6} m, the so-called *London penetration depth*[25], which is virtually frequency-independent until the skin depth δ_s—measured in the same material in its 'normal' state, i.e. the absence of superconductivity—becomes less than δ_L. (This crossover happens typically at frequencies $\sim 10^{13}$–10^{14} s^{-1}.) The smallness of δ_L on the human scale means that the magnetic field is pushed out of macroscopic samples at their transition into the superconducting state.

This *Meissner–Ochsenfeld effect*, discovered experimentally in 1933[26], may be partly understood using the following classical reasoning. The discussion of the Ohm

[22] A brief discussion of the mean free path may be found, for example, in *Part SM* chapter 6. In very clean metals at very low temperatures, δ_s may approach l at frequencies as low as ~1 GHz, but at room temperature the crossover from the normal to the anomalous skin affect takes place only at ~100 GHz.

[23] See, e.g. [1].

[24] Discovered experimentally in 1911 by H Kamerlingh Onnes.

[25] Named to acknowledge the pioneering theoretical work of brothers F and H London—see below.

[26] It is hardly fair to shorten the name to just the 'Meissner effect', as is frequently done, because of the reportedly crucial contribution by R Ochsenfeld, then W Meissner's student, to the discovery.

law in section 4.2 implied that the current's (and hence the electric field's) frequency ω is either zero or sufficiently low. In the classical Drude reasoning, this is acceptable while $\omega\tau \ll 1$, where τ is the effective carrier scattering time participating in Eqs. (4.12) and (4.13). If this condition is not satisfied, we should take into account the charge carrier inertia; moreover, in the opposite limit $\omega\tau \gg 1$ we may neglect the scattering at all. Classically, we can describe the charge carriers in such a 'perfect conductor' as particles with a non-zero mass, which are accelerated by the electric field in accordance with Newton's second law (4.11),

$$m\dot{\mathbf{v}} = \mathbf{F} = q\mathbf{E}, \tag{6.39}$$

so that the current density $\mathbf{j} = qn\mathbf{v}$ they create changes in time as

$$\dot{\mathbf{j}} = \frac{q^2 n}{m}\mathbf{E}. \tag{6.40}$$

In terms of the Fourier amplitudes of the functions $\mathbf{j}(t)$ and $\mathbf{E}(t)$, this means

$$-i\omega\mathbf{j}_\omega = \frac{q^2 n}{m}\mathbf{E}_\omega. \tag{6.41}$$

Comparing this formula with the relation $\mathbf{j}_\omega = \sigma\mathbf{E}_\omega$ implied in the last section, we see that we can use all its results with the following replacement:

$$\sigma \to i\frac{q^2 n}{m\omega}. \tag{6.42}$$

This change replaces the characteristic equation (6.29) with

$$-i\omega = \frac{\kappa^2 m\omega}{iq^2 n\mu}, \qquad \text{i.e. } \kappa^2 = \frac{\mu q^2 n}{m}, \tag{6.43}$$

i.e. replaces the skin effect with the field penetration by the following frequency-independent depth:

$$\delta \equiv \frac{1}{\kappa} = \left(\frac{m}{\mu q^2 n}\right)^{1/2}. \tag{6.44}$$

Superficially, this means that the field decay into the superconductor does not depend on frequency:

$$H(x, t) = H(0, t)e^{-x/\delta}, \tag{6.45}$$

thus explaining the Meissner–Ochsenfeld effect.

However, there are two problems with this result. First, for parameters typical for good metals ($q = -e$, $n \sim 10^{29}\ \text{m}^{-3}$, $m \sim m_e$, $\mu \approx \mu_0$), Eq. (6.44) gives $\delta \sim 10^{-8}$ m, a factor of $\sim$10 to $\sim 10^2$ lower than the typical experimental values of δ_L. Experiments also show that the penetration depth diverges at $T \to T_c$, which is not predicted by Eq. (6.44). Another, much more fundamental problem with Eq. (6.44) is that it has been derived for $\omega\tau \gg 1$. Even if we assume that somehow there are no collisions at all, i.e. $\tau = \infty$, at $\omega \to 0$ both parts of the characteristic equation (6.43) vanish, and we cannot make any conclusion about κ. This is not just a mathematical artifact we can ignore. For example, let us place a non-magnetic metal into a static external magnetic field at $T > T_c$. The field would completely penetrate into the sample. Now let us cool it. As soon as the temperature is decreased below T_c, the above calculations would become valid, forbidding penetration into the superconductor of any *change* of the field, so that the initial field would be 'frozen' inside the sample. Experiments shows something completely different: as T is lowered below T_c, the initial field is pushed out of the sample.

The resolution of these contradictions has been provided by quantum mechanics. As was explained in 1957 in a seminal work by J Bardeen, L Cooper, and J Schrieffer (commonly referred to as the *BSC theory*), superconductivity is due to the correlated motion of electron pairs, with opposite spins and nearly opposite momenta. Such *Cooper pairs*, each with the electric charge $q = -2e$ and zero spin, may form only in a narrow energy layer near the Fermi surface, of certain thickness $\Delta(T)$. Parameter $\Delta(T)$, which may also be considered as the binding energy of the pair, tends to zero at $T \to T_c$, while at $T \ll T_c$ it has a virtually constant value $\Delta(0) \approx 3.5\, k_B T_c$, of the order of a few meV for most superconductors. This fact readily explains the relatively low spatial density of the Cooper pairs: $n_p \sim n\Delta(T)/\varepsilon_F \sim 10^{26}\ \text{m}^{-3}$. With the correction $n \to n_p$, our Eq. (6.38) for the penetration depth becomes

$$\delta \to \delta_L = \left(\frac{m}{\mu q^2 n_p}\right)^{1/2}. \tag{6.46}$$

This expression diverges at $T \to T_c$ and generally fits the experimental data reasonably well, at least for so-called 'clean' superconductors (with the mean free path $l = v_F \tau$, where $v_F \sim (2m\varepsilon_F)^{1/2}$ is the r.m.s. velocity of electrons on the Fermi surface, much longer that the Cooper pair size ξ—see below).

The smallness of the coupling energy $\Delta(T)$ is also a key factor in the explanation of the Meissner–Ochsenfeld effect, as well as several *macroscopic quantum phenomena* in superconductors. Because of Heisenberg's quantum uncertainty relation $\delta r \delta p \sim \hbar$, the spatial extension of the Cooper pair's wavefunction (the so-called *coherence length* of the superconductor) is relatively large: $\xi \sim \delta r \sim \hbar/\delta p \sim \hbar v_F/\Delta(T) \sim 10^{-6}$ m. As a result, $n_p \xi^3 \gg 1$, meaning that the wavefunctions of the pairs are strongly overlapped in space. Now, due to their integer spin, Cooper pairs behave like bosons, which means in particular that at low temperature they exhibit the so-called *Bose–Einstein condensation* onto the same ground energy level ε_g.[27] This means that the frequency $\omega = \varepsilon_g/\hbar$ of the time evolution of each pair's wavefunction

[27] A qualitative discussion of the Bose–Einstein condensation of bosons may be found in *Part SM* section 3.4, although the full theory of superconductivity is more complex, because it describes the condensation taking place *simultaneously* with the formation of effective bosons (Cooper pairs) from fermions (single electrons). For a more detailed, but very readable coverage of physics of superconductors, I would refer the reader to the monograph [2].

$\Psi = \psi \exp\{-i\omega t\}$ is the same, i.e. that the phases φ of the wavefunctions, defined by the relation

$$\psi = |\psi| \ e^{i\varphi}, \tag{6.47}$$

become equal, so that the electric current is carried not by individual Cooper pairs but rather their *Bose–Einstein condensate* described by a single wavefunction. Due to this coherence, the quantum effects (which are, in the usual Fermi-liquids of single electrons, masked by the statistical spread of their energies and hence of their phases) become very explicit—'macroscopic'.

To illustrate this, let us write the well-known quantum-mechanical formula for the probability current density of a free, non-relativistic particle[28],

$$\mathbf{j}_w = \frac{i\hbar}{2m}(\psi\nabla\psi^* - \text{c.c.}) \equiv \frac{1}{2m}[\psi^*(-i\hbar\nabla)\psi - \text{c.c.}], \tag{6.48}$$

where c.c. means the complex conjugate of the previous expression. Now let me borrow one result that will be proved later in this course (in section 9.7) when we discuss the analytical mechanics of a charged particle moving in an electromagnetic field. Namely, in order to account for the magnetic field effects, the particle's *kinetic momentum* $\mathbf{p} \equiv m\mathbf{v}$ (where $\mathbf{v} \equiv d\mathbf{r}/dt$ is particle's velocity) has to be distinguished from its *canonical momentum*[29],

$$\mathbf{P} \equiv \mathbf{p} + q\mathbf{A}. \tag{6.49}$$

where $\mathbf{A}$ is the vector-potential of the field, defined by Eq. (5.27). In contrast with the Cartesian components $p_j = mv_j$ of the momentum $\mathbf{p}$, the canonical momentum's components are the generalized momenta corresponding to the Cartesian components r_j of the radius-vector $\mathbf{r}$, considered as generalized coordinates of the particle: $P_j = \partial\mathscr{L}/\partial v_j$, where $\mathscr{L}$ is the particle's Lagrangian function. According to the general rules of transfer from classical to quantum mechanics[30], it is the vector $\mathbf{P}$ whose operator (in the Schrödinger picture) equals $-i\hbar\nabla$, so that the operator of the kinetic momentum $\mathbf{p} = \mathbf{P} - q\mathbf{A}$ is equal to $-i\hbar\nabla - q\mathbf{A}$. Hence, the in order to account for the magnetic field[31] effects, we should make the following replacement,

$$-i\hbar\nabla \rightarrow -i\hbar\nabla - q\mathbf{A}, \tag{6.50}$$

in all field-free quantum-mechanical relations. In particular, Eq. (6.48) has to be generalized as

---

[28] See, e.g. *Part QM* section 1.4, in particular Eq. (1.47).

[29] I am sorry to use traditional notations $\mathbf{p}$ and $\mathbf{P}$ for the momenta—the same symbols which were used for the electric dipole moment and polarization in chapter 3. I hope there will be no confusion, because the latter notions are not used in this section.

[30] See, e.g. *Part CM* section 10.1, in particular Eq. (10.26).

[31] The account of the electric field is easier, because the related energy $q\phi$ of the particle may be directly included into the potential energy operator—not participating in our current discussion.

$$\mathbf{j}_w = \frac{1}{2m}[\psi^*(-i\hbar\nabla - q\mathbf{A})\psi - \text{c.c.}]. \tag{6.51}$$

This expression becomes more transparent if we take the wavefunction in the form (6.47):

$$\mathbf{j}_w = \frac{\hbar}{m}|\psi|^2\left(\nabla\varphi - \frac{q}{\hbar}\mathbf{A}\right). \tag{6.52}$$

This relation means, in particular, that in order to keep $\mathbf{j}_w$ gauge-invariant, the transformation (6.8)–(6.9) has to be accompanied by a simultaneous transformation of the wavefunction's phase:

$$\varphi \to \varphi + \frac{q}{\hbar}\chi. \tag{6.53}$$

It is fascinating that the quantum-mechanical wavefunction (or more exactly, its phase) is *not* gauge-invariant, meaning that you may change it in your mind—at will! Again, this does not change any observable (such as $\mathbf{j}_w$ or the probability density $\psi\psi^*$), i.e. any experimental results.

Now for the *electric* current density of the whole superconducting condensate, Eq. (6.52) yields the following constitutive relation:

$$\mathbf{j} \equiv \mathbf{j}_w q n_p = \frac{\hbar q n_p}{m}|\psi|^2\left(\nabla\varphi - \frac{q}{\hbar}\mathbf{A}\right), \tag{6.54}$$

This equation shows that this *supercurrent* may be induced by the dc magnetic field alone and does not require any electric field. Indeed, for the simplest, 1D geometry shown in figure 6.2a, $\mathbf{j}(\mathbf{r}) = j(x)\mathbf{n}_z$, $\mathbf{A}(\mathbf{r}) = A(x)\,\mathbf{n}_z$, and $\partial/\partial z = 0$, so that the Coulomb gauge condition Eq. (5.48) is satisfied for any choice of the gauge function $\chi(x)$ and, for the sake of simplicity, we can choose it to provide $\varphi(\mathbf{r}) \equiv \text{const}$[32], so that

$$\mathbf{j} = -\frac{q^2 n_p}{m}\mathbf{A} \equiv -\frac{1}{\mu\delta_L^2}\mathbf{A}. \tag{6.55}$$

where δ_L is given by Eq. (6.46), and the field is assumed to be small and hence not affecting the probability $|\psi|^2$ (normalized to 1 in the absence of the field). This is the so-called *London equation*, proposed (in a different form) by brothers F and H London in 1935 to explain the Meissner–Ochsenfeld effect. Combining it with Eq. (5.44), generalized for a linear magnetic medium by the replacement $\mu_0 \to \mu$, we obtain

$$\nabla^2\mathbf{A} = \frac{1}{\delta_L^2}\mathbf{A}, \tag{6.56}$$

[32] This is the so-called *London gauge*; for our simple geometry it is also the Coulomb gauge (5.48).

This simple differential equation, similar to Eq. (6.23), for our simple geometry has an exponential solution similar to Eq. (6.32):

$$A(x) = A(0)\exp\left\{-\frac{x}{\delta_L}\right\}, \quad B(x) = B(0)\exp\left\{-\frac{x}{\delta_L}\right\},$$
$$j(x) = j(0)\exp\left\{-\frac{x}{\delta_L}\right\}, \tag{6.57}$$

which shows that the magnetic field and supercurrent penetrate into a superconductor only by the London penetration depth δ_L, regardless of frequency[33]. By the way, integrating the last result through the penetration layer and using the vector-potential's definition $\mathbf{B} = \nabla \times \mathbf{A}$ (for our geometry, giving $B(x) = dA(x)/dx = -\delta_L A(x)$) we may readily check that the linear density $\mathbf{J}$ of the surface supercurrent still satisfies the universal coarse-grain relation (6.38).

This universality should bring to our attention the following common feature of the skin effect (in 'normal' conductors) and the Meissner–Ochsenfeld effect (in superconductors): if the linear size of a bulk sample is much larger than, respectively, δ_s or δ_L, then $\mathbf{B} = 0$ in the dominating part of its interior. According to Eq. (5.110), a formal description of such conductors (valid only on a coarse scale much larger than either δ_s or δ_L) may be achieved by formally treating the sample a as an *ideal diamagnet*, with $\mu = 0$. In particular, we can use this description and Eq. (5.124) to immediately obtain the magnetic field's distribution outside a bulk sphere:

$$\mathbf{B} = \mu_0\mathbf{H} = -\mu_0\nabla\phi_m, \qquad \text{with } \phi_m = H_0\left(-r - \frac{R^3}{2r^2}\right)\cos\theta, \qquad \text{for } r \geqslant R. \tag{6.58}$$

Figure 6.3 shows the corresponding surfaces of equal potential ϕ_m. It is evident that the magnetic field lines (which are normal to the equipotential surfaces) bend to become parallel to the surface near it. This pattern also helps to answer the question that might arise when making the assumption (6.24): what happens to bulk conductors placed into in a *normal* ac magnetic field (and to superconductors in a dc magnetic field)? The answer is: the field is deformed outside the conductor to sustain the 'coarse-grain' boundary condition[34]

$$B_n|_{\text{surface}} = 0, \tag{6.59}$$

which follows from Eq. (5.118) and the requirement $\mathbf{B}|_{\text{inside}} = 0$.

---

[33] Since not all electrons in a superconductor form Cooper pairs, at any frequency $\omega \neq 0$ the unpaired electrons provide energy-dissipating Ohmic currents, which are not described by Eq. (6.54). These losses become very substantial when frequency ω becomes so high that the skin-effect length δ_s of the material (as measured with superconductivity suppressed, say by a high magnetic field) becomes less than δ_L. For typical metallic superconductors this crossover takes place at frequencies of a few hundred GHz, so that even for microwaves Eq. (6.57) still gives a fairly good description of the field penetration.

[34] Sometimes this boundary condition, as well as the (compatible) Eq. (6.38), is called 'macroscopic'. However, this term may lead to confusion with the boundary conditions (5.117) and (5.118), which also ignore the atomic-scale microstructure of the 'effective currents' $\mathbf{j}_{ef} = \nabla \times \mathbf{M}$, but (as was shown in this section) still allow an explicit, detailed account of the skin currents (6.34) or supercurrents (6.55).

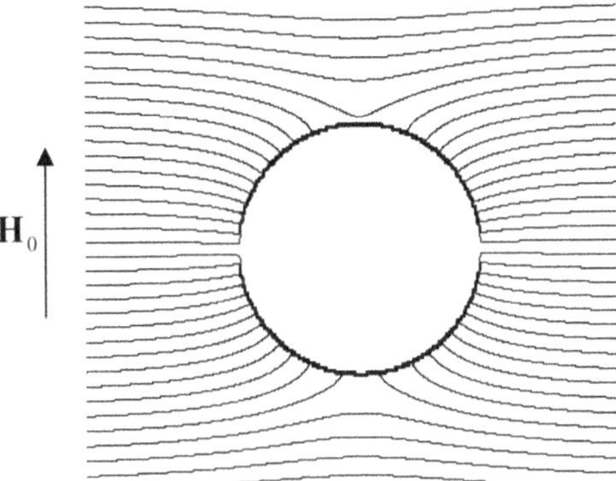

Figure 6.3. Equipotential surfaces ϕ_m = const around a (super)conducting sphere of radius $R \gg \delta_s$ (or δ_L), placed into a uniform magnetic field, within the coarse-grain model $\mu = 0$.

This answer should be taken with a grain of salt. First, for normal conductors it is only valid at sufficiently high frequencies, so that the skin depth (6.33) is sufficiently small: $\delta_s \ll a$, where a is the scale of the conductor's linear size—for a sphere, $a \sim R$. In superconductors, this simple picture is valid not only if $\delta_s \ll a$, but also only in sufficiently low magnetic fields, because strong fields *do* penetrate into superconductors, destroying superconductivity (completely or partly) and thus disrupting the Meissner–Ochsenfeld effect—see the next section.

6.5 Electrodynamics of macroscopic quantum phenomena[35]

Despite this superficial similarity of the skin effect and the Meissner–Ochsenfeld effect, the electrodynamics of superconductors is much richer. For example, let us use Eq. (6.54) to describe the fascinating effect of *magnetic flux quantization.* Consider a closed ring/loop (of any form) made of a superconducting 'wire' with a cross-section much larger than δ_L^2 (figure 6.4a).

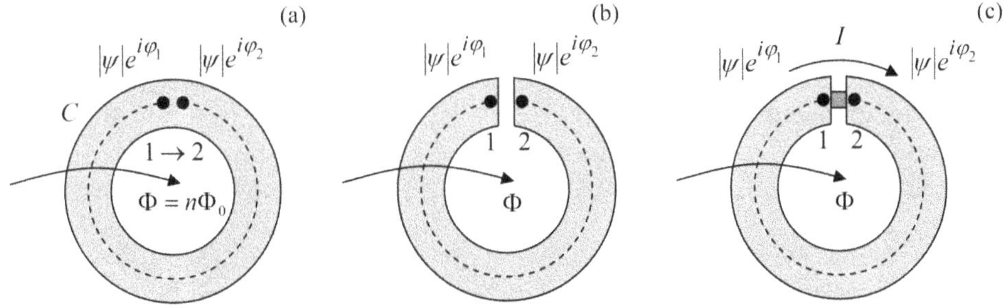

Figure 6.4. (a) A closed, flux-quantizing superconducting ring, (b) a ring with a narrow slit, and (c) a superconducting quantum interference device (SQUID).

[35] The material of this section is not covered in most E&M textbooks, and will not be used in later sections of this course. Thus the 'only' loss to the reader from skipping this section would be a lack of familiarity with one of the most fascinating fields of physics. Note also that we already have virtually all the tools necessary for its discussion, so that reading this section should not require much effort.

From the last section's discussion, we know that deep inside the wire the supercurrent is exponentially small. Integrating Eq. (6.54) along any closed contour C that does not approach the surface closer than a few δ_L at any point (see the dashed line in figure 6.4), so that $\mathbf{j} = 0$ at all its points, we obtain

$$\oint_C \nabla\varphi \cdot dr - \frac{q}{\hbar}\oint_C \mathbf{A} \cdot d\mathbf{r} = 0. \tag{6.60}$$

The first integral, i.e. the difference of φ in the initial and final points, has to be equal to either zero or an integer number of 2π, because the change $\varphi \rightarrow \varphi + 2\pi n$ does not change the Cooper pair condensate's wavefunction:

$$\psi' = |\psi|\, e^{i(\varphi+2\pi n)} = |\psi|\, e^{i\varphi} = \psi. \tag{6.61}$$

On the other hand, according to Eq. (5.65), the second integral in Eq. (6.60) is just the magnetic flux Φ through the contour[36]. As a result, we obtain a wonderful result:

$$\Phi = n\Phi_0, \quad \text{where } \Phi_0 \equiv \frac{2\pi\hbar}{|q|}, \quad \text{with} \quad n = 0, \pm1, \pm2, \ldots, \tag{6.62}$$

saying that the magnetic flux inside any superconducting loop can only take values multiple of the *flux quantum* Φ_0. This effect, predicted in 1950 by the same Fritz London (who expected q to be equal to the electron charge $-e$), was confirmed experimentally in 1961[37], but with $|q| = 2e$ (so that $\Phi_0 \approx 2.07 \times 10^{-15}$ Wb). Historically, this observation gave decisive support to the BSC theory of superconductivity based on Cooper pairs, with charge $q = -2e$, which had been put forward just 4 years earlier.

Note the truly macroscopic character of this quantum effect: it has been repeatedly observed in human-scale superconducting loops and, from what is known about superconductors, there is no doubt that if we made a giant superconducting wire loop extending, say, over the Earth's diameter, the magnetic flux through it would still be quantized—although with a very large flux quanta number n. This means that the coherence of the Bose–Einstein condensates may extend over (using H Casimir's famous expression) 'miles of dirty lead wire'. (Lead is a typical superconductor, with $T_c \approx 7.2$ K, and indeed retains its superconductivity even when highly contaminated by impurities.)

Moreover, hollow rings are not entirely necessary for flux quantization. In 1957, A Abrikosov explained the counter-intuitive high-field behavior of superconductors with $\delta_L > \xi\sqrt{2}$, known experimentally as their *mixed* (or 'Shubnikov') *phase* since the 1930s. He showed that a sufficiently high magnetic field may penetrate into such superconductors in the form of self-formed magnetic field 'threads' (or 'tubes') surrounded by vortex-shaped supercurrents—the so-called *Abrikosov vortices.* In the simplest case, the core of such a vortex is a straight line, on which

[36] Due to the Meissner–Ochsenfeld effect, the exact path of the contour is not important, and we may discuss Φ just as the magnetic flux through the ring.

[37] Independently and virtually simultaneously by two groups: B Deaver and W Fairbank, and R Doll and M Näbauer; their reports were published back-to-back in *Physical Review Letters*.

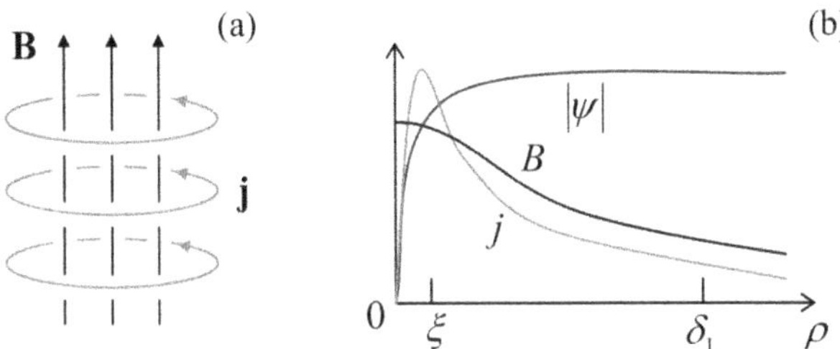

Figure 6.5. The Abrikosov vortex: (a) a 3D structure sketch and (b) the main variables as functions of the distance ρ from the axis (schematically).

the superconductivity is completely suppressed ($|\psi| = 0$), surrounded by circular, axially symmetric, persistent supercurrents $\mathbf{j}(\rho)$, where ρ is the distance from the vortex axis—see figure 6.5a. At the axis the current vanishes, and with distance it first rises and then falls, so that $\mathbf{j}(\infty) = 0$, reaching its maximum at $\rho \sim \xi$, while the magnetic field $\mathbf{B}(\rho)$, directed along the vortex axis, drops monotonically at distances $\sim\delta_L$ (figure 6.5b).

The total flux of the field equals exactly one flux quantum Φ_0, given by Eq. (6.62). Correspondingly, the wavefunction's phase φ performs just one $\pm 2\pi$ revolution along *any* contour drawn around the vortex's axis, so that $\nabla\varphi = \pm\mathbf{n}_\varphi/\rho$, where $\mathbf{n}_\varphi$ is the azimuthal unit vector[38]. This topological feature of the wavefunction's phase is sometimes called the *fluxoid quantization*—in order to distinguish it from the flux quantization, which is valid only for relatively large contours not approaching the axis by distances $\sim\delta_L$.

A quantitative analysis of the Abrikosov vortex requires, in addition to the equations we have discussed, one more constituent equation that would describe the changes of the number of Cooper pairs (quantified by $|\psi|^2$) by the magnetic field—or rather by the field-induced supercurrent. In his original work, Abrikosov used for this purpose the famous *Ginzburg–Landau* equation[39], which is qualitatively valid only at $T \approx T_c$. The equation may be conveniently represented using either of the following forms,

$$\begin{aligned} &\frac{1}{2m}(-i\hbar\nabla - q\mathbf{A})^2\psi = a\psi - b\psi\,|\psi|^2, \\ &\xi^2\psi^*\left(\nabla - i\frac{q}{\hbar}\mathbf{A}\right)^2\psi = (1 - |\psi|^2)\;|\psi|^2, \end{aligned} \qquad (6.63)$$

where a and b are certain temperature-dependent coefficients, with $a \to 0$ at $T \to T_c$. The first of these forms clearly shows that the Ginzburg–Landau equation (together with the similar *Gross–Pitaevskii* equation describing uncharged Bose–Einstein condensates) belongs to a broader class of *nonlinear Schrödinger equations*, differing

[38] The last (perhaps, evident) expression formally follows from Eq. (A.60) with $f = \pm\varphi$ + const.

[39] This equation was derived by V Ginzburg and L Landau from phenomenological arguments in 1950, i.e. before the advent of the 'microscopic' BSC theory, and may be used for simple analyses of a broad range of nonlinear effects in superconductors. The Ginzburg–Landau and Gross–Pitaevskii equations will be further discussed in *Part SM* section 4.3.

only by the additional nonlinear term from the usual Schrödinger equation, which is linear in ψ. The equivalent, second form of Eq. (6.63) is more convenient for applications, and shows more clearly that if the superconductor's condensate density, proportional to $|\psi|^2$, is suppressed only locally, it restores to its unperturbed value (with $|\psi|^2 = 1$) at distances of the order of the coherence length $\xi \equiv \hbar/(2ma)^{1/2}$.

This fact enables a simple quantitative analysis of the Abrikosov vortex in the limit $\xi \ll \delta_L$. Indeed, in this case (see figure 6.5) $|\psi|^2 = 1$ at most distances ($\rho \sim \delta_L$) where the field and current are distributed, so that these distributions may be readily calculated without any further involvement of Eq. (6.63), just from Eq. (6.54) with $\nabla\varphi = \pm\mathbf{n}_\varphi/\rho$, and the Maxwell equations (6.21) for the magnetic field, giving $\nabla\times\mathbf{B} = \mu\mathbf{j}$, and $\nabla\cdot\mathbf{B} = 0$. Indeed, combining these equations just as this was done at the derivation of Eq. (6.23), for the only Cartesian component of the vector $\mathbf{B}(\mathbf{r}) = B(\rho)\mathbf{n}_z$ (where the axis z is directed along the vortex' axis), we obtain a simple equation

$$\delta_L^2\nabla^2 B - B = -\frac{\hbar}{q}\nabla\times(\nabla\times\varphi) \equiv \mp\Phi_0\delta_2(\boldsymbol{\rho}), \qquad \text{at } \rho \gg \xi, \tag{6.64}$$

which coincides with Eq. (6.56) at all regular points $\rho \neq 0$. Spelling out the Laplace operator for our current case of axial symmetry[40], we obtain an ordinary differential equation,

$$\delta_L^2\frac{1}{\rho}\frac{d}{d\rho}\left(\rho\frac{dB}{d\rho}\right) - B = 0, \qquad \text{for } \rho \neq 0. \tag{6.65}$$

Comparing this equation with Eq. (2.155) with $\nu = 0$ and taking into account that we need the solution decreasing at $\rho \to \infty$, making any contribution from the function I_0 unacceptable, we obtain

$$B = CK_0\left(\frac{\rho}{\delta_L}\right) \tag{6.66}$$

—see the plot of this function (black line) in the right-hand panel of figure 2.22. The constant C should be calculated from fitting the 2D delta-function on the right-hand side of Eq. (6.64), i.e. by requiring

$$\int_{\text{vortex}} B(\rho)d^2\rho \equiv 2\pi\int_0^\infty B(\rho)\rho\, d\rho \equiv 2\pi\delta_L^2 C\int_0^\infty K_0(\zeta)\zeta\, d\zeta = \mp\Phi_0. \tag{6.67}$$

The last, dimensionless integral equals 1,[41] so that finally

$$B(\rho) = \frac{\Phi_0}{2\pi\delta_L^2}K_0\left(\frac{\rho}{\delta_L}\right), \qquad \text{at } \rho \gg \xi. \tag{6.68}$$

[40] See, e.g. Eq. (A.61) with $\partial/\partial\varphi = \partial/\partial z = 0$.

[41] This equality follows, for example, from the integration of both sides of Eq. (2.143) (which is valid for any Bessel functions, including K_n) with $n = 1$, from 0 to ∞, and then using the asymptotic values given by Eqs. (2.157) and (2.158): $K_1(\infty) = 0$, and $K_1(\zeta) \to 1/\zeta$ at $\zeta \to 0$.

The function K_0 (the modified Bessel function of the second kind), drops exponentially as its argument becomes larger than 1 (i.e. in our problem, at distances ρ much larger than δ_L), and diverges as its argument tends to zero—see, e.g. the second of Eqs. (2.157). However, this divergence is very slow (logarithmic) and, as was repeatedly discussed in this series, is avoided by the account of virtually any other factor. In our current case, this factor is the decrease of $|\psi|^2$ to zero at $\rho \sim \xi$ (see figure 6.5), not taken into account in Eq. (6.68). As a result, we may estimate the field on the axis of the vortex as

$$B(0) \approx \frac{\Phi_0}{2\pi\delta_L^2} \ln \frac{\delta_L}{\xi}; \tag{6.69}$$

the exact (much more involved) solution of the problem confirms this estimate with a minor correction: $\ln(\delta_L/\xi) \to [\ln(\delta_L/\xi) - 0.28]$.

The current density distribution may be now calculated using the Maxwell equation $\nabla \times \mathbf{B} = \mu \mathbf{j}$, giving $\mathbf{j} = j(\rho)\mathbf{n}_\varphi$, with[42]

$$j(\rho) = -\frac{1}{\mu}\frac{\partial B}{\partial \rho} = -\frac{\Phi_0}{2\pi\mu\delta_L^2}\frac{\partial}{\partial \rho}K_0\left(\frac{\rho}{\delta_L}\right) \equiv \frac{\Phi_0}{2\pi\mu\delta_L^3}K_1\left(\frac{\rho}{\delta_L}\right), \qquad \text{at } \rho \gg \xi, \tag{6.70}$$

where the same identity Eq. (2.158), with $J_n \to K_n$ and $n = 1$, was used. Now looking at Eqs. (2.157) and (2.158) with $n = 1$, we see that the supercurrent is exponentially small at $\rho \gg \delta_L$ (thus outlining the vortex's periphery), and is proportional to $1/\rho$ within the broad range $\xi \ll \rho \ll \delta_L$. This rise of the current at $\rho \to 0$ (which could be readily predicted directly from Eq. (6.54) with $\nabla\varphi = \pm\mathbf{n}_\varphi/\rho$, and the $\mathbf{A}$-term negligible at $\rho \ll \delta_L$) is quenched at $\rho \sim \xi$ by a rapid drop of the factor $|\psi|^2$ in the same Eq. (6.54), i.e. by the suppression of the superconductivity near the axis (by the same supercurrent!)—see figure 6.5 again.

This vortex structure may be used to calculate, in a straightforward way, its energy per unit length (i.e. its linear tension)

$$\mathscr{T} \equiv \frac{U}{l} \approx \frac{\Phi_0^2}{4\pi\mu\delta_L^2} \ln \frac{\delta_L}{\xi}, \tag{6.71}$$

and hence the 'first critical' value H_{c1} of the external magnetic field[43], at which the vortex formation becomes possible (in a long cylindrical sample parallel to the field):

$$H_{c1} = \frac{\mathscr{T}}{\Phi_0} \approx \frac{\Phi_0}{4\pi\mu\delta_L^2} \ln \frac{\delta_L}{\xi}. \tag{6.72}$$

Let me leave the proof of these two formulas for the reader's exercise.

---

[42] See, e.g. Eq. (A.63), with $f_\rho = f_\varphi = 0$, and $f_z = B(\rho)$.

[43] This term is used to distinguish H_{c1} from the higher 'second critical field' H_{c2}, at which the Abrikosov vortices are pressed to each other so tightly (to distances $d \sim \xi$) that they merge, and the remains of superconductivity vanish: $\psi \to 0$. Unfortunately, I do not have time/space to discuss these effects; the interested reader is referred to chapter 5 of [2].

The flux quantization and the Abrikosov vortices discussed above are just two of several *macroscopic quantum effects* in superconductivity. Let me discuss just one more, but perhaps the most interesting of such effects. Let us consider a superconducting ring/loop interrupted with a very narrow slit (figure 6.4b). Integrating Eq. (6.54) along any current-free path from point 1 to point 2 (see, e.g. dashed line in figure 6.4b), we obtain

$$0 = \int_1^2 \left(\nabla\varphi - \frac{q}{\hbar}\mathbf{A}\right) \cdot d\mathbf{r} = \varphi_2 - \varphi_1 - \frac{q}{\hbar}\Phi. \tag{6.73}$$

Using the flux quantum definition (6.62), this result may be rewritten as

$$\varphi \equiv \varphi_1 - \varphi_2 = \frac{2\pi}{\Phi_0}\Phi, \tag{6.74}$$

where φ is called the *Josephson phase difference*. Note that in contrast to each of the phases $\varphi_{1,2}$, their difference φ is gauge-invariant, because it is directly related to the gauge-invariant magnetic flux Φ.

Can this φ be measured? Yes, using the *Josephson effect*[44]. Let us consider two (for the argument simplicity, similar) superconductors, connected with some sort of *weak link*, for example a tunnel barrier, or a point contact, or a narrow thin-film bridge, through that a weak Cooper pair supercurrent can flow. (Such system of two weakly coupled superconductors is called a *Josephson junction.*) Let us think what this supercurrent I may be a function of. For that, reverse thinking is helpful: let us imagine we change the current; what parameter of the superconducting condensate can it affect? If the current is weak, it cannot perturb the superconducting condensate's density, proportional to $|\psi|^2$; hence it may only change the Cooper condensate phases $\varphi_{1,2}$. However, according to Eq. (6.53), the phases are not gauge-invariant, while the current should be. Hence the current may affect (or, if you like, may be a function of) only the phase difference φ defined by Eq. (6.74). Moreover, just has already been argued during the flux quantization discussion, a change of any of $\varphi_{1,2}$ (and hence of φ) by 2π or any of its multiples should not change the current. In addition, if the wavefunction is the same in both superconductors ($\varphi = 0$), the supercurrent should vanish due to the system's symmetry. Hence the function $I(\varphi)$ should satisfy conditions

$$I(0) = 0, \qquad I(\varphi + 2\pi) = I(\pi). \tag{6.75}$$

With these conditions on hand, we should not be terribly surprised by the following Josephson's result that for the weak link provided by tunneling[45],

$$I(\varphi) = I_c \sin\varphi, \tag{6.76}$$

where constant I_c, which depends on the weak link's strength and temperature, is called the *critical current*. Actually, Eqs. (6.54) and (6.63) enable not only a straightforward calculation of this relation, but even obtaining a simple expression

---

[44] It was predicted in 1961 by B Josephson (then a PhD student!), and observed experimentally by several groups soon after that.

[45] For some other types of weak links, the function $I(\varphi)$ may deviate from the sine form Eq. (6.76) rather considerably, still satisfying the general requirements (6.75).

of the critical current I_c via the link's normal-sate resistance—a task left for the (creative:-) reader's exercise.

Now let us see what happens if a Josephson junction is placed into the gap in a superconductor loop—see figure 6.4c. In this case, we may combine Eqs. (6.74) and (6.76), obtaining

$$I = I_c \sin 2\pi \frac{\Phi}{\Phi_0}. \quad (6.77)$$

This effect of a periodic dependence of the current on the magnetic flux is called the *macroscopic quantum interference*[46], while the system shown in figure 6.4c is a *superconducting quantum interference device*, abbreviated as SQUID (with all letters capital, please). The low value of the magnetic flux quantum Φ_0, and hence the high sensitivity of φ to external magnetic fields, allows the use of SQUIDs as ultra-sensitive magnetometers. Indeed, for a superconducting ring of area ~1 cm^2, one period of the change of the supercurrent (6.77) is produced by magnetic field change of the order of 10^{-11} T (10^{-7} Gs), while sensitive electronics allow us to measure a tiny fraction of this period—limited by thermal noise at a level of the order of a few fT. Such sensitivity allows measurements, for example, of the magnetic fields induced by the beating human heart, and even by brain activity, outside the body[47].

An important aspect of quantum interference is the so-called *Aharonov–Bohm* (AB) *effect* (which actually takes place for single quantum particles as well)[48]. Let the magnetic field lines be limited to the central part of the SQUID ring, so that no appreciable magnetic field ever touches the superconducting ring material. (This may be done experimentally with very good accuracy, for example using high-μ magnetic cores—see their discussion in section 5.6.) As predicted by Eq. (6.77), and confirmed by several careful experiments carried out in the mid-1960s[49], this restriction does not matter—the interference is observed anyway. This means that not only the magnetic field **B**, but also the vector-potential **A** represents physical reality, albeit in a quite peculiar way—remember the gauge transformation (5.46), which you may carry out off the top of your head, without changing any physical reality? (Fortunately, this transformation does not change the contour integral participating in Eq. (5.65), and hence the magnetic flux Φ, and hence the interference pattern.)

Actually, the magnetic flux quantization (6.62) and the macroscopic quantum interference (6.77) are not completely different effects, but just two manifestations of the inter-related macroscopic quantum phenomena. In order to show that, one

[46] The name is due to the deep analogy between this phenomenon and the interference between two coherent waves, to be discussed in detail in section 8.4.

[47] Other practical uses of SQUIDs include MRI signal detectors, high-sensitive measurements of magnetic properties of materials, and weak field detection in a broad variety of physical experiments—see, e.g. [3]. For a comparison of these devices with other sensitive magnetometers see, e.g. the review collection [4].

[48] For a more detailed discussion of the AB effect see, e.g. *Part QM* section 3.2.

[49] Similar experiments have been carried out with single (unpaired) electrons—moving either ballistically, in a vacuum, or in 'normal' (non-superconducting) conducting rings. In the last case, the effect is much harder to observe that in SQUIDs, because the ring size has to be very small and temperature very low to avoid the so-called *dephasing effects* due to unavoidable interactions of the electrons with their environment—see, e.g. *Part QM* chapter 7.

should note that if the critical current I_c (or rather its product by the loop's self-inductance L) is high enough, the flux Φ in the SQUID loop is due not only to the external magnetic field flux Φ_{ext}, but also has a self-field component—cf Eq. (5.68)[50]:

$$\Phi = \Phi_{ext} - LI, \quad \text{where } \Phi_{ext} \equiv \int_S (B_{ext})_n d^2r. \tag{6.78}$$

Now the relation between Φ and Φ_{ext} may be readily found by solving this equation together with Eq. (6.77). Figure 6.6 shows this relation for several values of the dimensionless parameter $\lambda \equiv 2\pi LI_c/\Phi_0$.

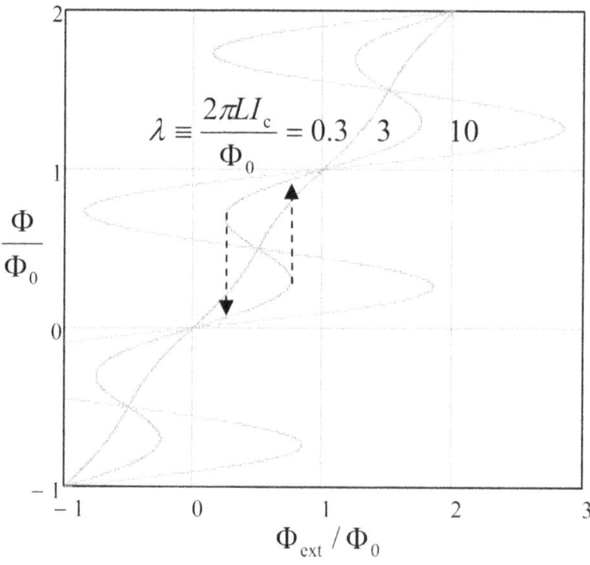

Figure 6.6. The function $\Phi(\Phi_{ext})$ for SQUIDs with various values of the normalized LI_c product. Dashed arrows show the flux leaps as the external field is changed. (The branches with $d\Phi/d\Phi_{ext} < 0$ are unstable.)

These plots show that if the critical current (or the inductance) is low, $\lambda \ll 1$, the self-field effects are negligible, and the total flux follows the external field (i.e. Φ_{ext}) faithfully. However, at $\lambda > 1$, the function $\Phi(\Phi_{ext})$ becomes hysteretic and at $\lambda \gg 1$ the stable (positive-slope) branches of this function are nearly flat, with the total flux values corresponding to Eq. (6.62). Thus, a superconducting ring closed by a high-I_c Josephson junction exhibits nearly perfect flux quantization.

The self-field effects described by Eq. (6.78) create certain technical problems for SQUID magnetometry, but they are the basis for one more application of these devices: ultrafast computing. Indeed, figure 6.6 shows that at the values of λ modestly above 1 (e.g. $\lambda \approx 3$), and within a certain range of applied field, the SQUID has two stable flux states that differ by $\Delta\Phi \approx \Phi_0$ and may be used for coding binary 0 and 1. For practical superconductors (such as Nb), the time of switching

[50] The sign before LI would be positive, as in Eq. (5.70), if I was the current flowing *into* the inductance. However, in order to keep the sign in Eq. (6.76) intact, I should mean the current flowing into the Josephson junction, i.e. *from* the inductance, thus changing the sign of the LI term in Eq. (6.78).

between these states (see the dashed arrows in figure 6.4) are of the order of a picosecond, while the energy dissipated at such an event may be as low as $\sim 10^{-19}$ J. (This bound is determined not by the device's physics, by the fundamental requirement for the energy barrier between the two states to be much higher than the thermal fluctuation energy scale $k_B T$, ensuring a sufficiently long information retention time.) While the picosecond switching speed may also be achieved with some semiconductor devices, the power consumption of the SQUID-based digital devices may be 5–6 orders of magnitude lower, enabling VLSI integrated circuits with 100 GHz scale clock frequencies. Unfortunately, the range of practical application of these *rapid single-flux-quantum* (RSFQ) digital circuits is still very narrow, due to the inconvenience of their deep refrigeration to temperatures below T_c.[51]

Since we have already obtained the basic relations (6.74) and (6.76) describing macroscopic quantum phenomena in superconductivity, let me mention in brief two other members of this group, called the *dc* and *ac Josephson effects*. Differentiating Eq. (6.74) over time, and using the Faraday induction law (6.2), we obtain[52]

$$\frac{d\varphi}{dt} = \frac{2e}{\hbar}V. \tag{6.79}$$

This famous *Josephson phase-to-voltage relation* should be valid regardless of the way how the voltage V has been created[53], so let us apply Eqs. (6.76) and (6.79) to the simplest circuit with a non-superconducting source of dc voltage—see figure 6.7.

If the current's magnitude is below the critical value, Eq. (6.76) allows the phase φ to have a time-independent value

$$\varphi = \sin^{-1}\frac{I}{I_c}, \qquad \text{if } -I_c < I < +I_c, \tag{6.80}$$

and hence, according to Eq. (6.79), a vanishing voltage drop across the junction: $V = 0$. This *dc Josephson effect* is not quite surprising—indeed, we have postulated from the very beginning that the Josephson junction may pass a certain

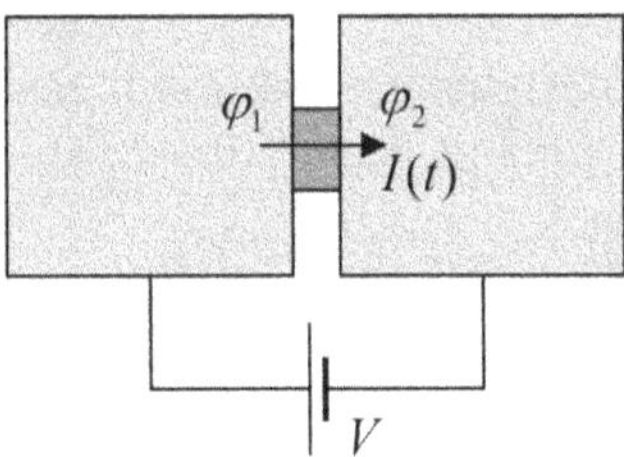

Figure 6.7. A dc-voltage-biased Josephson junction.

[51] For more on that technology see, e.g. the review paper [5] and references therein.

[52] Since the induced e.m.f. $\mathscr{V}_{\text{ind}}$ cannot drop on the superconducting path between the Josephson junction electrodes 1 and 2 (see figure 6.4c), it should be equal to $(-V)$, where V is the voltage across the junction.

[53] Indeed, it may also be obtained from simple Schrödinger equation arguments—see, e.g. *Part QM* section 1.6.

supercurrent. Much more fascinating is the so-called *ac Josephson effect*, which occurs if the voltage across the junction has a non-vanishing average (dc) component V_0. For simplicity, let us assume that this is the *only* voltage component: $V(t) = V_0 =$ const[54], then Eq. (6.79) may be easily integrated to give $\varphi = \omega_J t + \varphi_0$, where

$$\omega_J \equiv \frac{2e}{\hbar} V_0. \tag{6.81}$$

This result, plugged into Eq. (6.76), shows that the supercurrent oscillates,

$$I = I_c \sin(\omega_J t + \varphi_0), \tag{6.82}$$

with the so-called *Josephson frequency* ω_J (6.81), proportional to the applied dc voltage. For practicable voltages (above the typical noise level), the cyclic frequency $f_J = \omega_J/2\pi$ corresponds to the GHz or even THz ranges, because the proportionality coefficient in Eq. (6.81) is very high: $f_J/V_0 = e/\pi\hbar \approx 483$ MHz μV^{-1}.[55]

An important experimental fact is the universality of this coefficient. For example, in the mid-1980s, a Stony Brook group led by J Lukens proved that this factor is material-independent with a relative accuracy of at least 10^{-15}. Very few experiments, particularly in solid state physics, have ever reached such precision. This fundamental nature of the Josephson voltage-to-frequency relation (6.81) allows an important application of the ac Josephson effect in metrology. Namely, phase locking[56] the Josephson oscillations with an external microwave signal from atomic frequency standards, one can obtain a more precise dc voltage than from any other source. In NIST and other metrological institutions around the globe, this effect is used for the calibration of simpler 'secondary' voltage standards that can operate at room temperature.

6.6 Inductors, transformers, and ac Kirchhoff laws

Let a *wire coil* (meaning either a single loop as illustrated in figure 5.4b, or a series of such loops, such as one of the solenoids shown in figure 5.6) have a self-inductance L much larger than that of the wires connecting it to other components of our system: ac voltage sources, voltmeters, etc. (Since, according to Eq. (5.75), L scales as the number N of wire turns squared, this is condition is easier to satisfy at $N \gg 1$.) Then in a quasi-static system consisting of such *lumped induction coils* and external wires (and other circuit elements such as resistors, capacitances, etc), we may neglect the electromagnetic induction effects everywhere outside the coil, so that the electric field in those external regions is potential. Then the voltage V between coil's

[54] In experiments, this condition is hard to implement, due to relatively high inductance of the current leads providing the dc voltage supply. However, these complications do not change the main conclusion of the analysis.

[55] This 1962 prediction by B Josephson was confirmed experimentally—first implicitly (by phase locking of the oscillations with an external oscillator) in 1963, and then explicitly (by the detection of microwave radiation) in 1967.

[56] For a discussion of this very important (and general) effect, see, e.g. *Part CM* section 5.4.

terminals may be defined (as in electrostatics) as the difference of values of scalar potential ϕ between the terminals, i.e. as the integral

$$V = \int \mathbf{E} \cdot d\mathbf{r} \tag{6.83}$$

between the coil terminals along any path outside the coil. This voltage has to be balanced by the induction e.m.f. (6.2) in the coil, so that if the Ohmic resistance of the coil is negligible[57], we may write

$$V = \frac{d\Phi}{dt}, \tag{6.84}$$

where Φ is the magnetic flux in the coil. If the flux is due to the current I in the same coil only (i.e. if it is magnetically uncoupled from other coils), we may use Eq. (5.70) to obtain the well-known relation

$$V = L\frac{dI}{dt}, \tag{6.85}$$

where the compliance with the Lenz sign rule is achieved by selecting the relations between the assumed voltage polarity and the current direction as shown in figure 6.8a.

If similar conditions are satisfied for two magnetically coupled coils (figure 6.8b), then, in Eq. (6.84), we need to use Eqs. (5.69) instead, obtaining

$$V_1 = L_1\frac{dI_1}{dt} + M\frac{dI_2}{dt}, \qquad V_2 = L_2\frac{dI_2}{dt} + M\frac{dI_1}{dt}. \tag{6.86}$$

Such systems of inductively coupled coils have numerous applications in electrical engineering and physical experiments. Perhaps the most important of them is the *ac transformer*, in which the coils share a common soft-ferromagnetic core with the toroidal ('doughnut') topology—see figure 6.8c[58]. As we already know from the discussion in section 5.6, such cores with $\mu \gg \mu_0$ 'try' to suck in all magnetic field

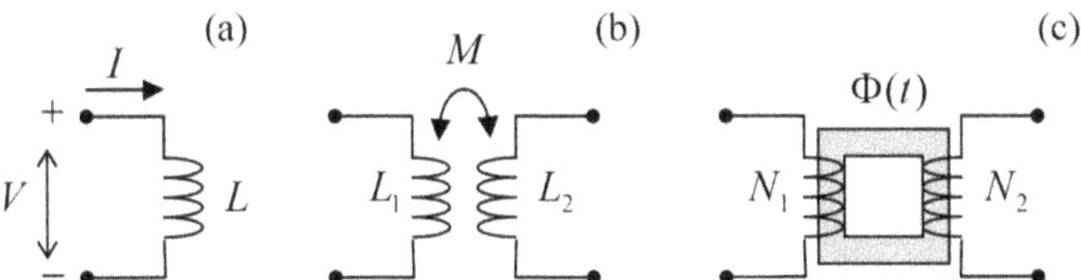

Figure 6.8. Some lumped ac circuit elements: (a) an induction coil, (b) two inductively coupled coils, and (c) an ac transformer.

---

[57] If the resistance is substantial, it may be represented by a separate lumped circuit element (resistor) connected in series with the coil.

[58] The first practically acceptable form of this device, called the *Stanley transformer*, in which multi-turn windings could be easily mounted onto a toroidal ferromagnetic (then a silicon–steel plate) core, was invented in 1886.

lines, so that the magnetic flux $\Phi(t)$ in the core is nearly the same in each of its cross-sections. With this, Eq. (6.84) yields

$$V_1 \approx N_1\frac{d\Phi}{dt}, \qquad V_2 \approx N_2\frac{d\Phi}{dt}, \tag{6.87}$$

where the voltage ratio is completely determined by the ratio N_1/N_2 of the number of wire turns.

Now we may generalize to the ac current case the Kirchhoff laws already discussed in chapter 4—see figure 4.3, reproduced in figure 6.9a. Let not only wire inductances but also the wire capacitances and resistances be negligible in comparison to those of the lumped (compact) circuit elements, whose list now would include not only resistors and current sources (as in the dc case), but also the induction coils (including magnetically coupled ones) and capacitors—see figure 6.9b. In the quasi-static limit, the current flowing in each wire is conserved, so that the 'node rule', i.e. the first Kirchhoff law (4.7*a*),

$$\sum_j I_j = 0, \tag{6.88a}$$

remains valid. Also, if the electromagnetic induction effect is restricted to the interior of lumped induction coils as discussed above, the voltage drops V_k across each circuit element may be still represented, just as in dc circuits, with differences of potentials of the adjacent nodes. As a result, the 'loop rule', i.e. the second Kirchhoff law (4.7*b*),

$$\sum_k V_k = 0, \tag{6.88b}$$

is also valid. In contrast to the dc case, Eqs. (6.88) are now the (ordinary) differential equations. However, if all circuit elements are linear (as in the examples presented in figure 6.9b), these equations may be readily reduced to linear algebraic equations using the Fourier expansion. (In the common case of sinusoidal ac sources, the final stage of the Fourier series summation is unnecessary.)

My experience suggests that the potential readers of this text are very familiar with the application of Eqs. (6.88) to such problems from their undergraduate studies, so I

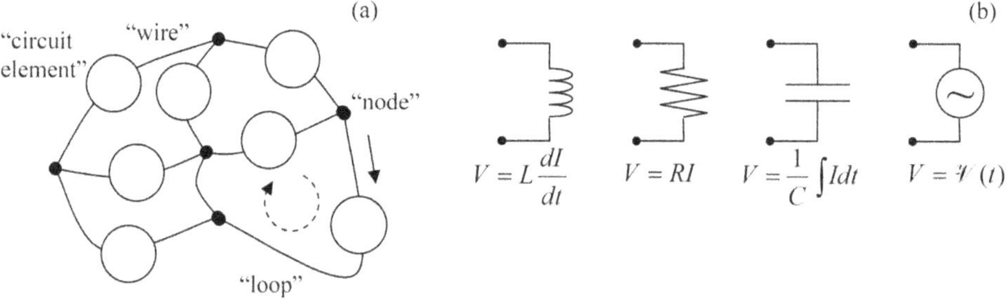

Figure 6.9. (a) A typical quasi-static ac circuit obeying the Kirchhoff laws and (b) the simplest lumped circuit elements.

would like to save time/space, skipping discussions of even the simplest examples of such circuits, such as *LC*, *LR*, *RC*, and *LRC* loops and periodic structures[59]. However, since these problems are very important in practice, my sincere advice to the reader is to carry out a self-test by solving a few problems of this type, provided at the end of this chapter, and if they cause difficulty, pursue remedial reading.

6.7 Displacement currents

Electromagnetic induction is not the only new effect arising in non-stationary electrodynamics. Indeed, although Eqs. (6.21) are adequate for the description of quasi-static phenomena, a deeper analysis shows that one of these equations, namely $\nabla \times \mathbf{H} = \mathbf{j}$, cannot be exact. To see that, let us take the divergence of both its sides:

$$\nabla \cdot (\nabla \times \mathbf{H}) = \nabla \cdot \mathbf{j}. \tag{6.89}$$

But, as for the divergence of any curl[60], the left-hand side should equal zero. Hence we obtain

$$\nabla \cdot \mathbf{j} = 0. \tag{6.90}$$

This is fine in statics, but in dynamics this equation forbids any charge accumulation, because according to the continuity relation (4.5)

$$\nabla \cdot \mathbf{j} = -\frac{\partial \rho}{\partial t}. \tag{6.91}$$

This discrepancy was recognized by James Clerk Maxwell who suggested, in the 1860s, a way out of this contradiction. If we generalize the equation for $\nabla \times \mathbf{H}$ by adding to $\mathbf{j}$ (that describes real currents) the so-called *displacement current* term,

$$\mathbf{j}_d \equiv \frac{\partial \mathbf{D}}{\partial t}, \tag{6.92}$$

(which of course vanishes in statics), then the equation takes the form

$$\nabla \times \mathbf{H} = \mathbf{j} + \mathbf{j}_d = \mathbf{j} + \frac{\partial \mathbf{D}}{\partial t}. \tag{6.93}$$

In this case, due to the equation (3.32), $\nabla \cdot \mathbf{D} = \rho$, the divergence of the right-hand side equals zero due to the continuity equation (6.92), and the discrepancy is removed. This incredible theoretical feat[61], confirmed by the 1886 experiments by Heinrich Hertz (see below) was perhaps the main triumph of the theoretical physics of the nineteenth century.

---

[59] Curiously enough, these effects include wave propagation in periodic *LC* circuits, even within the quasi-static approximation! However, the speed $1/(LC)^{1/2}$ of these waves in lumped circuits is much lower than the speed $1/(\varepsilon\mu)^{1/2}$ of electromagnetic waves in the surrounding medium—see section 6.8 below.

[60] Again, see Eq. (A.72)—if you need.

[61] It looks deceptively simple with current mathematical tools (much superior to those available to Maxwell), and after the fact.

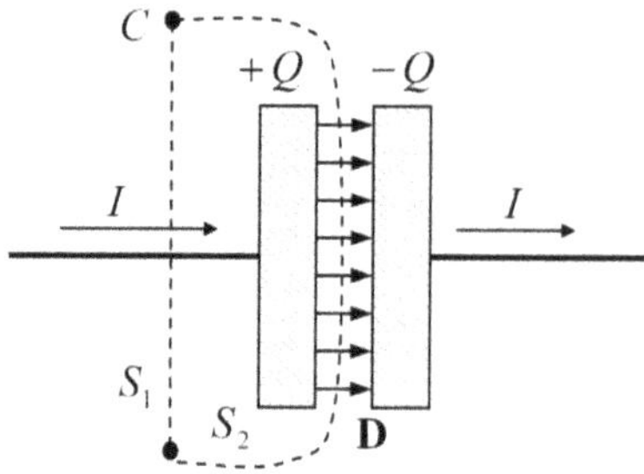

Figure 6.10. The Ampère law applied to a recharged capacitor.

The Maxwell's displacement current concept, expressed by Eq. (6.93), is so important that it is worth having one more look at its derivation using the particular model shown in figure 6.10[62]. Neglecting the fringe field effects, we may use Eq. (4.1) to describe the relation between the current I flowing through the wires and the electric charge Q of the capacitor[63]:

$$\frac{dQ}{dt} = I. \tag{6.94}$$

Now let us consider a closed contour C drawn around the wire. (The solid points in figure 6.10 show the places where the contour pierces the plane of drawing.) This contour may be seen as the line limiting either the surface S_1 (crossed by the wire) or the surface S_2 (avoiding such crossing by passing through capacitor's gap). Applying the macroscopic Ampère law (5.116) to the former surface we obtain

$$\oint_C \mathbf{H} \cdot d\mathbf{r} = \int_{S_1} j_n \, d^2r = I, \tag{6.95}$$

while for the latter surface the same law gives a different result,

$$\oint_C \mathbf{H} \cdot d\mathbf{r} = \int_{S_2} j_n \, d^2r = 0, \quad \text{(WRONG)}, \tag{6.96}$$

for the same integral. This is just an integral-form manifestation of the discrepancy outlined above, but it shows clearly how serious the problem is (or rather was—before Maxwell).

Now let us see how the introduction of the displacement currents saves the day, considering for the sake of simplicity a plane capacitor of area A with a constant electrode spacing. In this case, as we already know, the field inside it is uniform with

[62] No physicist should be ashamed of doing this. For example, J Maxwell's main book, *A Treatise of Electricity and Magnetism*, is full of drawings of plane capacitors, inductance coils, and voltmeters. More generally, the whole history of science teaches us that snobbishness toward particular examples and practical systems is a sure way toward producing nothing of either practical value or fundamental importance. In any productive science, all ways leading to novel, correct results should be welcome.

[63] This is of course just the integral form of the continuity equation (6.91).

$D = \sigma$, so that the total capacitor's charge $Q = A\sigma = AD$ and the current (6.94) may be represented as

$$I = \frac{dQ}{dt} = A\frac{dD}{dt}. \qquad (6.97)$$

So, instead of Eq. (6.96), the modified Ampère law gives

$$\oint_C \mathbf{H} \cdot d\mathbf{r} = \int_{S_2} (j_d)_n d^2r = \int_{S_2} \frac{\partial D_n}{\partial t} d^2r = \frac{dD}{dt} A = I, \qquad (6.98)$$

i.e. the Ampère integral becomes independent of the choice of the surface limited by the contour C—as it has to, because the surface exists only in our imagination.

6.8 Finally, the full Maxwell equation system

This is a very special moment in this course: with the displacement current inclusion, i.e. with the replacement of Eq. (5.107) with Eq. (6.93), we have finally arrived at the full set of macroscopic Maxwell equations for time-dependent fields[64],

$$\nabla \times \mathbf{E} + \frac{\partial \mathbf{B}}{\partial t} = 0, \qquad \nabla \times \mathbf{H} - \frac{\partial \mathbf{D}}{\partial t} = \mathbf{j}, \qquad (6.99a)$$

$$\nabla \cdot \mathbf{D} = \rho, \qquad \nabla \cdot \mathbf{B} = 0, \qquad (6.99b)$$

whose validity has been confirmed in by an enormous body of experimental data. Indeed, despite numerous efforts, no other corrections (e.g. additional terms) to the Maxwell equations have ever been found, and these equations are still considered exact within the range of their validity, i.e. while the electric and magnetic fields may be considered classically. Moreover, even in the quantum case, these equations are believed to be *strictly* valid as relations between the Heisenberg operators of the electric and magnetic field[65].

The most striking feature of these equations is that, even in the absence of stand-alone charges and currents, when all the equations become homogeneous,

$$\nabla \times \mathbf{E} = -\frac{\partial \mathbf{B}}{\partial t}, \qquad \nabla \times \mathbf{H} = \frac{\partial \mathbf{D}}{\partial t}, \qquad (6.100a)$$

$$\nabla \cdot \mathbf{D} = 0, \qquad \nabla \cdot \mathbf{B} = 0, \qquad (6.100b)$$

they still describe something very non-trivial: *electromagnetic waves*, including light. The physics of the waves may be clearly seen from Eq. (6.100*a*): according to the first

[64] This vector form of the equations, magnificent it its symmetry and simplicity, was developed in 1884–85 by O Heaviside, with substantial contributions by H Lorentz. (The original Maxwell's result looked like a system of 20 equations for Cartesian components of the vector and scalar potentials.) Note that the *microscopic* Maxwell equations for genuine ('microscopic') fields **E** and **B** may be formally obtained from Eqs. (6.99) by the substitutions $\mathbf{D} = \varepsilon_0\mathbf{E}$ and $\mathbf{H} = \mathbf{B}/\mu_0$, and the simultaneous replacement of the stand-alone charge and current densities in their right-hand sides with the full ones.

[65] See, e.g. *Part QM* chapter 9.

of them, the change of magnetic field creates a vortex-like (divergence-free) electric field. On the other hand, the second of Eqs. (6.100*a*) describes how the changing electric field, in turn, creates a vortex-like magnetic field. Thus exchanging energy, the electomagnetic field may propagate as waves.

We will carry out a detailed quantitative analysis of the waves in the next chapter; here I will only use this notion to fulfil the promise given in section 6.3, namely to establish the condition of validity of the quasi-static approximation (6.21). For simplicity, let us consider an electromagnetic wave with a time period $\mathcal{T}$, velocity v, and hence the wavelength $\lambda = v\mathcal{T}$ in a linear medium with $\mathbf{D} = \varepsilon\mathbf{E}$, $\mathbf{B} = \mu\mathbf{H}$, and $\mathbf{j} = 0$. Then the magnitude of the left-hand side of the first of Eqs. (6.100*a*) is of the order of $E/\lambda = E/v\mathcal{T}$, while that of its right-hand side may be estimated as $B/\mathcal{T} = \mu H/\mathcal{T}$. Using similar estimates for the second of Eqs. (6.100*a*), we arrive at the following two (approximate) requirements[66]:

$$\frac{E}{H} \sim \mu v \sim \frac{1}{\varepsilon v}. \tag{6.101}$$

In order to insure the compatibility of these two relations, the waves' speed should satisfy the estimate

$$v \sim \frac{1}{(\varepsilon\mu)^{1/2}}, \tag{6.102}$$

reduced to $v \sim 1/(\varepsilon_0\mu_0)^{1/2} \equiv c$ in free space, while the ratio of the electric and magnetic field amplitudes should be of the following order:

$$\frac{E}{H} \sim \mu v \sim \mu\frac{1}{(\varepsilon\mu)^{1/2}} = \left(\frac{\mu}{\varepsilon}\right)^{1/2}. \tag{6.103}$$

(In the next chapter we will see that these are indeed the *exact* results for a planar electromagnetic wave.)

Now, let a system of size $\sim a$ carry currents producing a certain magnetic field H. Then, according to Eq. (6.100*a*), their magnetic field Faraday-induces the electric field of magnitude $E \sim \mu Ha/\mathcal{T}$, whose displacement currents in turn produce an additional magnetic field with magnitude

$$H' \sim \frac{a\varepsilon}{\mathcal{T}}E \sim \frac{a\varepsilon}{\mathcal{T}}\frac{\mu a}{\mathcal{T}}H \sim \left(\frac{a\lambda}{v\mathcal{T}\lambda}\right)^2 H = \left(\frac{a}{\lambda}\right)^2 H. \tag{6.104}$$

Hence, the displacement current effects are negligible for a system of size $a \ll \lambda$.[67]

In particular, the quasi-static picture of the skin effect, which was discussed in section 6.3, is valid while the skin depth (6.33) remains much *smaller* than the corresponding wavelength,

[66] The fact that $\mathcal{T}$ cancels, shows that these estimates are valid for waves of arbitrary frequency.

[67] Let me emphasize that if this condition is *not* fulfilled, the lumped-circuit representation of the system (see figure 6.9 and its discussion) is typically inadequate—apart from some special cases, to be discussed in chapter 7.

$$\lambda = v\mathcal{T} = \frac{2\pi v}{\omega} = \left(\frac{4\pi^2}{\varepsilon\mu\omega^2}\right)^{1/2}. \tag{6.105}$$

The wavelength decreases with the frequency as $1/\omega$, i.e. faster than $\delta_s \propto 1/\omega^{1/2}$, so that they become comparable at the crossover frequency

$$\omega_r = \frac{\sigma}{\varepsilon} \equiv \frac{\sigma}{\kappa\varepsilon_0}, \tag{6.106}$$

which is nothing else than the reciprocal charge relaxation time (4.10). As was discussed in section 4.2, for good metals this frequency is extremely high (about 10^{18} s^{-1}), so the validity of Eq. (6.33) is typically limited by the anomalous skin effect (which was briefly discussed in section 6.3), rather than the wave effects.

Before going after the analysis of the full Maxwell equations in particular situations (that will be the main goal of all the next chapters of this course), let us have a look at the energy balance they yield for a certain volume V, which may include both charged particles and the electromagnetic field. Since according to Eq. (5.10) the magnetic field does no work on charged particles even if they move, the total power $\mathscr{P}$ being transferred from the field to the particles inside the volume is due to the electric field alone—see Eq. (4.38):

$$\mathscr{P} = \int_V \not{p}\, d^3r, \quad \not{p} = \mathbf{j} \cdot \mathbf{E}, \tag{6.107}$$

Expressing $\mathbf{j}$ from the corresponding Maxwell equation of the system (6.99), we obtain

$$\mathscr{P} = \int_V \left[\mathbf{E} \cdot (\nabla \times \mathbf{H}) - \mathbf{E} \cdot \frac{\partial \mathbf{D}}{\partial t}\right] d^3r. \tag{6.108}$$

Let us pause here for a second and transform the divergence of vector $\mathbf{E} \times \mathbf{H}$, using the well-known vector algebra identity[68]:

$$\nabla \cdot (\mathbf{E} \times \mathbf{H}) = \mathbf{H} \cdot (\nabla \times \mathbf{E}) - \mathbf{E} \cdot (\nabla \times \mathbf{H}). \tag{6.109}$$

The last term on the right-hand side of this equation is exactly the first term in the square brackets of Eq. (6.108), so that we can rewrite that formula as

$$\mathscr{P} = \int_V \left[-\nabla \cdot (\mathbf{E} \times \mathbf{H}) + \mathbf{H} \cdot (\nabla \times \mathbf{E}) - \mathbf{E} \cdot \frac{\partial \mathbf{D}}{\partial t}\right] d^3r. \tag{6.110}$$

However, according to the Maxwell equation for $\nabla \times \mathbf{E}$, it is equal to $-\partial\mathbf{B}/\partial t$, so that the second term in the square brackets of Eq. (6.110) equals $-\mathbf{H} \cdot \partial\mathbf{B}/\partial t$ and, according to Eq. (6.14), is just the (minus) time derivative of the magnetic energy per unit volume. Similarly, according to Eq. (3.76), the third term under the integral is the minus time derivative of the electric energy per unit volume. Finally, we can use

[68] See, e.g. Eq. (A.77) with $\mathbf{f} = \mathbf{E}$ and $\mathbf{g} = \mathbf{H}$.

the divergence theorem to transform the integral of the first term in the square brackets to a 2D integral over the surface S limiting the volume V. As a result, we obtain the so-called *Poynting theorem*[69] for the power balance in the system:

$$\int_V \left(\not{p} + \frac{\partial u}{\partial t} \right) d^3r + \oint_S S_n \, d^2r = 0. \tag{6.111}$$

Here u is the density of the total (electric plus magnetic) energy of the electromagnetic field, with

$$\delta u \equiv \mathbf{E} \cdot \delta \mathbf{D} + \mathbf{H} \cdot \delta \mathbf{B}, \tag{6.112}$$

so that for an isotropic, linear, and dispersion-free medium, with $\mathbf{D}(t) = \varepsilon \mathbf{E}(t)$, $\mathbf{B}(t) = \mu \mathbf{H}(t)$,

$$u = \frac{\mathbf{E} \cdot \mathbf{D}}{2} + \frac{\mathbf{H} \cdot \mathbf{B}}{2} \equiv \frac{\varepsilon E^2}{2} + \frac{B^2}{2\mu}, \tag{6.113}$$

and $\mathbf{S}$ is the *Poynting vector* defined as[70]

$$\mathbf{S} \equiv \mathbf{E} \times \mathbf{H}. \tag{6.114}$$

The first integral in Eq. (6.111) is evidently the net change of the energy of the system (particles + field) per unit time, so that the second (surface) integral has to be the power flowing out from the system through the surface. As a result, it is tempting to interpret the Poynting vector $\mathbf{S}$ locally, as the power flow density at the given point. In many cases such a local interpretation of the vector $\mathbf{S}$ is legitimate; however, in some cases it may lead to incorrect conclusions. Indeed, let us consider the simple system shown in figure 6.11: a charged plane capacitor placed into a static and uniform external magnetic field, so that the electric and magnetic fields are mutually perpendicular.

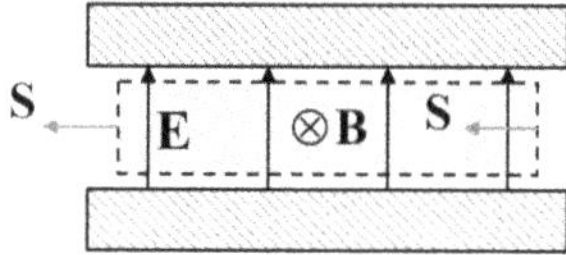

Figure 6.11. The Poynting vector paradox.

In this static situation, with no charges moving, both $\not{p}$ and $\partial/\partial t$ are equal to zero, and there should be no power flow in the system. However, Eq. (6.114) shows that the Poynting vector is not equal to zero inside the capacitor, being directed as the red arrows in figure 6.11 show. From the point of view of our only unambiguous corollary of the Maxwell equations, Eq. (6.111), there is no contradiction here,

[69] Named after J Poynting, although this fact was independently discovered by O Heaviside, while a similar expression for the intensity of mechanical elastic waves had been derived earlier by N Umov.

[70] Actually, an addition to $\mathbf{S}$ of the curl of an arbitrary vector function $\mathbf{f}(\mathbf{r}, t)$ does not change Eq. (6.111). Indeed, we may use the divergence theorem to transform the corresponding change of the surface integral in Eq. (6.111) to a volume integral of scalar function $\nabla \cdot (\nabla \times \mathbf{f})$ that equals zero at any point—see, e.g. Eq. (A.72).

because the fluxes of vector **S** through the side boundaries of the volume shaded in figure 6.11, are equal and opposite (and they are zero for other faces of this rectilinear volume), so that the total flux of the Poynting vector through the volume boundary equals zero, as it should. It is, however, useful to recall this example each time before giving a local interpretation to the vector **S**.

The paradox illustrated in figure 6.11 is closely related to the *radiation recoil effects*, due to the electromagnetic field's momentum—more exactly, it *linear momentum*. Indeed, acting as at the Poynting theorem derivation, it is straightforward to use the *microscopic* Maxwell equations[71] to prove that, neglecting the boundary effects, the vector sum of the mechanical linear momentum of the particles in an arbitrary volume and the integral of the following vector,

$$\mathbf{g} \equiv \frac{\mathbf{S}}{c^2}, \tag{6.115}$$

over the same volume is conserved, allowing us to interpret **g** as the density of the linear momentum of the electromagnetic field. (It will be more convenient for me to prove this relation and discuss the related issues in section 9.8, using the four-vector formalism of the special relativity.) Due to this conservation, if some static fields coupled to mechanical bodies are suddenly decoupled from them and are allowed to propagate in space, i.e. change their local integral of **g**, they give the bodies an opposite and equal impulse of force.

Finally, to complete our initial discussion of the Maxwell equations[72], let us rewrite them in terms of potentials **A** and ϕ, because this is more convenient for the solution of some (though not all!) problems. Even when dealing with the system (6.99) of the more general Maxwell equations than discussed before, Eqs. (6.7) are still used for the potential definitions. It is straightforward to verify that with these definitions, the two homogeneous Maxwell equations (6.99*b*) are satisfied automatically. Plugging Eq. (6.7) into the inhomogeneous equations (6.99*a*) and considering, for simplicity, a linear, uniform medium with frequency-independent ε and μ, we obtain

$$\nabla^2\phi + \frac{\partial}{\partial t}(\nabla \cdot \mathbf{A}) = -\frac{\rho}{\varepsilon}, \qquad \nabla^2\mathbf{A} - \varepsilon\mu\frac{\partial^2\mathbf{A}}{\partial t^2} - \nabla\left(\nabla \cdot \mathbf{A} + \varepsilon\mu\frac{\partial\phi}{\partial t}\right) = -\mu\mathbf{j}. \tag{6.116}$$

This is a more complex result than we would like to obtain. However, let us select a special gauge, which is frequently called (particularly for the free space case, when $v = c$) the *Lorenz gauge condition*[73]

[71] The situation with the *macroscopic* Maxwell equations is more complex, and is still the subject of some lingering discussions (usually called the *Abraham–Minkowski controversy*, despite contributions by many other scientists including A Einstein), because of the ambiguity of the momentum's division between the field and particle parts—see, e.g. the recent review paper [6].

[72] We will return to their general discussion (in particular, to the analytical mechanics of the electromagnetic field and its stress tensor) in section 9.8, after we have equipped ourselves with the theory of special relativity.

[73] This condition, named after *L Lorenz*, should not be confused with the so-called *Lorentz invariance condition* of the relativity theory, due to *H Lorentz* (note the last names' spellings), to be discussed in section 9.4.

$$\nabla \cdot \mathbf{A} + \varepsilon\mu \frac{\partial \phi}{\partial t} = 0, \tag{6.117}$$

which is a natural generalization of the Coulomb gauge (5.48) to time-dependent phenomena. With this condition, Eq. (6.107) are reduced to a simpler, beautifully symmetric form,

$$\nabla^2 \phi - \frac{1}{v^2}\frac{\partial^2 \phi}{\partial t^2} = -\frac{\rho}{\varepsilon}, \qquad \nabla^2 \mathbf{A} - \frac{1}{v^2}\frac{\partial^2 \mathbf{A}}{\partial t^2} = -\mu \mathbf{j}, \tag{6.118}$$

where $v^2 \equiv 1/\varepsilon\mu$. Note that these equations are essentially a set of four similar equations for four scalar functions (namely, ϕ and three Cartesian components of the vector $\mathbf{A}$) and thus clearly invite the four-component vector formalism of the theory of relativity; it will be discussed in chapter 9.[74]

If ϕ and $\mathbf{A}$ depend on just one spatial coordinate, say z, in a region without field sources $\rho = 0$, $\mathbf{j} = 0$, Eqs. (6.118) are reduced to the well-known 1D wave equations

$$\frac{\partial^2 \phi}{\partial^2 z} - \frac{1}{v^2}\frac{\partial^2 \phi}{\partial t^2} = 0, \qquad \frac{\partial^2 \mathbf{A}}{\partial^2 z} - \frac{1}{v^2}\frac{\partial^2 \mathbf{A}}{\partial t^2} = 0\,. \tag{6.119}$$

These equations describe waves with arbitrary waveforms, propagating with the same speed v in either of the directions of axis z. Due to the definitions of the constants ε_0 and μ_0, in the free space v is just the speed of light:

$$v = \frac{1}{(\varepsilon_0\mu_0)^{1/2}} \equiv c. \tag{6.120}$$

Historically, the experimental observation of relatively low-frequency (GHz-scale) electromagnetic waves and the proof that their speed in free space is equal to that of light was the decisive proof of Maxwell's theory. This was first accomplished in 1886 by H Hertz, using electronic circuits and antennas he had invented for this purpose.

Before proceeding to the detailed analysis of these waves in the following chapters, let me mention that the invariance of Eqs. (6.119) with respect to the wave propagation direction is not occasional; it is just a manifestation of one more general property of the Maxwell equations (6.99), called the *Lorentz reciprocity*. We have already met its simplest example, for time-independent electrostatic fields, in

[74] Here I have to mention in passing the so-called *Hertz vector potentials* $\mathbf{\Pi}_e$ and $\mathbf{\Pi}_m$ (whose introduction may be traced back to at least the 1904 work by E Whittaker). They may be defined by the following relations:

$$\mathrm{A} = \mu \frac{\partial \Pi_e}{\partial t} + \mu \nabla \cdot \Pi_m, \quad \varphi = -\frac{1}{\varepsilon}\nabla \cdot \Pi_e,$$

which make the Lorentz gauge condition (6.117) automatically satisfied. These potentials are particularly convenient for the solution of problems in which the electromagnetic field is excited by external sources characterized by externally fixed electric and magnetic polarizations $\mathbf{P}$ and $\mathbf{M}$—rather than fixed charge and current densities ρ and $\mathbf{j}$. Indeed, it is straightforward to check that both $\mathbf{\Pi}_e$ and $\mathbf{\Pi}_m$ satisfy the equations similar to Eq. (6.118), but with their right-hand sides equal to, respectively, $-\mathbf{P}$ and $-\mathbf{M}$. Unfortunately, I would not have time for a discussion of such problems.

one of the problems of chapter 1. Let us now consider a much more general case when two time-dependent electromagnetic fields, say $\{\mathbf{E}_1(\mathbf{r}, t), \mathbf{H}_1(\mathbf{r}, t)\}$ and $\{\mathbf{E}_2(\mathbf{r}, \mathrm{t}),$ $\mathbf{H}_2(\mathbf{r}, t)\}$, are induced, respectively, by spatially localized stand-alone currents $\mathbf{j}_1(\mathbf{r}, t)$ and $\mathbf{j}_2(\mathbf{r}, t)$. Then it may be proved[75] that if the medium is linear, and either isotropic or even anisotropic, but with symmetric tensors $\varepsilon_{jj'}$ and $\mu_{jj'}$, then for any volume V, limited by a closed surface S,

$$\int_V (\mathbf{j}_1 \cdot \mathbf{E}_2 - \mathbf{j}_2 \cdot \mathbf{E}_1)\, d^3r = \oint_S (\mathbf{E}_1 \times \mathbf{H}_2 - \mathbf{E}_2 \times \mathbf{H}_1)_n\, d^2r. \qquad (6.121)$$

This property implies, in particular, that the waves propagate similarly in two reciprocal directions even in situations much more general than the 1D case described by Eqs. (6.119). For some important practical applications (e.g. for low-noise amplifiers and detectors) such reciprocity is rather inconvenient. Fortunately, Eq. (6.121) may be violated in anisotropic media with asymmetric tensors $\varepsilon_{jj'}$ and/or $\mu_{jj'}$. The simplest, and most important case of such an anisotropy, the *Faraday rotation* of the wave polarization in plasma, will be discussed in the next chapter.

6.9 Problems

Problem 6.1. Prove that the electromagnetic induction e.m.f. $\mathscr{V}_{\text{ind}}$ in a conducting loop may be measured as shown on two panels of figure 6.1:

(i) by measuring the current $I = \mathscr{V}_{\text{ind}}/R$ induced in the closed loop with Ohmic resistance R, or
(ii) using a voltmeter inserted into the loop.

Problem 6.2. The flux Φ of the magnetic field that pierces a resistive ring is being changed in time, while the magnetic field outside the ring is negligibly low. A voltmeter is connected to a part of the ring, as shown in the figure below. What would the voltmeter show?

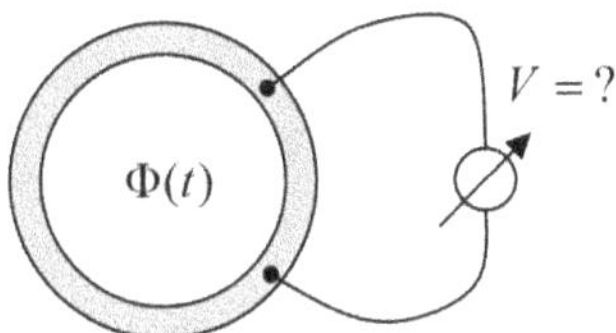

Problem 6.3. A weak, uniform magnetic field $\mathbf{B}$ is applied to an axially symmetric permanent magnet, with the dipole magnetic moment $\mathbf{m}$ directed along the symmetry axis, rapidly rotating about the same axis, with an angular momentum $\mathbf{L}$. Calculate the electric field resulting from the magnetic field's application and formulate the conditions of your result's validity.

[75] A warning: the proof of Eq. (6.121), given in many textbooks and online sites, is deficient.

Problem 6.4. The similarity of Eq. (5.53), obtained in section 5.3 without any use of the Faraday induction law, and Eq. (5.54), proved in section 6.2 using the law, implies that the law may be derived from magnetostatics. Prove that this is indeed true for a particular case of a current loop, being slowly deformed in a fixed magnetic field **B**.

Problem 6.5. Could problem 5.1 (i.e. the analysis of the mechanical stability of the system shown in the figure below) be solved using potential energy arguments?

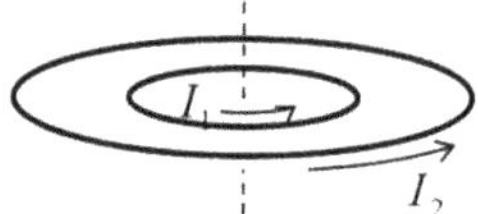

Problem 6.6. Use energy arguments to calculate the pressure exerted by the magnetic field **B** inside a long uniform solenoid of length l, and a cross-section of area $A \ll l^2$, with $N \gg l/A^{1/2} \gg 1$ turns, on its 'walls' (windings), and the force exerted by the field on the solenoid's ends, for two cases:

(i) the current through the solenoid is fixed by an external source, and
(ii) after the initial current setting, the ends of the solenoid's wire, with a negligible resistance, are connected so that it continues to carry a non-zero current.

Compare the results and give a physical interpretation of the direction of these forces.

Problem 6.7. The *electromagnetic railgun* is a projectile launch system consisting of two long, parallel conducting rails and a sliding conducting projectile, shorting the current I fed into the system by a powerful source—see panel (a) in the figure below. Calculate the force exerted on the projectile using two approaches:

(i) by a direct calculation, assuming that the cross-section of the system has the simple shape shown on panel (b) of the figure below, with $t \ll w$, l, and
(ii) using the energy balance (for simplicity, neglecting the Ohmic resistances in the system),

and compare the results.

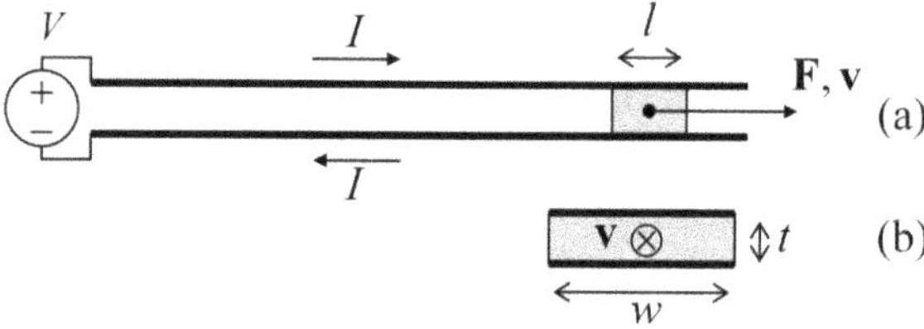

Problem 6.8. A uniform, static magnetic field **B** is applied along the axis of a long round pipe of a radius R, and a very small thickness τ, made of a material with Ohmic conductivity σ. A sphere of mass M and radius $R' < R$, made of a linear magnetic with permeability $\mu \gg \mu_0$, is launched with an initial velocity v_0 to fly ballistically along the pipe's axis—see the figure below. Use the quasi-static approximation to calculate the distance the sphere would pass before it stops. Formulate the conditions of validity of your result.

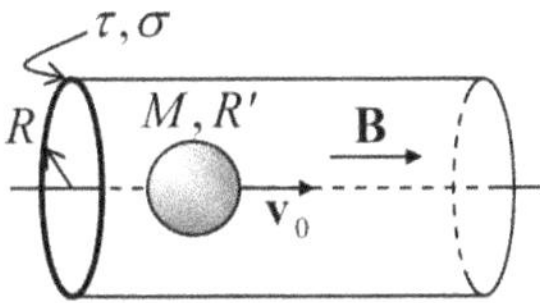

Problem 6.9. AC current of frequency ω is being passed through a long uniform wire with a round cross-section of radius R comparable to the skin depth δ_s. In the quasi-static approximation, find the current's distribution across the cross-section and analyze it in the limits $R \ll \delta_s$ and $\delta_s \ll R$. Calculate the effective ac resistance of the wire (per unit length) in these two limits.

Problem 6.10. A very long, round cylinder of radius R, made of a uniform conductor with Ohmic conductivity σ and magnetic permittivity μ, has been placed into a uniform ac magnetic field $\mathbf{H}_{ext}(t) = \mathbf{H}_0 \cos \omega t$ directed along its symmetry axis. Calculate the spatial distribution of the magnetic field's amplitude, and in particular its value on the cylinder's axis. Spell out the last result in the limits of relatively small and large R.

Problem 6.11.* Define and calculate an appropriate spatial-temporal Green's function for Eq. (6.25), and then use this function to analyze the dynamics of propagation of the external magnetic field, suddenly turned on at $t = 0$ and then left constant:

$$H(x < 0, t) = \begin{cases} 0, & \text{at } t < 0, \\ H_0, & \text{at } t > 0, \end{cases}$$

into an Ohmic conductor occupying the half-space $x > 0$—see figure 6.2.

Hint: Try to use a function proportional to $\exp\{-(x - x')^2/2(\delta x)^2\}$, with a suitable time dependence of the parameter δx, and a properly selected pre-exponential factor.

Problem 6.12. Solve the previous problem using the variable separation method, and compare the results.

Problem 6.13. A small, planar wire loop, carrying current I, is located far from a plane surface of a superconductor. Within the 'coarse-grain' (ideal-diamagnetic) description of the Meissner–Ochsenfeld effect, calculate:

(i) the energy of the loop-superconductor interaction,
(ii) the force and torque acting on the loop, and
(iii) the distribution of supercurrents on the superconductor surface.

Problem 6.14. A straight, uniform magnet of length 2 l, cross-section area $A \ll l^2$, and mass m, with a permanent longitudinal magnetization M_0, is placed over a horizontal surface of a superconductor—see the figure below. Within the ideal-diamagnet description of the Meissner–Ochsenfeld effect, find the stable equilibrium position of the magnet.

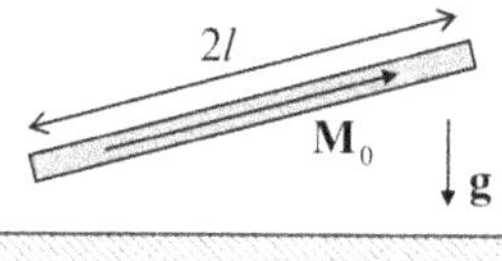

Problem 6.15. A planar superconducting wire loop of area A and inductance L may rotate, without friction, about a horizontal axis 0 (in the figure below, perpendicular to the plane of drawing) passing through its center of mass. Initially the loop was horizontal (with $\theta = 0$), and carried supercurrent I_0 in such direction that its magnetic dipole vector was directed down. Then a uniform magnetic field **B**, directed vertically up, was applied. Using the ideal-diamagnet description of the Meissner–Ochsenfeld effect, find all possible equilibrium positions of the loop, analyze their stability, and give a physical interpretation of the results.

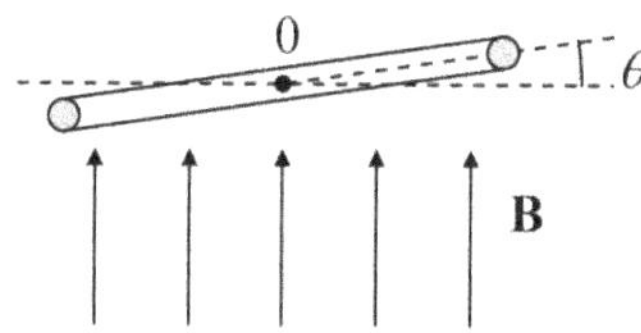

Problem 6.16. Use the London equation to analyze the penetration of a uniform external magnetic field into a thin ($t \sim \delta_L$), planar superconducting film, whose plane is parallel to the field.

Problem 6.17. Use the London equation to calculate the distribution of supercurrent density **j** inside a long, straight superconducting wire, with a circular cross-section of radius $R \sim \delta_L$, carrying dc current I.

Problem 6.18. Use the London equation to calculate the inductance (per unit length) of a long, uniform superconducting strip placed close to the surface of a similar superconductor—see the figure below, which shows the structure's cross-section.

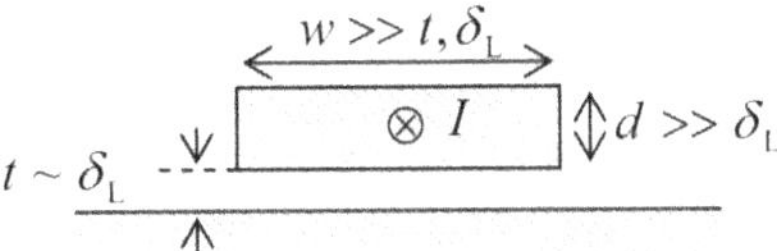

Problem 6.19. Calculate the inductance (per unit length) of a superconducting cable with a round cross-section, shown in the figure below, in the following limits:

(i) $\delta_L \ll a, b, c - b$, and
(ii) $a \ll \delta_L \ll b, c - b$.

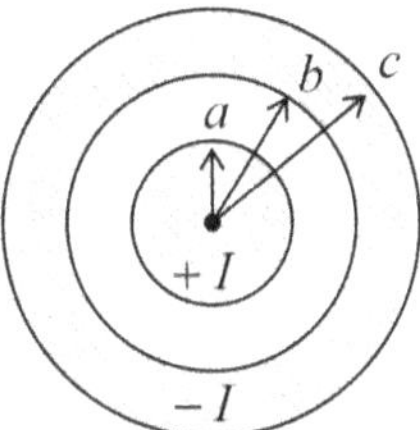

Problem 6.20. Use the London equation to analyze the magnetic field shielding by a superconducting thin film of thickness $t \ll \delta_L$, by calculating the penetration of the field induced by current I flowing in a thin wire which runs parallel to a wide, planar thin film, at distance $d \gg t$ from it, into the half-space behind the film.

Problem 6.21. Use the Ginzburg–Landau equations (6.54) and (6.63) to calculate the largest ('critical') value of the supercurrent in a uniform, long superconducting wire of a small cross-section $A_w \ll \delta_L{}^2$.

Problem 6.22. Use the discussion of a long, straight Abricosov vortex in the approximation $\xi \ll \delta_L$ in section 6.5 to prove Eqs. (6.71) and (6.72) for its energy per unit length and the first critical field.

Problem 6.23.* Use the Ginzburg–Landau equations (6.54) and (6.63) to prove the Josephson's relation (6.76) for a small superconducting weak link, and express its critical current I_c via the Ohmic resistance R_n of the same weak link in its normal state.

Problem 6.24. Use Eqs. (6.76) and (6.79) to calculate the coupling energy of a Josephson junction, and the full potential energy of the SQUID shown in figure 6.4c.

Problem 6.25. Analyze the possibility of wave propagation in a long, uniform chain of lumped inductances and capacitances—see the figure below.

Hint: Readers without prior experience with electromagnetic wave analysis may like to use a substantial analogy between this effect and mechanical waves in a 1D chain of elastically coupled particles[76].

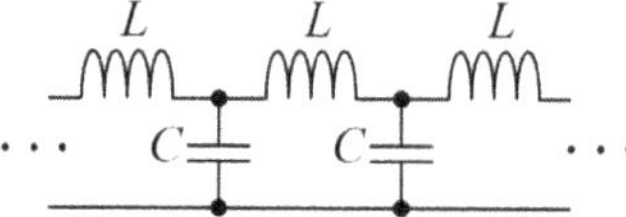

Problem 6.26. A sinusoidal e.m.f. of amplitude V_0 and frequency ω is applied to an end of a long chain of similar, lumped resistors and capacitors, shown in the figure below. Calculate the law of decay of the ac oscillation amplitude in the chain.

Problem 6.27. As was discussed in section 6.7, the displacement current concept allows one to generalize the Ampère law to time-dependent processes as

$$\oint_C \mathbf{H} \cdot d\mathbf{r} = I_S + \frac{\partial}{\partial t}\int_S D_n \, d^2r.$$

We also have seen that such generalization makes $\oint \mathbf{H} \cdot d\mathbf{r}$ over an external contour, such as the one shown in figure 6.10, independent of the choice of the surface S limited by the contour. However, it may look like the situation is different for a contour drawn inside the capacitor—see the figure below. Indeed, if the contour's size is much larger than the capacitor's thickness, the magnetic field $\mathbf{H}$ created by the linear current I on the contour's line is virtually the same as that of a continuous wire, and hence the integral $\oint \mathbf{H} \cdot d\mathbf{r}$ along the contour apparently does not depend on its area, while the magnetic flux $\int D_n \, d^2r$ does, so that the equation displayed above seems invalid. (The current I_S piercing this contour evidently equals zero.) Resolve the paradox, for simplicity considering an axially symmetric system.

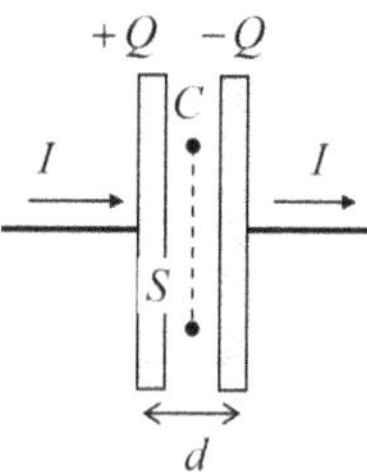

[76] See, e.g. *Part CM* section 6.3.

Problem 6.28. A straight, uniform, long wire with circular cross-section of radius R, is made of an Ohmic conductor with conductivity σ, and carries dc current I. Calculate the flux of the Poynting vector through its surface, and compare it with the Joule energy losses.

References

[1] Abrikosov A 1972 Introduction to the Theory of Normal *Metals* (New York: Academic)
[2] Tinkham M 1996 *Introduction to Superconductivity* 2nd edn (New York: McGraw-Hill)
[3] Clarke J and Braginski A (ed) *The SQUID Handbook* vol 2 (Wiley)
[4] Grosz A *et al* (ed) 2017 *High Sensitivity Magnetometers* (Springer)
[5] Bunyk P *et al* 2001 *Int. J. High Speed Electron. Syst.* **11** 257
[6] Pfeiffer R *et al* 2007 *Rev. Mod. Phys.* **79** 1197

Konstantin K Likharev

Chapter 7

Electromagnetic wave propagation

This (rather long) chapter focuses on the most important effect that follows from the time-dependent Maxwell equations, namely electromagnetic waves, at this stage avoiding discussion of their origin—radiation—which will the subject of chapters 8 *and* 10. *The discussion starts from the simplest plane waves in a uniform and isotropic medium, and then proceeds to non-uniform systems, in particular those with sharp boundaries between different materials, bringing up such effects as reflection and refraction. Then we will discuss the so-called guided waves, propagating along various long transmission lines—such as coaxial cables, waveguides, and optical fibers. Finally, the end of the chapter is devoted to final-length fragments of such lines serving as resonators and to the effects of energy dissipation in transmission lines and resonators.*

7.1 Plane waves

Let us start from considering a spatial region that does not contain field sources ($\rho = 0$, $\mathbf{j} = 0$), and is filled with a linear, uniform, isotropic medium, which obeys Eqs. (3.46) and (5.110):

$$\mathbf{D} = \varepsilon \mathbf{E}, \quad \mathbf{B} = \mu \mathbf{H}. \tag{7.1}$$

Moreover, let us assume, for a while, that these constitutive equations hold for all frequencies of interest. (Of course, these relations are *exactly* valid for the very important particular case of free space, where we may formally use the macroscopic Maxwell equations (6.100), but with $\varepsilon = \varepsilon_0$ and $\mu = \mu_0$.)

As was already shown in section 6.8, in this case the Lorenz gauge condition (6.117) allows the Maxwell equations to be recast into the wave equations (6.118) for the vector and scalar potentials. However, for most purposes it is more convenient to use directly the homogeneous Maxwell equations (6.100) for the electric and magnetic

doi:10.1088/978-0-7503-1404-6ch7

fields (which are independent of the gauge choice). After the elementary elimination of **D** and **B** using Eqs. (7.1)[1], these equations take a simple, symmetric form

$$\nabla \times \mathbf{E} + \mu \frac{\partial \mathbf{H}}{\partial t} = 0, \qquad \nabla \times \mathbf{H} - \varepsilon \frac{\partial \mathbf{E}}{\partial t} = 0, \tag{7.2a}$$

$$\nabla \cdot \mathbf{E} = 0, \qquad \nabla \cdot \mathbf{H} = 0. \tag{7.2b}$$

Now, acting by operator $\nabla \times$ on each of Eqs. (7.2*a*), i.e. taking their curl, and then using the vector algebra identity (5.31), whose first term for both **E** and **H** vanishes due to Eqs. (7.2*b*), we obtain similar wave equations for the electric and magnetic fields:

$$\left(\nabla^2 - \frac{1}{v^2}\frac{\partial^2}{\partial t^2}\right)\mathbf{E} = 0, \quad \left(\nabla^2 - \frac{1}{v^2}\frac{\partial^2}{\partial t^2}\right)\mathbf{H} = 0, \tag{7.3}$$

where the parameter v is defined by relation

$$v^2 \equiv \frac{1}{\varepsilon\mu}, \tag{7.4}$$

with $v^2 = 1/\varepsilon_0\mu_0 \equiv c^2$ in free space—see Eq. (6.120) again.

The two vector equation (7.3) are of course just a shorthand for six similar equations for the three Cartesian components of **E** and **H**. These equations allow, in particular, the following solution,

$$f = f(z - vt), \tag{7.5}$$

where z is the Cartesian coordinate along a certain (arbitrary) direction **n**. This solution describes a specific type of *wave*, i.e. a certain field pattern moving without deformation along axis z with velocity v. According to Eq. (7.5), each variable f has the same value in each plane perpendicular to the direction **n** of wave propagation, hence the name—*plane wave*.

According to Eqs. (7.2), the independence of the wave equations (7.3) for vectors **E** and **H** does not mean that their plane-wave *solutions* are independent. Indeed, plugging the solutions of the type (7.5) into Eqs. (7.2*a*), we obtain

$$\mathbf{H} = \frac{\mathbf{n} \times \mathbf{E}}{Z}, \qquad \text{i.e. } \mathbf{E} = Z\,\mathbf{H} \times \mathbf{n}, \tag{7.6}$$

where the constant Z is defined as

$$Z \equiv \frac{E}{H} = \left(\frac{\mu}{\varepsilon}\right)^{1/2}. \tag{7.7}$$

[1] Although **B** rather than **H** is the actual magnetic field, mathematically it is a bit more convenient (just as it was in section 6.2) to use the latter vector in the following discussion, because at sharp media boundaries, **H** obeys the boundary condition (5.117) similar to that for **E**—cf Eq. (3.37).

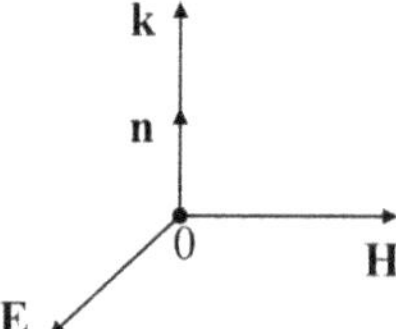

Figure 7.1. Field vectors in a plane electromagnetic wave propagating along direction **n**.

The vector relation (7.6) means, first of all, that the vectors **E** and **H** are perpendicular not only to vector **n** (such waves are called *transverse*), but also to each other (figure 7.1)—at any point of space and at any time instant. Second, this relation does not depend on the function f, meaning that the electric and magnetic fields increase and decrease *simultaneously*.

Finally, the field magnitudes are related by the constant Z, called the *wave impedance* of the medium. Very soon we will see that the impedance plays a pivotal role in many problems, in particular at the wave reflection from the interface between the two media. Since the dimensionality of E, in SI units, is V m^{-1}, and that of H is A m^{-1}, Eq. (7.7) shows that Z has the dimensionality of V/A, i.e. ohms (Ω).[2] In particular, in free space,

$$Z = Z_0 \equiv \left(\frac{\mu_0}{\varepsilon_0}\right)^{1/2} = 4\pi \times 10^{-7} c \approx 377 \ \Omega. \tag{7.8}$$

Now plugging Eq. (7.6) into Eqs. (6.113) and (6.114), we obtain

$$u = \varepsilon E^2 = \mu H^2, \tag{7.9a}$$

$$\mathbf{S} \equiv \mathbf{E} \times \mathbf{H} = \mathbf{n}\frac{E^2}{Z} = \mathbf{n}ZH^2, \tag{7.9b}$$

so that, according to Eqs. (7.4) and (7.7), the wave's energy and power densities are universally related as

$$\mathbf{S} = \mathbf{n}uv. \tag{7.9c}$$

In the view of the Poynting vector paradox discussed in section 6.8 (see figure 6.11), one may wonder whether this expression may be interpreted as the actual density of power flow. In contrast to the static situation shown in figure 6.11 that limits the electric and magnetic fields to a vicinity of their sources, waves may travel far from them. As a result, they can form *wave packets* of a finite length in free space—see figure 7.2. Let us apply the Poynting theorem (6.111) to the cylinder shown with dashed lines in figure 7.2, with one lid inside the wave packet and another lid in the region already passed by the wave. Then, according to Eq. (6.111), the rate of change of the full field energy E inside the volume is $d\mathscr{E}/dt = -SA$

[2] In the Gaussian units, E and H have a similar dimensionality (in particular, in a free-space wave, $E = H$), making the (very useful) notion of the wave impedance less manifestly exposed—so that in some older physics textbooks it is not mentioned at all!

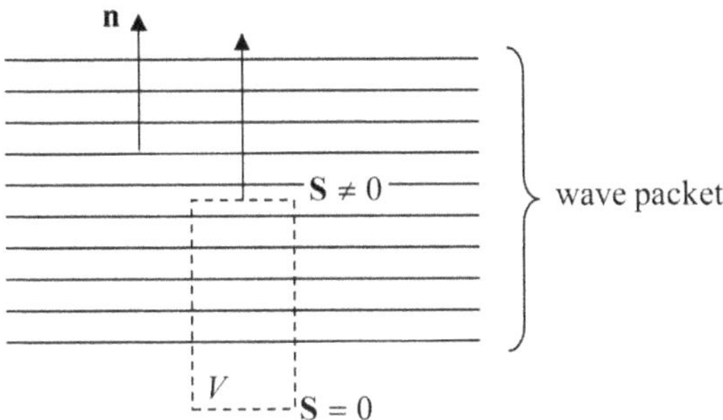

Figure 7.2. Interpreting the Poynting vector in an electromagnetic wave.

(where A is the lid area), so that S may be indeed interpreted as the power flow (per unit area) from the volume. Making a reasonable assumption that the finite length of a sufficiently long wave packet does not affect the physics inside it, we may indeed interpret the **S** given by Eqs. (7.9*b*) and (7.9*c*) as the power flow density inside a plane electromagnetic wave.

As we will see later in this chapter, the free-space value Z_0 of the wave impedance, given by Eq. (7.8), establishes the scale of Z of virtually all wave transmission lines, so we may use it, together with Eq. (7.9), to obtain a better feeling of how different the electric and magnetic field amplitudes in the waves are, on the scale of typical electrostatics and magnetostatics experiments. For example, according to Eqs. (7.9), a wave of a modest intensity $S = 1\ \text{W m}^{-2}$ (this is what we get from a usual electric bulb a few meters away from it) has $E \sim (SZ_0)^{1/2} \sim 20\ \text{V m}^{-1}$, quite comparable with the dc field created by a standard AA battery right outside it. On the other hand, the wave's magnetic field $H = (S/Z_0)^{1/2} \approx 0.05\ \text{A m}^{-1}$. For this particular case, the relation following from Eqs. (7.1), (7.4), and (7.7),

$$B = \mu H = \mu\frac{E}{Z} = \mu\frac{E}{(\mu/\varepsilon)^{1/2}} = (\varepsilon\mu)^{1/2}E = \frac{E}{v}, \tag{7.10}$$

gives $B = \mu_0 H = E/c \sim 7 \times 10^{-8}\ \text{T}$, i.e. a magnetic field thousand times lower than the Earth's field, and about eight orders of magnitude lower than the field of a typical permanent magnet. This huge difference may be interpreted as follows: the scale of magnetic fields $B \sim E/c$ in the waves is 'normal' for electromagnetism, while the permanent magnet fields are abnormally high, because they are due to the ferromagnetic alignment of electron spins, essentially relativistic objects—see the discussion in section 5.5.

As soon as ε and μ are simple constants, the wave speed v is also constant, and Eq. (7.5) is valid for an arbitrary function f—defined by the initial conditions. In plain English, a medium with frequency-independent ε and μ supports the propagation of plane waves with an arbitrary waveform—without either decay (*attenuation*) or deformation (*dispersion*). However, for any real medium but pure vacuum, this approximation is valid only within limited frequency intervals. We will discuss the effects of attenuation and dispersion in the next section and see that all our prior formulas remain valid even for an arbitrary linear medium, provided that we limit them to single-frequency (i.e. sinusoidal, frequently called *monochromatic*) waves. Such waves may be most conveniently represented as[3]

$$f = \text{Re}\,[f_\omega e^{i(kz-\omega t)}], \tag{7.11}$$

where f_ω is the *complex amplitude* of the wave, and k is its *wave number* (the magnitude of the *wave vector* $\mathbf{k} \equiv \mathbf{n}k$), sometimes also called the *spatial frequency*. The last term is justified by the fact, evident from Eq. (7.11), that k is related to the wavelength λ exactly as the usual ('temporal') frequency ω is related to the time period $\mathcal{T}$:

$$k = \frac{2\pi}{\lambda}, \qquad \omega = \frac{2\pi}{\mathcal{T}}. \tag{7.12}$$

Requiring Eq. (7.11) to be a particular form of Eq. (7.5), i.e. the argument $(kz - \omega t) \equiv k[z - (\omega/k)t]$ to be proportional to $(z - vt)$, so that $\omega/k = v$, we see that the wave number should equal

$$k = \frac{\omega}{v} = (\varepsilon\mu)^{1/2}\omega, \tag{7.13}$$

showing that in this 'dispersion-free' case the *dispersion relation* $\omega(k)$ is linear.

Now note that Eq. (7.6) does not mean that vectors **E** and **H** retain their direction in space. (The simple case when they do is called the *linear polarization* of the wave.) Indeed, nothing in the Maxwell equations prevents, for example, a joint rotation of this pair of vectors around the fixed vector **n**, while still keeping all these three vectors perpendicular to each other at any instant—see figure 7.1. An arbitrary rotation law, or even an arbitrary constant frequency of such rotation, however, would violate the single-frequency (monochromatic) character of the elementary sinusoidal wave (7.11). In order to understand what is the most general type of polarization the wave may have without violating that condition, let us represent two Cartesian components of one of these vectors (say, **E**) along any two fixed axes x and y, perpendicular to each other and axis z (i.e. to vector **n**), in the same form as used in Eq. (7.11):

$$E_x = \text{Re}\,[E_{\omega x} e^{i(kz-\omega t)}], \qquad E_y = \text{Re}\,[E_{\omega y} e^{i(kz-\omega t)}]. \tag{7.14}$$

In order to keep the wave monochromatic, the complex amplitudes $E_{\omega x}$ and $E_{\omega y}$ must be constant; however, they may have different magnitudes and an arbitrary phase shift between them.

In the simplest case when the arguments of the complex amplitudes are equal,

$$E_{\omega x,y} = \left|E_{\omega x,y}\right| e^{i\varphi}, \tag{7.15}$$

[3] As we have already seen in the previous chapter (see also *Part CM* section 5.1), such a complex-exponential representation of sinusoidally changing variables is more convenient for mathematical manipulation than using sine and cosine functions, particularly because in all linear relations, the operator Re may be omitted (implied) until the very end of the calculation. Note, however, that this is *not* valid for the quadratic forms such as Eqs. (7.9*a*) and (7.9*b*).

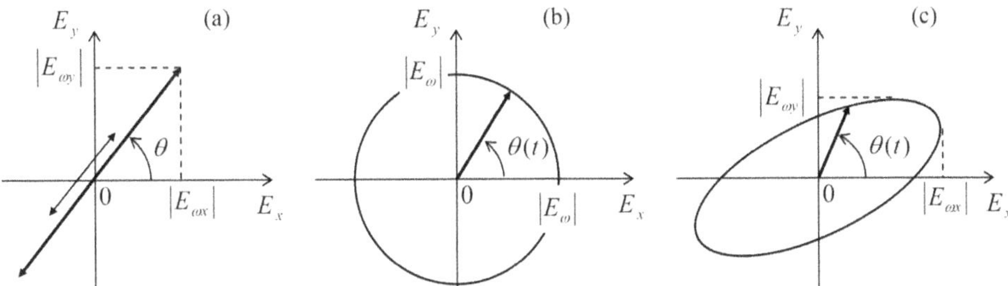

Figure 7.3. Time evolution of the electric field vector in monochromatic waves with: (a) a linear polarization, (b) the circular polarization, and (c) an elliptical polarization.

the real field components have the same phase,

$$E_x = |E_{\omega x}| \cos(kz - \omega t + \varphi), \qquad E_y = |E_{\omega y}| \cos(kz - \omega t + \varphi), \tag{7.16}$$

so that their ratio is constant in time—see figure 7.3a. This means that the wave is linearly polarized, within the polarization plane defined by the relation

$$\tan\theta = |E_{\omega y}|/|E_{\omega x}|. \tag{7.17}$$

Another simple case is when the moduli of the complex amplitudes $E_{\omega x}$ and $E_{\omega y}$ are equal, but their phases are shifted by $+\pi/2$ or $-\pi/2$:

$$E_{\omega x} = |E_\omega|\, e^{i\varphi}, \quad E_{\omega y} = |E_\omega| e^{i(\varphi \pm \pi/2)}. \tag{7.18}$$

In this case

$$\begin{aligned} E_x &= |E_\omega| \cos(kz - \omega t + \varphi), \\ E_y &= |E_\omega| \cos\left(kz - \omega t + \varphi \pm \frac{\pi}{2}\right) = \mp|E_\omega| \sin(kz - \omega t + \varphi). \end{aligned} \tag{7.19}$$

This means that on the $[x, y]$ plane, the end of the vector $\mathbf{E}$ measured at a fixed coordinate z moves with the wave's frequency ω either clockwise or counter-clockwise around a circle—see figure 7.3b:

$$\theta(t) = \mp(\omega t - \varphi). \tag{7.20}$$

Such waves are called *circularly polarized*[4]. These particular solutions of the Maxwell equations are very convenient for quantum electrodynamics, because single electromagnetic field quanta with a certain (positive or negative) spin direction may be considered as elementary excitations of the corresponding circularly polarized wave[5]. (This fact does not exclude, from the quantization scheme, waves

[4] In the dominant convention, the wave is called *right-polarized* (RP) if it is described by the lower sign in Eqs. (7.18)–(7.20) (i.e. if the vector $\boldsymbol{\omega}$ of the angular frequency of the field vector's rotation coincides with the wave propagation's direction $\mathbf{n}$), and *left-polarized* (LP) in the opposite case.

[5] This issue is closely related to that of the radiation's angular momentum; it will be more convenient for me to discuss it later in this chapter (in section 7.7).

of other polarizations, because any monochromatic wave may be represented as a linear combination of two opposite circularly polarized waves—just as Eqs. (7.14) represent it as a linear combination of two linearly polarized waves.)

Finally, in the general case of arbitrary complex amplitudes $E_{\omega x}$ and $E_{\omega y}$, the electric field vector end moves along an ellipse on the $[x, y]$ plane (figure 7.3c), such wave is called *elliptically polarized.* The eccentricity and orientation of the ellipse are completely described by one complex number, the ratio $E_{\omega x}/E_{\omega y}$, i.e. by two real numbers: $|E_{\omega x}/E_{\omega y}|$ and $\varphi = \arg(E_{\omega x}/E_{\omega y})$.[6]

7.2 Attenuation and dispersion

Let me start the discussion of the dispersion and attenuation effects by considering a particular case of the time evolution of the electric polarization of a dilute, non-polar medium, with negligible interaction between its elementary dipoles $\mathbf{p}(t)$. As was discussed in section 3.3, in this case the local electric field acting on each elementary dipole equals the macroscopic field $\mathbf{E}(t)$. Then, the polarization $\mathbf{p}(t) \neq 0$ may be caused only by the values of the field $\mathbf{E}$ at the same moment of time (t), or at the earlier moments of time, $t < t'$. Due to the linear superposition principle, the macroscopic polarization $\mathbf{P}(t) \equiv n\mathbf{p}(t)$ should be a linear sum (integral) of the values of $\mathbf{E}(t')$ at all previous moments of time, $t' < t$, weighed by some function of t and t':[7]

$$P(t) = \int_{-\infty}^{t} E(t')G(t, t')dt'. \tag{7.21}$$

The condition $t' < t$, which is implied by this relation, expresses a key principle of physics, the *causal relation* between a cause (in our case, the electric field applied to each dipole) and its effect (the polarization it creates). The function $G(t, t')$ is called the *temporal Green's function* for the electric polarization[8]. In order to understand its physical sense, let us consider the case when the applied field $E(t)$ is a very short pulse at the moment $t_0 < t$, that may be approximated with Dirac's delta-function:

$$E(t) = \delta(t - t_0). \tag{7.22}$$

[6] Note that the same information may be expressed via four so-called *Stokes parameters* s_0, s_1, s_2, s_3, which are popular in practical optics, because they may be used for the description of not only completely coherent waves that are discussed here, but also of party coherent or even fully incoherent waves—including the *natural light* emitted by thermal sources such as our Sun. (In contrast to the coherent waves (7.14), whose complex amplitudes are deterministic numbers, the amplitudes of incoherent waves should be treated as stochastic variables.) For more on the Stokes parameters, as well as about many optics topics I will not have time to cover (in particular geometrical optics), I can recommend the classical text [1].

[7] In an isotropic medium, vectors $\mathbf{E}$, $\mathbf{P}$, and hence $\mathbf{D} = \varepsilon_0\mathbf{E} + \mathbf{P}$, are all parallel, and for simplicity of notation I will drop the vector sign in the following formulas. I am also assuming that $\mathbf{P}$ at any point $\mathbf{r}$ is only dependent on the electric field at the same point, and hence drop the term ikz from the exponent's argument. This assumption is valid if the wavelength λ is much larger than the elementary media dipole's size a. In most systems of interest, the scale of a is atomic ($\sim 10^{-10}$ m), so that the last approximation is valid up to very high frequencies, $\omega \sim c/a \sim 10^{18}\ \mathrm{s}^{-1}$, corresponding to hard x-rays.

[8] The idea of these functions is very similar to that of the spatial Green's functions (see section 2.10), but with a new twist, due to the causality principle. A discussion of the temporal Green's functions in application to classical mechanics may be found in *Part CM* section 5.1.

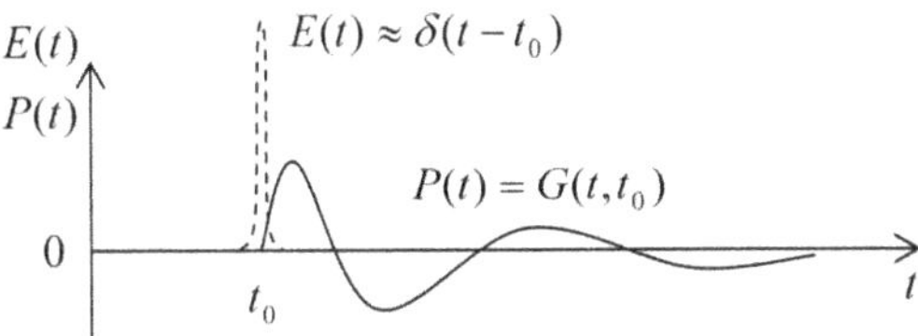

Figure 7.4. An example of the temporal Green's function for electric polarization.

Then Eq. (7.21) yields just $P(t) = G(t, t_0)$, showing that the Green's function is just the polarization at moment t, created by a unit δ-functional pulse of the applied field at moment t' (figure 7.4).

What are the general properties of the temporal Green's function? First, the function is evidently real, since the dipole moment $\mathbf{p}$ and hence polarization $\mathbf{P} = n\mathbf{p}$ are real by definition—see Eq. (3.6). Next, for systems without infinite internal 'memory', G should tend to zero at $t - t' \rightarrow \infty$, although the type of this approach (e.g. whether the function G oscillates approaching zero—see figure 7.4) depends on the elementary dipole's properties. Finally, if parameters of the medium do not change in time, the polarization response to an electric field pulse should be dependent not on its absolute timing, but only on the time difference $\theta \equiv t - t'$ between the pulse and observation instants, i.e. Eq. (7.21) is reduced to

$$P(t) = \int_{-\infty}^{t} E(t')G(t - t')dt' \equiv \int_{0}^{\infty} E(t - \theta)G(\theta)d\theta. \tag{7.23}$$

For a sinusoidal waveform, $E(t) = \mathrm{Re}\,[E_\omega e^{-i\omega t}]$, this equation yields

$$P(t) = \mathrm{Re}\int_{0}^{\infty} E_\omega e^{-i\omega(t-\theta)}G(\theta)d\theta = \mathrm{Re}\left\{\left[E_\omega \int_{0}^{\infty} G(\theta)e^{i\omega\theta}\,d\theta\right] e^{-i\omega t}\right\}. \tag{7.24}$$

The expression in square brackets is of course nothing more that the complex amplitude P_ω of the polarization. This means that even if the static linear relation (3.43), $P = \chi_e \varepsilon_0 E$, is invalid for an arbitrary time-dependent process, we may still keep its Fourier analog,

$$P_\omega = \chi_e(\omega)\varepsilon_0 E_\omega, \quad \text{with} \quad \chi_e(\omega) \equiv \frac{1}{\varepsilon_0}\int_{0}^{\infty} G(\theta)e^{i\omega\theta}\,d\theta, \tag{7.25}$$

for each sinusoidal component of the process, using it as the definition of the frequency-dependent electric susceptibility $\chi_e(\omega)$. Similarly, the frequency-dependent electric permittivity may be defined using the Fourier analog of Eq. (3.46):

$$D_\omega \equiv \varepsilon(\omega)E_\omega. \tag{7.26a}$$

Then, according to the definition (3.33), the permittivity is related to the temporal Green's function by the usual Fourier transform:

$$\varepsilon(\omega) \equiv \varepsilon_0 + \frac{P_\omega}{E_\omega} = \varepsilon_0 + \int_{0}^{\infty} G(\theta)e^{i\omega\theta}\,d\theta. \tag{7.26b}$$

This relation shows that $\varepsilon(\omega)$ may be complex,

$$\varepsilon(\omega) = \varepsilon'(\omega) + i\varepsilon''(\omega), \quad \text{with} \quad \varepsilon'(\omega) = \varepsilon_0 + \int_0^\infty G(\theta)\cos\omega\theta\, d\theta,$$
$$\varepsilon''(\omega) = \int_0^\infty G(\theta)\sin\omega\theta\, d\theta, \tag{7.27}$$

and that its real part $\varepsilon'(\omega)$ is always an even function of frequency, while the imaginary part $\varepsilon''(\omega)$ is an odd function of ω.

Although the particular causal relationship (7.21) between $P(t)$ and $E(t)$ is conditioned by the elementary dipole independence, the frequency-dependent complex electric permittivity $\varepsilon(\omega)$ may be introduced in a similar way if *any* two linear combinations of these variables are related by a similar formula. Absolutely similar arguments show that magnetic properties of a linear, isotropic medium may be characterized with frequency-dependent, complex permeability $\mu(\omega)$.

Now rewriting Eq. (7.1) for the complex amplitudes of the fields at a particular frequency, we may repeat all calculations of section 7.1, and verify that all its results are valid for monochromatic waves even for a dispersive (but necessarily linear!) medium. In particular, Eqs. (7.7) and (7.13) now become

$$Z(\omega) = \left(\frac{\mu(\omega)}{\varepsilon(\omega)}\right)^{1/2}, \quad k(\omega) = \omega[\varepsilon(\omega)\mu(\omega)]^{1/2}, \tag{7.28}$$

so that the wave impedance and the wave number may be both complex functions of frequency.

This fact has important consequences for the electromagnetic wave propagation. First, plugging the representation of the complex wave number as the sum of its real and imaginary parts, $k(\omega) \equiv k'(\omega) + ik''(\omega)$, into Eq. (7.11),

$$f = \text{Re}\,\{f_\omega e^{i[k(\omega)z-\omega t]}\} = e^{-k''(\omega)z}\,\text{Re}\,\{f_\omega e^{i[k'(\omega)z-\omega t]}\}, \tag{7.29}$$

we see that $k''(\omega)$ describes the rate of wave *attenuation* in the medium at frequency ω.[9] Second, if the waveform is not sinusoidal (and hence should be represented as a sum of several/many sinusoidal components), the frequency dependence of $k'(\omega)$ provides for wave *dispersion*, i.e. the waveform deformation at the propagation, because the propagation velocity (7.4) of component waves is now different[10].

---

[9] It may be tempting to attribute this effect to wave *absorption*, i.e. the dissipation of the wave's energy, but we will see very soon that wave attenuation may also be due to different effects.

[10] The reader is probably familiar with the most noticeable effect of the dispersion, namely the difference between that *group velocity* $v_{gr} \equiv d\omega/dk'$, giving the speed of the envelope of a wave packet with a narrow frequency spectrum, and the *phase velocity* $v_{ph} \equiv \omega/k'$ of the component waves. The second-order dispersion effect, proportional to $d^2\omega/d^2k'$, leads to the deformation (gradual broadening) of the envelope itself. Following tradition, these effects are discussed in more detail in the quantum-mechanics part of this series (*Part QM* section 2.1), because they are the crucial factor of Schrödinger's wave mechanics. (See also their brief discussion in *Part CM* section 6.3.)

As an example of such a dispersive medium, let us consider a simple but very representative *Lorentz oscillator model*[11]. In dilute atomic or molecular systems (e.g. gases), electrons respond to the external electric field particularly strongly when frequency ω is close to certain eigenfrequencies ω_j corresponding to the spectrum of quantum transitions of a single atom/molecule. An approximate, phenomenological description of this behavior may be obtained from a classical model of several externally driven harmonic oscillators, generally with non-zero damping. For a single oscillator, driven by electric field's force $F(t) = qE(t)$, we can write Newton's second law as

$$m\left(\ddot{x} + 2\delta_0\dot{x} + \omega_0^2 x\right) = qE(t), \tag{7.30}$$

where ω_0 is the own frequency of the oscillator, and δ_0 its damping coefficient. For the electric field of a monochromatic wave[12], $E(t) = \mathrm{Re}[E_\omega \exp\{-i\omega t\}]$, we may look for a particular, *forced-oscillation* solution of this equation in a similar form $x(t) = \mathrm{Re}[x_\omega \exp\{-i\omega t\}]$.[13] Plugging this solution into Eq. (7.30), we can readily find the complex amplitude of these oscillations:

$$x_\omega = \frac{q}{m}\frac{E_\omega}{(\omega_0^2 - \omega^2) - 2i\omega\delta_0}. \tag{7.31}$$

Using this result to calculate the complex amplitude of the dipole moment as $p_\omega = qx_\omega$, and then the electric polarization $P_\omega = np_\omega$ of a dilute medium with n independent oscillators for unit volume, for its frequency-dependent permittivity (7.26) we obtain

$$\varepsilon(\omega) = \varepsilon_0 + \frac{nq^2}{m}\frac{1}{\left(\omega_0^2 - \omega^2\right) - 2i\omega\delta_0}. \tag{7.32}$$

This result may be readily (and obviously) generalized to the case when the system has several types of oscillators with different eigenfrequencies:

$$\varepsilon(\omega) = \varepsilon_0 + n\frac{q^2}{m}\sum_j \frac{f_j}{\left(\omega_j^2 - \omega^2\right) - 2i\omega\delta_j}, \tag{7.33}$$

where $f_j \equiv n_j/n$ is the fraction of oscillators with eigenfrequency ω_j, so that the sum of all f_j equals 1. Figure 7.5 shows a typical behavior of the real and imaginary parts of the complex dielectric constant, described by Eq. (7.33), as functions of frequency.

---

[11] This example is focused on the frequency dependence of ε, because electromagnetic waves interact with 'usual' media via their electric field much more than via the magnetic field. However, as will be discussed in section 7.6, forgetting about the possible dispersion of $\mu(\omega)$ might result in missing some remarkable opportunities for manipulating the waves.

[12] According to Eq. (7.7), the magnetic field of the wave $B = \mu H = E/v \sim E/c$, so that the magnetic component of the Lorentz force (5.10), acting on a non-relativistic partice ($F_m \sim quB \sim (u/c)qE$), is much smaller than that of its electric component ($F_e = qE$), and may be neglected.

[13] If this point is not absolutely clear, please see *Part CM* section 5.1 for a more detailed discussion.

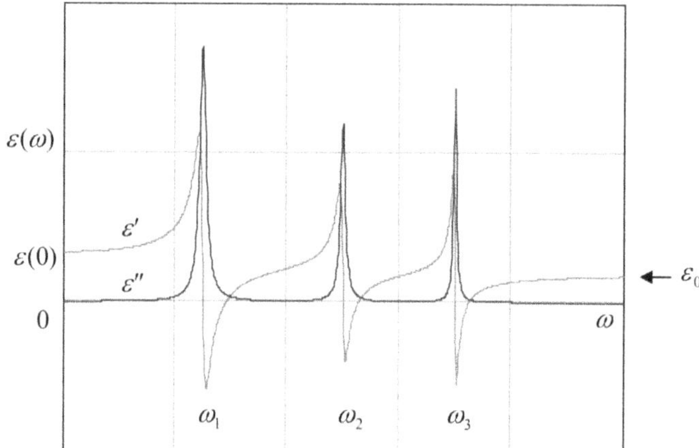

Figure 7.5. A typical frequency dependence of the real and imaginary parts of the electric permittivity of a complex dielectric constant according to the generalized Lorentz oscillator model.

The oscillator resonances' effect is clearly visible and dominates the medium's response at $\omega \approx \omega_j$, in particular in the case of low damping, $\delta_j \ll \omega_j$. Note that in the low-damping limit, the imaginary part of the dielectric constant ε'', and hence the wave attenuation k'', are negligibly small at all frequencies apart from small vicinities of frequencies ω_j, where derivative $d\varepsilon'(\omega)/d\omega$ is negative[14]. Thus, for a system of weakly damped oscillators, Eq. (7.33) may be approximated at most frequencies as a sum of odd singularities ('poles'):

$$\varepsilon(\omega) \approx \varepsilon_0 + n\frac{q^2}{2m}\sum_j \frac{f_j}{\omega_j - \omega}, \quad \text{for} \quad \delta_j \ll |\omega - \omega_j| \ll |\omega_j - \omega_{j'}|. \tag{7.34}$$

This result is particularly important because, according to quantum mechanics[15], Eq. (7.34) is also valid for a set of non-interacting, similar quantum systems (whose dynamics may be completely different from that of a harmonic oscillator!), provided that ω_j are replaced with frequencies of possible quantum interstate transitions, and coefficients f_j are replaced with the so-called *oscillator strengths* of the transitions—which obey the same *sum rule*, $\Sigma_j f_j = 1$. At $\omega \to 0$, the imaginary part of the permittivity also vanishes (for any δ_j), while its real part approaches its electrostatic ('dc') value

$$\varepsilon(0) = \varepsilon_0 + q^2\sum_j \frac{n_j}{m_j\omega_j^2}. \tag{7.35}$$

Note that according to Eq. (7.30), the denominator in Eq. (7.35) is just the effective spring constant $\kappa_j = m_j\omega_j^2$ of the jth oscillator, so that the oscillator masses m_j as such are actually (and quite naturally) not involved in the static dielectric response.

[14] In optics, such behavior is called the *anomalous dispersion*.

[15] See, e.g. *Part QM* chapters 5–7.

In the opposite limit of very high frequencies, $\omega \gg \omega_j$, δ_j, the permittivity (7.33) also becomes real and may be represented as

$$\varepsilon(\omega) = \varepsilon_0\left(1 - \frac{\omega_p^2}{\omega^2}\right), \quad \text{where} \quad \omega_p^2 \equiv \frac{q^2}{\varepsilon_0}\sum_j \frac{n_j}{m_j}. \tag{7.36}$$

The last result is very important because it is also valid at *all* frequencies if all ω_j and δ_j vanish, i.e. for a gas of free charged particles, in particular for *plasmas*—ionized atomic gases, provided that the ion collision effects are negligible. (This is why the parameter ω_p defined by Eq. (7.36) is called the *plasma frequency*.) Typically, the plasma as a whole is neutral, i.e. the density n of positive atomic ions is equal to that of the free electrons. Since the ratio n_j/m_j for electrons is much higher than that for ions, the general formula (7.36) for the plasma frequency is usually well approximated by the following simple expression:

$$\omega_p^2 \equiv \frac{ne^2}{\varepsilon_0 m_e}. \tag{7.37}$$

This expression has a simple physical sense: the effective spring constant $\kappa_{ef} = m_e\omega_p^2 = ne^2/\varepsilon_0$ describes the Coulomb force that appears when the electron subsystem of a plasma is shifted, as a whole, from its positive-ion subsystem, thus violating electroneutrality. (Indeed, let us consider such a small shift, Δx, perpendicular to the plane surface of a broad, plane slab filled with a plasma. The uncompensated ion charges, with equal and opposite surface densities $\sigma = \pm en\Delta x$, that appear at the slab surfaces, create inside it, according to Eq. (2.3), a uniform electric field with $E_x = en\Delta x/\varepsilon_0$. This field exerts force $-eE = -(ne^2/\varepsilon_0)\Delta x = -\kappa_{ef}\Delta x$ on each electron, pulling it back to its equilibrium position.) Hence, there is no surprise that the function $\varepsilon(\omega)$ vanishes at $\omega = \omega_p$: at this resonance frequency, the electric field $\mathbf{E}$ may oscillate, i.e. have a non-zero amplitude $E_\omega = D_\omega/\varepsilon(\omega)$, even in the absence of external forces induced by external (stand-alone) charges, i.e. in the absence of the field $\mathbf{D}$ these charges induce—see Eq. (3.32).

The behavior of electromagnetic waves in a medium that obeys Eq. (7.36), is very remarkable. If the wave frequency ω is above ω_p, the dielectric constant $\varepsilon(\omega)$, and hence the wave number (7.28) are positive and real, and waves propagate without attenuation, following the dispersion relation,

$$k(\omega) = \omega[\varepsilon(\omega)\mu_0]^{1/2} = \frac{1}{c}\left(\omega^2 - \omega_p^2\right)^{1/2}, \tag{7.38}$$

which is shown in figure 7.6. At $\omega \to \omega_p$ the wave number k tends to zero. Beyond that point (at $\omega < \omega_p$), we still can use Eq. (7.38), but it is instrumental to rewrite it in the mathematically equivalent form

$$k(\omega) = \frac{i}{c}\left(\omega_p^2 - \omega^2\right)^{1/2} = \frac{i}{\delta}, \quad \text{where } \delta \equiv \frac{c}{\left(\omega_p^2 - \omega^2\right)^{1/2}}. \tag{7.39}$$

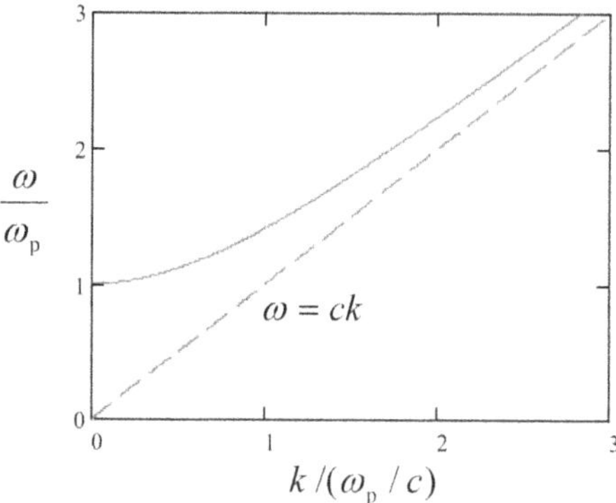

Figure 7.6. The plasma dispersion law (solid line) in comparison with the linear dispersion of the free space (dashed line).

According to Eq. (7.29), this means that the electromagnetic field exponentially decreases with distance:

$$f = \mathrm{Re} f_\omega e^{i(kz-\omega t)} \equiv \exp\left\{-\frac{z}{\delta}\right\} \mathrm{Re} f_\omega e^{-i\omega t}. \tag{7.40}$$

Does this mean that the wave is being absorbed in the plasma? Answering this question is a good pretext to calculate the time average of the Poynting vector $\mathbf{S} = \mathbf{E} \times \mathbf{H}$ of a monochromatic electromagnetic wave in an *arbitrary* dispersive (but still linear!) medium. First, let us spell out the fields' time dependences:

$$\begin{aligned} E(t) &= \mathrm{Re}\,[E_\omega e^{-i\omega t}] \equiv \frac{1}{2}[E_\omega e^{-i\omega t} + \text{c.c.}], \\ H(t) &= \mathrm{Re}\,[H_\omega e^{-i\omega t}] \equiv \frac{1}{2}\left[\frac{E_\omega}{Z(\omega)} e^{-i\omega t} + \text{c.c.}\right]. \end{aligned} \tag{7.41}$$

Now, a straightforward calculation yields[16]

$$\overline{S} = \overline{E(t)H(t)} = \frac{E_\omega E_\omega^*}{4}\left[\frac{1}{Z(\omega)} + \frac{1}{Z^*(\omega)}\right] \equiv \frac{E_\omega E_\omega^*}{2}\mathrm{Re}\frac{1}{Z(\omega)} \equiv \frac{|E_\omega|^2}{2}\mathrm{Re}\left[\frac{\varepsilon(\omega)}{\mu(\omega)}\right]^{1/2}. \tag{7.42}$$

Let us apply this important general formula to our simple model of plasma at $\omega < \omega_p$. In this case the magnetic permeability equals μ_0, i.e. $\mu(\omega) = \mu_0$ is positive and

[16] For an arbitrary plane wave the total average power flow may be calculated as an integral of Eq. (7.42) over all frequencies. By the way, combining this integral and the Poynting theorem (6.111), is it straightforward to prove the following interesting expression for the average electromagnetic energy density of a narrow ($\Delta\omega \ll \omega$) wave packet propagating in an arbitrary dispersive (but linear and isotropic) medium:

$$\overline{u} = \frac{1}{2}\int_{\text{packet}} \left\{\frac{d[\omega\varepsilon'(\omega)]}{d\omega} E_\omega E_\omega^* + \frac{d[\omega\mu'(\omega)]}{d\omega} H_\omega H_\omega^*\right\} d\omega$$

real, while $\varepsilon(\omega)$ is real and negative, so that $1/Z(\omega) = [\varepsilon(\omega)/\mu(\omega)]^{1/2}$ is purely imaginary, and the average Poynting vector (7.42) vanishes. This means that energy, on average, does not flow along axis z. However, this does not mean that the waves with $\omega < \omega_p$ are absorbed in the plasma. (Indeed, the Lorentz model with $\delta_j = 0$ does not describe any energy dissipation mechanism.) Instead, as we will see in the next section, the waves are instead *reflected* from the plasma's boundary.

Note also that in the limit $\omega \ll \omega_p$, Eq. (7.39) yields

$$\delta \to \frac{c}{\omega_p} = \left(\frac{c^2\varepsilon_0 m_e}{ne^2}\right)^{1/2} = \left(\frac{m_e}{\mu_0 ne^2}\right)^{1/2}. \tag{7.43}$$

But this is just a particular case (for $q = e$, $m = m_e$, and $\mu = \mu_0$) of the expression (6.44), which was derived in section 6.4 for the depth of the magnetic field's penetration into a lossless (collision-free) conductor in the quasi-static approximation. This fact shows again that, as was already discussed in section 6.7, that this approximation (in which the displacement currents are neglected) gives an adequate description of the time-dependent phenomena at $\omega \ll \omega_p$, i.e. at $\delta \ll c/\omega = 1/k = \lambda/2\pi$.[17]

There are two very important examples of plasmas. For the Earth's ionosphere, i.e. the upper part of its atmosphere, which is almost completely ionized by the UV and x-ray components of the Sun's radiation, the maximum value of n, reached at about 300 km over the Earth's surface, is between 10^{10} and 10^{12} m^{-3} (depending on the time of the day and the Sun's activity phase), so that that the maximum plasma frequency (7.37) is between 1 and 10 MHz. This is much higher than the particles' typical reciprocal collision time τ^{-1}, so that the first of Eq. (7.36) gives a good description of the plasma's electric polarization. The effect of the reflection of waves with $\omega < \omega_p$ from the ionosphere enables long-range (over-the-globe) radio communications and broadcasting at the so-called *short waves*, with frequencies of the order of 10 MHz: they may propagate in the flat channel formed by the Earth's surface and the ionosphere, reflected repeatedly by these 'walls'. Unfortunately, due to the random variations of the Sun's activity, and hence ω_p, such a natural communication channel is not very reliable and, in our age of trans-world fiber-optic cables, its practical importance has diminished.

Another important example of plasmas is free electrons in metals and other conductors. For a typical metal, n is of the order of 10^{23} cm^{-3} $\equiv 10^{29}$ m^{-3}, so that (7.37) yields $\omega_p \sim 10^{16}$ s^{-1}. Note that this value of ω_p is somewhat higher than the mid-optical frequencies ($\omega \sim 3 \times 10^{15}$ s^{-1}). This explains why planar, clean metallic surfaces, such as the aluminum and silver films used in mirrors, are so shiny: at these frequencies their complex permittivity $\varepsilon(\omega)$ is almost exactly real and negative,

[17] One more convenience of the simple model of a collision-free plasma, which has led us to Eq. (7.36), is that it may be readily generalized to the case of an additional strong dc magnetic field $\mathbf{B}_0$ (much higher that that of the wave) applied in the direction $\mathbf{n}$ of wave propagation. It is straightforward (and hence left for the reader) to show that such plasma exhibits the *Faraday effect* of the polarization plane's rotation, and hence gives an example of an anisotropic media that violates the Lorentz reciprocity relation (6.121).

leading to light reflection with very little absorption. However, the simple model (7.36), which neglects electron scattering, becomes inadequate at lower frequencies, $\omega\tau \sim 1$.

A phenomenological way of extending the model to account for scattering is to take, in Eq. (7.33), the lowest frequency ω_j to be equal zero (to describe the free electrons), while keeping the damping coefficient δ_0 of this mode finite to account for their energy loss due to scattering. Then Eq. (7.33) is reduced to

$$\varepsilon_{\text{ef}}(\omega) = \varepsilon_{\text{opt}}(\omega) + \frac{n_0 q^2}{m} \frac{1}{-\omega^2 - 2i\omega\delta_0} = \varepsilon_{\text{opt}}(\omega) + \frac{i}{\omega} \frac{n_0 q^2}{2\delta_0 m} \frac{1}{1 - i\omega/2\delta_0}, \tag{7.44}$$

where the response $\varepsilon_{\text{opt}}(\omega)$ at high (in practice, optical) frequencies is still given by Eq. (7.33), but now with $j \neq 0$. The result (7.44) allows for a simple interpretation. To show that, let us incorporate into our calculations the Ohmic conduction of the medium, generalizing Eq. (4.7) as $\mathbf{j}_\omega = \sigma(\omega)\mathbf{E}_\omega$ to account for the possible frequency dependence of the Ohmic conductivity. Plugging this relation into the Fourier image of the relevant macroscopic Maxwell equation, $\nabla \times \mathbf{H}_\omega = \mathbf{j}_\omega - i\omega\mathbf{D}_\omega \equiv \mathbf{j}_\omega - i\omega\varepsilon(\omega)\mathbf{E}_\omega$, we obtain

$$\nabla \times \mathbf{H}_\omega = [\sigma(\omega) - i\omega\varepsilon(\omega)]\mathbf{E}_\omega. \tag{7.45}$$

This relation shows that for a sinusoidal process, the addition of the Ohmic current density $\mathbf{j}_\omega$ to the displacement current density is equivalent to the addition of $\sigma(\omega)$ to $-i\omega\varepsilon(\omega)$, i.e. to the following change of the ac electric permittivity[18]:

$$\varepsilon(\omega) \rightarrow \varepsilon_{\text{ef}}(\omega) \equiv \varepsilon_{\text{opt}}(\omega) + i\frac{\sigma(\omega)}{\omega}. \tag{7.46}$$

Now the comparison of Eqs. (7.44) and (7.46) shows that they coincide if we take

$$\sigma(\omega) = \frac{n_0 q^2 \tau}{m_0} \frac{1}{1 - i\omega\tau} = \sigma(0)\frac{1}{1 - i\omega\tau}, \tag{7.47}$$

where the dc conductivity $\sigma(0)$ is described by the Drude formula (4.13) and the phenomenologically introduced coefficient δ_0 is associated with $1/2\tau$. Relation (7.47), which is frequently called the *generalized* (or 'ac', or 'rf') *Drude formula*[19], gives a very reasonable (semi-quantitative) description of the ac conductivity of many metals almost all the way up to optical frequencies.

Now returning to the discussion of the generalized Lorentz model (7.33), we see that the frequency dependences of the real (ε') and imaginary (ε'') parts of the complex permittivity it yields are not quite independent. For example, let us have one more look at the resonance peaks in figure 7.5. Each time the real part drops

[18] Alternatively, according to Eq. (7.45), it is possible (and in the field of infrared spectroscopy, conventional) to attribute the ac response of a medium at *all* frequencies to an effective complex conductivity: $\sigma_{\text{ef}}(\omega) \equiv \sigma(\omega) - i\omega\varepsilon(\omega) \equiv -i\omega\varepsilon_{\text{ef}}(\omega)$.

[19] It may also be derived from the Boltzmann kinetic equation in the so-called relaxation-time approximation (RTA)—see, e.g. *Part SM* section 6.2.

with frequency, $d\varepsilon'/d\omega < 0$, its imaginary part ε'' has a positive peak. R de L Kronig (in 1926) and H A Kramers (in 1927) independently showed that this is not an occasional coincidence pertinent only to this particular model. Moreover, full knowledge of the function $\varepsilon'(\omega)$ allows one to *calculate* the function $\varepsilon''(\omega)$ and vice versa. The reason is that both these functions are always related to a single real function $G(\theta)$ by Eq. (7.28).

To derive the Kramers–Kronig relations, let us consider Eq. (7.26b) on the complex frequency plane, $\omega \to \boldsymbol{\omega} \equiv \omega' + i\omega''$:

$$f(\boldsymbol{\omega}) \equiv \varepsilon(\boldsymbol{\omega}) - \varepsilon_0 = \int_0^\infty G(\theta) e^{i\boldsymbol{\omega}\theta}\, d\theta \equiv \int_0^\infty G(\theta) e^{i\omega'\theta} e^{-\omega''\theta}\, d\theta. \tag{7.48}$$

For all stable physical systems, $G(\theta)$ has to be finite for all important values of the real integration variable ($\theta > 0$), and tend to zero at $\theta \to 0$ and $\theta \to \infty$. (Indeed, according to Eq. (7.23), a non-vanishing $G(0)$ would mean an instantaneous response of the medium to the external force, while $G(\infty) \neq 0$ would mean that is has an infinitely long memory.) Because of that, and thanks to factor $e^{-\omega''\theta}$, the expression under the integral tends to zero at $|\boldsymbol{\omega}| \to \infty$ in all the upper half-plane ($\omega'' \geqslant 0$). As a result, we may claim that the complex function $f(\boldsymbol{\omega})$, defined by Eq. (7.48), is analytical in that half-plane. This fact allows us to use for it the general *Cauchy integral* formula[20]

$$f(\boldsymbol{\omega}) = \frac{1}{2\pi i} \oint_C f(\boldsymbol{\Omega}) \frac{d\boldsymbol{\Omega}}{\boldsymbol{\Omega} - \boldsymbol{\omega}}, \tag{7.49}$$

where $\boldsymbol{\Omega} \equiv \Omega' + i\Omega''$, with $\Omega' \equiv \Omega$, is a complex variable, for the case when $\boldsymbol{\omega} = \omega + i0 \equiv \omega$ is real. Let us take the integration contour shown in figure 7.7, with the radius R of the larger semicircle tending to infinity, and the radius r of the smaller semicircle (around the singular point $\boldsymbol{\Omega} = \omega$) tending to zero.

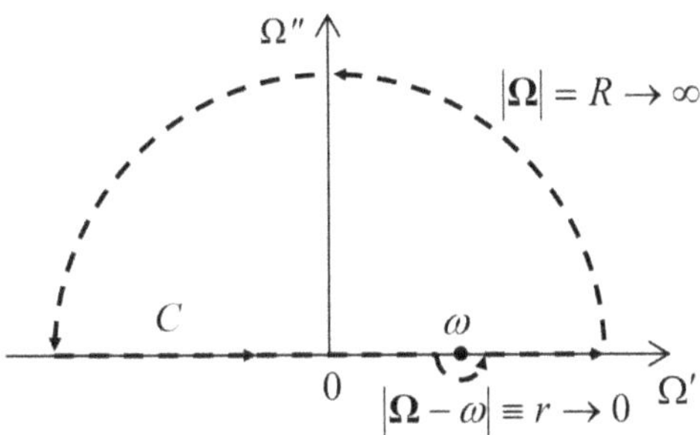

Figure 7.7. Deriving the Kramers–Kronig dispersion relations.

Due to the exponential decay of $|f(\boldsymbol{\Omega})|$ at $|\boldsymbol{\Omega}| \to \infty$, the contribution to the right-hand side of Eq. (7.49) from the larger semicircle vanishes[21], while the contribution from the small semicircle, where $\boldsymbol{\Omega} = \omega + r\exp\{i\varphi\}$ with $-\pi \leqslant \varphi \leqslant 0$, is

---

[20] See, e.g. Eq. (A.92).

[21] Strictly speaking, this also requires $|f(\Omega)|$ to decrease faster than Ω^{-1} at the real axis (at $\Omega'' = 0$), but due to the inertia of charged particles, this requirement is fulfilled for all realistic models of dispersion—see, e.g. Eq. (7.36).

$$\lim_{r\to 0}\frac{1}{2\pi i}\int_{\Omega=\omega+r\exp\{i\varphi\}} f(\Omega)\frac{d\Omega}{\Omega-\omega} = \frac{f(\omega)}{2\pi i}\int_{-\pi}^{0}\frac{ir\exp\{i\varphi\}d\varphi}{r\exp\{i\varphi\}} \equiv \frac{f(\omega)}{2\pi}\int_{-\pi}^{0}d\varphi \equiv \frac{1}{2}f(\omega). \tag{7.50}$$

As a result, for our contour C, Eq. (7.49) yields

$$f(\omega) = \lim_{r\to 0}\frac{1}{2\pi i}\left(\int_{-\infty}^{\omega-r} + \int_{\omega+r}^{+\infty}\right)f(\Omega)\frac{d\Omega}{\Omega-\omega} + \frac{1}{2}f(\omega). \tag{7.51}$$

Such an integral, excluding a symmetric infinitesimal vicinity of a pole singularity, is called the *principal value* of the (formally, diverging) integral from $-\infty$ to $+\infty$, and is denoted by letter P before it[22]. Using this notation, subtracting $f(\omega)/2$ from both parts of Eq. (7.51), and multiplying them by 2, we obtain

$$f(\omega) = \frac{1}{\pi i}\mathrm{P}\int_{-\infty}^{+\infty} f(\Omega)\frac{d\Omega}{\Omega-\omega}. \tag{7.52}$$

Now plugging into this complex equality the polarization-related difference $f(\omega) \equiv \varepsilon(\omega) - \varepsilon_0$ in the form $[\varepsilon'(\omega) - \varepsilon_0] + i[\varepsilon''(\omega)]$, and requiring both real and imaginary components of two sides of Eq. (7.52) to be equal separately, we obtain the famous *Kramers–Kronig dispersion relations*

$$\begin{aligned}\varepsilon'(\omega) &= \varepsilon_0 + \frac{1}{\pi}\mathrm{P}\int_{-\infty}^{+\infty}\varepsilon''(\Omega)\frac{d\Omega}{\Omega-\omega},\\ \varepsilon''(\omega) &= -\frac{1}{\pi}\mathrm{P}\int_{-\infty}^{+\infty}[\varepsilon'(\Omega)-\varepsilon_0]\frac{d\Omega}{\Omega-\omega}.\end{aligned} \tag{7.53}$$

We may use the already mentioned fact that $\varepsilon'(\omega)$ is always an even function, while $\varepsilon''(\omega)$ an odd function of frequency, to rewrite these relations in the following equivalent form,

$$\begin{aligned}\varepsilon'(\omega) &= \varepsilon_0 + \frac{2}{\pi}\mathrm{P}\int_{0}^{+\infty}\varepsilon''(\Omega)\frac{\Omega\, d\Omega}{\Omega^2-\omega^2},\\ \varepsilon''(\omega) &= -\frac{2\omega}{\pi}\mathrm{P}\int_{0}^{+\infty}[\varepsilon'(\Omega)-\varepsilon_0]\frac{d\Omega}{\Omega^2-\omega^2},\end{aligned} \tag{7.54}$$

which is more convenient for most applications, because it involves only physical (positive) frequencies.

Although the Kramers–Kronig relations are 'global' in frequency, in certain cases they allow an approximate calculation of dispersion from experimental data for absorption, collected even in a limited ('local') frequency range. Most importantly, if a medium has a sharp absorption peak at some frequency ω_j, we may describe it as

[22] I am typesetting this symbol in a Roman (upright) font, to avoid its confusion with the medium's polarization.

$$\varepsilon''(\omega) \approx c\delta(\omega - \omega_j) + \text{a more smooth function of } \omega, \tag{7.55}$$

and the first of Eq. (7.54) immediately gives

$$\varepsilon'(\omega) \approx \varepsilon_0 + \frac{2c}{\pi}\frac{\omega_j}{\omega_j^2 - \omega^2} + \text{another smooth function of } \omega, \tag{7.56}$$

thus predicting the anomalous dispersion near such a point. This calculation shows that such behavior observed in the Lorentz oscillator model (see figure 7.5) is by no means occasional or model-specific.

Let me emphasize again that the Kramers–Kronig relations (7.53) and (7.54) are much more general than the Lorentz model (7.33), hinging only on the causal, linear relation (7.21) between the polarization $P(t)$ with the electric field $E(t')$. Hence, these relations are also valid for the complex functions relating Fourier images of any cause/effect-related pair of variables. In particular, at a measurement of *any* linear response $r(t)$ of *any* experimental sample to *any* external applied field $f(t')$, whatever the nature of this response and the physics behind it, we may be confident that there is a causal relation between the variables r and f, so that the complex function $\chi(\omega) \equiv r_\omega/f_\omega$ does obey the Kramers–Kronig relations. However, it is still important to remember that a linear relation between the Fourier amplitudes of two variables does *not* necessarily imply the causal relationship between them[23].

7.3 Reflection

The most important new effect arising in non-uniform media is the wave *reflection*. Let us start its discussion from the simplest case of a plane electromagnetic wave that is normally incident on a sharp interface between two uniform, linear, isotropic media.

As the simplest example, let us assume that one of the two media (say, located at $z > 0$, see figure 7.8) cannot sustain any electric field at all[24]:

$$E|_{z\geqslant 0} = 0. \tag{7.57}$$

This condition is evidently incompatible with the single traveling wave (7.5). However, this solution may be readily corrected using the fact that the dispersion-free 1D wave equation,

$$\left(\frac{\partial^2}{\partial z^2} - \frac{1}{v^2}\frac{\partial^2}{\partial t^2}\right)E = 0, \tag{7.58}$$

[23] For example, the function $\varphi(\omega) \equiv E_\omega/P_\omega$ in the Lorentz oscillator model does not obey the Kramers–Kronig relations. This is evident not only from the fact that $E(t)$ is *not* a causal function of $P(t)$, but even mathematically. Indeed, the Green's function describing a causal relationship has to tend to zero at small time delays $\theta \equiv t - t'$, so that its Fourier image has to tend to zero at $\omega \to \pm\infty$. This is certainly true for the function $f(\omega)$ given by Eq. (7.32), but not for the reciprocal function $\varphi(\omega) \equiv 1/f(\omega) \propto (\omega^2 - \omega_0^2) - 2i\delta\omega$, which diverges at large frequencies.

[24] Such equality is given, in particular, by the macroscopic model of a good conductor—see Eq. (2.1).

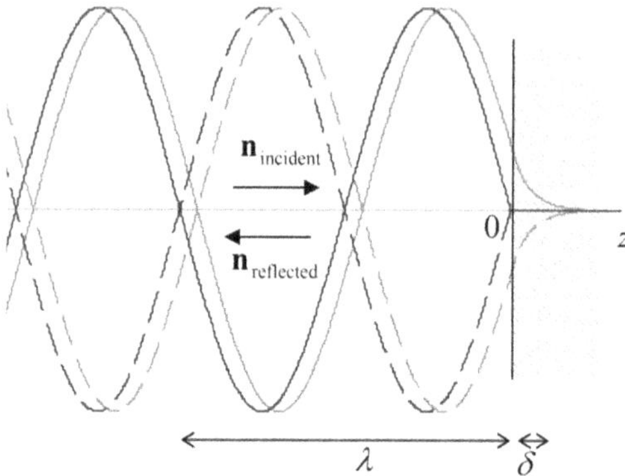

Figure 7.8. A snapshot of the electric field at the reflection of a sinusoidal wave from a perfect conductor: a realistic pattern (red lines) and its macroscopic, ideal-mirror approximation (blue lines). Dashed lines show the snapshots after a half-period time delay ($\omega\Delta t = \pi$).

supports waves propagating, with the same speed, in opposite directions. As a result, the following linear superposition of two such waves,

$$E|_{z\leqslant 0} = f(z - vt) - f(-z - vt), \tag{7.59}$$

satisfies both the equation and the boundary condition (7.57) for an arbitrary function f. The second term in Eq. (7.59) may be interpreted as the *total reflection* of the incident wave (described by its first term)—in this particular case, with the change of the electric field's sign. This means, in particular, that within the macroscopic model a conductor acts as a perfect mirror. By the way, since the vector **n** of the reflected wave is opposite to that incident one (see the arrows in figure 7.8), Eq. (7.6) shows that the magnetic field of the wave does *not* change its sign at the reflection:

$$H|_{z\leqslant 0} = \frac{1}{Z}[f(z - vt) + f(-z - vt)]. \tag{7.60}$$

The blue lines in figure 7.8 show the resulting pattern (7.59) for the simplest, monochromatic wave:

$$E|_{z\leqslant 0} = \text{Re}\,[E_\omega e^{i(kz-\omega t)} - E_\omega e^{i(-kz-\omega t)}]. \tag{7.61a}$$

Depending on convenience in a particular context, this pattern may be legitimately represented and interpreted either as the linear superposition (7.61a) of two *traveling* waves, or as a single *standing wave*:

$$\begin{aligned} E|_{z\leqslant 0} &= -\,2\text{Im}(E_\omega e^{-i\omega t})\sin kz \\ &\equiv 2\,\text{Re}\,(iE_\omega e^{-i\omega t})\sin kz \equiv 2\,\text{Re}\,[E_\omega e^{-i(\omega t-\pi/2)}]\sin kz, \end{aligned} \tag{7.61b}$$

in which the electric and magnetic field oscillate with phase shifts of $\pi/2$ both in time and space:

$$H|_{z\leqslant 0} = \mathrm{Re}\left[\frac{E_\omega}{Z}e^{i(kz-\omega t)} + \frac{E_\omega}{Z}e^{i(-kz-\omega t)}\right] \equiv 2\,\mathrm{Re}\left(\frac{E_\omega}{Z}e^{-i\omega t}\right)\cos kz. \tag{7.62}$$

As the result of this shift, the time average of the Poynting vector's magnitude,

$$S(z, t) = EH = \frac{1}{Z}\,\mathrm{Re}\,[E_\omega^2 e^{-2i\omega t}]\,\sin 2kz, \tag{7.63}$$

equals zero, showing that at the total reflection there is no *average* power flow. (This is natural, because a perfect mirror can neither transmit a wave nor absorb it.) However, Eq. (7.63) shows that the standing wave provides local oscillations of energy, transferring it periodically between the concentrations of the electric and magnetic fields, separated by the distance $\Delta z = \pi/2k = \lambda/4$.

In the case of sinusoidal waves, the reflection effects may be readily explored even for the more general case of dispersive and/or lossy (but still linear) media in which $\varepsilon(\omega)$ and $\mu(\omega)$, and hence the wave vector $k(\omega)$ and the wave impedance $Z(\omega)$, defined by Eq. (7.28), are certain complex functions of frequency. The 'only' new factors we have to account for is that in this case the reflection may not be full, and that inside the second medium we have to use the traveling-wave solution as well. Both these factors may be taken care of by looking for the solution of our boundary problem in the form

$$E|_{z\leqslant 0} = \mathrm{Re}\,[E_\omega(e^{ik_-z} + R\,e^{-ik_-z})\,e^{-i\omega t}], \quad E|_{z\geqslant 0} = \mathrm{Re}\,[E_\omega T e^{ik_+z}e^{-i\omega t}], \tag{7.64}$$

and hence, according to Eq. (7.6),

$$\begin{aligned} H|_{z\leqslant 0} &= \mathrm{Re}\left[\frac{E_\omega}{Z_-(\omega)}(e^{ik_-z} - R\,e^{-ik_-z})\,e^{-i\omega t}\right], \\ H|_{z\geqslant 0} &= \mathrm{Re}\left[\frac{E_\omega}{Z_+(\omega)}Te^{ik_+z}e^{-i\omega t}\right]. \end{aligned} \tag{7.65}$$

(Indices + and – correspond to, respectively, the media at $z > 0$ and $z < 0$.) Please note the following important features of these relations:

(i) Due to the problem's linearity, we could (and did) take the complex amplitudes of the reflected and transmitted wave proportional to (E_ω) of the incident wave, scaling them with dimensionless, generally complex coefficients R and T. As the comparison of Eqs. (7.64) and (7.65) with Eqs. (7.61) and (7.62) shows, the total reflection from an ideal mirror, that was discussed above, corresponds to the particular case $R = -1$ and $T = 0$.

(ii) Since the incident wave we are considering arrives from one side only (from $z = -\infty$), there is no need to include a term proportional to $\exp\{-ik_+z\}$ into Eqs. (7.64) and (7.65), in our current problem. However, we would need such a term if the medium at $z > 0$ had been non-uniform (e.g. had at least one more interface or any other inhomogeneity), because the wave reflected from that additional inhomogeneity would be incident on our interface (located at $z = 0$) from the right.

(iii) The solution (7.64) and (7.65) is sufficient even for the description of the cases when waves cannot propagate at $z \geqslant 0$, for example a conductor or a plasma with $\omega_p > \omega$. Indeed, the exponential drop of the field amplitude at $z > 0$ in such cases is automatically described by the imaginary part of the wave number k_+—see Eq. (7.29).

In order to calculate the coefficients R and T, we need to use boundary conditions at $z = 0$. Since the reflection does not change the transverse character of the partial waves, at the normal incidence both vectors $\mathbf{E}$ and $\mathbf{H}$ remain tangential to the interface plane (in our notation, $z = 0$). Reviewing the arguments that have led us, in statics, to the boundary conditions (3.37) and (5.117) for these components, we see that they remain valid for the time-dependent situation as well[25], so that for our current case of normal incidence we may write:

$$E|_{z=-0} = E|_{z=+0}, \qquad H|_{z=-0} = H|_{z=+0}. \tag{7.66}$$

Plugging equations (7.64) and (7.65) into these conditions, we obtain two equations for the coefficients R and T:

$$1 + R = T, \qquad \frac{1}{Z_-}(1 - R) = \frac{1}{Z_+}T. \tag{7.67}$$

Solving this simple system of linear equations, we obtain[26]

$$R = \frac{Z_+ - Z_-}{Z_+ + Z_-}, \qquad T = \frac{2Z_+}{Z_+ + Z_-}. \tag{7.68}$$

These formulas are very important, and much more general than one might think, because they are applicable for virtually any 1D waves—electromagnetic or not—if only the impedance Z is defined in a proper way[27]. Since in the general case the wave impedances $Z_\pm$, defined by Eq. (7.28) with the corresponding indices, are complex functions of frequency, Eq. (7.68) show that coefficients R and T may have imaginary parts as well. This fact has most important consequences at $z < 0$, where the reflected wave, proportional to R, mixes ('interferes') with the incident wave.

---

[25] For example, the first of conditions (7.66) may be obtained by integrating the full (time-dependent) Maxwell equation $\nabla \times \mathbf{E} + \partial\mathbf{B}/\partial t = 0$ over a narrow and long rectangular contour with dimensions l and d ($d \ll l$) stretched along the interface. In the Stokes theorem, the first term gives $\Delta E_\tau l$, while the contribution of the second term is proportional to product dl and vanishes as $d/l \to 0$. The proof of the second boundary condition is similar—as was already discussed in section 6.2.

[26] Please note that only the medium impedances (rather than wave velocities) are important for the reflection in this case! Unfortunately, this fact is not clearly emphasized in some textbooks that discuss only the case $\mu_\pm = \mu_0$, when $Z = (\mu_0/\varepsilon)^{1/2}$ and $v = 1/(\mu_0\varepsilon)^{1/2}$ are proportional to each other.

[27] See, e.g. the discussion of elastic waves of mechanical deformations in *Part CM* sections 6.3, 6.4, 7.7, and 7.8.

Indeed, with $R = |R|e^{i\varphi}$ (where $\varphi \equiv \arg R$ is a real phase shift), the expression in parentheses in the first of Eqs. (7.64) becomes

$$\begin{aligned} e^{ik_-z} + R\, e^{-ik_-z} &= (1 - |R| + |R|)e^{ik_-z} + |R|\, e^{i\varphi}\, e^{-ik_-z} \\ &= (1 - |R|)e^{ik_-z} + 2\,|R|\, e^{i\varphi/2} \sin[k_-(z - \delta_-)]\,, \\ \text{where}\;\; \delta_- &\equiv \frac{\varphi - \pi}{2k_-}. \end{aligned} \tag{7.69}$$

This means that the field may be represented as a sum of a traveling wave and a standing wave, with amplitude proportional to $|R|$, shifted by distance δ_- toward the interface, relative to the ideal-mirror pattern (7.61b). This effect is frequently used for the experimental measurement of an unknown impedance Z_+ of some medium, provided than Z_- is known (e.g. for the free space, $Z_- = Z_0$). For that, a small antenna (the *probe*), not disturbing the fields' distribution too much, is placed into the wave field, and the amplitude of the ac voltage induced in it by the wave in the probe is measured by a detector (e.g. a semiconductor diode with a quadratic *I–V* curve) as a function of z (figure 7.9). From the results of such a measurement, it is straightforward to find both $|R|$ and δ_-, and hence restore the complex R, and then use Eq. (7.68) to calculate both the modulus and the argument of Z_+. (Before the advent of computers, a specially lined paper, called the *Smith chart*, was commercially available for performing this recalculation graphically; it is occasionally used even now for the presentation of results.)

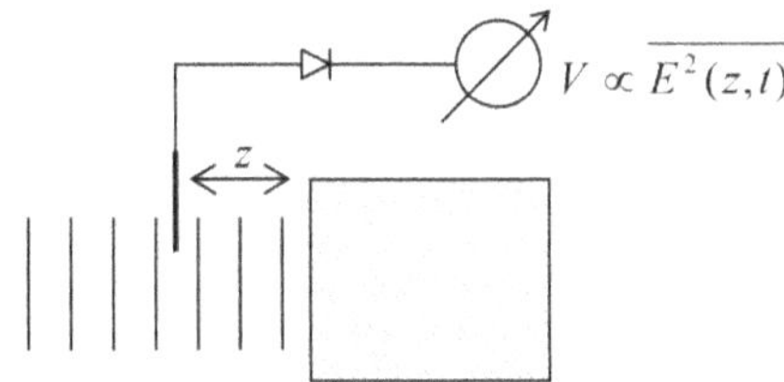

Figure 7.9. Measurement of the complex impedance of a medium (schematically).

Now let us discuss what these results give for waves incident from the free space ($Z_-(\omega) = Z_0 = \text{const}$, $k_- = k_0 = \omega/c$) onto the surface of two particular, important media.

(i) For a collision-free plasma (with negligible magnetization) we may use Eq. (7.36) with $\mu(\omega) = \mu_0$, to represent the impedance in either of two equivalent forms:

$$Z_+ = Z_0 \frac{\omega}{\left(\omega^2 - \omega_p^2\right)^{1/2}} \equiv -iZ_0 \frac{\omega}{\left(\omega_p^2 - \omega^2\right)^{1/2}}. \tag{7.70}$$

The former of these forms is more convenient in the case $\omega > \omega_p$, when the wave vector k_+ and the wave impedance Z_+ of the plasma are real, so that part of the incident wave does propagate into the plasma. Plugging this expression into the latter of Eqs. (7.68), we see that T is real:

$$T = \frac{2\omega}{\omega + \left(\omega^2 - \omega_p^2\right)^{1/2}}. \tag{7.71}$$

Note that according to this formula, and somewhat counter-intuitively, $T > 1$ for any frequency (above ω_p). How can the transmitted wave be more intense than the incident one that has induced it? For a better understanding of this result, let us compare the powers (rather than the electric field amplitudes) of these two waves, i.e. their average Poynting vectors (7.42):

$$\bar{S}_{\text{incident}} = \frac{|E_\omega|^2}{2Z_0}, \qquad \overline{S_+} = \frac{|TE_\omega|^2}{2Z_+} = \frac{|E_\omega|^2}{2Z_0}\frac{4\omega\left(\omega^2 - \omega_p^2\right)^{1/2}}{\left[\omega + \left(\omega^2 - \omega_p^2\right)^{1/2}\right]^2}. \tag{7.72}$$

It is easy to check that the ratio of these two values[28] is always below 1 (and tends to zero at $\omega \to \omega_p$), so that only a fraction of the incident wave power may be transmitted. Hence the result $T > 1$ may be interpreted as follows: the interface between two media also works as an *impedance transformer*. Although it can never transmit more *power* than the incident wave provides (i.e. can only decrease the product $S = EH$), since the ratio $Z = E/H$ changes at the interface, the amplitude of *one of the fields* may increase at the transmission.

Now let us proceed to case $\omega < \omega_p$, when the waves cannot propagate in the plasma. In this case, the latter of the expressions (7.70) is more convenient, because it immediately shows that Z_+ is purely imaginary, while $Z_- = Z_0$ is purely real. This means that $(Z_+ - Z_-) = (Z_+ + Z_-)^*$, i.e. according to the first equation of Eqs. (7.68), $|R| = 1$, so that the reflection is total, i.e. no incident power (on average) is transferred into the plasma—as was already discussed in section 7.2. However, the complex R has a finite argument,

$$\varphi \equiv \arg R = 2\arg(Z_+ - Z_0) = -2\tan^{-1}\frac{\omega}{\left(\omega_p^2 - \omega^2\right)^{1/2}}, \tag{7.73}$$

and hence provides a finite spatial shift (7.69) of the standing wave toward the plasma surface:

$$\delta_- = \frac{\varphi - \pi}{2k_0} = \frac{c}{\omega}\tan^{-1}\frac{\omega}{\left(\omega_p^2 - \omega^2\right)^{1/2}}. \tag{7.74}$$

On the other hand, we already know from Eq. (7.40) that the solution at $z > 0$ is exponential, with the decay length δ that is described by Eq. (7.39). Calculating from

[28] This ratio is sometimes also called the transmission coefficient, but in order to avoid its confusion with the T defined by Eq. (7.64), it is better to call it the *power* transmission coefficient.

the coefficient T the exact coefficient before this exponent, it is straightforward to verify that the electric and magnetic fields are indeed continuous at the interface, forming the pattern shown with red lines in figure 7.8. This penetration may be experimentally observed, for example, by bringing close to the interface another material transparent as frequency ω. Even without solving this problem exactly, it is evident that if the distance between these two interfaces becomes comparable to δ, a part of the exponential 'tail' of the field is picked up by the second material, and induces a propagating wave. This is an electromagnetic analog of the quantum-mechanical tunneling through a potential barrier[29].

Note that at low frequencies, both δ_- and δ tend to the same frequency-independent value,

$$\delta,\ \delta_- \to \frac{c}{\omega_p} = \left(\frac{c^2\varepsilon_0 m_e}{ne^2}\right)^{1/2} = \left(\frac{m_e}{\mu_0 ne^2}\right)^{1/2}, \qquad \text{at } \frac{\omega}{\omega_p} \to 0, \qquad (7.75)$$

which is just the field penetration depth δ (6.44) calculated for a perfect conductor model (assuming $m = m_e$ and $\mu = \mu_0$) in the quasi-static limit. This is natural, because the condition $\omega \ll \omega_p$ may be recast as $\lambda_0 \equiv 2\pi c/\omega \gg 2\pi c/\omega_p \equiv 2\pi\delta$, justifying the quasi-static approximation.

(ii) Now let us consider the electromagnetic wave reflection from an Ohmic, non-magnetic conductor. In the simplest low-frequency limit, when $\omega\tau$ is much less than 1, the conductor may be described by a frequency-independent conductivity σ.[30] According to Eq. (7.46), in this case we can take

$$Z_+ = \left(\frac{\mu_0}{\varepsilon_{\text{opt}}(\omega) + i\sigma/\omega}\right)^{1/2}. \qquad (7.76)$$

With this substitution, Eqs. (7.68) immediately give us all the results of interest. In particular, in the most important quasi-static limit (when $\delta_s \equiv (2/\mu_0\sigma\omega)^{1/2} \ll \lambda_0 \equiv 2\pi c/\omega$, i.e. $\sigma/\omega \gg \varepsilon_0 \sim \varepsilon_{\text{opt}}$), the conductor' impedance is low:

$$Z_+ \approx \left(\frac{\mu_0\omega}{i\sigma}\right)^{1/2} \equiv \pi\left(\frac{2}{i}\right)^{1/2}\frac{\delta_s}{\lambda_0}Z_0, \qquad \text{i.e.} \left|\frac{Z_+}{Z_0}\right| \ll 1. \qquad (7.77)$$

The impedance is complex, and hence some fraction $\mathscr{A}$ of the incident wave is absorbed by the conductor. This fraction may be found as the ratio of the dissipated power (either calculated, as was done above, from Eq. (7.68), or just taken from Eq. (6.36) with the magnetic field amplitude $|H_\omega| = 2|E_\omega|/Z_0$—see Eq. (7.62)) to the incident wave's power given by the first of Eq. (7.72). The result is

[29] See, e.g. *Part QM* section 2.3.

[30] In a typical metal, $\tau \sim 10^{-13}$ s, so that this approximation works well all the way up to $\omega \sim 10^{13}$ s^{-1}, i.e. up to far-infrared frequencies.

$$\mathscr{f} = \frac{2\omega\delta_s}{c} \equiv 4\pi\frac{\delta_s}{\lambda_0} \ll 1. \tag{7.78}$$

This important result is widely used for crude estimates of the energy dissipation in metallic-wall waveguides and resonators, and immediately shows that in order to keep the energy losses low, the characteristic size of such systems (which gives a scale of the free-space wavelengths λ_0 at which they are used) should be much larger than δ_s. A more detailed theory of these structures, and the effects of energy loss in them, will be discussed later in this chapter.

7.4 Refraction

Now let us consider the effects arising at a plane interface between two uniform media if the wave's incidence angle θ (figure 7.10) is arbitrary (rather than equal to zero as in our previous analysis), for the simplest case of fully transparent media with real $\varepsilon_\pm(\omega)$ and $\mu_\pm(\omega)$. (For the sake of notation simplicity, the argument of these functions will be dropped, i.e. just implied in most formulas below.)

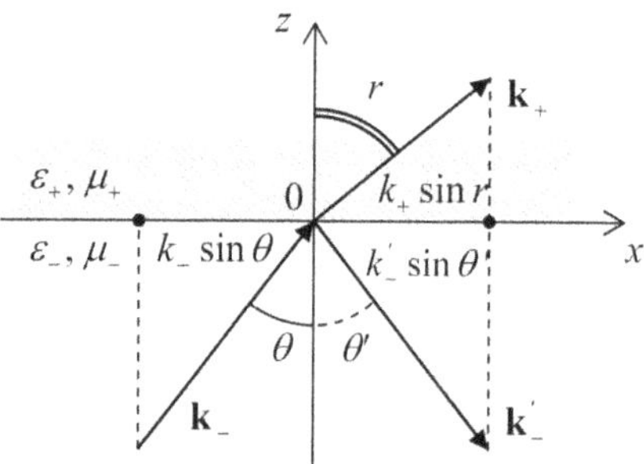

Figure 7.10. Plane wave reflection, transmission, and refraction at a plane interface. The plane of the drawing is selected to contain all three wave vectors: $\mathbf{k}_+$, $\mathbf{k}_-$, and $\mathbf{k}'_-$.

In contrast with the case of normal incidence, here the wave vectors $\mathbf{k}_-$, $\mathbf{k}'_-$, and $\mathbf{k}_+$ of the three component (incident, reflected, and transmitted) waves may have different directions. (This change of the transmitted wave's direction is called *refraction*.) Hence now we have to start our analysis by writing a general expression for a single-plane, monochromatic wave for the case when its wave vector $\mathbf{k}$ has all three Cartesian components, rather than one. An evident generalization of Eq. (7.11) for this case is

$$f(r, t) = \text{Re}\left[f_\omega e^{i(k_x x + k_y y + k_z z) - \omega t)}\right] = \text{Re}\left[f_\omega e^{i(\mathbf{k}\cdot\mathbf{r} - \omega t)}\right]. \tag{7.79}$$

This relation enables a ready analysis of 'kinematic' relations that are independent of the media impedances. Indeed, it is sufficient to notice that in order to satisfy *any* linear, homogeneous boundary conditions at the interface ($z = 0$), all waves must have the same temporal and spatial dependence on this plane. Hence if we select the plane xz so that the vector $\mathbf{k}_-$ lies in it, then $(k_-)_y = 0$, and $\mathbf{k}_+$ and $\mathbf{k}'_-$ cannot have any y-component either, i.e. all three vectors lie in the same plane—which is selected as the plane of drawing in figure 7.10. Moreover, due to the same reason their x-components should be equal:

$$k_- \sin\theta = k_-' \sin\theta' = k_+ \sin r. \tag{7.80}$$

From here we immediately obtain the well-known laws of reflection

$$\theta' = \theta, \tag{7.81}$$

and refraction[31]

$$\frac{\sin r}{\sin\theta} = \frac{k_-}{k_+}. \tag{7.82}$$

In this form, the laws are valid for plane waves of any nature. In optics, the Snell law (7.82) is frequently represented in the form

$$\frac{\sin r}{\sin\theta} = \frac{n_-}{n_+}, \tag{7.83}$$

where $n_\pm$ is the *index of refraction*, also called the 'refractive index' of the corresponding medium, defined as its wave number normalized to that of the free space (at the particular wave's frequency):

$$n_\pm \equiv \frac{k_\pm}{k_0} = \left(\frac{\varepsilon_\pm\mu_\pm}{\varepsilon_0\mu_0}\right)^{1/2}. \tag{7.84}$$

Perhaps the most famous corollary of the Snell law is that if a wave propagates from a medium with a higher index of refraction to that with a lower one (i.e. if $n_- > n_+$ in figure 7.10), for example from water into air, there is always a certain *critical* value θ_c of the incidence angle,

$$\theta_c = \sin^{-1}\frac{n_+}{n_-} \equiv \sin^{-1}\left(\frac{\varepsilon_+\mu_+}{\varepsilon_-\mu_-}\right)^{1/2}, \tag{7.85}$$

at which the refraction angle r (see figure 7.10 again) reaches $\pi/2$. At a larger θ, i.e. within the range $\theta_c < \theta < \pi/2$, the boundary conditions (7.80) cannot be satisfied by a refracted wave with a real wave vector, so that the wave experiences so-called *total internal reflection*. This effect is very important in practice, because it means that dielectric surfaces may be used as optical mirrors, in particular in optical fibers—to be discussed in more detail in section 7.7. This is very fortunate for telecommunication technology, because light's reflection from metals is rather imperfect. Indeed,

[31] This relation is traditionally called the *Snell law*, after seventeenth century author W Snellius, though it has been traced all the way back to a circa 984 manuscript by Abu Saad al-Ala ibn Sahl.

according to Eq. (7.78), in the optical range ($\lambda_0 \sim 0.5$ μm, i.e. $\omega \sim 10^{15}$ s^{-1}), even the best conductors (with $\sigma \sim 6 \times 10^8$ s m^{-1} and hence the normal skin depth $\delta_s \sim 1.5$ nm) suffer losses of at least a few percent at each reflection.

Note, however, that even within the range $\theta_c < \theta < \pi/2$, the field at $z > 0$ is not identically equal to zero: it penetrates into the less dense medium by a distance of the order of λ_0, exponentially decaying inside it, just as it does at normal incidence—see figure 7.8. However, at $\theta \neq 0$ the penetrating field still propagates, with the wave number (7.80), along the interface. Such a field, exponentially dropping in one direction but still propagating as a wave in another direction, is commonly called the *evanescent wave*.

One more remark: just as at the normal incidence, the field's penetration into another medium causes a phase shift of the reflected wave—see, e.g. Eq. (7.69) and its discussion. A new feature of this phase shift, arising at $\theta \neq 0$, is that it also has a component parallel to the interface—the so-called called the *Goos–Hänchen effect*. In the geometric optics, this effect leads to an image shift (relative to its position in a perfect mirror) with components both normal and parallel to the interface.

Now let us carry out an analysis of the 'dynamic' relations that determine the amplitudes of the refracted and reflected waves. For this we need to write explicitly the boundary conditions at the interface (i.e. the plane $z = 0$). Since now the electric and/or magnetic fields may have components normal to the plane, in addition to the continuity of their tangential components, which were repeatedly discussed above,

$$E_{x,y}\Big|_{z=-0} = E_{x,y}\Big|_{z=+0}, \quad H_{x,y}\Big|_{z=-0} = H_{x,y}\Big|_{z=+0}, \tag{7.86}$$

we also need relations for the normal components. As follows from the homogeneous macroscopic Maxwell equations (6.99*b*), these conditions are also the same as in statics, i.e. D_n = const and B_n = const, for our reference frame choice (figure 7.10) giving

$$\varepsilon_- E_z\Big|_{z=-0} = \varepsilon_+ E_z\Big|_{z=+0}, \quad \mu_- H_z\Big|_{z=-0} = \mu_+ H_z\Big|_{z=+0}. \tag{7.87}$$

The expressions of these components via the amplitudes E_ω, RE_ω, and TE_ω of the incident, reflected, and transmitted waves depend on the incident wave's polarization. For example, for a linearly polarized wave with the electric field vector *perpendicular* to the plane of incidence, i.e. *parallel* to the interface plane, the reflected and refracted waves are similarly polarized—see figure 7.11a. As a result, all E_z are equal to zero (so that the first of Eq. (7.87) is inconsequential), while the tangential components of the electric field are just equal to their full amplitudes, just like at the normal incidence, so we still can use Eq. (7.64), expressing these components via the coefficients R and T. However, at $\theta \neq 0$ the magnetic fields have not only tangential components

$$H_x\Bigg|_{z=-0} = \text{Re}\left[\frac{E_\omega}{Z_-}(1 - R)\cos\theta\, e^{-i\omega t}\right], \quad H_x\Bigg|_{z=+0} = \text{Re}\left[\frac{E_\omega}{Z_+}T\cos r\; e^{-i\omega t}\right], \tag{7.88}$$

but also normal components (see figure 7.11a):

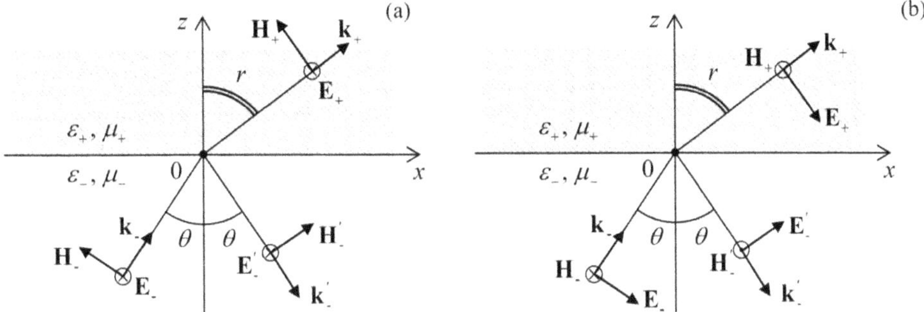

Figure 7.11. Reflection and refraction at two different linear polarizations of the incident wave.

$$H_z\Big|_{z=-0} = \text{Re}\left[\frac{E_\omega}{Z_-}(1 + R)\sin\theta\, e^{-i\omega t}\right], \quad H_z\Big|_{z=+0} = \text{Re}\left[\frac{E_\omega}{Z_+}T \sin r\, e^{-i\omega t}\right]. \quad (7.89)$$

Plugging these expressions into the boundary conditions expressed by Eq. (7.86) (in this case, for y components only) and the second of Eqs. (7.87), we obtain *three* equations for *two* unknown coefficients R and T. However, two of these equations duplicate each other because of the Snell law, and we obtain just two independent equations,

$$1 + R = T, \qquad \frac{1}{Z_-}(1 - R)\cos\theta = \frac{1}{Z_+}T\cos r, \qquad (7.90)$$

which are a very natural generalization of Eq. (7.67), with replacements $Z_- \to Z_-\cos r$, $Z_+ \to Z_+\cos\theta$. As a result, we can immediately use Eqs. (7.68) to write the solution of the system (7.90)[32]:

$$R = \frac{Z_+\cos\theta - Z_-\cos r}{Z_+\cos\theta + Z_-\cos r}, \qquad T = \frac{2Z_+\cos\theta}{Z_+\cos\theta + Z_-\cos r}. \qquad (7.91a)$$

If we want to express these coefficients via the angle of incidence alone, we should use the Snell law (7.82) to eliminate the angle r, obtaining the commonly used, more bulky expressions:

$$R = \frac{Z_+\cos\theta - Z_-[1 - (k_-/k_+)^2\sin^2\theta]^{1/2}}{Z_+\cos\theta + Z_-[1 - (k_-/k_+)^2\sin^2\theta]^{1/2}},$$
$$T = \frac{2Z_+\cos\theta}{Z_+\cos\theta + Z_-[1 - (k_-/k_+)^2\sin^2\theta]^{1/2}}. \qquad (7.91b)$$

[32] Note that we may calculate the reflection and transmission coefficients R' and T' for the wave traveling in the opposite direction just by making the following parameter swaps: $Z_+ \leftrightarrow Z_-$ and $\theta \leftrightarrow r$, and that the resulting coefficients satisfy the following *Stokes relations*: $R' = -R$, and $R^2 + TT' = 1$, for any $Z_\pm$.

However, my strong preference is to use the kinematic relation (7.82) and the dynamic relations (7.91a) separately, because Eq. (7.91b) obscures the very important physical fact that and the ratio of $k_\pm$, i.e. of the wave velocities of the two media, is only involved in the Snell law, while the dynamic relations essentially include only the ratio of wave impedances—just as in the case of normal incidence.

In the opposite case of linear polarization of the electric field within the plane of incidence (figure 7.11b), it is the magnetic field which does not have a normal component, so it is now the second of Eqs. (7.87) that does not participate in the solution. However, now the electric fields in two media have not only tangential components,

$$E_x|_{z=-0} = \mathrm{Re}\,[E_\omega(1+R)\cos\theta\; e^{-i\omega t}], \quad E_x|_{z=+0} = \mathrm{Re}\,[E_\omega T\cos r\; e^{-i\omega t}], \tag{7.92}$$

but also normal components (figure 7.11b):

$$E_z|_{z=-0} = E_\omega(-1+R)\sin\theta, \quad E_z|_{z=+0} = -E_\omega T\sin r. \tag{7.93}$$

As a result, instead of Eqs. (7.90), the reflection and transmission coefficients are related as

$$(1+R)\cos\theta = T\cos r, \qquad \frac{1}{Z_-}(1-R) = \frac{1}{Z_+}T. \tag{7.94}$$

Again, the solution of this system may be immediately written using the analogy with Eqs. (7.67):

$$R = \frac{Z_+\cos r - Z_-\cos\theta}{Z_+\cos r + Z_-\cos\theta}, \qquad T = \frac{2Z_+\cos\theta}{Z_+\cos r + Z_-\cos\theta}, \tag{7.95a}$$

or, alternatively, using the Snell law, in a bulkier form:

$$\begin{aligned} R &= \frac{Z_+[1-(k_-/k_+)^2\sin^2\theta]^{1/2} - Z_-\cos\theta}{Z_+[1-(k_-/k_+)^2\sin^2\theta]^{1/2} + Z_-\cos\theta}, \\ T &= \frac{2Z_+\cos\theta}{Z_+[1-(k_-/k_+)^2\sin^2\theta]^{1/2} + Z_-\cos\theta}. \end{aligned} \tag{7.95b}$$

For the particular case $\mu_+ = \mu_- = \mu_0$, when $Z_+/Z_- = (\varepsilon_-/\varepsilon_+)^{1/2} = k_-/k_+ = n_-/n_+$ (which is approximately correct for traditional optical media), Eqs. (7.91b) and (7.95b) are called the *Fresnel formulas*[33]. Most textbooks are quick to point out that there is a major difference between them: for the electric field polarization within the

[33] After A-J Fresnel (1788–1827), one of the pioneers of wave optics, who is credited, among many other contributions (see, in particular, the discussions in chapter 8), for the concept of light as a purely transverse wave.

plane of incidence (figure 7.11b), the reflected wave amplitude (proportional to the coefficient R) turns to zero[34] at a special value of θ (called the *Brewster angle*)[35]

$$\theta_B = \tan^{-1}\frac{n_+}{n_-}, \tag{7.96}$$

while there is no such angle in the opposite case (figure 7.11a). However, note that this statement, as well as Eq. (7.96), is true only for the case $\mu_+ = \mu_-$. In the general case of different ε and μ, Eqs. (7.91) and (7.95) show that the reflected wave vanishes at $\theta = \theta_B$ with

$$\tan^2\theta_B = \frac{\varepsilon_-\mu_+ - \varepsilon_+\mu_-}{\varepsilon_+\mu_+ - \varepsilon_-\mu_-} \times \begin{cases} (\mu_+/\mu_-), & \text{for } \mathbf{E}\perp\mathbf{n}_z \text{ (figure 11a)}, \\ (-\varepsilon_+/\varepsilon_-), & \text{for } \mathbf{H}\perp\mathbf{n}_z \text{ (figure 11b)}. \end{cases} \tag{7.97}$$

Note the natural $\varepsilon \leftrightarrow \mu$ symmetry of these relations, resulting from the $\mathbf{E} \leftrightarrow \mathbf{H}$ symmetry for these two polarization cases (figure 7.11). These formulas also show that for any set of parameters of the two media (with $\varepsilon_\pm$, $\mu_\pm > 0$), $\tan^2\theta_B$ is positive (and hence a real Brewster angle θ_B exists) only for one of these two polarizations. In particular, if the interface is due to the change of μ alone (i.e. if $\varepsilon_+ = \varepsilon_-$), the first of Eq. (7.97) is reduced to the simple form (7.96) again, while for the polarization shown in figure 7.11b there is no Brewster angle, i.e. the reflected wave has a non-vanishing amplitude for any θ.

Such an account of both media parameters, ε and μ, on an equal footing is in especially necessary to describe the so-called *negative refraction* effects[36]. As was shown in section 7.2, in a medium with electric-field-driven resonances, the function $\varepsilon(\omega)$ may be almost real and negative, at least within limited frequency intervals—see, in particular, Eq. (7.34) and figure 7.5. As has already been discussed, if the function $\mu(\omega)$ is real and positive at these frequencies, then $k^2(\omega) = \omega^2\varepsilon(\omega)\mu(\omega) < 0$ and k may be represented as i/δ with real δ, meaning that the exponential field decays into the medium. However, let us consider the case when both $\varepsilon(\omega) < 0$ *and* $\mu(\omega) < 0$ at a certain frequency. (This is evidently possible in a medium with both **E**-driven and **H**-driven resonances, at a proper choice of their resonant frequencies.) Since in this case $k^2(\omega) = \omega^2\varepsilon(\omega)\mu(\omega) > 0$, the wave vector is real, so that Eq. (7.79) describes a traveling wave, and one could think that there is nothing new in this case. Not quite so!

[34] This effect is used in practice to obtain linearly polarized light, with the electric field vector perpendicular to the plane of incidence, from the natural light with its random polarization. An even more practical application of the effect is a partial reduction of undesirable glare from wet surfaces (for the water/air interface, $n_+/n_- \approx 1.33$, giving $\theta_B \approx 50°$) by making car-light covers and sunglasses of vertically polarizing materials.

[35] A very simple interpretation of Eq. (7.96) is based on the fact that, together with the Snell law (7.82), it gives $r + \theta = \pi/2$. As a result, the vector $\mathbf{E}_+$ is parallel to the vector $\mathbf{k}'_-$, and hence oscillating electric dipoles of the medium at $z > 0$ do not have the component which could induce the transverse electric field $\mathbf{E}'_-$ of the potential reflected wave.

[36] Despite some important background theoretical work by A Schuster (1904), L Mandelstam (1945), D Sivikhin (1957), and in particular V Veselago (1966–67), the negative refractivity effects have only recently become a subject of intensive scientific research and engineering development.

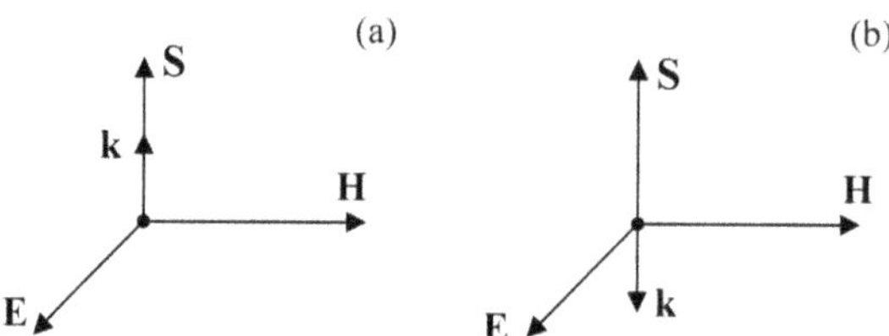

Figure 7.12. Directions of the main vectors of a plane wave inside a medium with (a) positive and (b) negative values of ε and μ.

First of all, for a sinusoidal, plane wave (7.79), the operator ∇ is equivalent to the multiplication by $i\mathbf{k}$. As the Maxwell equations (7.2a) show, this means that at a fixed direction of vectors **E** and **k**, the simultaneous reversal of the signs of ε and μ means the reversal of the direction of the vector **H**. Namely, if both ε and μ are positive, these equations are satisfied with mutually orthogonal vectors **E**, **H**, and **k** forming the usual, *right-hand* system (see figure 7.1 and figure 7.12a), the name stemming from the popular 'right-hand rule' used to determine the vector product's direction. However, if both ε and μ are negative, the vectors form a *left-hand* system —see figure 7.12b. (Due to this fact, the media with $\varepsilon < 0$ and $\mu < 0$ are frequently called *left-handed materials*, LHM for short.) According to the fundamental relation (6.114), which does not involve media parameters, this means that for a plane wave in a left-hand material, the Poynting vector $\mathbf{S} = \mathbf{E} \times \mathbf{H}$, i.e. the energy flow, is directed *opposite* to the wave vector **k**.

This fact may look strange, but is in no contradiction with any fundamental principle. Let me remind you that, according to the definition of vector **k**, its direction shows the direction of the *phase* velocity $v_{\text{ph}} = \omega/k$ of a sinusoidal (and hence infinitely long) wave that cannot be used, for example, for signaling. Such signaling (by sending wave packets—see figure 7.13) is possible with the *group* velocity $v_{\text{gr}} = d\omega/dk$. This velocity in left-hand materials is always positive (directed along the vector **S**).

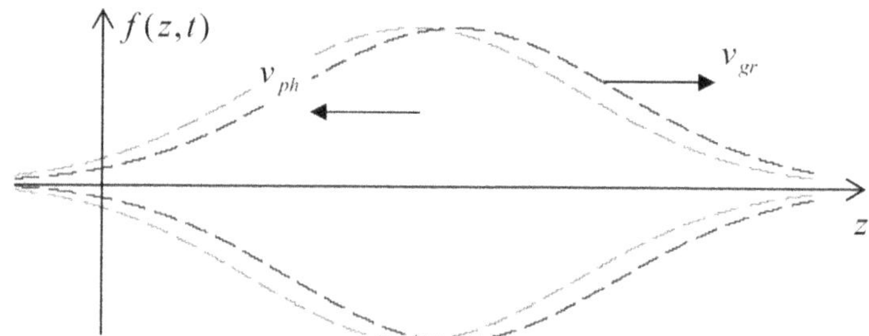

Figure 7.13. An example of a wave packet moving along axis z with a negative phase velocity, but positive group velocity. Blue lines show a packet snapshot a short time interval after the first snapshot (red lines).

Maybe the most fascinating effect possible with left-hand materials is the wave refraction at their interfaces with the usual, right-handed materials—first predicted by V Veselago in 1960. Consider the example shown in figure 7.14a. In the incident

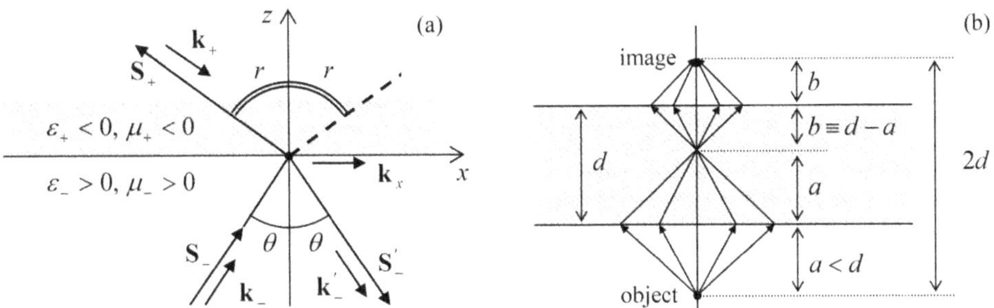

Figure 7.14. Negative refraction: (a) waves at the interface between media with positive and negative values of $\varepsilon\mu$, and (b) the hypothetical *perfect lens*: a parallel plate made of a material with $\varepsilon = -\varepsilon_0$ and $\mu = -\mu_0$.

wave, coming from the usual material, the directions of vectors $\mathbf{k}_-$ and $\mathbf{S}_-$ coincide, and so they are in the reflected wave with vectors $\mathbf{k}'_-$ and S'_-. This means that the electric and magnetic fields in the interface plane ($z = 0$) are, at our choice of coordinates, proportional to $\exp\{ik_x x\}$ with a positive component $k_x = k_- \cos\theta$. In order to satisfy any linear boundary conditions, the refracted wave, going into the left-handed material, should match that dependence, i.e. have a positive x-component of its wave vector $\mathbf{k}_+$. But in this medium, this vector has to be antiparallel to the vector $\mathbf{S}$ that, in turn, should be directed out of the interface, because it represents the power flow from the interface into the material bulk. These conditions cannot be reconciled by the refracted wave propagating along the usual Snell-law direction (shown with the dashed line in figure 7.14a), but are all satisfied at refraction in the direction given by Snell's angle with the opposite sign. (Hence the term 'negative refraction')[37].

In order to understand how unusual the results of the negative refraction may be, let us consider a parallel slab of thickness d, made of a hypothetical left-handed material with exactly selected values $\varepsilon = -\varepsilon_0$ and $\mu = -\mu_0$—see figure 7.14b. For such a material, placed in free space, the refraction angle $r = -\theta$, so that the rays from a point source located in free space, at a distance $a < d$ from the slab, propagate as shown in the figure, i.e. all meet again at the distance a inside the plate and then continue to propagate to the second surface of the slab. Repeating our discussion for this surface, we see that a point's image is also formed beyond the plate at distance $2a + 2b = 2a + 2(d - a) = 2d$ from the object.

Superficially, this system looks like the usual lens, but the well-known lens formula, which relates a and b with the focal length f, is *not* satisfied. (In particular, a parallel beam is *not* focused into a point at any finite distance.) As an additional difference to the usual lens, the system shown in figure 7.14b *does not reflect* any part of the incident light. Indeed, it is straightforward to check that in order for all above formulas for R and T to be valid, the sign of the wave impedance Z in left-handed

[37] Inspired by this fact, in some publications the left-handed materials are prescribed a negative index of refraction n. However, this prescription should be treated with care (for example, it complies with the first form of Eq. (7.84), but not its second form), and the sign of n, in contrast to that of the wave vector $\mathbf{k}$, is the matter of convention.

materials has to be kept positive. Thus, for our particular choice of parameters ($\varepsilon = -\varepsilon_0$, $\mu = -\mu_0$), Eqs. (7.91a) and (7.95a) are valid with $Z_+ = Z_- = Z_0$ and $\cos r = \cos \theta = 1$, giving $R = 0$ for any linear polarization, and hence for any other wave polarization—circular, elliptic, natural, etc.

The perfect lens suggestion has triggered a wave of efforts to implement left-handed materials experimentally. (Attempts to find such materials in nature have failed so far.) Most progress in this direction has been achieved using the so-called *metamaterials*, which are essentially quasi-periodic arrays of specially designed electromagnetic resonators, ideally with high density $n \gg \lambda^{-3}$. For example, figure 7.15 shows the metamaterial that was used for the first demonstration of negative refractivity in the microwave region, i.e. a few GHz frequencies[38]. It combines straight strips of a metallic film, working as lumped resonators with a large electric dipole moment (hence strongly coupled to the wave's electric field **E**), and several almost-closed film loops (so-called *split rings*), working as lumped resonators with large magnetic dipole moments, coupled to the field **H**. The negative refractivity is achieved by designing the resonance frequencies close to each other. More recently, metamaterials with negative refractivity were demonstrated in the optical range[39], although to the best of my knowledge, their relatively large absorption still prevents practical applications.

Figure 7.15. An artificial left-handed material providing negative refraction at microwave frequencies (~10 GHz). The original image by Jeffrey D Wilson (in the public domain) is available at https://en.wikipedia.org/wiki/Metamaterial.

This progress has stimulated the development of other potential uses of metamaterials (not necessarily the left-handed ones), in particular designs of non-uniform systems with engineered distributions $\varepsilon(\mathbf{r}, \omega)$ and $\mu(\mathbf{r}, \omega)$, which may provide electromagnetic wave propagation along desired paths, e.g. around a certain region

[38] See [2, 3].

[39] See, e.g. [4].

of space, making it virtually invisible for an external observer—so far, within a limited frequency range[40].

As was mentioned in section 5.5, another way to reach negative values of $\mu(\omega)$ is to place a ferromagnetic material into such an external dc magnetic field that the frequency ω_r of the ferromagnetic resonance is somewhat lower than ω. If thin layers of such a material (e.g. nickel) are interleaved with layers of a non-magnetic, very good conductor (such as copper), the resulting metamaterial has an average value of $\mu(\omega)$—say, positive, but substantially below μ_0. According to Eq. (6.33), the skin-depth δ_s of such a material may be larger than that of the good conductor alone, enforcing a more uniform distribution of the ac current flowing along the layers, and hence making the energy losses lower than in the good conductor alone. This effect may be useful, in particular, for electronic circuit interconnects[41].

7.5 Transmission lines: TEM waves

So far, we have analyzed plane electromagnetic waves, implying that their cross-section is infinite—evidently an unrealistic assumption. The cross-section may be limited, still sustaining wave propagation, using *wave transmission lines*[42]: cylindrically shaped structures made of either good conductors or dielectrics. Let us first discuss the first option, using the following simplifying assumptions:

(i) the structure is a cylinder (not necessarily with a round cross-section, see figure 7.16) filled with a usual (right-handed), uniform dielectric material with negligible energy losses ($\varepsilon'' = \mu'' = 0$), and

(ii) the wave attenuation due to the skin effect is also negligibly low. (As Eq. (7.78) indicates, for this to be the case the characteristic size a of the line's cross-section has to be much larger than the skin-depth δ_s of its wall material. The energy dissipation effects will be analyzed in section 7.9 below.)

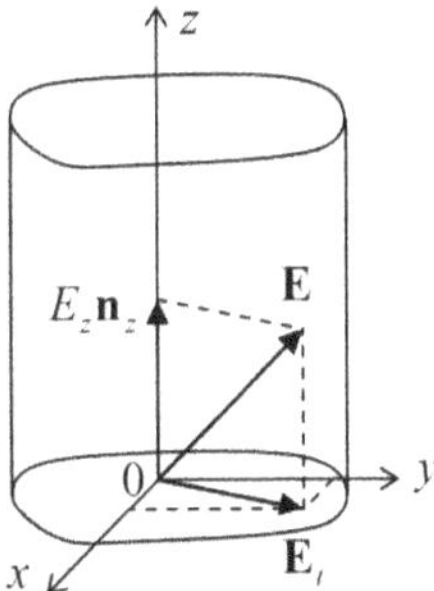

Figure 7.16. Electric field's decomposition in a transmission line.

[40] For a review of such 'invisibility cloaks', see, e.g. [5].

[41] See, for example, [6] and references therein.

[42] Another popular term is the *waveguide*, but it is typically reserved for transmission lines with a singly connected cross-section, to be analyzed in the next section.

With such exclusion of energy losses, we may look for a particular solution of the macroscopic Maxwell equations in the form of a monochromatic wave traveling along the line:

$$\mathbf{E}(\mathbf{r}, t) = \text{Re}\,[\mathbf{E}_\omega(x, y)e^{i(k_z z-\omega t)}], \quad \mathbf{H}(\mathbf{r}, t) = \text{Re}\,[\mathbf{H}_\omega(x, y)e^{i(k_z z-\omega t)}], \tag{7.98}$$

with real k_z, where the axis **z** is directed along the transmission line—see figure 7.16. Note that this form allows an account for a substantial coordinate dependence of the electric and magnetic field in the plane $\{x, y\}$ of the transmission line's cross-section, as well as for longitudinal components of the fields, so that the solution (7.98) is substantially more complex than the plane waves we have discussed above. We will see in a minute that as a result of this dependence, the parameter k_z may be very different from its plane-wave value (7.13), $k \equiv \omega(\varepsilon\mu)^{1/2}$, in the same material, at the same frequency.

In order to describe these effects qualitatively, let us decompose the complex amplitudes of the wave's fields into their longitudinal and transverse components (figure 7.16)[43]

$$\mathbf{E}_\omega = E_z\mathbf{n}_z + \mathbf{E}_t, \quad \mathbf{H}_\omega = H_z\mathbf{n}_z + \mathbf{H}_t. \tag{7.99}$$

Plugging Eqs. (7.98) and (7.99) into the source-free Maxwell equations (7.2), and requiring the longitudinal and transverse components to be balanced separately, we obtain

$$\begin{aligned}
&ik_z\mathbf{n}_z \times \mathbf{E}_t - i\omega\mu\mathbf{H}_t = -\nabla_t \times (E_z\mathbf{n}_z), \qquad && ik_z\mathbf{n}_z \times \mathbf{H}_t + i\omega\varepsilon\mathbf{E}_t = -\nabla_t \times (H_z\mathbf{n}_z),\\
&\nabla_t \times \mathbf{E}_t = i\omega\mu H_z\mathbf{n}_z, && \nabla_t \times \mathbf{H}_t = -i\varepsilon\omega E_z\mathbf{n}_z,\\
&\nabla_t \cdot \mathbf{E}_t = -ik_zE_z, && \nabla_t \cdot \mathbf{H}_t = -ik_zH_z.
\end{aligned} \tag{7.100}$$

where ∇_t is the 2D del operator acting in the transverse plane $[x, y]$ only (i.e. the usual ∇, but with $\partial/\partial z = 0$). These equations may look even more bulky than the original equations (7.2), but actually are much simpler for analysis. Indeed, eliminating the transverse components from these equations (or, even simpler, just plugging Eq. (7.99) into Eq. (7.3) and keeping only their z-components), we obtain a pair of self-consistent equations for the longitudinal components of the fields[44],

$$\left(\nabla_t^2 + k_t^2\right)E_z = 0, \qquad \left(\nabla_t^2 + k_t^2\right)H_z = 0, \tag{7.101}$$

where k is still defined by Eq. (7.13), $k = (\varepsilon\mu)^{1/2}\omega$, while

$$k_t^2 \equiv k^2 - k_z^2 = \omega^2\varepsilon\mu - k_z^2. \tag{7.102}$$

After the distributions $E_z(x,y)$ and $H_z(x,y)$ have been found from these equations, they provide the right-hand sides for rather simple, closed system of Eq. (7.100) for

[43] Note that for notation simplicity, I am dropping index ω in the complex amplitudes of the field components, and also have dropped the argument ω in k_z and Z, although these parameters of the wave may depend on its frequency rather substantially—see below.

[44] The wave equation represented in the form (7.101), even with the 3D Laplace operator, is called the *Helmholtz equation*, after H von Helmholtz (1821–94)—the mentor of H Hertz and M Planck, among many others.

the transverse components of field vectors. Moreover, as we will see below, each of the following three types of solutions,

(i) with $E_z = 0$ and $H_z = 0$ (called *transverse*, or *TEM waves*),

(ii) with $E_z = 0$, but $H_z \neq 0$ (called either *TE waves* or, more frequently, *H-modes*), and

(iii) with $E_z \neq 0$, but $H_z = 0$ (so-called *TM waves* or *E-modes*)

has its own dispersion law and hence its own wave propagation velocity; as a result, these *modes* (i.e. the field distribution patterns) may be considered separately.

In this section we will focus on the simplest, TEM waves, with *no longitudinal components* of either field. For them, the top two equations of the system (7.100) immediately give Eqs. (7.6) and (7.13), and $k_z = k$. In plain English, this means that $\mathbf{E} = \mathbf{E}_t$ and $\mathbf{H} = \mathbf{H}_t$ are proportional to each other and mutually perpendicular (just as in the plane wave) at each point of the cross-section, and that the TEM wave's impedance $Z \equiv E/H$ and dispersion law $\omega(k)$, and hence the propagation speed, are the same as in a plane wave in the same material. In particular, if ε and μ are frequency-independent within a certain frequency range, the dispersion law is linear, $\omega = k/(\varepsilon\mu)^{1/2}$, and the wave's speed does not depend on its frequency. For practical applications to telecommunications, this is a very important advantage of TEM waves over their TM and TE counterparts—to be discussed below.

Unfortunately, such waves cannot propagate in every transmission line. In order to show this, let us have a look at the two last lines of Eq. (7.100). For the TEM waves ($E_z = 0$, $H_z = 0$, $k_z = k$), they yield

$$\begin{aligned} \nabla_t \times \mathbf{E}_t &= 0, \qquad & \nabla_t \times \mathbf{H}_t &= 0, \\ \nabla_t \cdot \mathbf{E}_t &= 0, \qquad & \nabla_t \cdot \mathbf{H}_t &= 0. \end{aligned} \tag{7.103}$$

In the macroscopic approximation for the conducting walls of the line (i.e. completely neglecting the skin effect), we have to require that inside them, $\mathbf{E} = \mathbf{H} = 0$. Close to a wall but outside it, the normal component E_n of the electric field may be different from zero, because surface charges may sustain its jump—see section 2.1, in particular Eq. (2.3). Similarly, the tangential component H_τ of the magnetic field may have a finite jump at the surface due to skin currents—see section 6.3, in particular Eq. (6.38). However, the tangential component of the electric field and the normal component of magnetic field cannot experience such a jump, and in order to have them equal to zero inside the walls they have to equal zero outside the inside the walls as well:

$$\mathbf{E}_\tau = 0, \qquad H_n = 0. \tag{7.104}$$

But the left columns of Eqs. (7.103) and (7.104) coincide with the formulation of the 2D boundary problem of electrostatics for the electric field induced by electric charges of the conducting walls, with the only difference that in our current case the value of ε actually means $\varepsilon(\omega)$. Similarly, the right columns of those relations coincide with the formulation of the 2D boundary problem of magnetostatics for the magnetic field induced by currents in the walls, with $\mu \to \mu(\omega)$. The only difference is that in our current ac case the magnetic fields cannot penetrate inside the conductors.

Now we immediately see that in waveguides with a singly connected wall, for example a hollow conducting tube (see, e.g. the example shown in figure 7.16), TEM waves are impossible because there is no way to create a finite electrostatic field inside a conductor with such cross-section. However, such fields (and hence TEM waves) are possible in structures with cross-sections consisting of two or more disconnected (galvanically insulated) parts—see, e.g. Figure 7.17.

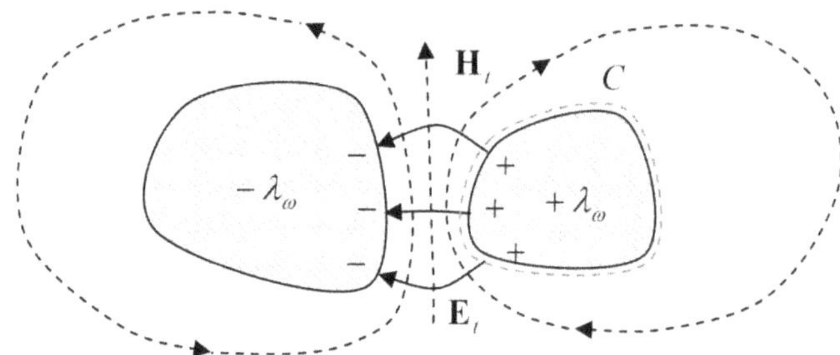

Figure 7.17. An example of the cross-section of a transmission line that may support TEM wave propagation.

In order to derive 'global' relations for such a line, let us consider the contour C drawn very close to the surface of one of its conductors—see, e.g. the red dashed line in figure 7.17. We can consider it, on one hand, as the cross-section of a cylindrical Gaussian volume of a certain elementary length $dz \ll \lambda \equiv 2\pi/k$. Using the generalized Gauss law (3.34), we obtain

$$\oint_C (\mathbf{E}_t)_n \, dr = \frac{\lambda_\omega}{\varepsilon}, \tag{7.105}$$

where λ_ω (not to be confused with the wavelength λ!) is the complex amplitude of the linear density of electric charge of the conductor. On the other hand, the same contour C may be used in the generalized Ampère law (5.116) to write

$$\oint_C (\mathbf{H}_t)_\tau \, dr = I_\omega, \tag{7.106}$$

where I_ω is the total current flowing along the conductor (or rather its complex amplitude). But, as was mentioned above, in the TEM wave the ratio E_t/H_t of the field components participating in these two integrals is constant and equal to $Z = (\mu/\varepsilon)^{1/2}$, so that Eqs. (7.105) and (7.106) give the following simple relation between the 'global' variables of the conductor:

$$I_\omega = \frac{\lambda_\omega/\varepsilon}{Z} \equiv \frac{\lambda_\omega}{(\varepsilon\mu)^{1/2}} \equiv \frac{\omega}{k}\lambda_\omega. \tag{7.107}$$

This important relation may be also obtained by a different means; let me describe it as well, because it has an independent value. Let us consider a small segment $dz \ll \lambda = 2\pi/k$ of the line's conductor, and apply the electric charge conservation law (4.1) to the *instant* values of the linear charge density and current. The cancellation of dz in both parts yields

$$\frac{\partial \lambda(z, t)}{\partial t} = -\frac{\partial I(z, t)}{\partial z}. \tag{7.108}$$

If we accept the sinusoidal waveform, $\exp\{i(kz - \omega t)\}$, for both these variables, we immediately recover Eq. (7.107) for their complex amplitudes, showing that this relation expresses just the charge continuity law.

The global Eq. (7.108) may be made more specific in the case when the frequency dependence of ε and μ is negligible, and the transmission line consists of just two isolated conductors (see, e.g. Figure 7.17). In this case, in order to have the wave localized in the space near the two conductors, we need a sufficiently fast decrease of its electric field at large distances. For that, their linear charge densities for each value of z should be equal and opposite, and we can simply relate them to the potential difference V between the conductors:

$$\frac{\lambda(z, t)}{V(z, t)} = C_0, \tag{7.109}$$

where C_0 is the mutual capacitance of the conductors (per unit length)—which was repeatedly discussed in chapter 2. Then Eq. (7.108) takes the form

$$C_0\frac{\partial V(z, t)}{\partial t} = -\frac{\partial I(z, t)}{\partial z}. \tag{7.110}$$

Next, let us consider the contour shown with the red dashed line in figure 7.18, which shows a cross-section of the transmission line by a plane containing the wave propagation axis z, and apply to it the Faraday induction law (6.3).

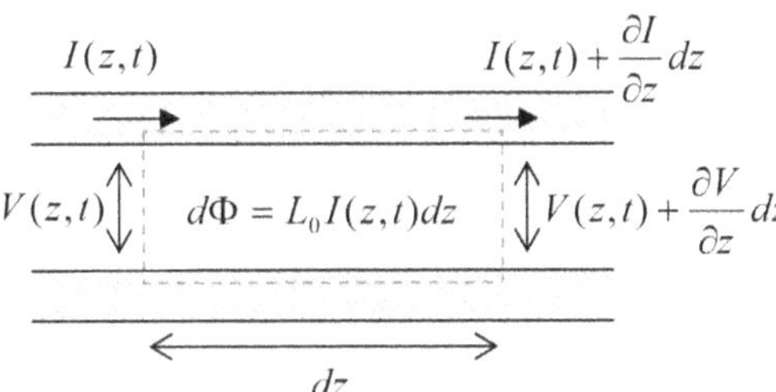

Figure 7.18. Electric current, magnetic flux, and voltage in a two-conductor transmission line.

Since the electric field is zero inside the conductors (in figure 7.18, on the horizontal segments of the contour), the total e.m.f. equals the difference of voltages V at the end of the segment dz, while the only sources of magnetic flux through the area limited by the contour are the (equal and opposite) currents $\pm I$ in the conductors, we can use Eq. (5.70) to express it. As a result, canceling dz in both parts of the equation, we obtain

$$L_0 \frac{\partial I(z, t)}{\partial t} = -\frac{\partial V(z, t)}{\partial z}, \tag{7.111}$$

where L_0 is the mutual inductance of the conductors per unit length. The only difference between this L_0 and the dc mutual inductances discussed in chapter 5 is that at the high frequencies we are analyzing now, L_0 should be calculated neglecting its penetration into the conductors. (In the dc case, we had the same situation for superconductor electrodes, within their coarse-grain, ideal-diamagnetic description.)

The system of Eqs. (7.110) and (7.111) is frequently called the *telegrapher's equations*. Combined, they give for any 'global' variable f (either V, or I, or λ) the usual 1D wave equation,

$$\frac{\partial^2 f}{\partial z^2} - L_0 C_0 \frac{\partial^2 f}{\partial t^2} = 0, \tag{7.112}$$

which describes the dispersion-free TEM wave's propagation.

Again, this equation is only valid within the frequency range where the frequency dependence of both ε and μ is negligible. If it is not so, the global approach may still be used for sinusoidal waves $f = \text{Re}[f_\omega \exp\{i(kz - \omega t)\}]$. Repeating the above arguments, instead of Eqs. (7.110) and (7.111) we obtain a system of two algebraic equations

$$\omega C_0 V_\omega = k I_\omega, \qquad \omega L_0 I_\omega = k V_\omega, \tag{7.113}$$

in which $L_0 \propto \mu$ and $C_0 \propto \varepsilon$ may now depend on frequency. These equations are consistent only if

$$L_0 C_0 = \frac{k^2}{\omega^2} \equiv \frac{1}{v^2} \equiv \varepsilon\mu. \tag{7.114}$$

In addition to the fact we have already learned (that the TEM wave's speed is the same as that of the plane wave), Eq. (7.114) gives us a result that I confess was not emphasized enough in chapter 5: the product $L_0 C_0$ does not depend on the shape or size of the line's cross-section (provided that the magnetic field penetration into the conductors is negligible). Hence, if we have calculated the mutual capacitance C_0 of a system of two cylindrical conductors, the result immediately gives us their mutual inductance: $L_0 = \varepsilon\mu / C_0$. This relation stems from the fact that both the electric and magnetic fields may be expressed via the solution of a 2D Laplace equation for system's cross-section.

With Eq. (7.114) satisfied, any of Eqs. (7.113) gives the same result for the following ratio:

$$Z_W \equiv \frac{V_\omega}{I_\omega} = \left(\frac{L_0}{C_0}\right)^{1/2}, \tag{7.115}$$

which is called the *transmission line's impedance*. This parameter has the same dimensionality (in SI units, ohms) as the wave impedance (7.7),

$$Z \equiv \frac{E_\omega}{H_\omega} = \left(\frac{\mu}{\varepsilon}\right)^{1/2}, \tag{7.116}$$

but these parameters should not be confused, because Z_W depends on the cross-section's geometry, while Z does not. In particular, Z_W is the only relevant parameter of a transmission line for matching with a lumped load circuit (figure 7.19), in the important case when both the cable cross-section's size and the load's linear dimensions are much smaller than the wavelength[45].

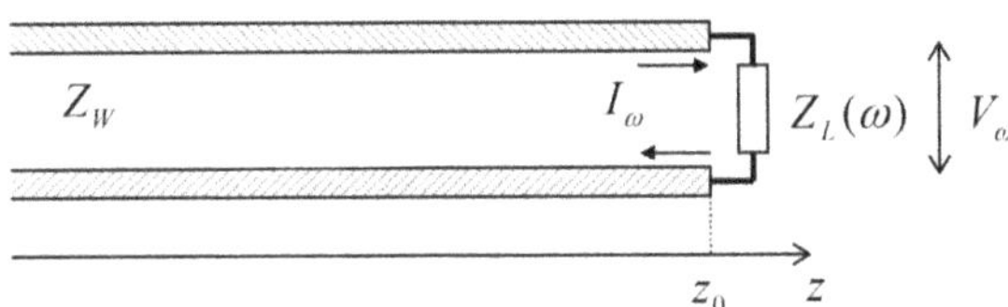

Figure 7.19. Passive, lumped termination of a TEM transmission line.

Indeed, in this case we may consider the load in the quasi-static limit and write

$$V_\omega(z_0) = Z_L(\omega)I_\omega(z_0), \tag{7.117}$$

where $Z_L(\omega)$ is the (generally complex) impedance of the load. Taking $V(z, t)$ and $I(z, t)$ in the form similar to Eqs. (7.61) and (7.62), and writing the two Kirchhoff's laws for the point $z = z_0$, we obtain for the reflection coefficient a result similar to Eq. (7.68):

$$R = \frac{Z_L(\omega) - Z_W}{Z_L(\omega) + Z_W}. \tag{7.118}$$

This formula shows that for perfect matching (i.e. the total wave absorption in the load), the load's impedance $Z_L(\omega)$ should be real and equal to Z_W—but not necessarily to Z.

As an example, let us consider one of the simplest (and the most important) transmission lines: the coaxial cable (figure 7.20)[46]. For this geometry, we already know the expressions for both C_0 and L_0[47], although they have to be modified for the dielectric and magnetic constants, and the magnetic field's non-penetration into the conductors. As a result of such modification, we obtain the formulas,

$$C_0 = \frac{2\pi\varepsilon}{\ln(b/a)}, \qquad L_0 = \frac{\mu}{2\pi}\ln(b/a), \tag{7.119}$$

illustrating that the universal relation (7.114) is indeed valid. In contrast, for the cable's impedance (7.115), Eqs. (7.119) yield a geometry-dependent value

[45] The ability of TEM lines to have such a small cross-section is their other important practical advantage.

[46] It was invented by O Heaviside in 1880.

[47] See, respectively, Eqs. (2.49) and (5.79).

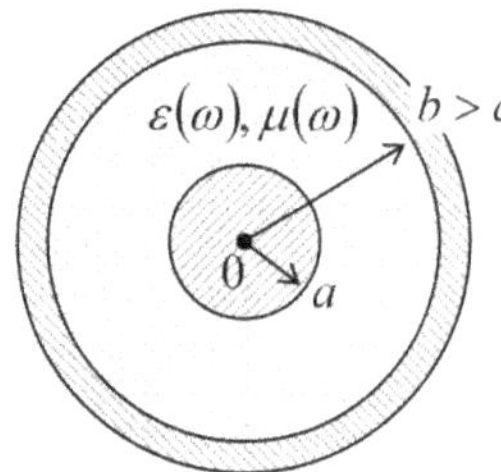

Figure 7.20. The cross-section of a coaxial cable with a (possibly, dispersive) dielectric filling.

$$Z_W = \left(\frac{\mu}{\varepsilon}\right)^{1/2} \frac{\ln(b/a)}{2\pi} \equiv Z\frac{\ln(b/a)}{2\pi} \neq Z. \qquad (7.120)$$

For the standard TV antenna cables (such as RG-6/U, with $b/a \sim 3$, $\varepsilon/\varepsilon_0 \approx 2.2$), $Z_W = 75$ ohms, while for most computer component connections, cables with $Z_W = 50$ ohms (such as RG-58/U) are prescribed by electronic engineering standards. Such cables are broadly used for the transmission of electromagnetic waves with frequencies up to 1 GHz over distances of a few km, and up to ~20 GHz on the tabletop scale (a few meters), limited by wave attenuation—see section 7.9.

Moreover, the following two facts enable a broad application in electrical engineering and physical experiments of the coaxial-cable-*like* systems. First, as Eq. (5.78) shows, in a cable with $a \ll b$, most wave energy is localized near the internal conductor. Second, the theory to be discussed in the next section shows that waves of other (H- and E-) modes in the cable are impossible until the wavelength λ becomes smaller than $\sim\pi(a + b)$. As a result, the TEM mode propagation in a cable with $a \ll b < \lambda/\pi$ is not much affected even if the internal conductor is not straight, but is bent—for example, into a helix—see, e.g. Figure 7.21.

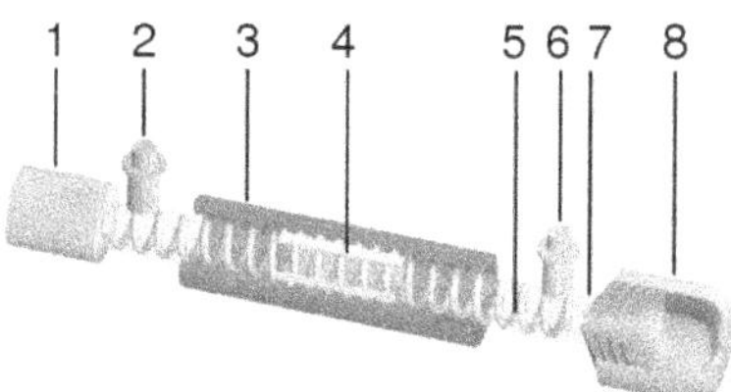

Figure 7.21. A typical traveling-wave tube: (1) electron gun, (2) RF input, (3) beam-focusing magnets, (4) wave attenuator, (5) helix coil, (6) RF output, (7) vacuum tube, (8) electron collector. Adapted from https://en.wikipedia.org/wiki/Traveling-wave_tube under the Creative Commons BY-SA 3.0 license.

In such a system, called the *traveling-wave tube* (TWT), the quasi-TEM wave propagates with velocity $v = 1/(\varepsilon\mu)^{1/2} \sim c$ along the helix's length, so that the velocity's component along the cable's axis may be made close to the velocity $u \ll c$ of the electron beam moving ballistically along the tube's axis, enabling their effective (length-accumulating) interaction and, as a result, a broadband amplification of the wave[48].

Another important example of a TEM transmission line is a set of two parallel wires. In the form of *twisted pairs*[49], they allow communications, in particular long-range telephone and DSL Internet connections, at frequencies up to a few hundred kHz, as well as relatively short, multi-line Ethernet and TV cables at frequencies up to ~1 GHz, limited mostly by the mutual interference ('crosstalk') between the individual lines of the same cable and the unintended radiation of the wave into the environment.

7.6 Waveguides: *H* and *E* waves

Let us now return to Eqs. (7.100) and explore the *H*- and *E*-waves—with, respectively, either H_z or E_z different from zero. At first sight they may seem more complex. However, Eqs. (7.101), which determine the distribution of these longitudinal components over the cross-section, are just 2D Helmholtz equations for scalar functions. For simple cross-section geometries, they may be solved using the methods discussed for the Laplace equation in chapter 2, in particular variable separation. After the solution of such an equation has been found, the transverse components of the fields may be calculated by differentiation, using the simple formulas

$$\mathbf{E}_t = \frac{i}{k_t^2}[k_z \nabla_t E_z - kZ(\mathbf{n}_z \times \nabla_t H_z)], \quad \mathbf{H}_t = \frac{i}{k_t^2}\left[k_z \nabla_t H_z + \frac{k}{Z}(\mathbf{n}_z \times \nabla_t E_z)\right], \qquad (7.121)$$

which follow from the two equations in the first line of Eqs. (7.100)[50].

In comparison with the boundary problems of electro- and magnetostatics, the only conceptually new feature of Eqs. (7.101) is that they form the so-called *eigenproblems*, with typically many solutions (*eigenfunctions*), each describing a specific wave mode and corresponding to a specific *eigenvalue* of the parameter k_t. The good news here is that these values of k_t are determined by this 2D boundary problem and hence do not depend on k_z. As a result, the dispersion law $\omega(k_z)$ of any mode, which follows from the last form of Eq. (7.102),

$$\omega = \left(\frac{k_z^2 + k_t^2}{\varepsilon\mu}\right)^{1/2} = (v^2 k_z^2 + \omega_c^2)^{1/2}, \qquad (7.122)$$

is functionally the same and is also absolutely similar to that of plane waves in a plasma—see Eq. (7.38), figure 7.6, and their discussion in section 7.2. The only

[48] Very unfortunately, in this course I will not have time/space to discuss the (rather elegant) theory of such devices. The reader seriously interested in this field may be referred, for example, to the detailed monograph [7].

[49] Such twisting, around the line's axis, reduces mutual induction ('crosstalk') between adjacent lines, and the parasitic radiation at their bends.

[50] For that, one of these two linear equations should first be vector-multiplied by $\mathbf{n}_z$. Note that this approach could not be used to analyze the TEM waves, because for them $k_t = 0$, $E_z = 0$, $H_z = 0$, and Eqs. (7.121) yield uncertainty.

differences are that the speed in light c is now replaced with $v = 1/(\varepsilon\mu)^{1/2}$, the speed of plane (or any TEM) waves in the medium filling the waveguide, and ω_p is replaced with the so-called *cutoff frequency*

$$\omega_c \equiv v k_t, \tag{7.123}$$

specific for each mode. (As Eq. (7.101) implies, and as we will see from several examples below, k_t has the order of $1/a$, where a is the characteristic dimension of waveguide's cross-section, so that the critical value of the free-space wavelength $\lambda \equiv 2\pi c/\omega$ is of the order of a.) Below the cutoff frequency of each particular mode, such wave cannot propagate in the waveguide[51]. As a result, the modes with the *lowest* values of ω_c present special practical interest, because the choice of the signal frequency ω between the two lowest values of the cutoff frequency (7.123) guarantees that the waves propagate in the form of only one mode, with the lowest k_t. Such a choice allows one to simplify the excitation of the desired mode by wave generators, and to avoid the parasitic transfer of electromagnetic wave energy to undesirable modes by (virtually unavoidable) small inhomogeneities of the system.

The boundary conditions for the Helmholtz equation (7.101) depend on the propagating wave type. For the E-modes, with $H_z = 0$ but $E_z \neq 0$, the condition $E_\tau = 0$ immediately gives

$$E_z|_C = 0, \tag{7.124}$$

where C is the contour limiting the conducting wall's cross-section. For the H-modes, with $E_z = 0$ but $H_z \neq 0$, the boundary condition is slightly less obvious and may be obtained using, for example, the second equation of the system (7.100), vector-multiplied by $\mathbf{n}_z$. Indeed, for the component perpendicular to the conductor surface the result of such multiplication is

$$ik_z(\mathbf{H}_t)_n - i\frac{k}{Z}(\mathbf{n}_z \times \mathbf{E}_t)_n = \frac{\partial H_z}{\partial n}. \tag{7.125}$$

But the first term on the left-hand side of this relation must be zero on the wall surface, because of the second of Eqs. (7.104), while according to the first of Eqs. (7.104), the vector $\mathbf{E}_t$ in the second term cannot have a component tangential to the wall. As a result, the vector product in that term cannot have a normal component, so that the term should equal zero as well, and Eq. (7.125) is reduced to

$$\left.\frac{\partial H_z}{\partial n}\right|_C = 0. \tag{7.126}$$

Let us see how all this machinery works for a simple but practically important case of a metallic-wall waveguide with a rectangular cross-section—see figure 7.22.

[51] An interesting recent twist in the ideas of electromagnetic metamaterials (mentioned in section 7.5) is the so-called *ε-near-zero* materials, designed to have an effective product $\varepsilon\mu$ much lower than $\varepsilon_0\mu_0$ within certain frequency ranges. Since at these frequencies the speed v (7.4) becomes much lower than c, the cutoff frequency (7.123) virtually vanishes. As a result, the waves may 'tunnel' through very narrow sections of metallic waveguides filled with such materials—see, e.g. [8].

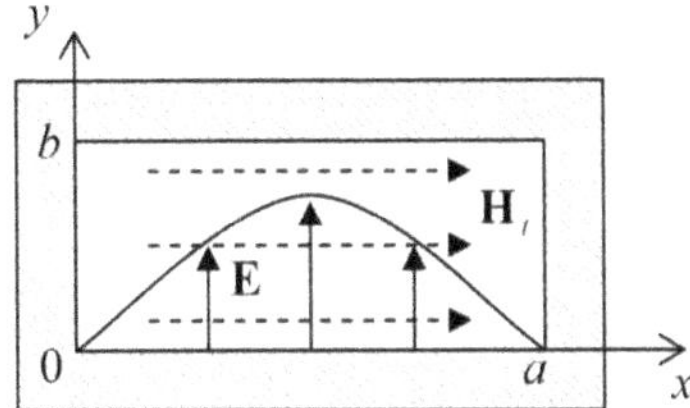

Figure 7.22. A schematic representation of a rectangular waveguide and the transverse field distribution in its fundamental mode H_{10}.

In the natural Cartesian coordinates shown in this figure, both equations (7.101) take the simple form

$$\left(\frac{\partial^2}{\partial x^2}+\frac{\partial^2}{\partial y^2}+k_t^2\right)f=0,\quad \text{where } f=\begin{cases}E_z, & \text{for } E\text{-modes},\\ H_z, & \text{for } H\text{-modes}.\end{cases}\tag{7.127}$$

From chapter 2 we know that the most effective way of solving of such equations in a rectangular region is variable separation, in which the general solution is represented as a sum of partial solutions of the type

$$f=X(x)Y(y).\tag{7.128}$$

Plugging this expression into Eq. (7.127) and dividing each term by XY, we obtain the equation

$$\frac{1}{X}\frac{d^2X}{dx^2}+\frac{1}{Y}\frac{d^2Y}{dy^2}+k_t^2=0,\tag{7.129}$$

which should be satisfied for all values of x and y within the waveguide's interior. This is only possible if each term of the sum equals a constant. Taking the X-term and Y-term constants in the form $(-k_x^2)$ and $(-k_y^2)$, respectfully, and solving the corresponding ordinary differential equations[52], for the eigenfunction (7.128) we obtain

$$f=(c_x\cos k_x x+s_x\sin k_x x)(c_y\cos k_y y+s_y\sin k_y y)\ ,\quad \text{with } k_x^2+k_y^2=k_t^2,\tag{7.130}$$

where the constants c and s should be found from the boundary conditions. Here the difference between the H-modes and E-modes pitches in.

For the H-modes, Eq. (7.130) is valid for H_z, and we should use the boundary condition (7.126) on all metallic walls of the waveguide ($x = 0$ and a; $y = 0$ and b—see figure 7.22). As a result, we obtain very simple expressions for eigenfunctions and eigenvalues:

$$(H_z)_{nm}=H_l\cos\frac{\pi nx}{a}\cos\frac{\pi my}{b},\tag{7.131}$$

[52] Let me hope that the solution of equations of the type $d^2X/dx^2+k_x^2X=0$ does not present a problem for the reader, due to his or her prior experience with problems such as standing waves on a guitar string, wavefunctions in a flat 1D quantum well, or (with the replacement $x\to t$) a classical harmonic oscillator.

$$k_x = \frac{\pi n}{a}, \quad k_y = \frac{\pi m}{b}, \quad (k_t)_{nm} = \left(k_x^2 + k_y^2\right)^{1/2} = \pi\left[\left(\frac{n}{a}\right)^2 + \left(\frac{m}{b}\right)^2\right]^{1/2}, \tag{7.132}$$

where H_l is the longitudinal field's amplitude, and n and m are two arbitrary integer numbers, except that they cannot be equal to zero simultaneously. (Otherwise, the function $H_z(x,y)$ would be constant, so that according to Eq. (7.121) the transverse components of the electric and magnetic field would equal zero. As a result, as the last two lines of Eq. (7.100) show, the whole field would be zero for any $k_z \neq 0$.) Assuming for certainty that $a \geqslant b$ (as shown in figure 7.22), we see that the lowest eigenvalue of k_t, and hence the lowest cutoff frequency (7.123), is achieved for the so-called H_{10} *mode* with $n = 1$ and $m = 0$, and hence with

$$(k_t)_{10} = \frac{\pi}{a} \tag{7.133}$$

(thus confirming our prior estimate of k_t).

Depending on the a/b ratio, the second lowest k_t (and hence ω_c) belongs to either the H_{11} mode with $n = 1$ and $m = 1$,

$$(k_t)_{11} = \pi\left(\frac{1}{a^2} + \frac{1}{b^2}\right)^{1/2} = \left[1 + \left(\frac{a}{b}\right)^2\right]^{1/2}(k_t)_{10}, \tag{7.134}$$

or to the H_{20} mode with $n = 2$ and $m = 0$,

$$(k_t)_{20} = \frac{2\pi}{a} = 2(k_t)_{10}. \tag{7.135}$$

These values become equal at $a/b = \sqrt{3} \approx 1.7$; in practical waveguides, the a/b ratio is made not too far from this value. For example, in the standard X-band (~10 GHz) waveguide WR90, $a \approx 2.3$ cm ($f_c \equiv \omega_c/2\pi \approx 6.5$ GHz), and $b \approx 1.0$ cm.

Now let us have a fast look at the alternative E-modes. For them, we still should use the general solution (7.130) with $f = E_z$, but now with the boundary condition (7.124). This gives us the eigenfunctions

$$(E_z)_{nm} = E_l \sin\frac{\pi n x}{a} \sin\frac{\pi m y}{b}, \tag{7.136}$$

and the same eigenvalue spectrum (7.132) as for the H modes. However, now neither n nor m can be equal to zero; otherwise Eq. (7.136) would give the trivial solution $E_z(x,y) = 0$. Hence the lowest cutoff frequency of TM waves is achieved in the so-called E_{11} *mode* with $n = 1$, $m = 1$, and with the eigenvalue given by Eq. (7.134), always higher than $(k_t)_{10}$.

Thus the fundamental H_{10} mode is certainly the most important wave in rectangular waveguides; let us have a better look at its field distribution. Plugging the corresponding solution (7.131) with $n = 1$ and $m = 0$ into the general relation (7.121), we easily obtain

$$(H_x)_{10} = -i\frac{k_z a}{\pi}H_l \sin\frac{\pi x}{a}, \qquad (H_y)_{10} = 0, \tag{7.137}$$

$$(E_x)_{10} = 0, \qquad (E_y)_{10} = i\frac{ka}{\pi}ZH_l \sin\frac{\pi x}{a}. \tag{7.138}$$

This field distribution is (schematically) shown in figure 7.22. Neither of the fields depends on the coordinate y—the feature very convenient, in particular, for microwave experiments with small samples. The electric field has only one (in figure 7.22, vertical) component that vanishes at the side walls and reaches its maximum at the waveguide's center; its field lines are straight, starting and ending on wall surface charges (whose distribution propagates along the waveguide together with the wave). In contrast, the magnetic field has two non-vanishing components (H_x and H_z) and its field lines are shaped as horizontal loops wrapped around the electric field maxima.

An important question is whether the H_{10} wave may be usefully characterized by a unique impedance introduced similarly to Z_W of the TEM modes—see Eq. (7.115). The answer is 'no', because the main value of Z_W is a convenient description of the impedance matching of a transmission line with a lumped load—see figure 7.19 and Eq. (7.118). As was discussed above, such simple description is possible (i.e. does not depend on the exact geometry of the connection) only if both dimensions of the line's cross-section are much less than λ. But for the H_{10} wave (and more generally, any non-TEM mode) this is impossible—see, e.g. Eq. (7.129): its lowest frequency corresponds to the TEM wavelength $\lambda_{\max} = 2\pi/(k_t)_{\min} = 2\pi/(k_t)_{10} = 2a$.[53]

Now let us consider metallic-wall waveguides with a round cross-section (figure 7.23a). In this single-connected geometry, the TEM waves are impossible again, while for the analysis of H-modes and E-modes, the polar coordinates $\{\rho, \varphi\}$ are most natural. In these coordinates, the 2D Helmholtz equation (7.101) takes the following form:

$$\left[\frac{1}{\rho}\frac{\partial}{\partial\rho}\left(\rho\frac{\partial}{\partial\rho}\right) + \frac{1}{\rho^2}\frac{\partial^2}{\partial\varphi^2} + k_t^2\right]f = 0, \qquad f = \begin{cases} E_z, & \text{for } E\text{-modes}, \\ H_z, & \text{for } H\text{-modes}. \end{cases} \tag{7.139}$$

Separating the variables as $f = \mathcal{R}(\rho)\mathcal{F}(\varphi)$, we obtain

$$\frac{1}{\rho\mathcal{R}}\frac{d}{d\rho}\left(\rho\frac{d\mathcal{R}}{d\rho}\right) + \frac{1}{\rho^2\mathcal{F}}\frac{d^2\mathcal{F}}{d\varphi^2} + k_t^2 = 0. \tag{7.140}$$

[53] The reader is encouraged to find a simple interpretation of this equality.

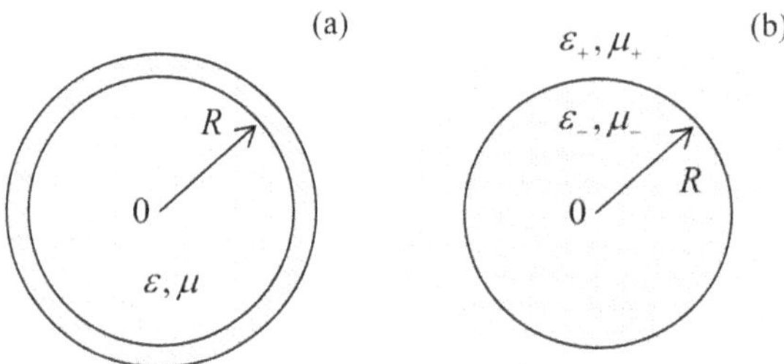

Figure 7.23. (a) Metallic and (b) dielectric waveguides with circular cross-sections.

But this is exactly the Eq. (2.127) that was studied in section 2.7 in the context of electrostatics, just with a replacement of notation: $\gamma \to k_t$. So we already know that in order to have 2π-periodic functions $\mathcal{F}(\varphi)$, and finite values $\mathcal{R}(0)$ (which are evidently necessary for our current case—see figure 7.23a), the general solution must have the form given by Eq. (2.136), i.e. the eigenfunctions are expressed via integer-order Bessel functions of the first kind:

$$f_{nm} = J_n(k_{nm}\rho)(c_n \cos n\varphi + s_n \sin n\varphi) \equiv \text{const} \times J_n(k_{nm}\rho) \cos n(\varphi - \varphi_0), \quad (7.141)$$

with the eigenvalues k_{nm} of the transverse wave number k_t to be determined from appropriate boundary conditions, and an arbitrary constant φ_0.

As for the rectangular waveguide, let us start from the H-modes ($f = H_z$). Then the boundary condition on the wall surface ($\rho = R$) is given by Eq. (7.126), which for the solution (7.141) takes the form

$$\frac{d}{d\xi}J_n(\xi) = 0, \quad \text{where } \xi \equiv kR. \quad (7.142)$$

This means that eigenvalues of Eq. (7.139) are

$$k_t = k_{nm} = \frac{\xi'_{nm}}{R}, \quad (7.143)$$

where ξ'_{nm} is the mth root of the function $dJ_n(\xi)/d\xi$. The approximate values of these roots for several lowest n and m may be read out from the plots in figure 2.18; their more accurate values are given in table 7.1. The table shows, in particular, that the lowest of the roots is $\xi'_{11} \approx 1.84$.[54] Thus, a bit counter-intuitively, the fundamental mode, providing the lowest cutoff frequency $\omega_c = vk_{nm}$, is H_{11}, corresponding to $n = 1$ rather than $n = 0$:

$$H_z = H_l J_1\left(\xi'_{11}\frac{\rho}{R}\right)\cos(\varphi - \varphi_0). \quad (7.144)$$

[54] Mathematically, the lowest root of Eq. (7.142) with $n = 0$ equals 0. However, it would yield $k = 0$ and hence a constant field H_z, which, according to the first of Eqs. (7.121), would give zero electric field.

Table 7.1. Roots ξ'_{nm} of the function $dJ_n(\xi)/d\xi$ for a few lowest values of the Bessel function's index n and the root's number m.

	$m = 1$	2	3
$n = 0$	3.83171	7.015587	10.1735
1	1.84118	5.33144	8.53632
2	3.05424	6.70613	9.96947
3	4.20119	8.01524	11.34592

It has the transverse wave number $k_t = k_{11} = \xi'_{11}/R \approx 1.84/R$, and hence the cutoff frequency corresponding to the TEM wavelength $\lambda_{\max} = 2\pi/k_{11} \approx 3.41\ R$. Thus the ratio of $\lambda_{\max}$ to the waveguide diameter $2R$ is about 1.7, i.e. is close to the ratio $\lambda_{\max}/a = 2$ for the rectangular waveguide. The origin of this proximity is clear from figure 7.24, which shows the transverse field distribution in the H_{11} mode. (It may be readily calculated from Eq. (7.121) with $E_z = 0$, and H_z given by Eq. (7.144).)

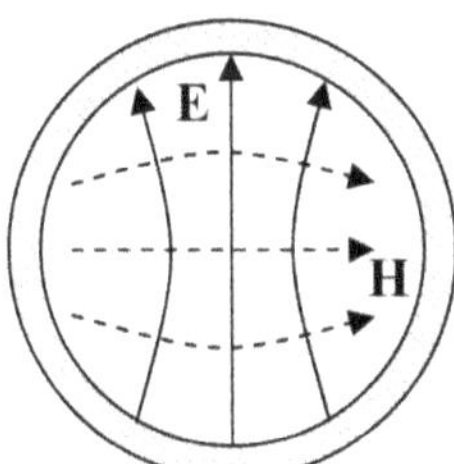

Figure 7.24. Schematic representation of transverse field components in the fundamental H_{11} mode of a metallic, circular waveguide.

One can see that the field structure is actually very similar to that of the fundamental mode in the rectangular waveguide, shown in figure 7.22, despite the different nomenclature (which is due to the different coordinates used for the solution). However, note the arbitrary constant angle φ_0, indicating that in circular waveguides the transverse field's polarization is arbitrary. For some practical applications, such degeneracy of these 'quasi-linearly polarized' waves creates problems; they may be avoided by using waves with circular polarization.

As table 7.1 shows, the next lowest H-mode is H_{21}, for which $k_t = k_{21} = \xi'_{21}/R \approx 3.05/R$, almost twice larger than that of the fundamental mode, and only then comes the first mode with no angular dependence of any field, H_{01}, with $k_t = k_{01} = \xi'_{01}/R \approx 3.83/R$.[55]

[55] The electric field lines in the H_{01} mode (as well as all higher H_{0m} modes) are directed straight from the symmetry axis to the walls, recalling those of the TEM waves in a coaxial cable. Due to this property, these modes provide, at $\omega \gg \omega_c$, much lower energy losses (see section 7.9 below) than the fundamental H_{11} mode, and are sometimes used in practice, despite the inconvenience of working in the multimode frequency range.

For the E modes, we may still use Eq. (7.141) (with $f = E_z$), but with the boundary condition (7.124) at $\rho = R$. This gives the following equation for the problem eigenvalues:

$$J_n(k_{nm}R) = 0, \qquad \text{i. e. } k_{nm} = \frac{\xi_{nm}}{R}, \tag{7.145}$$

where ξ_{nm} is the mth root of function $J_n(\xi)$—see table 2.1. The table shows that the lowest k_t equals to $\xi_{01}/R \approx 2.405/R$. Hence the corresponding mode (E_{01}), with

$$E_z = E_l J_0\left(\xi_{01}\frac{\rho}{R}\right), \tag{7.146}$$

has the second lowest cutoff frequency, approximately 30% higher than that of the fundamental mode H_{11}.

Finally, let us discuss one more topic of general importance—the number N of electromagnetic modes that may propagate in a waveguide within a certain range of relatively large frequencies $\omega \gg \omega_c$. It is easy to calculate for a rectangular waveguide, with its simple expressions (7.132) for the eigenvalues of $\{k_x, k_y\}$. Indeed, these expressions describe a rectangular mesh on the $[k_x, k_y]$ plane, so that each point corresponds to the plane area $\Delta A_k = (\pi/a)(\pi/b)$, and the number of modes in a large k-plane area $A_k \gg \Delta A_k$ is $N = A_k/\Delta A_k = abA_k/\pi^2 = AA_k/\pi^2$, where A is the waveguide's cross-section area[56]. However, it is frequently more convenient to discuss transverse wave vectors $\mathbf{k}_t$ of arbitrary direction, i.e. with an arbitrary sign of their components k_x and k_y. Taking into account that the opposite values of each component actually give the same wave, the actual number of different modes of each type (E- or H-) is a factor of 4 lower than was calculated above. This means that the number of modes of *both* types is

$$N = 2\frac{A_k A}{(2\pi)^2}. \tag{7.147}$$

Let me leave it for the reader to give hand-waving (but convincing) arguments that this *mode counting rule* is valid for waveguides with a cross-section of any shape, and any boundary conditions on the walls, provided that $N \gg 1$.

7.7 Dielectric waveguides, optical fibers, and paraxial beams

Now let us discuss electromagnetic wave propagation in *dielectric waveguides*. The conceptually simplest, *step-index* waveguide (see figures 7.23b and 7.25) consists of an inner *core* and an outer shell (in the optical fiber technology lingo, called the *cladding*) with a higher wave propagation speed, i.e. lower index of refraction:

[56] This formula ignores the fact that, according to our analysis, some modes (with $n = 0$ *and* $m = 0$ for H modes, and $n = 0$ *or* $m = 0$ for E modes), are forbidden. However, for $N \gg 1$, the associated corrections of Eq. (7.147) are negligible.

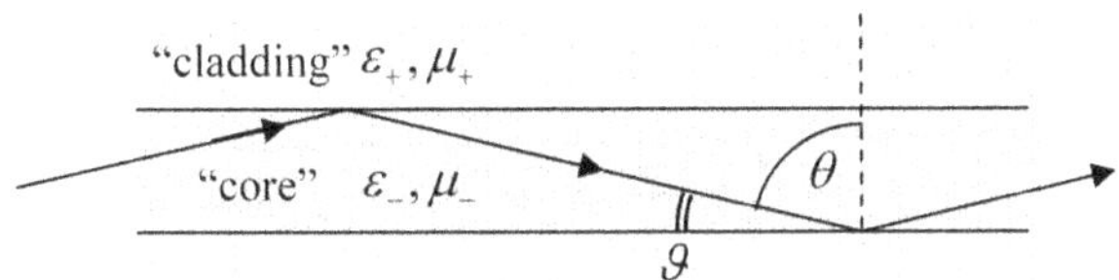

Figure 7.25. Wave propagation in a thick optical fiber.

$$v_+ > v_-, \quad \text{i.e. } n_+ < n_-, \quad k_+ < k_-, \quad \varepsilon_+\mu_+ < \varepsilon_-\mu_-. \tag{7.148}$$

at the same frequency. (In most cases the difference is achieved due to that in the electric permittivity, $\varepsilon_+ < \varepsilon_-$, while magnetically both materials are virtually passive: $\mu_- \approx \mu_+ \approx \mu_0$, so that their refraction indices $n_\pm$, defined by Eq. (7.84), are very close to $(\varepsilon_\pm/\varepsilon_0)^{1/2}$; I will limit my discussion to this approximation.) The idea of the waveguide operation may be readily understood in the limit when the wavelength λ is much smaller than the characteristic size R of the core's cross-section. If this 'geometric optics' limit, at the distances of the order of λ from the core-to-cladding interface, which determines the wave reflection, we can approximate the interface with a plane. As we know from section 7.4, if the angle θ of the wave's incidence on such an interface is larger than the critical value θ_c specified by Eq. (7.85), the wave is totally reflected. As a result, the waves launched into the fiber core at such 'grazing' angles propagate inside the core, repeatedly reflected from the cladding—see figure 7.25.

The most important type of dielectric waveguides are *optical fibers*[57]. Due to a heroic technological effort, in about three decades starting from the mid-1960s the attenuation of glass fibers has been decreased from the values of the order of 20 db km^{-1} (typical for window glass) to the fantastically low values of about 0.2 db km^{-1} (meaning virtually perfect transparency of 10 km long fiber segments!), combined with extremely low plane-wave (so-called *chromatic*) *dispersion* below 10 ps km^{-1} · nm^{-1}.[58] In conjunction with the development of inexpensive erbium-based quantum amplifiers, this breakthrough has enabled inter-continental (undersea), broadband[59] optical cables, which are the backbone of all the modern telecommunication infrastructure.

The only bad news is that these breakthroughs were achieved for just one kind of material (silica-based glasses)[60] within a very narrow range of chemical composition. As a result, the dielectric constants $\kappa_\pm \equiv \varepsilon_\pm/\varepsilon_0$ of the cladding and core of practical

---

[57] For a comprehensive discussion of this vital technology see, e.g. [9].

[58] Both these parameters are best not in the visible light range (from 380 to 740 nm), but in the near-infrared, with the attenuation lowest between approximately 1500 and 1650 nm, so that two windows—the so-called C-band (1530–1565 nm) and L-band (1570–1610 nm) are used in modern optical communication systems.

[59] Each of the frequency bands mentioned above, at a typical signal-to-noise ratio $S/N > 10^5$ (50 db), corresponds to the Shannon bandwidth $\Delta f \log_2(S/N)$ exceeding 10^{14} bits per second, five orders of magnitude (!) higher than that of a modern Ethernet cable. The practical bandwidth of a fiber is somewhat lower, but an optical cable, with many fibers in parallel, has a proportionately higher aggregate bandwidth. A recent (circa 2017) example is the transatlantic (6600 km long) cable *Marea*, with eight fiber pairs and an aggregate useable bandwidth of 160 Tbits per second.

[60] The silica-based fibers were suggested in 1966 by C Kao (awarded with the 2009 Nobel Prize in physics), but the idea of using optical fibers for communications may be traced back to at least the 1963 work by J Nishizawa.

optical fibers are both close to 2.2 ($n_\pm \approx 1.5$) and hence very close to each other, so that the relative difference of the refraction indices,

$$\Delta \equiv \frac{n_- - n_+}{n_-} = \frac{\varepsilon_-^{1/2} - \varepsilon_+^{1/2}}{\varepsilon_-^{1/2}} \approx \frac{\varepsilon_- - \varepsilon_+}{2\varepsilon_\pm}, \qquad (7.149)$$

is typically below 0.5%. This factor limits the fiber bandwidth. Indeed, let us use the geometric-optics picture to calculate the number of quasi-plane-wave modes that may propagate in the fiber. For the complementary angle (figure 7.25)

$$\vartheta \equiv \frac{\pi}{2} - \theta, \qquad \text{so that } \sin\theta = \cos\vartheta, \qquad (7.150)$$

Eq. (7.85) gives the following propagation condition:

$$\cos\vartheta > \frac{n_+}{n_-} = 1 - \Delta. \qquad (7.151)$$

In the limit $\Delta \ll 1$, when the incidence angles $\theta > \theta_c$ of all propagating waves are very close to $\pi/2$ and hence the complimentary angles are small, we can keep only the two first terms in the Taylor expansion of the left-hand side of Eq. (7.151) and obtain

$$\vartheta_{\max}^2 \approx 2\Delta. \qquad (7.152)$$

(Even for the higher-end value $\Delta = 0.005$, this critical angle is only ~0.1 radian, i.e. close to 5°.) Due to this smallness, we can approximate the maximum transverse component of the wave vector as

$$(k_t)_{\max} = k(\sin\vartheta)_{\max} \approx k\vartheta_{\max} \approx \sqrt{2}k\Delta, \qquad (7.153)$$

and use Eq. (7.147) to calculate the number N of propagating modes:

$$N \approx 2\frac{(\pi R^2)\left(\pi k^2 \vartheta_{\max}^2\right)}{(2\pi)^2} = (kR)^2\Delta. \qquad (7.154)$$

For typical values $k = 0.73 \times 10^7$ m^{-1} (corresponding to the free-space wavelength $\lambda_0 = n\lambda = 2\pi n/k \approx 1.3$ μm), $R = 25$ μm, and $\Delta = 0.005$, this formula gives $N \approx 150$.

Now we can calculate the *geometric dispersion* of such a fiber, i.e. the difference of the mode propagation speed, which is commonly characterized in terms of the difference between the wave delay times (traditionally measured in picoseconds per kilometer) of the fastest and the slowest mode. Within the geometric optics approximation, the difference of time delays of the fastest mode (with $k_z = k$) and the slowest mode (with $k_z = k \sin\theta_c$) at distance l is

$$\Delta t = \Delta\left(\frac{l}{v_z}\right) = \Delta\left(\frac{k_z l}{\omega}\right) = \frac{l}{\omega}\Delta k_z = \frac{l}{v}(1 - \sin\theta_c) = \frac{l}{v}\left(1 - \frac{n_+}{n_-}\right) = \frac{l}{v}\Delta. \qquad (7.155)$$

For the example considered above, the TEM wave speed in the glass, $v = c/n \approx 2 \times 10^8$ m s^{-1}, and the geometric dispersion $\Delta t/l$ is close to 25 ps m^{-1}, i.e. 25 000 ps km^{-1}.

(This means, for example, that a 1 ns pulse, being distributed between the modes, would spread to a ~25 ns pulse after passing through a just 1 km fiber segment.) This result should be compared with the chromatic dispersion mentioned above, below 10 ps $\text{km}^{-1} \cdot \text{nm}^{-1}$, which gives dt/l of the order of only 1000 ps km^{-1} in the whole communication band $d\lambda \sim 100$ nm. For this reason, such relatively thick ($2R \sim$ 50 nm) *multi-mode fibers* with high geometric dispersion are used for the transfer of signals/power over only short distances below ~100 m. (In return, they may carry relatively large power, beyond 10 mW.)

Long-range telecommunications are based on *single-mode fibers*, with thin cores (typically with diameters $2R \sim 5$ μm, i.e. of the order of $\lambda/\Delta^{1/2}$). For such structures, Eq. (7.154) yields $N \sim 1$, but in this case the geometric optics approximation is not quantitatively valid, and for the fiber analysis, we should get back to the Maxwell equations. In particular, this analysis should take into an explicit account the evanescent wave in the cladding, because its penetration depth may be comparable to R.[61]

Since the cross-section of an optical fiber lacks metallic walls, the Maxwell equations describing them cannot be exactly satisfied with TEM-wave, H-mode, or E-mode solutions. Instead, the fibers can carry the so-called *HE* and *EH modes*, with both vectors **H** and **E** having longitudinal components simultaneously. In such modes, both E_z and H_z inside the core ($\rho \leqslant R$) have a form similar to Eq. (7.141):

$$f_- = f_l J_n(k_t\rho)\cos n(\varphi - \varphi_0), \quad \text{where } k_t^2 = k_-^2 - k_z^2 > 0, \ \text{and } k_-^2 \equiv \omega^2\varepsilon_-\mu_-, \tag{7.156}$$

where the constant angles φ_0 may be different for each field. On the other hand, for the evanescent wave in the cladding, we may rewrite Eqs. (7.101)–(7.102) as

$$\left(\nabla^2 - \kappa_t^2\right)f_+ = 0, \quad \text{where} \quad \kappa_t^2 \equiv k_z^2 - k_+^2 > 0, \ \text{ and } k_+^2 \equiv \omega^2\varepsilon_+\mu_+. \tag{7.157}$$

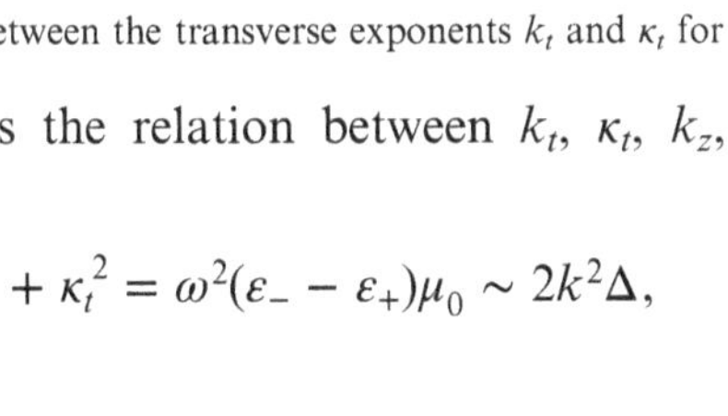

Figure 7.26. Relation between the transverse exponents k_t and κ_t for waves in optical fibers.

Figure 7.26 illustrates the relation between k_t, κ_t, k_z, and $k_\pm$; note that the following sum,

$$k_t^2 + \kappa_t^2 = \omega^2(\varepsilon_- - \varepsilon_+)\mu_0 \sim 2k^2\Delta, \tag{7.158}$$

[61] The following quantitative analysis of the single-mode fibers is very valuable—both in practice and as a very good example of the solution of the Maxwell equations. However, its results will not be used in the following parts of the course, so that if the reader is not interested in this topic, he or she may safely jump to the text following Eq. (7.181). (I believe that the discussion of the angular momentum of electromagnetic radiation, starting at that point, is a compulsory reading for every professional physicist.)

is fixed (at fixed frequency) and, for typical fibers, is very small ($\ll k^2$). In particular, figure 7.26 shows that neither k_t nor κ_t can be larger than $\omega[(\varepsilon_- - \varepsilon_+)\mu_0]^{1/2} = (2\Delta)^{1/2}\,k$. This means that the depth $\delta = 1/\kappa_t$ of the wave penetration into the cladding is at least $1/k(2\Delta)^{1/2} = \lambda/2\pi(2\Delta)^{1/2} \gg \lambda/2\pi$. This is why the cladding layers in practical optical fibers are made as thick as ~50 μm, so that only a negligibly small tail of this evanescent wave field reaches their outer surfaces.

In the polar coordinates, Eq. (7.157) becomes

$$\left[\frac{1}{\rho}\frac{\partial}{\partial\rho}\left(\rho\frac{\partial}{\partial\rho}\right) + \frac{1}{\rho^2}\frac{\partial^2}{\partial\varphi^2} - \kappa_t^2\right]f_+ = 0, \tag{7.159}$$

the equation to be compared with Eq. (7.139) for the circular metallic-wall waveguide. From section 2.7 we know that the eigenfunctions of Eq. (7.159) are the products of the sine and cosine functions of $n\varphi$ by a linear combination of the modified Bessel functions I_n and K_n, shown in figure 2.22, now of argument $\kappa_t\rho$. The fields have to vanish at $\rho \to \infty$, so that only the latter functions (of the second kind) can participate:

$$f_+ \propto K_n(\kappa_t\rho)\cos n(\varphi - \varphi_0). \tag{7.160}$$

Now we have to reconcile Eqs. (7.156) and (7.160) using the boundary conditions at $\rho = R$ for both longitudinal and transverse components of both fields, with the latter components first calculated using Eq. (7.121). Such a conceptually simple, but somewhat bulky calculation (which I am leaving for the reader's exercise), yields a system of two linear, homogeneous equations for complex amplitudes E_l and H_l, which are compatible if

$$\left(\frac{k_-^2}{k_t}\frac{J_n'}{J_n} + \frac{k_+^2}{\kappa_t}\frac{K_n'}{K_n}\right)\left(\frac{1}{k_t}\frac{J_n'}{J_n} + \frac{1}{\kappa_t}\frac{K_n'}{K_n}\right) = \frac{n^2}{R^2}\left(\frac{k_-^2}{k_t^2} + \frac{k_+^2}{\kappa_t^2}\right)\left(\frac{1}{k_t^2} + \frac{1}{\kappa_t^2}\right), \tag{7.161}$$

where the prime sign denotes the derivative of each function over its full argument: $k_t\rho$ for J_n, and $\kappa_t\rho$ for K_n.

For any given frequency ω, the system of equations (7.158) and (7.161) determines the values of k_t and κ_t, and hence k_z. Actually, for any $n > 0$, this system provides two different solutions: one corresponding to the so-called *HE* wave, with a larger ratio E_z/H_z, and the *EH* wave, with a smaller value of that ratio. For angular-symmetric modes with $n = 0$ (for which we might naively expect the lowest cutoff frequency), the equations may be satisfied by the fields having just one finite longitudinal component (either E_z or H_z), so that the *HE* modes are the usual *E* waves, while the *EH* modes are the *H* waves. For the *H* modes, the characteristic equation is reduced to the requirement that the second parentheses on the left-hand side of Eq. (7.161) equal zero. Using the Bessel function identities $J'_0 = -J_1$ and $K'_0 = -K_1$, this equation may be rewritten in a simpler form:

$$\frac{1}{k_t}\frac{J_1(k_tR)}{J_0(k_tR)} = -\frac{1}{\kappa_t}\frac{K_1(\kappa_tR)}{K_0(\kappa_tR)}. \tag{7.162}$$

Using the simple relation between k_t and κ_t, given by Eq. (7.158), we may plot both parts of Eq. (7.162) as a function of the same argument, say, $\xi \equiv k_t R$—see figure 7.27.

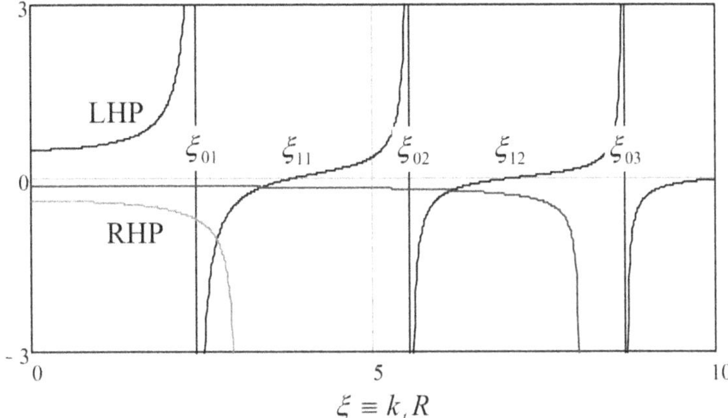

Figure 7.27. Two sides of the characteristic equation (7.162), plotted as functions of $k_t R$, for two values of its dimensionless parameter: $\mathscr{V} = 8$ (blue line) and $\mathscr{V} = 3$ (red line). Note that according to Eq. (7.158), the argument of the functions K_0 and K_1 is $\kappa_t R = [\mathscr{V}^2 - (k_t R)^2]^{1/2} \equiv (\mathscr{V}^2 - \xi^2)^{1/2}$.

The right-hand side of Eq. (7.162) depends not only on ξ but also on the dimensionless parameter $\mathscr{V}$ defined as the normalized right-hand side of Eq. (7.158):

$$\mathscr{V}^2 \equiv \omega^2(\varepsilon_- - \varepsilon_+)\mu_0 R^2 \approx 2\Delta k_\pm^2 R^2. \tag{7.163}$$

(According to Eq. (7.154), if $\mathscr{V} \gg 1$, it gives twice the number N of the fiber modes —the conclusion confirmed by figure 7.27, taking into account that it describes only the H modes.) Since the ratio K_1/K_0 is positive for all values of the functions' argument (see, e.g. the right-hand panel of figure 2.22), the right-hand side of Eq. (7.162) is always negative, so that the equation may have solutions only in the intervals where the ratio J_1/J_0 is negative, i.e. at

$$\xi_{01} < k_t R < \xi_{11}, \quad \xi_{02} < k_t R < \xi_{12}, \ldots, \tag{7.164}$$

where ξ_{nm} is the mth zero of function $J_n(\xi)$—see table 2.1. The right-hand side of the characteristic equation (7.162) diverges at $\kappa_t R \to 0$, i.e. at $k_t R \to \mathscr{V}$, so that no solutions are possible if $\mathscr{V}$ is below the critical value $\mathscr{V}_c = \xi_{01} \approx 2.405$. At this cutoff point, Eq. (7.163) yields $k_\pm \approx \xi_{01}/R(2\Delta)^{1/2}$. Hence, the cutoff frequency of the lowest H mode corresponds to the TEM wavelength

$$\lambda_{\max} = \frac{2\pi R}{\xi_{01}}(2\Delta)^{1/2} \approx 3.7 R\Delta^{1/2}. \tag{7.165}$$

For typical parameters $\Delta = 0.005$ and $R = 2.5$ μm, this result yields $\lambda_{\max} \sim 0.65$ μm, corresponding to the free-space wavelength $\lambda_0 \sim 1$ μm. A similar analysis of the first parentheses on the left-hand side of Eq. (7.161) shows that at $\Delta \to 0$, the cutoff frequency for the E modes is similar.

This situation may look exactly like that in metallic-wall waveguides, with no waves possible at frequencies below ω_c, but this is not so. The basic reason for the difference is that in the metallic waveguides, the approach to ω_c results in the divergence of the longitudinal wavelength $\lambda_z \equiv 2\pi/k_z$. On the other hand, in dielectric waveguides this approach leaves λ_z finite ($k_z \to k_+$). Due to this difference, a certain linear superposition of *HE* and *EH* modes with $n = 1$ can propagate at frequencies well below the cutoff frequency for $n = 0$, which we have just calculated[62]. This mode, in the limit $\varepsilon_+ \approx \varepsilon_-$ (i.e. $\Delta \ll 1$) allows a very interesting and simple description using the *Cartesian* (rather than polar) components of the fields, but still expressed as functions of the *polar* coordinates ρ and φ. The reason is that this mode is very close to a linearly polarized TEM wave. (For this reason, this mode is referred to as LP_{01}.)

Let us select axis x parallel to the transverse component of the magnetic field vector, so that $E_x|_{\rho=0} = 0$, but $E_y|_{\rho=0} \neq 0$, and $H_x|_{\rho=0} \neq 0$, but $H_y|_{\rho=0} = 0$. The only suitable solutions of the 2D Helmholtz equation (which should be obeyed not only by z-components of the field, but also the x- and y-components) are proportional to $J_0(k_t\rho)$ with zero coefficients for E_x and H_y:

$$E_x = 0, \quad E_y = E_0 J_0(k_t\rho), \quad H_x = H_0 J_0(k_t\rho), \quad H_y = 0, \quad \text{for } \rho \leqslant R. \tag{7.166}$$

Now we can use the last two of Eqs. (7.100) to calculate the longitudinal components of the fields:

$$\begin{aligned} E_z &= \frac{1}{-ik_z}\frac{\partial E_y}{\partial y} = -i\frac{k_t}{k_z}E_0 J_1(k_t\rho)\sin\varphi, \\ H_z &= \frac{1}{-ik_z}\frac{\partial H_x}{\partial x} = -i\frac{k_t}{k_z}H_0 J_1(k_t\rho)\cos\varphi, \end{aligned} \tag{7.167}$$

where I have used the mathematical identities $J'_0 = -J_1$, $\partial\rho/\partial x = x/\rho = \cos\varphi$, and $\partial\rho/\partial y = y/\rho = \sin\varphi$. As a sanity check, we see that the longitudinal component or each field is a (legitimate!) eigenfunction of the type (7.141), with $n = 1$. Note also that if $k_t \ll k_z$ (this relation is always true if $\Delta \ll 1$—see either Eq. (7.158) or figure 7.26), the longitudinal components of the fields are much smaller than their transverse counterparts, so that the wave is indeed very close to the TEM one. Because of that, the ratio of the electric and magnetic field amplitudes is also close to that in the TEM wave: $E_0/H_0 \approx Z_- \approx Z_+$.

Now in order to satisfy the boundary conditions at the core-to-cladding interface ($\rho = R$), we need to have a similar angular dependence of these components at $\rho \geqslant R$. The longitudinal components of the fields are tangential to the interface and thus should be continuous. Using solutions similar to Eq. (7.160) with $n = 1$, we obtain

[62] This fact becomes less surprising if we recall that in the circular metallic waveguide, discussed in section 7.6, the fundamental mode (H_{11}, see figure 7.23) also corresponded to $n = 1$ rather than $n = 0$.

$$E_z = -i\frac{k_t}{k_z}\frac{J_1(k_tR)}{K_1(\kappa_tR)}E_0K_1(\kappa_t\rho)\sin\varphi,$$
$$H_z = -i\frac{k_t}{k_z}\frac{J_1(k_tR)}{K_1(\kappa_tR)}H_0K_1(\kappa_t\rho)\cos\varphi, \quad \text{for } \rho \geqslant R. \tag{7.168}$$

For the transverse components, we should require the continuity of the normal magnetic field μH_n, for our simple field structure equal to just $\mu H_x\cos\varphi$, of the tangential electric field $E_\tau = E_y\sin\varphi$, and of the normal component of $D_n = \varepsilon E_n = \varepsilon E_y\cos\varphi$. Assuming that $\mu_- = \mu_+ = \mu_0$, and $\varepsilon_+ \approx \varepsilon_-$,[63] we can satisfy these conditions with the following solutions:

$$E_x = 0, \quad E_y = \frac{J_0(k_tR)}{K_0(\kappa_tR)}E_0K_0(\kappa_t\rho),$$
$$H_x = \frac{J_0(k_tR)}{K_0(k_tR)}H_0K_0(\kappa_t\rho), \quad H_y = 0, \quad \text{for } \rho \geqslant R. \tag{7.169}$$

From here, we can calculate components from E_z and H_z, using the same approach as for $\rho \leqslant R$:

$$E_z = \frac{1}{-ik_z}\frac{\partial E_y}{\partial y} = -i\frac{\kappa_t}{k_z}\frac{J_0(k_tR)}{K_0(\kappa_tR)}E_0K_1(\kappa_t\rho)\sin\varphi,$$
$$H_z = \frac{1}{-ik_z}\frac{\partial H_x}{\partial x} = -i\frac{\kappa_t}{k_z}\frac{J_0(k_tR)}{K_0(\kappa_tR)}H_0K_1(\kappa_t\rho)\cos\varphi, \quad \text{for } \rho \geqslant R. \tag{7.170}$$

We see that this relation provides the same functional dependence of the fields as Eq. (7.167), i.e. the internal and external fields are compatible, but their amplitudes at the interface coincide only if

$$k_t\frac{J_1(k_tR)}{J_0(k_tR)} = \kappa_t\frac{K_1(\kappa_tR)}{K_0(\kappa_tR)}. \tag{7.171}$$

This characteristic equation (which may be also derived from Eq. (7.161) with $n = 1$ in the limit $\Delta \to 0$) looks similar to Eq. (7.162), but functionally is very different from it—see figure 7.28. Indeed, its right-hand side is always positive, and the left-hand side tends to zero at $k_tR \to 0$. As a result, Eq. (7.171) may have a solution for arbitrary small values of the parameter $\mathscr{V}$, defined by Eq. (7.163), i.e. for *arbitrary low frequencies*. This is why this mode is used in practical single-mode fibers: there are no other modes that can propagate at $\omega < \omega_c$, so that they cannot be unintentionally excited on small inhomogeneities of the fiber.

It is easy to use the Bessel function approximations by the first terms of the Taylor expansions (2.132) and (2.157) to show that in the limit $\mathscr{V} \to 0$, κ_tR tends to zero

[63] This is the core assumption of this approximate theory, which accounts only for the most important effect of the difference of dielectric constants ε_+ and ε_-: the opposite signs of the differences $(k_+^2 - k_z^2) = k_t^2$ and $(k_-^2 - k_z^2) = -\kappa_t^2$. For more discussion of the accuracy of this approximation and some exact results, let me refer the interested reader either to the monograph [10], or to chapter 3 and appendix B in [9].

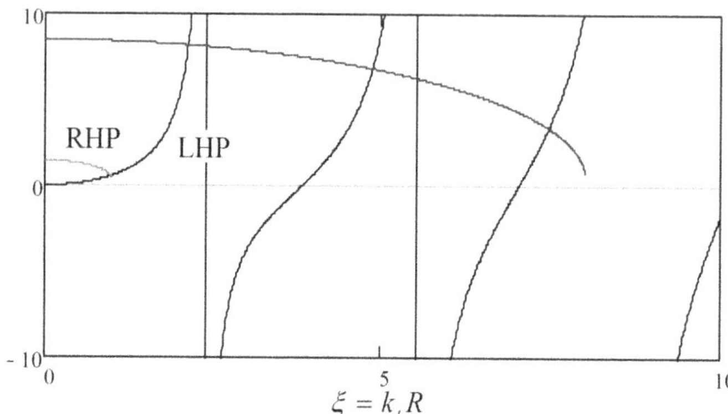

Figure 7.28. Two sides of the characteristic equation (7.171) for the LP_{01} mode, plotted as a function of $k_t R$, for two values of the dimensionless parameter: $\mathscr{V} = 8$ (blue line) and $\mathscr{V} = 1$ (red line).

much faster than $k_t R \approx \mathscr{V}$: $\kappa_t R \to 2\exp\{-1/\mathscr{V}\} \ll \mathscr{V}$. This means that the scale $\rho_c \equiv 1/\kappa_t$ of the radial distribution of the LP_{01} wave's fields in the cladding becomes very large. In this limit, this mode may be interpreted as a virtually TEM wave propagating in the cladding, just slightly deformed (and guided) by the fiber's core. The drawback of this feature is that it requires a very thick cladding, in order to avoid energy losses in outer ('buffer' and 'jacket') layers that defend the silica components from the elements, but lack their low optical absorption. Due to this reason, the core radius is usually selected so that the parameter $\mathscr{V}$ is just slightly less than the critical value $\mathscr{V}_c = \xi_{01} \approx 2.4$ for higher modes, thus ensuring the single-mode operation and eliminating the geometric dispersion problem.

In order to reduce the field spread into the cladding, the step-index fibers discussed above may be replaced with *graded-index* fibers whose dielectric constant ε is gradually and slowly decreased from the center to the periphery[64]. Keeping only the main two terms in the Taylor expansion of the function $\varepsilon(\rho)$ at $\rho = 0$, we may approximate such reduction as

$$\varepsilon(\rho) \approx \varepsilon(0)\left(1 - \frac{\zeta}{2}\rho^2\right), \tag{7.172}$$

where $\zeta \equiv -[(d^2\varepsilon/d\rho^2)/\varepsilon]_{\rho=0}$ is a positive constant characterizing the fiber composition gradient[65]. Moreover, if this constant is sufficiently small ($\zeta \ll k^2$), the field distribution across the fiber's cross-section may be described by the same 2D Helmholtz equation (7.101), but with the space-dependent transverse wave vector[66]:

[64] Due to the technological difficulties of achieving wave attenuation below a few dm km^{-1}, the graded-index fibers are still not used as broadly as the step-index ones.

[65] For an axially symmetric smooth function $\varepsilon(\rho)$, the *first* derivative $d\varepsilon/d\rho$ always vanishes at $\rho = 0$.

[66] Such approach is invalid at arbitrary (large) ζ. Indeed, in the macroscopic Maxwell equations, $\varepsilon(\mathbf{r})$ is under the differentiation sign, and the exact Helmholtz-type equations for fields have additional terms containing $\nabla\varepsilon$.

$$\left[\nabla_t^2 + k_t^2(\rho)\right]f = 0, \quad \text{where}$$
$$k_t^2(\rho) \equiv k^2(\rho) - k_z^2 = \omega^2\varepsilon(\rho)\mu_0 - k_z^2 = k_t^2(0)\left(1 - \frac{\zeta}{2}\rho^2\right). \tag{7.173}$$

Surprisingly for such an axially symmetric problem, because of its special dependence on the radius, this equation may be most readily solved in Cartesian coordinates. Indeed, rewriting it as

$$\left[\frac{\partial^2}{\partial x^2} + \frac{\partial^2}{\partial y^2} + k_t^2(0)\left(1 - \frac{\zeta}{2}x^2 - \frac{\zeta}{2}y^2\right)\right]f = 0, \tag{7.174}$$

and separating variables as $f = X(x)Y(y)$, we obtain

$$\frac{d^2X}{Xdx^2} + \frac{d^2Y}{Ydy^2} + k_t^2(0)\left(1 - \frac{\zeta}{2}x^2 - \frac{\zeta}{2}y^2\right) = 0, \tag{7.175}$$

so that the functions X and Y obey similar differential equations, for example

$$\frac{d^2X}{dx^2} + \left[k_x^2 - k_t^2(0)\frac{\zeta}{2}x^2\right]X = 0, \tag{7.176}$$

with the separation constants satisfying the following relation:

$$k_x^2 + k_y^2 = k_t^2(0) = \omega^2\varepsilon(0)\mu_0 - k_z^2. \tag{7.177}$$

The ordinary differential equation (7.176) is well known from the elementary quantum mechanics, because the Schrödinger equation for the perhaps most important quantum system, a 1D harmonic oscillator, may be rewritten in this form. Its eigenvalues are described by a simple formula

$$\begin{aligned}(k_x^2)_n &= k_t(0)\left(\frac{\zeta}{2}\right)^{1/2}(2n+1),\\ \left(k_y^2\right)_m &= k_t(0)\left(\frac{\zeta}{2}\right)^{1/2}(2m+1), \qquad n, m = 0, 1, 2, \ldots\end{aligned} \tag{7.178}$$

but eigenfunctions $X_n(x)$ and $Y_m(y)$ have to be expressed via not quite elementary functions—the Hermite polynomials[67]. For most practical purposes, however, the lowest eigenfunctions $X_0(x)$ and $Y_0(y)$ are sufficient, because they correspond to the lowest $k_{x,y}$ and hence the lowest

$$[k_t^2(0)]_{\min} = (k_x^2)_0 + \left(k_y^2\right)_0 = 2[k_t(0)]_{\min}\left(\frac{\zeta}{2}\right)^{1/2}, \qquad \text{i.e. } [k_t^2(0)]_{\min} = 2\zeta, \tag{7.179}$$

and hence the highest propagation speed. The eigenfunctions corresponding to these lowest eigenvalues are simple:

[67] See, e.g. *Part QM* section 2.6.

$$X_0(x) = \text{const} \times \exp\left\{-\frac{\zeta x^2}{2}\right\}, \tag{7.180}$$

and similarly for $Y_0(y)$, so that the field distribution follows the Gaussian ('bell curve') function

$$\begin{aligned} f_0(\rho) &= f_0(0)\exp\left\{-\frac{\zeta(x^2+y^2)}{2}\right\} \\ &\equiv f_0(0)\exp\left\{-\frac{\zeta\rho^2}{2}\right\} \equiv f_0(0)\exp\left\{-\frac{\rho^2}{2a^2}\right\}, \end{aligned} \tag{7.181}$$

where $a \equiv 1/\zeta^{1/2} \gg 1/k$ is the effective width of the field's extension in the radial direction, normal to the wave propagation axis z. This is the so-called *Gaussian beam*, very convenient for some applications.

The Gaussian beam (7.181) is just one example of the so-called *paraxial beams*, which may be represented as a result of modulation of a plane wave with a wave number k with an axially symmetric *envelope function* $f(\rho)$, where $\boldsymbol{\rho} \equiv \{x, y\}$, with a relatively large effective radius $a \gg 1/k$.[68] Such beams give me a convenient opportunity to deliver on the promise made in section 7.1: calculate the angular momentum $\mathbf{L}$ of a circularly polarized wave, propagating in free space, and prove its fundamental relation to the wave's energy U. Let us start from the calculation of U for a paraxial beam (with an arbitrary, but spatially limited envelope f) of the circularly polarized waves, with the transverse electric field components given by Eq. (7.19):

$$E_x = E_0 f(\rho)\cos\psi, \qquad E_y = \mp E_0 f(\rho)\sin\psi, \tag{7.182a}$$

where E_0 is the real amplitude of the wave's electric field at the propagation axis, $\psi \equiv kz - \omega t + \varphi$ is its total phase, and the two signs correspond to two possible directions of the circular polarization[69]. According to Eq. (7.6), the corresponding transverse components of the magnetic field are

$$H_x = \pm\frac{E_0}{Z_0}f(\rho)\sin\psi, \qquad H_y = \frac{E_0}{Z_0}f(\rho)\cos\psi. \tag{7.182b}$$

These expressions are sufficient to calculate the energy density (6.113) of the wave[70],

[68] Note that propagating in a uniform medium, i.e. outside grade-index fibers or other focusing systems, such beams gradually increase their width a due to diffraction—to be analyzed in the next chapter.

[69] For our task of calculation of two *quadratic* forms of the fields ($\mathbf{L}$ and U), their real representation (7.182) is more convenient then the complex-exponent one. However, for *linear* manipulations, the latter representation of the circularly polarized waves, $\mathbf{E}_t = E_0 f(\rho)\text{Re}[(\mathbf{n}_x \pm i\mathbf{n}_y)\exp\{i\psi\}]$, $\mathbf{H}_t = (E_0/Z_0)f(\rho)\text{Re}[(\mp i\mathbf{n}_x + \mathbf{n}_y)\exp\{i\psi\}]$, is usually more convenient, and is broadly used.

[70] Note that, in contrast to a linearly polarized wave (7.16), the energy density of a circularly polarized wave does not depend on the full phase ψ—in particular, on t at fixed z, or vice versa. This is natural, because its field vectors rotate (keeping their magnitude) rather than oscillate—see figure 7.3b.

$$u = \frac{\varepsilon_0\left(E_x^2 + E_y^2\right)}{2} + \frac{\mu_0\left(H_x^2 + H_y^2\right)}{2} = \frac{\varepsilon_0 E_0^2 f^2}{2} + \frac{\mu_0 E_0^2 f^2}{2Z_0^2} \equiv \varepsilon_0 E_0^2 f^2, \qquad (7.183)$$

and hence the full energy (per unit length in the direction z of the wave's propagation) of the beam:

$$U = \int u\, d^2r \equiv 2\pi \int_0^\infty u\rho\, d\rho = 2\pi\varepsilon_0 E_0^2 \int_0^\infty f^2 \rho\, d\rho. \qquad (7.184)$$

However, the transverse fields (7.182) are insufficient to calculate a non-vanishing average of **L**. Indeed, following the angular moment's definition in mechanics[71], $\mathbf{L} \equiv \mathbf{r} \times \mathbf{p}$, where **p** is a particle's (linear) momentum, we may use Eq. (6.115) for the electromagnetic field momentum's density **g** in free space, to define the field's angular momentum's density as

$$\mathbf{l} \equiv \mathbf{r} \times \mathbf{g} \equiv \frac{1}{c^2}\mathbf{r} \times \mathbf{S} \equiv \frac{1}{c^2}\mathbf{r} \times (\mathbf{E} \times \mathbf{H}). \qquad (7.185)$$

Let us use the familiar *bac minus cab* rule of the vector algebra[72] to transform this expression to

$$\begin{aligned} \mathbf{l} &= \frac{1}{c^2}[\mathbf{E}(\mathbf{r} \cdot \mathbf{H}) - \mathbf{H}(\mathbf{r} \cdot \mathbf{E})] \\ &\equiv \frac{1}{c^2}\{\mathbf{n}_z[E_z(\mathbf{r} \cdot \mathbf{H}) - H_z(\mathbf{r} \cdot \mathbf{E})] + [\mathbf{E}_t(\mathbf{r} \cdot \mathbf{H}) - \mathbf{H}_t(\mathbf{r} \cdot \mathbf{E})]\}. \end{aligned} \qquad (7.186)$$

If the field is purely transverse ($E_z = H_z = 0$), as it is in a strictly plane wave, the first square brackets in the last expression vanish, while the second bracket gives an azimuthal component of **l**, which oscillates in time, and vanishes at its time averaging. (This is exactly the reason why I have not tried to calculate **L** at our first discussion of the circularly polarized waves in section 7.1.)

Fortunately, our discussion of optical fibers, in particular, the derivation of Eqs. (7.167), (7.168), and (7.170), gives us a very clear clue how to solve this paradox. If the envelope function $f(\rho)$ differs from a constant, the transverse wave components (7.182) alone do *not* satisfy the Maxwell equations (7.2b), which necessitate longitudinal components E_z and H_z of the fields, with[73]

$$\frac{\partial E_z}{\partial z} = -\frac{\partial E_x}{\partial x} - \frac{\partial E_y}{\partial y}, \qquad \frac{\partial H_z}{\partial z} = -\frac{\partial H_x}{\partial x} - \frac{\partial H_y}{\partial y}. \qquad (7.187)$$

However, as these expressions show, if the envelope function f changes very slowly in the sense $df/d\rho \sim f/a \ll kf$, the longitudinal components are very small and do not

[71] See, e.g. *Part CM* Eq. (1.31).

[72] See, e.g. Eq. (A.47).

[73] The complex-exponential versions of these equalities are given by the bottom line of Eq. (7.100).

have a back effect on the transverse components, so that the above calculation of U is still valid (asymptotically, at $ka \to \infty$). Hence, we may still use Eqs. (7.182) on the right-hand side of Eq. (7.187),

$$\frac{\partial E_z}{\partial z} = E_0\left(-\frac{\partial f}{\partial x}\cos\psi \pm \frac{\partial f}{\partial x}\sin\psi\right),$$
$$\frac{\partial H_z}{\partial z} = \frac{E_0}{Z_0}\left(\mp\frac{\partial f}{\partial x}\sin\psi - \frac{\partial f}{\partial x}\cos\psi\right), \qquad (7.188)$$

and integrate them over z as

$$E_z = E_0\int\left(-\frac{\partial f}{\partial x}\cos\psi \pm \frac{\partial f}{\partial x}\sin\psi\right)dz$$
$$= \frac{E_0}{k}\left(-\frac{\partial f}{\partial x}\int\cos\psi\, d\psi \pm \frac{\partial f}{\partial x}\int\sin\psi\, d\psi\right) \qquad (7.189a)$$
$$\equiv \frac{E_0}{k}\left(-\frac{\partial f}{\partial x}\sin\psi \mp \frac{\partial f}{\partial x}\cos\psi\right).$$

Here the integration constant is taken for zero, because evidently no wave field component may have a time-independent part. Integrating, absolutely similarly, the second of Eq. (7.188), we obtain

$$H_z = \frac{E_0}{kZ_0}\left(\pm\frac{\partial f}{\partial x}\cos\psi - \frac{\partial f}{\partial y}\sin\psi\right). \qquad (7.189b)$$

With the same approximation we may calculate the longitudinal (z-)component of $\mathbf{l}$, given by the first term of Eq. (7.186), keeping only the dominant, transverse fields (7.182) in the scalar products:

$$l_z = E_z(\mathbf{r}\cdot\mathbf{H}_t) - H_z(\mathbf{r}\cdot\mathbf{E}_t) \equiv E_z(xH_x + yH_y) - H_z(xE_x + yE_y). \qquad (7.190)$$

Plugging in Eqs. (7.182) and (7.189), and taking into account that in free space, $k = \omega/c$, and hence $1/Z_0c^2k = \varepsilon_0/\omega$, we obtain:

$$l_z = \mp\frac{\varepsilon_0 E_0^2}{\omega}\left(xf\frac{\partial f}{\partial x} + y\frac{\partial f}{\partial y}\right)$$
$$\equiv \mp\frac{\varepsilon_0 E_0^2}{2\omega}\left[x\frac{\partial(f^2)}{\partial x} + y\frac{\partial(f^2)}{\partial y}\right] \equiv \mp\frac{\varepsilon_0 E_0^2}{2\omega}\boldsymbol{\rho}\cdot\nabla(f^2) \equiv \mp\frac{\varepsilon_0 E_0^2}{2\omega}\rho\frac{d(f^2)}{d\rho}. \qquad (7.191)$$

Hence the total angular momentum of the beam (per unit length), is

$$L_z = \int l_z \, d^2r \equiv 2\pi \int_0^\infty l_z \rho \, d\rho$$
$$= \mp \pi \frac{\varepsilon_0 E_0^2}{\omega} \int_0^\infty \rho^2 \frac{d(f^2)}{d\rho} d\rho \equiv \mp \pi \frac{\varepsilon_0 E_0^2}{\omega} \int_{\rho=0}^{\rho=\infty} \rho^2 d(f^2). \tag{7.192}$$

Taking this integral by parts, with the assumption that $\rho f \to 0$ at $\rho \to 0$ and $\rho \to \infty$ (at is true for the Gaussian beam (7.181) and all realistic paraxial beams), we finally obtain

$$L_z = \pm \pi \frac{\varepsilon_0 E_0^2}{\omega} \int_0^\infty f^2 d(\rho^2) \equiv \pm 2\pi \frac{\varepsilon_0 E_0^2}{\omega} \int_0^\infty f^2 \rho \, d\rho. \tag{7.193}$$

Now comparing this expression with Eq. (7.184), we see that remarkably, the ratio L_z/U does not depend on the shape and the width of the beam (and of course on the wave's amplitude E_0), so these parameters are very simply and universally related:

$$L_z = \pm \frac{U}{\omega}. \tag{7.194}$$

Since this relation is valid in the plane-wave limit $a \to \infty$, it may be attributed to plane waves as well, with the understanding that in real life they always have some kind of the wave width ('aperture') restriction.

As the reader certainly knows, in quantum mechanics the energy excitations of any harmonic oscillator of frequency ω are quantized in units of $\hbar\omega$, while the components of the internal angular momentum of a particle are quantized in units of $s\hbar$, where s is its spin. In this context, the classical relation (7.194) is used in quantum electrodynamics as the basis for treating the electromagnetic field excitation quanta (*photons*) as a sort of quantum particles with spin $s = 1$. (Such integer spin also fits the Bose–Einstein statistics of the electromagnetic radiation.)

Unfortunately, I do not have time for a further discussion of the (very interesting) physics of paraxial beams, but cannot help noticing, at least in passing, the very curious effect of *helical waves*—the beams carrying not only the 'spin' momentum (7.194), but also an additional 'orbital' angular momentum. The distribution of their energy in space is not monotonic, and it is in the Gaussian beam (7.181), but resembles several threads twisted around the propagation axis—hence the term 'helical'[74]. Mathematically, this structure is described by the *associate Laguerre polynomials*—the same special functions that are used for the quantum-mechanical description of hydrogen-like atoms[75]. Presently there are efforts to use such beams for the so-called *orbital angular momentum* (OAM) multiplexing for high-rate information transmission[76].

[74] Note that such solutions of the Maxwell equations may be traced back to at least the 1943 theoretical work by J Humblet; however, this issue had not been much discussed in the literature until the spectacular 1992 experiments by L Allen *et al* who demonstrated a simple way of generating helical optical beams—see, e.g. [11]. For a review of later work see, e.g. [12], and references therein.

[75] See, e.g. *Part QM* section 3.7.

[76] See, e.g. [13].

7.8 Resonators

Resonators are distributed oscillators, i.e. structures that may sustain standing waves (in electrodynamics, oscillations of the electromagnetic field) even without a source, until the oscillation amplitude slowly decreases in time due to unavoidable energy losses. If the resonator quality (described by the so-called *Q-factor*, which will be defined and discussed in the next section) is high, $Q \gg 1$, this decay takes many oscillation periods. Alternatively, high-Q resonators may sustain oscillating fields permanently, if fed with a relatively weak incident wave.

Conceptually the simplest resonator is the *Fabry–Pérot interferometer*[77] that may be obtained by placing two well-conducting planes parallel to each other[78]. Indeed, in section 7.3 we have seen that if a plane wave is normally incident on such a 'perfect mirror', located at $z = 0$, its reflection, at negligible skin depth, results in a standing wave described by Eq. (7.61*b*):

$$E(z, t) = \text{Re}\,(2E_\omega e^{-i\omega t + i\pi/2}) \sin kz. \tag{7.195}$$

Hence the wave would not change if we had suddenly put the second mirror (isolating the segment of length l from the external wave source) at any position $z = l$ with $\sin kl = 0$, i.e.

$$kl = p\pi, \quad \text{where } p = 1, 2, ... \tag{7.196}$$

This condition, which determines the *eigen-* (or *resonance-*)*frequency spectrum* of the resonator of fixed length l,

$$\omega_p = vk_p = \frac{\pi v}{a}p, \quad \text{with} \quad v = \frac{1}{(\varepsilon\mu)^{1/2}}, \tag{7.197}$$

has a simple physical sense: the resonator's length l equals exactly p half-waves of frequency ω_p. Though this is all very simple, please note a considerable change of philosophy from what we have been doing in the previous sections: the main task of resonator's analysis is finding its eigenfrequencies ω_p that are now determined by the system geometry rather than by an external wave source.

Before we move to more complex resonators, let us use Eq. (7.62) to represent the magnetic field in the Fabry–Pérot interferometer:

$$H(z, t) = \text{Re}\left(2\frac{E_\omega}{Z}e^{-i\omega t}\right)\cos kz \ . \tag{7.198}$$

Expressions (7.195) and (7.198) show that in contrast to traveling waves, each field of the standing wave changes simultaneously (proportionately) at all points of the Fabry–Pérot resonator, turning to zero everywhere twice a period. At the instants when the energy of the corresponding field vanishes, the total energy of oscillations

---

[77] The device is named after its inventors, M Fabry and A Pérot; it is also called the *Fabry–Pérot etalon* (meaning 'gauge'), because of its initial use for light wavelength measurements.

[78] The resonators formed by well conducting (usually, metallic) walls are frequently called *resonant cavities*.

stays constant, because the counterpart field oscillates with the phase shift $\pi/2$. Such behavior is typical for all electromagnetic resonators.

Another, more technical remark is that we can readily get the same results (7.195)–(7.198) by solving the Maxwell equations from scratch. For example, we already know that in the absence of dispersion, losses, and sources, they are reduced to the wave equations (7.3) for any field components. For the Fabry–Pérot resonator's analysis, we can use the 1D form of these equations, say, for the transverse component of the electric field:

$$\left(\frac{\partial^2}{\partial z^2} - \frac{1}{v^2}\frac{\partial^2}{\partial t^2}\right)E = 0, \tag{7.199}$$

and solve it as a part of an eigenvalue problem with the corresponding boundary conditions. Indeed, separating time and space variables as $E(z, t) = Z(z)\mathcal{T}(t)$, we obtain

$$\frac{1}{Z}\frac{d^2Z}{dz^2} - \frac{1}{v^2}\frac{1}{\mathcal{T}}\frac{d^2\mathcal{T}}{dt^2} = 0. \tag{7.200}$$

Calling the separation constant k^2, we obtain two similar ordinary differential equations,

$$\frac{d^2Z}{dz^2} + k^2Z = 0, \qquad \frac{d^2\mathcal{T}}{dt^2} + k^2v^2\mathcal{T} = 0, \tag{7.201}$$

both with sinusoidal solutions, so that the product $Z(z)\mathcal{T}(t)$ is a standing wave with the wave vector k and frequency $\omega = kv$.[79] Now using the boundary conditions $E(0, t) = E(l, t) = 0$,[80] we obtain the eigenvalue spectrum for k_p and hence for $\omega_p = vk_p$, given by Eqs. (7.196) and (7.197).

Lessons from this simple case study may be readily generalized for an arbitrary resonator: there are (at least) two approaches to finding the eigenfrequency spectrum:

(i) We may look at a traveling wave solution and find where reflecting mirrors may be inserted without affecting the wave's structure. Unfortunately, this method is limited to simple geometries.

(ii) We may solve the general 3D wave equations,

$$\left(\nabla^2 - \frac{1}{v^2}\frac{\partial^2}{\partial t^2}\right)f(\mathbf{r}, t) = 0, \tag{7.202}$$

[79] In this form, the equations are valid even in the presence of dispersion, but with a frequency-dependent wave speed: $v^2 = 1/\varepsilon(\omega)\mu(\omega)$.

[80] This is of course the expression of the first of the general boundary conditions (7.104). The second of these conditions (for the magnetic field) is satisfied automatically for the transverse waves we are considering.

for field components as an eigenvalue problem with appropriate boundary conditions. If the system parameters (and hence the coefficient v) do not change in time, the spatial and temporal variables of Eq. (7.202) may be *always* separated by taking

$$f(\mathbf{r}, t) = \mathcal{R}(\mathbf{r})\mathcal{T}(t), \tag{7.203}$$

where the function $\mathcal{T}(t)$ *always* obeys the same equation as in Eq. (7.201), having the sinusoidal solution of frequency $\omega = vk$. Plugging this solution back into Eq. (7.201), for the spatial distribution of the field we obtain the *3D Helmholtz equation*,

$$(\nabla^2 + k^2)\mathcal{R}(\mathbf{r}) = 0, \tag{7.204}$$

whose solution (for non-symmetric geometries) may be much more complex.

Let us use these approaches to find the eigenfrequency spectrum of a few simple, but practically important resonators. First of all, the first method is completely sufficient for the analysis of any resonator formed as a fragment of a uniform TEM transmission line (e.g. a coaxial cable), confined with two conducting lids perpendicular to the line direction. Indeed, since in such lines $k_z = k = \omega/v$, and the electric field is perpendicular to the propagation axis, e.g. parallel to the lid surface, the boundary conditions are exactly the same as in the Fabry–Pérot resonator, and we again arrive at the eigenfrequency spectrum (7.197).

Now let us analyze a slightly more complex system: a rectangular metallic-wall cavity of volume $a \times b \times l$—see figure 7.29. In order to use the first approach outlined above, let us consider the resonator as a finite-length ($\Delta z = l$) section of the rectangular waveguide stretched along axis z, which was analyzed in detail in section 7.6. As a reminder, at $a < b$, in the fundamental H_{10} traveling wave mode, both vectors $\mathbf{E}$ and $\mathbf{H}$ do not depend on y, with $\mathbf{E}$ having only an y-component. In contrast, $\mathbf{H}$ has two components, H_x and H_z, with the phase shift $\pi/2$ between them, with H_x having the same phase as E_y—see Eqs. (7.131), (7.137), and (7.138). Hence, if a plane, perpendicular to axis z, is placed so that the electric field vanishes on it, H_x also vanishes, so that both boundary conditions (7.104) pertinent to a perfect metallic wall are fulfilled simultaneously.

As a result, the H_{10} wave would not be perturbed by two metallic walls separated by an integer number of half-wavelength $\lambda_z/2$ corresponding to the wave number given by the combination of Eqs. (7.102) and (7.133):

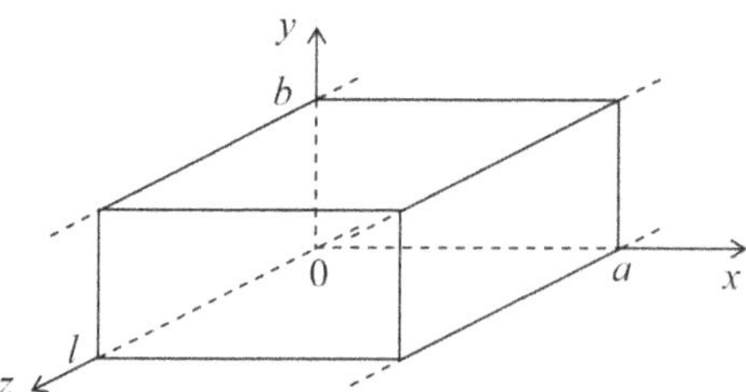

Figure 7.29. A rectangular metallic-wall resonator as a finite section of a waveguide with the cross-section shown in figure 7.22.

$$k_z = \left(k^2 - k_t^2\right)^{1/2} = \left(\frac{\omega^2}{v^2} - \frac{\pi^2}{a^2}\right). \tag{7.205}$$

Using this expression, we see that the smallest of these distances, $l = \lambda_z/2 = \pi/k_z$, gives the resonance frequency[81]

$$\omega_{101} = v\left[\left(\frac{\pi}{a}\right)^2 + \left(\frac{\pi}{l}\right)^2\right]^{1/2}, \tag{7.206}$$

where the indices of ω show the numbers of half-waves along each dimension of the system. This is the lowest (fundamental) eigenfrequency of the resonator (if $b < a, l$).

The field distribution in this mode is close to that in the corresponding waveguide mode H_{10} (figure 7.22), with the important difference that the magnetic and electric fields are shifted by phase $\pi/2$ both in space and time, just as in the Fabry–Pérot resonator—see Eqs. (7.195) and (7.198). Such a time shift allows for a very simple interpretation of the H_{101} mode that is particularly adequate for very flat resonators, with $b \ll a, l$. At the instant when the electric field reaches its maximum (figure 7.30a), i.e. when the magnetic field vanishes in the whole volume, the surface electric charge of the walls (with the areal density $\sigma = E_n/\varepsilon$) is largest, being localized mostly in the middle of the broadest (in figure 7.30, horizontal) faces of the resonator. At immediate later times, the walls start to recharge via surface currents whose density J is largest in the side walls, and reaches its maximal value in a quarter period of the oscillation period of frequency ω_{101}—see figure 7.30b. The currents generate the vortex magnetic field, with looped field lines in the plane of the broadest face of the resonator. The surface currents continue to flow in this direction until (in one more quarter period) the broader walls of the resonator are fully recharged in the polarity opposite to that shown in figure 7.30a. After that, the surface currents start to flow in the direction opposite to that shown in figure 7.30b. This process, which repeats again and again, is conceptually similar to the well-known oscillations in a lumped LC circuit, with the role of (now, distributed) capacitance played mostly by the broadest faces of the resonator, and that of (now, distributed) inductance, mostly by its narrower walls.

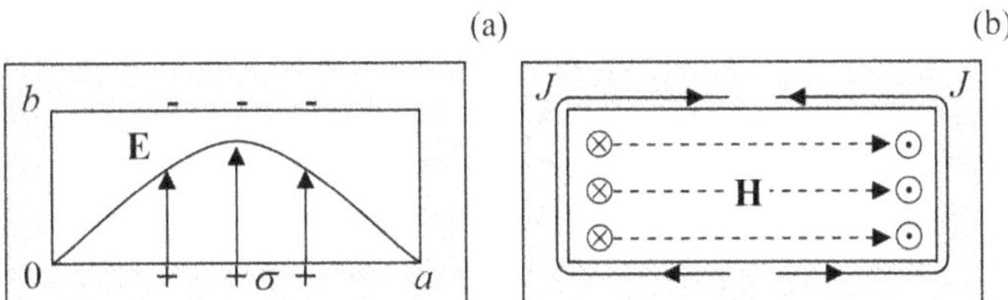

Figure 7.30. A schematic representation of the fields, charges, and currents in the fundamental (H_{101}) mode of a rectangular metallic resonator, at two instants separated by $\Delta t = \pi/2\omega_{101}$.

[81] In most electrical engineering handbooks, the index corresponding to the shortest side of the resonator is listed last, so that the fundamental mode is nominated as H_{110} and its eigenfrequency as ω_{110}.

In order to generalize the result (7.206) to higher oscillation modes, the second of the approaches discussed above is more prudent. Separating variables as $\mathcal{R}(\mathbf{r}) = X(x)Y(y)Z(z)$ in the Helmholtz equation (7.204), we see that X, Y, and Z have to be either sinusoidal or cosinusoidal functions of their arguments, with wave vector components satisfying the characteristic equation

$$k_x^2 + k_y^2 + k_z^2 = k^2 \equiv \frac{\omega^2}{v^2}. \qquad (7.207)$$

In contrast to the wave propagation problem, now we are dealing with standing waves along all three dimensions and have to satisfy the macroscopic boundary conditions (7.104) on all sets of parallel walls. It is straightforward to check that these conditions ($E_\tau = 0$, $H_n = 0$) are fulfilled at the following field component distribution,

$$\begin{aligned} E_x &= E_1 \cos k_x x \ \sin k_y y \ \sin k_z z, & H_x &= H_1 \sin k_x x \ \cos k_y y \ \cos k_z z, \\ E_y &= E_2 \sin k_x x \ \cos k_y y \ \sin k_z z, & H_y &= H_2 \cos k_x x \ \sin k_y y \ \cos k_z z, \\ E_z &= E_3 \sin k_x x \ \sin k_y y \ \cos k_z z, & H_z &= H_3 \cos k_x x \ \cos k_y y \ \sin k_z z, \end{aligned} \qquad (7.208)$$

with each of the wave vector components having an equidistant spectrum, similar to Eq. (7.196),

$$k_x = \frac{\pi n}{a}, \quad k_y = \frac{\pi m}{b}, \quad k_z = \frac{\pi p}{l}, \qquad (7.209)$$

so that the full spectrum of eigenfrequencies is given by the following formula,

$$\omega_{nmp} = vk = v\left[\left(\frac{\pi n}{a}\right)^2 + \left(\frac{\pi m}{b}\right)^2 + \left(\frac{\pi p}{l}\right)^2\right]^{1/2}, \qquad (7.210)$$

which is a natural generalization of Eq. (7.206). Note, however, that of the three integers m, n, and p, at least two have to be different from zero in order to keep the fields (7.206) from vanishing at all points.

We may use Eq. (7.210), in particular, to evaluate the number of different modes in a relatively small range $d^3k \ll k^3$ of the wave vector space, which is still much larger than the reciprocal volume, $1/V = 1/abl$, of the resonator. Taking into account that each eigenfrequency (7.210), with $nml \neq 0$, corresponds to two field modes with different polarizations[82], argumentation absolutely similar to that used at the end of section 7.7 for the 2D case yields

$$dN = 2V\frac{d^3k}{(2\pi)^3}. \qquad (7.211)$$

[82] This fact becomes evident from plugging Eqs. (7.208) into the Maxwell equation $\nabla \cdot \mathbf{E} = 0$. The resulting equation, $k_xE_1 + k_yE_2 + k_zE_3 = 0$, with the discrete, equidistant spectrum (7.209) for each wave vector component, may be satisfied by two linearly independent sets of the constants $E_{1,2,3}$.

This property, valid for resonators of arbitrary shape, is broadly used in classical and quantum statistical physics[83] in the following form. If some electromagnetic mode functional $f(\mathbf{k})$ is a smooth function of the wave vector $\mathbf{k}$ and the volume V is large enough, then Eq. (7.211) may be used to approximate the sum of the functional's values over the modes by an integral:

$$\sum_{\mathbf{k}} f(\mathbf{k}) \approx \int_N f(\mathbf{k})dN \equiv \int_{\mathbf{k}} f(\mathbf{k})\frac{dN}{d^3k}d^3k = 2\frac{V}{(2\pi)^3}\int_{\mathbf{k}} f(\mathbf{k})d^3k. \tag{7.212}$$

Leaving the similar analyses of resonant cavities of other shapes for reader's exercises, let me finish this section by noting that low-loss resonators may be also formed by finite-length sections of not only metallic-wall waveguides with different cross-sections, but also of the dielectric waveguides. Moreover, even a simple slab of a dielectric material with a μ/ε ratio substantially different from that of its environment (say, the free space) may be used as a high-Q Fabry–Pérot interferometer (figure 7.31), due to an effective wave reflection from its surfaces at normal

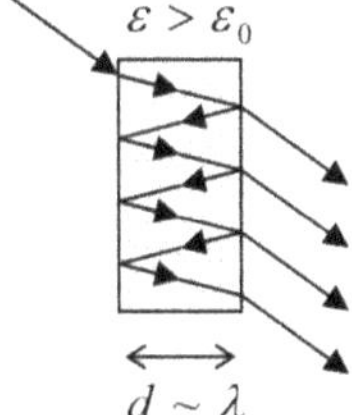

Figure 7.31. A dielectric Fabry–Pérot interferometer.

and in particular inclined incidence—see, respectively, Eq. (7.68), and Eqs. (7.91) and (7.95).

Actually, such dielectric Fabry–Pérot interferometers are frequently more convenient for practical purposes than metallic-wall resonators, not only due to possibly lower losses (particularly in the optical range), but also due to a natural coupling to the environment that enables a ready method of wave insertion and extraction—see figure 7.31 again. However, this coupling to environment provides an additional mechanism of power losses, limiting the resonance quality—see the next section.

7.9 Energy loss effects

The inevitable energy losses ('dissipation') in passive media lead, in two different situations, to two different effects. In a long transmission line fed by a constant wave source at one end, the losses lead to a gradual *attenuation* of the wave, i.e. to a decrease of its amplitude, and hence its power $\mathscr{P}$, with the distance z along the line. In linear materials, the time-averaged losses are proportional to the wave amplitude squared, i.e. to the time-averaged power $\mathscr{P}$ of the wave itself, so that the energy balance on a small segment dz takes the form

[83] See, e.g. *Part QM* section 1.1 and *Part SM* section 2.6.

$$d\mathscr{P} = -\frac{d\mathscr{P}_{\text{loss}}}{dz}dz = -\alpha\mathscr{P}\,dz. \tag{7.213}$$

The coefficient α, participating in the last form of Eq. (7.213) and defined by that relation,

$$\alpha \equiv \frac{d\mathscr{P}_{\text{loss}}/dz}{\mathscr{P}}, \tag{7.214}$$

is called the *attenuation constant*[84]. Comparing the solution of Eq. (7.213),

$$\mathscr{P}(z) = \mathscr{P}(0)e^{-\alpha z}, \tag{7.215}$$

with Eq. (7.29), where k is replaced with k_z, we see that α may expressed as

$$\alpha = 2\text{Im}\,k_z, \tag{7.216}$$

where k_z is the component of the wave vector along the transmission line. In the most important limit when the losses are low in the sense $\alpha \ll |k_z| \approx \text{Re}\,k_z$, its effects on the field distribution along the line's cross-section are negligible, making the calculation of α rather straightforward. In particular, in this limit the contributions to attenuation from two major sources, energy losses in the filling dielectric and the skin effect-losses in conducting walls, are independent and additive.

The dielectric losses are especially simple to describe. Indeed, a review of our calculations in sections 7.5–7.7 shows that all of them remain valid if either $\varepsilon(\omega)$ or $\mu(\omega)$, or both, and hence $k(\omega)$, have small imaginary parts:

$$k'' = \omega\text{Im}[\varepsilon^{1/2}(\omega)\mu^{1/2}(\omega)] \ll k'. \tag{7.217}$$

In TEM transmission lines $k_z = k$ and hence Eq. (7.216) yields

$$\alpha_{\text{filling}} = 2k'' = 2\omega\text{Im}[\varepsilon^{1/2}(\omega)\mu^{1/2}(\omega)]. \tag{7.218}$$

For dielectric waveguides, in particular optical fibers, these losses are the main attenuation mechanism. As we already know from section 7.7, in practical optical fibers $\kappa_t R \gg 1$, i.e. most of the field propagates (as an evanescent wave) in the cladding, with a field distribution very close to the TEM wave. This is why Eq. (7.218) is approximately valid if it is applied to the cladding material alone. In waveguides with non-TEM waves, we can readily use the relations between k_z and k, derived in the previous sections, to re-calculate k'' into Im k_z. (Note that at such a re-calculation, the values of k_t have to be kept real, because they are just the eigenvalues of the Helmholtz equation (7.101), which does not include the filling media parameters.).

[84] In engineering, attenuation is frequently measured in *decibels per meter* (acronymed as db m^{-1} or just dbm):

$$\alpha\Big|_{\text{db}/m} \equiv 10\log_{10}\frac{\mathscr{P}(z=0)}{\mathscr{P}(z=1\text{m})} = 10\log_{10}e^{\alpha[1/\text{m}]} = \frac{10}{\ln 10}\alpha[\text{m}^{-1}] \approx 4.34\alpha[\text{m}^{-1}].$$

In transmission lines and waveguides and with metallic walls, much higher energy losses may come from the skin effect. If the wavelength λ is much larger than δ_s, as it usually is[85], we may use Eq. (6.36)[86]:

$$\frac{d\mathscr{P}_{\text{loss}}}{dA} = H_{\text{wall}}^2 \frac{\mu\omega\delta_s}{4}, \tag{7.219}$$

where H_{wall} is the real amplitude of the tangential component of the magnetic field at the wall's surface. The total power loss $d\mathscr{P}_{loss}/dz$ per unit length of a waveguide, i.e. the right-hand side of Eq. (7.213), now may be calculated by the integration of the ratio $d\mathscr{P}_{loss}/dA$ along the contour(s) limiting the cross-section of all conducting walls. Since our calculation is only valid for low losses, we may ignore their effect on the field distribution, so that the unperturbed distributions may be used both in Eq. (7.219), i.e. in the nominator of Eq. (7.214), and also for the calculation of the average propagating power, i.e. the denominator of Eq. (7.214)—as the integral of the Poynting vector over the cross-section of the waveguide.

Let us see how this approach works for the TEM mode in one of the simplest transmission lines, the coaxial cable (figure 7.20). As we already know from section 7.5, in the absence of losses, the distribution of TEM mode fields is the same as in statics, namely:

$$H_z = 0, \quad H_\rho = 0, \quad H_\varphi(\rho) = H_0\frac{a}{\rho}, \tag{7.220}$$

where H_0 is the field's amplitude on the surface of the inner conductor, and

$$E_z = 0, \quad E_\rho(\rho) = ZH_\varphi(\rho) = ZH_0\frac{a}{\rho}, \quad E_\varphi = 0, \quad Z \equiv \left(\frac{\mu}{\varepsilon}\right)^{1/2}. \tag{7.221}$$

Now we can, neglecting losses for now, use Eq. (7.42) to calculate the time-averaged Poynting vector

$$\overline{S} = \frac{Z\left|H_\varphi(\rho)\right|^2}{2} = \frac{Z\left|H_0\right|^2}{2}\left(\frac{a}{\rho}\right)^2, \tag{7.222}$$

and from it, the total power propagating through the cross-section:

$$\mathscr{P} = \int_A \overline{S}\, d^2r = \frac{Z\left|H_0\right|^2 a^2}{2} 2\pi \int_a^b \frac{\rho d\rho}{\rho^2} = \pi Z\left|H_0\right|^2 a^2 \ln\frac{b}{a}. \tag{7.223}$$

For the particular case of the coaxial cable (figure 7.20), the contours limiting the wall cross-sections are circles of radii $\rho = a$ (where the surface field amplitude H_{walls} equals, in our notation, H_0), and $\rho = b$ (where, according to Eq. (7.214), the field is a factor of b/a lower). As a result, for the power loss per unit length, Eq. (7.219) yields

[85] As follows from Eq. (7.78), which may be used for crude estimates even in cases of arbitrary incidence, this condition is necessary for low attenuation: $\alpha \ll k$ only if $\not{f} \ll 1$.

[86] For a normally incident plane wave, this formula would bring us back to Eq. (7.78).

$$\frac{d\mathscr{P}_{\text{loss}}}{dz} = \left(2\pi a\,|H_0|^2 + 2\pi b\left|H_0\,\frac{a}{b}\right|^2\right)\frac{\mu_0\omega\delta_s}{4} = \frac{\pi}{2}a\left(1+\frac{a}{b}\right)\mu\omega\delta_s\,|H_0|^2. \qquad (7.224)$$

Note that at $a \ll b$ the losses in the inner conductor dominate, despite its smaller surface, because of the higher surface field. Now we may plug Eqs. (7.223) and (7.224) into the definition (7.214) of α to calculate the skin-effect contribution to the attenuation constant:

$$\alpha_{\text{skin}} \equiv \frac{d\mathscr{P}_{\text{loss}}/dz}{\mathscr{P}} = \frac{1}{2\ln(b/a)}\left(\frac{1}{a}+\frac{1}{b}\right)\frac{\mu\omega\delta_s}{Z} = \frac{k\delta_s}{2\ln(b/a)}\left(\frac{1}{a}+\frac{1}{b}\right). \qquad (7.225)$$

We see that the relative (dimensionless) attenuation, α/k, scales approximately as the ratio $\delta_s/\min[a, b]$. This result has to be compared with Eq. (7.78) for the normal incidence of plane waves on a conducting surface.

Let us use this result to evaluate α for the standard TV cable RG-6/U (with copper conductors of diameters $2a = 1$ mm, $2b = 4.7$ mm, and $\varepsilon \approx 2.2\varepsilon_0$, $\mu \approx \mu_0$). According to Eq. (6.33), for frequency $f = 100$ MHz ($\omega \approx 6.3 \times 10^8$ s^{-1}) the skin depth of pure copper at room temperature (with $\sigma \approx 6.0 \times 10^7$ s m^{-1}) is close to 6.5×10^{-6} m, while $k = \omega(\varepsilon\mu)^{1/2} = (\varepsilon/\varepsilon_0)^{1/2}(\omega/c) \approx 3.1$ m^{-1}. As a result, the attenuation is rather low: $\alpha_{\text{skin}} \approx 0.016$ m^{-1}, so that the attenuation length scale $l_d \equiv 1/\alpha$ is about 60 m. Hence the attenuation in a cable connecting a roof TV antenna to a TV set in the same house is not a big problem, although using a worse conductor, e.g. steel, would make the losses rather noticeable. (Hence the current worldwide shortage of copper.) However, an attempt to use the same cable in the X-band ($f \sim 10$ GHz) is more problematic. Indeed, though the skin depth $\delta_s \propto \omega^{-1/2}$ decreases with frequency, the wavelength drops, i.e. k increases even faster ($k \propto \omega$), so that the attenuation $\alpha_{\text{skin}} \propto \omega^{1/2}$ becomes close to 0.16 m^{-1}, i.e. l_d to ~6 m. This is why at such frequencies, it may be necessary to use rectangular waveguides, with their larger internal dimensions $a, b \sim 1/k$, and hence lower attenuation. Let me leave the calculation of this attenuation, using Eq. (7.219) and the results derived in section 7.7, for the reader's exercise.

The power loss effect on free oscillations *in resonators* is different: here it leads to a gradual decay of the oscillating fields' energy U in time. A useful dimensionless measure of this decay, called the *Q-factor*, may be introduced by writing the temporal analog of Eq. (7.213)[87]:

$$dU = -\mathscr{P}_{\text{loss}}\,dt = -\frac{\omega}{Q}U\,dt, \qquad (7.226)$$

where ω is the eigenfrequency in the loss-free limit and

[87] As losses grow, the oscillation waveform deviates from sinusoidal, and the very notion of 'oscillation frequency' becomes vague. As a result, the parameter Q is well defined only if it is much higher than 1.

$$\frac{\omega}{Q} \equiv \frac{\mathscr{P}_{\text{loss}}}{U} \tag{7.227}$$

is the temporal analog of Eq. (7.214). The solution of Eq. (7.226),

$$U(t) = U(0)e^{-t/\tau}, \quad \text{with } \tau \equiv \frac{Q}{\omega} = \frac{Q/2\pi}{\omega/2\pi} = \frac{Q\mathcal{T}}{2\pi}, \tag{7.228}$$

which is the temporal analog of Eq. (7.215), shows the physical meaning of the Q-factor: the characteristic time τ of the oscillation energy's decay is $(Q/2\pi)$ times longer than the oscillation period $\mathcal{T} = 2\pi/\omega$. (Another useful interpretation of Q comes from the relation[88]

$$Q = \frac{\omega}{\Delta\omega}, \tag{7.229}$$

where $\Delta\omega$ is the so-called FWHM[89] bandwidth of the resonance, namely the difference between the two values of the external signal frequency, one above and one below ω, at which the energy of the forced oscillations induced in the resonator by an input signal is twice lower than its resonance value.)

In the important particular case of resonators formed by insertion of metallic walls into a TEM transmission line of small cross-section (with the linear size scale a much less than the wavelength λ), there is no need to calculate the Q-factor directly, provided that the line attenuation coefficient α is already known. Indeed, as was discussed in section 7.8 above, the standing waves in such a resonator, of the length given by Eq. (7.196): $l = p(\lambda/2)$ with $p = 1, 2,\ldots$, may be understood as an overlap of two TEM waves running in opposite directions or, in other words, a traveling wave plus its reflection from one of the ends, the whole roundtrip taking time $\Delta t = 2\,l/v = p\lambda/v = 2\pi p/\omega = p\,\mathcal{T}$. According to Eq. (7.215), at this distance the wave's power drops by the factor of $\exp\{-2\alpha l\} = \exp\{-p\alpha\lambda\}$. On the other hand, the same decay may be viewed as happening in time and, according to Eq. (7.228), results in the drop by $\exp\{-\Delta t/\tau\} = \exp\{-(p\mathcal{T})/(Q/\omega)\} = \exp\{-2\pi p/Q\}$. Comparing these two exponents, we obtain

$$Q = \frac{2\pi}{\alpha\lambda} = \frac{k}{\alpha}. \tag{7.230}$$

This simple relation neglects the losses at the wave reflection from the walls limiting the resonator length. Such approximation is indeed legitimate at $a \ll \lambda$; if this relation is violated, or if we are dealing with more complex resonator modes (such as those based on the reflection of E or H waves), the Q-factor may be smaller than that given by Eq. (7.230) and needs to be calculated directly. A substantial relief for such a direct calculation is that, just at the calculation of small attenuation in waveguides, in the low-loss limit ($Q \gg 1$) both the nominator and denominator of the right-hand side of Eq. (7.227) may be calculated neglecting the effects of the

[88] See, e.g. *Part CM* section 5.1.

[89] This is the acronym for 'full width at half-maximum'.

power loss on the field distribution in the resonator. I am leaving such a calculation, for the simplest (rectangular and circular) resonators, for the reader's exercise.

To conclude this chapter, let me make a final remark: in some resonators (including certain dielectric resonators and metallic resonators with holes in their walls), additional losses due to the wave radiation into the environment are also possible. In some simple cases (say, the Fabry–Pérot interferometer shown in figure 7.31) the calculation of these *radiative losses* is straightforward, but sometimes it requires more elaborated approaches, which will be discussed in the next chapter.

7.10 Problems

Problem 7.1.* Find the temporal Green's function of a medium whose complex dielectric constant obeys the Lorentz oscillator model, given by Eq. (7.32), using:

(i) the Fourier transform, and
(ii) the direct solution of Eq. (7.30).

Hint: For the Fourier-transform approach, you may like to use the Cauchy integral[90].

Problem 7.2. The electric polarization of a material responds in the following way to an electric field step[91]:

$$P(t) = \varepsilon_1 E_0 (1 - e^{-t/\tau}), \qquad \text{if } E(t) = E_0 \times \begin{cases} 0, & \text{for } t < 0, \\ 1, & \text{for } 0 < t, \end{cases}$$

where τ is a positive constant. Calculate the complex permittivity $\varepsilon(\omega)$ of this material and discuss a possible simple physical model giving such dielectric response.

Problem 7.3. Calculate the complex dielectric constant $\varepsilon(\omega)$ for a material whose dielectric-response Green's function, defined by Eq. (7.23), is

$$G(\theta) = G_0(1 - e^{-\theta/\tau}),$$

with some positive constants G_0 and τ. What is the difference between this dielectric response and the apparently similar one considered in the previous problem?

Problem 7.4. Use the Lorentz oscillator model of an atom, given by Eq. (7.30), to calculate the average potential energy of the atom in a uniform, sinusoidal ac electric field, and use the result to calculate the potential profile created for the atom by a standing electromagnetic wave with the electric field amplitude $E_\omega(\mathbf{r})$.

Problem 7.5. The solution of the previous problem shows that a standing plane wave exerts a time-averaged force on a non-relativistic charged particle. Reveal the

[90] See, e.g. Eq. (A.92).

[91] This function $E(t)$ is of course proportional to the well-known step function $\theta(t)$—see, e.g. Eq. (A.87). I am not using this notation just to avoid possible confusion between two different uses of the Greek letter θ.

physics of this force by writing and solving the equations of motion of a free, charged particle in:

(i) a linearly polarized, monochromatic, plane traveling wave, and
(ii) a similar but standing wave.

Problem 7.6. Calculate, sketch, and discuss the dispersion relation for electromagnetic waves propagating in a medium described by the Lorentz oscillator model (7.32), for the case of negligible damping.

Problem 7.7. As was briefly discussed in section 7.2[92], a wave pulse of a finite but relatively large spatial extension $\Delta z \gg \lambda \equiv 2\pi/k$ may be represented with a *wave packet*—a sum of sinusoidal waves with wave vectors $\mathbf{k}$ within a relatively narrow interval. Consider an electromagnetic plane wave packet of this type, with the electric field distribution

$$\mathbf{E}(\mathbf{r}, t) = \mathrm{Re}\int_{-\infty}^{+\infty} \mathbf{E}_k e^{i(kz-\omega_k t)}dk, \quad \text{with } \omega_k[\varepsilon(\omega_k)\mu(\omega_k)]^{1/2} \equiv |k|,$$

propagating along axis z in an isotropic, linear, and loss-free (but not necessarily dispersion-free) medium. Express the full energy of the packet (per unit area of wave's front) via the complex amplitudes $\mathbf{E}_k$, and discuss its dependence of time.

Problem 7.8.* Analyze the effect of a constant, uniform magnetic field $\mathbf{B}_0$, parallel to the direction $\mathbf{n}$ of electromagnetic wave propagation, on the wave's dispersion in plasma, within the same simple model that was used in section 7.2 for the derivation of Eq. (7.38). (Limit your analysis to relatively weak waves, whose magnetic field is negligible in comparison with $\mathbf{B}_0$.)

Hint: You may like to represent the incident wave as a linear superposition of two circularly polarized waves, with opposite polarization directions.

Problem 7.9. A monochromatic, plane electromagnetic wave is normally incident, from the free space, on a uniform slab of a material with electric permittivity ε and magnetic permeability μ, with the slab thickness d comparable with the wavelength.

(i) Calculate the power transmission coefficient $\mathcal{T}$, i.e. the fraction of the incident power that is transmitted through the slab.
(ii) Assuming that ε and μ are frequency-independent and positive, analyze in detail the frequency dependence of $\mathcal{T}$. In particular, how does the function $\mathcal{T}(\omega)$ depend on the slab's thickness d and the wave impedance $Z = (\mu/\varepsilon)^{1/2}$ of its material?

Problem 7.10. A monochromatic, plane electromagnetic wave, with free-space wave number k_0, is normally incident on a planar, conducting film of thickness $d \sim \delta_s \ll 1/k_0$. Calculate the power transmission coefficient of the system, i.e. the fraction of

[92] And in more detail in *Part CM* section 5.3, and particularly in *Part QM* section 2.2.

incident wave's power propagating beyond the film. Analyze the result in the limits of small and large ratio d/δ_s.

Problem 7.11. A plane wave of frequency ω is normally incident, from free space, on a planar surface of a material with real values of electric permittivity ε' and magnetic permeability μ'. To minimize the wave reflection from the surface, you may cover it with a layer of thickness d of another transparent material—see the figure below. Calculate the optimal values for ε, μ, and d.

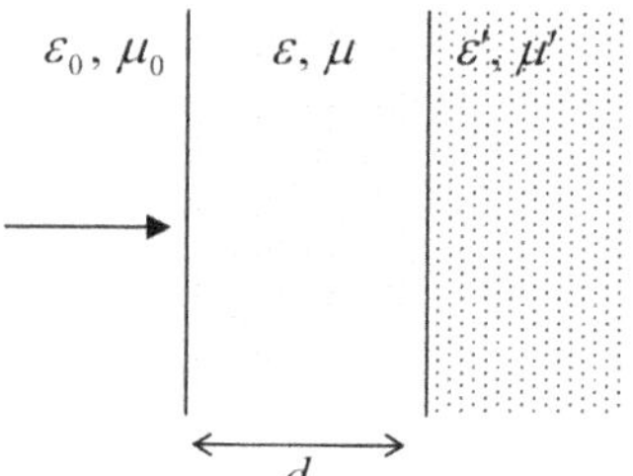

Problem 7.12. A monochromatic, plane wave is incident from inside a medium with $\varepsilon\mu > \varepsilon_0\mu_0$ onto its plane surface, at an angle of incidence θ larger than the critical angle $\theta_c = \sin^{-1}(\varepsilon_0\mu_0/\varepsilon\mu)^{1/2}$. Calculate the depth δ of the evanescent wave penetration into the free space, and analyze its dependence on θ. Does the result depend on the wave's polarization?

Problem 7.13. Analyze the possibility of propagation of surface electromagnetic waves along a plane boundary between a plasma and the free space. In particular, calculate and analyze the dispersion relation of the waves.

Hint: Assume that the magnetic field of the wave is parallel to the boundary and perpendicular to the wave's propagation direction. (After solving the problem, justify this mode choice.)

Problem 7.14. Light from a very distant source arrives to an observer through a planar layer of a non-uniform medium with a certain refraction index distribution, $n(z)$, at angle θ_0—see the figure below. What is the actual direction θ_i to the source, if $n(z) \to 1$ at $z \to \infty$? (This problem is obviously important for high-precision astronomical measurements from the Earth's surface.)

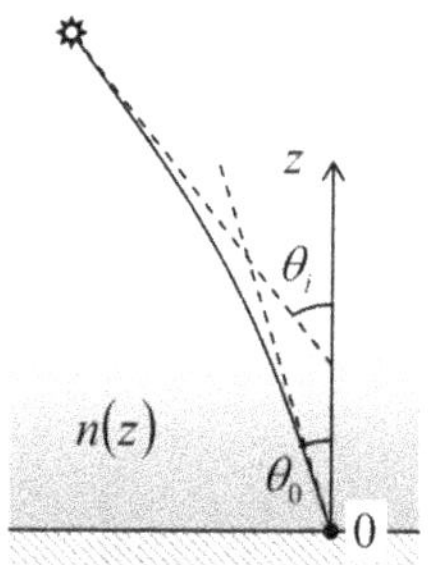

Problem 7.15. Calculate the impedance Z_W of the long, straight TEM transmission lines formed by metallic electrodes with the cross-sections shown in the figure below:

(i) two round, parallel wires, separated by distance $d \gg R$,
(ii) a *microstrip line* of width $w \gg d$,
(iii) a *stripline* with $w \gg d_1 \sim d_2$,

in all cases using the macroscopic boundary conditions on metallic surfaces. Assume that the conductors are embedded into a linear dielectric with constant ε and μ.

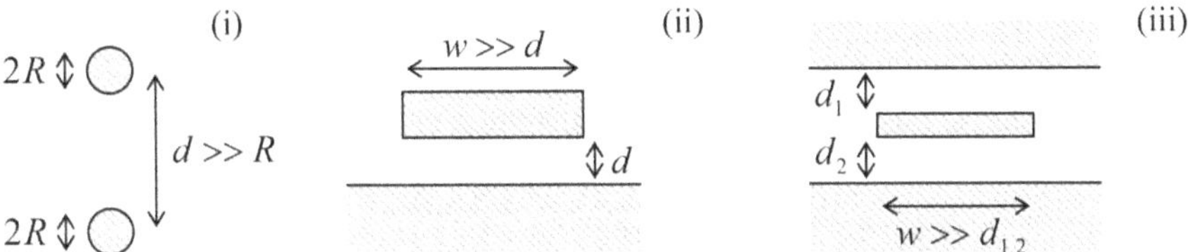

Problem 7.16. Modify the solution of task (ii) of the previous problem for a superconductor microstrip line, taking into account the magnetic field penetration into both the strip and the ground plane.

Problem 7.17.* What lumped ac circuit would be equivalent to the TEM-line system shown in figure 7.19, with an incident wave's power $\mathscr{P}_i$? Assume that the wave reflected from the lumped load circuit does not return to it.

Problem 7.18. Find the lumped ac circuit equivalent to a loss-free TEM transmission line of length $l \sim \lambda$, with a small cross-section area $A \ll \lambda^2$, as 'seen' (measured) from one end, if the line's conductors are galvanically connected ('shortened') at the other end—see the figure below. Discuss the result's dependence on the signal frequency.

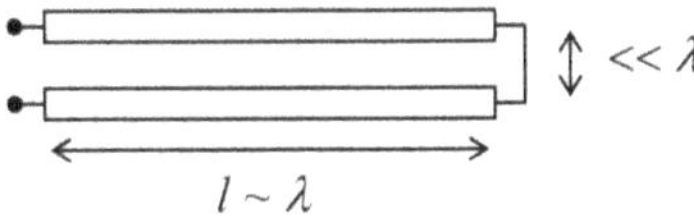

Problem 7.19. Represent the fundamental H_{10} wave in a rectangular waveguide (figure 7.22) with a sum of two plane waves, and discuss the physics behind such a representation.

Problem 7.20.* For a metallic coaxial cable with the circular cross-section (figure 7.20), find the lowest non-TEM mode and calculate its cutoff frequency.

Problem 7.21. Two coaxial cable sections are connected coaxially—see the figure below, which shows the cut along the system's symmetry axis. Relations (7.118) and (7.120) seem to imply that if the ratios b/a of these sections are equal, their impedance matching is perfect, i.e. a TEM wave incident from one side on the connection would pass it without any reflection at all: $R = 0$. Is this statement correct?

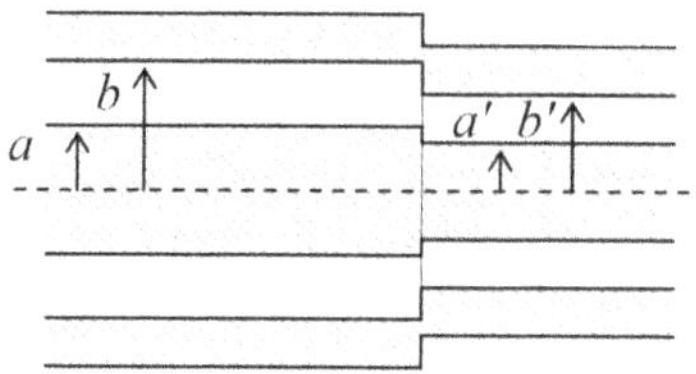

Problem 7.22. Prove that TEM-like waves may propagate, in the radial direction, in the free space between two coaxial, round, metallic cones—see the figure below. Can this system be characterized by a certain transmission line impedance Z_W, as defined by Eq. (7.115)?

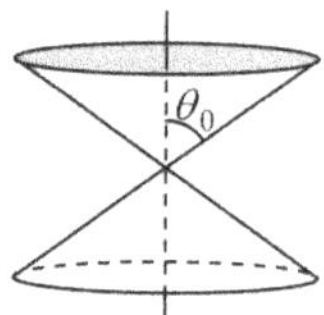

Problem 7.23.* Use the recipe outlined in section 7.7 to prove the characteristic equation (7.161) for the *HE* and *EH* modes in a round, step-index optical fiber.

Problem 7.24. Neglecting the skin-effect depth δ_s, find the lowest eigenfrequencies, and the corresponding field distributions, of the standing electromagnetic waves inside a round cylindrical resonant cavity—see the figure below.

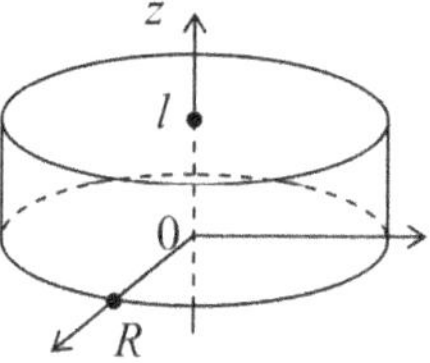

Problem 7.25. A plane, monochromatic wave propagates through a medium with an Ohmic conductivity σ, and negligible electric and magnetic polarization effects. Calculate the wave's attenuation, and relate the result to a certain calculation carried out in chapter 6.

Problem 7.26. Generalize the telegrapher's equations (7.110) and (7.111) by accounting for small energy losses:

(i) in the transmission line's conductors, and
(ii) in the medium separating the conductors,

using their simplest (Ohmic) models. Formulate the conditions of validity of the resulting equations.

Problem 7.27. Calculate the skin-effect contribution to the attenuation coefficient α, defined by Eq. (7.214), for the fundamental (H_{10}) mode propagating in a metallic-wall waveguide with a rectangular cross-section—see figure 7.22. Use the results to evaluate the wave decay length $l_d \equiv 1/\alpha$ for a 10 GHz wave in the standard X-band waveguide WR-90 (with copper walls, a = 23 mm, b = 10 mm, and no dielectric filling), at room temperature. Compare the result with that, made in section 7.9, for the standard TV coaxial cable, at same frequency.

Problem 7.28.* Calculate the skin-effect contribution to the attenuation coefficient α of

(i) the fundamental (H_{11}) wave, and
(ii) the H_{01} wave,

in a metallic-wall waveguide with the circular cross-section (see figure 7.23a), and analyze the low-frequency ($\omega \to \omega_c$) and high-frequency ($\omega \gg \omega_c$) behaviors of α for each of these modes.

Problem 7.29. For a rectangular metallic-wall resonator with dimensions $a \times b \times l$ ($b \leqslant a, l$), calculate the Q-factor in the fundamental oscillation mode, due to the skin-effect losses in the walls. Evaluate the factor for a 23 × 23 × 10 mm^3 resonator with copper walls, at room temperature.

Problem 7.30.* Calculate the lowest eigenfrequency and the Q-factor (due to the skin-effect losses) of the toroidal (axially-symmetric) resonator with metallic walls, and interior's cross-section shown in the figure below, in the case when $d \ll r, R$.

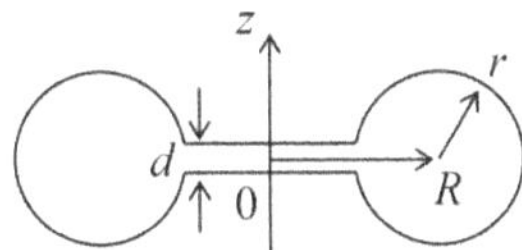

Problem 7.31. Express the contribution to the damping coefficient (the reciprocal Q-factor) of a resonator, from small energy losses in the dielectric that fills it, via the complex functions $\varepsilon(\omega)$ and $\mu(\omega)$ of the material.

Problem 7.32. For the dielectric Fabry–Pérot resonator (figure 7.31) with the normal wave incidence, calculate the Q-factor due to radiation losses, in the limit of a strong impedance mismatch ($Z \gg Z_0$), using two approaches:

(i) from the energy balance, using Eq. (7.227), and
(ii) from the frequency dependence of the power transmission coefficient, using Eq. (7.229).

Compare the results.

References

[1] Born M *et al* 1999 *Principles of Optics* 7th edn (Cambridge University Press)
[2] Shelby R *et al* 2001 *Science* **292** 77
[3] Wilson J and Schwartz Z 2005 *Appl. Phys. Lett.* **86** 021113
[4] Valentine J *et al* 2008 *Nature* **455** 376
[5] Wood B 2009 *C. R. Phys.* **10** 379
[6] Sato N *et al* 2012 *J. Appl. Phys.* **111** 07A501
[7] Whitaker J 2012 *Power Vacuum Tubes Handbook* 3rd edn (Boca Raton, FL: CRC Press)
[8] Silveirinha M and Engheta N 2006 *Phys. Rev. Lett.* **97** 157403
[9] Yariv A and Yeh P 2007 *Photonics* 6th edn (Oxford University Press)
[10] Snyder A and Love D 1983 *Optical Waveguide Theory* (Chapman and Hall)
[11] Allen L *et al* 2003 *Optical Angular Momentum* (Bristol: IOP Publishing)
[12] Marrucchi L *et al* 2011 *J. Opt.* **13** 064001
[13] Wang J *et al* 2012 *Nat. Photon.* **6** 488

IOP Publishing

Classical Electrodynamics
Lecture notes
Konstantin K Likharev

Chapter 8

Radiation, scattering, interference, and diffraction

This chapter continues the discussion of electromagnetic wave propagation, now focusing on the results of wave incidence on various objects with more complex shapes. Depending on the shape, the result of this interaction is called either scattering, *or* diffraction, *or* interference. *However, as the reader will see, the boundaries between these effects are blurry, and their mathematical description may be conveniently based on a single key calculation—the electric dipole radiation of a spherical wave by a localized source. Naturally, I will start the chapter from this calculation, deriving it from an even more general result—the 'retarded potentials' solution of the Maxwell equations.*

8.1 Retarded potentials

Let us start from finding the general solution of the macroscopic Maxwell equations (6.99) in a dispersion-free, linear, uniform, isotropic medium, characterized by frequency-independent, real ε and μ.[1] The easiest way to perform this calculation is to use the scalar (ϕ) and vector ($\mathbf{A}$) potentials of electromagnetic field, defined via the electric and magnetic fields by Eqs (6.7):

$$\mathbf{E} = -\nabla\phi - \frac{\partial \mathbf{A}}{\partial t}, \qquad \mathbf{B} = \nabla \times \mathbf{A}\,. \tag{8.1}$$

As was discussed in section 6.8, by imposing on the potentials the Lorenz gauge condition (6.117),

[1] When necessary (e.g. in the discussion of Cherenkov radiation in section 10.4), it will not be too difficult to generalize the results to a dispersive medium.

doi:10.1088/978-0-7503-1404-6ch8

$$\nabla \cdot \mathbf{A} + \frac{1}{v^2}\frac{\partial \phi}{\partial t} = 0, \quad \text{with } v^2 \equiv \frac{1}{\varepsilon\mu}, \tag{8.2}$$

which does not affect the fields **E** and **B**, the Maxwell equations may be used to obtain a pair of very similar, simple equations (6.118) for the potentials:

$$\nabla^2 \phi - \frac{1}{v^2}\frac{\partial^2 \phi}{\partial t^2} = -\frac{\rho}{\varepsilon}, \tag{8.3a}$$

$$\nabla^2 \mathbf{A} - \frac{1}{v^2}\frac{\partial^2 \mathbf{A}}{\partial t^2} = -\mu \mathbf{j}. \tag{8.3b}$$

Let us find the general solution of these equations, assuming that the densities $\rho(\mathbf{r}, t)$ and $\mathbf{j}(\mathbf{r}, t)$ of the stand-alone charges and currents are known[2]. The idea of such a solution is borrowed from electro- and magnetostatics. Indeed, for the stationary case ($\partial/\partial t = 0$), the solutions of Eqs. (8.3) are given by the ready generalization of, respectively, Eqs. (1.38) and (5.28) to a uniform, linear medium:

$$\phi(\mathbf{r}) = \frac{1}{4\pi\varepsilon}\int \rho(\mathbf{r}')\frac{d^3r'}{|\mathbf{r} - \mathbf{r}'|}, \tag{8.4a}$$

$$\mathbf{A}(\mathbf{r}) \equiv \frac{\mu}{4\pi}\int \mathbf{j}(\mathbf{r}')\frac{d^3r'}{|\mathbf{r} - \mathbf{r}'|}. \tag{8.4b}$$

As we know, these expressions may be derived by first calculating the potential of a point source and then using the linear superposition principle for a system of such sources.

Let us do the same for the time-dependent case, starting from the field induced by a time-dependent point charge at the origin[3]:

$$\rho(\mathbf{r}, t) = q(t)\delta(\mathbf{r}). \tag{8.5}$$

In this case Eq. (8.3a) is homogeneous everywhere but the origin:

$$\nabla^2 \phi - \frac{1}{v^2}\frac{\partial^2 \phi}{\partial t^2} = 0, \quad \text{at } r \neq 0. \tag{8.6}$$

[2] This assumption will not prevent the results from being valid for the case when $\rho(\mathbf{r}, t)$ and $\mathbf{j}(\mathbf{r}, t)$ should be calculated self-consistently.

[3] Admittedly, this expression does *not* satisfy the continuity equation (4.5), but this deficiency will be corrected imminently, at the linear superposition stage—see Eqs. (8.17) below.

Due to the spherical symmetry of the problem, it is natural to look for a spherically symmetric solution to this equation[4]. Thus, we may simplify the Laplace operator[5] correspondingly and reduce Eq. (8.6) to

$$\left[\frac{1}{r^2}\frac{\partial}{\partial r}\left(r^2\frac{\partial}{\partial r}\right) - \frac{1}{v^2}\frac{\partial^2}{\partial t^2}\right]\phi = 0, \quad \text{at } r \neq 0. \tag{8.7}$$

If we now introduce a new variable $\chi \equiv r\phi$, Eq. (8.7) is reduced to a 1D wave equation

$$\left(\frac{\partial^2}{\partial r^2} - \frac{1}{v^2}\frac{\partial^2}{\partial t^2}\right)\chi = 0, \quad \text{at } r \neq 0. \tag{8.8}$$

From discussions in chapter 7,[6] we know that its general solution may be represented as

$$\chi(r, t) = \chi_{\text{out}}\left(t - \frac{r}{v}\right) + \chi_{\text{in}}\left(t + \frac{r}{v}\right), \tag{8.9}$$

where χ_{in} and χ_{out} are (so far) arbitrary functions of one variable. The physical sense of $\phi_{\text{out}} = \chi_{\text{out}}/r$ is a spherical wave propagating from our source (at $r = 0$) to outer space, i.e. exactly the solution we are looking for. On the other hand, $\phi_{\text{in}} = \chi_{\text{in}}/r$ describes a spherical wave that could be created by some distant spherically symmetric source that converges exactly on our charge located at the origin—evidently not the effect we want to consider here. Discarding this term, and returning to $\phi = \chi/r$, we can write the solution (8.9) as

$$\phi(r, t) = \frac{1}{r}\chi_{\text{out}}\left(t - \frac{r}{v}\right). \tag{8.10}$$

In order to calculate the function χ_{out}, let us consider the solution (8.10) at distances r so small that the time derivative in Eq. (8.3a), with the right-hand side (8.5),

$$\nabla^2\phi - \frac{1}{v^2}\frac{\partial^2\phi}{\partial t^2} = -\frac{q(t)}{\varepsilon}\delta(\mathbf{r}), \tag{8.11}$$

is much smaller that the spatial derivative (which diverges at $r \to 0$). Then Eq. (8.11) is reduced to the electrostatic equation, whose solution (8.4a) for the source (8.5) is

$$\phi(r \to 0, t) = \frac{q(t)}{4\pi\varepsilon\, r}. \tag{8.12}$$

[4] Let me confess that this is *not* the general solution to Eq. (8.6). For example, it does nor describe the possible waves created by other sources that pass by the considered charge $q(t)$. However, such fields are irrelevant for our current task: to calculate the field *created* by the charge $q(t)$. The solution becomes general when it is integrated (as it will be) over all charges of interest.

[5] See, e.g. Eq. (A.67).

[6] See also *Part CM* section 6.3.

Now requiring the two solutions, (8.10) and (8.12), to coincide at $r \ll vt$, we obtain $\chi_{out}(t) = q(t)/4\pi\varepsilon r$, so that Eq. (8.10) becomes

$$\phi(r, t) = \frac{1}{4\pi\varepsilon\, r} q\left(t - \frac{r}{v}\right). \tag{8.13}$$

Just as was done in statics, this result may be readily generalized for the arbitrary position $\mathbf{r}'$ of the point charge,

$$\rho(\mathbf{r}, t) = q(t)\delta(\mathbf{r} - \mathbf{r}') \equiv q(t)\delta(\mathbf{R}), \tag{8.14}$$

where R is the distance between the field observation point $\mathbf{r}$ and the source position point $\mathbf{r}'$, i.e. the length of the vector,

$$\mathbf{R} \equiv \mathbf{r} - \mathbf{r}', \tag{8.15}$$

connecting these points—see figure 8.1.

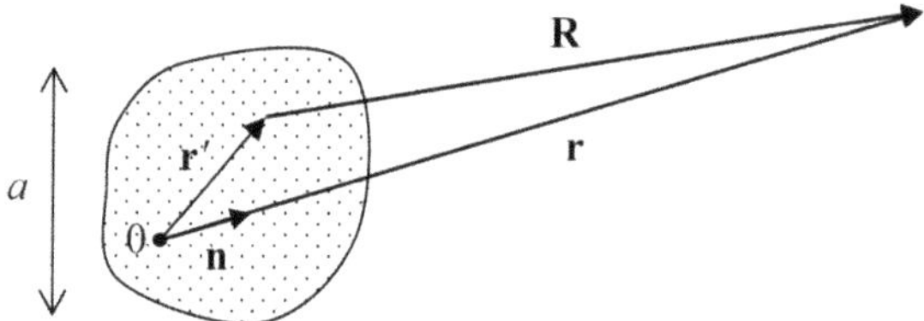

Figure 8.1. Calculating the retarded potentials of a localized source.

Obviously, now Eq. (8.13) becomes

$$\phi(\mathbf{r}, t) = \frac{1}{4\pi\varepsilon\, R} q\left(t - \frac{R}{v}\right). \tag{8.16}$$

Finally, we may use the linear superposition principle to write, for the arbitrary charge distribution,

$$\phi(\mathbf{r}, t) = \frac{1}{4\pi\varepsilon} \int \rho\left(\mathbf{r}', t - \frac{R}{v}\right) \frac{d^3r'}{R}, \tag{8.17a}$$

where integration is extended over all charges of the system under analysis. Acting absolutely similarly, for the vector-potential we obtain[7]

$$\mathbf{A}(\mathbf{r}, t) = \frac{\mu}{4\pi} \int \mathbf{j}\left(\mathbf{r}', t - \frac{R}{v}\right) \frac{d^3r'}{R}. \tag{8.17b}$$

The solutions (8.17) are called the *retarded potentials*[8], the name signifying that the observed fields are 'retarded' (delayed) in time by $\Delta t = R/v$ relative to the source

[7] Now nothing prevents functions $\rho(\mathbf{r}, t)$ and $\mathbf{j}(\mathbf{r}, t)$ from satisfying the continuity relation.

[8] As should be clear from the analogy of Eqs. (8.17) with their stationary forms (8.4), which were discussed, respectively, in chapters 1 and 5, in the Gaussian units the retarded potential formulas are valid with the coefficient $1/4\pi$ dropped in Eq. (8.17a), and replaced with the coefficient $1/c$ in Eq. (8.17b).

variations, due to the finite speed v of the electromagnetic wave propagation. These solutions are so important that they deserve at least a couple of general remarks.

First, very remarkably, these simple expressions are *exact* solutions of the macroscopic Maxwell equations (in a uniform, linear, dispersion-free) medium for an *arbitrary* distribution of stand-alone charges and currents. They also may be considered as the *general* solutions of these equations, provided that the integration is extended over all field sources in the Universe—or at least in its part that affects our observations.

Second, due to the mathematical similarity of the microscopic and macroscopic Maxwell equations, Eqs. (8.17) are valid, with the coefficient replacement $\varepsilon \to \varepsilon_0$ and $\mu \to \mu_0$, for the exact, rather than the macroscopic fields, provided that the functions $\rho(\mathbf{r}, t)$ and $\mathbf{j}(\mathbf{r}, t)$ describe not only stand-alone but *all* charges and currents in the system. (Alternatively, this statement may be formulated as the validity of Eqs. (8.17), with the same coefficient replacement, in the free space.)

Finally, Eqs. (8.17) may be plugged into Eq. (8.1), giving (after an explicit differentiation) the so-called *Jefimenko equations* for fields **E** and **B**—similar in structure to Eqs. (8.17), but more cumbersome. Conceptually, the existence of such equations is good news, because they are free from the gauge ambiguity pertinent to the potentials ϕ and **A**. However, the practical value of these explicit expressions for the fields is not too high: for all applications I am aware of, it is easier to use Eqs. (8.17) to calculate the particular expressions for the potentials first, and only then calculate the fields from Eq. (8.1). Let me now present an (apparently, the most important) example of this approach.

8.2 Electric dipole radiation

Consider again the problem that was discussed in electrostatics (section 3.1), namely the field of a localized source with linear dimensions $a \ll r$ (see figure 8.1 again), but now with time-dependent charge and/or current distributions. Using the arguments of that discussion, in particular the condition expressed by Eq. (3.1), $r' \ll r$, we may apply the Taylor expansion (3.3), truncated to two leading terms,

$$f(\mathbf{R}) = f(\mathbf{r}) - \mathbf{r}' \cdot \nabla f(\mathbf{r}) + \cdots, \qquad (8.18)$$

to the function $f(\mathbf{R}) \equiv R$ (for which $\nabla f(\mathbf{r}) = \nabla R = \mathbf{n}$, where $\mathbf{n} \equiv \mathbf{r}/r$ is the unit vector directed toward the observation point—see figure 8.1) to approximate the distance R as

$$R \approx r - \mathbf{r}' \cdot \mathbf{n}. \qquad (8.19)$$

In each of the retarded potential formulas (8.17), R participates in two places: in the denominator and in the source's time argument. If ρ and $\mathbf{j}$ change in time on scale $\sim 1/\omega$, where ω is some characteristic frequency, then any change of the argument $(t - R/v)$ on that time scale, for example due to a change of R on the spatial scale $\sim v/\omega = 1/k$, may substantially change these functions. Thus, the expansion (8.19) may be applied to R in the argument $(t - R/v)$ only if $ka \ll 1$, i.e. if the system's size a is much smaller than the radiation wavelength $\lambda = 2\pi/k$. On the other hand, the function $1/R$ changes relatively slowly, and for it even the first term of the expansion (8.19) gives a good approximation as soon as $a \ll r,\ R$. In this approximation, Eq. (8.17a) yields

$$\phi(\mathbf{r},\ t) \approx \frac{1}{4\pi\varepsilon r}\int\rho\left(\mathbf{r}',\ t - \frac{R}{v}\right)d^3r' \equiv \frac{1}{4\pi\varepsilon r}Q\left(t - \frac{R}{v}\right), \tag{8.20}$$

where $Q(t)$ is the net electric charge of the localized system. Due to charge conservation, this charge cannot change with time, so that the approximation (8.20) describes just a static Coulomb field of our localized source, rather than a radiated wave.

Let us, however, apply the similar approximation to the vector potential (8.17b):

$$\mathbf{A}(\mathbf{r},\ t) \approx \frac{\mu}{4\pi\ r}\int\mathbf{j}\left(\mathbf{r}',\ t - \frac{R}{v}\right)d^3r'. \tag{8.21}$$

According to Eq. (5.87), in statics the right-hand side of this expression would vanish, but in dynamics this is no longer true. For example, if the current is due to a non-relativistic motion[9] of a system of point charges q_k, we can write

$$\int\mathbf{j}(\mathbf{r}',\ t)d^3r' = \sum_k q_k\dot{\mathbf{r}}_k(t) = \frac{d}{dt}\sum_k q_k\mathbf{r}_k(t) \equiv \dot{\mathbf{p}}(t), \tag{8.22}$$

where $\mathbf{p}(t)$ is the dipole moment of the localized system, defined by Eq. (3.6). Now, after the integration, we may keep only the first term of the approximation (8.19) in the argument $(t - R/v)$ as well, obtaining

$$\mathbf{A}(\mathbf{r},\ t) \approx \frac{\mu}{4\pi\ r}\dot{\mathbf{p}}\left(t - \frac{r}{v}\right). \tag{8.23}$$

Let us analyze exactly what this result, valid in the limit $ka \ll 1$, describes. The second of Eqs. (8.1) allows us to calculate the magnetic field by the spatial differentiation of $\mathbf{A}$. At large distances $r \gg \lambda$ (i.e. in the so-called *far field zone*), where Eq. (8.23) describes a virtually plane wave, the main contribution into this derivative is given by the dipole moment factor:

$$\mathbf{B}(\mathbf{r},\ t) = \frac{\mu}{4\pi\ r}\nabla\times\dot{\mathbf{p}}\left(t - \frac{r}{v}\right) = -\frac{\mu}{4\pi\ rv}\mathbf{n}\times\ddot{\mathbf{p}}\left(t - \frac{r}{v}\right). \tag{8.24}$$

This expression means that the magnetic field, at the observation point, is perpendicular to the vectors $\mathbf{n}$ and (the retarded value of) $\ddot{\mathbf{p}}$, and its magnitude is

[9] For relativistic particles, moving with velocities of the order of speed of light, one has to be more careful. As a result, I will postpone the discussion of their radiation until chapter 10, i.e. until after the detailed discussion of special relativity in chapter 9.

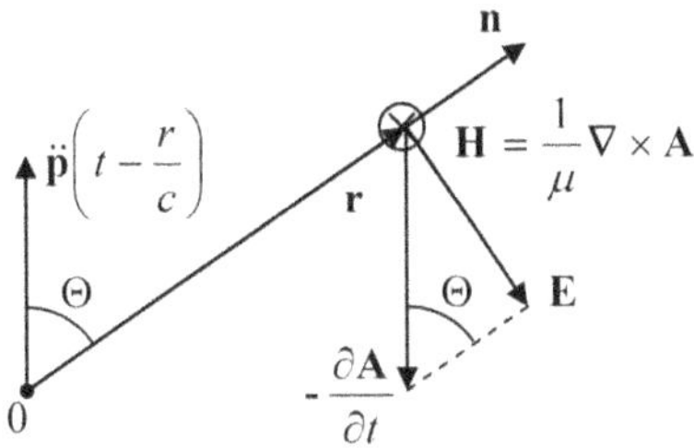

Figure 8.2. Far-zone fields of a localized source, contributing into its electric dipole radiation.

$$B = \frac{\mu}{4\pi r v}\ddot{p}\left(t - \frac{r}{v}\right)\sin\Theta, \quad \text{i.e. } H = \frac{1}{4\pi r v}\ddot{p}\left(t - \frac{r}{v}\right)\sin\Theta, \tag{8.25}$$

where Θ is the angle between those two vectors—see figure 8.2[10].

The most important feature of this result is that the time-dependent field decreases very slowly (only as $1/r$) with the distance from the source, so that the radial component of the corresponding Poynting vector (7.9b)[11],

$$S_r = ZH^2 = \frac{Z}{(4\pi\ vr)^2}\left[\ddot{p}\left(t - \frac{r}{v}\right)\right]^2 \sin^2\Theta \tag{8.26}$$

drops as $1/r^2$, i.e. the full instant power $\mathscr{P}$ of the emitted wave,

$$\mathscr{P} \equiv \oint_{r=\text{const}} S_r\, d^2r = \frac{Z}{(4\pi\ v)^2}\ddot{p}^2 2\pi\int_0^{\pi} \sin^3\Theta\, d\Theta = \frac{Z}{6\pi\ v^2}\ddot{p}^2, \tag{8.27}$$

does not depend on the distance from the source—as it should for radiation[12].

This is the famous *Larmor formula*[13] for the *electric dipole radiation*; it is the dominant component of radiation by a localized system of charges—unless $\ddot{\mathbf{p}} = 0$. Please notice its angular dependence: the radiation vanishes at the axis of the retarded vector $\ddot{\mathbf{p}}$ (where $\Theta = 0$), and reaches its maximum in the plane perpendicular to that axis.

In order to find the average power, Eq. (8.27) has to be averaged over a sufficiently long time. In particular, if the source is monochromatic, $\mathbf{p}(t) = \text{Re}\,[\mathbf{p}_\omega \exp\{-i\omega t\}]$, with a time-independent vector $\mathbf{p}_\omega$, such averaging may be carried out just over one period, giving an extra factor 2 in the denominator:

$$\overline{\mathscr{P}} = \frac{Z\omega^4}{12\pi\ v^2}\,|p_\omega|^2. \tag{8.28}$$

[10] From the first of Eqs. (8.1), for the electric field in the first approximation (8.23) we would obtain $-\partial\mathbf{A}/\partial t = -(1/4\pi\varepsilon vr)\,\ddot{\mathbf{p}}(t - r/v) = -(Z/4\pi r)\ddot{\mathbf{p}}(t - r/v)$. The transverse component of this vector (see figure 8.2) is the proper electric field $\mathbf{E} = Z\mathbf{H} \times \mathbf{n}$ of the radiated wave, while its longitudinal component is exactly compensated by $(-\nabla\phi)$ in the *next* term of the expansion of Eq. (8.17a) with respect to the small parameter $r/\lambda \ll 1$.

[11] Note the 'doughnut' dependence of S_r on the direction $\mathbf{n}$, frequently used to visualize the dipole radiation.

[12] In the Gaussian units, for free space ($v = c$), Eq. (8.27) reads $\mathscr{P} = (2/3c^3)\ddot{p}^2$.

[13] After J Larmor, who was first to derive it (in 1897) for the particular case of a single point charge q moving with acceleration $\ddot{\mathbf{r}}$, when $\ddot{\mathbf{p}} = q\ddot{\mathbf{r}}$.

The easiest example of application of the formula is to a point charge oscillating, with frequency ω, along a straight line (which we may take for axis z), with amplitude a. In this case, $\mathbf{p} = q\mathbf{n}_z z(t) = \mathbf{n}_z qa \operatorname{Re}[\exp\{-i\omega t\}]$, and if the charge velocity amplitude, $a\omega$, is much less than the wave speed v, we may use Eq. (8.28) with $p_\omega = qa$, giving

$$\overline{\mathscr{P}} = \frac{Zq^2a^2\omega^4}{12\pi v^2}. \tag{8.29}$$

Applied to an electron ($q = -e \approx -1.6 \times 10^{-19}$ C), initially rotating about a nucleus at an atomic distance $a \sim 10^{-10}$ m, the Larmor formula shows[14] that the energy loss due to the dipole radiation is so large that it would cause the electron to collapse on the atom's nucleus in just $\sim 10^{-10}$ s. In the beginning of the 1900 s, this classical result was one of the main arguments for the development of quantum mechanics, which prevents such collapse of electrons in their lowest-energy (ground) quantum state.

Another example of a very useful application of Eq. (8.28) is the radio wave radiation by a short, straight, symmetric antenna which is fed, for example, by a TEM transmission line such as a coaxial cable—see figure 8.3.

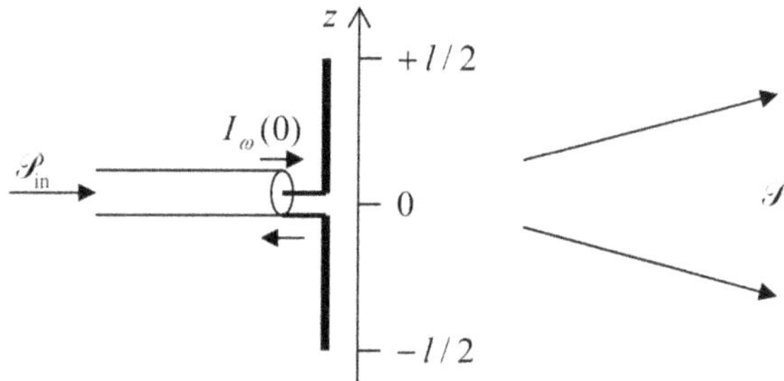

Figure 8.3. The dipole antenna.

The exact solution of this problem is rather complex, because the law $I_\omega(z)$ of the current variation along antenna's length should be calculated self-consistently with the distribution of the electromagnetic field induced by the current in the surrounding space. (This fact is unfortunately ignored in some textbooks.) However, one may argue that at $l \ll \lambda$, the current should be largest in the feeding point (in figure 8.3, taken for $z = 0$), vanish at antenna's ends ($z = \pm l/2$), and that the only possible scale of the current variation in the antenna is l itself, so that the linear function,

$$I_\omega(z) = I_\omega(0)\left(1 - \frac{2}{l}|z|\right), \tag{8.30}$$

gives a good approximation to the actual distribution—as it indeed does. Now we can use the continuity equation $\partial Q/\partial t = I$, i.e. $-i\omega Q_\omega = I_\omega$, to calculate the complex

[14] Actually, the formula needs a numerical coefficient adjustment to account for the electron's orbital (rather than linear) motion—the task left for reader's exercise. However, this adjustment does not affect the order-of-magnitude estimate given above.

amplitude $Q_\omega(z) = iI_\omega(z)\text{sgn}(z)/\omega$ of the electric charge $Q(z, t) = \text{Re}[Q_\omega \exp\{-i\omega t\}]$ of the wire beyond point z, and from it, the amplitude of the linear density of charge

$$\lambda_\omega(z) \equiv \frac{dQ_\omega(z)}{d\,|z|} = -i\frac{2I_\omega(0)}{\omega l}\,\text{sgn}\,z. \tag{8.31}$$

From here, the dipole moment's amplitude is

$$p_\omega = 2\int_0^{l/2} \lambda_\omega(z) z\, dz = -i\frac{I_\omega(0)}{2\omega}l, \tag{8.32}$$

so that Eq. (8.28) yields

$$\overline{\mathscr{P}} = Z\frac{\omega^4}{12\pi v^2}\frac{|I_\omega(0)|^2}{4\omega^2}l^2 = \frac{Z(kl)^2}{24\pi}\frac{|I_\omega(0)|^2}{2}, \tag{8.33}$$

where $k = \omega/v$. The analogy between this result and the dissipation power, $\mathscr{P} = \text{Re}\, Z|I_\omega{}^2/2|$, in a lumped linear circuit element, allows the interpretation of the first fraction in the last form of Eq. (8.33) as the real part of the antenna's impedance:

$$\text{Re}\, Z_A = Z\frac{(kl)^2}{24\pi}, \tag{8.34}$$

as felt by the transmission line.

According to Eq. (7.118), the wave traveling along the line toward the antenna is fully radiated, i.e. not reflected back, only if Z_A equals Z_W of the line. As we know from section 7.5 (and the solution of related problems), for typical TEM lines $Z_W \sim Z_0$, while Eq. (8.34), which is only valid in the limit $kl \ll 1$, shows that for radiation into the free space ($Z = Z_0$), $\text{Re}Z_A$ is much less than Z_0. Hence in order to reach the impedance matching condition $Z_W = Z_A$, the antenna's length should be increased—as a more involved theory shows, to $l \sim \lambda/2$. However, in many cases, practical considerations make short antennas necessary. The example most frequently encountered nowadays is the cell phone antennas, which use frequencies close to 1 or 2 GHz, with free-space wavelengths λ between 15 and 30 cm, i.e. much larger than the phone size[15]. The quadratic dependence of the antenna's efficiency on l, following from Eq. (8.34), explains why every millimeter counts in the design of such antennas, and why the designs are carefully optimized using software packages for (virtually exact) numerical solution of time-dependent Maxwell equations for the specific shape of the antenna and other phone parts[16].

To conclude this section, let me note that if the wave source is not monochromatic, so that $\mathbf{p}(t)$ should be represented as a Fourier series,

[15] The situation will be partly remedied by the planned transfer of the wireless mobile technology to its next (5 G) generation, with the frequencies moving to the 28 GHz, 37–39 GHz, and possibly even the 64–71 GHz bands.

[16] A partial list of popular software packages of this kind includes both publicly available codes such as NEC-2 (whose various versions are available online, e.g. at http://alioth.debian.org/projects/necpp/ and http://www.qsl.net/4nec2/), and proprietary packages—such as *Momentum* from Aglient Technologies (now owned by Hewlett-Packard), *FEKO* from EM Software & Systems, and *XFdtd* from Remcom.

$$\mathbf{p}(t) = \text{Re}\sum_{\omega}\mathbf{p}_\omega e^{-i\omega t} \tag{8.35}$$

the terms corresponding to interference of spectral components with different frequencies ω are averaged out at the time averaging of the Poynting vector, so that the *average* radiated power is just a sum of contributions (8.28) from all substantial frequency components.

8.3 Wave scattering

The formalism described above may be immediately used in the theory of *scattering*—the phenomenon illustrated by figure 8.4. Generally, scattering is a complex problem. However, in many cases it allows the so-called *Born approximation*[17], in which the scattered wave field's effect on the scattering object is assumed to be much weaker than that of the incident wave, and is neglected.

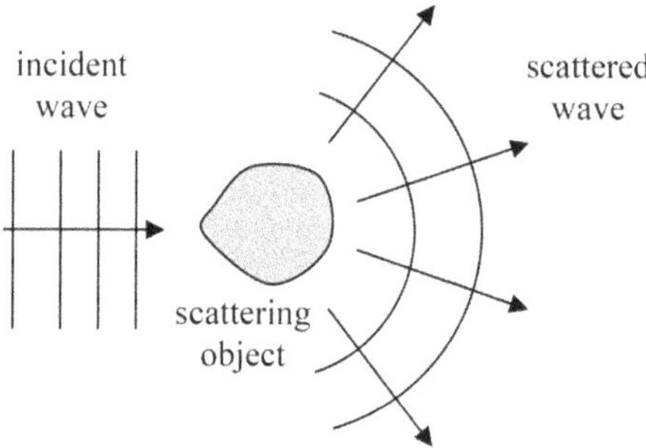

Figure 8.4. Schematic representation of scattering.

As the first example of this approach, let us consider the scattering of a plane wave, propagating in free space ($Z = Z_0$, $v = c$), by a free[18] charged particle whose motion may be described by non-relativistic classical mechanics. (This requires, in particular, the incident wave not to be too powerful, so that the speed of the induced charge motion remains much lower than the speed of light.) As was already discussed at the derivation of Eq. (7.32), in this case the magnetic component of the Lorentz force (5.10) is negligible in comparison with the force $\mathbf{F}_e = q\mathbf{E}$ exerted by its electric field. Thus, assuming that the incident wave is linearly polarized along some axis x, the equation of the particle's motion in the Born approximation is just $m\ddot{x} = qE(t)$, so that for the x-component $p_x = qx$ of its dipole moment we can write

$$\ddot{p} = q\ddot{x} = \frac{q^2}{m}E(t). \tag{8.36}$$

[17] Named after M Born, one of the founding fathers of quantum mechanics. Note, however, the basic idea of this approach was developed in electromagnetic theory much earlier (1881) by Lord Rayleigh (born J Stuff, 1842–1919), whose numerous contributions to science include the discovery of argon.

[18] As Eq. (7.30) shows, this calculation is also valid for an oscillator with its own frequency $\omega_0 \ll \omega$.

As we already know from section 8.2, oscillations of the dipole moment lead to radiation of a wave with a wide angular distribution of intensity; in our case this is the scattered wave—see figure 8.4. Its full power may be found by plugging Eq. (8.36) into Eq. (8.27),

$$\mathscr{P} = \frac{Z_0}{6\pi\, c^2}\ddot{p}^2 = \frac{Z_0 q^4}{6\pi c^2 m^2}E^2(t), \tag{8.37}$$

so that for the average power we obtain

$$\overline{\mathscr{P}} = \frac{Z_0 q^4}{12\pi c^2 m^2}\,|E_\omega|^2. \tag{8.38}$$

Since the power is proportional to incident wave's intensity S, it is customary to characterize the scattering ability of the object by the ratio,

$$\sigma \equiv \frac{\overline{\mathscr{P}}}{\overline{S_{\text{incident}}}} \equiv \frac{\overline{\mathscr{P}}}{|E_\omega|^2/2Z_0}, \tag{8.39}$$

which has the dimension of area and is called the *full cross-section* of scattering[19]. For this measure, Eq. (8.38) yields the famous result

$$\sigma = \frac{Z_0^2 q^4}{6\pi c^2 m^2} = \frac{\mu_0^2 q^4}{6\pi m^2}, \tag{8.40}$$

which is called the *Thomson scattering formula*[20], in particular when applied to an electron. This relation is most frequently represented in the form[21]

$$\sigma = \frac{8\pi}{3}r_c^2, \quad \text{with } r_c \equiv \frac{q^2}{4\pi\varepsilon_0}\cdot\frac{1}{mc^2} = 10^{-7}\frac{q^2}{m}. \tag{8.41}$$

This constant r_c is called the *classical radius of the particle* (or sometimes the 'Thomson scattering length'); for an electron ($q = -e$, $m = m_e$) it is close to 2.82×10^{-15} m. Its possible interpretation is evident from the first form of Eq. (8.41) for r_c: at that distance between two similar particles, the potential energy $q^2/4\pi\varepsilon_0 r$ of their electrostatic interaction is equal to the particle's rest-mass energy mc^2.[22]

[19] This definition parallels those accepted in the classical and quantum theories of *particle* scattering—see, e.g. respectively, *Part CM* section 3.5 and *Part QM* section 3.3.

[20] Named after Sir J J Thomson (1856–1940), the discoverer of the electron—and of isotopes as well! He should not be confused with his son, G P Thomson, who discovered (simultaneously with C Davisson and L Germer) the quantum-mechanical wave properties of the same electron.

[21] In Gaussian units, this formula looks like $r_c = q^2/mc^2$ (giving, of course, the same numerical value: for the electron, $r_c \approx 2.82 \times 10^{-13}$ cm). This *classical* quantity should not be confused with the particle's *Compton wavelength* $\lambda_c \equiv 2\pi\hbar/mc$ (for the electron, close to 2.24×10^{-12} cm), which naturally arises in *quantum* electrodynamics—see a brief discussion in the next chapter, and also *Part QM* section 1.1.

[22] It is fascinating how smartly has the *relativistic* expression mc^2 sneaked into the result (8.40)–(8.41), which was obtained using a *non-relativistic* equation of particle motion. This was possible because the calculation engaged electromagnetic waves, which propagate with the speed of light, and whose quanta (*photons*), as a result, may be frequently treated as relativistic (moreover, ultra-relativistic) particles—see the next chapter.

Now we have to go back and establish the conditions under which the Born approximation, when the field of the scattered wave is negligible, is indeed valid for a point-object scattering. Since the scattered wave's intensity, described by Eq. (8.26), diverges as $1/r^2$, according to the definition (8.39) of the cross-section, it may become comparable to S_{incident} at $r^2 \sim \sigma$. However, Eq. (8.38) itself is only valid if $r \gg \lambda$, so the Born approximation does not lead to a contradiction only if

$$\sigma \ll \lambda^2. \tag{8.42}$$

For the Thompson scattering by an electron, this condition means $\lambda \gg r_c \sim 3 \times 10^{-15}$ m and is fulfilled for all frequencies up to very hard γ-rays with energies ~100 MeV.

Possibly the most notable feature of result (8.40) is its independence of the wave frequency. As follows from its derivation, particularly from Eq. (8.37), this independence is intimately related to the unbound character of charge motion. For bound charges, say for electrons in gas molecules, this result is only valid if the wave frequency ω is much higher than all eigenfrequencies ω_j of molecular resonances. In the opposite limit, $\omega \ll \omega_j$, the result is dramatically different. Indeed, in this limit we can approximate the molecule's dipole moment by its static value (3.48)

$$\mathbf{p} = \alpha \mathbf{E}. \tag{8.43}$$

In the Born approximation, and in the absence of the molecular field effects discussed in section 3.3, **E** in this expression is just the incident wave's field, and we can use Eq. (8.28) to calculate the power of the wave scattered by a single molecule:

$$\overline{\mathscr{P}} = \frac{Z_0 \omega^4}{4\pi\, c^2} \alpha^2 \,|E_\omega|^2. \tag{8.44}$$

Now, using the last form of the definition (8.39) of the cross-section, we obtain a very simple result,

$$\sigma = \frac{Z_0^2 \omega^4}{6\pi c^2} \alpha^2, \tag{8.45}$$

showing that in contrast to Eq. (8.40), at low frequencies σ grows as fast as ω^4.

Now let us explore the effect of such *Rayleigh scattering* on wave propagation in a gas, with a relatively low volumic density n. We may expect (and will prove in the next section) that due to the randomness of molecule positions, the waves scattered by individual molecules may be treated as *incoherent* ones, so that the total scattering power may be calculated just as the sum of those scattered by each molecule. We can use this fact to write the balance of the incident's wave intensity in a small volume dV of length (along the incident wave direction) dz, and area A across it. Since such a segment includes $ndV = nA\, dz$ molecules and, according to definition (8.39), each of them scatters power $S\sigma = \mathscr{P}\sigma/A$, the total scattered power is $n\mathscr{P}\sigma\, dz$; hence the incident power's change is

$$d\mathscr{P} \equiv -n\sigma\mathscr{P}\, dz. \tag{8.46}$$

Comparing this equation with the definition (7.213) of the wave attenuation constant, applied to scattering[23],

$$d\mathscr{P} \equiv -\alpha_{\text{scat}}\mathscr{P}\, dz. \tag{8.47}$$

we see that the scattering gives the following contribution to attenuation: $\alpha_{\text{scat}} = n\sigma$. From here, using Eq. (3.50) to write $\alpha = \varepsilon_0(\kappa - 1)/n$, where κ is the dielectric constant, and Eq. (8.45) for σ, we obtain

$$\alpha_{\text{scat}} = \frac{k^2}{6\pi\, n}(\kappa - 1)^2, \quad \text{where } k \equiv \frac{2\pi}{\lambda_0} = \frac{\omega}{c}. \tag{8.48}$$

This is the famous *Rayleigh scattering formula*, which in particular explains the colors of blue skies and red sunsets. Indeed, through the visible light spectrum, ω changes almost two-fold; as a result, the scattering of blue components of sunlight is an order of magnitude higher than that of its red components. More qualitatively, for the air near the Earth's surface, $\kappa - 1 \approx 6 \times 10^{-4}$, and $n \sim 2.5 \times 10^{25}$ m^{-3}—see section 3.3. Plugging these numbers into (8.47), we see that the characteristic length $l_{\text{scat}} \equiv 1/\alpha_{\text{scat}}$ of scattering is ~30 km for blue light and ~200 km for red light[24]. The effective thickness h of the Earth's atmosphere is ~10 km, so that the Sun looks just a bit yellowish during most of the day. However, elementary geometry shows that at the sunset, the light should travel the length $l \sim (R_E h)^{1/2} \approx 300$ km to reach an Earth-surface observer; as a result, the blue components of Sun's light spectrum are almost completely scattered out, and even the red components are weakened substantially.

Now let us discuss scattering by objects with size of the order of, or even larger than λ. For such extended objects, the phase difference factors (neglected above) step in, leading in particular to the important effects of *interference* and *diffraction*, to whose discussion we now proceed.

8.4 Interference and diffraction

These effects show up not as much in the total power of the scattered radiation, as in its angular distribution. It is traditional to characterize this distribution by the *differential cross-section* defined as

$$\frac{d\sigma}{d\Omega} \equiv \frac{\overline{S}_r r^2}{S_{\text{incident}}}, \tag{8.49}$$

where r is the distance from the scatterer, at which the scattered wave is observed[25]. Both the definition and notation may become more clear if we notice that according to Eq. (8.26), at large distances ($r \gg a$), the nominator on the right-hand side of

[23] Sorry for using the same letter (α) for both the molecular polarizability and the wave attenuation, but both notations are traditional. Hopefully, the subscript 'scat', marking α in the latter meaning, excludes any possibility of confusion.

[24] These values are approximate, because both n and ($\kappa - 1$) vary through the atmosphere's thickness.

[25] Just as in the case of the full cross-section, this definition is also similar to that accepted at the *particle* scattering—see, e.g. *Part CM* section 3.5 and *Part QM* section 3.3.

Eq. (8.49), and hence the differential cross-section as the whole, does not depend on r, and that its integral over the total solid angle $\Omega = 4\pi$ coincides with the full cross-section defined by Eq. (8.39):

$$\begin{aligned}\oint_{4\pi} \frac{d\sigma}{d\Omega} d\Omega &= \frac{1}{\overline{S}_{\text{incident}}} r^2 \oint_{4\pi} \overline{S}_r d\Omega = \frac{1}{\overline{S}_{\text{incident}}} \oint_{r=\text{const}} \overline{S}_r d^2r \\ &= \frac{\overline{\mathscr{P}}}{\overline{S}_{\text{incident}}} \equiv \sigma.\end{aligned} \tag{8.50}$$

For example, according to Eq. (8.26), the angular distribution of the radiation scattered by a single dipole is rather broad; in particular, in the quasi-static case (8.43) and in the Born approximation,

$$\frac{d\sigma}{d\Omega} = \left(\frac{\alpha k^2}{4\pi\varepsilon_0}\right)^2 \sin^2\Theta. \tag{8.51}$$

If the wave is scattered by a small dielectric body, with a characteristic size $a \ll \lambda$ (i.e. $ka \ll 1$), then all its parts re-radiate the incident wave coherently. Hence, we can calculate it in the similar way, just replacing the molecular dipole moment (8.43) with the total dipole moment of the object—see Eq. (3.45):

$$\mathbf{p} = \mathbf{P}V = (\kappa - 1)\,\varepsilon_0 \mathbf{E} V, \tag{8.52}$$

where $V \sim a^3$ is the body's volume. As a result, the differential cross-section may be obtained from Eq. (8.51) with the replacement $\alpha_{\text{mol}} \rightarrow (\kappa - 1)\varepsilon_0 V$:

$$\frac{d\sigma}{d\Omega} = \left(\frac{k^2 V}{4\pi}\right)^2 (\kappa - 1)^2 \sin^2\Theta, \tag{8.53}$$

i.e. follows the same $\sin^2\Theta$ law. The situation for extended objects, with at least one dimension of the order of or larger than the wavelength, is different: here we have to take into account the phase shifts introduced by various parts of the body. Let us analyze this issue first for an arbitrary collection of similar point scatterers located at points $\mathbf{r}_j$.

If the wave vector of the incident plane wave is $\mathbf{k}_0$, the wave's field has the phase factor $\exp\{i\mathbf{k}_0 \cdot \mathbf{r}\}$—see (7.79). At the location of the jth scattering center, the factor equals $\exp\{i\mathbf{k}_0 \cdot \mathbf{r}_j\}$, so that the local dipole vectors $\mathbf{p}$ and the scattered wave they create are proportional to this factor. On its way to the observation point $\mathbf{r}$, the scattered wave, with the wave vector $\mathbf{k}$ (with $k = k_0$), acquires an additional phase factor $\exp\{\mathbf{i}\mathbf{k} \cdot (\mathbf{r} - \mathbf{r}_j)\}$, so that the scattered wave field is proportional to

$$\begin{aligned}\exp\{i\mathbf{k}_0 \cdot \mathbf{r}_j + i\mathbf{k}(\mathbf{r} - \mathbf{r}_j)\} &= \exp\{i(\mathbf{k}_0 - \mathbf{k}) \cdot \mathbf{r}_j + i\mathbf{k} \cdot \mathbf{r}\} \\ &= e^{i\mathbf{k}\cdot\mathbf{r}} \exp\{-i(\mathbf{k} - \mathbf{k}_0) \cdot \mathbf{r}_j\}.\end{aligned} \tag{8.54}$$

Since the first factor in the last expression does not depend on $\mathbf{r}_j$, in order to calculate the total scattering wave it is sufficient to sum up the last phase factors, $\exp\{-\mathbf{i}\mathbf{q} \cdot \mathbf{r}_j\}$, where the vector

$$\mathbf{q} \equiv \mathbf{k} - \mathbf{k}_0 \tag{8.55}$$

has the physical sense of the wave vector change at scattering[26]. It may look like the phase factor depends on the choice of the reference frame. However, according to Eq. (7.42), the average intensity of the scattered wave is proportional to $E_\omega E_\omega^*$, i.e. to the following real scalar function of vector $\mathbf{q}$:

$$\begin{aligned} F(\mathbf{q}) &= \left(\sum_j \exp\{-i\mathbf{q}\cdot\mathbf{r}_j\}\right)\left(\sum_{j'} \exp\{-i\mathbf{q}\cdot\mathbf{r}_{j'}\}\right)^* \\ &= \sum_{j,j'} \exp\{i\mathbf{q}\cdot(\mathbf{r}_j - \mathbf{r}_{j'})\} \equiv |I(\mathbf{q})|^2, \end{aligned} \tag{8.56}$$

where the complex function

$$I(q) \equiv \sum_j \exp\{-iq\cdot r_j\} \tag{8.57}$$

is called the *phase sum*, and may be calculated in any reference frame, without affecting the final result (8.56).

So, in addition to the $\sin^2\Theta$ factor, the differential cross-section (8.49) of scattering by an extended object is also proportional to the scattering function (8.56). Its double-sum form is convenient to notice that for a system of *many* ($N \gg 1$) of similar but randomly located scatterers, only the terms with $j = j'$ accumulate at summation, so that $F(\mathbf{q})$ and hence $d\sigma/d\Omega$ scale as N rather than N^2—thus justifying again our treatment of the Rayleigh scattering problem in the previous section.

Let us start using Eq. (8.56) by applying it to a simple problem of just *two* similar small scatterers, separated by a fixed distance a:

$$\begin{aligned} F(\mathbf{q}) &= \sum_{j,j'=1}^{2} \exp\{i\mathbf{q}\cdot(\mathbf{r}_j - \mathbf{r}_{j'})\} = 2 + \exp\{-iq_a a\} + \exp\{iq_a a\} \\ &= 2(1 + \cos q_a a) = 4\cos^2\frac{q_a a}{2}, \end{aligned} \tag{8.58}$$

where $q_a \equiv \mathbf{q}\cdot\mathbf{a}/a$ is the component of the vector $\mathbf{q}$ along the vector $\mathbf{a}$ connecting the scatterers. The apparent simplicity of this result may be a bit misleading, because the mutual plane of the vectors $\mathbf{k}$ and $\mathbf{k}_0$ (and hence of the vector $\mathbf{q}$) does not necessarily coincide with the mutual plane of vectors $\mathbf{k}_0$ and $\mathbf{E}_\omega$, so that the *scattering angle* θ between $\mathbf{k}$ and $\mathbf{k}_0$ is generally different from $(\pi/2 - \Theta)$—see figure 8.5.

Moreover, the angle between the vectors $\mathbf{q}$ and $\mathbf{a}$ (within their common plane) is one more parameter independent of both θ and Θ. As a result, the angular dependence[27] of the scattered wave's intensity (and hence $d\sigma/d\Omega$) that depends on all three angles may be rather involved, but some of its details do not affect the basic physics of the interference/diffraction.

[26] In quantum mechanics, $\hbar\mathbf{q}$ has a very clear sense of the momentum transferred from the scattering object to the scattered particle (for example, a photon) and this terminology is sometimes smuggled even into classical electrodynamics texts.

[27] In optics, such patterns, observed as dark and bright spots on a screen, are called *interference fringes*.

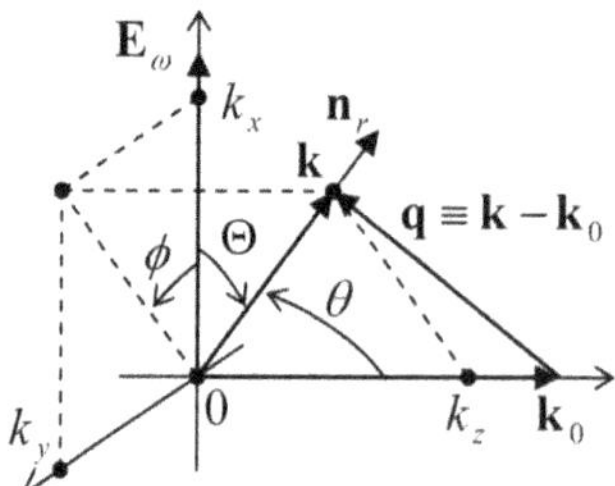

Figure 8.5. The angles important for the general scattering problem.

Thus let me consider only the simple case when the vectors **k**, $\mathbf{k}_0$, and **a** are all in the same plane (figure 8.6a), with $\mathbf{k}_0$ perpendicular to **a**. Then, with our choice of coordinates, $q_a = q_x = k\sin\theta$, and Eq. (8.58) is reduced to

$$F(\mathbf{q}) = 4\cos^2\frac{ka\sin\theta}{2}. \tag{8.59}$$

This function always has two maxima, at $\theta = 0$ and $\theta = \pi$, and, if the product ka is large enough, also other maxima at special angles θ_n that satisfy the simple condition

$$ka\sin\theta_n = 2\pi n, \quad \text{i.e. } a\sin\theta_n = n\lambda. \tag{8.60}$$

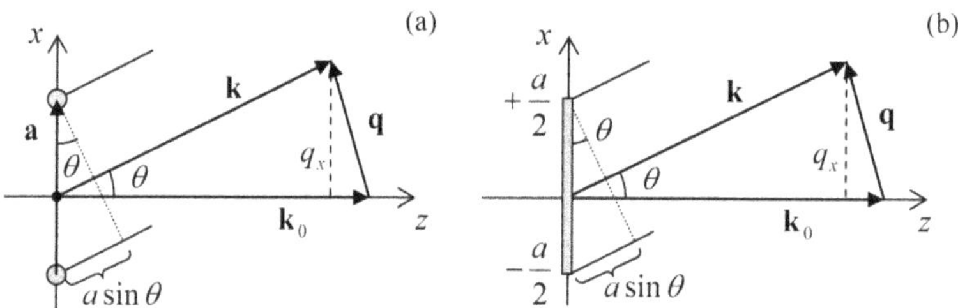

Figure 8.6. The simplest cases of (a) interference and (b) diffraction.

As figure 8.6a shows, this condition may be readily understood as that of the in-phase addition (the *constructive interference*) of two coherent waves scattered from the two points, when the difference between their paths toward the observer, $a\sin\theta$, equals an integer number of wavelengths. At each such maximum, $F = 4$, due to the doubling of the wave amplitude and hence quadrupling its power.

If the distance between the point scatterers is large ($ka \gg 1$), the first maxima (8.60) correspond to small scattering angles, $\theta \ll 1$. For this region, Eq. (8.59) is reduced to a simple sinusoidal dependence of function F on the angle θ. Moreover, within the range of small θ, the wave polarization factor $\sin^2\Theta$ is virtually constant, so that the scattered wave intensity and hence the differential cross-section

$$\frac{d\sigma}{d\Omega} \propto F(\mathbf{q}) = 4\cos^2\frac{ka\theta}{2}. \tag{8.61}$$

This is of course the simple *interference pattern*, well known from Young's two-slit experiment[28]. (As will be discussed in the next section, the theoretical description of the two-slit experiment is more complex than that of the Born scattering, but is preferable experimentally, because at such scattering, the wave of intensity (8.61) has to be observed on the backdrop of a stronger incident wave that propagates in almost the same direction, $\theta = 0$.)

A very similar analysis of scattering from $N > 2$ similar, equidistant scatterers, located along the same straight line (left for the reader's exercise), shows that the positions (8.60) of the constructive interference maxima do not change (because the derivation of this condition is still applicable to each pair of adjacent scatterers), but the increase of N makes these peaks sharper and sharper.

Now let me jump to the limit $N \to 0$, in which we may ignore the scatterers' discreteness. The resulting pattern is similar to that at scattering by a continuous thin rod (see figure 8.6b), so let us first spell out the Born scattering formula by an arbitrary an extended, continuous, uniform dielectric body. Transferring Eq. (8.56) from the sum to an integral, for the differential cross-section we obtain

$$\frac{d\sigma}{d\Omega} = \left(\frac{k^2}{4\pi}\right)^2 (\kappa - 1)^2 F(\mathbf{q}) \sin^2\Theta \equiv \left(\frac{k^2}{4\pi}\right)^2 (\kappa - 1)^2 |I(\mathbf{q})|^2 \ \sin^2\Theta, \qquad (8.62)$$

where $I(\mathbf{q})$ now becomes the *phase integral*[29],

$$I(\mathbf{q}) = \int_V \exp\{-i\mathbf{q}\cdot\mathbf{r}'\} d^3r', \qquad (8.63)$$

with the dimensionality of volume.

Now we may return to the particular case of a thin rod (with both dimensions of the cross-section's area much smaller than λ, but an arbitrary length a), otherwise keeping the same simple geometry as for two point scatterers—see figure 8.6b. In this case the phase integral is just

$$I(\mathbf{q}) = A\int_{-a/2}^{+a/2} \exp\{-iq_x x'\} dx' = A\frac{\exp\{-iq_x a/2\} - \exp\{-iq_x a/2\}}{-iq} = V\frac{\sin\xi}{\xi}, \qquad (8.64)$$

where $V = Aa$ is the volume of the rod, and ξ is the dimensionless parameter defined as

$$\xi \equiv \frac{q_x a}{2} \equiv \frac{ka\sin\theta}{2}. \qquad (8.65)$$

[28] This experiment was described as early as in 1803 by T Young—one more universal genius of science, who has also introduced the Young modulus in the elasticity theory (see, e.g. *Part CM* chapter 7), in addition to numerous other achievements—including deciphering Egyptian hieroglyphs! The two-slit experiment firmly established the wave picture of light, only to be replaced by the dualistic photon-versus-wave picture, formalized by quantum electrodynamics 100+ years later.

[29] Since the observation point's position $\mathbf{r}$ does not participate in this formula explicitly, the prime sign in $\mathbf{r}'$ could be dropped, but I keep it as a reminder that the integral is taken over points $\mathbf{r}'$ of the *scattering object*.

The fraction participating in Eq. (8.64) is met in physics so frequently that is has deserved the special name *sinc* (not 'sync', please!) *function*:

$$\operatorname{sinc} \xi \equiv \frac{\sin \xi}{\xi}. \tag{8.66}$$

Obviously, this function, plotted in figure 8.7, vanishes at all points $\xi_n = \pi n$, with integer n, apart from the point with $n = 0$: $\operatorname{sinc} \xi_0 = \operatorname{sinc} 0 = 1$.

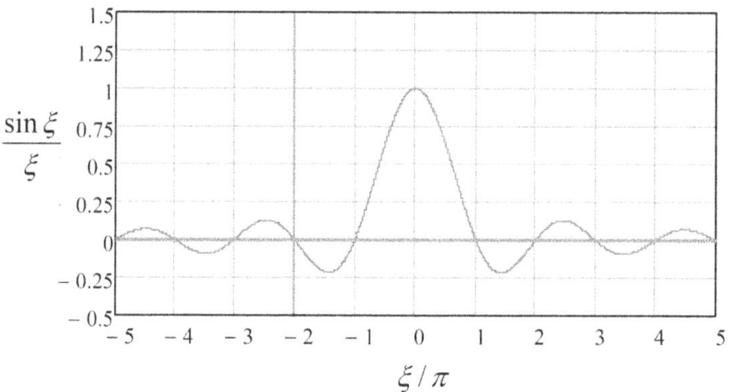

Figure 8.7. The sinc function.

The function $F(\mathbf{q}) = V^2 \operatorname{sinc}^2 \xi$, resulting from Eq. (8.64), is plotted with the red line in figure 8.8 and is called the *Fraunhofer diffraction pattern*.

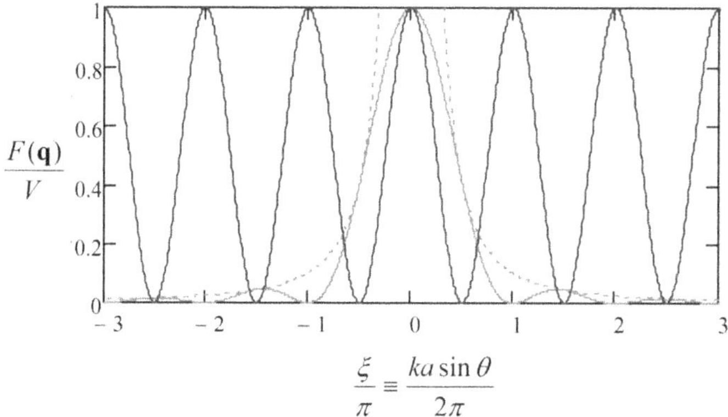

Figure 8.8. The Fraunhofer diffraction pattern (solid red line) and its envelope $1/\xi^2$ (dashed red line). For comparison, the blue line shows the standard interference pattern $\cos^2 \xi$ —cf equation (8.59).

Note that it oscillates with the same argument period $\Delta(ka\sin\theta) = 2\pi/ka$ as the interference pattern (8.59) from two point scatterers (shown with the blue line in figure 8.8). However, at the interference, the scattered wave intensity vanishes at angles θ_n' that satisfy the condition

$$\frac{ka \sin \theta_n'}{2\pi} = n + \frac{1}{2}, \tag{8.67}$$

i.e. when the optical paths difference $a\sin\alpha$ equals to a semi-integer number of wavelengths $\lambda/2 = \pi/k$, and hence the two waves from the scatterers arrive to the observer in anti-phase—the so-called *destructive interference*. On the other hand, for the diffraction from a continuous rod the minima occur at a different set of scattering angles:

$$\frac{ka\sin\theta_n}{2\pi} = n, \tag{8.68}$$

i.e. exactly where the two-point interference pattern has its maxima—see figure 8.8 again. The reason for this relation is that the wave diffraction on the rod may be considered as a simultaneous interference of waves from all its elementary fragments, and exactly at the observation angles when the rod edges give waves with phases shifted by $2\pi n$, the interior points of the rod give waves with all possible phases, with their algebraic sum equal to zero. As is even more visible in figure 8.8, at the diffraction, the intensity oscillations are limited by a rapidly decreasing envelope function $1/\xi^2$ (while at the two-point interference, the oscillations retain the same amplitude). The reason for this fast decrease is that with each Fraunhofer diffraction period, a smaller and smaller fraction of the road gives an unbalanced contribution to the scattered wave.

If the rod's length is small ($ka \ll 1$, i.e. $a \ll \lambda$), then the sinc function's argument ξ is small at all scattering angles θ, so $I(\mathbf{q}) \approx V$, and Eq. (8.62) is reduced to Eq. (8.53). In the opposite limit, $a \gg \lambda$, the first zeros of the function $I(\mathbf{q})$ correspond to very small angles θ, for which $\sin\Theta \approx 1$, so that the differential cross-section is just

$$\frac{d\sigma}{d\Omega} = \left(\frac{k^2}{4\pi}\right)^2 (\kappa - 1)^2 \text{sinc}^2\frac{ka\theta}{2}, \tag{8.69}$$

i.e. figure 8.8 shows the scattering intensity as a function of the direction toward the observation point—if this point is within the plane containing the rod.

Finally, let us discuss a problem of a large importance for applications: calculate the positions of the maxima of the interference pattern arising at the incidence of a plane wave on a very large 3D periodic system of point scatterers. For that, first of all, let us quantify the notion of 3D periodicity. The periodicity in one dimension is simple: the system we are considering (say, the positions of point scatterers) should be invariant with respect to the linear translation by some period a, and hence by any multiple sa of this period. (Here s is any integer.) Anticipating the 3D generalization, we may require any of the possible *translation vectors* $\mathbf{R}$ to be equal $s\mathbf{a}$, where the *primitive vector* $\mathbf{a}$ is directed along the (only) axis of the 1D system.

Now we are ready for the common definition of the 3D periodicity—as the invariance of the system with respect to the translation by any vector of the following set:

$$\mathbf{R} = \sum_{l=1}^{3} s_l \mathbf{a}_l, \tag{8.70}$$

where s_l are three independent integers and $\{\mathbf{a}_l\}$ is a set of three linearly independent primitive vectors. The set of geometric points described by Eq. (8.70) is called the *Bravais lattice*; perhaps the most non-trivial feature of this relation is that the vectors $\mathbf{a}_l$ should not necessarily be orthogonal to each other. (That requirement would severely restrict the set of possible lattices, and make it unsuitable for the description, for example, of many solid-state crystals.) For the scattering problem we are considering, let us assume that the position $\mathbf{r}_j$ of each scatterer coincides with one of the points $\mathbf{R}$ of some Bravais lattice, with a given set of primitive vectors $\mathbf{a}_l$, so that the index j is coding the set of integers $\{s_1, s_2, s_3\}$.

Now let us consider a similar Bravais lattice, but in the reciprocal (wave-number) space, numbered by independent integers $\{t_1, t_2, t_3\}$:

$$\mathbf{Q} = \sum_{m=1}^{3} t_m \mathbf{b}_m, \quad \text{with } \mathbf{b}_m = 2\pi \frac{\mathbf{a}_{m''} \times \mathbf{a}_{m'}}{\mathbf{a}_m \cdot (\mathbf{a}_{m''} \times \mathbf{a}_{m'})}, \tag{8.71}$$

where the indices m, m', and m'' are all different. This is the so-called *reciprocal lattice*, which plays an important role in all physics of periodic structures, in particular in the quantum energy-band theory[30]. To reveal its most important property, and thus justify the above definition of the primitive vectors $\mathbf{b}_m$, let us calculate the scalar product

$$\begin{aligned} \mathbf{R} \cdot \mathbf{Q} &\equiv \sum_{l,m=1}^{3} s_l t_m \mathbf{a}_l \cdot \mathbf{b}_m \equiv 2\pi \sum_{l,m=1}^{3} s_l t_m \mathbf{a}_l \cdot \frac{\mathbf{a}_{m''} \times \mathbf{a}_{m'}}{\mathbf{a}_m \cdot (\mathbf{a}_{m''} \times \mathbf{a}_{m'})} \\ &\equiv 2\pi \sum_{l,m=1}^{3} s_l t_k \frac{\mathbf{a}_l \cdot (\mathbf{a}_{m''} \times \mathbf{a}_{m'})}{\mathbf{a}_m \cdot (\mathbf{a}_{m''} \times \mathbf{a}_{m'})}. \end{aligned} \tag{8.72}$$

Applying to the nominator of the last fraction the *operand rotation rule* of vector algebra[31], we see that it is equal to zero if $l \neq m$, while for $l = m$ the whole fraction is evidently equal to 1. Thus the double sum (8.72) is reduced to a single sum,

$$\mathbf{R} \cdot \mathbf{Q} = 2\pi \sum_{l=1}^{3} s_l t_l = 2\pi \sum_{l=1}^{3} n_l, \tag{8.73}$$

where each of the products $n_l \equiv s_l t_l$ is an integer and hence their sum,

$$n \equiv \sum_{l=1}^{3} n_l \equiv s_1 t_1 + s_2 t_2 + s_3 t_3, \tag{8.74}$$

is an integer as well, so that the main property of the direct/reciprocal lattice couple is very simple:

---

[30] See, e.g. *Part QM* section 3.4, where several particular Bravais lattices **R**, and their reciprocals **Q**, are considered.

[31] See, e.g. Eq. (A.48).

$$\mathbf{R} \cdot \mathbf{Q} = 2\pi n, \quad \text{and } \exp\{-i\mathbf{R} \cdot \mathbf{Q}\} = 1. \tag{8.75}$$

Now returning to the scattering function (8.56), we see that if the vector $\mathbf{q} \equiv \mathbf{k} - \mathbf{k}_0$ coincides with *any* vector $\mathbf{Q}$ of the reciprocal lattice, then all terms of the phase sum (8.57) take their largest possible values (equal to 1), and hence the sum as the whole is largest as well, giving a constructive interference maximum. This equality, $\mathbf{q} = \mathbf{Q}$, where $\mathbf{Q}$ is given by Eq. (8.71), is called the *von Laue condition* of the constructive interference; it is, in particular, the basis of all field of the x-ray crystallography of solids and polymers—the main tool for revealing their atomic/molecular structure[32].

In order to recast the von Laue condition in a more vivid, geometric form, let us consider one of the vectors $\mathbf{Q}$ of the reciprocal lattice, corresponding to a certain integer n in Eq. (8.75), and notice that if that relation is satisfied for one point $\mathbf{R}$ of the direct Bravais lattice (8.70), i.e. for one set of integers $\{s_1, s_2, s_3\}$, it is also satisfied for a 2D system of other integer sets, which may be parameterized, for example, by two integers S_1 and S_2:

$$s_1' = s_1 + S_1 t_3, \quad s_2' = s_2 + S_2 t_3, \quad s_3' = s_3 - S_1 t_1 - S_2 t_2. \tag{8.76}$$

Indeed, each of these sets has the same value of the integer n, defined by Eq. (8.74), as the original one:

$$n' \equiv s_1' t_1 + s_2' t_2 + s_3' t_3 \equiv (s_1 + S_1 t_3)\, t_1 + (s_2 + S_2 t_3)\, t_2 + (s_3 - S_1 t_1 - S_2 t_2)\, t_3 = n. \tag{8.77}$$

Since according to Eq. (8.75), the vector of distance between any pair of the corresponding points of the direct Bravais lattice (8.70),

$$\Delta\mathbf{R} = \Delta S_1 t_3 \mathbf{a}_1 + \Delta S_2 t_3 \mathbf{a}_2 - (\Delta S_1 t_1 + \Delta S_2 t_2)\mathbf{a}_3, \tag{8.78}$$

satisfies the condition $\Delta\mathbf{R} \cdot \mathbf{Q} = 2\pi\Delta n = 0$, this vector is normal to the (fixed) vector $\mathbf{Q}$. Hence, all the points corresponding to the 2D set (8.76), with arbitrary integers S_1 and S_2, are located on one geometric plane, called the *crystal* (or 'lattice') *plane*. In a 3D system of $N \gg 1$ scatterers (such as $N \sim 10^{23}$ atoms in a $\sim$1-cm^3 solid crystal), with all linear dimensions comparable, such a plane contains $\sim N^{2/3} \gg 1$ points. As a result, the constructive interference peaks are very sharp.

Now rewriting (8.75) as a relation for the vector $\mathbf{R}$'s component along the vector $\mathbf{Q}$,

$$R_Q = \frac{2\pi}{Q} n, \quad \text{where } R_Q \equiv \mathbf{R} \cdot \mathbf{n}_Q \equiv \mathbf{R} \cdot \frac{\mathbf{Q}}{Q}, \quad \text{and } Q \equiv |\mathbf{Q}|, \tag{8.79}$$

we see that the parallel crystal planes, corresponding to different numbers n (but the same $\mathbf{Q}$) are located in space periodically, with the smallest distance

[32] For more reading on this important topic I can recommend, for example, the classical monograph [1]. Note that it uses the alternative popular name of the field, once again illustrating how blurry the boundary between interference and diffraction is.

$$d = \frac{2\pi}{Q}, \tag{8.80}$$

so that the von Laue condition $\mathbf{q} = \mathbf{Q}$ may be rewritten as the following rule for the possible magnitudes of the scattering vector $\mathbf{q} \equiv \mathbf{k} - \mathbf{k}_0$:

$$q = \frac{2\pi n}{d}. \tag{8.81}$$

Figure 8.9a shows the diagram of the three wave vectors $\mathbf{k}$, $\mathbf{k}_0$, and $\mathbf{q}$, taking into account the elastic scattering condition $|\mathbf{k}| = |\mathbf{k}_0| = k \equiv 2\pi/\lambda$:

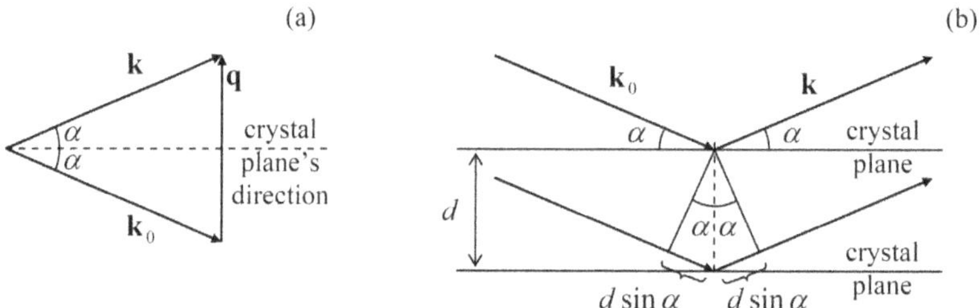

Figure 8.9. Deriving the Bragg rule: (a) from the von Laue condition (in the reciprocal space) and (b) from a direct-space diagram. Note that the scattering angle θ equals 2α.

From the diagram, we immediately obtain the famous *Bragg rule*[33] for the (equal) angles $\alpha \equiv \theta/2$ between the crystal plane and each of the vectors $\mathbf{k}$ and $\mathbf{k}_0$:

$$k \sin \alpha = \frac{q}{2} = \frac{\pi n}{d}, \quad \text{i.e. } 2d \sin \alpha = n\lambda. \tag{8.82}$$

The physical sense of this relation is very simple—see figure 8.9b (drawn in the 'direct' space of the radius-vectors $\mathbf{r}$, rather than in the reciprocal space of the wave vectors as figure 8.9a). It shows that if the Bragg condition (8.82) is satisfied, the total difference $2d\sin \alpha$ of the optical paths of two waves, partly reflected from the adjacent crystal planes, is equal to an integer number of wavelengths, so these waves interfere constructively.

Finally, note that the von Laue and Bragg rules, as well as the similar condition (8.60) for the 1D system of scatterers, are valid not only in the Born approximation, but also follow from any adequate theory of scattering. This is because the phase sum (8.57) does not depend on the magnitude of the wave propagating from an elementary scatterer, provided that they are all similar.

[33] Named after Sir William Bragg and his son Sir William Lawrence Bragg who were the first to demonstrate (in 1912) x-ray diffraction by atoms in crystals. The Braggs' experiments made the existence of atoms (before that, a hypothetical notion, which had been ignored by many physicists) indisputable.

8.5 The Huygens principle

The Born approximation is very convenient for tracing the basic features of (and the difference between) the phenomena of interference and diffraction. Unfortunately, this approximation, based on the relative weakness of the scattered wave, cannot be used for more typical experimental implementations of these phenomena, for example, the Young's two-slit experiment or the diffraction on a single slit or orifice—see, e.g. Figure 8.10. Indeed, in such experiments, the orifice size a is typically much larger than the light's wavelength and, as a result, no clear decomposition of the fields to the 'incident' and 'scattered' waves is possible.

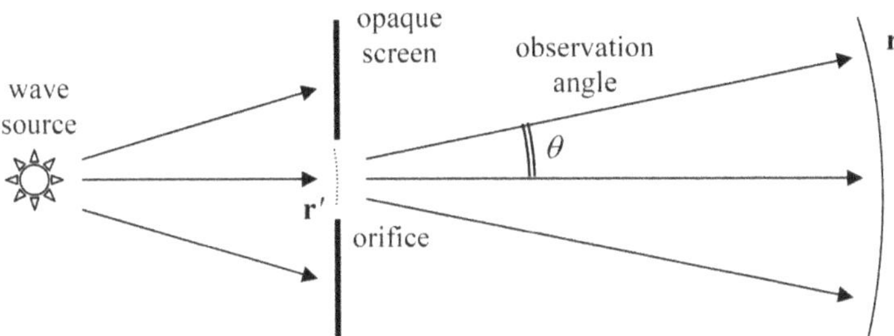

Figure 8.10. Deriving the Huygens principle.

However, for such experiments, another approximation called the *Huygens* (or 'Huygens–Fresnel') *principle*[34] is very instrumental. In this approach, the wave beyond the screen is represented as a linear superposition of spherical waves of the type (8.17), as if they were emitted by every point of the incident wave's front that has arrived at the orifice. This approximation is valid if the following strong conditions are satisfied:

$$\lambda \ll a \ll r, \tag{8.83}$$

where r is the distance of the observation point from the orifice. In addition, as we have seen in the last section, at small λ/a the diffraction phenomena are confined to angles $\theta \sim 1/ka \sim \lambda/a \ll 1$. For observation at such small angles, the mathematical expression of the Huygens principle, for the complex amplitude $f_\omega(\mathbf{r})$ of a monochromatic wave $f(\mathbf{r}, t) = \text{Re}[f_\omega e^{-i\omega t}]$, is given by the following simple formula

$$f_\omega(\mathbf{r}) = C\int_{\text{orifice}} f_\omega(\mathbf{r}')\frac{e^{ikR}}{R}d^2r'. \tag{8.84}$$

[34] Named after C Huygens (1629–95) who conjectured the wave nature of light (which remained controversial for more than a century, until T Young's experiments) and A-J Fresnel (1788–1827) who developed a quantitative theory of diffraction, and in particular gave a mathematical formulation of the Huygens principle. (Note that Eq. (8.91), sufficient for the purposes of this course, is not its most general form.)

Here f is any transverse component of any of wave's fields (either **E** or **H**)[35], R is the distance between point $\mathbf{r}'$ at the orifice and the observation point $\mathbf{r}$ (i.e. the magnitude of vector $\mathbf{R} \equiv \mathbf{r} - \mathbf{r}'$), and C is a complex constant.

Before describing the proof of Eq. (8.84), let me carry out its sanity check—which also will give us the constant C. Let us see what the Huygens principle gives for the case when the field under the integral is a plane wave with the complex amplitude $f_\omega(z)$, propagating along axis z, with an unlimited x–y front (i.e. when there is no opaque screen at all), so we should take the whole $[x, y]$ plane, say with $z' = 0$, as the integration area in Eq. (8.84)—see figure 8.11.

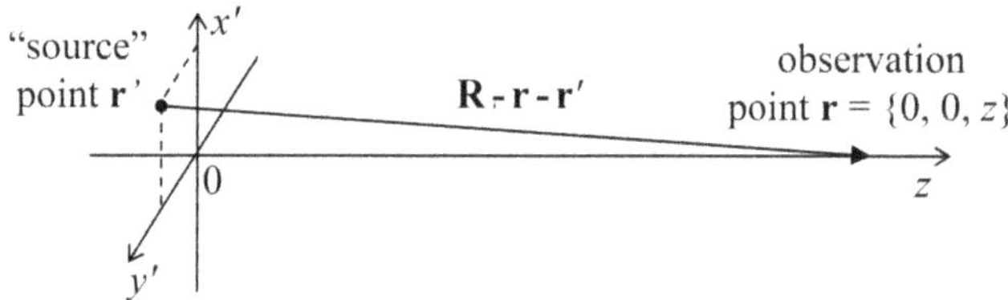

Figure 8.11. Applying the Huygens principle to a plane incident wave.

Then, for the observation point with coordinates $x = 0$, $y = 0$, and $z \gg \lambda$, Eq. (8.84) yields

$$f_\omega(z) = Cf_\omega(0)\int dx'\int dy' \frac{\exp\{ik(x'^2 + y'^2 + z^2)^{1/2}\}}{(x'^2 + y'^2 + z^2)^{1/2}}. \tag{8.85}$$

Before specifying the integration limits, let us consider the range $|x'|, |y'| \ll z$. In this range the square root, participating in Eq. (8.85) twice, may be approximated as

$$\begin{aligned}(x'^2 + y'^2 + z^2)^{1/2} &\equiv z\left(1 + \frac{x'^2 + y'^2}{z^2}\right)^{1/2} \approx z\left(1 + \frac{x'^2 + y'^2}{2z^2}\right)\\ &\equiv z + \frac{x'^2 + y'^2}{2z}.\end{aligned} \tag{8.86}$$

The denominator of Eq. (8.85) is a much slower function of x' and y' than the exponent, and in it (as we will check *a posteriori*), it is sufficient to keep just the main, first term of the expansion (8.86). With that, Eq. (8.85) becomes

$$f_\omega(z) = Cf_\omega(0)\frac{e^{ikz}}{z}\int dx'\int dy' \exp\frac{ik(x'^2 + y'^2)}{2z} = Cf_\omega(0)\frac{e^{ikz}}{z}I_x I_y, \tag{8.87}$$

where I_x and I_y are two similar integrals; for example,

[35] The fact that the Huygens principle is valid for any field component should not be too surprising. Due to condition $a \gg \lambda$, the real boundary conditions at the orifice edges are not important; what is important is only that the screen that limits the orifice is opaque. Because of this, the Huygens principle's expression (8.84) is part of the so-called *scalar theory of diffraction*. (In this course, I will not have time to pursue the more accurate, vector theory of these effects - see, e.g. chapter 11 of the classical monograph [2].)

$$I_x = \int \exp \frac{ikx'^2}{2z} dx' = \left(\frac{2z}{k}\right)^{1/2} \int \exp \{i\xi^2\} d\xi$$
$$= \left(\frac{2z}{k}\right)^{1/2} \left[\int \cos (\xi^2) d\xi + i \int \sin (\xi^2) d\xi\right], \tag{8.88}$$

where $\xi \equiv (k/2z)^{1/2}x'$. These are the so-called *Fresnel integrals.* I will discuss them in more detail in the next section (see, in particular, figure 8.13), and for now, only one property[36] of these integrals is important for us: if taken in symmetric limits $[-\xi_0, +\xi_0]$, both of them rapidly converge to the same value, $(\pi/2)^{1/2}$, as soon as ξ_0 becomes much larger than 1. This means that even if we do not impose any exact limits on the integration area in Eq. (8.85) this integral converges to the value

$$f_\omega(z) = Cf_\omega(0)\frac{e^{ikz}}{z}\left\{\left(\frac{2z}{k}\right)^{1/2}\left[\left(\frac{\pi}{2}\right)^{1/2} + i\left(\frac{\pi}{2}\right)^{1/2}\right]\right\}^2 = \left(C\frac{2\pi\, i}{k}\right)f_\omega(0)e^{ikz}, \tag{8.89}$$

due to contributions from the central area with linear size of the order of $\Delta\xi \sim 1$, i.e.

$$\Delta x \sim \Delta y \sim \left(\frac{z}{k}\right)^{1/2} \sim (\lambda z)^{1/2}, \tag{8.90}$$

so that the contributions from the front points $\mathbf{r}'$ well beyond the range (8.90) are negligible[37]. (Within our assumptions (8.83), which in particular require λ to be much less than z, the *diffraction angle* $\Delta x/z \sim \Delta y/z \sim (\lambda/z)^{1/2}$, corresponding to the important area of the front, is small.) According to Eq. (8.89), in order to sustain the unperturbed plane wave propagation, $f_\omega(z) = f_\omega(0)e^{ikz}$, the constant C has to be taken equal to $k/2\pi i$. Thus, the Huygens principle's prediction (8.84), in its final form, reads

$$f_\omega(\mathbf{r}) = \frac{k}{2\pi i}\int_{\text{orifice}} f_\omega(\mathbf{r}')\frac{e^{ikR}}{R} d^2r', \tag{8.91}$$

and describes, in particular, the straight propagation of the plane wave (in a uniform media).

Let me pause to emphasize how non-trivial this result is. It would be a natural corollary of Eqs. (8.25) (and the linear superposition principle) if all points of the orifice were filled with point scatterers that re-emit all the incident waves into spherical waves. However, as it follows from the above proof, the Huygens principle is also valid if there is nothing in the orifice but the free space!

[36] See, e.g. (A.37).

[37] This result very is natural, because the function $\exp\{ikR\}$ oscillates fast with the change of $\mathbf{r}'$, so that the contributions from various front points are averaged out. Indeed, the only reason why the central part of plane $[x', y']$ gives a non-zero contribution (8.89) to $f_\omega(z)$ is that the phase exponents stops oscillating as $(x'^2 + y'^2)$ is reduced below $\sim z/k$—see Eq. (8.86).

This is why let us discuss a proof of the principle,[38] based on Green's theorem (2.207). Let us apply this theorem to the function $f = f_\omega$, where f_ω is the complex amplitude of a scalar component of one of the wave's fields, which satisfies the Helmholtz equation (7.204),

$$(\nabla^2 + k^2)f_\omega(\mathbf{r}) = 0, \tag{8.92}$$

and the function $g = G_\omega$, which is the time Fourier image of the corresponding Green's function. The latter function may be defined, as usual, as the solution to the same equation, but with the added delta-functional right-hand side with an arbitrary coefficient, for example,

$$(\nabla^2 + k^2)\, G_\omega(\mathbf{r}, \mathbf{r}') = -4\pi\delta(\mathbf{r} - \mathbf{r}'). \tag{8.93}$$

Using Eqs. (8.92) and (8.93) to express the Laplace operators of the functions f_ω and G_ω, we may rewrite Eq. (2.207) as

$$\begin{aligned}&\int_V \left\{f_\omega[-k^2G_\omega(\mathbf{r}, \mathbf{r}') - 4\pi\delta(\mathbf{r} - \mathbf{r}')] - G_\omega(\mathbf{r}, \mathbf{r}')\left[-k^2 f_\omega\right]\right\} d^3r \\ &= \oint_S \left[f_\omega \frac{\partial G_\omega(\mathbf{r}, \mathbf{r}')}{\partial n} - G_\omega(\mathbf{r}, \mathbf{r}')\frac{\partial f_\omega}{\partial n}\right] d^2r,\end{aligned} \tag{8.94}$$

where $\mathbf{n}$ is the outward normal to the surface S limiting the integration volume V. Two terms on the left-hand side of this relation cancel, so that after swapping the arguments $\mathbf{r}$ and $\mathbf{r}'$ we obtain

$$-4\pi f_\omega(\mathbf{r}) = \oint_S \left[f_\omega(\mathbf{r}')\frac{\partial G_\omega(\mathbf{r}', \mathbf{r})}{\partial n'} - g_\omega(\mathbf{r}', \mathbf{r})\frac{\partial G_\omega(\mathbf{r}')}{\partial n'}\right] d^2r'. \tag{8.95}$$

This relation is only correct if the selected volume V includes the point $\mathbf{r}$ (otherwise we would not obtain its left-hand side from the integration of the delta-function), but does not include the genuine source of the wave—otherwise Eq. (8.92) would have a non-zero right-hand side. Let $\mathbf{r}$ be the field observation point, V be all the source-free half-space (for example, the half-space right of the screen in figure 8.10), so that S is the surface of the screen, including the orifice. Then the right-hand side of Eq. (8.95) describes the field at the observation point $\mathbf{r}$, induced by the wave passing through the orifice points $\mathbf{r}'$. Since no waves are emitted by the opaque parts of the screen, we can limit the integration by the orifice area[39]. Assuming also that the opaque parts of the screen do not re-emit the waves

[38] This proof was given in 1882 by G Kirchhoff.

[39] Actually, this is a non-trivial point of the proof. Indeed, it may be shown that the solution of Eq. (8.94) identically equals zero if $f(\mathbf{r}')$ and $\partial f(\mathbf{r}')/\partial n'$ vanish together at any part of the boundary. A more careful analysis of this issue (which is the task of the formal vector theory of diffraction, which I will not have time to pursue) confirms our intuition-based conclusion.

'radiated' by the orifice, we can take the solution of Eq. (8.93) to be the retarded potential for the free space[40]:

$$G_\omega(\mathbf{r}, \mathbf{r}') = \frac{e^{ikR}}{R}. \tag{8.96}$$

Plugging this expression into Eq. (8.82) we obtain

$$-4\pi f_\omega(\mathbf{r}) = \oint_{\text{orifice}} \left[f_\omega(\mathbf{r}') \frac{\partial}{\partial n'} \left(\frac{e^{ikR}}{R} \right) - \left(\frac{e^{ikR}}{R} \right) \frac{\partial f_\omega(\mathbf{r}')}{\partial n'} \right] d^2 r'. \tag{8.97}$$

This is the so-called *Kirchhoff* (or 'Fresnel–Kirchhoff') *integral.* (Again, with the integration extended over *all* boundaries of the volume V, this would be an exact mathematical result.) Now, let us make two additional approximations. The first of them stems from Eq. (8.83): at $ka \gg 1$, the wave's spatial dependence in the orifice area may be represented as

$$f_\omega(\mathbf{r}') = (\text{a slow function of } \mathbf{r}') \times \exp\{i\mathbf{k}_0 \cdot \mathbf{r}'\}, \tag{8.98}$$

where 'slow' means a function that changes on the scale of a rather than λ. If, also, $kR \gg 1$, then the differentiation in Eq. (8.97) may be, in both instances, limited to the rapidly changing exponents, giving

$$-4\pi f_\omega(\mathbf{r}) = \oint_{\text{orifice}} i(\mathbf{k} + \mathbf{k}_0) \cdot \mathbf{n}' \frac{e^{ikR}}{R} f(\mathbf{r}') d^2 r', \tag{8.99}$$

Second, if all observation angles are small, we can take $\mathbf{k} \cdot \mathbf{n}' \approx \mathbf{k}_0 \cdot \mathbf{n}' \approx -k$. With that, Eq. (8.99) is reduced to Eq. (8.91) expressing the Huygens principle.

It is clear that the principle immediately gives a very simple description of the interference of waves passing through two small holes in the screen. Indeed, if the holes' sizes are negligible in comparison with the distance a between them (though still much larger than the wavelength!), Eq. (8.91) yields

$$f_\omega(\mathbf{r}) = c_1 e^{ikR_1} + c_2 e^{ikR_2}, \quad \text{with } c_{1,2} \equiv \frac{k f_{1,2} A_{1,2}}{2\pi i R_{1,2}}, \tag{8.100}$$

where $R_{1,2}$ are the distances between the holes and the observation point, and $A_{1,2}$ are the hole areas. For the wave intensity, Eq. (8.100) yields

$$\overline{S} \propto f_\omega f_\omega^* = |c_1|^2 + |c_2|^2 + 2\,|c_1||c_2| \cos[k(R_1 - R_2) + \phi], \quad \text{where} \quad \phi \equiv \arg c_1 - \arg c_2. \tag{8.101}$$

The first two terms in this relation clearly represent the intensities of the partial waves passed through each hole, while the last one is the result of their interference. The interference pattern's *contrast ratio*

[40] It follows, e.g. from Eq. (8.16) with a monochromatic source $q(t) = q_\omega \exp\{-i\omega t\}$, with the amplitude $q_\omega = 4\pi\varepsilon$ that fits the right-hand side of Eq. (8.93).

$$\mathcal{R} \equiv \frac{\overline{S}_{\max}}{\overline{S}_{\min}} = \left(\frac{|c_1| + |c_2|}{|c_1| - |c_2|}\right)^2, \tag{8.102}$$

is largest (infinite) when both waves have equal amplitudes.

The analysis of the interference pattern is simple if the line connecting the holes is perpendicular to wave vector $\mathbf{k} \approx \mathbf{k}_0$—see figure 8.6a. Selecting the coordinate axes as shown in that figure, and using for distances $R_{1,2}$ the same expansion as in Eq. (8.86), for the interference term in Eq. (8.101) we obtain

$$\cos[k(R_1 - R_2) + \phi] \approx \cos\left(\frac{kxa}{z} + \phi\right). \tag{8.103}$$

This means that the intensity does not depend on y, i.e. the interference pattern in the plane of constant z is a set of straight, parallel strips, perpendicular to the vector $\mathbf{a}$, with the period given by Eq. (8.60), i.e. by the Bragg law[41]. Note that this result is strictly valid only at $(x^2 + y^2) \ll z^2$; it is straightforward to use the next term in the Taylor expansion (8.73) to show that farther from the interference pattern center the strips start to diverge.

8.6 Fresnel and Fraunhofer diffraction patterns

Now let us use the Huygens principle to analyze a more complex problem: a plane wave's diffraction on a long, straight slit of a constant width a (figure 8.12).

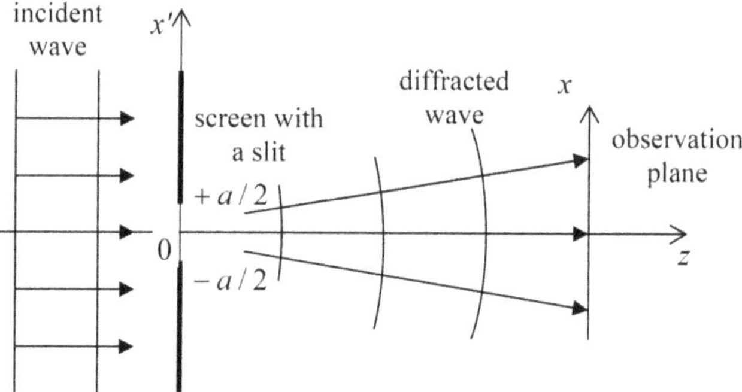

Figure 8.12. Diffraction on a slit.

According to Eq. (8.83), in order to use the Huygens principle for the problem analysis we need to have $\lambda \ll a \ll z$. Moreover, the simple version (8.91) of the principle is only valid for small observation angles, $|x| \ll z$. Note, however, that the relation between two dimensionless numbers, z/a and a/λ, both much less than 1, is

[41] The phase shift ϕ vanishes at the normal incidence of a plane wave on the holes. Note, however, that the spatial shift of the interference pattern following from Eq. (8.103), $\Delta x = -(z/ka)\phi$, is extremely convenient for the experimental measurement of the phase shift between two waves, in particular if it is induced by some factor (such as insertion of a transparent object into one of the interferometer's arms, etc) that may be turned on/off at will.

so far arbitrary. As we will see in a minute, this relation determines the type of observed diffraction pattern.

Let us apply Eq. (8.91) to our current problem (figure 8.12), for the sake of simplicity assuming normal wave incidence, and taking $z = 0$ at the screen plane:

$$f_\omega(x, z) = f_0 \frac{k}{2\pi i} \int_{-a}^{+a} dx' \int_{-\infty}^{+\infty} dy' \frac{\exp\{ik[(x - x')^2 + y'^2 + z^2]^{1/2}\}}{[(x - x')^2 + y'^2 + z^2]^{1/2}}, \tag{8.104}$$

where $f_0 \equiv f_\omega(x', 0) = \text{const}$ is the incident wave's amplitude. This is the same integral as in Eq. (8.85), except for the finite limits for the integration variable x', and may be simplified similarly, using the small-angle condition $(x - x')^2 + y'^2 \ll z^2$:

$$\begin{aligned} f_\omega(x, z) &\approx f_0 \frac{k}{2\pi i} \frac{e^{ikz}}{z} \int_{-a/2}^{+a/2} dx' \int_{-\infty}^{+\infty} dy' \exp \frac{ik[(x - x')^2 + y'^2]}{2z} \\ &\equiv f_0 \frac{k}{2\pi i} \frac{e^{ikz}}{z} I_x I_y. \end{aligned} \tag{8.105}$$

The integral over y' is the same as in the last section,

$$I_y \equiv \int_{-\infty}^{+\infty} \exp \frac{iky'^2}{2z} dy' = \left(\frac{2\pi iz}{k}\right)^{1/2}, \tag{8.106}$$

but the integral over x' is more complicated, because of its finite limits:

$$I_x \equiv \int_{-a/2}^{+a/2} \exp \frac{ik(x - x')^2}{2z} dx'. \tag{8.107}$$

It may be simplified in the following two (opposite) limits.

(i) *Fraunhofer diffraction* takes place when $z/a \gg a/\lambda$—the relation which may be rewritten either as $a \ll (z\lambda)^{1/2}$, or as $ka^2 \ll z$. In this limit, the ratio kx'^2/z is negligibly small for all values of x' under the integral, and we can approximate it as

$$\begin{aligned} I_x &= \int_{-a/2}^{+a/2} \exp \frac{ik(x^2 - 2xx' + x'^2)}{2z} dx' \approx \int_{-a/2}^{+a/2} \exp \frac{ik(x^2 - 2xx')}{2z} dx' \\ &\equiv \exp \frac{ikx^2}{2z} \int_{-a/2}^{+a/2} \exp\left\{-\frac{ikxx'}{z}\right\} dx' = \frac{2z}{kx} \exp\left\{\frac{ikx^2}{2z}\right\} \sin \frac{kxa}{2z}, \end{aligned} \tag{8.108}$$

so that Eq. (8.105) yields

$$f_\omega(x, z) \approx f_0 \frac{k}{2\pi i} \frac{e^{ikz}}{z} \frac{2z}{kx} \left(\frac{2\pi i z}{k}\right)^{1/2} \exp\left\{\frac{ikx^2}{2z}\right\} \sin \frac{kxa}{2z}, \tag{8.109}$$

and hence the relative wave intensity is

$$\frac{\overline{S}(x, z)}{S_0} = \left|\frac{f_\omega(x, z)}{f_0}\right|^2 = \frac{8z}{\pi k x^2}\sin^2\frac{kxa}{2z} \equiv \frac{2}{\pi}\frac{ka^2}{z}\text{sinc}^2\left(\frac{ka\theta}{2}\right), \tag{8.110}$$

where S_0 is the (average) intensity of the incident wave and $\theta \equiv x/z \ll 1$ is the observation angle. Comparing this expression with Eq. (8.69), we see that this diffraction pattern is exactly the same as that of a similar (uniform, 1D) object in the Born approximation—see the red line in figure 8.8. Note again that the angular width $\delta\theta$ of the Fraunhofer pattern is of the order of $1/ka$, so that its linear width $\delta x = z\ \delta\theta$ is of the order of $z/ka \sim z\lambda/a$.[42] Hence the condition of the Fraunhofer approximation's validity may be also represented as $a \ll \delta x$.

(ii) *Fresnel diffraction*. In the opposite limit of a relatively wide slit, with $a \gg \delta x = z\ \delta\theta \sim z/ka \sim z\lambda/a$, i.e. $ka^2 \gg z$, the diffraction patterns at two slit edges are well separated. Hence, near each edge (for example, near $x' = -a/2$) we may simplify Eq. (8.107) as

$$I_x(x) \approx \int_{-a/2}^{+\infty} \exp\frac{ik(x - x')^2}{2z} dx' \equiv \left(\frac{2z}{k}\right)^{1/2} \int_{(k/2z)^{1/2}(x+a/2)}^{+\infty} \exp\{i\zeta^2\} d\zeta, \tag{8.111}$$

and express it via the special functions called the Fresnel integrals[43],

$$\mathscr{C}(\xi) \equiv \left(\frac{2}{\pi}\right)^{1/2} \int_0^\xi \cos(\zeta^2) d\zeta, \qquad \mathscr{S}(\xi) \equiv \left(\frac{2}{\pi}\right)^{1/2} \int_0^\xi \sin(\zeta^2) d\zeta, \tag{8.112}$$

whose plots are shown in figure 8.13a. As was mentioned above, at large values of their argument (ξ), both functions tend to ½.

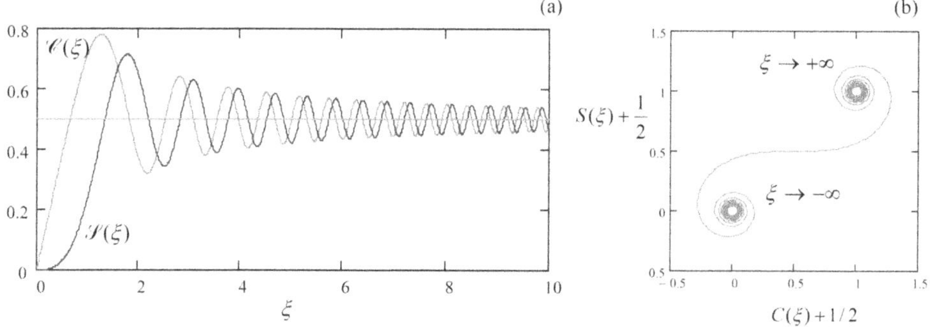

Figure 8.13. (a) The Fresnel integrals and (b) their parametric representation.

[42] Note also that since in this limit $ka^2 \ll z$, Eq. (8.97) shows that even the maximum value $S(0, z)$ of the diffracted wave intensity is much less than the intensity S_0 of the incident wave. This is natural, because the incident power $S_0 a$ per unit length of the slit is now distributed over a much larger width $\delta x \gg a$, so that $S(0, z) \sim S_0\ (a/\delta x) \ll S_0$.

[43] Slightly different definitions of these functions, mostly affecting the constant factors, may also be met in literature.

Plugging this expression into Eqs. (8.105) and (8.111), for the diffracted wave intensity in the Fresnel limit (i.e. at $| x + a/2 | \ll a$) we obtain

$$\frac{\overline{S}(x, z)}{S_0} = \frac{1}{2}\left\{\left[\mathscr{C}\left(\left(\frac{k}{2z}\right)^{1/2}\left(x + \frac{a}{2}\right)\right) + \frac{1}{2}\right]^2 + \left[\mathscr{S}\left(\left(\frac{k}{2z}\right)^{1/2}\left(x + \frac{a}{2}\right)\right) + \frac{1}{2}\right]^2\right\}. \quad (8.113)$$

A plot of this function (figure 8.14) shows that the diffraction pattern is very peculiar: while in the 'shade' region $x < -a/2$ the wave intensity fades monotonically, the transition to the 'light' region within the gap ($x > -a/2$) is accompanied by intensity oscillations, just as at the Fraunhofer diffraction—cf figure 8.8.

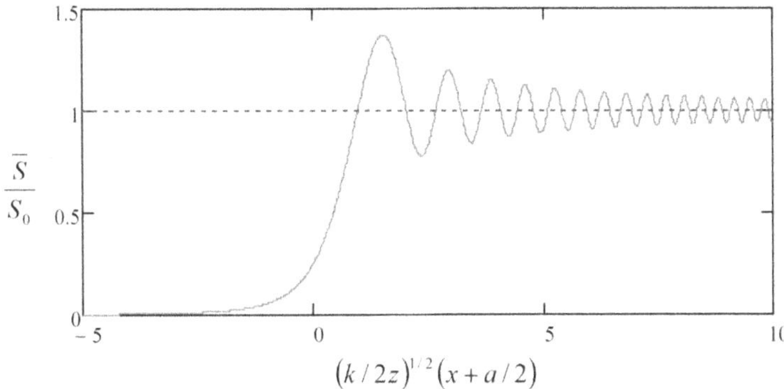

Figure 8.14. The Fresnel diffraction pattern.

This behavior, which is described by the following asymptotes,

$$\frac{\overline{S}}{S_0} \to \begin{cases} 1 + \dfrac{1}{\sqrt{\pi}}\dfrac{\sin(\xi^2 - \pi/4)}{\xi}, & \text{for } \xi \equiv \left(\dfrac{k}{2z}\right)^{1/2}(x + a/2) \to +\infty, \\ \dfrac{1}{4\pi\xi^2}, & \text{for } \xi \to -\infty\,, \end{cases} \quad (8.114)$$

is essentially an artifact of observing just the wave intensity (i.e. its real amplitude) rather than its phase as well. Indeed, as may be seen even more clearly from the parametric representation of the Fresnel integrals, shown in figure 8.13b, these functions oscillate similarly at large positive and negative values of their argument. (This famous pattern is called either the *Euler spiral* or the *Cornu spiral*.) Physically, this means that the wave diffraction at the slit edge leads to similar oscillations of its phase at $x < -a/2$ and $x > -a/2$; however, in the latter region (i.e. inside the slit) the diffracted wave overlaps the incident wave passing through the slit directly and their interference reveals the phase oscillations, making them visible in the measured intensity as well.

Note that according to Eq. (8.113) the linear scale δx of the Fresnel diffraction pattern is of the order of $(2z/k)^{1/2}$, i.e. is complies with the estimate (8.90). If the slit is gradually narrowed so that its width a becomes comparable to δx,[44] the Fresnel diffraction patterns from both edges start to 'collide' (interfere). The resulting wave, fully described by Eq. (8.107), is just a sum of two contributions of the type (8.111) from both edges of the slit. The resulting interference pattern is somewhat complicated, and only when a becomes substantially less than δx, it is reduced to the simple Fraunhofer pattern (8.110). Of course, this crossover from the Fresnel to Fraunhofer diffraction may be also observed at fixed wavelength λ and slit width a by increasing z, i.e. by measuring the diffraction pattern farther and farther away from the slit.

Note also that the Fraunhofer limit is always valid if the diffraction is measured as a function of the diffraction angle θ alone, i.e. effectively at infinity, $z \to \infty$. This may be done, for example, by collecting the diffracted wave with a 'positive' (converging) lens, and observing the diffraction pattern in its focal plane.

8.7 Geometrical optics placeholder

Behind all these details, I would not like the reader to miss the main feature of the Fresnel diffraction, which has an overwhelming practical significance. Namely, apart from narrow diffraction 'cones' (actually, parabolic-shaped regions) with the lateral scale $\delta x \sim (\lambda z)^{1/2}$, the wave far behind a slit of width $a \gg \lambda$, δx, repeats the field just behind the slit, i.e. reproduces the unperturbed incident wave inside the slit, and has negligible intensity in the shade regions outside it. An evident generalization of this fact is that when a plane wave (in particular an electromagnetic wave) passes any opaque object of a large size $a \gg \lambda$, it propagates around it by distances z up to $\sim a^2/\lambda$ along straight lines with virtually negligible diffraction effects. This fact gives the strict foundation for the very notion of the wave *ray* (or *beam*), as the line perpendicular to the local front of a quasi-plane wave. In a uniform medium such a ray follows a straight line, but it refracts in accordance with the Snell law at the interface of two media with different wave speeds v, i.e. different values of the refraction index. The notion of rays enables the whole field of *geometric (or 'geometrical') optics*, devoted mostly to ray tracing in various (sometimes very complex) systems.

This is why, at this point, an E&M course that followed scientific logic more faithfully than this one would give an extended discussion of the geometric and quasi-geometric optics, including (as a minimum[45]) such vital topics as

- the so-called *lensmaker's equation* expressing the focus length f of a lens via the curvature radii of its spherical surfaces and the refraction index of the lens material,
- the *thin lens formula* relating the image distance from the lens via f and the source distance,
- the concepts of basic optical instruments such as *telescopes* and *microscopes*, and

[44] Note that this condition may be also rewritten as $a \sim \delta x$, i.e. $z/a \sim a/\lambda$.

[45] Admittedly, even this list leaves aside several spectacular effects, including such a beauty as *conical refraction* in biaxial crystals—see, e.g. chapter 15 of the textbook [2].

- the concepts of the spherical, angular, and chromatic *aberrations* (image distortions).

However, since I have made a (possibly, incorrect) decision to follow the common tradition in selecting the main topics for this course, I do not have time left for such discussion. Still, I am placing this 'placeholder' pseudo-section to relay my deep conviction that any educated physicist has to know the basics of geometric optics. If the reader has not had exposure to this subject during his or her undergraduate studies, I highly recommend at least browsing one of the available textbooks[46].

8.8 Fraunhofer diffraction from more complex scatterers

So far, our discussion of diffraction has been limited to a very simple geometry—a single slit in an otherwise opaque screen (figure 8.12). However, in the most important Fraunhofer limit, $z \gg ka^2$, it is easy to obtain a very simple expression for the plane wave diffraction/interference by a plane orifice (with a linear size $\sim a$) of an arbitrary shape. Indeed, the evident 2D generalization of the approximation (8.106) and (8.107) is

$$\begin{aligned} I_x I_y &= \int_{\text{orifice}} \exp \frac{ik[(x-x')^2 + (y-y')^2]}{2z} dx'dy' \\ &\approx \exp\left\{\frac{ik(x^2+y^2)}{2z}\right\} \int_{\text{orifice}} \exp\left\{-i\frac{kxx'}{z} - i\frac{kyy'}{z}\right\} dx'dy', \end{aligned} \tag{8.115}$$

so that apart from the inconsequential total phase factor, Eq. (8.105) is reduced to

$$f(\boldsymbol{\rho}) \propto f_0 \int_{\text{orifice}} \exp\{-i\boldsymbol{\kappa}\cdot\boldsymbol{\rho}'\} d^2\rho' \equiv f_0 \int_{\text{all screen}} T(\boldsymbol{\rho}')\exp\{-i\boldsymbol{\kappa}\cdot\boldsymbol{\rho}'\} d^2\rho', \tag{8.116}$$

where the 2D vector $\boldsymbol{\kappa}$ (not to be confused with wave vector $\mathbf{k}$ that is virtually perpendicular to $\boldsymbol{\kappa}$!) is defined as

$$\boldsymbol{\kappa} \equiv k\frac{\boldsymbol{\rho}}{z} \approx \mathbf{q} \equiv \mathbf{k} - \mathbf{k}_0, \tag{8.117}$$

$\boldsymbol{\rho} = \{x, y\}$ and $\boldsymbol{\rho}' = \{x', y'\}$ are 2D radius-vectors in the, respectively, observation and screen planes (both nearly normal to vectors $\mathbf{k}$ and $\mathbf{k}_0$). In the last form of Eq. (8.116), the function $T(\boldsymbol{\rho}')$ describes the screen's transparency at point $\boldsymbol{\rho}'$, and the integral is over the whole screen plane $z' = 0$. (Although the strict equivalence of the two forms of Eq. (8.116) is only valid if $T(\boldsymbol{\rho}')$ equals either 1 or 0, its last form may be readily obtained from Eq. (8.91) with $f(\mathbf{r}') = T(\boldsymbol{\rho}')f_0$ for any transparency profile, provided that $T(\boldsymbol{\rho}')$ is an arbitrary function, but changes only at distances much larger than $\lambda \equiv 2\pi/k$.)

[46] My top recommendation for that purpose would be chapters 3–6 and section 8.6 in [2]. A simpler alternative is chapter 10 in [3]. Note also that the venerable field of optical microscopy is currently revitalized by holographic/tomographic methods, using the scattered wave's phase information. These methods are particularly productive in biology and medicine—see, e.g. [4, 5].

From the mathematical point of view, the last form of Eq. (8.116) is just the 2D spatial Fourier transform of the function $T(\boldsymbol{\rho}')$, with the variable $\boldsymbol{\kappa}$ defined by the observation point's position: $\boldsymbol{\rho} = (z/k)\boldsymbol{\kappa} = (z\lambda/2\pi)\boldsymbol{\kappa}$. This interpretation is useful because of the experience we all have with the Fourier transform, mostly in the context of its time/frequency applications. For example, if the orifice is a single small hole, $T(\boldsymbol{\rho}')$ may be approximated by a delta-function, so that Eq. (8.116) yields $|f(\boldsymbol{\rho})| \approx$ const. This corresponds (at least for the small diffraction angles $\theta \equiv \rho/z$ for which the Huygens approximation is valid) to a spherical wave spreading from the point-like orifice. Next, for two small holes, Eq. (8.116) immediately gives the Young interference pattern (8.103). Let me now use Eq. (8.116) to analyze the simplest (and most important) 1D transparency profiles, leaving 2D cases for the reader's exercise.

(i) A single slit of width a (figure 8.12) may be described by transparency

$$T(\boldsymbol{\rho}') = \begin{cases} 1, & \text{for } |x'| < a/2, \\ 0, & \text{otherwise.} \end{cases} \tag{8.118}$$

Its substitution into Eq. (8.116) yields

$$\begin{aligned} f(\boldsymbol{\rho}) \propto f_0 \int_{-a/2}^{+a/2} \exp\{-i\kappa_x x'\} dx' &= f_0 \frac{\exp\{-i\kappa_x a/2\} - \exp\{i\kappa_x a/2\}}{-i\kappa_x} \\ &\propto \operatorname{sinc}\left(\frac{\kappa_x a}{2}\right) = \operatorname{sinc}\left(\frac{kxa}{2z}\right), \end{aligned} \tag{8.119}$$

naturally returning us to Eqs (8.64) and (8.110), and hence to the red lines in figure 8.8 for the wave intensity. (Please note again that Eq. (8.116) describes only the Fraunhofer, but not the Fresnel diffraction!)

(ii) Two narrow, similar, parallel slits with a much larger distance a between them, may be described by taking

$$T(\boldsymbol{\rho}') \propto \delta(x' - a/2) + \delta(x' + a/2), \tag{8.120}$$

so that Eq. (8.116) yields the generic 1D interference pattern,

$$f(\boldsymbol{\rho}) \propto f_0 \left[\exp\left\{-\frac{i\kappa_x a}{2}\right\} + \exp\left\{\frac{i\kappa_x a}{2}\right\}\right] \propto \cos\frac{\kappa_x a}{2} = \cos\frac{kxa}{2z}, \tag{8.121}$$

whose intensity is shown with the blue line in figure 8.8.

(iii) In a more realistic version of the Young-type two-slit experiment, each slit has a width (say, w) which is much larger than light wavelength λ, but still much smaller than the slit spacing a. This situation may be described by the following transparency function

$$T(\boldsymbol{\rho}') = \sum_{\pm} \begin{cases} 1, & \text{for } |x' \pm a/2| < w/2, \\ 0, & \text{otherwise,} \end{cases} \tag{8.122}$$

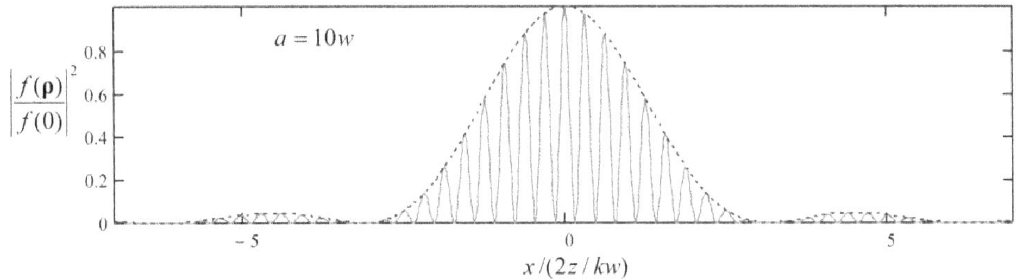

Figure 8.15. Young's double-slit interference pattern for a finite-width slit.

for which Eq. (8.116) yields a natural combination of results (8.119) (with a replaced with w) and (8.121):

$$f(\mathbf{r}) \propto \ \text{sinc}\left(\frac{kxw}{2z}\right)\cos\left(\frac{kxa}{2z}\right). \tag{8.123}$$

This is the usual interference pattern, modulated with a Fraunhofer-diffraction envelope (shown with the dashed blue line in figure 8.15). Since the function $\text{sinc}^2\,\xi$ decreases very fast beyond its first zeros at $\xi = \pm\pi$, the practical number of observable interference fringes is close to $2a/w$.

(iv) A structure very useful for experimental and engineering practice is a set of many parallel, similar slits, called a *diffraction grating*[47]. Indeed, if the slit width is much smaller than the grating period d, then the transparency function may be approximated as

$$T(\boldsymbol{\rho}') \propto \sum_{n=-\infty}^{+\infty} \delta(x' - nd) \tag{8.124}$$

and Eq. (8.116) yields

$$f(\boldsymbol{\rho}) \propto \sum_{n=-\infty}^{n=+\infty} \exp\{-in\kappa_x d\} = \sum_{n=-\infty}^{n=+\infty} \exp\left\{-i\frac{nkxd}{z}\right\}. \tag{8.125}$$

This sum vanishes for all values of $\kappa_x d$ that are not multiples of 2π, so that the result describes sharp intensity peaks at the following diffraction angles:

$$\theta_m \equiv \left(\frac{x}{z}\right)_m = \left(\frac{\kappa_x}{k}\right)_m = \frac{2\pi}{kd}m = \frac{\lambda}{d}m. \tag{8.126}$$

Taking into account that this result is only valid for small angles $|\theta_m| \ll 1$, it may be interpreted exactly as Eq. (8.59)—see figure 8.6a. However, in contrast with the interference (8.121) from two slits, the destructive interference from many slits kills the net wave as soon as the angle is even slightly different from each value (8.60). This is very convenient for spectroscopic purposes, because the diffraction lines

[47] The rudimentary diffraction grating effect, produced by the parallel fibers of a bird's feather, was discovered as early as in 1673 by J Gregory—who also invented the reflecting ('Gregorian') telescope.

produced by multi-frequency waves do not overlap even if the frequencies of their adjacent components are very close.

Two unavoidable features of practical diffraction gratings make their properties different from this simple, ideal picture. First, the finite number N of slits, which may be described by limiting the sum (8.125) to the interval $n = [-N/2, +N/2]$, results in a non-zero spread, $\delta\theta/\theta \sim 1/N$, of each diffraction peak, and hence in the reduction of the grating's spectral resolution. (Unintentional variations of the inter-slit distance d have a similar effect, so that before the advent of high-resolution photolithography, special high-precision mechanical tools have been used for grating fabrication.)

Second, a finite slit width w leads to the diffraction peak pattern modulation by a $\text{sinc}^2(kw\theta/2)$ envelope, similarly to the pattern shown in figure 8.15. Actually, for spectroscopic purposes such modulation is sometimes a plus, because only one diffraction peak (say, with $m = \pm 1$) is practically used, and if the frequency spectrum of the analyzed wave is very broad (covers more than one octave), the higher peaks produce undesirable hindrance. Because of this reason, w is frequently selected to be equal exactly to $d/2$, thus suppressing each other diffraction maximum. Moreover, sometimes semi-transparent films are used to make the transparency function $T(\mathbf{r}')$ continuous and close to the sinusoidal one:

$$T(\boldsymbol{\rho}') \approx T_0 + T_1 \cos\frac{2\pi x'}{d} \equiv T_0 + \frac{T_1}{2}\left(\exp\left\{i\frac{2\pi x'}{d}\right\} + \exp\left\{-i\frac{2\pi x'}{d}\right\}\right). \quad (8.127)$$

Plugging the last expression into Eq. (8.116) and integrating, we see that the output wave consists of just three components: the direct-passing wave (proportional to T_0) and two diffracted waves (proportional to T_1) propagating in the directions of the two lowest Bragg angles, $\theta_{\pm 1} = \pm\lambda/d$.

The same relation (8.116) may be also used to obtain one more general (and rather curious) result, called the *Babinet principle*. Consider two experiments with diffraction of similar plane waves on two 'complementary' screens who together would cover the whole plane, without a hole or an overlap. (Think, for example, about an opaque disk of radius R and a large opaque screen with a round orifice of the same radius.) Then, according to the Babinet principle, the diffracted wave patterns produced by these two screens in all directions with $\theta \neq 0$ are *identical*. The proof of this principle is straightforward: since the transparency functions produced by the screens are complementary in the following sense:

$$T(\boldsymbol{\rho}') \equiv T_1(\boldsymbol{\rho}') + T_2(\boldsymbol{\rho}') = 1, \quad (8.128)$$

and (in the Fraunhofer approximation only!) the diffracted wave is a Fourier transform of $T(\boldsymbol{\rho}')$, which is a linear operation, we obtain

$$f_1(\boldsymbol{\rho}) + f_2(\boldsymbol{\rho}) = f_0(\boldsymbol{\rho}), \quad (8.129)$$

where f_0 is the wave 'scattered' by the composite screen with $T_0(\boldsymbol{\rho}') \equiv 1$, i.e. the unperturbed initial wave propagating in the initial direction ($\theta = 0$). In all other directions, $f_1 = -f_2$, i.e. the diffracted waves are indeed similar apart from the difference in sign—which is equivalent to a phase shift by $\pm\pi$. However, it is important to remember that the Babinet principle notwithstanding, in real experiments, with screens at finite distances, the diffracted waves may interfere with the unperturbed plane wave $f_0(\boldsymbol{\rho})$, leading to different diffraction patterns in the cases 1 and 2—see, e.g. Figure 8.14 and its discussion.

8.9 Magnetic dipole and electric quadrupole radiation

Throughout this chapter, we have seen how many important results may be obtained from Eq. (8.26) for the electric dipole radiation by a small-size source (figure 8.1). Only in rare cases when this radiation is absent, for example if the dipole moment **p** of the source equals zero (or does not change at time—either at all, or at the frequency of our interest), higher-order effects may be important. I will discuss the main two of them, the *quadrupole electric radiation* and the *dipole magnetic radiation.*

In section 8.2 above, the electric dipole radiation was calculated by plugging the expansion (8.19) into the exact formula (8.17b) for the retarded vector-potential $\mathbf{A}(\mathbf{r}, t)$. Let us make a more exact calculation, by keeping the second term of that expansion as well:

$$\mathbf{j}\left(\mathbf{r}', t - \frac{R}{v}\right) \approx \mathbf{j}\left(\mathbf{r}', t - \frac{r}{v} + \frac{\mathbf{r}' \cdot \mathbf{n}}{v}\right) \equiv \mathbf{j}\left(\mathbf{r}', t' + \frac{\mathbf{r}' \cdot \mathbf{n}}{v}\right), \quad \text{where } t' \equiv t - \frac{r}{v}. \tag{8.130}$$

Since the expansion is only valid if the last term in the time argument of **j** is relatively small, in the Taylor expansion of **j** with respect to that argument we may keep just two leading terms:

$$\mathbf{j}\left(\mathbf{r}', t' + \frac{\mathbf{r}' \cdot \mathbf{n}}{v}\right) \approx \mathbf{j}(\mathbf{r}', t') + \frac{1}{v}\frac{\partial}{\partial t'}\mathbf{j}(\mathbf{r}', t')(\mathbf{r}' \cdot \mathbf{n}), \tag{8.131}$$

so that Eq. (8.17b) yields $\mathbf{A} = \mathbf{A}_e + \mathbf{A}'$, where $\mathbf{A}_e$ is the electric dipole contribution as given by Eq. (8.23), and $\mathbf{A}'$ is the new term of the next order in the small parameter $r' \ll r$:

$$\mathbf{A}'(\mathbf{r}, t) = \frac{\mu}{4\pi\, rv}\frac{\partial}{\partial t'}\int \mathbf{j}(\mathbf{r}', t')(\mathbf{r}' \cdot \mathbf{n})d^3r'. \tag{8.132}$$

Just as it was done in section 8.2, let us evaluate this term for a system of non-relativistic particles with electric charges q_k and radius-vectors $\mathbf{r}_k(t)$:

$$\mathbf{A}'(\mathbf{r}, t) = \frac{\mu}{4\pi\, rv}\left[\frac{d}{dt}\sum_k q_k \dot{\mathbf{r}}_k(\mathbf{r}_k \cdot \mathbf{n})\right]_{t=t'}. \tag{8.133}$$

Using the 'bac minus cab' identity of the vector algebra again[48], the vector operand of Eq. (8.133) may be rewritten as

[48] If you need, see, e.g. Eq. (A.47).

$$
\begin{aligned}
\dot{\mathbf{r}}_k(\mathbf{r}_k \cdot \mathbf{n}) &\equiv \frac{1}{2}\dot{\mathbf{r}}_k(\mathbf{r}_k \cdot \mathbf{n}) + \frac{1}{2}\dot{\mathbf{r}}_k(\mathbf{n} \cdot \mathbf{r}_k) \\
&= \frac{1}{2}(\mathbf{r}_k \times \dot{\mathbf{r}}_k) \times \mathbf{n} + \frac{1}{2}\mathbf{r}_k(\mathbf{n} \cdot \dot{\mathbf{r}}_k) + \frac{1}{2}\dot{\mathbf{r}}_k(\mathbf{n} \cdot \mathbf{r}_k) \\
&\equiv \frac{1}{2}(\mathbf{r}_k \times \dot{\mathbf{r}}_k) \times \mathbf{n} + \frac{1}{2}\frac{d}{dt}[\mathbf{r}_k(\mathbf{n} \cdot \mathbf{r}_k)],
\end{aligned}
\tag{8.134}
$$

so that the right-hand side of Eq. (8.133) may be represented as a sum of two terms, $\mathbf{A}' = \mathbf{A}_{\mathrm{m}} + \mathbf{A}_{\mathrm{q}}$, where

$$
\begin{aligned}
\mathbf{A}_{\mathrm{m}}(\mathbf{r}, t) &= \frac{\mu}{4\pi\, rv}\dot{\mathbf{m}}(t') \times \mathbf{n} = \frac{\mu}{4\pi\, rv}\dot{\mathbf{m}}\left(t - \frac{r}{v}\right) \times \mathbf{n}, \\
&\text{with } \mathbf{m}(t) \equiv \frac{1}{2}\sum_k \mathbf{r}_k(t) \times q_k\dot{\mathbf{r}}_k(t),
\end{aligned}
\tag{8.135}
$$

$$
\mathbf{A}_{\mathrm{q}}(\mathbf{r}, t) = \frac{\mu}{8\pi\, rv}\left[\frac{d^2}{dt^2}\sum_k q_k\mathbf{r}_k(\mathbf{n} \cdot \mathbf{r}_k)\right]_{t=t'}. \tag{8.136}
$$

Comparing the second of Eqs. (8.135) with Eq. (5.91), we see that $\mathbf{m}$ is just the total magnetic moment of the source. On the other hand, the first of Eqs. (8.135) is absolutely similar in structure to Eq. (8.23), with $\mathbf{p}$ replaced with $(\mathbf{m} \times \mathbf{n})/v$, so that for the corresponding component of the magnetic field it gives (in the same approximation $r \gg \lambda$) a result similar to Eq. (8.24):

$$
\mathbf{B}_{\mathrm{m}}(\mathbf{r}, t) = \frac{\mu}{4\pi\, rv}\nabla \times \left[\dot{\mathbf{m}}\left(t - \frac{r}{v}\right) \times \mathbf{n}\right] = -\frac{\mu}{4\pi\, rv^2}\mathbf{n} \times \left[\ddot{\mathbf{m}}\left(t - \frac{r}{v}\right) \times \mathbf{n}\right]. \tag{8.137}
$$

According to this expression, just as at the electric dipole radiation, the vector $\mathbf{B}$ is perpendicular to the vector $\mathbf{n}$, and its magnitude is also proportional to $\sin\Theta$, where Θ is now the angle between the direction toward the observation point and the second time derivative of the vector $\mathbf{m}$—rather than $\mathbf{p}$:

$$
B_{\mathrm{m}} = \frac{\mu}{4\pi\, rv^2}\ddot{m}\left(t - \frac{r}{v}\right)\sin\Theta. \tag{8.138}
$$

As the result, the intensity of this *magnetic dipole radiation* has a similar angular distribution:

$$
S_r = ZH^2 = \frac{Z}{(4\pi\, v^2 r)^2}\left[\ddot{m}\left(t - \frac{r}{v}\right)\right]^2 \sin^2\Theta, \tag{8.139}
$$

—cf Eq. (8.26), except for the (generally) different meaning of the angle Θ.

Note, however, that this radiation is usually much weaker than its electric-dipole counterpart. For example, for a non-relativistic particle with electric charge q, moving on a trajectory of size $\sim a$, the electric dipole moment is of the order of qa, while its magnetic moment scales as $qa^2\omega$, where ω is the motion frequency. As a result, the ratio of the magnetic and electric dipole radiation intensities is of the order of $(a\omega/v)^2$, i.e. the squared ratio of the particle's speed to the speed of emitted waves—that has to be much smaller than 1 for our non-relativistic calculation to be valid.

The angular distribution of the *electric quadrupole radiation*, described by Eq. (8.136), is more complicated. In order to show this, we may add to $\mathbf{A}_q$ a vector parallel to $\mathbf{n}$ (i.e. along the wave's propagation), obtaining

$$\mathbf{A}_q(\mathbf{r}, t) \to \frac{\mu}{24\pi\, rv}\ddot{\mathscr{Q}}\left(t - \frac{r}{v}\right), \quad \text{where } \mathscr{Q} \equiv \sum_k q_k\left\{3\mathbf{r}_k(\mathbf{n}\cdot\mathbf{r}_k) - \mathbf{n}r_k^2\right\}, \tag{8.140}$$

because this addition does not give any contribution to the transverse component of the electric and magnetic fields, i.e. to the radiated wave. According to the above definition of the vector $\mathscr{Q}$, its Cartesian components may be represented as

$$\mathscr{Q}_j = \sum_{j'=1}^{3} \mathscr{Q}_{jj'} n_{j'}, \tag{8.141}$$

where $\mathscr{Q}_{jj'}$ are the elements of the electric quadrupole tensor of the system—see the last of Eqs. (3.4):

$$\mathscr{Q}_{jj'} = \sum_k q_k(3r_j r_{j'} - r^2\delta_{jj'})_k. \tag{8.142}$$

Let me hope that the reader has already obtained some expertise in the calculation of this tensor's components for the simple systems listed in problems 3.2 and 3.3.

Now taking the curl of the first of Eq. (8.140) at $r \gg \lambda$, we obtain

$$\mathbf{B}_q(\mathbf{r}, t) = -\frac{\mu}{24\pi\, rv^2}\mathbf{n} \times \dddot{\mathscr{Q}}\left(t - \frac{r}{v}\right). \tag{8.143}$$

This expression is similar to Eqs (8.24) or (8.137), but according to Eqs (8.140) and (8.142) the components of the vector $\mathscr{Q}$ do depend on the direction of the vector $\mathbf{n}$, leading to a different angular dependence of S_r.

As the simplest example, let us consider a system of two equal point electric charges moving symmetrically, at equal distances $d(t) \ll \lambda$ from a stationary center—see figure 8.16.

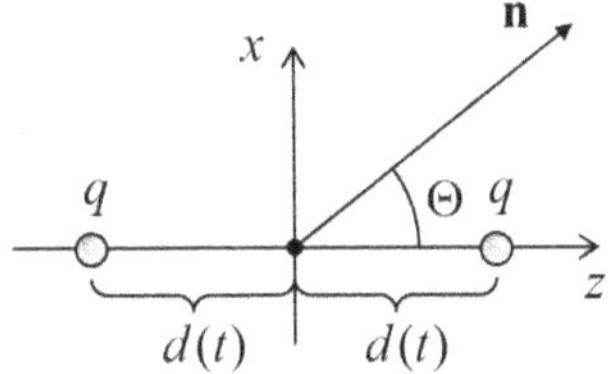

Figure 8.16. The simplest system emitting electric quadrupole radiation.

Due to the symmetry of the system, its dipole moments **p** and **m** (and hence its electric and magnetic dipole radiation) vanish, but the quadrupole tensor (8.142) still has non-zero components. With the coordinate choice shown in figure 8.16, these components are diagonal:

$$\mathscr{Q}_{xx} = \mathscr{Q}_{yy} = -2qd^2, \quad \mathscr{Q}_{zz} = 4qd^2. \tag{8.144}$$

With the axis x selected within the common plane of the axis z and the direction **n** toward the source (figure 8.16), so that $n_x = \sin\Theta$, $n_y = 0$, and $n_z = \cos\Theta$, Eq. (8.141) yields

$$\mathscr{Q}_x = -2qd^2 \sin\Theta, \quad \mathscr{Q}_y = 0, \quad \mathscr{Q}_z = 4qd^2 \cos\Theta, \tag{8.145}$$

and the vector product in Eq. (8.143) has only one non-vanishing Cartesian component:

$$(\mathbf{n} \times \dddot{\mathscr{Q}})_y = n_z\dddot{\mathscr{Q}}_x - n_x\dddot{\mathscr{Q}}_z = -6q \sin\Theta \cos\Theta \frac{d^3}{dt^3}[d^2(t)]. \tag{8.146}$$

As a result, the quadrupole radiation intensity, $S \propto B_q^{\;2}$, is proportional to $\sin^2\Theta$ $\cos^2\Theta$, i.e. vanishes not only along the symmetry axis of the system (as the electric-dipole and the magnetic-dipole radiations would do), but also in all directions perpendicular to this axis, reaching its maxima at $\Theta = \pm\pi/4$.

For more complex systems, the angular distribution of the electric quadrupole radiation may be different, but it may be proved that its total (instant) power always obeys the simple formula

$$\mathscr{P}_q = \frac{Z}{1440\pi v^4} \sum_{j,j'=1}^{3} (\dddot{\mathscr{Q}}_{jj'})^2. \tag{8.147}$$

Let me finish this section by giving, also without proof, one more fact important for some applications: due to their different spatial structures, the magnetic-dipole and electric-quadrupole radiation fields do not interfere, i.e. the total power of radiation (neglecting the electric-dipole and higher multipole terms) may be found as the sum of these components, calculated independently. In contrast, if the electric-dipole and magnetic-dipole radiations of the same system are comparable, they typically interfere coherently, so that their radiation fields (rather than powers) should be added.

8.10 Problems

Problem 8.1. In the electric-dipole approximation, calculate the angular distribution and the total power of electromagnetic radiation using the following classical model of the hydrogen atom: an electron rotating, at a constant distance R, about a much heavier proton. Use the latter result to evaluate the classical lifetime of the atom, borrowing the initial value of R from quantum mechanics: $R(0) = r_B \approx 0.53 \times 10^{-10}$ m.

Problem 8.2. A non-relativistic particle of mass m, with electric charge q, is placed into a uniform magnetic field **B**. Derive the law of decrease of the particle's kinetic energy due to its electromagnetic radiation at the *cyclotron frequency* $\omega_c = qB/m$. Evaluate the rate of such radiation cooling for electrons in a magnetic field of 1 T, and estimate the electron energy interval in which this result is qualitatively correct.

Hint: The cyclotron motion will be discussed in detail (for arbitrary particle velocities $v \sim c$) in section 9.6 below, but I hope that the reader knows that in the non-relativistic case ($v \ll c$) the above formula for ω_c may be readily obtained by combining the Newton's second law $mv_\perp^2/R = qv_\perp B$ for the circular motion of the particle under the effect of the magnetic component of the Lorentz force (5.10), and the geometric relation $v_\perp = R\omega_c$. (Here $\mathbf{v}_\perp$ is particle's velocity within the plane normal to the vector **B**.)

Problem 8.3. Solve the dipole antenna radiation problem discussed in section 8.2 (see figure 8.3) for the optimal length $l = \lambda/2$, assuming[49] that the current distribution in each of its arms is sinusoidal:

$$I(z, t) = I_0 \cos(\pi z/l) \cos \omega t.$$

Problem 8.4. Use the Lorentz oscillator model of a bound charge, given by Eq. (7.30), to explore the transition between the two scattering limits discussed in section 8.3, in particular the *resonant scattering* taking place at $\omega \approx \omega_0$. In this context, discuss the contribution of scattering into the oscillator's damping.

Problem 8.5.* A sphere of radius R, made of a material with a uniform permanent electric polarization $\mathbf{P}_0$ and a constant mass density ρ, is free to rotate about its center. Calculate the average total cross-section of scattering, by the sphere, of a linearly polarized electromagnetic wave of frequency $\omega \ll R/c$, propagating in free space, in the limit of a small wave amplitude, assuming that the initial orientation of the polarization vector $\mathbf{P}_0$ is random.

Problem 8.6. Use Eq. (8.56) to analyze the interference/diffraction pattern produced by a plane wave's scattering on a set of N similar, equidistant points on a straight line normal to the direction of the incident wave's propagation—see the figure below. Discuss the trend(s) of the pattern in the limit $N \to \infty$.

Problem 8.7. Use the Born approximation to calculate the differential cross-section of the plane wave scattering by a non-magnetic, uniform dielectric sphere of an

[49] As was emphasized in section 8.2, this is a reasonable guess rather than a controllable approximation. The exact (rather involved!) theory shows that this assumption gives errors ~5%.

arbitrary radius R. In the limits $kR \ll 1$ and $1 \ll kR$ (where k is the wave number), analyze the angular dependence of the differential cross-section, and calculate the full cross-section of scattering.

Problem 8.8. A sphere of radius R is made of a uniform, non-magnetic, linear dielectric material, with an arbitrary dielectric constant. Derive an exact expression for its full cross-section of scattering of a low-frequency monochromatic wave, with $k \ll 1/R$, and compare the result with the solution of the previous problem.

Problem 8.9. Use the Born approximation to calculate the differential cross-section of the plane wave scattering on a right, circular cylinder of length l and radius R, for an arbitrary angle of incidence.

Problem 8.10. Formulate the quantitative condition of the Born approximation's validity for a uniform linear-dielectric scatterer, with all linear dimensions of the order of the same scale a.

Problem 8.11. If a scatterer absorbs some part of the incident wave's power, it may be characterized by an *absorption cross-section* σ_a defined similarly to Eq. (8.39) for the scattering cross-section:

$$\sigma_a \equiv \frac{\overline{\mathscr{P}_a}}{|E_\omega|^2/2Z_0},$$

where the nominator is the time-averaged power absorbed in the scatterer. Calculate σ_a for a very small sphere of radius $R \ll k^{-1}, \delta_s$, made of a non-magnetic material with Ohmic conductivity σ, and with high-frequency permittivity $\varepsilon_{opt} = \varepsilon_0$. Can σ_a of such a sphere be larger than its geometric cross-section πR^2?

Problem 8.12. Use the Huygens principle to calculate the wave's intensity on the symmetry plane of the slit diffraction experiment (i.e. at $x = 0$ in figure 8.12), for an arbitrary ratio z/ka^2.

Problem 8.13. A plane wave with wavelength λ is normally incident on an opaque, plane screen, with a round orifice of radius $R \gg \lambda$. Use the Huygens principle to calculate the passing wave's intensity distribution along the system's symmetry axis, at distances $z \gg R$ from the screen (see the figure below) and analyze the result.

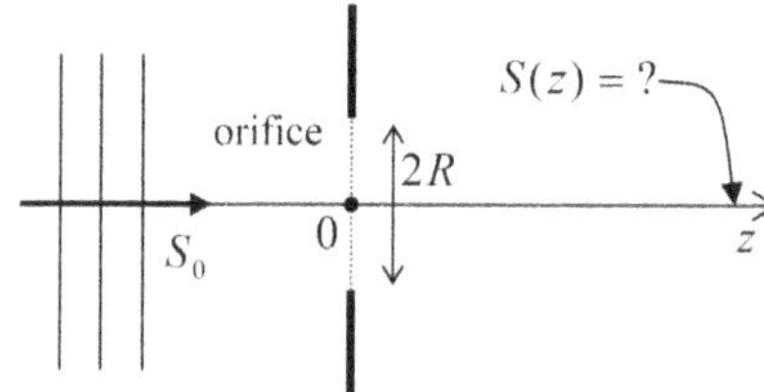

Problem 8.14. A planar monochromatic wave is now normally incident on an opaque circular disk of radius $R \gg \lambda$. Use the Huygens principle to calculate the wave's intensity at distance $z \gg R$ behind the disk's center (see the figure below). Discuss the result.

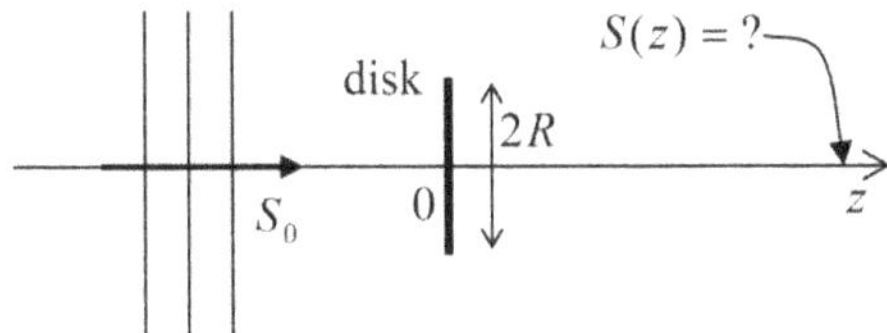

Problem 8.15. Use the Huygens principle to analyze the Fraunhofer diffraction of a plane wave normally incident on a square-shaped hole, of size $a \times a$, in an opaque screen. Sketch the diffraction pattern you would observe at a sufficiently large distance, and quantify the expression 'sufficiently large' for this case.

Problem 8.16. Use the Huygens principle to analyze the propagation of a monochromatic Gaussian beam described by Eq. (7.181), with the initial characteristic width $a_0 \gg \lambda$, in a uniform, isotropic medium. Use the result for a semi-quantitative derivation of the so-called *Abbe limit*[50] for the spatial resolution of an optical system,

$$w_{\min} = \lambda/2 \sin \theta,$$

where θ is the half-angle of the wave cone propagating from the object, and captured by the system.

Problem 8.17. Within the Fraunhofer approximation, analyze the pattern produced by a 1D diffraction grating with the periodic transparency profile shown in the figure below, for the normal incidence of a plane, monochromatic wave.

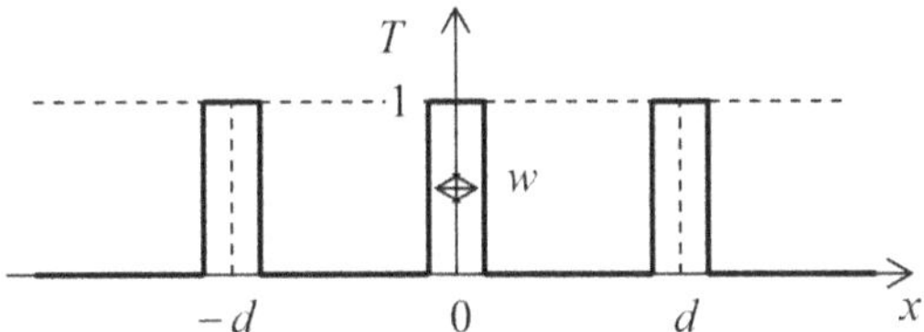

Problem 8.18. N equal point charges are attached, at equal intervals, to a circle rotating with a constant angular velocity about its center—see the figure below. For what values of N does the system emit:

(i) the electric dipole radiation?
(ii) the magnetic dipole radiation?
(iii) the electric quadrupole radiation?

---

[50] Reportedly, due to not only E Abbe (1873), but also to H von Helmholtz (1874).

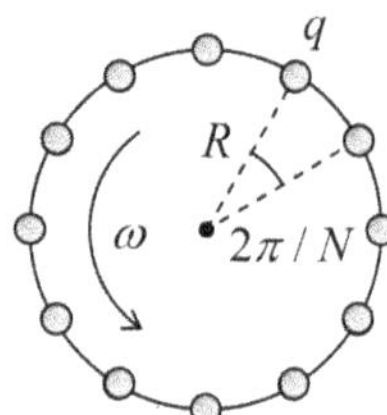

Problem 8.19. What general statements can you made about:

(i) the electric dipole radiation, and
(ii) the magnetic dipole radiation

due to a collision of an arbitrary number of similar classical, non-relativistic particles?

Problem 8.20. Calculate the angular distribution and the total power radiated by a small round loop antenna with radius R, fed by ac current $I(t)$ with frequency ω and amplitude I_0, into the free space.

Problem 8.21. The orientation of a magnetic dipole $\mathbf{m}$, of a fixed magnitude, is rotating about a certain axis with angular velocity ω, with angle α between them staying constant. Calculate the angular distribution and the average power of its radiation (into the free space).

Problem 8.22. Solve problem 8.8 (also in the low-frequency limit $kR \ll 1$), for the case when the sphere's material has a frequency-independent Ohmic conductivity σ, and $\varepsilon_{\text{opt}} = \varepsilon_0$, in two limits:

(i) of a very large skin depth ($\delta_s \gg R$), and
(ii) of a very small skin depth ($\delta_s \ll R$).

Problem 8.23. Complete the solution of the problem started in section 8.9, by calculating the full power of radiation of the system of two charges oscillating in antiphase along the same straight line—see figure 8.16. Also, calculate the average radiation power for the case of harmonic oscillations, $d(t) = a\cos\omega t$, compare it with the case of a single charge performing similar oscillations, and interpret the difference.

References

[1] Cullity B 1978 *Elements of x-ray Diffraction* 2nd edn (New York: Addison-Wesley)
[2] Born M *et al* 1999 *Principles of Optics* 7th edn (Cambridge University Press)
[3] Fowles G 1989 *Introduction to Modern Optics* 2nd edn (New York: Dover)
[4] Brezinski M 2006 *Optical Coherence Tomography* (New York: Academic)
[5] Popescu G 2011 *Quantitative Phase Imaging of Cells and Tissues* (New York: McGraw-Hill)

IOP Publishing

Chapter 9

Special relativity

This chapter starts with a review of the basics of special relativity, including the very convenient four-vector formalism. This background is then used for the analysis of the relation between the electromagnetic field's values measured in different reference frames moving relative to each other. The results of this discussion enable the analysis of the relativistic particle dynamics in the electric and magnetic fields, the analytical mechanics of the particles, and of the electromagnetic field as such.

9.1 Einstein postulates and the Lorentz transform

As was emphasized at the derivation of expressions for the dipole and quadrupole radiations in the previous chapter, they are only valid for systems of non-relativistic particles. Thus, these results cannot be used for the description of such important phenomena as the Cherenkov radiation or synchrotron radiation, in which relativistic effects are essential. Moreover, an analysis of the motion of charged relativistic particles in electric and magnetic fields is also a natural part of electrodynamics. This is why I will follow the tradition of using this course for a (by necessity, brief) introduction to the theory of special relativity. This theory is based on the fundamental idea that measurements of physical variables (including the spatial and even temporal intervals between two events) may give different results in different reference frames, in particular in two inertial frames moving relative to each other translationally (i.e. without rotation), with a certain constant velocity $\mathbf{v}$ (figure 9.1).

In non-relativistic (Newtonian) mechanics the problem of transfer between such reference frames has a simple solution, at least in the limit $v \ll c$, because the basic equation of particle dynamics (Newton's second law)[1]

[1] Let me hope that the reader does not need a reminder that in order for Eq. (9.1) to be valid, the reference frames 0 and 0′ have to be inertial—see, e.g. *Part CM* section 1.2.

doi:10.1088/978-0-7503-1404-6ch9

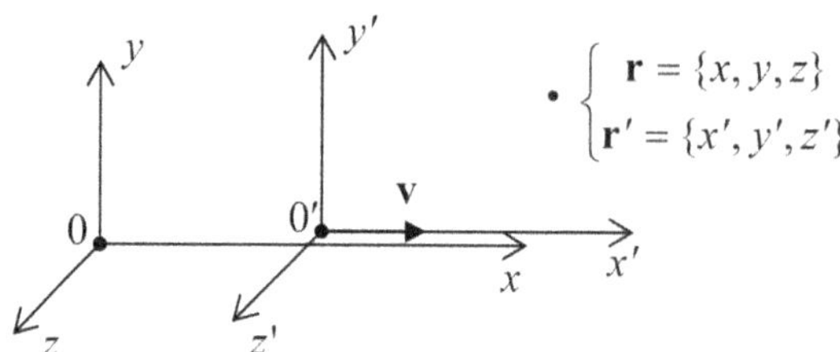

Figure 9.1. The translational, uniform mutual motion of two reference frames.

$$m_k \ddot{\mathbf{r}}_k = -\nabla_k \sum_{k'} U(\mathbf{r}_k - \mathbf{r}_{k'}), \tag{9.1}$$

where U, the potential energy of inter-particle interactions, is invariant with respect to the so-called *Galilean transformation* (or just 'transform' for short)[2]. Choosing the coordinate axes of both frames so that the axes x and x' are parallel to the vector $\mathbf{v}$ (figure 9.1), the transform[3] may be represented as

$$x = x' + vt', \quad y = y', \quad z = z', \quad t = t', \tag{9.2a}$$

and plugging Eq. (9.2*a*) into Eq. (9.1) we obtain an absolutely similar looking equation of motion in the 'moving' reference frame 0′. Since the reciprocal transform,

$$x' = x - vt, \quad y = y', \quad z' = z, \quad t' = t, \tag{9.2b}$$

is similar to the direct one, with the replacement of $(+v)$ by $(-v)$, we may say that the Galilean invariance means that there is no 'master' (*absolute*) spatial reference frame in classical mechanics, although the spatial and temporal intervals between different instant events are absolute, i.e. reference-frame invariant: $\Delta x = \Delta x', \ldots, \Delta t = \Delta t'$.

However, it is straightforward to use Eq. (9.2) to check that the form of the wave equation

$$\left(\frac{\partial^2}{\partial x^2} + \frac{\partial^2}{\partial y^2} + \frac{\partial^2}{\partial z^2} - \frac{1}{c^2}\frac{\partial^2}{\partial t^2} \right) f = 0, \tag{9.3}$$

describing in particular the electromagnetic wave propagation in free space[4], is *not* Galilean-invariant[5]. For the 'usual' (say, elastic) waves, which obey a similar

[2] It was first formulated by G Galilei as early as in 1638—4 years before Newton was *born*!

[3] Note the very unfortunate term 'boost' sometimes used for the description of the transfer between reference frames. (It is particularly unnatural in special relativity, which does not describe any accelerations.) In these notes, this term is avoided.

[4] The discussions in this chapter and most of the next chapter will be restricted to the free space (and hence dispersion-free) case; some media effects on the radiation by relativistic particles will be discussed in section 10.4.

[5] It is interesting that the usual Schrödinger equation, whose fundamental solution for a free particle is a similar monochromatic wave (albeit with a different dispersion law), *is* Galilean-invariant, with a certain addition to the wavefunction's phase—see, e.g. *Part QM* chapter 1. This is natural, because this equation is non-relativistic.

equation albeit with a different speed[6], this lack of Galilean invariance is natural and is compatible with the invariance of Eq. (9.1), from which the wave equation originates. This is because elastic waves are essentially the oscillations of interacting particles of a certain medium (e.g. an elastic solid), which makes the reference frame, connected to this medium, special. So, if the electromagnetic waves were oscillations of a certain special medium (which was first called the 'luminiferous aether'[7] and later the *aether*, or just 'ether'), similar arguments might be applicable to reconcile Eqs. (9.2) and (9.3).

The detection of such a medium was the goal of the Michelson–Morley measurements (carried out between 1881 and 1887 with better and better precision), that are sometimes called 'the most famous failed experiment in physics'. Figure 9.2 shows a crude scheme of their experiments.

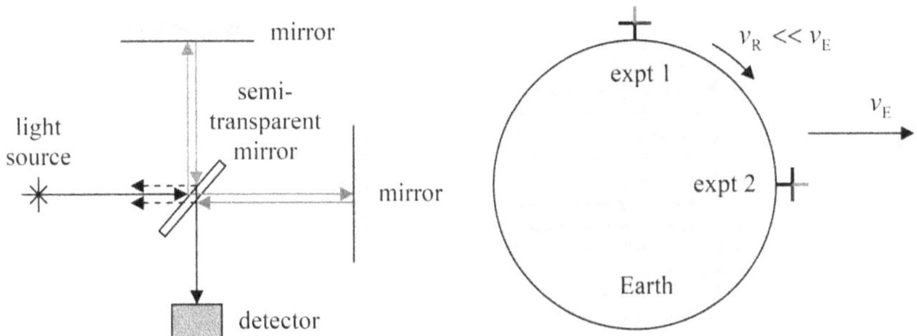

Figure 9.2. The Michelson–Morley experiment.

A nearly monochromatic wave from a light source is split in two parts (nominally, of equal intensity), using a semi-transparent mirror tilted by 45° to the incident wave direction. These two partial waves are reflected back by two full-reflection mirrors and arrive at the same semi-transparent mirror again. Here a half of each wave is returned to the light source area (where they vanish without affecting the source), but the other half travels toward the detector, forming, with its counterpart, an interference pattern similar to that in the Young experiment. Thus each of the interfering waves has traveled twice (back and forth) along each of the two mutually perpendicular 'arms' of the interferometer. Assuming that the aether, in which light propagates with speed c, moves with speed $v < c$ along one of the arms, of length l_l, it is straightforward (and hence left for the reader's exercise) to obtain the following expression for the difference between the light roundtrip times:

$$\Delta t = \frac{2}{c}\left[\frac{l_t}{(1 - v^2/c^2)^{1/2}} - \frac{l_l}{1 - v^2/c^2}\right] \approx \frac{l}{c}\left(\frac{v}{c}\right)^2, \tag{9.4}$$

where l_t is the length of the second, 'transverse' arm of the interferometer (perpendicular to $\mathbf{v}$), and the last, approximate expression is valid at $l_t \approx l_l$ and $v \ll c$.

[6] See, e.g. *Part CM* sections 6.5 and 7.7.

[7] In the ancient Greek mythology, aether is the clean air breathed by the gods residing on Mount Olympus.

Since the Earth moves around the Sun with a speed $v_E \approx 30$ km s$^{-1} \approx 10^{-4}$ c, the arm positions relative to this motion alternate, due to the Earth's rotation about its axis, every 6 h—see the right-hand panel of figure 9.2. Hence if we assume that the aether rests in the Sun's reference frame, Δt (and the corresponding shift of the interference fringes) has to change its sign with this half-period as well. The same alternation may be achieved, at a smaller time scale, by a deliberate rotation of the instrument by $\pi/2$. In the most precise version of the Michelson–Morley experiment (1887), this shift was expected to be close to 0.4 of the fringe pattern period. The results of the search for such a shift were negative, with the error bar about 0.01 of the fringe period[8].

The most prominent immediate explanation of this zero result[9] was suggested in 1889 by G FitzGerald and (independently and more qualitatively) by H Lorentz in 1892: as is evident from Eq. (9.4), if the longitudinal arm of the interferometer itself experiences so-called *length contraction*,

$$l_l(v) = l_l(0)\left(1 - \frac{v^2}{c^2}\right)^{1/2}, \qquad (9.5)$$

while the transverse arm's length is not affected by its motion through the aether, this kills the shift Δt. This extremely radical idea received strong support from the proof, in 1887–1905, that the Maxwell equations, and hence the wave Eq. (9.3), are form-invariant under the so-called *Lorentz transform*[10]. For the choice of coordinates shown in figure 9.1, the transform reads

$$x = \frac{x' + vt'}{(1 - v^2/c^2)^{1/2}}, \quad y = y', \quad z = z', \quad t = \frac{t' + (v/c^2)x'}{(1 - v^2/c^2)^{1/2}}. \qquad (9.6a)$$

It is elementary to solve these equations for the primed coordinates to obtain the reciprocal transform

$$x' = \frac{x - vt}{(1 - v^2/c^2)^{1/2}}, \quad y' = y, \quad z' = z, \quad t' = \frac{t - (v/c^2)x}{(1 - v^2/c^2)^{1/2}}. \qquad (9.6b)$$

(I will soon represent Eqs. (9.6) in a more elegant form.)

The Lorentz transform relations (9.6) are evidently reduced to the Galilean transform formulas (9.2) at $v^2 \ll c^2$. As will be proved in the next section, Eqs. (9.6) also yield the Lorentz length contraction (9.5). However, all attempts to give a

[8] Through the twentieth century, Michelson–Morley-type experiments were repeated using more and more refined experimental techniques, always with the zero result for the apparent aether motion speed. For example, recent experiments, using cryogenically cooled optical resonators, have reduced the upper limit for such a speed to just 3×10^{-15} c—see [1].

[9] The zero result of a slightly later experiment, namely precise measurements of the torque which should be exerted by the moving aether on a charged capacitor, carried out in 1903 by F Trouton and H Noble (following G FitzGerald's suggestion) seconded Michelson and Morley's conclusions.

[10] The theoretical work toward this goal (which I do not have time to review in detail) included important contributions by W Voigt (in 1887), H Lorentz (1892–1904), J Larmor (1897 and 1900), and H Poincaré (1900 and 1905).

reasonable interpretation of these equations while retaining the notion of the aether have failed, in particular because of the restrictions imposed by the results of earlier experiments carried out in 1851 and 1853 by H Fizeau—which were repeated with higher accuracy by the same Michelson and Morley in 1886. These experiments have shown that if one sticks to the aether concept, this hypothetical medium should be partially 'dragged' by any moving dielectric material with a speed proportional to $(\kappa - 1)$. Such a local drag is irreconcilable with the assumed continuity of the aether.

In his famous 1905 paper Albert Einstein suggested a bold resolution of this contradiction, essentially removing the concept of the aether altogether. Moreover, he argued that the Lorentz transform is a general property of time and space, rather than of the electromagnetic field alone. He has started with two postulates, the first one essentially repeating the principle of relativity, formulated earlier (1904) by H Poincaré in the following form:

> ...the laws of physical phenomena should be the same, whether for an observer fixed, or for an observer carried along in a uniform movement of translation; so that we have not and could not have any means of discerning whether or not we are carried along in such a motion.[11]

The second Einstein's postulate was that the speed of light c, in free space, should be constant in all reference frames. (This is essentially a denial of the aether's existence.)

Then, Einstein showed how naturally the Lorenz transform relations (9.6) follow from his postulates, with a few (very natural) additional assumptions. Let a point source emit a short flash of light, at the moment $t = t' = 0$ when the origins of the reference frames shown in figure 9.1 coincide. Then, according to the second of Einstein's postulates, in each of the frames the spherical wave propagates with the same speed c, i.e. the coordinates of points of its front, measured in the two frames, have to obey equations

$$\begin{aligned}(ct)^2 - (x^2 + y^2 + z^2) = 0,\\ (ct')^2 - (x'^2 + y'^2 + z'^2) = 0.\end{aligned} \tag{9.7}$$

What might be the general relation between the combinations on the left-hand side of these equations—not for this particular wave's front, but in general? A very natural (essentially, the only justifiable) choice is

$$[(ct)^2 - (x^2 + y^2 + z^2)] = f(v^2)[(ct')^2 - (x'^2 + y'^2 + z'^2)]. \tag{9.8}$$

Now, according to the first postulate, the same relation should be valid if we swap the reference frames ($x \leftrightarrow x'$, etc) and replace v with $(-v)$. This is only possible

[11] Note that although the relativity principle excludes the notion of the special ('absolute') spatial reference frame, its verbal formulation still leaves the possibility of the Galilean 'absolute time' open. The quantitative relativity theory kills this option—see Eq. (9.6) and their discussion below.

if $f^2 = 1$, so that excluding the option $f = -1$ (which is incompatible with the Galilean transform in the limit $v/c \to 0$), we obtain

$$(ct)^2 - (x^2 + y^2 + z^2) = (ct')^2 - (x'^2 + y'^2 + z'^2). \tag{9.9}$$

For the line with $y = y' = 0$, $z = z' = 0$, Eq. (9.9) is reduced to

$$(ct)^2 - x^2 = (ct')^2 - x'^2. \tag{9.10}$$

It is very illuminating to interpret this relation as the one resulting from a mutual rotation of the reference frames (that now have to include clocks to measure time) on the plane of the coordinate x and the so-called *Euclidian time* $\tau \equiv ict$—see figure 9.3.

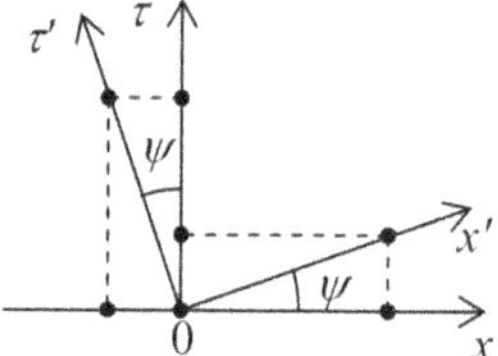

Figure 9.3. The Lorentz transform as a mutual rotation of two reference frames on the $[x, \tau]$ plane.

Indeed, rewriting Eq. (9.10) as

$$\tau^2 + x^2 = \tau'^2 + x'^2, \tag{9.11}$$

we may consider it as the invariance of the squared radius at the rotation that is shown in figure 9.3 and described by the evident geometric relations

$$\begin{aligned} x &= x' \cos\psi - \tau' \sin\psi, \\ \tau &= x' \sin\psi + \tau' \cos\psi, \end{aligned} \tag{9.12a}$$

with the reciprocal relations

$$\begin{aligned} x' &= x \cos\psi + \tau \sin\psi, \\ \tau' &= -x \sin\psi + \tau \cos\psi. \end{aligned} \tag{9.12b}$$

So far, the angle ψ has been arbitrary. In the spirit of Eq. (9.8), a natural choice is $\psi = \psi(v)$, with the requirement $\psi(0) = 0$. In order to find this function, let us write the definition of the velocity v of the reference frame 0′, as measured in the frame 0 (which was implied above): for $x' = 0$, $x = vt$. In the variables x and τ, this means

$$\left.\frac{x}{\tau}\right|_{x'=0} \equiv \left.\frac{x}{ict}\right|_{x'=0} = \frac{v}{ic}. \tag{9.13}$$

On the other hand, for the same point $x' = 0$, Eqs. (9.12*a*) yield

$$\left.\frac{x}{\tau}\right|_{x'=0} = -\tan\psi. \tag{9.14}$$

These two expressions are compatible only if

$$\tan\psi = \frac{iv}{c}, \tag{9.15}$$

so that

$$\begin{aligned} \sin\psi &\equiv \frac{\tan\psi}{(1+\tan^2\psi)^{1/2}} = \frac{iv/c}{(1-v^2/c^2)^{1/2}} \equiv i\beta\gamma, \\ \cos\psi &\equiv \frac{1}{(1+\tan^2\psi)^{1/2}} = \frac{1}{(1-v^2/c^2)^{1/2}} \equiv \gamma, \end{aligned} \tag{9.16}$$

where β and γ are two very convenient and commonly used dimensionless parameters defined as

$$\boldsymbol{\beta} \equiv \frac{\mathbf{v}}{c}, \quad \gamma \equiv \frac{1}{(1-v^2/c^2)^{1/2}} = \frac{1}{(1-\beta^2)^{1/2}}. \tag{9.17}$$

(The vector $\boldsymbol{\beta}$ is called the *normalized velocity*, while the scalar γ is the *Lorentz factor*[12].)

Using the relations for ψ, Eqs. (9.12) become

$$x = \gamma(x' - i\beta\tau'), \qquad \tau = \gamma(i\beta x' + \tau'), \tag{9.18a}$$

$$x' = \gamma(x + i\beta\tau), \qquad \tau' = \gamma(-i\beta x + \tau). \tag{9.18b}$$

Now returning to the real variables $[x, ct]$, we obtain the Lorentz transform relations (9.6), in a more compact form:

$$x = \gamma(x' + \beta\ ct'), \quad y = y', \quad z = z', \quad ct = \gamma(ct' + \beta\ x'), \tag{9.19a}$$

$$x' = \gamma(x - \beta\ ct), \quad y' = y, \quad z' = z, \quad ct' = \gamma(ct - \beta\ x). \tag{9.19b}$$

An immediate corollary of Eqs. (9.19) is that for γ to stay real, we need $v^2 \leqslant c^2$, i.e. that the speed of any physical body (to which we could connect a meaningful reference frame) cannot exceed the speed of light, as measured in *any* other meaningful reference frame[13].

9.2 Relativistic kinematic effects

In order to discuss other corollaries of Eqs. (9.19), we need to spend a few minutes discussing what these relations actually mean. Evidently, they are trying to tell us that the spatial and temporal intervals are not absolute (as they are in Newtonian space), but do depend on the reference frame they are measured in. So, we have to

[12] Note the following identities: $\gamma^2 \equiv 1/(1-\beta^2)$ and $(\gamma^2 - 1) \equiv \beta^2/(1-\beta^2) \equiv \gamma^2\beta^2$, which are frequently handy for relativity-related algebra. One more function of β, the *rapidity* $\varphi \equiv \tanh^{-1}\beta$ (so that $\psi = i\varphi$), is also useful for some calculations.

[13] All attempts to rationally conjecture particles moving with $v > c$, called *tachyons*, have failed—so far, at least. Possibly the strongest objection against their existence is the fact that tachyons could be used to communicate backwards in time, thus violating the causality principle—see, e.g. [2].

understand very clearly what exactly may be measured—and thus may be discussed in a physics theory. Recognizing this necessity, Einstein has introduced the notion of numerous imaginary *observers* that may be distributed all over each reference frame. Each observer has a clock and may use it to measure the instants of *local* events. He also conjectured, very reasonably, that:

(i) all observers within the same reference frame may agree on a common length measure ('a scale'), i.e. on their relative positions in that frame, and synchronize their clocks[14], and

(ii) the observers belonging to different reference frames may agree on the nomenclature of *world events* (e.g. short flashes of light) to which their respective measurements refer.

Actually, these additional postulates have been already implied in our 'derivation' of the Lorentz transform in section 9.1. For example, by $\{x, y, z, \text{and } t\}$ we mean the results of space and time measurements of a certain world event, about which all observers belonging to the frame 0 agree. Similarly, all observers of the frame 0′ have to agree about the results $\{x', y', z', t'\}$. Finally, when the origin of frame 0′ passes some sequential points x_k of frame 0, observers in that frame may measure its passage times t_k without fundamental error, and know that all these times belong to $x' = 0$.

Now we can analyze the major corollaries of the Lorentz transform, which are rather striking from the point of view of our everyday (rather non-relativistic) experience.

(i) *Length contraction.* Let us consider a rigid rod, stretched along axis x, with its length $l \equiv x_2 - x_1$, where $x_{1,2}$ are the coordinates of the rod's ends as measured in its rest frame 0, at any instant t (figure 9.4). What would be the rod's length l' measured by the Einstein observers in the moving frame 0′?

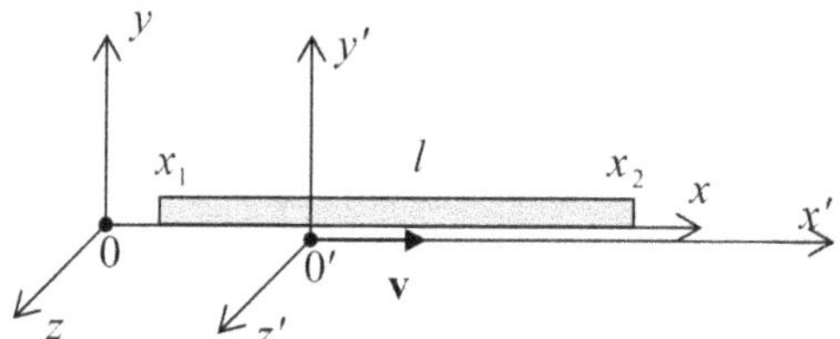

Figure 9.4. The relativistic length contraction.

At a time instant t' agreed upon in advance, the observers who find themselves exactly at the rod's ends may register that fact and then subtract their coordinates

[14] *A posteriori*, the Lorenz transform may be used to show that consensus-creating procedures (such as clock synchronization) are indeed possible. The basic idea of the proof is that at $v \ll c$, the relativistic corrections to space and time intervals are of the order of $(v/c)^2$, they have negligible effects on clocks being brought together into the same point for synchronization very slowly. The reader interested in detailed discussion of this and other fine points of special relativity is referred to, e.g. either [3] or [4].

$x'_{1,2}$ to calculate the apparent rod length $l' \equiv x'_2 - x'_1$ in the moving frame. According to Eq. (9.19*a*), l may be expressed via l' as

$$l \equiv x_2 - x_1 = \gamma(x'_2 + \beta ct') - \gamma(x'_1 + \beta ct') = \gamma(x'_2 - x'_1) \equiv \gamma\, l'. \tag{9.20a}$$

Hence, the rod's length, as measured in the *moving* reference frame is

$$l' = \frac{l}{\gamma} = l\left(1 - \frac{v^2}{c^2}\right)^{1/2} \leqslant l, \tag{9.20b}$$

in accordance with the FitzGerald–Lorentz hypothesis (9.5). This is the *relativistic length contraction* effect: an object is always the longest (has the so-called *proper length l*) if measured in its *rest frame*. Note that according to Eq. (9.19), the length contraction takes place only in the direction of the relative motion of two reference frames. As has been noted in section 9.1, this result immediately explains the zero result of the Michelson–Morley-type experiments, so that they provide very convincing evidence (if not irrefutable proof) for Eq. (9.20).

(ii) *Time dilation.* Now let us use Eq. (9.19*a*) to find the time interval Δt as measured in the frame 0 between two world events—say, two ticks of a clock moving with the frame 0′ (figure 9.5), i.e. having constant values of x', y', and z'.

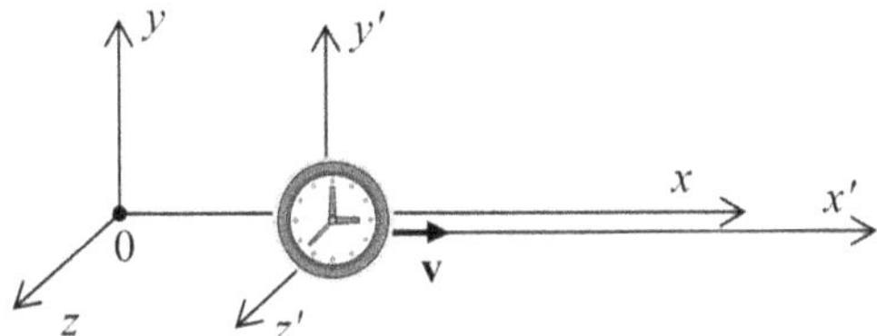

Figure 9.5. The relativistic time dilation.

Let the time interval between these two events, measured in the clock's rest frame 0′, be $\Delta t' \equiv t'_2 - t'_1$. At these two moments, the clock would fly by two certain Einstein observers at rest in the frame 0, so that they can record the corresponding moments $t_{1,2}$ shown by their clocks, and then calculate Δt as their difference. According to the second of Eqs. (9.19*a*),

$$\Delta t \equiv t_2 - t_1 = \frac{\gamma}{c}[(ct'_2 + \beta x') - (ct'_1 + \beta x')] \equiv \gamma \Delta t', \tag{9.21a}$$

so that, finally,

$$\Delta t = \gamma \Delta t' \equiv \frac{\Delta t'}{(1 - v^2/c^2)^{1/2}} \geqslant \Delta t'. \tag{9.21b}$$

This is the famous *relativistic time dilation* (or 'dilatation') effect: a time interval is *longer* if measured in a frame (in our case, frame 0) *moving relatively to the clock*, while that in the rest frame is the shortest—the so-called *proper time interval.*

This counter-intuitive effect is the everyday reality in experiments with high-energy elementary particles. For example, in a typical (and by no means record-breaking) experiment carried out in Fermilab, a beam of charged 200 GeV pions with $\gamma \approx 1400$ travelled a distance $l = 300$ m with a measured loss of only 3% of the initial beam intensity due to pion decay (mostly, into muon–neutrino pairs) with the proper lifetime $t_0 \approx 2.56 \times 10^{-8}$ s. Without time dilation, only an $\exp\{-l/ct_0\} \sim 10^{-17}$ fraction of the initial pions would survive, while the relativity-corrected number $\exp\{-l/ct\} = \exp\{-l/c\gamma t_0\} \approx 0.97$ was in a full accordance with experimental measurements.

As another example, global positioning systems (say, GPS) are designed to take into account the time dilation due to the velocity of their satellites (and also some gravity-induced, i.e. general-relativity corrections, which I do not have time to discuss) and would give large errors without such corrections. So, there is no doubt that time dilation (9.21) is a reality, although the precision of all its experimental tests I am aware of has been limited to a few percent, because of the almost unavoidable involvement of less controllable gravity effects[15].

Before the first reliable observation of time dilation (by B Rossi and D Hall in 1940), there had been serious doubts on the reality of this effect, the most famous being the *twin paradox* first posed (together with an immediate suggestion of its resolution) by P Langevin in 1911. Let us send one of two twins on a long star journey with a speed v approaching c. Upon his return to Earth, which of the twins would be older? The naïve approach is to say that due to the relativity principle, neither one would be older (and hence there is no time dilation), because each twin could claim that his counterpart, rather than himself, was moving, with the same speed, just in the opposite direction. The resolution of the paradox in the general theory of relativity (which can handle gravity and acceleration effects) is that one of the twins had to be accelerated to be brought back, and hence the reference frames have to be dissimilar: only one of them may stay inertial all the time. Because of that, the twin who had been accelerated ('actually traveling') would be younger than his or her sibling, when they finally come together.

(iii) *Velocity transformation.* Now let us calculate velocity $\mathbf{u}$ of a particle, as observed in the reference frame 0, provided that its velocity, as measured in the frame 0′, is $\mathbf{u}'$ (figure 9.6).

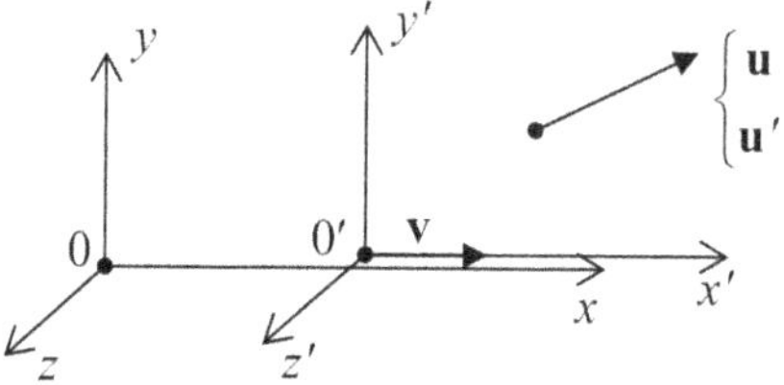

Figure 9.6. The relativistic velocity addition.

[15] See, e.g. [5].

Keeping the usual definition of velocity, but with due attention to the relativity of not only spatial but also temporal intervals, we may write

$$\mathbf{u} \equiv \frac{d\mathbf{r}}{dt}, \qquad \mathbf{u}' \equiv \frac{d\mathbf{r}'}{dt'}. \tag{9.22}$$

Plugging in the differentials of the Lorentz transform relations (9.6*a*) into these definitions, we obtain

$$\begin{aligned} u_x &\equiv \frac{dx}{dt} = \frac{dx' + vdt'}{dt' + vdx'/c^2} = \frac{u_x' + v}{1 + u_x'v/c^2}, \\ u_y &\equiv \frac{dy}{dt} = \frac{1}{\gamma}\frac{dy'}{dt' + vdx'/c^2} = \frac{1}{\gamma}\frac{u_y'}{1 + u_x'v/c^2}, \end{aligned} \tag{9.23}$$

and a similar formula for u_z. In the classical limit $v/c \to 0$, these relations are reduced to

$$u_x = u_x' + v, \qquad u_y = u_y', \qquad u_z = u_z', \tag{9.24a}$$

and may be merged into the familiar Galilean vector form

$$\mathbf{u} = \mathbf{u}' + \mathbf{v}, \qquad \text{for} \quad v \ll c. \tag{9.24b}$$

In order to see how unusual the full relativistic rules (9.23) are, let us first consider a purely longitudinal motion, $u_y = u_z = 0$; then[16]

$$u = \frac{u' + v}{1 + u'v/c^2}, \tag{9.25}$$

where $u \equiv u_x$ and $u' \equiv u_x'$. Figure 9.7 shows u as the function of u', given by this formula, for several values of the reference frames' relative velocity v. The first sanity check is that if $v = 0$, i.e. the reference frames are at rest relative to each other, then $u = u'$, as it should be—see the diagonal straight line. Next, if the magnitudes of u' and v are both below c, so is the magnitude of u. (Also good, because otherwise ordinary particles in one frame would be tachyons in the other one and the theory would be in big trouble.) Now strange things begin: even as u' and v are both approaching c, then u is also close to c, but does not exceed it. As an example, if we fired ahead a bullet with the relative speed $0.9c$ from a spaceship moving from the Earth also at $0.9c$, Eq. (9.25) predicts the speed of the bullet relative to Earth to be just $[(0.9 + 0.9)/(1 + 0.9 \times 0.9)]c \approx 0.994c < c$, rather than

[16] With an account of the well-known trigonometric identity $\tan(a + b) = (\tan a + \tan b)/(1 - \tan a \tan b)$ and Eq. (9.15), Eq. (9.25) shows that that the rapidities ψ add up exactly as the longitudinal velocities in non-relativistic motion, making that notion very convenient for the analysis of transfer between several frames.

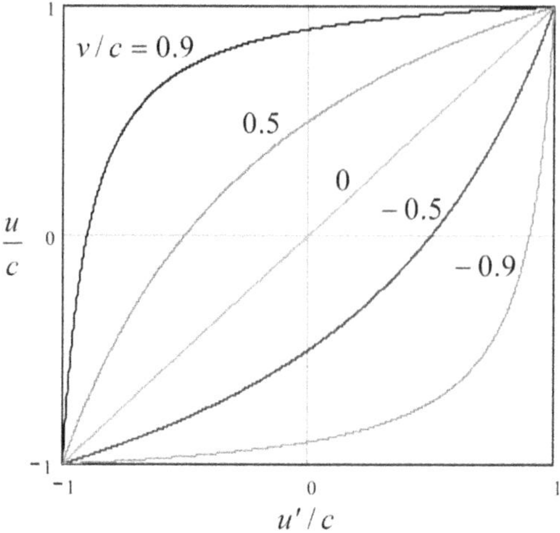

Figure 9.7. The addition of longitudinal velocities.

$(0.9 + 0.9)c = 1.8c > c$ as in the Galilean kinematics. Actually, we could expect this strangeness, because it is necessary to fulfil Einstein's second postulate: the independence of the speed of light in any reference frame. Indeed, for $u' = \pm c$, Eq. (9.25) yields $u = \pm c$, regardless of v.

In the opposite case of purely transverse motion, when a particle moves across the relative motion of the frames (for example, at our choice of coordinates, $u'_x = u'_z = 0$), Eqs. (9.23) yield a much less spectacular result

$$u_y = \frac{u'_y}{\gamma} \leqslant u'_y. \tag{9.26}$$

This effect comes purely from the time dilation, because the transverse coordinates are Lorentz-invariant.

In the case when both $u_{x'}$ and $u_{y'}$ are substantial (but $u_{z'}$ is still zero), we may divide Eqs. (9.23) by each other to relate the angles θ of particle propagation, as observed in the two reference frames:

$$\tan\theta \equiv \frac{u_y}{u_x} = \frac{u'_y}{\gamma(u'_x + v)} = \frac{\sin\theta'}{\gamma(\cos\theta' + v/u')}. \tag{9.27}$$

This expression describes, in particular, the so-called *stellar aberration* effect, the dependence of the observed direction θ toward a star on the speed v of the telescope's motion relative to the star—see figure 9.8. (The effect is readily observable experimentally as the *annual aberration* due to the periodic change of speed v by $2v_E \approx 60\ \text{km s}^{-1}$ because of the Earth's rotation about the Sun. Since the aberration's main part is of the first order in $v_E/c \sim 10^{-4}$, the effect is very significant and has been known since the early 1700s.)

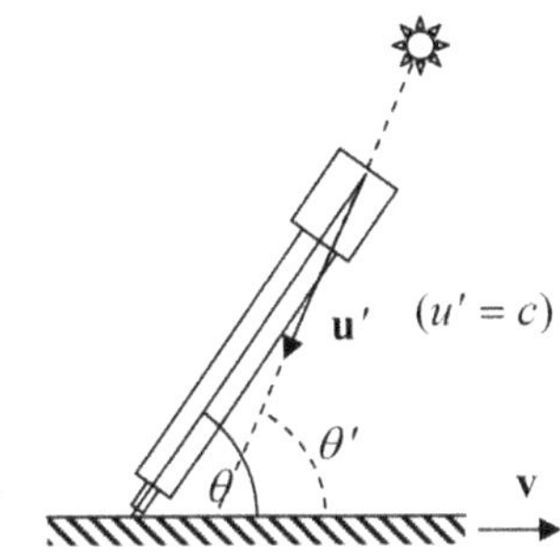

Figure 9.8. The stellar aberration.

For the analysis of this effect, it is sufficient to take, in Eq. (9.27), $u' = c$, i.e. $v/u' = \beta$, and interpret θ' as the 'proper' direction to the star that would be measured at $v = 0$.[17] At $\beta \ll 1$, both Eq. (9.27) and the Galilean result (which the reader is invited to derive directly from figure 9.8),

$$\tan\theta = \frac{\sin\theta'}{\cos\theta' + \beta}, \tag{9.28}$$

may be well approximated by the first-order term

$$\Delta\theta \equiv \theta' - \theta \approx \beta\sin\theta. \tag{9.29}$$

Unfortunately, it is not easy to use the difference between Eqs. (9.27) and (9.28), of second order in β, for the special relativity confirmation, because other components of the Earth's motion, such as its rotation, nutation, and torque-induced precession[18], provide masking first-order contributions to the aberration.

Finally, at a completely arbitrary direction of vector $\mathbf{u}'$, Eqs. (9.23) may be readily used to calculate the velocity magnitude. The most popular form of the resulting expression is for the square of the relative velocity (or rather the relative reduced velocity $\boldsymbol{\beta}$) of two particles,

$$\beta^2 = \frac{(\boldsymbol{\beta}_1 - \boldsymbol{\beta}_2)^2 - |\boldsymbol{\beta}_1 \cdot \boldsymbol{\beta}_2|}{(1 - \boldsymbol{\beta}_1 \cdot \boldsymbol{\beta}_2)^2} \leqslant 1. \tag{9.30}$$

where $\boldsymbol{\beta}_{1,2} \equiv \mathbf{v}_{1,2}/c$ are their normalized velocities as measured in the same reference frame.

[17] Strictly speaking, in order to reconcile the geometries shown in figure 9.1 (for which all our formulas, including Eq. (9.27), are valid) and figure 9.8 (giving the traditional scheme of the stellar aberration), it is necessary to invert the signs of $\mathbf{u}$ (and hence $\sin\theta'$ and $\cos\theta'$) and $\mathbf{v}$, but as is evident from Eq. (9.27), all the minus signs cancel and the formula is valid as it is.

[18] See, e.g. *Part CM* sections 4.4 and 4.5.

(iv) *The Doppler effect.* Now let us consider a plane, monochromatic wave moving along axis x:

$$f = \mathrm{Re}\,[f_\omega \exp\{i(kx - \omega t)\}] \equiv |f_\omega| \cos(kx - \omega t + \arg f_\omega). \tag{9.31}$$

Its total phase $\Psi \equiv kx - \omega t + \arg f_\omega$ (in contrast to the real amplitude $|f_\omega|$!) cannot depend on the observer's reference frame, because all fields of a traveling wave vanish simultaneously at $\Psi = 2\pi n$ (for all integer n), and such 'world events' should take place in all reference frames. The only way to keep $\Psi = \Psi'$ at all times is to have[19]

$$kx - \omega t = k'x' - \omega' t'. \tag{9.32}$$

First, let us use this general relation to consider the Doppler effect in the usual non-relativistic waves, e.g. oscillations of particles of a certain medium. Using the Galilean transform (9.2), we may rewrite Eq. (9.32) as

$$k(x' + vt) - \omega t = k'x' - \omega' t. \tag{9.33}$$

Since this transform leaves all space intervals (including wavelength $\lambda = 2\pi/k$) intact, we can take $k = k'$ so that Eq. (9.33) yields

$$\omega' = \omega - kv. \tag{9.34}$$

For a dispersion-free medium, the wave number k is the ratio of its frequency ω, as measured in the reference frame bound to the medium, and the wave velocity v_W. In particular, if the wave source rests in the medium, we may bind the reference frame 0 to the medium as well, and the frame 0′ to wave's receiver (so that $v = v_\mathrm{r}$), so that

$$k = \frac{\omega}{v_\mathrm{W}}, \tag{9.35}$$

and for the frequency perceived by the receiver, Eq. (9.34) yields

$$\omega' = \omega \frac{v_\mathrm{W} - v_\mathrm{r}}{v_\mathrm{W}}. \tag{9.36}$$

On the other hand, if the receiver and the medium are at rest in the reference frame 0′, while the wave source is bound to the frame 0 (so that $v = -v_\mathrm{S}$), Eq. (9.35) should be replaced with

$$k = k' = \frac{\omega'}{v_\mathrm{W}}, \tag{9.37}$$

and Eq. (9.34) yields a different result:

[19] Strictly speaking, Eq. (9.32) is valid to an additive constant, but for simplicity of notation it may always be made equal to zero by selecting (as has already been done in all relations of section 9.1) the reference frame origins and/or clock turn-on times so that at $t = 0$ and $x = 0$, $t' = 0$, and $x' = 0$ as well.

$$\omega' = \omega \frac{v_\text{w}}{v_\text{w} - v_\text{s}}. \tag{9.38}$$

Finally, if both the source and detector are moving, it is straightforward to combine these two results to obtain the general relation

$$\omega' = \omega \frac{v_\text{w} - v_\text{r}}{v_\text{w} - v_\text{s}}. \tag{9.39}$$

At low speeds of both the source and the receiver this result simplifies,

$$\omega' \approx \omega(1 - \beta), \quad \beta \equiv \frac{v_\text{r} - v_\text{s}}{v_\text{w}}, \tag{9.40}$$

but at speeds comparable to v_w we have to use the more general Eq. (9.39). Thus, the usual Doppler effect is generally affected not only by the relative speed $(v_\text{r} - v_\text{s})$ of the wave's source and detector, but also of their speeds relative to the medium in which the waves propagate.

Somewhat counter-intuitively, for electromagnetic waves the calculations are simpler, because for them the propagation medium (aether) does not exist, the wave velocity equals $\pm c$ in any reference frame, and there are not two separate cases: we can always take $k = \pm\omega/c$ and $k' = \pm\omega'/c$. Plugging these relations, together with the Lorentz transform (9.19*a*), into the phase-invariance condition (9.32), we obtain

$$\pm\frac{\omega}{c}\gamma(x' + \beta\, ct') - \omega\gamma\frac{ct' + \beta\, x'}{c} = \pm\frac{\omega'}{c}x' - \omega' t'. \tag{9.41}$$

This relation has to hold for any x' and t', so we may require that the net coefficients before these variables vanish. These two requirements yield the same equality:

$$\omega' = \omega\gamma(1 \mp \beta). \tag{9.42}$$

This result is already quite simple, but may be transformed further to be even more illuminating:

$$\omega' = \omega\frac{1 \mp \beta}{(1 - \beta^2)^{1/2}} \equiv \omega\left[\frac{(1 \mp \beta)(1 \mp \beta)}{(1 + \beta)(1 - \beta)}\right]^{1/2}. \tag{9.43}$$

At any sign before β, one pair of parentheses cancel, so that

$$\omega' = \omega\left(\frac{1 \mp \beta}{1 \pm \beta}\right)^{1/2}. \tag{9.44}$$

(It may look like the reciprocal expression of ω via ω' is different, violating the relativity principle. However, in this case we have to change the sign of β, because the relative velocity of the system is opposite, so we return to Eq. (9.44) again.)

Thus the Doppler effect for electromagnetic waves depends only on the relative velocity $v = \beta c$ between the wave source and detector—as it should be, given the aether's absence. At velocities much below c, Eq. (9.43) may be approximated as

$$\omega' \approx \omega \frac{1 \mp \beta/2}{1 \pm \beta/2} \approx \omega\,(1 \mp \beta), \tag{9.45}$$

i.e. in the first approximation in $\beta \equiv v/c$ it coincides with the corresponding limit (9.40) of the usual Doppler effect. However, even at $v \ll c$ there is still a difference, of the order of $(v/c)^2$, between the Galilean and Lorentzian relations.

If the wave vector $\mathbf{k}$ is tilted by angle θ to the vector $\mathbf{v}$ (as measured in frame 0), then we have to repeat the calculations, with k replaced by k_x, and components k_y and k_z left intact at the Lorentz transform. As a result, Eq. (9.42) is generalized as

$$\omega' = \omega\gamma(1 - \beta\cos\theta). \tag{9.46}$$

For the cases $\cos\theta = \pm 1$, Eq. (9.46) reduces to our previous result (9.42). However, at $\theta = \pi/2$ (i.e. $\cos\theta = 0$), the relation is rather different:

$$\omega' = \gamma\omega = \frac{\omega}{(1 - \beta^2)^{1/2}}. \tag{9.47}$$

This is the *transverse Doppler effect*—which is completely absent in non-relativistic physics. Its first experimental evidence was obtained using electron beams (as suggested in 1906 by J Stark) by H Ives and G Stilwell in 1938 and 1941. Later, similar experiments were repeated several times, but the first unambiguous measurements were only performed in 1979 by D Hasselkamp *et al* who confirmed Eq. (9.47) with a relative accuracy of about 10%. This precision may not look too spectacular, but in addition to the special tests discussed above, the Lorentz transform formulas have been also confirmed, less directly, by a huge body of other experimental data, especially in high energy physics, in agreement with calculations incorporating the transform as a part. This is why, with all respect to the spirit of challenging authority, I should warn the reader: if you decide to challenge the theory of relativity (which is called 'theory' by tradition only), you would also need to explain all of these data[20]. Best of luck with that!

[20] The same fact, ignored by crackpots, is also valid for other favorite points of their attacks, including the expansion of the Universe and quantum mechanics in physics, and the theory of evolution in biology.

9.3 Four-vectors, momentum, mass, and energy

Before proceeding to relativistic dynamics, let us discuss a mathematical formalism which makes all the calculations more compact—and more beautiful. We have already seen that the three spatial coordinates $\{x, y, z\}$ and the product ct are Lorentz-transformed similarly—see Eqs. (9.19). So it is natural to consider them as components of a single four-component vector (or, for short, *four-vector*),

$$\{x_0, x_1, x_2, x_3\} \equiv \{ct, \mathbf{r}\}, \tag{9.48}$$

with components

$$x_0 \equiv ct, \quad x_1 \equiv x, \quad x_2 \equiv y, \quad x_3 \equiv z. \tag{9.49}$$

According to Eqs. (9.19), its components are Lorentz-transformed as

$$x_j = \sum_{j'=0}^{3} L_{jj'} x'_{j'}, \tag{9.50}$$

where $L_{jj'}$ are the elements of the following 4×4 *Lorentz transform matrix*

$$\begin{pmatrix} \gamma & \beta\gamma & 0 & 0 \\ \beta\gamma & \gamma & 0 & 0 \\ 0 & 0 & 1 & 0 \\ 0 & 0 & 0 & 1 \end{pmatrix}. \tag{9.51}$$

Since four-vectors are a new notion for our course, and are used for many more aims than the just the space–time transform, we need to discuss the mathematical rules they obey. Indeed, as was mentioned in section 8.9, the usual (three-component) vector is not just any ordered set (*string*) of three scalars $\{A_x, A_y, A_z\}$; if we want it to represent a reference-frame-independent physical reality, the vector's components have to obey certain rules at transfer from one reference frame to another. In particular, the vector's *norm* (its magnitude squared),

$$A^2 = A_x^2 + A_y^2 + A_z^2, \tag{9.52}$$

should be an invariant at the Galilean transform (9.2). However, a naïve extension of this formula to four-vectors would not work because, according to the calculations of section 9.1, the Lorentz transform keeps intact combinations of the type (9.7), with one sign negative, rather than the sum of all components squared. Hence for the four-vector all the rules of the game have to be reviewed and adjusted—or rather redefined from the very beginning.

An arbitrary four-vector is a string of four scalars,

$$\{A_0, A_1, A_2, A_3\}, \tag{9.53}$$

defined in 4D *Minkowski space*[21], whose components A_j, as measured in systems 0 and 0′, shown in figure 9.1, obey a Lorentz transform similar to Eq. (9.50):

$$A_j = \sum_{j'=0}^{3} L_{jj'} A'_{j'}. \tag{9.54}$$

As we have already seen on the example of the space–time four-vector (9.48), this means in particular that

$$A_0^2 - \sum_{j=1}^{3} A_j^2 = (A'_0)^2 - \sum_{j=1}^{3} (A'_j)^2. \tag{9.55}$$

This is the so-called *Lorentz invariance* condition of the *norm* of the four-vector. (The difference between this relation and Eq. (9.52), pertaining to the Euclidian geometry, is the reason why the Minkowski space is called *pseudo-Euclidian.*) It is also straightforward to use Eqs. (9.51) and (9.54) to check that an evident generalization of the norm, the *scalar product* of two arbitrary four-vectors,

$$A_0 B_0 - \sum_{j=1}^{3} A_j B_j, \tag{9.56}$$

is also Lorentz-invariant.

Now consider the four-vector corresponding to a small *interval* between two close world events:

$$\{dx_0, dx_1, dx_2, dx_3\} = \{c\,dt, d\mathbf{r}\}; \tag{9.57}$$

its norm,

$$(ds)^2 \equiv dx_0^2 - \sum_{j=1}^{3} dx_j^2 = c^2(dt)^2 - (dr)^2, \tag{9.58}$$

is of course also Lorentz-invariant. Since the speed of any particle (or signal) cannot be larger than c, for any pair of world events that are in a causal relation with each other, dr cannot be larger than cdt, i.e. such *time-like* interval $(ds)^2$ cannot be negative. The 4D surface separating such intervals from *space-like* intervals $(ds)^2 < 0$ is called the light cone (figure 9.9).

Now let us assume that these two close world events happen with the same particle that moves with velocity $\mathbf{u}$. Then in the frame moving with a particle ($\mathbf{v} = \mathbf{u}$), the last term on the right-hand side of Eq. (9.58) equals zero, so that

[21] After H Minkowski, who was the first to recast (in 1907) the relations of special relativity in a form in which the spatial coordinates and time (or rather ct) are treated on an equal footing.

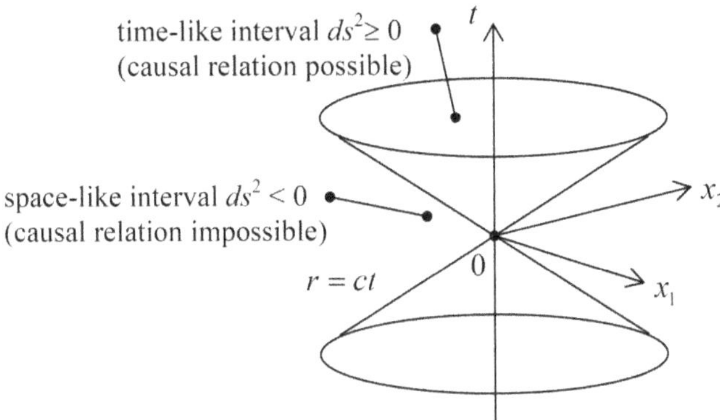

Figure 9.9. A 2+1-dimensional image of the light cone (which is actually 3+1-dimensional).

$$ds = c\, d\tau, \tag{9.59}$$

where $d\tau$ is the proper time interval. But according to Eq. (9.21) this means that we can write

$$d\tau = \frac{dt}{\gamma}, \tag{9.60}$$

where dt is the time interval in an *arbitrary* (besides being inertial) reference frame, while

$$\boldsymbol{\beta} \equiv \frac{\mathbf{u}}{c} \quad \text{and} \quad \gamma \equiv \frac{1}{(1-\beta^2)^{1/2}} = \frac{1}{(1-u^2/c^2)^{1/2}} \tag{9.61}$$

are the parameters (9.17) corresponding to the *particle's* velocity ($\mathbf{u}$) in that frame, so that $ds = c\, dt/\gamma$.[22]

Let us use Eq. (9.60) to explore whether a four-vector may be formed using the spatial components of particle's velocity

$$\mathbf{u} = \left\{\frac{dx}{dt}, \frac{dy}{dt}, \frac{dz}{dt}\right\}. \tag{9.62}$$

Here we have a slight problem: as Eqs. (9.23) show, these components do not obey the Lorentz transform (9.54). However, let us use $d\tau \equiv dt/\gamma$, the proper time interval of the particle, to form the following string:

$$\left\{\frac{dx_0}{d\tau}, \frac{dx_1}{d\tau}, \frac{dx_2}{d\tau}, \frac{dx_3}{d\tau}\right\} \equiv \gamma\left\{c, \frac{dx}{dt}, \frac{dy}{dt}, \frac{dz}{dt}\right\} \equiv \gamma\{c, \mathbf{u}\}. \tag{9.63}$$

[22] I have opted against using special indices (e.g. $\boldsymbol{\beta}_u$, γ_u) to distinguish Eqs. (9.17) and (9.61) here and below, in hope that the suitable velocity (of either a reference frame or a particle) will be always clear from the context.

As follows from comparison of the first form of this expression with Eq. (9.48), since the time–space vector obeys the Lorentz transform, and τ is Lorentz-invariant, the string (9.63) is a legitimate four-vector. It is called the *four-velocity* of the particle.

Now we are well equipped to proceed to relativistic dynamics. Let us start with such basic notions as the momentum **p** and the energy $\mathscr{E}$—so far, for a free particle[23]. Perhaps the most elegant way to 'derive' (or rather guess[24]) the expressions for **p** and $\mathscr{E}$ as functions of the particle's velocity **u** is based on analytical mechanics. Due to the conservation of **v**, the trajectory of a free particle in the 4D Minkowski space is always a straight line. Hence, from the Hamilton principle[25], we may expect its action $\mathscr{S}$, between points 1 and 2, to be a linear function of the space–time interval (9.59):

$$\mathscr{S} = \alpha \int_1^2 ds \equiv \alpha c \int_1^2 d\tau \equiv \alpha c \int_{t_1}^{t_2} \frac{dt}{\gamma}, \tag{9.64}$$

where α is some constant. On the other hand, in analytical mechanics the action is defined as

$$\mathscr{S} = \int_{t_1}^{t_2} \mathscr{L}\, dt, \tag{9.65}$$

where $\mathscr{L}$ is the particle's Lagrangian function[26]. Comparing these two expressions, we obtain

$$\mathscr{L} = \frac{\alpha c}{\gamma} \equiv \alpha c \left(1 - \frac{u^2}{c^2}\right)^{1/2}. \tag{9.66}$$

In the non-relativistic limit ($u \ll c$), this function tends to

$$\mathscr{L} \approx \alpha c \left(1 - \frac{u^2}{2c^2}\right) = \alpha c - \frac{\alpha u^2}{2c}. \tag{9.67}$$

In order to correspond to Newtonian mechanics[27], the last (velocity-dependent) term should equal $mu^2/2$. From here we find $\alpha = -mc$, so that, finally,

$$\mathscr{L} = -mc^2 \left(1 - \frac{u^2}{c^2}\right)^{1/2}. \tag{9.68}$$

[23] I am sorry for using, as in section 6.3, for the particle's momentum the same traditional notation (**p**) as had been used for the electric dipole moment. However, since the latter notion will be virtually unused in the balance of the notes, this is unlikely to lead to confusion.

[24] Indeed, such a derivation uses additional assumptions, however natural (such as the Lorentz-invariance of $\mathscr{S}$), so it can hardly be considered as a real proof of the final results, so that they require experimental confirmation. Fortunately, such confirmations have been numerous—see below.

[25] See, e.g. *Part CM* section 10.3.

[26] See, e.g. *Part CM* section 2.1.

[27] See, e.g. *Part CM* Eq. (2.19b).

Now we can find the Cartesian components p_j of the particle's momentum as the generalized momenta corresponding to the components r_j (j = 1, 2, 3) of the 3D radius-vector **r**:[28]

$$p_j = \frac{\partial \mathscr{L}}{\partial \dot{r}_j} = \frac{\partial \mathscr{L}}{\partial u_j} = -mc^2 \frac{\partial}{\partial u_j}\left(1 - \frac{u_1^2 + u_2^2 + u_3^2}{c^2}\right)^{1/2} = \frac{mu_j}{(1 - u^2/c^2)^{1/2}} \tag{9.69}$$
$$= m\gamma\, u_j.$$

Thus for the 3D vector of momentum, we can write the result in the same form as in non-relativistic mechanics,

$$\mathbf{p} = m\gamma\,\mathbf{u} \equiv M\mathbf{u}, \tag{9.70}$$

where the reference-frame-dependent scalar M (called the *relativistic mass*) is defined as

$$M \equiv m\gamma = \frac{m}{(1 - u^2/c^2)^{1/2}} \geqslant m, \tag{9.71}$$

m being the non-relativistic mass of the particle. (It is also called the *rest mass*, because in the reference frame in which the particle rests, Eq. (9.71) yields $M = m$.)

Next, let us return to the analytical mechanics to calculate the particle's energy $\mathscr{E}$ (which for a free particle coincides with the Hamiltonian function $\mathscr{H}$)[29]:

$$\begin{aligned} \mathscr{E} = \mathscr{H} &= \sum_{j=1}^{3} p_j u_j - \mathscr{L} = \mathbf{p}\cdot\mathbf{u} - \mathscr{L} \\ &= \frac{mu^2}{(1 - u^2/c^2)^{1/2}} + mc^2\left(1 - \frac{u^2}{c^2}\right) \equiv \frac{mc^2}{(1 - u^2/c^2)^{1/2}}. \end{aligned} \tag{9.72}$$

Thus, we have arrived at the most famous of Einstein's formulas (and probably the most famous formula of physics as a whole),

$$\mathscr{E} = m\gamma\, c^2 = Mc^2, \tag{9.73}$$

which expresses the relation between the free particle's mass and its energy[30]. In the non-relativistic limit, it reduces to

$$\mathscr{E} = \frac{mc^2}{(1 - u^2/c^2)^{1/2}} \approx mc^2\left(1 + \frac{u^2}{2c^2}\right) = mc^2 + \frac{mu^2}{2}, \tag{9.74}$$

the first term mc^2 being called the *rest energy* of a particle.

---

[28] See, e.g. *Part CM* section 2.3, in particular Eq. (2.31).

[29] See, e.g. *Part CM* Eq. (2.32).

[30] Let me hope that the reader understands that all the layman talk about the 'mass to energy conversion' is only valid in a very limited sense of the word. While the Einstein relation (9.73) does allow the conversion of 'massive' particles (with $m \neq 0$) into particles with $m = 0$, such as photons, each of the latter particles also has a non-zero relativistic mass M, and *simultaneously* the energy $\mathscr{E}$ related to M by Eq. (9.73).

Now let us consider the following string of four scalars:

$$\left\{\frac{\mathscr{E}}{c}, p_1, p_2, p_3\right\} = \left\{\frac{\mathscr{E}}{c}, \mathbf{p}\right\}. \tag{9.75}$$

Using Eqs. (9.70) and (9.73) to represent this expression as

$$\left\{\frac{\mathscr{E}}{c}, \mathbf{p}\right\} = m\gamma\{c, \mathbf{u}\}, \tag{9.76}$$

and comparing the result with Eq. (9.63), we immediately see that, since m is a Lorentz-invariant constant, this string is a legitimate four-vector of *energy–momentum*. As a result, its norm,

$$\left(\frac{\mathscr{E}}{c}\right)^2 - p^2, \tag{9.77a}$$

is Lorentz-invariant, and in particular has to be equal to the norm in the particle-bound frame. But in that frame $p = 0$ and, according to Eq. (9.73), $\mathscr{E} = mc^2$, and so the norm is just

$$\left(\frac{\mathscr{E}}{c}\right)^2 = \left(\frac{mc^2}{c}\right)^2 \equiv (mc)^2, \tag{9.77b}$$

so that in an arbitrary frame

$$\left(\frac{\mathscr{E}}{c}\right)^2 - p^2 = (mc)^2. \tag{9.78a}$$

This very important relation[31] between the relativistic energy and momentum (valid for free particles only!) is usually represented in the form[32]

$$\mathscr{E}^2 = (mc^2)^2 + (pc)^2. \tag{9.78b}$$

According to Eq. (9.70), in the *ultra-relativistic limit* $u \to c$, p tends to infinity, while mc^2 stays constant, so that $pc \gg mc^2$. As follows from Eq. (9.78), in this limit $\mathscr{E} \approx pc$. Although the above discussion was for particles with finite m, the four-vector formalism allows us to consider particles with zero rest mass as ultra-relativistic particles for which the above energy-to-moment relation,

$$\mathscr{E} = pc, \qquad \text{for} \qquad m = 0, \tag{9.79}$$

is exact. Quantum electrodynamics[33] tells us that under certain conditions, the electromagnetic field quanta (photons) may also be considered as such *massless*

[31] Please note one more simple and useful relation following from Eqs. (9.70) and (9.73): $\mathbf{p} = (\mathscr{E}/c^2)\mathbf{u}$.

[32] It may be tempting to interpret this relation as the perpendicular-vector-like addition of the rest energy mc^2 and the 'kinetic energy' pc, but from the point of view of the total energy conservation (see below), a better definition of the kinetic energy is $T(u) \equiv \mathscr{E}(u) - \mathscr{E}(0)$.

[33] Briefly reviewed in *Part QM* chapter 9.

particles, with momentum $\mathbf{p} = \hbar\mathbf{k}$. Plugging (the modulus of) the last relation into Eq. (9.78), for the photon's energy we obtain $\mathscr{E} = pc = \hbar kc = \hbar\omega$. Please note again that according to Eq. (9.73), the relativistic mass of a photon is not equal to zero: $M = \mathscr{E}/c^2 = \hbar\omega/c^2$, so that the term 'massless particle' has a limited meaning: $m = 0$. For example, the relativistic mass of an optical phonon is of the order of 10^{-36} kg. This is not too much, but still a noticeable (approximately one-millionth) part of the rest mass m_e of an electron.

The fundamental relations (9.70) and (9.73) have been repeatedly verified in numerous particle collision experiments in which the total energy and momentum of a system of particles are conserved—at the same conditions as in non-relativistic dynamics. (For the momentum this is the absence of external forces, and for the energy this is the elasticity of particle interactions—in other words, the absence of alternative channels of energy escape.) Of course, generally, the total energy of the system is conserved, including the potential energy of particle interactions. However, at typical high-energy particle collisions, the potential energy vanishes so rapidly with the distance between them that we can use the momentum and energy conservation laws using Eq. (9.73).

As an example, let us calculate the minimum energy $\mathscr{E}_{\text{min}}$ of a proton (p_a), necessary for the well-known high-energy reaction that generates a new proton–antiproton pair, $\text{p}_a + \text{p}_b \rightarrow \text{p} + \text{p} + \text{p} + \bar{\text{p}}$, provided that before the collision the proton p_b has been at rest in the lab frame. This minimum evidently corresponds to the vanishing relative velocity of the reaction products, i.e. their motion with virtually the same velocity ($\mathbf{u}_{\text{fin}}$), as seen from the lab frame—see figure 9.10.

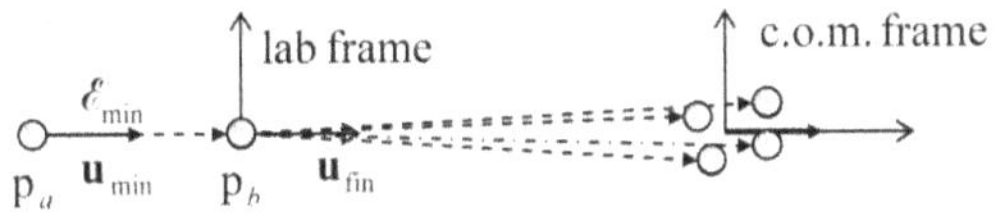

Figure 9.10. Schematic representation of a high-energy proton reaction at $\mathscr{E} \approx \mathscr{E}_{\text{min}}$.

Due to the momentum conservation, this velocity should have the same direction as the initial velocity ($\mathbf{u}_{\text{min}}$) of proton p_a. This is why two scalar equations, for the energy conservation

$$\frac{mc^2}{(1 - u_{\text{min}}^2/c^2)^{1/2}} + mc^2 = \frac{4mc^2}{(1 - u_{\text{fin}}^2/c^2)^{1/2}} \tag{9.80a}$$

and for the momentum conservation

$$\frac{mu}{(1 - u_{\text{min}}^2/c^2)^{1/2}} + 0 = \frac{4mu_{\text{fin}}}{(1 - u_{\text{fin}}^2/c^2)^{1/2}}, \tag{9.80b}$$

are sufficient to find both u_{min} and u_{fin}. After a conceptually simple but rather tedious solution of this system of two nonlinear equations, we obtain

$$u_{\text{min}} = \frac{4\sqrt{3}}{7}c, \quad u_{\text{fin}} = \frac{\sqrt{3}}{2}c. \tag{9.81}$$

Finally, we can use Eq. (9.73) to calculate the required energy; the result is $\mathscr{E}_{\min} = 7mc^2$. (Note that of the kinetic energy of the initial moving particle, $6mc^2$, only $2mc^2$ goes into the 'useful' proton–antiproton pair production.) The proton's rest mass, $m_\text{p} \approx 1.67 \times 10^{-27}$ kg, corresponds to $m_\text{p}c^2 \approx 1.502 \times 10^{-10}$ J ≈ 0.938 GeV, so that $\mathscr{E}_{\min} \approx 6.57$ GeV.

The second, more intelligent way to solve the same problem is to use the center-of-mass (c.o.m.) reference frame that, in relativity, is defined as the frame in which the total momentum of the system vanishes[34]. In this frame, at $\mathscr{E} = \mathscr{E}_{\min}$, the velocity and momenta of all reaction products are equal to zero, while the velocities of protons p_a and p_b before the collision are equal and opposite, with some magnitude u'. Hence the energy conservation law becomes

$$\frac{2mc^2}{(1 - u'^2/c^2)^{1/2}} = 4mc^2, \tag{9.82}$$

readily giving $u' = (\sqrt{3}/2)\, c$. (This is of course the same result as Eq. (9.81) gives for u_fin.) Now we can use the fact that the velocity of the proton p_b in the c.o.m. frame is $(-u')$, and hence the velocity of the proton p_a is $(+u')$. Hence we may find the lab-frame speed of the proton p_a using the velocity transform formula (9.25):

$$u_{\min} = \frac{2u'}{1 + u'^2/c^2}. \tag{9.83}$$

With the above result for u', this relation gives the same result as the first method, $u_{\min} = (4\sqrt{3}/7)c$, but in a much simpler way.

9.4 More on four-vectors and four-tensors

This is a good moment to describe a formalism that will allow us, in particular, to solve the same proton collision problem in one more (and arguably, the most elegant) way. Much more importantly, this formalism will be virtually necessary for the description of the Lorentz transform of the electromagnetic field, and its interaction with relativistic particles—otherwise the formulas would be too cumbersome. Let us call the four-vectors we have used before,

$$A^\alpha \equiv \{A_0, \mathbf{A}\}, \tag{9.84}$$

contravariant, and denote them with the top index, and introduce also *covariant* vectors,

$$A_\alpha \equiv \{A_0, -\mathbf{A}\}, \tag{9.85}$$

marked by the lower index. Now if we form a scalar product of these vectors using the *standard* (3D-like) rule, just as a sum of the products of the corresponding components, we immediately obtain

[34] Note that according to this definition, the c.o.m.'s radius-vector is $\mathbf{R} = \Sigma_k M_k \mathbf{r}_k / \Sigma_k M_k \equiv \Sigma_k \gamma_k m_k \mathbf{r}_k / \Sigma_k \gamma_k m_k$, i.e. is generally different from the well-known expression $\mathbf{R} = \Sigma_k m_k \mathbf{r}_k / \Sigma_k m_k$ of the non-relativistic mechanics.

$$A_\alpha A^\alpha \equiv A^\alpha A_\alpha \equiv A_0^2 - A^2. \tag{9.86}$$

Here and below the sign of the sum of four components of the product has been dropped[35].

The scalar product (9.86) is just the norm of the four-vector in our former definition, and as we already know, is Lorentz-invariant. Moreover, the scalar product of two different vectors (also a Lorentz invariant), may be written in any of two similar forms[36]:

$$A_0B_0 - \mathbf{A} \cdot \mathbf{B} \equiv A_\alpha B^\alpha = A^\alpha B_\alpha; \tag{9.87}$$

again, the only caveat is to take one vector in the covariant, and another in the contravariant form.

Now let us return to our sample problem (figure 9.10). Since all the components ($\mathscr{E}/c$ and $\mathbf{p}$) of the total four-momentum of our system are conserved at the collision, its norm is conserved as well:

$$(p_a + p_b)_\alpha (p_a + p_b)^\alpha = (4p)_\alpha (4p)^\alpha. \tag{9.88}$$

Since now the vector product is the usual math construct, we know that the parentheses on the left-hand side of this equation may be multiplied as usual. We may also swap the operands and move constant factors around as convenient. As a result, we obtain

$$(p_a)_\alpha (p_a)^\alpha + (p_b)_\alpha (p_b)^\alpha + 2(p_a)_\alpha (p_b)^\alpha = 16 p_\alpha p^\alpha. \tag{9.89}$$

Thanks to the Lorentz-invariance of each of the terms, we may calculate it in the reference frame we like. For the first two terms on the left-hand side, as well as for the right-hand side term, it is beneficial to use the frames in which that particular proton is at rest. As a result, according to Eq. (9.77*b*), each of the left-hand side terms equals $(mc)^2$, while the right-hand side equals $16(mc)^2$. In contrast, the last term of the left-hand side is more easily evaluated in the lab frame, because in that frame the three spatial components of the four-momentum p_b vanish, and the scalar product is the just the product of scalars $\mathscr{E}/c$ for protons a and b. For the latter proton this ratio is just mc, so that we obtain a simple equation,

$$(mc)^2 + (mc)^2 + 2\frac{\mathscr{E}_{\min}}{c} mc = 16(mc)^2, \tag{9.90}$$

immediately giving the final result $\mathscr{E}_{\min} = 7mc^2$ we had already obtained in two more complex ways.

[35] This compact notation may take some time to become accustomed to, but is very convenient (compact) and can hardly lead to any confusion, due to the following rule: the summation is implied when, and only when an index is repeated twice, one on the top and another at the bottom. In these notes, this shorthand notation will be used only for four-vectors, but not for the usual (spatial) vectors.

[36] Note also that, by definition, for any two four-vectors, $A_\alpha B^\alpha = B^\alpha A_\alpha$.

Let me hope that this example was a convincing demonstration of the convenience of representing four-vectors in the contravariant (9.84) and covariant (9.85) forms[37] with Lorentz-invariant norms (9.86). To be useful for more complex tasks, the formalism should be developed a little bit further. In particular, it is crucial to know how the four-vectors change under the Lorentz transform. For contravariant vectors, we already know the answer (9.54); let us rewrite it in the new notation:

$$A^{\alpha} = L^{\alpha}_{\beta} A'^{\beta}. \tag{9.91}$$

where L^{α}_{β} is the matrix (9.51), generally called the *mixed Lorentz tensor*[38]:

$$L^{\alpha}_{\beta} = \begin{pmatrix} \gamma & \beta\gamma & 0 & 0 \\ \beta\gamma & \gamma & 0 & 0 \\ 0 & 0 & 1 & 0 \\ 0 & 0 & 0 & 1 \end{pmatrix}, \tag{9.92}$$

Note that although the position of indices α and β in the Lorentz tensor notation is not crucial, because it is symmetric, it is convenient to place them using the general *index balance rule*: the difference of the numbers of the upper and lower indices should be the same in both parts of any four-vector/tensor equality. (Check yourself that all the formulas above do satisfy this rule.)

In order to rewrite Eq. (9.91) in a more general form that would not depend on the particular orientation of the coordinate axes (figure 9.1), let us use the contravariant and covariant forms of the four-vector of the time-space interval (9.57),

$$dx^{\alpha} = \{c\,dt,\, d\mathbf{r}\}, \quad dx_{\alpha} = \{c\,dt,\, -d\mathbf{r}\}; \tag{9.93}$$

then its norm (9.58) may be represented as[39]

$$(ds)^2 \equiv (c\,dt)^2 - (dr)^2 = dx^{\alpha}\,dx_{\alpha} = dx_{\alpha}\,dx^{\alpha}. \tag{9.94}$$

---

[37] These forms are four-vector extensions of the notions of contravariance and covariance (introduced in the 1850s by J Sylvester) for the description of the change of the usual (three-component) vectors at the transfer between different reference frames—e.g. resulting from the frame rotation. In this case, the contravariance or covariance of a vector is uniquely determined by its nature: if the Cartesian coordinates of a vector (such as the non-relativistic velocity $\mathbf{v} = d\mathbf{r}/dt$) are transformed similarly to the radius-vector $\mathbf{r}$, it is called contravariant, while the vectors (such as ∇f) that require a reciprocal transform are called covariant. In the Minkowski space, both forms may be used for any four-vector.

[38] Just as the four-vectors, the four-tensors with two top indices are called contravariant, and those with two bottom indices, covariant. The tensors with one top and one bottom index are called mixed.

[39] Another way to write this relation is $(ds)^2 = g_{\alpha\beta}\,dx^{\alpha}dx^{\beta} = g^{\alpha\beta}dx_{\alpha}\,dx_{\beta}$, where double summation over indices α and β is implied, and g is the so-called *metric tensor*,

$$g^{\alpha\beta} \equiv g_{\alpha\beta} \equiv \begin{pmatrix} 1 & 0 & 0 & 0 \\ 0 & -1 & 0 & 0 \\ 0 & 0 & -1 & 0 \\ 0 & 0 & 0 & -1 \end{pmatrix},$$

which may be used, in particular, to a transfer a covariant vector into the corresponding contravariant one and back: $A^{\alpha} = g^{\alpha\beta}A_{\beta}$, $A_{\alpha} = g_{\alpha\beta}\,A^{\beta}$. The metric tensor plays a key role in general relativity, in which it is affected by gravity—'curved' by particles' masses.

Applying Eq. (9.91) to the contravariant form of vector (9.93), we obtain

$$dx^{\alpha} = L^{\alpha}_{\beta} dx'^{\beta}. \tag{9.95}$$

But, with our new shorthand notation, we can also write the usual rule of differentiation of each component x^{α}, considering it as a (in our case, linear) function of four arguments x'^{β}, as follows[40]:

$$dx^{\alpha} = \frac{\partial x^{\alpha}}{\partial x'^{\beta}} dx'^{\beta}. \tag{9.96}$$

Comparing Eqs. (9.95) and (9.96), we can rewrite the general Lorentz transform rule (9.92) in the new form,

$$A^{\alpha} = \frac{\partial x^{\alpha}}{\partial x'^{\beta}} A'^{\beta}. \tag{9.97a}$$

which evidently does not depend on the orientation of the coordinate axes.

It is straightforward to verify that the reciprocal transform may be represented as

$$A'^{\alpha} = \frac{\partial x'^{\alpha}}{\partial x^{\beta}} A^{\beta}. \tag{9.97b}$$

However, the reciprocal transform has to differ from the direct one only by the sign of the relative velocity of the frames, so that the transform is given by the inverse matrix $\partial x'^{\alpha}/\partial x^{\beta}$; for the coordinate choice shown in figure 9.1 the matrix is

$$\frac{\partial x'^{\alpha}}{\partial x^{\beta}} = \begin{pmatrix} \gamma & -\beta\gamma & 0 & 0 \\ -\beta\gamma & \gamma & 0 & 0 \\ 0 & 0 & 1 & 0 \\ 0 & 0 & 0 & 1 \end{pmatrix}. \tag{9.98}$$

Since, according to Eqs. (9.84) and (9.85), covariant four-vectors differ from the contravariant ones by the sign of the spatial components, their direct transform is given by the matrix (9.98). Hence their direct and reciprocal transforms may be represented, respectively, as

$$A_{\alpha} = \frac{\partial x'^{\beta}}{\partial x^{\alpha}} A'_{\beta}, \quad A'_{\alpha} = \frac{\partial x^{\beta}}{\partial x'^{\alpha}} A_{\beta}, \tag{9.99}$$

evidently satisfying the index balance rule. (Note that primed quantities are now multiplied, rather than divided as in the contravariant case.) As a sanity check, let us apply this formalism to the scalar product $A_{\alpha}A^{\alpha}$. As Eq. (9.96) shows, the implicit summation notation allows us to multiply and divide any equality by the same partial differential of a coordinate, so that we can write:

$$A_{\alpha}A^{\alpha} = \frac{\partial x'^{\beta}}{\partial x^{\alpha}} \frac{\partial x^{\alpha}}{\partial x'^{\gamma}} A'_{\beta}A'^{\gamma} = \frac{\partial x'^{\beta}}{\partial x'^{\gamma}} A'_{\beta}A'^{\gamma} = \delta_{\beta\gamma}A'_{\beta}A'^{\gamma} = A'_{\gamma}A'^{\gamma}, \tag{9.100}$$

i.e. the scalar product $A_{\alpha}A^{\alpha}$ (as well as $A^{\alpha}A_{\alpha}$) is Lorentz-invariant, as it should be.

[40] Note that in the index balance rule, the top index in the denominator of a fraction is counted as a bottom index in the nominator, and vice versa.

Now, let us consider the four-vectors of derivatives. Here we should be very careful. Consider, for example, the following vector operator

$$\frac{\partial}{\partial x^\alpha} \equiv \left\{ \frac{\partial}{\partial(ct)}, \nabla \right\}. \tag{9.101}$$

As was discussed above, the operator is not changed by its multiplication and division by another differential, e.g. $\partial x'^\beta$ (with the corresponding implied summation over β), so that

$$\frac{\partial}{\partial x^\alpha} = \frac{\partial x'^\beta}{\partial x^\alpha} \frac{\partial}{\partial x'^\beta}. \tag{9.102}$$

But, according to the first of Eqs. (9.99), this is exactly how the covariant vectors are Lorentz-transformed! Hence, we have to consider the derivative over a *contravariant* space–time interval as a *covariant* four-vector, and vice versa[41]. (This result might be also expected from the index balance rule.) In particular, this means that the scalar product

$$\frac{\partial}{\partial x^\alpha} A^\alpha \equiv \frac{\partial A_0}{\partial(ct)} + \nabla \cdot \mathbf{A} \tag{9.103}$$

should be Lorentz-invariant for any legitimate four-vector. A convenient shorthand for the covariant derivative, which complies with the index balance rule, is

$$\frac{\partial}{\partial x^\alpha} \equiv \partial_\alpha, \tag{9.104}$$

so that the invariant scalar product may be written just as $\partial_\alpha A^\alpha$. A similar definition of the contravariant derivative,

$$\partial^\alpha \equiv \frac{\partial}{\partial x_\alpha} = \left\{ \frac{\partial}{\partial(ct)}, -\nabla \right\}, \tag{9.105}$$

allows us to write the Lorentz-invariant scalar product (9.103) in any of two forms:

$$\frac{\partial A_0}{\partial(ct)} + \nabla \cdot \mathbf{A} = \partial^\alpha A_\alpha = \partial_\alpha A^\alpha. \tag{9.106}$$

Finally, let us see how does the general Lorentz transform change four-tensors. A second-rank 4×4 matrix is a legitimate four-tensor if the both four-vectors it relates obey the Lorentz transform. For example, if two legitimate four-vectors are related as

$$A^\alpha = T^{\alpha\beta} B_\beta, \tag{9.107}$$

we should require that

[41] As was mentioned above, this is also a property of the 'usual' reference-frame transform of 3D vectors.

$$A'^{\alpha} = T'^{\alpha\beta}B'_{\beta}, \tag{9.108}$$

where A^{α} and A'^{α} are related by Eqs. (9.97), while B_{β} and B'_{β} are related by Eqs. (9.99). This requirement immediately yields

$$T^{\alpha\beta} = \frac{\partial x^{\alpha}}{\partial x'^{\gamma}}\frac{\partial x^{\beta}}{\partial x'^{\delta}}T'^{\gamma\delta}, \quad T'^{\alpha\beta} = \frac{\partial x'^{\alpha}}{\partial x^{\gamma}}\frac{\partial x'^{\beta}}{\partial x^{\delta}}T^{\gamma\delta}, \tag{9.109}$$

with the implied summation over two indices, γ and δ. The rules for the covariant and mixed tensors are similar[42].

9.5 Maxwell equations in the four-form

This four-vector formalism background is already sufficient to analyze the Lorentz transform of the electromagnetic field. Just to warm up, let us consider the continuity equation (4.5),

$$\frac{\partial \rho}{\partial t} + \nabla \cdot \mathbf{j} = 0, \tag{9.110}$$

which expresses the electric charge conservation, and, as we already know, is compatible with the Maxwell equations. If we now define the contravariant and covariant *four-vectors of electric current* as

$$j^{\alpha} \equiv \{\rho c, \mathbf{j}\}, \qquad j_{\alpha} \equiv \{\rho c, -\mathbf{j}\}, \tag{9.111}$$

then Eq. (9.110) may be represented in the form

$$\partial^{\alpha} j_{\alpha} = \partial_{\alpha} j^{\alpha} = 0, \tag{9.112}$$

showing that the continuity equation is *form-invariant*[43] with respect to the Lorentz transform.

Of course, such a *form's* invariance of a relation does not mean that all component *values* of the four-vectors participating in it are the same in both frames. For example, let us have some static charge density ρ in the frame 0; then Eq. (9.97*b*), applied to the contravariant form of the four-vector (9.111), reads

$$j'^{\alpha} = \frac{\partial x'^{\alpha}}{\partial x^{\beta}} j^{\beta}, \qquad \text{with } j^{\beta} = \{\rho c, 0, 0, 0\}. \tag{9.113}$$

Using the particular form (9.98) of the reciprocal Lorentz matrix for the coordinate choice shown in figure 9.1, we see that this relation yields

$$\rho' = \gamma\rho, \quad j'_x = -\gamma\beta\rho c = -\gamma v \rho, \quad j'_y = j'_z = 0. \tag{9.114}$$

---

[42] It is straightforward to check that transfer between the contravariant and covariant forms of the same tensor may be readily achieved using the metric tensor g: $T_{\alpha\beta} = g_{\alpha\gamma}T^{\gamma\delta}g_{\delta\beta}$, $T^{\alpha\beta} = g^{\alpha\gamma}T_{\gamma\delta}\, g^{\delta\beta}$.

[43] In some texts, the equations preserving their form at a transform are called 'covariant', creating a possibility for confusion with the covariant vectors and tensors. On the other hand, calling such *equations* 'invariant' would not distinguish them properly from invariant *quantities*, such as the scalar products of four-vectors.

Since the charge velocity, as observed from frame 0′, is $(-\mathbf{v})$, the non-relativistic result would be $\mathbf{j} = -\mathbf{v}\rho$. The additional γ factor in the relativistic results (for both the charge density and the current) is caused by the length contraction $dx' = dx/\gamma$, so that in order to keep the total charge $dQ = \rho\, d^3r = \rho\, dx\, dy\, dz$ inside the elementary volume $d^3r = dx\, dy\, dz$ intact, ρ (and hence j_x) should increase proportionally.

Next, at the end of chapter 6 we have seen that Maxwell equations for potentials ϕ and $\mathbf{A}$ may be represented in similar forms (6.118), under the Lorenz (again, not 'Lorentz' please!) gauge condition (6.117). For the free space, this condition takes the form

$$\nabla \cdot \mathbf{A} + \frac{1}{c^2}\frac{\partial \phi}{\partial t} = 0. \tag{9.115}$$

This expression gives us a hint as to how to form the four-vector of potentials[44]:

$$A^\alpha \equiv \left\{\frac{\phi}{c}, \mathbf{A}\right\}, \qquad A_\alpha \equiv \left\{\frac{\phi}{c}, -\mathbf{A}\right\}; \tag{9.116}$$

indeed, such vector satisfies Eq. (9.115) in its four-vector form:

$$\partial^\alpha A_\alpha = \partial_\alpha A^\alpha = 0. \tag{9.117}$$

Since this scalar product is Lorentz-invariant, and the derivatives (9.104) and (9.105) are legitimate four-vectors, this implies that the four-vector (9.116) is also legitimate, i.e. obeys the Lorentz transform formulas (9.97) and (9.99). More convincing evidence of this fact may be obtained from the Maxwell equation (6.118) for the potentials. In the free space they may be rewritten as

$$\left[\frac{\partial^2}{\partial(ct)^2} - \nabla^2\right]\frac{\phi}{c} = \frac{(\rho c)}{\varepsilon_0 c^2} \equiv \mu_0(\rho c), \qquad \left[\frac{\partial^2}{\partial(ct)^2} - \nabla^2\right]\mathbf{A} = \mu_0\mathbf{j}. \tag{9.118}$$

Using the definition (9.116), these equations may be merged into one[45]:

$$\Box A^\alpha = \mu_0 j^\alpha, \tag{9.119}$$

where $\Box$ is the *d'Alembert operator*,[46] which may be represented as either of two scalar products,

$$\Box \equiv \frac{\partial^2}{\partial(ct)^2} - \nabla^2 = \partial^\beta\partial_\beta = \partial_\beta\partial^\beta, \tag{9.120}$$

and hence is Lorentz-invariant. Because of that, and the fact that the Lorentz transform changes both four-vectors A^α and j^α in a similar way, Eq. (9.119) does not depend on the reference frame choice. Thus we have arrived at a key point of this chapter: we see that the Maxwell equations are indeed form-invariant with respect to

[44] In the Gaussian units, the scalar potential should not be divided by c in this relation.

[45] In the Gaussian units, the coefficient μ_0 in Eq. (9.119) should be replaced, as usual, with $4\pi/c$.

[46] Named after J-B d'Alembert (1717–83). (Some older textbooks use notation $\Box^2$ for this operator.)

the Lorentz transform. As a by-product, the four-vector form (9.119) of these equations (for potentials) is extremely simple—and beautiful.

However, as we have seen in chapter 7, for many applications the Maxwell equations for the field vectors are more convenient, so let us represent them in the four-form as well. For that, we may express all Cartesian components of the usual (3D) field vectors

$$\mathbf{E} = -\nabla\phi - \frac{\partial \mathbf{A}}{\partial t}, \qquad \mathbf{B} = \nabla \times \mathbf{A}, \tag{9.121}$$

via those of the potential four-vector A^α. For example,

$$E_x = -\frac{\partial \phi}{\partial x} - \frac{\partial A_x}{\partial t} = -c\left(\frac{\partial}{\partial x}\frac{\phi}{c} + \frac{\partial A_x}{\partial (ct)}\right) \equiv -c(\partial^0 A^1 - \partial^1 A^0), \tag{9.122}$$

$$B_x = \frac{\partial A_z}{\partial y} - \frac{\partial A_y}{\partial z} \equiv -(\partial^2 A^3 - \partial^3 A^2). \tag{9.123}$$

Completing similar calculations for other field components (or just generating them by the appropriate index shifts), we find that the following asymmetric, contravariant *field-strength tensor*,

$$F^{\alpha\beta} \equiv \partial^\alpha A^\beta - \partial^\beta A^\alpha, \tag{9.124}$$

may be expressed via the field components as follows[47]:

$$F^{\alpha\beta} = \begin{pmatrix} 0 & -E_x/c & -E_y/c & -E_z/c \\ E_x/c & 0 & -B_z & B_y \\ E_y/c & B_z & 0 & -B_x \\ E_z/c & -B_y & B_x & 0 \end{pmatrix}, \tag{9.125a}$$

so that the covariant form of the tensor is

$$F_{\alpha\beta} \equiv g_{\alpha\gamma} F^{\gamma\delta} g_{\delta\beta} = \begin{pmatrix} 0 & E_x/c & E_y/c & E_z/c \\ -E_x/c & 0 & -B_z & B_y \\ -E_y/c & B_z & 0 & -B_x \\ -E_z/c & -B_y & B_x & 0 \end{pmatrix}. \tag{9.125b}$$

If this expression looks a bit too bulky, please note that as a reward, the pair of *inhomogeneous* Maxwell equations, i.e. the two first equations of the system (6.99), which in free space ($\mathbf{D} = \varepsilon_0\mathbf{E}$, $\mathbf{B} = \mu_0\mathbf{H}$) may be rewritten as

[47] In Gaussian units, this formula, as well as Eq. (9.131) for $G^{\alpha\beta}$, does not have the factors c in all the denominators.

$$\nabla \cdot \frac{\mathbf{E}}{c} = \mu_0 c\rho, \qquad \nabla \times \mathbf{B} - \frac{\partial}{\partial(ct)}\frac{\mathbf{E}}{c} = \mu_0 j, \tag{9.126}$$

may now be represented in a very simple (and manifestly form-invariant) way,

$$\partial_\alpha F^{\alpha\beta} = \mu_0 j^\beta, \tag{9.127}$$

which is comparable with Eq. (9.119) in its beauty and simplicity. Somewhat counter-intuitively, the pair of *homogeneous* Maxwell equations (6.99),

$$\nabla \times \mathbf{E} + \frac{\partial \mathbf{B}}{\partial t} = 0, \qquad \nabla \cdot \mathbf{B} = 0, \tag{9.128}$$

look, in the four-vector notation, a bit more complicated[48]:

$$\partial_\alpha F_{\beta\gamma} + \partial_\beta F_{\gamma\alpha} + \partial_\gamma F_{\alpha\beta} = 0. \tag{9.129}$$

Note, however, that Eqs. (9.128) may be also represented in a much simpler form,

$$\partial_\alpha G^{\alpha\beta} = 0, \tag{9.130}$$

using the so-called *dual tensor*

$$G^{\alpha\beta} = \begin{pmatrix} 0 & B_x & B_y & B_z \\ -B_x & 0 & -E_z/c & E_y/c \\ -B_y & E_z/c & 0 & -E_x/c \\ -B_z & -E_y/c & E_x/c & 0 \end{pmatrix}, \tag{9.131}$$

which may be obtained from $F^{\alpha\beta}$, given by Eq. (9.125), with the following replacements:

$$\frac{\mathbf{E}}{c} \to -\mathbf{B}, \quad \mathbf{B} \to \frac{\mathbf{E}}{c}. \tag{9.132}$$

In addition to the proof of the form-invariance of the Maxwell equations with respect to the Lorentz transform, the four-vector formalism allows us to achieve our initial goal: find out how do the electric and magnetic field components change at the transfer between (inertial) reference frames. Let us apply to the tensor $F^{\alpha\beta}$ the reciprocal Lorentz transform described by the second of Eqs. (9.109). Generally, it gives, for each field component, a sum of 16 terms, but since (for our choice of coordinates, shown in figure 9.1) there are many zeros in the Lorentz transform matrix, and the diagonal components of $F^{\gamma\delta}$ equal zero as well, the calculations are rather doable. Let us calculate, for example, $E'_x \equiv -cF'^{01}$. The only non-vanishing terms on the right-hand side are

[48] To be fair, note that just as Eq. (9.127), Eq. (9.129) is also a set of four scalar equations—in the latter case with indices α, β, and γ taking any three *different* values of the set $\{0, 1, 2, 3\}$.

$$E'_x = -cF'^{01} = -c\left(\frac{\partial x'^0}{\partial x^1}\frac{\partial x'^1}{\partial x^0}F^{10} + \frac{\partial x'^0}{\partial x^0}\frac{\partial x'^1}{\partial x^1}F^{01}\right) \equiv -c\gamma^2(\beta^2 - 1)\frac{E_x}{c} \equiv E_x. \quad (9.133)$$

Repeating the calculation for the other five components of the fields, we obtain the very important relations

$$\begin{aligned} E'_x &= E_x, & B'_x &= B_x, \\ E'_y &= \gamma(E_y - vB_z), & B'_y &= \gamma(B_y + vE_z/c^2), \\ E'_z &= \gamma(E_z + vB_y), & B'_z &= \gamma(B_z - vE_y/c^2), \end{aligned} \quad (9.134)$$

whose more compact 'semi-vector' form is

$$\begin{aligned} \mathbf{E}'_{\|} &= \mathbf{E}, & \mathbf{B}'_{\|} &= \mathbf{B}, \\ \mathbf{E}'_{\perp} &= \gamma(\mathbf{E} + \mathbf{v} \times \mathbf{B})_{\perp}, & \mathbf{B}'_{\perp} &= \gamma(\mathbf{B} - \mathbf{v} \times \mathbf{E}/c^2)_{\perp}, \end{aligned} \quad (9.135)$$

where the indices $_{\|}$ and $_{\perp}$ stand, respectively, for the field components parallel and perpendicular to the relative velocity $\mathbf{v}$ of the two reference frames. In the non-relativistic limit, the Lorentz factor γ tends to 1, and Eqs. (9.135) acquire an even simpler form

$$\mathbf{E}' \to \mathbf{E} + \mathbf{v} \times \mathbf{B}, \quad \mathbf{B}' \to \mathbf{B} - \frac{1}{c^2}\mathbf{v} \times \mathbf{E}. \quad (9.136)$$

Thus we see that the electric and magnetic fields actually transform to each other even in the first order of the v/c ratio. For example, if we fly across the field lines of a uniform, static, purely electric field $\mathbf{E}$ (e.g. the one in a plane capacitor) we will see not only the electric field's re-normalization (in the second order of the v/c ratio), but also a non-vanishing dc magnetic field $\mathbf{B}'$ perpendicular to both the vector $\mathbf{E}$ and the vector $\mathbf{v}$, i.e. to the direction of our motion. This is of course what might be expected from the relativity principle: from the point of view of the moving observer (which is as legitimate as that of a stationary observer), the surface charges of the capacitor's plates, which create the field $\mathbf{E}$, move back creating the dc currents (9.114), which induce the magnetic field $\mathbf{B}'$. Similarly, motion across a magnetic field creates, from the point of view of the moving observer, an electric field.

This fact is very important philosophically. One can say there is no such thing in Mother Nature as an electric field (or a magnetic field) all by itself. Not only can the electric field induce the magnetic field (and vice versa) in dynamics, but even in an apparently static configuration, what exactly we measure depends on our speed relative to the field sources—hence the very appropriate term for the whole field we are studying: *electromagnetism*.

Another simple but very important application of Eqs. (9.134) and (9.135) is the calculation of the fields created by a charged particle moving in free space by inertia, i.e. along a straight line with constant velocity $\mathbf{u}$, at the *impact parameter*[49] (the closest distance) b from the observer. Selecting the frame 0′ to move with the particle

[49] This term is very popular in the theory of particle scattering—see, e.g. *Part CM* section 3.7.

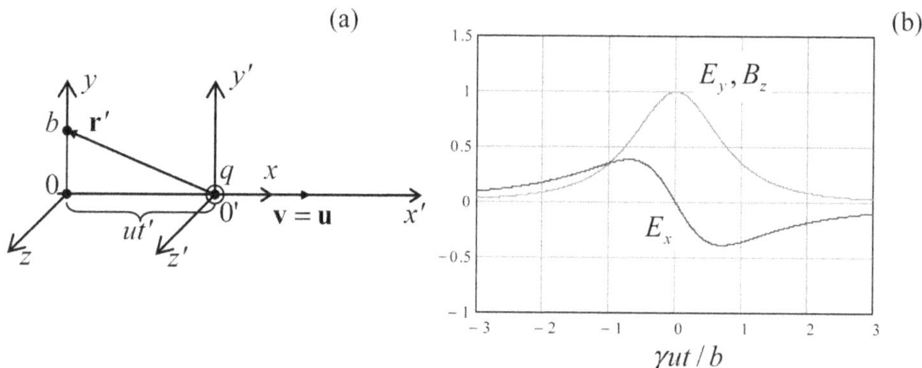

Figure 9.11. The field pulses induced by a uniformly moving charge.

in its origin, and the frame 0 to reside in the 'lab' (in that the fields **E** and **B** are measured), we can use the above formulas with $\mathbf{v} = \mathbf{u}$. In this case the fields $\mathbf{E}'$ and $\mathbf{B}'$ may be calculated from, respectively, electro- and magnetostatics,

$$\mathbf{E}' = \frac{q}{4\pi\varepsilon_0}\frac{\mathbf{r}'}{r'^3}, \qquad \mathbf{B}' = 0, \tag{9.137}$$

because in the frame $0'$ the particle does not move. Selecting the coordinate axes so that at the measurement point $x = 0$, $y = b$, $z = 0$ (figure 9.11a), for this point we may write $x' = -ut'$, $y' = b$, $z' = 0$, so that $r' = (u^2t'^2 + b^2)^{1/2}$, and the Cartesian components of the fields (9.137) are:

$$E'_x = -\frac{q}{4\pi\varepsilon_0}\frac{ut'}{(u^2t'^2 + b^2)^{3/2}}, \quad E'_y = \frac{q}{4\pi\varepsilon_0}\frac{b}{(u^2t'^2 + b^2)^{3/2}},$$
$$E'_z = 0, \ B'_x = B'_y = B'_z = 0. \tag{9.138}$$

Now using the last of Eqs. (9.19*b*), with $x = 0$, for the time transform, and the equations reciprocal to Eq. (9.134) for the field transform (it is evident that they are similar to the direct transform, with v replaced with $-v = -u$), in the lab frame we obtain

$$E_x = E'_x = -\frac{q}{4\pi\varepsilon_0}\frac{u\gamma\, t}{(u^2\gamma^2t^2 + b^2)^{3/2}},$$
$$E_y = \gamma E'_y = \frac{q}{4\pi\varepsilon_0}\frac{\gamma\, b}{(u^2\gamma^2t^2 + b^2)^{3/2}}, \qquad E_z = 0, \tag{9.139}$$

$$B_x = 0, \quad B_y = 0, \quad B_z = \frac{\gamma u}{c^2}E'_y = \frac{u}{c^2}\frac{q}{4\pi\varepsilon_0}\frac{\gamma b}{(u^2\gamma^2t^2 + b^2)^{3/2}} \equiv \frac{u}{c^2}E_y. \tag{9.140}$$

These results[50], plotted in figure 9.11b, reveal two major effects. First, the charge passage by the observer generates not only an electric field pulse, but also a magnetic field pulse. This is natural, because, as was repeatedly discussed in chapter 5, any charge motion is essentially an electric current[51]. Second, Eqs. (9.139) and (9.140) show that the pulse duration scale is

$$\Delta t = \frac{b}{\gamma u} = \frac{b}{u}\left(1 - \frac{u^2}{c^2}\right)^{1/2}, \tag{9.141}$$

i.e. shrinks to zero as the charge's velocity u approaches the speed of light. This is of course a direct corollary of the relativistic length contraction: in the frame 0′ moving with the charge, the longitudinal spread of its electric field at distance b from the motion line is of the order of $\Delta x' = b$. When observed from the lab frame 0, this interval, in accordance with (9.20), shrinks to $\Delta x = \Delta x'/\gamma = b/\gamma$, and so does the pulse duration scale $\Delta t = \Delta x/u = b/\gamma u$.

9.6 Relativistic particles in electric and magnetic fields

Now let us analyze the dynamics of charged particles in electric and magnetic fields. Inspired by 'our' success in forming the four-vector (9.75) of the energy–momentum,

$$p^\alpha = \left\{\frac{\mathscr{E}}{c}, \mathbf{p}\right\} = \gamma\{mc, \mathbf{p}\} = m\frac{dx^\alpha}{d\tau} \equiv mu^\alpha, \tag{9.142}$$

where u^α is the contravariant form of the four-velocity (9.63) of the particle,

$$u^\alpha \equiv \frac{dx^\alpha}{d\tau}, \qquad u_\alpha \equiv \frac{dx_\alpha}{d\tau}, \tag{9.143}$$

we may notice that the non-relativistic equation of motion, resulting from the Lorentz-force formula (5.10) for the three spatial components of p^α, at a charged particle's motion in an electromagnetic field,

$$\frac{d\mathbf{p}}{dt} = q(\mathbf{E} + \mathbf{u} \times \mathbf{B}), \tag{9.144}$$

is fully consistent with the following four-vector equality (which is evidently form-invariant with respect to the Lorentz transform):

$$\frac{dp^\alpha}{d\tau} = qF^{\alpha\beta}u_\beta. \tag{9.145}$$

For example, according to Eq. (9.125), the $\alpha = 1$ component of this equation reads

[50] In the next chapter, we will re-derive them in a different way.

[51] It is straightforward to use Eq. (9.140) and the linear superposition principle to calculate, for example, the magnetic field of a string of charges moving along the same line, and separated by equal distances $\Delta x = a$ (so that the average current, as measured in frame 0, is qu/a), and to show that the time-average of the magnetic field is given by the familiar Eq. (5.20) of magnetostatics, with b instead of ρ.

$$\frac{dp^1}{d\tau} = qF^{1\beta}u_\beta = q\left[\frac{E_x}{c}\gamma c + 0 \cdot (-\gamma u_x) + (-B_z)(-\gamma u_y) + B_y(-\gamma u_z)\right] \\ = q\gamma[\mathbf{E} + \mathbf{u} \times \mathbf{B}]_x, \tag{9.146}$$

and similarly for two other spatial components ($\alpha = 2$ and $\alpha = 3$). It may look like these expressions differ from the Newton's second law (9.144) by the extra factor γ. However, plugging into Eq. (9.146) the definition of the proper time interval, $d\tau = dt/\gamma$, and canceling γ in both parts, we recover Eq. (9.144) exactly—for *any* velocity of the particle! The only caveat is that if u is comparable with c, the vector $\mathbf{p}$ in Eq. (9.144) has to be understood as the relativistic momentum (9.70) proportional to the velocity-dependent mass $M = \gamma m \geqslant m$ rather than to the rest mass m.

The only remaining task is to examine the meaning of the zeroth component of Eq. (9.145). Let us spell it out:

$$\frac{dp^0}{d\tau} = qF^{0\beta}u_\beta = q\left[0 \cdot \gamma c + \left(-\frac{E_x}{c}\right)(-\gamma u_x) + \left(-\frac{E_y}{c}\right)(-\gamma u_y) + \left(-\frac{E_z}{c}\right)(-\gamma u_z)\right] \\ = q\gamma\frac{\mathbf{E}}{c} \cdot \mathbf{u}. \tag{9.147}$$

Recalling that $p^0 = \mathscr{E}/c$, and using the basic relation $d\tau = dt/\gamma$ again, we see that Eq. (9.147) looks exactly like the non-relativistic relation for the kinetic energy change[52],

$$\frac{d\mathscr{E}}{dt} = q\mathbf{E} \cdot \mathbf{u}, \tag{9.148}$$

except that in the relativistic case the energy has to be taken in the general form (9.73).

There is no question that the four-component equation (9.145) of the relativistic dynamics is beautiful in its simplicity. However, for the solution of particular problems, Eqs. (9.144) and (9.148) are frequently more convenient. As an illustration of this point, let us now use these equations to explore the relativistic effects at charged particle motion in uniform, time-independent electric and magnetic fields. In doing that, we will, for the time being, neglect the contributions into the field by the particle itself[53].

(i) *Uniform magnetic field.* Let the magnetic field be constant and uniform in the 'lab' reference frame 0. Then in this frame, Eqs. (9.144) and (9.148) yield

$$\frac{d\mathbf{p}}{dt} = q\mathbf{u} \times \mathbf{B}, \qquad \frac{d\mathscr{E}}{dt} = 0. \tag{9.149}$$

[52] See, e.g. *Part CM* Eq. (1.20) with $d\mathbf{p}/dt = \mathbf{F} = q\mathbf{E}$. (As a reminder, the magnetic field cannot affect the particle's energy, because the magnetic component of the Lorentz force is perpendicular to its velocity.)

[53] As was emphasized earlier in this course, in statics this contribution has to be ignored. In dynamics, this is generally not true; these *self-action effects* will be discussed in section 10.6.

From the second equation, $\mathscr{E}$ = const, we obtain u = const, $\beta \equiv u/c$ = const, $\gamma \equiv (1 - \beta^2)^{-1/2}$ = const, and $M \equiv \gamma m$ = const, so that the first of Eqs. (9.149) may be rewritten as

$$\frac{d\mathbf{u}}{dt} = \mathbf{u} \times \boldsymbol{\omega}_c, \tag{9.150}$$

where $\boldsymbol{\omega}_c$ is the vector directed along the magnetic field **B**, with a magnitude equal to the *cyclotron frequency* (sometimes called 'gyrofrequency')

$$\omega_c \equiv \frac{qB}{M} = \frac{qB}{\gamma m} = \frac{qc^2B}{\mathscr{E}}. \tag{9.151}$$

If the particle's initial velocity $\mathbf{u}_0$ is perpendicular to the magnetic field, Eq. (9.150) describes its circular motion with a constant speed $u = u_0$ in a plane perpendicular to **B**, with the angular velocity (9.151). In the non-relativistic limit $u \ll c$, when $\gamma \to 1$, i.e. $M \to m$, the cyclotron frequency $\omega_c = qB/m$, i.e. is independent of the speed. However, as the kinetic energy is increased to become comparable with the rest energy mc^2 of the particle, the frequency decreases, and in the ultra-relativistic limit

$$\omega_c \approx qc\frac{B}{p} \ll \frac{qB}{m}, \quad \text{at} \quad u \approx c. \tag{9.152}$$

The cyclotron motion's radius may be calculated as $R = u/\omega_c$; in the non-relativistic limit it is proportional to the particle's speed, i.e. to the square root of its kinetic energy. However, as Eq. (9.151) shows, in the general case the radius is proportional to the particle's relativistic momentum rather than its speed:

$$R = \frac{u}{\omega_c} = \frac{Mu}{qB} = \frac{m\gamma u}{qB} = \frac{1}{q}\frac{p}{B}, \tag{9.153}$$

so that in the ultra-relativistic limit, when $p \approx \mathscr{E}/c$, R is proportional to the kinetic energy.

This dependence of ω_c and R on energy are the major factors in the design of circular accelerators of charged particles. In the simplest of these machines (the *cyclotron*, invented in 1929 by E Lawrence), the frequency ω of the accelerating ac electric field is constant, so that even if it is tuned to the ω_c of the initially injected particles, the drop of the cyclotron frequency with energy eventually violates this tuning. For to this reason, the largest achievable particle speed is limited to just ~0.1c (for protons, corresponding to the kinetic energy of just ~15 MeV). This problem may be addressed in several ways. In particular, in *synchrotrons* (such as Fermilab's Tevatron and the CERN's Large Hadron Collider, LHC[54]) the magnetic field is gradually increased in time to compensate the momentum increase ($B \propto p$), so that both R (9.148) and ω_c (9.147) stay constant, enabling the proton acceleration to energies as high as ~7 TeV, i.e. ~2000 mc^2.[55]

[54] https://home.cern/topics/large-hadron-collider

[55] For more on this topic, I have to refer the interested reader to special literature, for example either [6] or [7].

Returning to our initial problem, if the particle's initial velocity has a component $u_{||}$ along the magnetic field, then it is conserved in time, so that the trajectory is a spiral around the magnetic field lines. As Eqs. (9.149) show, in this case Eq. (9.150) remains valid, but in Eqs. (9.151) and (9.153) the full speed and momentum have to be replaced with magnitudes of their (also time-conserved) components, $u_\perp$ and $p_\perp$, normal to **B**, while the Lorentz factor γ in those formulas still includes the full speed of the particle.

Finally, in the special case when particle's initial velocity is directed *exactly* along the magnetic field's direction, it continues to move in a straight line along vector **B**. In this case, the cyclotron frequency still has its non-zero value (9.151), but does not correspond to any real motion, because $R = 0$.

(ii) *Uniform electric field.* This problem is (technically) more complex than the previous one, because in the electric field, the particle's energy may change. Directing the axis z along the field **E**, from Eq. (9.144) we obtain

$$\frac{dp_z}{dt} = qE, \qquad \frac{d\mathbf{p}_\perp}{dt} = 0. \tag{9.154}$$

If the field does not change in time, the first integration of these equations is elementary,

$$p_z(t) = p_z(0) + qEt, \qquad \mathbf{p}_\perp(t) = \text{const} = \mathbf{p}_\perp(0), \tag{9.155}$$

but the further integration requires care, because the effective mass $M = \gamma m$ of the particle depends on its full speed u, with

$$u^2 = u_z^2 + u_\perp^2, \tag{9.156}$$

making the two motions, along and across the field, mutually dependent.

If the initial velocity is perpendicular to the field **E**, i.e. if $p_z(0) = 0$, $p_\perp(0) = p(0) \equiv p_0$, the easiest way to proceed is to calculate the kinetic energy first:

$$\mathscr{E}^2 = (mc^2)^2 + c^2p^2(t) = \mathscr{E}_0^2 + c^2(qEt)^2, \quad \text{where} \quad \mathscr{E}_0 \equiv [(mc^2)^2 + c^2p_0^2]^{1/2}. \tag{9.157}$$

On the other hand, we can calculate the same energy by integrating Eq. (9.148),

$$\frac{d\mathscr{E}}{dt} = q\mathbf{E}\cdot\mathbf{u} \equiv qE\frac{dz}{dt}, \tag{9.158}$$

over time, with a simple result:

$$\mathscr{E} = \mathscr{E}_0 + qEz(t), \tag{9.159}$$

where (for simplicity of notation) I took $z(0) = 0$. Requiring Eq. (9.159) to give the same $\mathscr{E}^2$ as Eq. (9.157), we obtain a quadratic equation for $z(t)$,

$$\mathscr{E}_0^2 + c^2(qEt)^2 = [\mathscr{E}_0 + qEz(t)]^2, \tag{9.160}$$

whose solution (with the sign before the square root corresponding to $E > 0$, i.e. $z \geqslant 0$) is

$$z(t) = \frac{\mathscr{E}_0}{qE}\left\{\left[1 + \left(\frac{cqEt}{\mathscr{E}_0}\right)^2\right]^{1/2} - 1\right\}. \tag{9.161}$$

Now let us find the particle's trajectory. Selecting the axis x so that the initial velocity vector (and hence the velocity vector at any further instant) is within the $[x, z]$ plane, i.e. that $y(t) \equiv 0$, we may use Eqs. (9.155) to calculate the trajectory's slope, at its arbitrary point, as

$$\frac{dz}{dx} \equiv \frac{dz/dt}{dx/dt} \equiv \frac{Mu_z}{Mu_x} \equiv \frac{p_z}{p_x} = \frac{qEt}{p_0}. \tag{9.162}$$

Now let us use Eq. (9.160) to express the nominator of this fraction, qEt, as a function of z:

$$qEt = \frac{1}{c}[(\mathscr{E}_0 + qEz)^2 - \mathscr{E}_0^2]^{1/2}. \tag{9.163}$$

Plugging this expression into Eq. (9.161) we obtain

$$\frac{dz}{dx} = \frac{1}{cp_0}[(\mathscr{E}_0 + qEz)^2 - \mathscr{E}_0^2]^{1/2}. \tag{9.164}$$

This differential equation may be readily integrated, separating the variables z and x, and using the following substitution: $\xi \equiv \cosh^{-1}(qEz/\mathscr{E}_0 + 1)$. Selecting the origin of axis x at the initial point, so that $x(0) = 0$, we finally obtain the trajectory

$$z = \frac{\mathscr{E}_0}{qE}\left(\cosh\frac{qEx}{cp_0} - 1\right). \tag{9.165}$$

This curve is usually called *catenary*, but sometimes the *'chainette'*, because it (with the proper constant replacement) describes the stationary shape of a heavy, uniform chain in a uniform gravity field, directed along axis z. At the initial part of the trajectory, where $qEx \ll cp_0(0)$, this expression may be approximated by the first non-zero term of its Taylor expansion in small x, giving a parabola,

$$z = \frac{\mathscr{E}_0 qE}{2}\left(\frac{x}{cp_0}\right)^2, \tag{9.166}$$

so that if the initial velocity of the particle is much less than c (i.e. $p_0 \approx mu_0$, $\mathscr{E}_0 \approx mc^2$), we obtain the familiar non-relativistic formula:

$$z = \frac{qE}{2mu_0^2}x^2 \equiv \frac{a}{2}t^2, \quad \text{where } a = \frac{F}{m} = \frac{qE}{m}. \tag{9.167}$$

The straightforward generalization of this solution to the case of an arbitrary direction of the particle's initial velocity is left for the reader's exercise.

(iii) *Crossed uniform magnetic and electric fields* (**E** ⊥ **B**). In the view of the somewhat bulky solution of the previous problem (i.e. the particular case of the current problem for $\mathbf{B} = 0$), one might think that the new problem should be forbiddingly complex for an analytical solution. Counter-intuitively, this is not the case, due to the help from the field transform relations (9.135). Let us consider two possible cases.

Case 1: $E/c < B$. Let us consider an inertial reference frame moving (relatively the 'lab' reference frame 0 in which fields **E** and **B** are defined) with the following velocity:

$$\mathbf{v} = \frac{\mathbf{E} \times \mathbf{B}}{B^2}, \tag{9.168}$$

whose magnitude $v = c \times (E/c)/B < c$. Selecting the coordinate axes as shown in figure 9.12, so that

$$E_x = 0, \quad E_y = E, \quad E_z = 0; \quad B_x = 0, \quad B_y = 0, \quad B_z = 0, \tag{9.169}$$

we see that the Cartesian components of this velocity are $v_x = v$, $v_y = v_z = 0$.

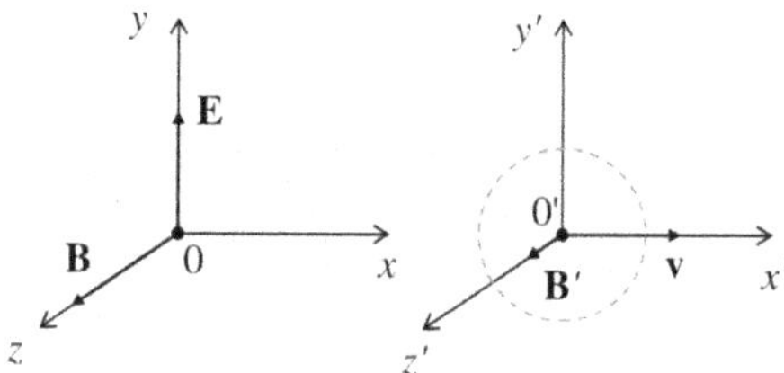

Figure 9.12. The particle's trajectory in crossed electric and magnetic fields (at $E/c < B$).

Since this choice of the coordinates complies with the one used to derive Eqs. (9.134), we can readily use that simple form of the Lorentz transform to calculate the field components in the moving reference frame:

$$E'_x = 0, \quad E'_y = \gamma(E - vB) = \gamma\left(E - \frac{E}{B}B\right) = 0, \quad E'_z = 0, \tag{9.170}$$

$$B'_x = 0, \; B'_y = 0,$$
$$B'_z = \gamma\left(B - \frac{vE}{c^2}\right) = \gamma B\left(1 - \frac{vE}{Bc^2}\right) = \gamma B\left(1 - \frac{v^2}{c^2}\right) = \frac{B}{\gamma} \leqslant B, \tag{9.171}$$

where the Lorentz parameter $\gamma \equiv (1 - v^2/c^2)^{-1/2}$ corresponds to the velocity (9.168) rather than that of the particle. These relations show that in this special reference frame the particle only sees a (re-normalized) uniform magnetic field $B' \leqslant B$, parallel to the initial field, i.e. perpendicular to the velocity (9.168). Using the result of the above example (i), we see that in this frame the particle moves along either a circle or a spiral winding about the direction of the magnetic field, with the angular speed (9.151),

$$\omega_c' = \frac{qB'}{\mathscr{E}'/c^2}, \tag{9.172}$$

and radius (9.153):

$$R' = \frac{p'_\perp}{qB'}. \tag{9.173}$$

Hence in the lab frame, the particle performs this orbital motion plus a 'drift' with the constant velocity **v** (figure 9.12). As a result, the lab-frame trajectory of the particle (or rater its projection onto the plane perpendicular to the magnetic field) is a *trochoid*-like curve[56] that, depending on the initial velocity, may be either *prolate* (self-crossing), as in figure 9.12, or *curtate* (stretched so much that it is not self-crossing).

Such looped motion of electrons (in practice, with $v \ll c$) is used, in particular, in *magnetrons*—very popular generators of microwave radiation. In such device (figure 9.13), the magnetic field, usually created by specially shaped permanent magnets, is nearly uniform (in the region of electron motion) and directed along the magnetron's axis (in figure 9.13, normal to the plane of drawing), while the electric field of magnitude $E \ll cB$, created by the dc voltage applied between the anode and cathode, is virtually radial.

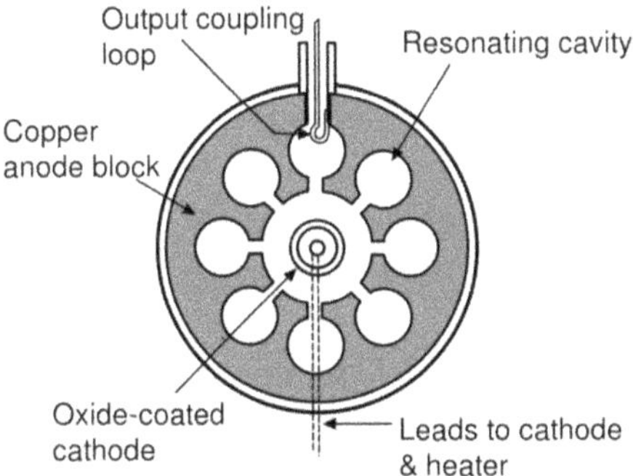

Figure 9.13. The cross-section of a typical magnetron. Figure adapted from https://en.wikipedia.org/wiki/Cavity_magnetron under the Free GNU Documentation License.

As a result, the above simple theory is only approximately valid, and the electron trajectories are close to *epicycloids* rather than trochoids. The applied electric field is adjusted so that these looped trajectories pass close to the anode's surface, and hence to the gap openings of the cylindrical microwave cavities drilled in the anode's bulk.

[56] As a reminder, a trochoid may be described as the trajectory of a point on a rigid disk rolled along a straight line. Its canonical parametric representation is $x = \Theta + a\cos\Theta$, $y = a\sin\Theta$. (For $a > 1$, the trochoid is *prolate*, if $a < 1$, it is *curtate*, and if $a = 1$, it is called the *cycloid*.) Note, however, that for our problem, the trajectory in the lab frame is exactly trochoidal only in the non-relativistic limit $v \ll c$ (i.e. $E/c \ll B$), because otherwise the Lorentz contraction in the drift direction squeezes the cyclotron orbit from a circle into an ellipse.

The fundamental mode of such a cavity is quasi-lumped, with the cylindrical walls working mostly as lumped inductances, and the gap openings as lumped capacitances, with the microwave electric field concentrated in these openings. This is why the mode is strongly coupled to the electrons passing nearby, and their interaction creates a large positive feedback (equivalent to a negative damping), which results in intensive microwave self-oscillations at the cavities' own frequency[57]. The oscillation energy, of course, is taken from the dc-field-accelerated electrons; due to the energy loss, the looped trajectory of each electron gradually moves closer to the anode and finally lands on its surface. The widespread use of such generators (in particular, in microwave ovens, which operate in a narrow frequency band around 2.45 GHz, allocated for these devices to avoid their interference with wireless communication systems) is due to their simplicity and high (up to 65%) efficiency.

Case 2: $E/c > B$. In this case, the speed given by Eq. (9.168) would be above the speed of light, so let us introduce a reference frame moving with a different velocity,

$$\mathbf{v} = \frac{\mathbf{E} \times \mathbf{B}}{(E/c)^2}, \tag{9.174}$$

whose direction is the same as before (figure 9.12), and magnitude $v = c \times B/(E/c)$ is again below c. A calculation absolutely similar to the one performed above for case 1, yields

$$\begin{aligned} E'_x = 0, \;\; E'_y &= \gamma(E - vB) \\ &= \gamma E\left(1 - \frac{vB}{E}\right) = \gamma E\left(1 - \frac{v^2}{c^2}\right) = \frac{E}{\gamma} \leqslant E, \quad E'_z = 0, \end{aligned} \tag{9.175}$$

$$B'_x = 0, \quad B'_y = 0, \quad B'_z = \gamma\left(B - \frac{vE}{c^2}\right) = \gamma\left(B - \frac{EB}{E}\right) = 0. \tag{9.176}$$

so that in the moving frame the particle sees only an electric field $E' \leqslant E$. According to the solution of our previous problem (ii), the trajectory of the particle in the moving frame is the catenary (9.165), so that in the lab frame it has an 'open', hyperbolic character as well.

To conclude this section, let me note that if the electric and magnetic fields are non-uniform, the particle motion may be much more complex, and in most cases the integration of Eqs. (9.144) and (9.148) may be carried out only numerically. However, if the field's non-uniformity is small, approximate analytical methods may be very effective. For example, if the magnetic field has a small *transverse* gradient ∇B in a direction perpendicular to the vector **B** itself, such that

$$\eta \equiv \frac{|\nabla B|}{B} \ll \frac{1}{R}, \tag{9.177}$$

[57] See, e.g. *Part CM* section 5.4.

where R is the cyclotron radius (9.153), then it is straightforward to use Eq. (9.150) to show[58] that the cyclotron orbit drifts perpendicular to both $\mathbf{B}$ and ∇B, with the drift speed

$$v_\mathrm{d} \approx \frac{\eta}{\omega_\mathrm{c}}\left(\frac{1}{2}u_\perp^2 + u_\parallel^2\right) \ll u. \tag{9.178}$$

The physics of this drift is rather simple: according to Eq. (9.153), the instant curvature of the cyclotron orbit is proportional to the local value of the field. Hence if the field is non-uniform, the trajectory bends slightly more on its parts passing through stronger field, thus acquiring a shape close to a curate trochoid.

For experimental physics and engineering practice, the effects of *longitudinal* gradients of magnetic field on the charged particle motion are much more important, but let me postpone their discussion until we have developed a few more analytical tools in the next section.

9.7 Analytical mechanics of charged particles

Eq. (9.145) gives a full description of relativistic particle dynamics in electric and magnetic fields, just as Newton's second law (9.1) does in the non-relativistic limit. However, we know that in the latter case the Lagrange formalism of analytical mechanics allows an easier solution of many problems[59]. We can fully expect that to be true in relativistic mechanics as well, so let us expand the analysis of section 9.3 to particles in the field.

Let recall that for a free particle, our main result was Eq. (9.68), which may be rewritten as

$$\gamma\mathscr{L} = -mc^2, \tag{9.179}$$

with $\gamma \equiv (1 - u^2/c^2)^{-1/2}$, showing that the product on the left-hand side is Lorentz-invariant. How can the electromagnetic field affect this relation? In non-relativistic electrostatics, we could write

$$\mathscr{L} = T - U = T - q\phi. \tag{9.180}$$

However, in relativity the scalar potential ϕ is just one component of the potential four-vector (9.116). The only way to obtain a Lorentz-invariant contribution to $\gamma\mathscr{L}$ from the full four-vector (9.116), that would also be proportional to the Lorentz force, i.e. to the first power of the particle's velocity (to account for the magnetic component of the Lorentz force), is evidently

$$\gamma\mathscr{L} = -mc^2 + \text{const} \times u^\alpha A_\alpha, \tag{9.181}$$

[58] See, e.g. section 12.4 in [8].

[59] See, e.g. *Part CM* section 2.2 and beyond.

where u^α is the four-velocity (9.63). In order to comply with Eq. (9.180) in electrostatics, the constant factor should be equal to $(-qc)$, so that Eq. (9.181) becomes

$$\gamma\mathscr{L} = -mc^2 - qu^\alpha A_\alpha \tag{9.182}$$

and, finally, we obtain an extremely important equality,

$$\mathscr{L} = -\frac{mc^2}{\gamma} - q\phi + q\mathbf{u}\cdot A, \tag{9.183}$$

whose Cartesian form is

$$\mathscr{L} = -mc^2\left(1 - \frac{u_x^2 + u_y^2 + u_z^2}{c^2}\right)^{1/2} - q\phi + q(u_xA_x + u_yA_y + u_zA_z). \tag{9.184}$$

Let us see whether this relation (which admittedly was obtained by an educated guess rather than by a strict derivation) passes a natural sanity check. For the case of unconstrained motion of a particle, we can select its three Cartesian coordinates r_j $(j = 1, 2, 3)$ as the generalized coordinates, and its linear velocity components u_j as the corresponding generalized velocities. In this case, the Lagrange equations of motion are[60]

$$\frac{d}{dt}\frac{\partial\mathscr{L}}{\partial u_j} - \frac{\partial\mathscr{L}}{\partial r_j} = 0. \tag{9.185}$$

For example, for $r_1 = x$, Eq. (9.184) yields

$$\frac{\partial\mathscr{L}}{\partial u_x} = \frac{mu_x}{(1 - u^2/c^2)^{1/2}} + qA_x \equiv p_x + qA_x, \qquad \frac{\partial\mathscr{L}}{\partial x} = -q\frac{\partial\phi}{\partial x} + q\mathbf{u}\cdot\frac{\partial\mathbf{A}}{\partial x}, \tag{9.186}$$

so that Eq. (9.185) takes the form

$$\frac{dp_x}{dt} = -q\frac{\partial\phi}{\partial x} + q\mathbf{u}\cdot\frac{\partial\mathbf{A}}{\partial x} - q\frac{dA_x}{dt}. \tag{9.187}$$

In the equations of motion, the field values have to be taken at the instant position of the particle, so that the last (full) derivative has components due to both the actual field's change (at a fixed point of space) and the particle's motion. Such addition is described by the so-called *convective derivative*[61]

$$\frac{d}{dt} = \frac{\partial}{\partial t} + \mathbf{u}\cdot\nabla. \tag{9.188}$$

[60] See, e.g. *Part CM* section 2.1.

[61] Alternatively called the 'Lagrangian derivative'; for its (rather simple) derivation see, e.g. *Part CM* section 8.3.

Spelling out both scalar products, we may group the terms remaining after cancellations as follows:

$$\frac{dp_x}{dt} = q\left[\left(-\frac{\partial \phi}{\partial x} - \frac{\partial A_x}{\partial t}\right) + u_y\left(\frac{\partial A_y}{\partial x} - \frac{\partial A_x}{\partial y}\right) - u_z\left(\frac{\partial A_x}{\partial z} - \frac{\partial A_z}{\partial x}\right)\right]. \tag{9.189}$$

But taking into account the relations (9.121) between the electric and magnetic fields and potentials, this expression is nothing more than

$$\frac{dp_x}{dt} = q(E_x + u_y B_z - u_z B_y) = q(\mathbf{E} + \mathbf{u} \times \mathbf{B})_x, \tag{9.190}$$

i.e. the x-component of Eq. (9.144). Since other Cartesian coordinates participate in (9.184) in a similar way, it is evident that the Lagrangian equations of motion along other coordinates yield other components of the same vector equation of motion.

So, Eq. (9.183) does indeed give the correct Lagrangian function, and we can use it for further analysis, in particular to discuss the first of Eqs. (9.186). This relation shows that in the electromagnetic field, the generalized momentum corresponding to particle's coordinate x is *not* $p_x = m\gamma u_x$, but[62]

$$P_x \equiv \frac{\partial \mathscr{L}}{\partial u_x} = p_x + qA_x. \tag{9.191}$$

Thus, as was already discussed (at that point, without a proof) in section 6.4, the particle's motion in a magnetic field may be described by two different momentum vectors: the *kinetic momentum* $\mathbf{p}$, defined by Eq. (9.70), and the *canonical* (or 'conjugate') *momentum*[63]

$$\mathbf{P} = \mathbf{p} + q\mathbf{A}. \tag{9.192}$$

In order to facilitate the discussion of this notion, let us generalize expression (9.72) for the Hamiltonian function $\mathscr{H}$ of a free particle to the case of a particle in the field:

$$\begin{aligned} \mathscr{H} &\equiv \mathbf{P} \cdot \mathbf{u} - \mathscr{L} = (\mathbf{p} + q\mathbf{A}) \cdot \mathbf{u} - \left(-\frac{mc^2}{\gamma} + q\mathbf{u} \cdot \mathbf{A} - q\phi\right) \\ &= \mathbf{p} \cdot \mathbf{u} + \frac{mc^2}{\gamma} + q\phi. \end{aligned} \tag{9.193}$$

Merging the first two terms exactly as it was done in Eq. (9.72), we obtain an extremely simple result,

$$\mathscr{H} = \gamma mc^2 + q\phi, \tag{9.194}$$

which may leave us wondering: where is the vector-potential $\mathbf{A}$ here—and the magnetic field effects it has to describe? The resolution of this puzzle is easy: for a practical use (e.g. for the alternative derivation of the equations of motion), $\mathscr{H}$ has to be represented as a function of the particle's generalized coordinates (in the case of unconstrained motion, these may be the Cartesian components of vector $\mathbf{r}$ that

[62] With regret, I have to use the same (common) notation as was used earlier for the electric polarization—which is not discussed below.

[63] In Gaussian units, Eq. (9.192) has the form $\mathbf{P} = \mathbf{p} + q\mathbf{A}/c$.

serves as an argument for potentials **A** and ϕ), and the generalized momenta, i.e. the Cartesian components of the vector **P** (plus, generally, time). Hence, the velocity u and the factor γ should be eliminated from (9.194). This may be done using the relation (9.192), $\gamma m\mathbf{u} = \mathbf{P} - q\mathbf{A}$. For such an elimination, it is sufficient to notice that according to Eq. (9.194), the difference $(\mathscr{H} - q\phi)$ is equal to the right-hand side of Eq. (9.72), so that the generalization of Eq. (9.78) is[64]

$$(\mathscr{H} - q\phi)^2 = (mc^2)^2 + c^2(\mathbf{P} - q\mathbf{A})^2, \quad \text{giving}$$
$$\mathscr{H} = mc^2\left[1 + \left(\frac{p}{mc}\right)^2\right]^{1/2} + q\phi. \tag{9.195}$$

It is straightforward to verify that the Hamilton equations of motion for three Cartesian coordinates of the particle, obtained in the regular way[65] from this $\mathscr{H}$, may be merged into the same vector equation (9.144). In the non-relativistic limit, performing the Taylor expansion of the latter of Eqs. (9.195) in p^2, and limiting it to two leading terms, we obtain the following generalization of Eq. (9.74):

$$\mathscr{H} \approx mc^2 + \frac{p^2}{2m} + q\phi, \quad \text{i.e.} \quad \mathscr{H} - mc^2 \approx \frac{1}{2m}(\mathbf{P} - q\mathbf{A})^2 + U, \tag{9.196}$$
with $U = q\phi$.

These expressions for $\mathscr{H}$, and Eq. (9.183) for $\mathscr{L}$, give a clear view of the electromagnetic field effect account in analytical mechanics. The electric part of the total Lorentz force $q(\mathbf{E} + \mathbf{u} \times \mathbf{B})$ can perform work on the particle, i.e. change its kinetic energy—see Eq. (9.148) and its discussion. As a result, the scalar potential ϕ, whose gradient gives a contribution into **E**, may be directly associated with the potential energy $U = q\phi$ of the particle. In contrast, the magnetic component $q\mathbf{u} \times \mathbf{B}$ of the Lorentz force is always perpendicular to the particle's velocity **u**, and cannot perform non-zero work on it, and as a result cannot be described by a contribution to U. However, if **A** did not participate in the functions $\mathscr{L}$ and/or $\mathscr{H}$ at all, the analytical mechanics would be unable to describe the effects of magnetic field $\mathbf{B} = \nabla \times \mathbf{A}$ on the particle's motion. The relations (9.183), (9.195) and (9.196) show the wonderful way in which physics (or Mother Nature herself?) solves this problem: the vector-potential gives such contributions to the functions $\mathscr{L}$ and $\mathscr{H}$ (if the latter is considered, as it should be, a function of **P** rather than **p**) that cannot be uniquely attributed to either kinetic or potential energy, but ensure the correct equation of motion (9.144) in both the Lagrange and Hamilton formalisms.

I believe I still owe the reader a discussion of the physical sense of the canonical momentum **P**. For that, let us consider a particle moving near a region of localized magnetic field $\mathbf{B}(\mathbf{r}, t)$, but not entering this region (see figure 9.14), so that on its trajectory $\mathbf{B} \equiv \nabla \times \mathbf{A} = 0$. If there is no electrostatic field (no other electric charges

[64] Alternatively, this relation may be obtained from the expression for the Lorentz-invariant norm, $p^\alpha p_\alpha = (mc)^2$, of the four-momentum (9.75), $p^\alpha = \{\mathscr{E}/c, \mathbf{p}\} = \{(\mathscr{H} - q\phi)/c, \mathbf{P} - q\mathbf{A}\}$.

[65] See, e.g. *Part CM* section 10.1.

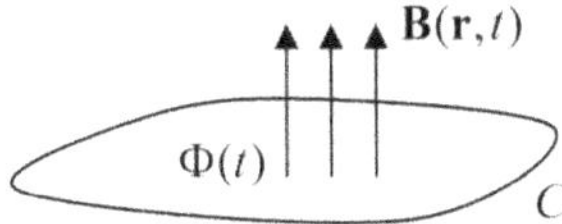

Figure 9.14. A particle's motion around a localized magnetic field with a time-dependent flux.

nearby), we may select such a local gauge that $\phi(\mathbf{r}, t) = 0$ and $\mathbf{A} = \mathbf{A}(t)$, so that Eq. (9.144) is reduced to

$$\frac{d\mathbf{p}}{dt} = q\mathbf{E} = -q\frac{d\mathbf{A}}{dt}, \tag{9.197}$$

so that Eq. (9.192) immediately gives

$$\frac{d\mathbf{P}}{dt} \equiv \frac{d\mathbf{p}}{dt} + q\frac{d\mathbf{A}}{dt} = 0. \tag{9.198}$$

Hence, even if the magnetic field is changed in time, so that the induced electric field does accelerate the particle, its conjugate momentum does not change. Hence $\mathbf{P}$ is a variable more stable to magnetic field changes than its kinetic counterpart $\mathbf{p}$. This conclusion may be criticized because it relies on a specific gauge, and generally $\mathbf{P} \equiv \mathbf{p} + q\mathbf{A}$ is not gauge-invariant, because the vector-potential $\mathbf{A}$ is not[66]. However, as was already discussed in section 5.3, the integral $\int \mathbf{A} \cdot d\mathbf{r}$ over a closed contour does not depend on the chosen gauge and equals the magnetic flux Φ through the area limited by the contour—see Eq. (5.65). So, integrating Eq. (9.197) over a closed trajectory of a particle (figure 9.14), and over the time of one orbit, we obtain

$$\Delta\oint_C \mathbf{p} \cdot d\mathbf{r} = -q\Delta\Phi, \quad \text{so that } \Delta\oint_C \mathbf{P} \cdot d\mathbf{r} = 0, \tag{9.199}$$

where $\Delta\Phi$ is the change of flux during that time. This gauge-invariant result confirms the above conclusion about the stability of the canonical momentum to magnetic field variations.

Generally, Eq. (9.199) is invalid if a particle moves inside a magnetic field and/or changes its trajectory at the field variation. However, if the field is almost uniform, i.e. its gradient small in the sense of Eq. (9.177), this result is (approximately) applicable. Indeed, analytical mechanics[67] tells us that for any canonical coordinate-momentum pair $\{q_j, p_j\}$, the corresponding *action variable*,

$$J_j \equiv \frac{1}{2\pi}\oint p_j\, dq_j, \tag{9.200}$$

remains virtually constant at slow variations of motion conditions. According to Eq. (9.191), for a particle in a magnetic field, the generalized momentum,

[66] The kinetic momentum $\mathbf{p} = M\mathbf{u}$ is just the usual $m\mathbf{u}$ product modified for relativistic effects, so that this variable is evidently gauge- (although not Lorentz-)invariant.

[67] See, e.g. *Part CM* section 10.2.

corresponding to the Cartesian coordinate r_j, is P_j rather than p_j. Thus forming the net action variable $J \equiv J_x + J_y + J_z$, we may write

$$2\pi J = \oint \mathbf{P} \cdot d\mathbf{r} = \oint \mathbf{p} \cdot d\mathbf{r} + q\Phi = \text{const.} \tag{9.201}$$

Let us apply this relation to the motion of a non-relativistic particle in an almost uniform magnetic field, with a small longitudinal velocity, $u_{||}/u_\perp \to 0$ (figure 9.15).

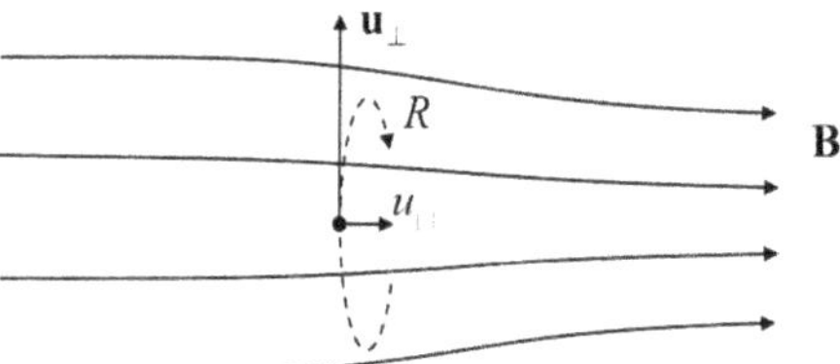

Figure 9.15. A particle in a magnetic field with a small longitudinal gradient $\nabla B || \mathbf{B}$.

In this case, Φ in Eq. (9.201) is the flux encircled by a cyclotron orbit, equal to $(-\pi R^2 B)$, where R is its radius given by Eq. (9.153) and the negative sign accounts for the fact that the 'correct' direction of the normal vector $\mathbf{n}$ in the definition of flux, $\Phi = \int \mathbf{B} \cdot \mathbf{n}\, d^2r$, is antiparallel to the vector $\mathbf{B}$. At $u \ll c$, the kinetic momentum is just $p_\perp = mu_\perp$, while Eq. (9.153) yields

$$mu_\perp = qBR. \tag{9.202}$$

Plugging these relations into Eq. (9.201), we obtain

$$2\pi J = mu_\perp 2\pi R - q\pi R^2 B = m\frac{qRB}{m}2\pi R - q\pi R^2 B \equiv (2-1)q\pi R^2 B \equiv -q\Phi. \tag{9.203}$$

This means that even if the circular orbit slowly moves in the magnetic field, the flux encircled by the cyclotron orbit should remain virtually constant. One manifestation of this effect is the result already mentioned at the end of section 9.6: if a small gradient of the magnetic field is perpendicular to the field itself, then the particle orbit's drift is perpendicular to ∇B, so that Φ stays constant. Now let us analyze the case of a small longitudinal gradient, $\nabla B || \mathbf{B}$ (figure 9.15). If the small initial longitudinal velocity $u_{||}$ is directed toward the higher field region, in order to keep Φ constant, the cyclotron orbit has to gradually shrink. Rewriting Eq. (9.202) as

$$mu_\perp = q\frac{\pi R^2 B}{\pi R} = q\frac{|\Phi|}{\pi R}, \tag{9.204}$$

we see that this reduction of R (at constant Φ) should increase the orbiting speed $u_\perp$. But since the magnetic field cannot perform any work on the particle, its kinetic energy,

$$\mathscr{E} = \frac{m}{2}\left(u_{||}^2 + u_\perp^2\right), \tag{9.205}$$

should stay constant, so that the longitudinal velocity $u_{||}$ has to decrease. Hence eventually the orbit's drift has to stop, and then the orbit has to start moving back toward the region of lower fields, being essentially repulsed from the high-field region. This effect is very important, in particular, for plasma confinement systems. In the simplest of such systems, two coaxial magnetic coils inducing magnetic fields of the same direction (figure 9.16) naturally form a 'magnetic bottle' which traps charged particles injected, with sufficiently low longitudinal velocities, into the region between the coils. More complex systems of this type, but working on the same basic principle, are the most essential components of the persisting large-scale efforts to achieve controllable nuclear fusion[68].

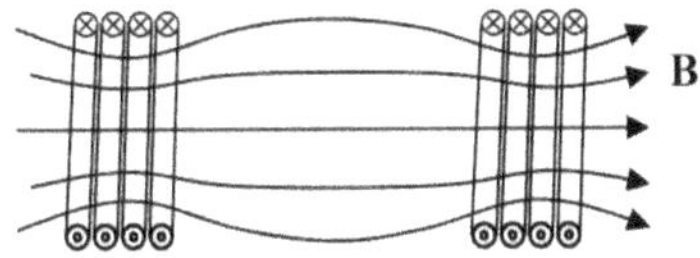

Figure 9.16. A magnetic bottle (schematically).

Returning to the constancy of the magnetic flux encircled by free particles, it reminds us of the Meissner–Ochsenfeld effect, which was discussed in section 6.4, and gives a motivation for a brief revisit of the electrodynamics of superconductivity. As was emphasized in that section, superconductivity is a substantially quantum phenomenon; nevertheless the classical notion of the conjugate momentum **P** helps us understand its theoretical description. Indeed, the general rule of quantization of physical systems[69] is that each canonical pair $\{q_j, p_j\}$ of a generalized coordinate and the corresponding momentum are described by quantum-mechanical operators that obey the following commutation relation

$$[\hat{q}_j, \hat{p}_{j'}] = i\hbar\delta_{jj'}. \tag{9.206}$$

According to Eq. (9.191), for the Cartesian coordinates r_j of a particle in the magnetic field the corresponding generalized momenta are P_j, so that their operators should obey the following commutation relations:

$$[\hat{r}_j, \hat{P}_{j'}] = i\hbar\delta_{jj'}. \tag{9.207}$$

In the coordinate representation of quantum mechanics, the canonical operators of the linear momentum are described by Cartesian components of the vector operator $-i\hbar\nabla$. As a result, ignoring the rest energy mc^2 (which gives an inconsequential phase factor $\exp\{-imc^2t/\hbar\}$ in the wave function), we can use Eq. (9.196) to rewrite the usual non-relativistic Schrödinger equation,

$$i\hbar\frac{\partial\psi}{\partial t} = \mathscr{H}\psi, \tag{9.208}$$

as follows:

---

[68] For further reading on this technology the reader is referred, for example, to a simple monograph [9], and/or a graduate-level theoretical treatment in [10].

[69] See, e.g. *Part CM* section 10.1.

$$i\hbar\frac{\partial\psi}{\partial t} = \left(\frac{\hat{p}^2}{2m} + U\right)\psi = \left[\frac{1}{2m}(-i\hbar\nabla - q\mathbf{A})^2 + q\varphi\right]\psi. \tag{9.209}$$

Thus, I believe I have finally delivered on my promise to justify the replacement (6.50), which was used in sections 6.4 and 6.5 to discuss the electrodynamics of superconductors, including the Meissner–Ochsenfeld effect[70].

9.8 Analytical mechanics of the electromagnetic field

We have just seen that the analytical mechanics of a *particle* in an electromagnetic field may be used to obtain some important results. The same is true for the analytical mechanics of the *field* as such, and the *field–particle system* as a whole. For such a space-distributed system as the field, governed by local dynamics laws (Maxwell equations), we need to apply analytical mechanics to the *local densities* ℓ and $\hbar$ of the Lagrangian and Hamiltonian functions, defined by relations

$$\mathscr{L} = \int \ell\, d^3r, \quad \mathscr{H} = \int \hbar\, d^3r. \tag{9.210}$$

Let us start, as usual, from the Lagrange formalism. Some clue toward the possible structure of the Lagrangian density ℓ may be obtained from that of the description of the particle–field interaction in this formalism, discussed in the previous section. As we have seen, for the case of a single particle, the interaction is described by the last two terms of Eq. (9.183):

$$\mathscr{L}_{\text{int}} = -q\phi - q\mathbf{u}\cdot\mathbf{A}. \tag{9.211}$$

It is virtually obvious that if charge q is continuously distributed over some volume, we may represent $\mathscr{L}$ as a volume integral of the following Lagrangian density:

$$\ell_{\text{int}} = -\rho\phi + \mathbf{j}\cdot\mathbf{A} \equiv -j_\alpha A^\alpha. \tag{9.212}$$

Notice that the density (in contrast to $\mathscr{L}_{\text{int}}$ itself) is Lorentz-invariant. (This is due to the contraction of the longitudinal coordinate, and hence volume, at the Lorentz transform.) Hence we may expect the density of the *field's* Lagrangian to be Lorentz-invariant as well. Moreover, in the view of the simple, local structure of the Maxwell equations (containing only the first spatial and temporal derivatives of the fields), ℓ_{field} should be a simple function of the potential's four-vector and its four-derivative:

$$\ell_{\text{field}} = \ell_{\text{field}}(A^\alpha, \partial_\alpha A^\beta). \tag{9.213}$$

Also, the density should be selected in such a way that the four-vector analog of the Lagrangian equations of motion,

[70] Eq. (9.209) is also the basis for discussion of numerous other magnetic field phenomena, including the AB and quantum Hall effects—see, e.g. *Part QM* sections 3.1 and 3.2.

$$\partial_\alpha \frac{\partial \ell_{\text{field}}}{\partial(\partial_\alpha A^\beta)} - \frac{\partial \ell_{\text{field}}}{\partial A^\beta} = 0, \tag{9.214}$$

gave us the correct inhomogeneous Maxwell equation (9.127)[71]. It is clear that the field part ℓ_{field} of the total Lagrangian density ℓ should be a scalar, and a quadratic form of the field strength, i.e. of $F^{\alpha\beta}$, so that the natural choice is

$$\ell_{\text{field}} = \text{const} \times F_{\alpha\beta}F^{\alpha\beta}, \tag{9.215}$$

with implied summation over both indices. Indeed, adding to this expression the interaction Lagrangian (9.212),

$$\ell = \ell_{\text{field}} + \ell_{\text{int}} = \text{const} \times F_{\alpha\beta}F^{\alpha\beta} - j_\alpha A^\alpha, \tag{9.216}$$

and performing the differentiation (9.214), we see that the relations (9.214) and (9.215) indeed yield Eqs. (9.127), provided that the constant factor equals $(-1/4\mu_0)$.[72] So, the field's Lagrangian density is

$$\ell_{\text{field}} = -\frac{1}{4\mu_0}F_{\alpha\beta}F^{\alpha\beta} = \frac{1}{2\mu_0}\left(\frac{E^2}{c^2} - B^2\right) \equiv \frac{\varepsilon_0}{2}E^2 - \frac{B^2}{2\mu_0} \equiv u_e - u_m, \tag{9.217}$$

where u_e is the electric field energy density (1.65), and u_m is the magnetic field energy density (5.57). Let me hope the reader agrees that Eq. (9.217) is a wonderful result, because the Lagrangian function has a structure absolutely similar to the well-known expression $\mathscr{L} = T - U$ of the classical mechanics. So, for the field alone, the 'potential' and 'kinetic' energies are separable again[73].

Now let us explore whether we can calculate a four-form of the field's Hamiltonian function $\mathscr{H}$. In the generic analytical mechanics,

$$\mathscr{H} = \sum_j \frac{\partial \mathscr{L}}{\partial \dot{q}_j}\dot{q}_j - \mathscr{L}. \tag{9.218}$$

However, just as for the Lagrangian function, for a field we should find the spatial density $\hbar$ of the Hamiltonian, defined by the second of Eqs. (9.210), for which a natural four-form of Eq. (9.218) is

$$\hbar^{\alpha\beta} = \frac{\partial \ell}{\partial(\partial_\alpha A^\gamma)}\partial^\beta A^\gamma - g^{\alpha\beta}\ell. \tag{9.219}$$

Calculated for the field alone, i.e. using Eq. (9.217) for ℓ_{field}, this definition yields

[71] Here the implicit summation over index α plays a role similar to the convective derivative (9.188) in replacing the full derivative over time, in a way that reflects the symmetry of time and space in special relativity. I do not want to spend more time justifying Eq. (9.214) for reasons that will be clear imminently.

[72] In the Gaussian units, the coefficient is $(-1/16\pi)$.

[73] Since the Lagrange equations of motion are homogeneous, the simultaneous change of the signs of T and U does not change them. Thus, it is not important which of the two energy densities, u_e or u_m, we count as the potential, and which as the kinetic energy. (Actually, such duality of the two field energy components is typical for all analytical mechanics, and was discussed already in *Part CM* section 2.2.)

$$\mathscr{h}^{\alpha\beta}_{\text{field}} = \theta^{\alpha\beta} - \tau^{\alpha\beta}_D, \tag{9.220}$$

where the tensor

$$\theta^{\alpha\beta} \equiv \frac{1}{\mu_0}\left(g^{\alpha\gamma}F_{\gamma\delta}F^{\delta\beta} + \frac{1}{4}g^{\alpha\beta}F_{\gamma\delta}F^{\gamma\delta}\right) \tag{9.221}$$

is gauge-invariant, while the remaining term,

$$\tau^{\alpha\beta}_D \equiv \frac{1}{\mu_0}g^{\alpha\gamma}F_{\gamma\delta}\partial^{\delta}A^{\beta}, \tag{9.222}$$

is not, so that it cannot correspond to any measurable variables. Fortunately, it is straightforward to verify that the last tensor may be represented in the form

$$\tau^{\alpha\beta}_D = \frac{1}{\mu_0}\partial_{\gamma}(F^{\gamma\alpha}A^{\beta}) \tag{9.223}$$

and as a result obeys the following relations,

$$\partial_{\alpha}\tau^{\alpha\beta}_D = 0, \qquad \int \tau^{0\beta}_D\, d^3r = 0, \tag{9.224}$$

so it does not interfere with the conservation properties of the gauge-invariant, symmetric *energy–momentum tensor* (also called the *symmetric stress tensor*) $\theta^{\alpha\beta}$, to be discussed below.

Let us use Eq. (9.125) to express the components of the latter tensor for the electric and magnetic fields. For $\alpha = \beta = 0$, we obtain

$$\theta^{00} = \frac{\varepsilon_0}{2}E^2 + \frac{B^2}{2\mu_0} = u_{\mathrm{e}} + u_{\mathrm{m}} \equiv u, \tag{9.225}$$

i.e. the expression for the total energy density u—see Eq. (6.113). The other three components of the same row/column turn out to be just the Cartesian components of the Poynting vector (6.114), divided by c:

$$\theta^{j0} = \frac{1}{\mu_0}\left(\frac{\mathbf{E}}{c} \times \mathbf{B}\right)_j = \left(\frac{\mathbf{E}}{c} \times \mathbf{H}\right)_j \equiv \frac{S_j}{c}, \quad \text{for } j = 1,\ 2,\ 3. \tag{9.226}$$

The remaining nine components $\theta_{jj'}$ of the tensor, with $j' = 1, 2, 3$, are usually represented as

$$\theta^{jj'} = -\tau^{(M)}_{jj'}, \tag{9.227}$$

where $\tau^{(M)}$ is the so-called *Maxwell stress tensor*:

$$\tau_{jj'}^{(M)} = \varepsilon_0\left(E_jE_{j'} - \frac{\delta_{jj'}}{2}E^2\right) + \frac{1}{\mu_0}\left(B_jB_{j'} - \frac{\delta_{jj'}}{2}B^2\right), \tag{9.228}$$

so that the whole symmetric energy–momentum tensor may be conveniently represented in the following symbolic way:

$$\theta^{\alpha\beta} = \begin{pmatrix} u \leftarrow & \mathbf{S}/c & \rightarrow \\ \uparrow & & \\ \dfrac{\mathbf{S}}{c} & & -\tau_{jj'}^{(M)} \\ \downarrow & & \end{pmatrix}. \tag{9.229}$$

The physical meaning of this tensor may be revealed in the following way. Considering Eq. (9.221) just as the *definition* of tensor $\theta^{\alpha\beta}$,[74] and using the four-vector form of the Maxwell equations given by Eqs. (9.127) and (9.129), it is straightforward to verify an extremely simple result for the four-derivative of the symmetric tensor:

$$\partial_\alpha \theta^{\alpha\beta} = -F^{\beta\gamma}j_\gamma. \tag{9.230}$$

This expression is valid in the presence of electromagnetic field sources, e.g. for any system of charged particles and the fields they have created. Of these four equations (for four values of the index β), the temporal one (with $\beta = 0$) may be simply expressed via the energy density (9.225) and the Poynting vector (9.226),

$$\frac{\partial u}{\partial t} + \nabla \cdot \mathbf{S} = -\mathbf{j} \cdot \mathbf{E}, \tag{9.231}$$

while three spatial equations (with $\beta = j = 1, 2, 3$) may be represented in the form

$$\frac{\partial}{\partial t}\frac{S_j}{c^2} - \sum_{j'=1}^{3}\frac{\partial}{\partial r_{j'}}\tau_{jj'}^{(M)} = -(\rho\mathbf{E} + \mathbf{j} \times \mathbf{B})_j. \tag{9.232}$$

Integrated over a volume V limited by surface S, with the account of the divergence theorem, Eq. (9.231) returns us to the Poynting theorem (6.111),

$$\int_V \left(\frac{\partial u}{\partial t} + \mathbf{j} \cdot \mathbf{E}\right) d^3r + \oint_S S_n\, d^2r = 0, \tag{9.233}$$

while Eq. (9.232) yields[75]

[74] In this way, we are using Eq. (9.219) just as a useful guess, which has led us to the definition of $\theta^{\alpha\beta}$, and may leave its strict justification for more in-depth field theory courses.

[75] Just like the Poynting theorem Eq. (9.233), Eq. (9.234) may be obtained directly from the Maxwell equations, without resorting to the four-vector formalism — see, e.g. section 8.2.2 in [11]. However, the derivation discussed above is preferable, because it shows the wonderful unity between the laws of conservation of energy and momentum.

$$\int_V \left[\frac{\partial}{\partial t}\frac{\mathbf{S}}{c^2} + \mathbf{f}\right]_j d^3r = \sum_{j'=1}^{3} \oint_S \tau_{jj'}^{(M)} dA_{j'}, \quad \text{with } \mathbf{f} \equiv \rho\mathbf{E} + \mathbf{j} \times \mathbf{B}, \tag{9.234}$$

where $dA_j = n_j\, dA = n_j\, d^2r$ is the jth component of the elementary area vector $d\mathbf{A} = \mathbf{n}\, dA = \mathbf{n}d^2\mathrm{r}$ that is normal to volume's surface, and directed out of the volume—see figure 9.17[76].

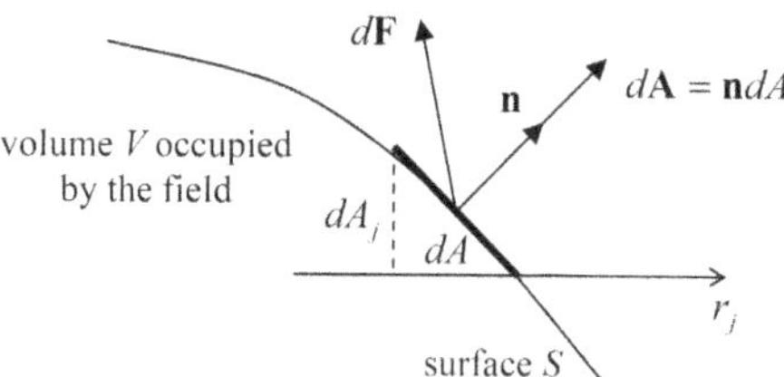

Figure 9.17. The force $d\mathbf{F}$ exerted on a boundary element $d\mathbf{A}$ of the volume V occupied by the field.

Since, according to Eq. (5.10), the vector $\mathbf{f}$ is nothing else than the density of volume-distributed Lorentz forces exerted by the field on the charged particles, we can use Newton's second law, in its relativistic form (9.144), to rewrite Eq. (9.234) for a stationary volume V as

$$\frac{d}{dt}\left[\int_V \frac{\mathbf{S}}{c^2} d^3r + \mathbf{p}_{\text{part}}\right] = \mathbf{F}, \tag{9.235}$$

where $\mathbf{p}_{\text{part}}$ is the total mechanical (relativistic) momentum of all particles in the volume V, and the vector $\mathbf{F}$ is defined by its Cartesian components:

$$F_j = \sum_{j'=1}^{3} \oint_A \tau_{jj'}^{(M)} dA_{j'}. \tag{9.236}$$

Relations (9.235) and (9.236) are our main new results. The first of them shows that the vector

$$\mathbf{g} \equiv \frac{\mathbf{S}}{c^2}, \tag{9.237}$$

already discussed in section 6.8 without derivation, may be indeed interpreted as the density of momentum of the electromagnetic field (per unit volume). This classical relation is consistent with the quantum-mechanical picture of photons of ultra-relativistic particles, with the momentum's magnitude $\mathscr{E}/c$, because then the total flux of the momentum carried by photons through a unit normal area per unit time may be represented either as S_n/c or as g_nc. It also allows us to revisit the Poynting vector paradox that was discussed in section 6.8—see figure 6.11 and its discussion. As has been emphasized at this discussion, the vector $\mathbf{S} = \mathbf{E} \times \mathbf{H}$ in this case does not correspond to any measurable energy flow. However, the corresponding momentum

[76] The same notions are used in the mechanical stress theory—see, e.g. *Part CM* section 7.2.

of the field, equal to the integral of the density (9.237) over a volume of interest, is not only real, but may be measured by the recoil impulse[77] it gives to the field sources—say, to a magnetic coil inducing the field **H**, or to the capacitor plates creating the field **E**.

Now let us turn to our second result, Eq. (9.236). It tells us that the 3 × 3-element Maxwell stress tensor complies with the general definition of the stress tensor[78] characterizing the force **F** exerted by external forces on the boundary of a volume, in this case occupied by the electromagnetic field (figure 9.17). Let us use this important result to analyze two simple examples of static fields.

(i) *Electrostatic field's* effect on a perfect conductor. Since Eq. (9.235) has been derived for a free space region, we have to select volume V outside the conductor, but we may align one of its faces with the conductor's surface (figure 9.18).

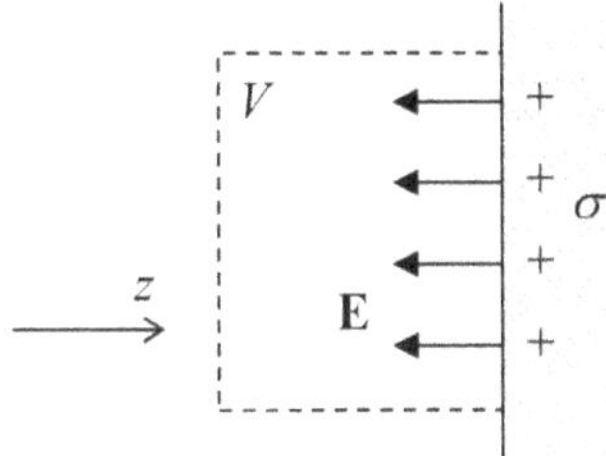

Figure 9.18. The electrostatic field near a conductor's surface.

From chapter 2, we know that the electrostatic field has to be perpendicular to the conductor's surface. Selecting axis z in this direction, we have $E_x = E_y = 0$, $E_z = \pm E$, so that only diagonal components of the tensor (9.228) are not equal to zero:

$$\tau_{xx}^{(M)} = \tau_{yy}^{(M)} = -\frac{\varepsilon_0}{2}E^2, \quad \tau_{zz}^{(M)} = \frac{\varepsilon_0}{2}E^2, \tag{9.238}$$

Since the elementary surface area vector has just one non-vanishing component, dA_z, according to Eq. (9.236), only the last component (which is positive regardless of the sign of E) gives a contribution to the surface force **F**. We see that the force exerted *by the conductor* (and eventually by external forces that hold the conductor in its equilibrium position) on the field is normal to the conductor and directed out of the field volume: $dF_z \geqslant 0$. Hence, by Newton's third law, the force exerted *by the field* on conductor's surface is directed toward the field-filled space:

$$dF_{\text{surface}} = -dF_z = -\frac{\varepsilon_0}{2}E^2\, dA. \tag{9.239}$$

This important result could be obtained by simpler means as well. (Actually, this was the task of one of the problems given in chapter 2.) For example, one could argue quite convincingly that the local relation between the force and the field

[77] This impulse is sometimes called the *hidden momentum*; this term makes sense if the field sources have finite masses, so that their velocity change at the field variation is measurable.

[78] See, e.g. *Part CM* section 7.2.

should not depend on the global configuration creating the field, and thus consider the simplest configuration, a planar capacitor (see e.g. figure 2.3) with the surfaces of both plates charged by equal and opposite charges of density $\sigma = \pm\varepsilon_0 E$. According to the Coulomb law, the charges should attract each other, pulling each plate toward the field region, so that the Maxwell-tensor result gives the correct direction of the force. The force's magnitude (9.239) can be verified either by the direct integration of the Coulomb law, or by the following simple reasoning. In the plane capacitor, the field $E_z = \sigma/\varepsilon_0$ is equally contributed by two surface charges; hence the field created by the negative charge of the counterpart plate (not shown in figure 9.18) is $E_- = \sigma/2\varepsilon_0$, and the force it exerts on the elementary surface charge $dQ = \sigma dA$ of the positively charged plate is $dF = E\, dQ = \sigma^2\, dA/2\varepsilon_0 = \varepsilon_0 E^2\, dA/2$, in accordance with Eq. (9.239)[79].

Quantitatively, even for such a high electric field as E = 3 MV m^{-1} (close to electric breakdown in air), the 'negative pressure' (dF/dA) given by Eq. (9.239) is of the order of 500 Pa (N m^{-2}), i.e. below one thousandth of the ambient atmospheric pressure (1 bar $\approx 10^5$ Pa). Still, this negative pressure may be substantial (above 1 bar) in some cases, for example in good dielectrics (such as high-quality SiO_2, grown at high temperature, which is broadly used in integrated circuits) that can withstand electric fields up to ~10^9 V m^{-1}.

(ii) *Static magnetic field's* effect on its source[80]—say, a solenoid's wall or a superconductor's surface (figure 9.19).

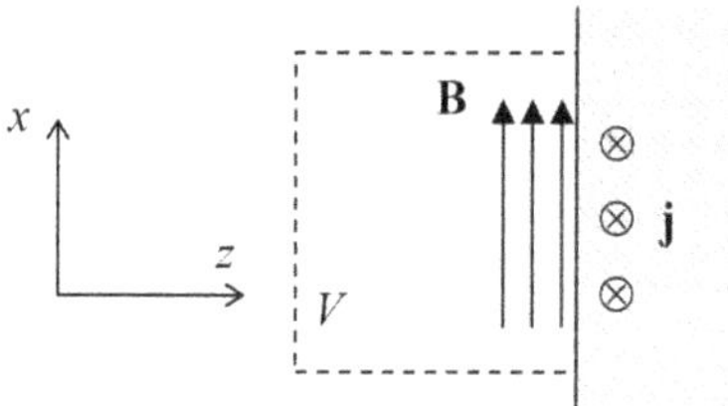

Figure 9.19. The static magnetic field near a current-carrying surface.

With the choice of coordinates shown in figure 9.19, we have $B_x = \pm B$, $B_y = B_z = 0$, so that the Maxwell stress tensor (9.228) is diagonal again:

$$\tau_{xx}^{(M)} = \frac{1}{2\mu_0}B^2, \quad \tau_{yy}^{(M)} = \tau_{zz}^{(M)} = -\frac{1}{2\mu_0}B^2. \tag{9.240}$$

However, since for this geometry only dA_z differs from 0 in Eq. (9.236), the sign of the resulting force is opposite to that in electrostatics, $dF_z \leqslant 0$, and the force exerted by the magnetic field upon the conductor's surface,

[79] By the way, repeating these arguments for a plane capacitor filled with a linear dielectric, we may readily see that Eq. (9.239) may be generalized for this case by replacing ε_0 for ε. The similar replacement ($\mu_0 \to \mu$) is valid for Eq. (9.241) in a linear magnetic medium.

[80] The causal relation is not important here. Especially in the case of a superconductor, the magnetic field may be induced by another source, with the surface supercurrent **j** just shielding the superconductor's bulk from its penetration—see section 9.6.

$$dF_{\text{surface}} = -dF_z = \frac{1}{2\mu_0} B^2\, dA, \tag{9.241}$$

corresponds to a positive pressure. For good laboratory magnets ($B \sim 10$ T), this pressure is of the order of 4×10^7 Pa ≈ 400 bars, i.e. is very substantial, so the magnets require a solid mechanical design.

The direction of the force (9.241) could also be readily predicted by elementary magnetostatics arguments. Indeed, we can imagine the magnetic field volume limited by another, parallel wall with the opposite direction of surface current. According to the starting point of magnetostatics, Eq. (5.1), such surface currents of opposite directions have to repulse each other—doing that via the magnetic field.

Another explanation of the fundamental sign difference between the electric and magnetic field pressures may be provided on the electric circuit language. As we know from chapter 2, the potential energy of the electric field stored in a capacitor may be represented in two equivalent forms,

$$U_{\text{e}} = \frac{CV^2}{2} = \frac{Q^2}{2C}. \tag{9.242}$$

Similarly, the magnetic field energy in an inductive coil is

$$U_{\text{m}} = \frac{LI^2}{2} = \frac{\Phi^2}{2L}. \tag{9.243}$$

If we do not want to consider the work of external sources at a virtual change of the system dimensions, we should use the latter forms of these relations, i.e. consider a galvanically detached capacitor (Q = const) and an externally shorted inductance (Φ = const)[81]. Now if we let the electric field forces (9.239) drag the capacitor's plates in the direction they 'want', i.e. toward each other, this would lead to a *reduction* of the capacitor thickness, and hence to an *increase* of its capacitance C, and to a *decrease* of U_{e}. Similarly, for a solenoid, allowing the positive pressure (9.241) to move its walls away from each other would lead to an *increase* of the solenoid volume, and hence of its inductance L, so that the potential energy U_{m} would be also *reduced*—as it should be. It is remarkable (actually, beautiful) how the local field formulas (9.239) and (9.241) 'know' about these global circumstances.

Finally, let us see whether the major results (9.237) and (9.241), obtained in this section, match each other. For that, let us return to the normal incidence of a plane, monochromatic wave from the free space upon the plane surface of a perfect conductor (see, e.g. figure 7.8 and its discussion), and use those results to calculate the time average of the pressure dF_{surface}/dA imposed by the wave on the surface. At elastic reflection from the conductor's surface, the electromagnetic field's momentum retains its amplitude but changes its sign, so that the momentum transferred to a unit area of the surface (i.e. average pressure) is

[81] Of course, this condition may hold 'forever' only for solenoids with superconducting wiring, but even in normal-metal solenoids with practicable inductances, the flux relaxation constants L/R may be rather large (practically, up to a few minutes), quite sufficient to carry out the force measurement.

$$\frac{\overline{dF_{\text{surface}}}}{dA} = 2c\overline{g_{\text{incident}}} = 2c\frac{\overline{S_{\text{incident}}}}{c^2} = 2c\frac{\overline{EH}}{c^2} = \frac{E_\omega H_\omega^*}{c}, \tag{9.244}$$

where E_ω and H_ω are complex amplitudes of the incident wave. Using the relation (7.7) between these amplitudes (for $\varepsilon = \varepsilon_0$ and $\mu = \mu_0$ giving $E_\omega = cB_\omega$) we obtain

$$\frac{\overline{dF_{\text{surface}}}}{dA} = \frac{1}{c}cB_\omega\frac{B_\omega^*}{\mu_0} = \frac{|B_\omega|^2}{\mu_0}. \tag{9.245}$$

On the other hand, as was discussed in section 7.3, at the surface of a perfect mirror the electric field vanishes while the magnetic field doubles, so that we can use Eq. (9.241) with $B \to B(t) = 2\text{Re}[B_\omega \exp\{-i\omega t\}]$. Averaging the pressure given by Eq. (9.241) over time, we obtain

$$\frac{\overline{dF_{\text{surface}}}}{dA} = \frac{1}{2\mu_0}\overline{(2\,\text{Re}\,[B_\omega e^{-i\omega t}])^2} = \frac{|B_\omega|^2}{\mu_0}, \tag{9.246}$$

i.e. the same result as Eq. (9.245).

For the development of physics intuition, it is useful to estimate the electromagnetic radiation pressure's magnitude. Even for a relatively high wave intensity S_n of 1 kW m^{-2} (close to that of the direct sunlight at the Earth's surface), the pressure $2cg_n = 2S_n/c$ is somewhat below 10^{-5} Pa ~ 10^{-10} bar. Still, this extremely small effect was experimentally observed (by P Lebedev) as early as in 1899, giving one more confirmation of the Maxwell's theory.

9.9 Problems

Problem 9.1. Use the non-relativistic picture of the Doppler effect, in which light propagates with velocity c in a Sun-bound aether, to derive Eq. (9.4).

Problem 9.2. Show that two successive Lorentz space/time transforms in the same direction, with velocities u' and v, are equivalent to a single transform, with the velocity u given by Eq. (9.25).

Problem 9.3. $N + 1$ reference frames, numbered by index n (taking values 0, 1, …, N), move in the same direction as a particle. Express the particle's velocity in the frame number 0 via its velocity u_N in the frame number N, and the set of velocities v_n of the frame number n relative to the frame number $(n - 1)$.

Problem 9.4. A spaceship, moving with constant velocity v directly from the Earth, sends back brief flashes of light with a period Δt_S—as measured by the spaceship's clock. Calculate the period with which Earth's observers receive the signals—as measured by the Earth's clock.

Problem 9.5. From the point of view of observers in a 'moving' reference frame 0′, a straight thin rod, parallel to axis x', is moving, without rotation, with a constant velocity $\mathbf{u}'$ directed along axis y'. The reference frame 0′ is itself moving relative to another ('lab') reference frame 0 with a constant velocity $\mathbf{v}$ along axis x, also without rotation—see the figure below. Calculate:

(i) the direction of the rod's velocity, and
(ii) the orientation of the rod on the $[x, y]$ plane,

both as observed from the lab reference frame. Is the velocity, in this frame, perpendicular to the rod?

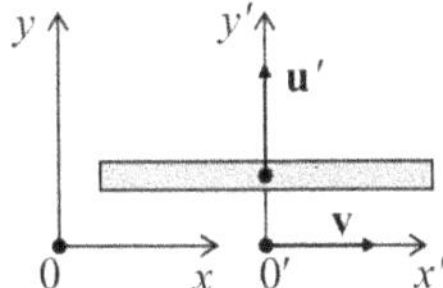

Problem 9.6. Starting from the rest at $t = 0$, a spaceship moves directly from the Earth with a constant acceleration as measured in its instantaneous rest frame. Find its displacement $x(t)$ from the Earth as measured from the Earth's reference frame and interpret the result.

Hint: The instantaneous rest frame of a moving particle is the inertial reference frame that, at the considered moment of time, has the same velocity as the particle.

Problem 9.7. Calculate the first relativistic correction to the frequency of a harmonic oscillator as a function of its amplitude.

Problem 9.8. An atom, with an initial rest mass m, has been excited to an internal state with an additional energy $\Delta\mathscr{E}$, still being at rest. Next, it returns into its initial state, emitting a photon. Calculate the photon's frequency, taking into account the relativistic recoil of the atom.

Hint: In this problem, and in problems 9.11–9.13, treat photons as ultra-relativistic point particles with zero rest mass, energy $\mathscr{E} = \hbar\omega$, and momentum $\mathbf{p} = \hbar\mathbf{k}$.

Problem 9.9. A particle of mass m, initially at rest, decays into two particles, with rest masses m_1 and m_2. Calculate the total energy of the first decay product, in the reference frame moving with that particle.

Problem 9.10. A relativistic particle with a rest mass m, moving with velocity u, decays into two particles with zero rest mass.

(i) Calculate the smallest possible angle between the decay product velocities (in the lab frame, in which the velocity u is measured).
(ii) What is the largest possible energy of one product particle?

Problem 9.11. A relativistic particle, propagating with velocity **u**, in the absence of external fields, decays into two photons[82]. Calculate the angular dependence of the probability of photon detection, as measured in the lab frame.

Problem 9.12. A photon with wavelength λ is scattered by an electron, initially at rest. Calculate the wavelength λ' of the scattered photon as a function of the scattering angle α—see the figure below[83].

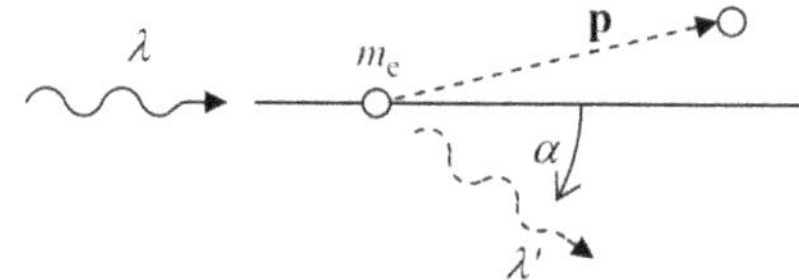

Problem 9.13. Calculate the threshold energy of a γ-photon for the reaction

$$\gamma + \mathrm{p} \to \mathrm{p} + \pi^0,$$

if the proton was initially at rest.

Hint: For protons $m_\mathrm{p}c^2 \approx 938$ MeV, while for neutral pions $m_\pi c^2 \approx 135$ MeV.

Problem 9.14. A relativistic particle with energy $\mathscr{E}$ and rest mass m collides with a similar particle, initially at rest in the laboratory reference frame. Calculate:

(i) the final velocity of the center of mass of the system, in the lab frame,
(ii) the total energy of the system, in the center-of-mass frame, and
(iii) the final velocities of both particles (in the lab frame), if they move along the same direction.

Problem 9.15. A 'primed' reference frame moves, with the reduced velocity $\boldsymbol{\beta} \equiv \mathbf{v}/c = \mathbf{n}_x\beta$, relative to the 'lab' frame. Use Eq. (9.109) to express the components T'^{00} and T'^{0j} (with $j = 1, 2, 3$) of an arbitrary contravariant four-tensor $T^{\gamma\delta}$ via its components in the lab frame.

Problem 9.16. Static fields **E** and **B** are uniform but arbitrary (both in magnitude and in direction). What should be the velocity of an inertial reference frame to have the vectors **E**′ and **B**′, observed from that frame, parallel? Is this solution unique?

Problem 9.17. Two charged particles, moving with the same constant velocity **u**, are offset by distance $\mathbf{R} = \{a, b\}$ (see the figure below), as measured in the lab frame. Calculate the forces between the particles—also in the lab frame.

---

[82] Such a decay may happen, for example, with a neutral pion.

[83] This the famous *Compton scattering* problem.

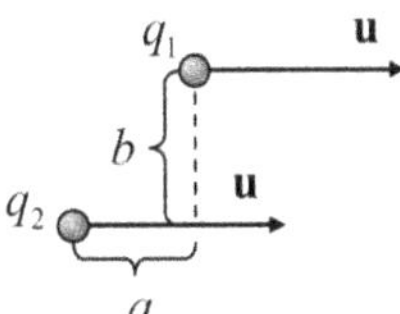

Problem 9.18. Each of two thin, long, parallel particle beams of the same velocity **u**, separated by distance d, carries electric charge with a constant density λ per unit length, as measured in the coordinate frame moving with the particles.

(i) Calculate the distribution of the electric and magnetic fields in the system (outside the beams), as measured in the lab frame.
(ii) Calculate the interaction force between the beams (per particle) and the resulting acceleration, both in the lab frame and in the system moving with the electrons. Compare the results and give a brief discussion of the comparison.

Problem 9.19. Spell out the Lorentz transform of the scalar potential and the vector potential components, and use the result to calculate the potentials of a point charge q moving with a constant velocity **u** as measured in the lab reference frame.

Problem 9.20. Calculate the scalar and vector potentials created by a time-independent electric dipole **p**, as measured in a reference frame that moves relatively to the dipole with a constant velocity **v**, with the shortest distance ('impact parameter') equal to b.

Problem 9.21. Solve the previous problem in the limit $v \ll c$ for a time-independent magnetic dipole **m**.

Problem 9.22. Assuming that the magnetic monopole does exist and has magnetic charge g, calculate the change $\Delta\Phi$ of magnetic flux in a superconductor ring due to the passage of a single monopole through it. Evaluate $\Delta\Phi$ for the monopole charge conjectured by P Dirac, $g = ng_0 \equiv n(2\pi\hbar/e)$, where n is an integer; compare the result with the magnetic flux quantum Φ_0 (6.62) with $|q| = e$, and discuss their relation.

Hint: For simplicity, you may consider the monopole's passage along the symmetry axis of a thin, round superconducting ring, in the otherwise free space.

Problem 9.23.* Calculate the trajectory of a relativistic particle in a uniform electrostatic field **E**, for arbitrary direction of its initial velocity **u**(0), using two different approaches—at least one of them different from the approach used in section 9.6 for the case $\mathbf{u}(0) \perp \mathbf{E}$.

Problem 9.24. A charged relativistic particle with velocity u performs planar cyclotron rotation in a uniform external magnetic field B. How much would the velocity and orbit radius change at a slow change of the field to a new magnitude B'?

Problem 9.25.* Analyze the motion of a relativistic particle in uniform, mutually perpendicular fields **E** and **B** for the particular case when E is *exactly* equal to cB.

Problem 9.26.* Find the law of motion of a relativistic particle in uniform, parallel, static fields **E** and **B**.

Problem 9.27. Neglecting relativistic effects, calculate the smallest voltage V that has to be applied between the anode and cathode of a magnetron (see figure 9.13 and its discussion) to enable electrons to reach the anode, at negligible electron–electron interactions (including the space-charge effects), and collisions with the residual gas molecules. You may:

(i) model the cathode and anode as two coaxial round cylinders, of radii R_1 and R_2, respectively;
(ii) assume that the magnetic field **B** is uniform and directed along their common axis; and
(iii) neglect the initial velocity of the electrons emitted by the cathode.

(After the solution, estimate the validity of the last assumption and of the non-relativistic approximation for reasonable values of parameters.)

Problem 9.28. A charged, relativistic particle has been injected into a uniform electric field whose magnitude oscillates in time with frequency ω. Calculate the time dependence of the particle's velocity, as observed from the lab reference frame.

Problem 9.29. Analyze the motion of a non-relativistic particle in a region where the electric and magnetic fields are both uniform and constant in time, but not necessarily parallel or perpendicular to each other.

Problem 9.30. A static distribution of electric charge in otherwise free space has created a time-independent distribution **E**(**r**) of the electric field. Use two different approaches to express the energy density u' and the Poynting vector $\mathbf{S}'$ as observed from a reference frame moving with constant velocity **v**, via components of the vector **E**. In particular, is $\mathbf{S}'$ equal to $(-\mathbf{v}u')$?

Problem 9.31. A plane wave, of frequency ω and intensity S, is normally incident on a perfect mirror, moving with velocity v in the same direction as the wave.

(i) Calculate the reflected wave's frequency, and
(ii) use the Lorentz transform of the fields to calculate the reflected wave's intensity,

both as observed from the lab reference frame.

Problem 9.32. Carry out the second task of the previous problem by using the relations between the wave's energy, power, and momentum.

Hint: As a by-product, this approach should also give you the pressure exerted by the wave on the moving mirror.

Problem 9.33. Consider the simple model of capacitor charging by a lumped current source, shown in the figure below, and prove that the momentum given by the constant, uniform external magnetic field **B** to the current-carrying conductor is equal and opposite to the momentum of the electromagnetic field that current $I(t)$ builds up in the capacitor. (You may assume that the capacitor is planar and very broad, and neglect the fringe field effects.)

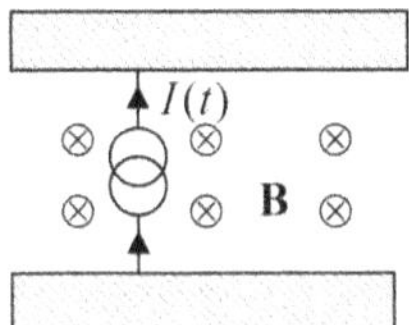

Problem 9.34. Consider an electromagnetic plane wave packet propagating in free space, with the electric field represented as the Fourier integral

$$\mathbf{E}(\mathbf{r}, t) = \text{Re} \int_{-\infty}^{+\infty} \mathbf{E}_k e^{i\psi_k} \, dk, \quad \text{with } \psi_k \equiv kz - \omega_k t, \ \text{ and } \omega_k \equiv c\,|\,k\,|.$$

Express the full linear momentum (per unit area of wave's front) of the packet via the complex amplitudes $\mathbf{E}_k$ of its Fourier components. Does the momentum depend on time? (In contrast with problem 7.7, in this case the wave packet is not necessarily narrow.)

Problem 9.35. Calculate the pressure exerted on well-conducting walls of a waveguide with a rectangular ($a \times b$) cross-section, by a wave propagating along it in the fundamental (H_{10}) mode. Give an interpretation of the result.

References

[1] Müller H *et al* 2003 *Phys. Rev. Lett.* **91** 020401
[2] Benford G *et al* 1970 *Phys. Rev.* D **2** 263
[3] Arzeliès H 1966 *Relativistic Kinematics* (Pergamon)
[4] Rindler W 1991 *Introduction to Special Relativity* 2nd edn (Oxford University Press)
[5] Hafele J and Keating R 1972 *Science* **177** 166
[6] Lee S 2004 *Accelerator Physics* 2nd edn (Singapore: World Scientific)
[7] Wilson E 2001 *An Introduction to Particle Accelerators* (Oxford University Press)
[8] Jackson J 1999 *Classical Electrodynamics* 3rd edn (Wiley)
[9] Chen F 1984 *Introduction to Plasma Physics and Controllable Fusion* vol 1 2nd edn (Springer)
[10] Hazeltine R and Meiss J 2003 *Plasma Confinement* (New York: Dover)
[11] Griffiths D 1999 *Introduction to Electrodynamics* 3rd edn (Prentice-Hall)

Chapter 10

Radiation by relativistic charges

The discussion of special relativity in the previous chapter enables us to revisit the analysis of electromagnetic radiation by charged particles, now for arbitrary speed. For a single particle, we will be able to calculate the radiated wave fields in an explicit form, and analyze them for such important specific cases as synchrotron radiation and 'Bremsstrahlung' (brake radiation). After that we will discuss the apparently unrelated effect of the so-called Coulomb losses of energy by a particle moving in condensed matter, because this discussion will naturally lead us to such important phenomena as the Cherenkov radiation and transitional radiation. At the end of the chapter, I will briefly review the effects of back action of the emitted radiation on the emitting particle, whose analysis reveals some limitations of classical electrodynamics.

10.1 Liénard–Wiechert potentials

A convenient starting point for the discussion of radiation by relativistic charges is provided by Eqs. (8.17) for the retarded potentials. In the free space, these formulas (with the integration variable changed from $\mathbf{r}'$ to $\mathbf{r}''$ for clarity in what follows) are reduced to

$$\begin{aligned}\phi(\mathbf{r}, t) &= \frac{1}{4\pi\varepsilon_0}\int\frac{\rho(\mathbf{r}'', t - R/c)}{R}d^3r'',\\ \mathbf{A}(\mathbf{r}, t) &= \frac{\mu_0}{4\pi}\int\frac{\mathbf{j}(\mathbf{r}'', t - R/c)}{R}d^3r'', \quad \text{with } \mathbf{R} \equiv \mathbf{r} - \mathbf{r}''.\end{aligned} \tag{10.1a}$$

As a reminder, Eqs. (10.1*a*) were derived from the Maxwell equations without any restrictions, and are very natural for situations with continuous distributions of the electric charge and/or current. However, for a single point charge, with

$$\rho(\mathbf{r}, t) = q\delta(\mathbf{r} - \mathbf{r}'), \qquad \mathbf{j}(\mathbf{r}, t) = q\mathbf{u}\delta(\mathbf{r} - \mathbf{r}'), \qquad \text{with } \mathbf{u} = \dot{\mathbf{r}}', \tag{10.1b}$$

doi:10.1088/978-0-7503-1404-6ch10

where $\mathbf{r}'$ is the instantaneous position of the charge, it is more convenient to recast Eqs. (10.1*a*) into an explicit form that would not require the integration in each particular case. Indeed, as Eqs. (10.1) show, the potentials at the observation point $\{\mathbf{r}, t\}$ are contributed by only one specific point $\{\mathbf{r}'(t_{\text{ret}}), t_{\text{ret}}\}$ of the particle's 4D trajectory (called its *world line*), which satisfies the following condition:

$$t_{\text{ret}} \equiv t - \frac{R_{\text{ret}}}{c}, \tag{10.2}$$

where t_{ret} is called the *retarded time*, and R_{ret} is the length of the following distance vector

$$\mathbf{R}_{\text{ret}} \equiv \mathbf{r}(t) - \mathbf{r}'(t_{\text{ret}}) \tag{10.3}$$

—physically, the distance covered by the electromagnetic wave from its emission to observation.

The reduction of Eqs. (10.1) to such a simpler form, however, requires care. Indeed, their naïve integration would yield the following apparent, but incorrect results:

$$\begin{aligned} \phi(\mathbf{r}, t) &= \frac{1}{4\pi\varepsilon_0}\frac{q}{R_{\text{ret}}}, \quad \text{i.e.} \quad \frac{\phi(\mathbf{r}, t)}{c} = \frac{\mu_0}{4\pi}\frac{qc}{R_{\text{ret}}}; \\ \mathbf{A}(\mathbf{r}, t) &= \frac{\mu_0}{4\pi}\frac{q\mathbf{u}_{\text{ret}}}{R_{\text{ret}}}, \quad \text{(WRONG!)} \end{aligned} \tag{10.4}$$

where $\mathbf{u}_{\text{ret}}$ is the particle's velocity at the retarded point $\mathbf{r}'(t_{\text{ret}})$. Eq. (10.4) is a good example how the theory of relativity (even the special theory) cannot be taken too lightly. Indeed, the strings (9.84) and (9.85), formed from the apparent potentials (10.4), would not obey the Lorentz transform rule (9.91), because according to Eqs. (10.2) and (10.3) the distance R_{ret} also depends on the reference frame it is measured in.

In order to correct the error, we need, first of all, to discuss the conditions (10.2) and (10.3). Combining them, we obtain the following equation for t_{ret}:

$$c(t - t_{\text{ret}}) = |\mathbf{r}(t) - \mathbf{r}'(t_{\text{ret}})|. \tag{10.5}$$

Figure 10.1 depicts the graphical solution of this self-consistency equation as the only[1] point of intersection of the light cone of the observation point (see figure 9.9 and its discussion) and the particle's word line. In Eq. (10.5), just as in Eqs. (10.1)–(10.3), all variables have to be measured in the same 'lab' reference frame, in which the observation point $\mathbf{r}$ rests. Now let us write Eqs. (10.1) for a point charge in another inertial reference frame 0′, whose velocity (as measured in the lab frame)

[1] As figure 10.1 shows, there is always another, 'advanced' point $\{\mathbf{r}'(t_{\text{adv}}), t_{\text{adv}}\}$ of the particle's world line, with $t_{\text{a}} > t$, which is also a solution of Eq. (10.5), but it does not fit Eq. (10.1), because the observation of the field, induced at the advanced point, at the point $\{\mathbf{r}, t < t_{\text{adv}}\}$ would violate the causality principle.

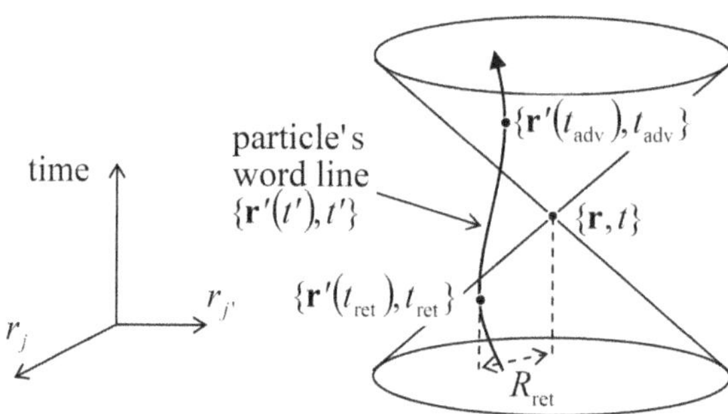

Figure 10.1. Graphical solution of Eq. (10.5).

coincides, at the moment $t' = t_{\text{ret}}$, with the velocity $\mathbf{u}_{\text{ret}}$ of the charge. In that frame the charge rests, so that, as we know from the electro- and magnetostatics,

$$\phi' = \frac{q}{4\pi\varepsilon_0 R'}, \qquad \mathbf{A}' = 0. \tag{10.6a}$$

(Remember that this R' may not be equal to R, because the latter distance is measured in the 'lab' reference frame.) Let us use the identity $1/\varepsilon_0 \equiv \mu_0 c^2$ to rewrite Eq. (10.6a) in the form of components of a four-vector similar in structure to Eq. (10.4):

$$\frac{\phi'}{c} = \frac{\mu_0}{4\pi}q\frac{c}{R'}, \qquad \mathbf{A}' = 0. \tag{10.6b}$$

Now it is easy to guess the correct answer for the whole four-potential:

$$A^\alpha = \frac{\mu_0}{4\pi}q\frac{cu^\alpha}{u_\beta R^\beta}, \tag{10.7}$$

where (just as a reminder), $A^\alpha \equiv \{\phi/c, \mathbf{A}\}$, $u^\alpha \equiv \gamma\{c, \mathbf{u}\}$,and R^α is a four-vector of the inter-event distance, formed similarly to that of a single event—cf Eq. (9.48):

$$R^\alpha \equiv \{c(t - t'), \mathbf{R}'\} \equiv \{c(t - t'), \mathbf{r} - \mathbf{r}'\}. \tag{10.8}$$

Indeed, we needed the four-vector A^α that would:

(i) obey the Lorentz transform,
(ii) have its spatial components A_j scaling at low velocity as u_j, and
(iii) be reduced to the correct result (10.6) in the reference frame moving with the charge.

Formula (10.7) evidently satisfies all these requirements, because the scalar product in its denominator is just

$$\begin{aligned} u_\beta R^\beta &= \gamma\{c, -\mathbf{u}\} \cdot \{c(t - t'), \mathbf{R}\} \equiv \gamma[c^2(t - t') - \mathbf{u} \cdot \mathbf{R}] \\ &\equiv \gamma c(R - \boldsymbol{\beta} \cdot \mathbf{R}) \equiv \gamma c R(1 - \boldsymbol{\beta} \cdot \mathbf{n}), \end{aligned} \tag{10.9}$$

where $\mathbf{n} \equiv \mathbf{R}/R$ is a unit vector in the observer's direction, $\boldsymbol{\beta} \equiv \mathbf{u}/c$ is the normalized velocity of the particle, and $\gamma \equiv 1/(1 - u^2/c^2)^{1/2}$. In the reference frame of the charge (in which $\boldsymbol{\beta} = 0$ and $\gamma = 1$), the expression (10.9) is reduced to cR, so that Eq. (10.7) is correctly reduced to Eq. (10.6*b*). Now let us spell out components of Eq. (10.7) in the lab frame (in which $t' = t_{\text{ret}}$ and $R = R_{\text{ret}}$):

$$\phi(\mathbf{r}, t) = \frac{1}{4\pi\varepsilon_0}\frac{q}{(R - \boldsymbol{\beta}\cdot\mathbf{R})_{\text{ret}}} = \frac{1}{4\pi\varepsilon_0}q\left[\frac{1}{R(1 - \boldsymbol{\beta}\cdot\mathbf{n})}\right]_{\text{ret}}, \qquad (10.10a)$$

$$\mathbf{A}(\mathbf{r}, t) = \frac{\mu_0}{4\pi}q\left(\frac{\mathbf{u}}{R - \boldsymbol{\beta}\cdot\mathbf{R}}\right)_{\text{ret}} = \frac{\mu_0}{4\pi}qc\left[\frac{\boldsymbol{\beta}}{R(1 - \boldsymbol{\beta}\cdot\mathbf{n})}\right]_{\text{ret}} \equiv \phi(\mathbf{r}, t)\frac{\mathbf{u}_{\text{ret}}}{c^2}. \qquad (10.10b)$$

These formulas are called the *Liénard–Wiechert potentials*[2]. In the non-relativistic limit, they coincide with the naïve guess (10.4), but in the general case include an additional factor $1/(1 - \boldsymbol{\beta}\cdot\mathbf{n})$. The explanation of its physical origin may be facilitated by one more formal calculation—which we will need anyway. Let us differentiate the geometric relation (10.5), rewritten as

$$R_{\text{ret}} = c(t - t_{\text{ret}}), \qquad (10.11)$$

over t_{ret} and then, independently, over t, assuming that $\mathbf{r}$ is fixed. For that let us first differentiate, over t_{ret}, both sides of the identity $R_{\text{ret}}^2 = \mathbf{R}_{\text{ret}}\cdot\mathbf{R}_{\text{ret}}$:

$$2R_{\text{ret}}\frac{\partial R_{\text{ret}}}{\partial t_{\text{ret}}} = 2\mathbf{R}_{\text{ret}}\cdot\frac{\partial \mathbf{R}_{\text{ret}}}{\partial t_{\text{ret}}}. \qquad (10.12)$$

If $\mathbf{r}$ is fixed, $\partial\mathbf{R}_{\text{ret}}/\partial t_{\text{ret}} \equiv \partial(\mathbf{r} - \mathbf{r}')/\partial t_{\text{ret}} = -\partial\mathbf{r}'/\partial t_{\text{ret}} \equiv -\mathbf{u}$, and Eq. (10.12) yields

$$\frac{\partial R_{\text{ret}}}{\partial t_{\text{ret}}} = \frac{\mathbf{R}_{\text{ret}}}{R_{\text{ret}}}\cdot\frac{\partial \mathbf{R}_{\text{ret}}}{\partial t_{\text{ret}}} = -(\mathbf{n}\cdot\mathbf{u})_{\text{ret}}. \qquad (10.13)$$

Now let us differentiate the same R_{ret} over t. On one hand, Eq. (10.11) yields

$$\frac{\partial R_{\text{ret}}}{\partial t} = c - c\frac{\partial t_{\text{ret}}}{\partial t}. \qquad (10.14)$$

On the other hand, according to Eq. (10.5), at the partial differentiation over time, i.e. if $\mathbf{r}$ is fixed, t_{ret} is a function of t alone so that (using Eq. (10.13) at the second step) we may write

$$\frac{\partial R_{\text{ret}}}{\partial t_{\text{ret}}} = \frac{\partial R_{\text{ret}}}{\partial t_{\text{ret}}}\frac{\partial t_{\text{ret}}}{\partial t} = -(\mathbf{n}\cdot\mathbf{u})_{\text{ret}}\frac{\partial t_{\text{ret}}}{\partial t}. \qquad (10.15)$$

Now requiring Eqs. (10.14) and (10.15) to give the same result, we obtain[3]:

[2] They were derived in 1898 by A-M Liénard and (apparently, independently) in 1900 by E Wiechert.
[3] This relation may be used for an alternative derivation of Eq. (10.10) directly from Eqs. (10.1)—an exercise highly recommended to the reader.

$$\frac{\partial t_{\text{ret}}}{\partial t} = \frac{c}{c - (\mathbf{n} \cdot \mathbf{u})_{\text{ret}}} = \left(\frac{1}{1 - \boldsymbol{\beta} \cdot \mathbf{n}}\right)_{\text{ret}}. \tag{10.16}$$

This relation may be readily re-derived (and more clearly understood) for the simple particular case when the charge's velocity is directed straight toward the observation point. In this case its vector **u** resides in the same space–time plane as the observation point's world line **r** = const—say, plane $[x, t]$, shown in figure 10.2.

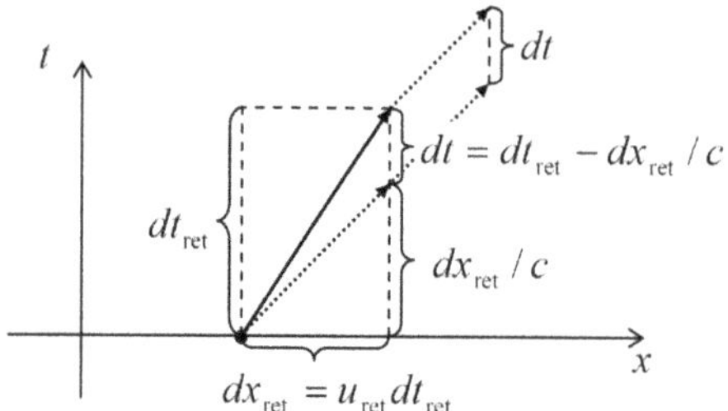

Figure 10.2. Deriving Eq. (10.16) for the case $\boldsymbol{\beta} \cdot \mathbf{n} = \beta$.

Let us consider an elementary time interval $dt_{\text{ret}} \equiv dt'$, during which the particle would travel the space interval $dx_{\text{ret}} = u_{\text{ret}}dt_{\text{ret}}$. In figure 10.2, the corresponding segment of its world line is shown with a solid vector. The dotted vectors in this figure show the world lines of the radiation emitted by the particle at the beginning and end of this interval, and propagating with the speed of light c. As follows from the drawing, the time interval dt between the instants of arrival of the radiation from these two points to any time-independent spatial point of observation is

$$\begin{aligned} &dt = dt_{\text{ret}} - \frac{dx_{\text{ret}}}{c} = dt_{\text{ret}} - \frac{u_{\text{ret}}}{c}dt_{\text{ret}}, \\ &\text{so that} \quad \frac{dt_{\text{ret}}}{dt} = \frac{1}{1 - u_{\text{ret}}/c} \equiv \frac{1}{1 - \beta_{\text{ret}}}. \end{aligned} \tag{10.17}$$

This expression coincides with (10.16), because in our particular case the directions of the vectors $\boldsymbol{\beta} \equiv \mathbf{u}/c$ and $\mathbf{n} \equiv \mathbf{R}/R$ (both taken at t_{ret}) coincide, and hence $(\boldsymbol{\beta} \cdot \mathbf{n})_{\text{ret}} = \beta_{\text{ret}}$. Now the general Eq. (10.16) may be interpreted by saying that the particle's velocity in the transverse directions (normal to the vector **n**) is not important for this kinematic effect[4]—a fact almost evident from figure 10.1.

So, the additional factor in the Liénard–Wiechert potentials is just the derivative $\partial t_{\text{ret}}/\partial t$. The reason for its appearance in Eq. (10.10) is usually interpreted along the following lines. Let the charge q be spread along the direction of vector $\mathbf{R}_{\text{ret}}$ (in figure 10.2, along axis x) by an infinitesimal speed-independent interval δx_{ret}, so that the linear density λ of its charge is proportional to $1/\delta x_{\text{ret}}$. Then the time rate of the

[4] Note that this effect (linear in β) has nothing to do with the Lorentz time dilation (9.21), which is quadratic in β. (Indeed, all our arguments above referred to the same, lab frame.) Rather, it is close in nature to the Doppler effect.

charge's arrival at some spatial point is $\lambda u_{\text{ret}} = \lambda dx_{\text{ret}}/dt_{\text{ret}}$, i.e. scales as $1/dt_{\text{ret}}$. However, the rate of the *radiation's* arrival at the observation point scales as $1/dt$, so that due to the non-vanishing velocity $\mathbf{u}_{\text{ret}}$ of the particle, this rate differs from the charge arrival rate by the factor of dt_{ret}/dt, given by Eq. (10.16). (If the particle moves toward the observation point, $(\boldsymbol{\beta} \cdot \mathbf{n})_{\text{ret}} > 0$, as shown in figure 10.2, this factor is larger than 1.) This radiation compression effect leads to the field change (at $(\boldsymbol{\beta} \cdot \mathbf{n})_{\text{ret}} > 0$, its enhancement) by the same factor (10.16)—as described by Eqs. (10.10).

So, the four-vector formalism was instrumental in the calculation of field potentials. It may be also used to calculate the fields **E** and **B**—by plugging Eq. (10.7) into Eq. (9.124) to calculate the field strength tensor. This calculation yields

$$F^{\alpha\beta} = \frac{\mu_0 q}{4\pi} \frac{1}{u_\gamma R^\gamma} \frac{d}{d\tau}\left[\frac{R^\alpha u^\beta - R^\beta u^\alpha}{u_\delta R^\delta}\right]. \tag{10.18}$$

Now using Eq. (9.125) to identify the elements of this tensor with the field components, we may bring the result to the following vector form[5]:

$$\mathbf{E} = \frac{q}{4\pi\varepsilon_0}\left[\frac{\mathbf{n} - \boldsymbol{\beta}}{\gamma^2(1 - \boldsymbol{\beta} \cdot \mathbf{n})^3 R^2} + \frac{\mathbf{n} \times \{(\mathbf{n} - \boldsymbol{\beta}) \times \dot{\boldsymbol{\beta}}\}}{(1 - \boldsymbol{\beta} \cdot \mathbf{n})^3 cR}\right]_{\text{ret}}. \tag{10.19}$$

$$\mathbf{B} = \frac{\mathbf{n}_{\text{ret}} \times \mathbf{E}}{c}, \quad \text{i.e. } \mathbf{H} = \frac{\mathbf{n}_{\text{ret}} \times \mathbf{E}}{Z_0}. \tag{10.20}$$

Thus the magnetic and electric fields of a relativistic particle are always proportional and perpendicular to each other, and related just as in a plane wave—cf Eq. (7.6), with the only difference that now the vector $\mathbf{n}_{\text{ret}}$ may be a function of time. Superficially, this result contradicts the electro- and magnetostatics, because for a particle at rest **B** should vanish while **E** stays finite. However, note that according to the Coulomb law for a point charge, in this case $\mathbf{E} = E\mathbf{n}_{\text{ret}}$, so that $\mathbf{B} \propto \mathbf{n}_{\text{ret}} \times \mathbf{E} \propto \mathbf{n}_{\text{ret}} \times \mathbf{n}_{\text{ret}} = 0$. (Actually, in these relations, the index 'ret' is unnecessary.)

As a sanity check, let us use Eq. (10.19) as an alternative way to find the electric field of a charge moving without acceleration, i.e. uniformly, along a straight line—see figure 9.11a reproduced, with minor changes, in figure 10.3. (This calculation will also illustrate very well the technical challenges of practical applications of the Liénard–Wiechert formulas for even simple cases.) In this case, the vector $\boldsymbol{\beta}$ does not change in time, so that the second term in Eq. (10.19) vanishes and all we need to do is to spell out the Cartesian components of the first term.

[5] An alternative way of deriving these formulas (highly recommended to the reader as an exercise) is to plug Eqs. (10.10) into the general relations (9.121), and carry out the required temporal and spatial differentiations directly, using Eq. (10.16) and its spatial counterpart (which may be derived absolutely similarly): $\nabla t_{\text{ret}} = -\left[\frac{\mathbf{n}}{c(1 - \boldsymbol{\beta} \cdot \mathbf{n})}\right]_{\text{ret}}$.

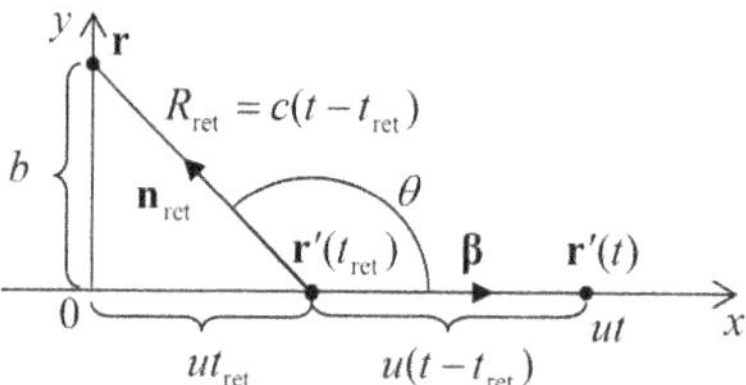

Figure 10.3. The linearly moving charge problem.

Let us select the coordinate axes and time origin in the way shown in figure 10.3, and make a clear distinction between the actual position, $\mathbf{r}'(t) = \{ut, 0, 0\}$ of the charged particle at the instant t we are considering, and its position $\mathbf{r}'(t_{ret})$ at the retarded instant defined by Eq. (10.5), i.e. the moment when the particle's field, propagating with the speed of light, had to be radiated to reach the observation point $\mathbf{r}$ at the time t. In these coordinates

$$\begin{aligned}&\boldsymbol{\beta} = \{\beta, 0, 0\}, \quad \mathbf{r} = \{0, b, 0\}, \quad \mathbf{r}'(t_{ret}) = \{ut_{ret}, 0, 0\},\\ &\mathbf{n}_{ret} = \{\cos\theta, \sin\theta, 0\},\end{aligned} \tag{10.21}$$

with $\cos\theta = -ut_{ret}/R_{ret}$, so that $[(\mathbf{n} - \boldsymbol{\beta})_x]_{ret} = -ut_{ret}/R_{ret} - \beta$, and Eq. (10.19) yields, in particular,

$$E_x = \frac{q}{4\pi\varepsilon_0}\frac{-ut_{ret}/R_{ret} - \beta}{\gamma^2[(1 - \boldsymbol{\beta}\cdot\mathbf{n})^3 R^2]_{ret}} \equiv \frac{q}{4\pi\varepsilon_0}\frac{-ut_{ret} - \beta R_{ret}}{\gamma^2[(1 - \boldsymbol{\beta}\cdot\mathbf{n})^3 R^3]_{ret}}. \tag{10.22}$$

But according to Eq. (10.5), the product βR_{ret} may be represented as $\beta c(t - t_{ret}) \equiv u(t - t_{ret})$. Plugging this expression into Eq. (10.22) we may eliminate the explicit dependence of E_x on time t_{ret}:

$$E_x = \frac{q}{4\pi\varepsilon_0}\frac{-ut}{\gamma^2[(1 - \boldsymbol{\beta}\cdot\mathbf{n})R]^3_{ret}}. \tag{10.23}$$

The non-vanishing transverse component of the field also has a similar form:

$$E_y = \frac{q}{4\pi\varepsilon_0}\left[\frac{\sin\theta}{\gamma^2(1 - \boldsymbol{\beta}\cdot\mathbf{n})^3 R^2}\right]_{ret} = \frac{q}{4\pi\varepsilon_0}\frac{b}{\gamma^2[(1 - \boldsymbol{\beta}\cdot\mathbf{n})R]^3_{ret}}, \tag{10.24}$$

while $E_z = 0$. From figure 10.3, $\boldsymbol{\beta}\cdot\mathbf{n}_{ret} = \beta\cos\theta = -\beta ut_{ret}/R_{ret}$, so that $(1 - \boldsymbol{\beta}\cdot\mathbf{n})R_{ret} \equiv R_{ret} + \beta ut_{ret}$, and we may again use Eq. (10.5) to obtain $(1 - \boldsymbol{\beta}\cdot\mathbf{n})R_{ret} = c(t - t_{ret}) + \beta ut_{ret} \equiv ct - ct_{ret}/\gamma^2$. What remains is to calculate t_{ret} from the self-consistency equation (10.5), whose square in our current case (figure 10.3) takes the form

$$R^2_{ret} \equiv b^2 + (ut_{ret})^2 = c^2(t - t_{ret})^2. \tag{10.25}$$

This is a simple quadratic equation for t_{ret}, which (with the appropriate negative sign before the square root, in order to obtain $t_{ret} < t$) yields

$$t_{ret} = \gamma^2 t - [(\gamma^2 t)^2 - \gamma^2(t^2 - b^2/c^2)]^{1/2} = \gamma^2 t - \frac{\gamma}{c}(u^2\gamma^2 t^2 + b^2)^{1/2}, \tag{10.26}$$

so that the only retarded-function combination that participates in Eqs. (10.23) and (10.24) is

$$[(1 - \boldsymbol{\beta} \cdot \mathbf{n})R]_{\text{ret}} = \frac{c}{\gamma^2}(u^2\gamma^2 t^2 + b^2)^{1/2} \tag{10.27}$$

and, finally, the electric field components are

$$E_x = -\frac{q}{4\pi\varepsilon_0}\frac{\gamma ut}{(b^2 + \gamma^2 u^2 t^2)^{3/2}}, \qquad E_y = \frac{q}{4\pi\varepsilon_0}\frac{\gamma b}{(b^2 + \gamma^2 u^2 t^2)^{3/2}}, \qquad E_z = 0. \tag{10.28}$$

These are exactly Eqs. (9.139)[6], which had been obtained in section 9.5 by much simpler means, without the necessity to solve the self-consistency equation (10.5). However, that alternative approach was essentially based on the inertial motion of the particle, and cannot be used in problems in which the particle moves with acceleration. In those problems, the second term in Eq. (10.19), dropping with distance as $1/R_{\text{ret}}$ and hence describing the wave radiation, is essential and most important.

10.2 Radiation power

Let us calculate the angular distribution of a particle's radiation. For that, we need to return to Eqs. (10.19) and (10.20) to find the Poynting vector $\mathbf{S} = \mathbf{E} \times \mathbf{H}$, and in particular its component $S_n = \mathbf{S} \cdot \mathbf{n}_{\text{ret}}$, at large distances R from the particle. Following tradition[7], let us express the result as the energy radiated into unit solid angle per unit time interval dt_{rad} of the *radiation*, rather than that (dt) of its *measurement*. (We will need to return to the measurement time t in the next section, in order to calculate the observed radiation spectrum.) Using Eq. (10.16), we obtain

$$\frac{d\mathscr{P}}{d\Omega} \equiv -\frac{d\mathscr{E}}{d\Omega\, dt_{\text{ret}}} = (R^2 S_n)_{\text{ret}}\frac{\partial t}{\partial t_{\text{ret}}} = (\mathbf{E} \times \mathbf{H}) \cdot [R^2\mathbf{n}\ (1 - \boldsymbol{\beta} \cdot \mathbf{n})]_{\text{ret}}. \tag{10.29}$$

At sufficiently large distances from the particle, i.e. in the limit $R_{\text{ret}} \to \infty$ (in the *radiation zone*), the contribution of the first (essentially, the Coulomb-field) term in the square brackets of Eq. (10.19) vanishes as $1/R^2$, and the substitution of the remaining term into Eqs. (10.20) and then (10.29) yields the following formula, valid for an arbitrary law of particle motion[8]:

$$\frac{d\mathscr{P}}{d\Omega} = \frac{Z_0 q^2}{(4\pi)^2}\frac{|\mathbf{n} \times [(\mathbf{n} - \boldsymbol{\beta}) \times \dot{\boldsymbol{\beta}}]|^2}{(1 - \mathbf{n} \cdot \boldsymbol{\beta})^5}. \tag{10.30}$$

Now, let us apply this important result to some simple cases. First of all, Eq. (10.30) says that a charge moving with a constant velocity $\boldsymbol{\beta}$ does not radiate

[6] A similar calculation of magnetic field components from Eq. (10.20) gives results identical to Eq. (9.140).

[7] This tradition may be reasonably justified. Indeed, we may say that the radiation field 'detaches' from the particle at times close to t_{ret}, while the observation time t depends on the detector's position, and hence is less relevant for the radiation process as such.

[8] If the direction of radiation, $\mathbf{n}$, does not change in time, this formula does not depend on the observer's position $\mathbf{R}$. Hence, from this point on, the index 'ret' may be safely dropped for brevity, though we should always remember that $\boldsymbol{\beta}$ in Eq. (10.30) is the reduced velocity of the particle at the instant of the radiation's *emission*, not its observation.

at all. This might be expected from our analysis of this case in section 9.5, because in the reference frame moving with the charge it produces only the Coulomb electrostatic field, i.e. no radiation.

Next, let us consider a linear motion of a point charge with a non-vanishing acceleration — evidently directed along the straight line of motion of the coordinate axes directed as shown in figure 10.4a. Each of the vectors involved in Eq. (10.30) has at most two non-vanishing Cartesian components,

$$\mathbf{n} = \{\sin\theta, 0, \cos\theta\}, \quad \boldsymbol{\beta} = \{0, 0, \beta\}, \quad \dot{\boldsymbol{\beta}} = \{0, 0, \dot{\beta}\}, \tag{10.31}$$

where θ is the angle between the directions of the particle's motion and radiation propagation. Plugging these expressions into Eq. (10.30) and performing the vector multiplications, we obtain

$$\frac{d\mathscr{P}}{d\Omega} = \frac{Z_0 q^2}{(4\pi)^2}\dot{\beta}^2 \frac{\sin^2\theta}{(1-\beta\cos\theta)^5}. \tag{10.32}$$

Figure 10.4b shows the angular distribution of this radiation, for three values of the particle's speed u.

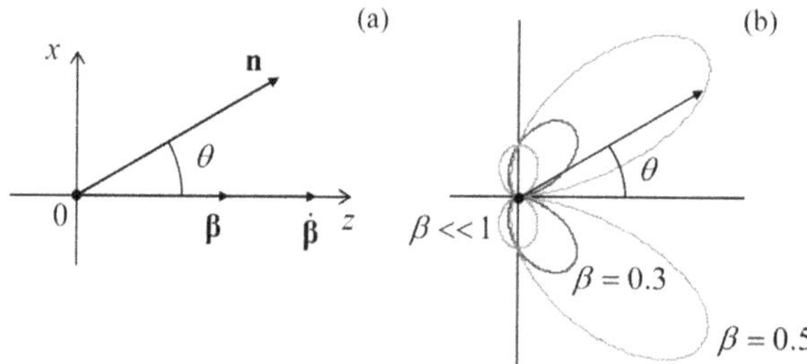

Figure 10.4. Particle's radiation at linear acceleration: (a) the problem's geometry, and (b) the last fraction of Eq. (10.32) as a function of the angle θ.

If the speed is relatively low ($u \ll c$, i.e. $\beta \ll 1$), the denominator in Eq. (10.32) is very close to 1 for all observation angles θ, so that the angular distribution of the radiation power is close to $\sin^2\theta$—just as follows from the general non-relativistic Larmor formula (8.27). However, as the velocity is increased, the denominator becomes less than 1 for $\theta < \pi/2$, i.e. for the forward-looking directions, and larger than 1 for backward directions. As a result, the radiation toward the particle's velocity is increased (somewhat counter-intuitively, regardless of the acceleration's sign!), while that in the back direction is suppressed. For ultra-relativistic particles ($\beta \to 1$) this trend is enormously exacerbated, and radiation to very small forward angles dominates. To describe this main part of the angular distribution we may expand the trigonometric functions of θ participating in Eq. (10.32) into the Taylor series in small θ, and keep only their leading terms: $\sin\theta \approx \theta$, $\cos\theta \approx 1 - \theta^2/2$, so that $(1 - \beta\cos\theta) \approx (1 + \gamma^2\theta^2)/2\gamma^2$. The resulting expression,

$$\frac{d\mathscr{P}}{d\Omega} \approx \frac{2Z_0 q^2}{\pi^2}\dot{\beta}^2\gamma^8 \frac{(\gamma\theta)^2}{(1+\gamma^2\theta^2)^5}, \quad \text{for } \gamma \gg 1, \tag{10.33}$$

describes a narrow 'hollow cone' distribution of radiation, with a maximum at the angle

$$\theta_0 = \frac{1}{2\gamma} \ll 1. \tag{10.34}$$

Another important aspect of Eq. (10.33) is how fast (as γ^8) the radiation density grows with the Lorentz factor γ, i.e. with the particle's energy $\mathscr{E} = \gamma mc^2$.

Still, the total radiated power $\mathscr{P}$ (into all observation angles) at linear acceleration is not too high for any practicable values of parameters. In order to show this, let us calculate $\mathscr{P}$ for an arbitrary motion of the particle. First, let me demonstrate how $\mathscr{P}$ may be found (or rather guessed) from the general relativistic arguments. In section 8.2, we have derived Eq. (8.27) for the power of the electric dipole radiation for non-relativistic particle motion. That result is valid, in particular, for one charged particle whose electric dipole moment's derivative over time may be expressed as $d(q\mathbf{r})/dt = (q/m)\mathbf{p}$, where $\mathbf{p}$ is the particle's mechanical momentum (*not* its electric dipole moment). As the result, the Larmor formula (8.27) in free space, i.e. with $v = c$, reduces to

$$\mathscr{P} = \frac{Z_0}{6\pi c^2}\left(\frac{q}{m}\frac{dp}{dt}\right)^2 \equiv \frac{Z_0 q^2}{6\pi m^2 c^2}\left(\frac{d\mathbf{p}}{dt}\cdot\frac{d\mathbf{p}}{dt}\right). \tag{10.35}$$

This is evidently not a Lorentz-invariant result, but it gives a clear hint as to how such an invariant, that would be reduced to Eq. (10.35) in the non-relativistic limit, may be formed:

$$\mathscr{P} = -\frac{Z_0 q^2}{6\pi m^2 c^2}\left(\frac{dp_\alpha}{d\tau}\cdot\frac{dp^\alpha}{d\tau}\right) \equiv \frac{Z_0 q^2}{6\pi m^2 c^2}\left[\left(\frac{d\mathbf{p}}{d\tau}\right)^2 - \frac{1}{c^2}\left(\frac{d\mathscr{E}}{d\tau}\right)^2\right]. \tag{10.36}$$

Plugging in the relativistic expressions, $\mathbf{p} = \gamma mc\boldsymbol{\beta}$, $\mathscr{E} = \gamma mc^2$, and $d\tau = dt/\gamma$, the last formula may be recast into the so-called *Liénard extension* of the Larmor formula[9]:

$$\mathscr{P} = \frac{Z_0 q^2}{6\pi}\gamma^6[(\dot{\boldsymbol{\beta}})^2 - (\boldsymbol{\beta}\times\dot{\boldsymbol{\beta}})^2] \equiv \frac{Z_0 q^2}{6\pi}\gamma^4[(\dot{\boldsymbol{\beta}})^2 + \gamma^2(\boldsymbol{\beta}\cdot\dot{\boldsymbol{\beta}})^2], \tag{10.37}$$

which may be also obtained by a direct integration of Eq. (10.30) over the full solid angle, thus confirming our guess.

However, for some applications, it is beneficial to express $\mathscr{P}$ the via the time evolution of particle's momentum alone. For that, we may differentiate the fundamental relativistic relation (9.78), $\mathscr{E}^2 = (mc^2)^2 + (pc)^2$, over the proper time τ to obtain

$$2\mathscr{E}\frac{d\mathscr{E}}{d\tau} = 2c^2 p\frac{dp}{d\tau}, \qquad \text{i.e.}\ \frac{d\mathscr{E}}{d\tau} = \frac{c^2 p}{\mathscr{E}}\frac{dp}{d\tau} = u\frac{dp}{d\tau}, \tag{10.38}$$

[9] The second form of Eq. (10.37), which is frequently more convenient for applications, may be readily obtained from the first one by applying Eq. (A.49*a*) to the vector product.

where, at the last step, the magnitude of the relativistic vector relation $c^2\mathbf{p}/\mathscr{E} = \mathbf{u}$, mentioned in section 9.3, has been used. Plugging Eq. (10.38) into Eq. (10.36), we may rewrite it as

$$\mathscr{P} = \frac{Z_0 q^2}{6\pi m^2 c^2}\left[\left(\frac{d\mathbf{p}}{d\tau}\right)^2 - \beta^2\left(\frac{dp}{d\tau}\right)^2\right]. \tag{10.39}$$

Note the difference between the squared derivatives in this expression: in the first of them we have to differentiate the momentum vector $\mathbf{p}$, and only then form a scalar by squaring the resulting vector derivative, while in the second case, only the magnitude of the vector has to be differentiated. For example, for a circular motion with a constant speed (to be analyzed in detail in the next section), the second term is zero, while the first one is not.

However, if we return to the simplest case of linear acceleration (figure 10.4), then $(d\mathbf{p}/d\tau)^2 = (dp/d\tau)^2$, and Eq. (10.39) is reduced to

$$\mathscr{P} = \frac{Z_0 q^2}{6\pi m^2 c^2}\left(\frac{dp}{d\tau}\right)^2(1 - \beta^2) \equiv \frac{Z_0 q^2}{6\pi m^2 c^2}\left(\frac{dp}{d\tau}\right)^2\frac{1}{\gamma^2} \equiv \frac{Z_0 q^2}{6\pi m^2 c^2}\left(\frac{dp}{dt_{\text{ret}}}\right)^2, \tag{10.40}$$

i.e. formally coincides with the non-relativistic relation (10.35). In order to obtain a better feeling of the magnitude of this radiation, we may use the fact that $dp/dt_{\text{ret}} = d\mathscr{E}/dz'$, where z' is the particle's coordinate at moment t_{ret}.[10] It allows us to rewrite Eq. (10.40) in the following form:

$$\mathscr{P} = \frac{Z_0 q^2}{6\pi m^2 c^2}\left(\frac{d\mathscr{E}}{dz}\right)^2 \equiv \frac{Z_0 q^2}{6\pi m^2 c^2}\frac{d\mathscr{E}}{dz'}\frac{d\mathscr{E}}{dt_{\text{ret}}}\frac{dt_{\text{ret}}}{dz'} \equiv \frac{Z_0 q^2}{6\pi m^2 c^2 u}\frac{d\mathscr{E}}{dz'}\frac{d\mathscr{E}}{dt_{\text{ret}}}. \tag{10.41}$$

For the most important case of ultra-relativistic motion ($u \to c$), this result reduces to

$$\frac{\mathscr{P}}{d\mathscr{E}/dt_{\text{ret}}} \approx \frac{2}{3}\frac{d(\mathscr{E}/mc^2)}{d(z'/r_c)}, \tag{10.42}$$

where r_c is the classical radius of the particle, defined by Eq. (8.41). This formula shows that the radiated power, i.e. the change of particle's energy due to radiation, is much smaller than that due to the accelerating field, unless an energy as large as $\sim mc^2$ is gained on the classical radius of the particle. For example, for an electron, with $r_c \approx 3 \times 10^{-15}$ m and $mc^2 \approx 0.5$ MeV, such an acceleration would require an accelerating electric field of the order of (0.5 MV)/(3×10^{-15} m) ~ 10^{14} MV m^{-1}, while practicable accelerating fields are below 10^3 MV m^{-1}—limited by the electric breakdown effects. (As described by the factor m^2 in the denominator of Eq. (10.41), for heavier particles such as protons, the relative losses are even lower.) Such negligible radiative losses of energy are actually a large advantage of linear accelerators—such as the famous 2 mile long SLAC[11], which can accelerate

[10] This relation follows, for example, from comparison of Eq. (9.144) with $\mathbf{B} = 0$, and Eq. (9.148) with $\mathbf{E} \parallel \mathbf{u}$.
[11] See, e.g. https://www6.slac.stanford.edu/.

electrons or positrons to energies up to 50 GeV, i.e. to $\gamma \approx 10^5$. If obtaining radiation from the accelerated particles is the goal, it may be readily achieved by bending their trajectories using additional magnetic fields—see the next section.

10.3 Synchrotron radiation

Consider a charged particle being accelerated in the direction perpendicular to its velocity $\mathbf{u}$ (for example by the magnetic component of the Lorentz force), so that its speed u, and hence the magnitude p of its momentum, do not change. In this case, the second term in the square brackets of Eq. (10.39) vanishes, and it yields

$$\mathscr{P} = \frac{Z_0 q^2}{6\pi m^2 c^2}\left(\frac{d\mathbf{p}}{d\tau}\right)^2 = \frac{Z_0 q^2}{6\pi m^2 c^2}\left(\frac{d\mathbf{p}}{dt_{\text{ret}}}\right)^2 \gamma^2. \tag{10.43}$$

Comparing this expression with Eq. (10.40), we see that for the same acceleration magnitude, the electromagnetic radiation is a factor of γ^2 larger. For modern accelerators, with $\gamma \sim 10^4$–10^5, such a factor creates an enormous difference. For example, if a particle is on a cyclotron orbit in a constant magnetic field (as was analyzed in section 9.6), both $\mathbf{u}$ and $\mathbf{p} = \gamma m\mathbf{u}$ obey Eq. (9.150), so that

$$\left|\frac{d\mathbf{p}}{dt_{\text{ret}}}\right| = \omega_c p = \frac{u}{R}p = \beta^2\gamma\frac{mc^2}{R}, \tag{10.44}$$

(where R is the orbit's radius), so that for the power of this *synchrotron radiation*, Eq. (10.43) yields

$$\mathscr{P} = \frac{Z_0 q^2}{6\pi}\beta^4\gamma^4\frac{c^2}{R^2}. \tag{10.45}$$

According to Eq. (9.153), at a fixed magnetic field (in particle accelerators, limited to a few tesla produced by the beam-bending magnets), the synchrotron orbit radius R scales as γ, so that according to Eq. (10.45), $\mathscr{P}$ scales as γ^2, i.e. grows as the square of the particle's energy $\mathscr{E} \propto \gamma$. For example, for typical parameters of the first electron cyclotrons (such as the General Electric's machine in which the synchrotron radiation was first noticed in 1947), $R \sim 1$ m, $\mathscr{E} \sim 0.3$ GeV ($\gamma \sim 600$), Eq. (10.45) gives a very modest electron energy loss per one revolution: $\mathscr{P}\mathscr{T} \equiv \mathscr{P}(2\pi\mathscr{P}R/u) \approx 2\pi R/c \sim 1$ keV. However, already by the mid-1970 s, electron accelerators, with $R \sim 100$ m, could give each particle an energy $\mathscr{E} \sim 10$ GeV, and the energy loss per revolution grew to ~ 10 MeV, becoming the major energy loss mechanism. For proton accelerators, such energy loss is much less of a problem, because γ of an ultra-relativistic particle (at fixed $\mathscr{E}$) is proportional to $1/m$, so that the estimates at the same R should be scaled back by $(m_p/m_e)^4 \sim 10^{13}$. Nevertheless, in the giant modern accelerators such as the already mentioned LHC (with $R \approx 4.3$ km and $\mathscr{E}$ up to 7 TeV), the synchrotron radiation loss per revolution is rather noticeable ($\mathscr{P}\Delta t_{\text{ret}} \sim 6$ keV), leading not as much to particle deceleration as to a substantial photoelectron emission from the beam tube's walls, creating harmful defocusing effects.

However, what is bad for particle accelerators and storage rings is good for the so-called *synchrotron light sources*—the electron accelerators designed specifically for the generation of intensive synchrotron radiation—with a spectrum extending well beyond the visible light range. Let us now analyze the angular and spectral distributions of such radiation. To calculate the angular distribution, let us select the coordinate axes as shown in figure 10.5, with the origin at the current location of the orbiting particle, axis z directed along its instant velocity (i.e. the vector $\boldsymbol{\beta}$), and axis x toward the orbit center.

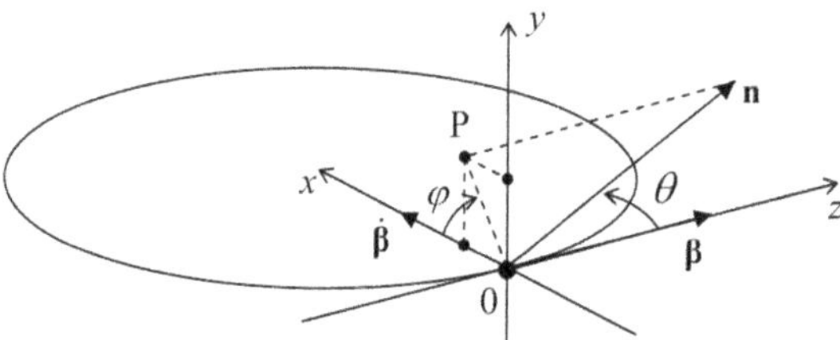

Figure 10.5. The synchrotron radiation problem's geometry.

In the general case, the unit vector $\mathbf{n}$ toward the radiation's observer is not within any of the coordinate planes, and hence should be described by two angles—the polar angle θ, and the azimuthal angle φ between the x axis and the projection OP of the vector $\mathbf{n}$ to the plane $[x, y]$. Since the length of the segment OP is $\sin\theta$, the Cartesian components of the relevant vectors are as follows:

$$\begin{aligned} \mathbf{n} &= \{\sin\theta\cos\phi,\quad \sin\theta\sin\phi,\quad \cos\theta\}, \\ \boldsymbol{\beta} &= \{0,\quad 0,\quad \beta\}, \qquad \text{and } \dot{\boldsymbol{\beta}} = \{\dot{\beta},\quad 0,\quad 0\}. \end{aligned} \tag{10.46}$$

Plugging these expressions into the general Eq. (10.30), we obtain

$$\begin{aligned} \frac{d\mathscr{P}}{d\Omega} &= \frac{2Z_0 q^2}{\pi^2}\,|\dot{\boldsymbol{\beta}}|^2\,\gamma^6 f(\theta, \phi), \\ \text{with}\quad f(\theta, \phi) &\equiv \frac{1}{8\gamma^6(1-\beta\cos\theta)^3}\left[1 - \frac{\sin^2\theta\cos^2\phi}{\gamma^2(1-\beta\cos\theta)^2}\right]. \end{aligned} \tag{10.47}$$

According to this result, just as at the linear acceleration, in the ultra-relativistic limit most radiation goes into a narrow cone (of a width $\Delta\theta \sim \gamma^{-1} \ll 1$) around the vector $\boldsymbol{\beta}$, i.e. around the instant direction of particle's propagation. For such small angles and $\gamma \gg 1$:

$$f(\theta, \phi) \approx \frac{1}{(1+\gamma^2\theta^2)^3}\left[1 - \frac{4\gamma^2\theta^2\cos^2\phi}{(1+\gamma^2\theta^2)^2}\right]. \tag{10.48}$$

The left-hand panel of figure 10.6 shows a color-coded contour map of the angular distribution $f(\theta, \varphi)$ of the radiation, as observed on a distant plane normal to the particle's instant velocity (in figure 10.5, parallel to the plane $[x, y]$), while the right-

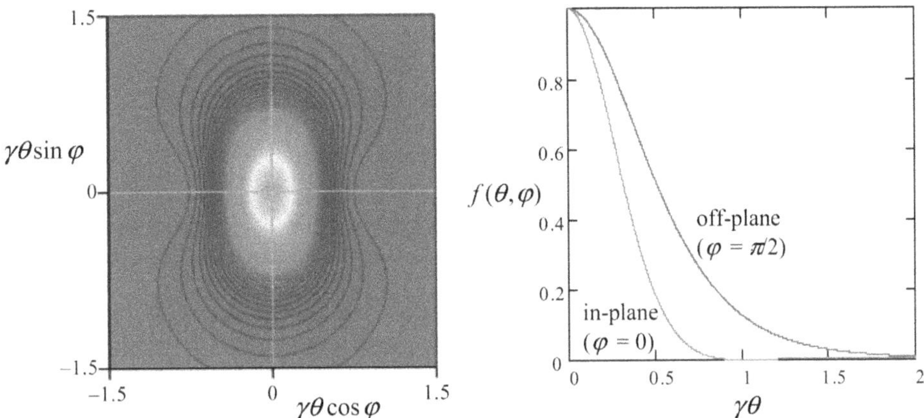

Figure 10.6. The angular distribution of the synchrotron radiation at $\gamma \gg 1$.

hand panel shows it as a function of θ in two perpendicular directions: within the particle rotation plane (in the direction parallel along axis x, i.e. at $\varphi = 0$) and perpendicular to this plane (along axis y, i.e. at $\varphi = \pm\pi/2$). The result shows, first of all, that, in contrast to the case of linear acceleration, the narrow radiation cone is now not hollow: the intensity maximum is reached at $\theta = 0$, i.e. exactly in the direction of the particle's motion. Second, the radiation cone is not axially symmetric: within the particle rotation plane, the intensity drops faster (and even has nodes at $\theta = \pm 1/\gamma$).

As figure 10.5 shows, the calculated angular distribution (10.47) of the synchrotron radiation is in the (inertial) reference frame whose origin coincides with the particle's position at this a particular instant, i.e. its radiation pattern is time-independent only in a frame moving with the particle. This pattern enables a semi-qualitative description of the radiation by an ultra-relativistic particle from the point of view of a stationary observer: if the observation point is on (or close to) the rotation plane[12], it is being 'struck' by the narrow radiation cone once each rotation period $\mathcal{T} \approx 2\pi R/c$, each 'strike' giving a pulse of a short duration $\Delta t_{\text{ret}} \ll 1/\omega_c$—see figure 10.7[13].

The evaluation of the time duration Δt of each pulse requires some care: its estimate $\Delta t_{\text{ret}} \sim 1/\gamma\omega_c$ is correct for the duration of the time during which its cone is aimed at the observer. However, due to the time compression effect, discussed in detail in section 10.1 and described by Eq. (10.12), the pulse duration as seen by the observer is a factor of $1/(1-\beta)$ shorter, so that

$$\Delta t = (1-\beta)\Delta t_{\text{ret}} \sim \frac{1-\beta}{\gamma\omega_c} \sim \frac{1}{\gamma^3\omega_c} \sim \gamma^{-3}\mathcal{T}, \qquad \text{for } \gamma \gg 1. \tag{10.49}$$

From the Fourier theorem, we can expect the frequency spectrum of such radiation to consist of numerous ($N \sim \gamma^3 \gg 1$) harmonics of the particle rotation frequency ω_c,

[12] It is easy (and hence is left as a reader's exercise) to show that if the observation point is greatly off-plane (say, is located on the particle orbit's axis), the radiation is virtually monochromatic, with frequency ω_c. (As we know from section 8.2, in the non-relativistic limit $u \ll c$ this is true for *any* observation point.)

[13] The fact that the in-plane component of each electric field's pulse $\mathbf{E}(t)$ is asymmetric with respect to its central point, and hence vanishes at that point (as figure 10.7b shows), readily follows from Eq. (10.19).

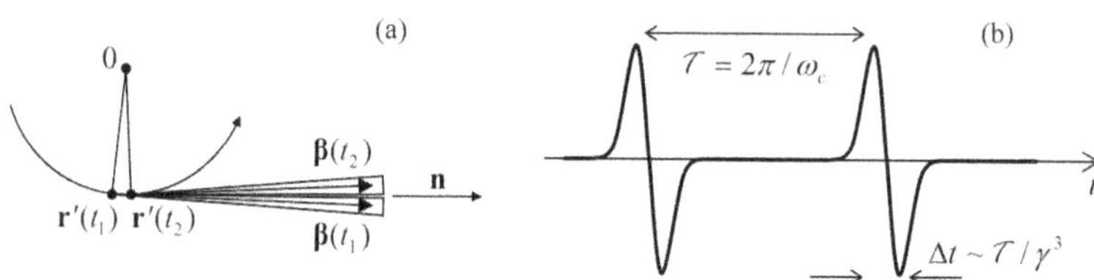

Figure 10.7. Schematic representations of (a) the synchrotron radiation cones (at $\gamma \gg 1$) for two close values of t_{ret}, and (b) the in-plane component of the electric field, observed in the rotation plane, as a function of time t.

with comparable amplitudes. However, if the orbital frequency fluctuates even slightly ($\delta\omega_c/\omega_c > 1/N \sim 1/\gamma^3$), as happens in most practical systems, the radiation pulses are not coherent, so that the average radiation power spectrum may be calculated as that of one pulse multiplied by the number of pulses per second. In this case, the spectrum is continuous, extending from low frequencies all the way to approximately

$$\omega_{\max} \sim 1/\Delta t \sim \gamma^3 \omega_c. \tag{10.50}$$

In order to verify and quantify this result, let us calculate the spectrum of radiation due to a single pulse. For that, we should first make the general notion of the radiation spectrum quantitative. Let us represent an arbitrary electric field (say that of the synchrotron radiation we are studying now), observed at a fixed point **r** as a function of the observation time t, as a Fourier integral[14]:

$$\mathbf{E}(t) = \int_{-\infty}^{+\infty} \mathbf{E}_\omega e^{-i\omega t}\, dt. \tag{10.51}$$

This expression may be plugged into the formula for the total energy of the radiation pulse (i.e. of the loss of particle's energy $\mathscr{E}$) per unit solid angle[15]:

$$-\frac{d\mathscr{E}}{d\Omega} \equiv \int_{-\infty}^{+\infty} S_n(t) R^2\, dt = \frac{R^2}{Z_0} \int_{-\infty}^{+\infty} |\mathbf{E}(t)|^2\, dt. \tag{10.52}$$

This substitution, plus a natural change of the integration order, yield

$$-\frac{d\mathscr{E}}{d\Omega} \equiv \frac{R^2}{Z_0} \int_{-\infty}^{+\infty} d\omega \int_{-\infty}^{+\infty} d\omega' \mathbf{E}_\omega \cdot \mathbf{E}_{\omega'} \int_{-\infty}^{+\infty} dt e^{-i(\omega+\omega')t}. \tag{10.53}$$

---

[14] In contrast to the single-frequency case (i.e. a monochromatic wave), we may avoid taking the real part of the complex function ($\mathbf{E}_\omega e^{-i\omega t}$) by requiring that in Eq. (10.51), $\mathbf{E}_{-\omega} = \mathbf{E}_\omega^*$. However, it is important to remember the factor ½ required for the transition to a monochromatic wave of frequency ω_0 and real amplitude E_0: $\mathbf{E}_\omega = \mathbf{E}_0 [\delta(\omega - \omega_0) + \delta(\omega + \omega_0)]/2$.

[15] Note that the expression under the integral differs from $d\mathscr{P}/d\Omega$ defined by Eq. (10.29) by the absence of term $(1 - \boldsymbol{\beta} \cdot \mathbf{n}) = \partial t_{\text{ret}}/\partial t$—see Eq. (10.16). This is natural, because this is the wave energy arriving at the observation point **r** during time interval dt rather than dt_{ret}.

But the inner integral (over t) is just $2\pi\delta(\omega + \omega')$.[16] This delta-function kills one of the frequency integrals (say, one over ω') and Eq. (10.53) gives us a result that may be recast as

$$-\frac{d\mathscr{E}}{d\Omega} = \int_0^{+\infty} I(\omega)d\omega, \qquad \text{with } I(\omega) \equiv \frac{4\pi R^2}{Z_0}\mathbf{E}_\omega \cdot \mathbf{E}_{-\omega} \equiv \frac{4\pi R^2}{Z_0}\mathbf{E}_\omega \mathbf{E}^*_\omega, \tag{10.54}$$

where the evident frequency symmetry of the scalar product $\mathbf{E}_\omega \cdot \mathbf{E}_{-\omega}$ has been utilized to fold the integral of $I(\omega)$ to positive frequencies only. The first of Eqs. (10.54) makes the physical sense of the function $I(\omega)$ clear: this is the so-called *spectral density* of the electromagnetic radiation (per unit solid angle)[17].

In order to calculate the spectral density, we need to express the function $\mathbf{E}_\omega$ via $\mathbf{E}(t)$ using the Fourier transform reciprocal to Eq. (10.51):

$$\mathbf{E}_\omega = \frac{1}{2\pi}\int_{-\infty}^{+\infty} \mathbf{E}(t)e^{i\omega t}\, dt. \tag{10.55}$$

In the particular case of radiation by a single point charge, we can use the second (radiative) term of Eq. (10.19):

$$\mathbf{E}_\omega = \frac{1}{2\pi}\frac{q}{4\pi\varepsilon_0}\frac{1}{cR}\int_{-\infty}^{+\infty}\left[\frac{\mathbf{n} \times \{(\mathbf{n} - \boldsymbol{\beta}) \times \dot{\boldsymbol{\beta}}\}}{(1 - \boldsymbol{\beta}\cdot\mathbf{n})^3}\right]_{\text{ret}} e^{i\omega t}\, dt. \tag{10.56}$$

Since the vectors $\mathbf{n}$ and $\boldsymbol{\beta}$ are more natural functions of the radiation (retarded) time t_{ret}, let us use Eqs. (10.5) and (10.16) to exclude the observation time t from this integral:

$$\mathbf{E}_\omega = \frac{q}{4\pi\varepsilon_0}\frac{1}{2\pi}\frac{1}{cR}\int_{-\infty}^{+\infty}\left[\frac{\mathbf{n} \times \{(\mathbf{n} - \boldsymbol{\beta}) \times \dot{\boldsymbol{\beta}}\}}{(1 - \boldsymbol{\beta}\cdot\mathbf{n})^2}\right]_{\text{ret}} \exp\left\{i\omega\left(t_{\text{ret}} + \frac{R_{\text{ret}}}{c}\right)\right\}dt_{\text{ret}}. \tag{10.57}$$

Assuming that the observer is sufficiently far from the particle[18], we may treat the unit vector $\mathbf{n}$ as a constant and also use the approximation (8.19) to reduce Eq. (10.57) to

$$\begin{aligned}\mathbf{E}_\omega = &\frac{1}{2\pi}\frac{q}{4\pi\varepsilon_0}\frac{1}{cR}\exp\left\{\frac{i\omega r}{c}\right\}\\ &\times\int_{-\infty}^{+\infty}\left[\frac{\mathbf{n} \times \{(\mathbf{n} - \boldsymbol{\beta}) \times \dot{\boldsymbol{\beta}}\}}{(1 - \boldsymbol{\beta}\cdot\mathbf{n})^2}\exp\left\{i\omega\left(t - \frac{\mathbf{n}\cdot\mathbf{r}'}{c}\right)\right\}\right]_{\text{ret}} dt_{\text{ret}}.\end{aligned} \tag{10.58}$$

---

[16] See, e.g., Eq. (A.88).

[17] The notion of spectral density may be readily generalized to random processes—see, e.g. *Part QM* section 7.4 and *Part SM* section 5.4.

[18] According to the estimate Eq. (10.49), for a synchrotron radiation's pulse this restriction requires the observer to be much farther than $\Delta r' \sim c\Delta t \sim R/\gamma^3$ from the particle. With the values $R \sim 10^4$ m and $\gamma \sim 10^5$ mentioned above, $\Delta r' \sim 10^{-11}$ m, so this requirement is satisfied for any realistic radiation detector.

Plugging this expression into Eq. (10.54), and using the definitions $c \equiv 1/(\varepsilon_0\mu_0)^{1/2}$ and $Z_0 \equiv (\mu_0/\varepsilon_0)^{1/2}$, we obtain[19]

$$I(\omega) = \frac{Z_0 q^2}{16\pi^3}\left|\int_{-\infty}^{+\infty}\left[\frac{\mathbf{n}\times\{(\mathbf{n}-\boldsymbol{\beta})\times\dot{\boldsymbol{\beta}}\}}{(1-\boldsymbol{\beta}\cdot\mathbf{n})^2}\exp\left\{i\omega\left(t-\frac{\mathbf{n}\cdot\mathbf{r}'}{c}\right)\right\}\right]_{\text{ret}} dt_{\text{ret}}\right|^2. \quad (10.59)$$

This result may be further simplified by noticing that the fraction before the exponent may be represented as a full derivative over t_{ret},

$$\begin{aligned}\left[\frac{\mathbf{n}\times\{(\mathbf{n}-\boldsymbol{\beta})\times\dot{\boldsymbol{\beta}}\}}{(1-\boldsymbol{\beta}\cdot\mathbf{n})^2}\right]_{\text{ret}} &\equiv \left[\frac{\mathbf{n}\times\{(\mathbf{n}-\boldsymbol{\beta})\times d\boldsymbol{\beta}/dt\}}{(1-\boldsymbol{\beta}\cdot\mathbf{n})^2}\right]_{\text{ret}} \\ &\equiv \frac{d}{dt}\left[\frac{\mathbf{n}\times(\mathbf{n}\times\boldsymbol{\beta})}{1-\boldsymbol{\beta}\cdot\mathbf{n}}\right]_{\text{ret}},\end{aligned} \quad (10.60)$$

and working out the resulting integral by parts. At this operation, the time differentiation of the parentheses in the exponent gives $d[t_{\text{ret}} - \mathbf{n}\cdot\mathbf{r}'(t_{\text{ret}})/c]/dt_{\text{ret}} = (1-\mathbf{n}\cdot\mathbf{u}/c)_{\text{ret}} \equiv (1-\boldsymbol{\beta}\cdot\mathbf{n})_{\text{ret}}$, leading to the cancellation of the remaining factor in the denominator and hence to a very simple general result[20]:

$$I(\omega) = \frac{Z_0 q^2\omega^2}{16\pi^3}\left|\int_{-\infty}^{+\infty}\left[\mathbf{n}\times(\mathbf{n}\times\boldsymbol{\beta})\exp\left\{i\omega\left(t-\frac{\mathbf{n}\cdot\mathbf{r}'}{c}\right)\right\}\right]_{\text{ret}} dt_{\text{ret}}\right|^2. \quad (10.61)$$

Now returning to the particular case of the synchrotron radiation, it is beneficial to choose the origin of time t_{ret} so that at $t_{\text{ret}} = 0$ the angle θ between the vectors $\mathbf{n}$ and $\boldsymbol{\beta}$ takes its smallest value θ_0, i.e. in terms of figure 10.5 the vector $\mathbf{n}$ is within the plane $[y, z]$. Fixing this direction of axes so that they do not move in further times, we can redraw that figure as shown in figure 10.8.

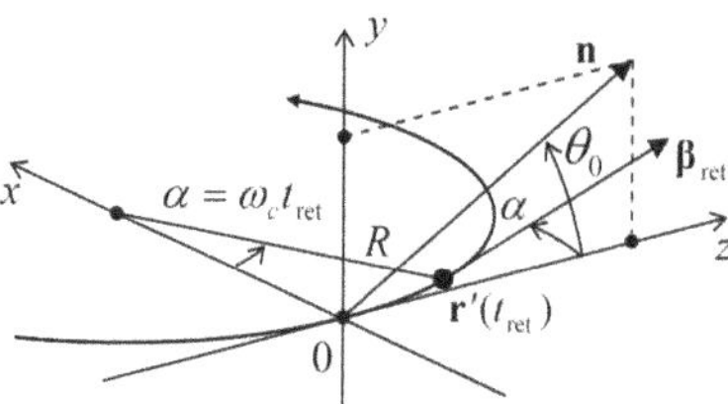

Figure 10.8. Deriving the spectral density of synchrotron radiation. The vector $\mathbf{n}$ is static within the plane $[y, z]$, while the vectors $\mathbf{r}'(t_{\text{ret}})$ and $\boldsymbol{\beta}_{\text{ret}}$ rotate within the plane $[x, y]$ with the angular velocity ω_c of the particle.

[19] Note that for our current purposes of calculation of the spectral density of radiation by a single particle, the factor $\exp\{i\omega r/c\}$ has been cancelled. However, as we have seen in chapter 8, this factor plays the central role at interference of radiation from several (many) sources. Such interference is important, in particular, in *undulators* and *free-electron lasers*—the devices to be (qualitatively) discussed below.

[20] Actually, this simplification is not occasional. According to Eq. (10.10b), the expression under the derivative in the last form of Eq. (10.60) is just the transverse component of the vector-potential **A** (give or take a constant factor), and from the discussion in section 8.2 we know that this component determines the electric dipole radiation of the particle, which dominates the radiation field in our current case of a particle with uncompensated electric charge.

In this 'lab' reference frame, the vector $\mathbf{n}$ does not depend on time, while the vectors $\mathbf{r}'(t_{\text{ret}})$ and $\boldsymbol{\beta}_{\text{ret}}$ do depend on it via angle $\alpha \equiv \omega_c t_{\text{ret}}$:

$$\begin{aligned}
&\mathbf{n} = \{0, \ \sin\theta_0, \ \cos\theta_0\}, \ \mathbf{r}'(t_{\text{ret}}) = \{R(1 - \cos\alpha), \ 0, \ R\sin\alpha\}, \\
&\boldsymbol{\beta}_{\text{ret}} \equiv \{\beta\sin\alpha, \ 0, \ \beta\cos\alpha\}.
\end{aligned} \tag{10.62}$$

Now an easy multiplication yields

$$[\mathbf{n}\times(\mathbf{n}\times\boldsymbol{\beta})]_{\text{ret}} = \beta\{\sin\alpha, \ \sin\theta_0\cos\theta_0\cos\alpha, \ -\sin^2\theta_0\sin\alpha\}, \tag{10.63}$$

$$\left[\exp\left\{i\omega\left(t - \frac{\mathbf{n}\cdot\mathbf{r}'}{c}\right)\right\}\right]_{\text{ret}} = \exp\left\{i\omega\left(t_{\text{ret}} - \frac{R}{c}\cos\theta_0\sin\alpha\right)\right\}. \tag{10.64}$$

As we already know, in the (most interesting) ultra-relativistic limit $\gamma \gg 1$, most radiation is confined to short pulses, so that only small angles $\alpha \sim \omega_c \Delta t_{\text{ret}} \sim \gamma^{-1}$ may contribute to the integral in Eq. (10.61). Moreover, since most radiation goes to small angles $\theta \sim \theta_0 \sim \gamma^{-1}$, it makes sense to consider only such small angles. Expanding both trigonometric functions of these small angles participating in parentheses of Eq. (10.64) into the Taylor series, and keeping only the leading terms, we obtain

$$t_{\text{ret}} - \frac{R}{c}\cos\theta_0\sin\alpha \approx t_{\text{ret}} - \frac{R}{c}\omega_c t_{\text{ret}} + \frac{R}{c}\frac{\theta_0^2}{2}\omega_c t_{\text{ret}} + \frac{R}{c}\frac{\omega_c^3}{6}t_{\text{ret}}^3. \tag{10.65}$$

Since $(R/c)\omega_c = u/c = \beta \approx 1$, in the two last terms we may approximate this parameter by 1. However, it is crucial to distinguish the difference of the two first terms, proportional to $(1 - \beta)t_{\text{ret}}$, from zero; as we have done before, we may approximate it with $t_{\text{ret}}/2\gamma^2$. On the right-hand side of Eq. (10.63), which does not have such a critical difference, we may be bolder, taking[21]

$$\begin{aligned}
\beta\{\sin\alpha, \ \sin\theta_0\cos\theta_0\cos\alpha, \ -\sin^2\theta_0\sin\alpha\} &\approx \{\alpha, \theta_0, 0\} \\
&\equiv \{\omega_c t_{\text{ret}}, \theta_0, 0\}.
\end{aligned} \tag{10.66}$$

As a result, Eq. (10.61) is reduced to

$$I(\omega) = \frac{Z_0 q^2}{16\pi^3}\,|a_x\mathbf{n}_x + a_y\mathbf{n}_y|^2 \equiv \frac{Z_0 q^2}{16\pi^3}\left(|a_x|^2 + |a_y|^2\right), \tag{10.67}$$

where a_x and a_y are dimensionless factors,

[21] This expression confirms that the in-plane (x) component of the electric field is an odd function of t_{ret} and hence of $t - t_0$ (see its sketch in figure 10.7b), while the normal (y) component is an even function of this difference. Also note that for an observer exactly in the rotation plane ($\theta_0 = 0$) the latter component equals zero for all times—the fact which could be predicted from the very beginning because of the evident mirror symmetry of the problem with respect to the particle rotation plane.

$$a_x \equiv \omega \int_{-\infty}^{+\infty} \omega_c t_{\text{ret}} \exp\left\{\frac{i\omega}{2}\left((\theta_0^2 + \gamma^{-2})t_{\text{ret}} + \frac{\omega_c^2}{3}t_{\text{ret}}^3\right)\right\} dt_{\text{ret}},$$
$$a_y \equiv \omega \int_{-\infty}^{+\infty} \theta_0 \exp\left\{\frac{i\omega}{2}\left((\theta_0^2 + \gamma^{-2})t_{\text{ret}} + \frac{\omega_c^2}{3}t_{\text{ret}}^3\right)\right\} dt_{\text{ret}}, \tag{10.68}$$

which describe the frequency spectra of two components of the synchrotron radiation, with mutually perpendicular directions of polarization. Defining the following dimensionless parameter

$$\nu \equiv \frac{\omega}{3\omega_c}(\theta_0^2 + \gamma^{-2})^{3/2}, \tag{10.69}$$

which is proportional to the observation frequency, and changing the integration variable to $\xi \equiv \omega_c t_{\text{ret}}/(\theta_0{}^2 + \gamma^{-2})^{1/2}$, the integrals (10.68) may be reduced to the modified Bessel functions of the second kind, but with fractional indices:

$$\begin{aligned} a_x &= \frac{\omega}{\omega_c}(\theta_0^2 + \gamma^{-2}) \int_{-\infty}^{+\infty} \xi \exp\left\{\frac{3}{2}i\nu\left(\xi + \frac{\xi^3}{3}\right)\right\} d\xi \\ &= \frac{2\sqrt{3}\ i}{(\theta_0^2 + \gamma^{-2})^{1/2}} \nu K_{2/3}(\nu), \\ a_y &= \frac{\omega}{\omega_c}\theta_0(\theta_0^2 + \gamma^{-2})^{1/2} \int_{-\infty}^{+\infty} \exp\left\{\frac{3}{2}i\nu\left(\xi + \frac{\xi^3}{3}\right)\right\} d\xi \\ &= \frac{2\sqrt{3}\ \theta_0}{\theta_0^2 + \gamma^{-2}} \nu K_{1/3}(\nu). \end{aligned} \tag{10.70}$$

Figure 10.9a shows the dependence of the Bessel factors, defining the amplitudes a_x and a_y, on the normalized observation frequency ν. It shows that the radiation intensity changes with frequency rather slowly (note the log–log scale of the plot) until the normalized frequency, defined by Eq. (10.69), is increased beyond ~1. For most important observation angles $\theta_0 \sim \gamma$ this means that our estimate (10.50) is indeed correct, although formally the frequency spectrum extends to infinity[22].

Naturally, the spectral density integrated over the full solid angle exhibits a similar frequency behavior. Without performing the integration[23], let me just give the result (also valid for $\gamma \gg 1$ only) for the reader's reference:

$$\oint_{4\pi} I(\omega)d\Omega = \frac{\sqrt{3}}{4\pi}q^2\gamma\zeta \int_{\zeta}^{\infty} K_{5/3}(\xi)d\xi, \qquad \text{where } \zeta \equiv \frac{2}{3}\frac{\omega}{\omega_c\gamma^3}. \tag{10.71}$$

[22] The law of the spectral density decrease at large ν may be readily obtained from the second of Eqs. (2.158), which is valid even for any (even non-integer) Bessel function index n: $a_x \propto a_y \propto \nu^{-1/2}\exp\{-\nu\}$. Here the exponential factor is certainly most important.

[23] For that, and many other details, the interested reader is referred, for example, to the fundamental review collection [1] or the more concise monograph [2].

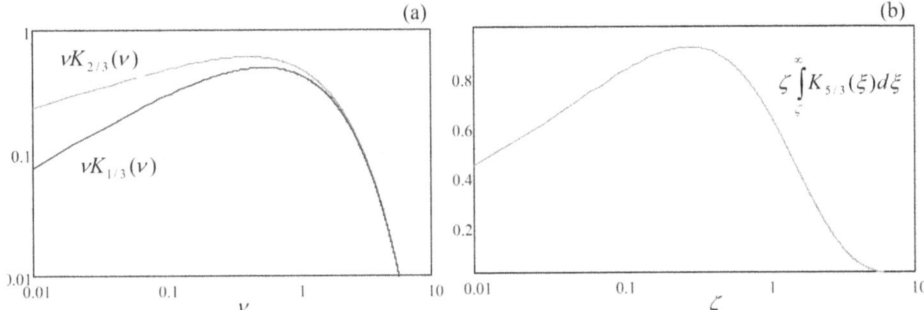

Figure 10.9. The frequency spectra of (a) two components of the synchrotron radiation, at a fixed angle θ_0, and (b) its total (polarization- and angle-averaged) intensity.

Figure 10.9b shows the dependence of this integral on the normalized frequency ζ. (This plot is sometimes called the 'universal flux curve'.) In accordance with estimate (10.50), it reaches maximum at

$$\zeta_{\max} \approx 0.3, \quad \text{i.e. } \omega_{\max} \approx \frac{\omega_c}{2}\gamma^3. \tag{10.72}$$

For example, in the new National Synchrotron Light Source (NSLS-II) in the Brookhaven National Laboratory near the SBU campus, with a ring circumference of 792 m, the electron revolution period $\mathcal{T}$ is 2.64 μs. Calculating ω_c as $2\pi/\mathcal{T} \approx 2.4 \times 10^6\ \text{s}^{-1}$, for the achieved $\gamma \approx 6 \times 10^3$ ($\mathscr{E} \approx 3$ GeV)[24] we obtain $\omega_{\max} \sim 3 \times 10^{17}\ \text{s}^{-1}$ (the photon energy $\hbar\omega_{\max} \sim 200$ eV), corresponding to soft x-rays. In the light of this estimate, the reader may be surprised by figure 10.10, which shows the calculated spectra of radiation that this facility was designed to produce, with the intensity maxima at photon energies up to a few keV.

The reason for this discrepancy is that in NLLS-II, and in all modern synchrotron light sources, most radiation is produced not by the circular orbit itself (which is, by the way, not exactly circular, but consists of a series of straight and bend-magnet sections), but by such bend sections, and also devices called *wigglers* and *undulators*: strings of several strong magnets with alternating field direction (figure 10.11), that induce periodic bending ('wiggling') of the electron's trajectory, with the synchrotron radiation emitted at each bend.

The difference between the wigglers and the undulators is more quantitative than qualitative: the former devices have a larger spatial period λ_u (the distance between the adjacent magnets of the same polarity, see figure 10.11), giving enough space for the electron beam to bend by an angle larger than γ^{-1}, i.e. larger than the radiation cone's width. As a result, the radiation arrives to an in-plane observer as a periodic sequence of individual pulses (figure 10.12a).

[24] By modern standards, this *energy* is not too high. The distinguished feature of NSLS-II is its unprecedented electron beam *intensity* (planned average beam current up to 500 mA) which should allow an extremely high synchrotron 'brightness'—see figure 10.10.

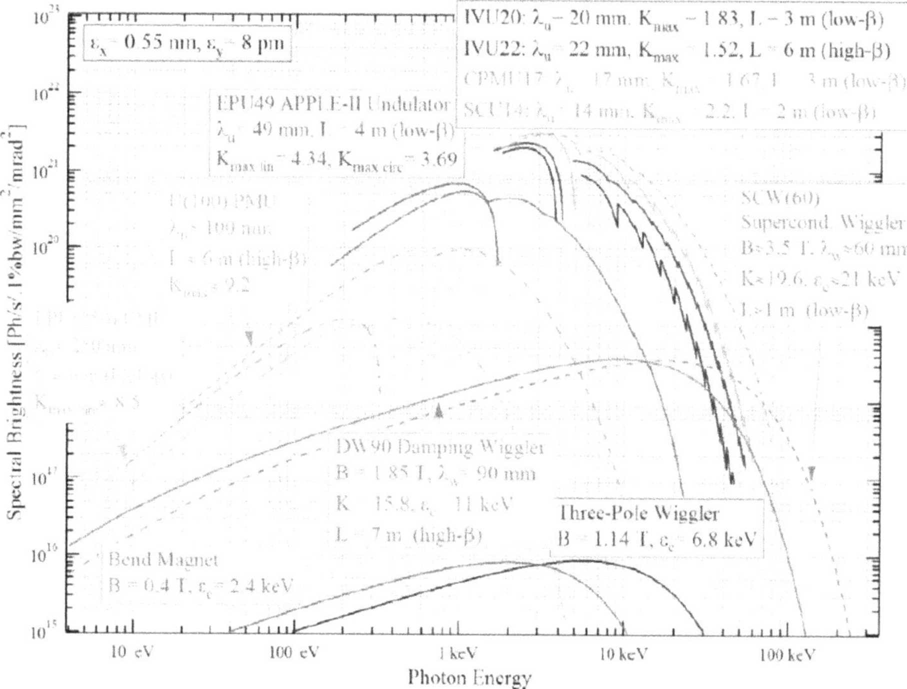

Figure 10.10. Design brightness of various synchrotron radiation sources of the NSLS-II facility. For bend magnets and wigglers, the 'brightness' may be obtained by multiplication of the spectral density $I(\omega)$ from one electron pulse, calculated above, by the number of electrons passing the source per second. (Note the non-SI units, commonly used by the synchrotron radiation community.) However, for undulators, there is an additional factor due to the partial coherence of radiation—see below. (Adapted from *NSLS-II Source Properties and Floor Layout*, available online at https://www.bnl.gov/ps/docs/pdf/SourceProperties.pdf.)

The shape of each pulse, and hence its frequency spectrum, is essentially similar to those discussed above[25], but with much higher local values of ω_c and hence ω_{max}—see figure 10.10. Another difference is a much higher frequency of the pulses. Indeed, the fundamental Eq. (10.16) allows us to calculate the time distance between them, for the observer, as

$$\Delta t \approx \frac{\partial t}{\partial t_{ret}} \Delta t_{ret} \approx (1 - \beta)\frac{\lambda_u}{u} \approx \frac{1}{2\gamma^2}\frac{\lambda_u}{c} \ll \frac{\lambda_u}{c}, \tag{10.73}$$

where the first two relations are valid at $\lambda_u \ll R$ (a relation that is typically satisfied very well, see the numbers in figure 10.10), and the last two relations assume the ultra-relativistic limit. As a result, the radiation intensity, which is proportional to

[25] Indeed, the period λ_u is typically a few centimeters (see the numbers in figure 10.10), i.e. is much larger than the interval $\Delta r' \sim R/\gamma^3$ estimated above. Hence the synchrotron radiation results may be applied locally, to each electron beam's bend. (In this context, a simple problem for the reader: use Eqs. (10.19) and (10.63) to explain the difference between the shapes of the in-plane electric field pulses emitted at opposite magnetic poles of the wiggler, which is schematically shown in figure 10.11a.)

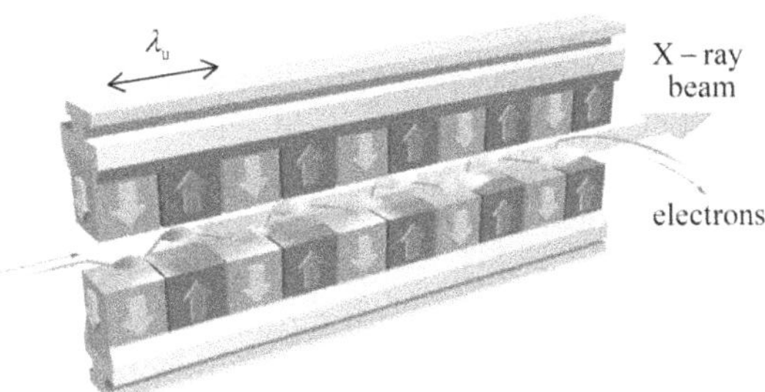

Figure 10.11. The generic structure of wigglers, undulators, and free-electron lasers. (Adapted from http://www.xfel.eu/overview/how_does_it_work/.)

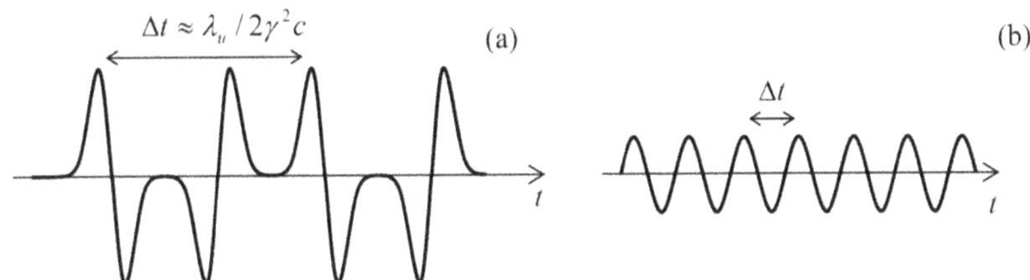

Figure 10.12. Schematic representation of waveforms of the radiation emitted by (a) a wiggler and (b) an undulator.

the number of the magnetic poles, is much higher than that from the bend magnets—see figure 10.10 again.

The situation is different in undulators—similar structures with a smaller spatial period λ_u, in which the electron's velocity vector oscillates with an angular amplitude smaller than γ^{-1}. As a result, the radiation pulses overlap (figure 10.12b) and the radiation waveform is closer to a sinusoidal one. As a result, the radiation spectrum narrows to the central frequency[26]

$$\omega_0 = \frac{2\pi}{\Delta t} \approx 2\gamma^2 \frac{2\pi c}{\lambda_u}. \tag{10.74}$$

For example, for the LSNL-II undulators with $\lambda_u = 2$ cm, this formula predicts the radiation peak at phonon energy $\hbar\omega_0 \approx 4$ keV, in reasonable agreement with the results of quantitative calculations, shown in figure 10.10.[27] Due to the spectrum narrowing, the intensity of undulators radiation is higher than that of wigglers using the same electron beam.

[26] This important formula may be also derived in the following way. Due to the relativistic length contraction (9.20), the undulator structure period as perceived by beam electrons is $\lambda' = \lambda_u/\gamma$, so that the central frequency of the radiation in the reference frame moving with the electrons is $\omega_0' = 2\pi c/\lambda' = 2\pi c\gamma/\lambda_u$. For the lab-frame observer, this frequency is Doppler-upshifted in accordance with Eq. (9.44): $\omega_0 = \omega_0'[(1+\beta)/(1-\beta)]^{1/2} \approx 2\gamma\omega_0'$, giving the same result as Eq. (10.74).

[27] Some of the difference is due to the fact that those plots show the spectral density of the number of *photons* $n = \mathscr{E}/\hbar\omega$ per second, which peaks at a frequency below that of the density of power, i.e. of the energy $\mathscr{E}$ per second.

This spectrum-narrowing trend is brought to its logical conclusion in the so-called *free-electron lasers*,[28] whose basic structure is the same as that of wigglers and undulators (figure 10.11), but the radiation at each beam bend is so intense and narrow-focused that it affects the electron motion downstream the radiation cone. As a result, the radiation spectrum narrows around the central frequency (10.74), and its power grows as a square of the number N of electrons in the structure (rather than proportionately to N in wigglers and undulators).

Finally, note that wigglers, undulators, and free-electron lasers may also be used at the end of linear electron accelerators (such as SLAC) which, as was noted above, may provide extremely high values of γ, and hence radiation frequencies, due to the absence of the radiation energy losses at the electron acceleration stage. Very unfortunately, I do not have time/space to discuss (the very interesting) physics of these devices in more detail[29].

10.4 Bremsstrahlung and Coulomb losses

Surprisingly, a very similar mechanism of radiation by charged particles works at much lower spatial scale, namely at their scattering by charged particles of the propagation medium. This effect, traditionally called by its German name *bremsstrahlung* ('brake radiation'), is responsible, in particular, for the continuous part of the frequency spectrum of the radiation produced in standard vacuum x-ray tubes, at the electrons' collisions with a metallic 'anti-cathode'[30].

The bremsstrahlung in condensed matter is generally a rather complicated phenomenon, because of the simultaneous involvement of many particles and (frequently) some quantum electrodynamic effects. This is why I will provide only a very brief glimpse of the theoretical description of this effect, for the simplest case when the scattering of incoming, relatively light charged particles (such as electrons, protons, α-particles, etc) is produced by atomic nuclei that remain virtually immobile during the scattering event (figure 10.13). This is a reasonable approximation if the energy of incoming particles is not too low; otherwise most scattering

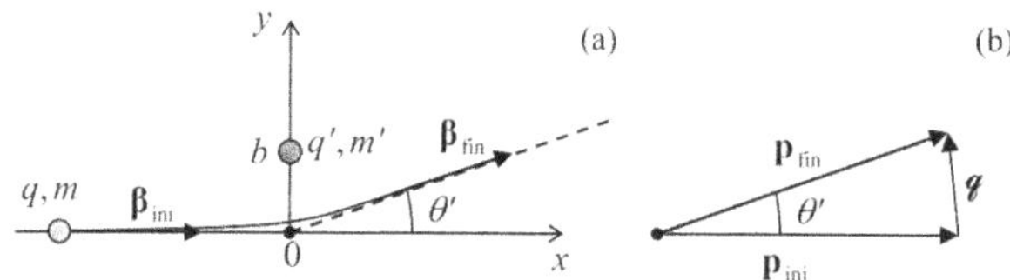

Figure 10.13. The basic geometry of the bremsstrahlung and the Coulomb loss problems in (a) direct and (b) reciprocal spaces.

[28] This name is somewhat misleading, because in contrast to the usual ('quantum') lasers, the free-electron laser is essentially a classical device, and the dynamics of electrons in it is very similar to that in other vacuum microwave generators, such as the magnetrons briefly discussed in section 9.6 and especially klystrons (see chapter 7).

[29] The interested reader may be referred, for example, to either [3] or [4].

[30] Such x-ray radiation was first observed experimentally, although not correctly interpreted, by N Tesla in 1887, i.e. before the radiation was re-discovered (in 1895) and studied in detail by W Röntgen.

is produced by atomic electrons whose dynamics is substantially quantum—see below.

To calculate the frequency spectrum of radiation emitted during a single scattering event, it is convenient to use a by-product of the last section's analysis, namely Eq. (10.59) with the replacement (10.60)[31]:

$$I(\omega) = \frac{1}{4\pi^2 c}\frac{q^2}{4\pi\varepsilon_0}\left|\int_{-\infty}^{+\infty}\left[\frac{d}{dt}\frac{\mathbf{n}\times(\mathbf{n}\times\boldsymbol{\beta})}{1-\boldsymbol{\beta}\cdot\mathbf{n}}\exp\left\{i\omega\left(t-\frac{\mathbf{n}\cdot\mathbf{r}'}{c}\right)\right\}\right]_{\text{ret}} dt_{\text{ret}}\right|^2. \quad (10.75)$$

A typical duration τ of a single scattering event we are discussing is of the order of $a_0/c \sim (10^{-10}\ \text{m})/(3\times 10^8\ \text{m s}^{-1}) \sim 10^{-18}$ s in solids, and only an order of magnitude longer in gases at ambient conditions. This is why for most frequencies of interest, from zero all the way up to *at least* soft x-rays[32], we can use the so-called *low-frequency approximation*, taking the exponent in Eq. (10.75) for 1 through the whole collision event, i.e. the integration interval. This approximation immediately yields

$$I(\omega) = \frac{1}{4\pi^2 c}\frac{q^2}{4\pi\varepsilon_0}\left|\frac{\mathbf{n}\times(\mathbf{n}\times\boldsymbol{\beta}_{\text{fin}})}{1-\boldsymbol{\beta}_{\text{fin}}\cdot\mathbf{n}} - \frac{\mathbf{n}\times(\mathbf{n}\times\boldsymbol{\beta}_{\text{ini}})}{1-\boldsymbol{\beta}_{\text{ini}}\cdot\mathbf{n}}\right|^2. \quad (10.76)$$

In the non-relativistic limit ($\beta_{\text{ini}}, \beta_{\text{fin}} \ll 1$), this formula is reduced to the result

$$I(\omega) = \frac{1}{4\pi^2 c}\frac{q^2}{4\pi\varepsilon_0}\frac{\mathscr{q}^2}{m^2c^{-2}}\sin^2\theta \quad (10.77)$$

(which may be derived from Eq. (8.27) as well), where $\mathscr{q}$ is the momentum transferred from the scattering center to the scattered charge (figure 10.13b)[33]:

$$\mathscr{q} \equiv \mathbf{p}_{\text{fin}} - \mathbf{p}_{\text{ini}} = m\Delta\mathbf{u} = mc\Delta\boldsymbol{\beta} = mc(\boldsymbol{\beta}_{\text{fin}} - \boldsymbol{\beta}_{\text{ini}}), \quad (10.78)$$

and θ is the angle between the vector $\mathscr{q}$ and the direction $\mathbf{n}$ toward the observer (at the collision moment).

[31] In publications on this topic (whose development peak was in the 1920–30 s) Gaussian units are more common, and the letter Z is usually reserved for expressing charges as multiples of the fundamental charge e, rather than for the wave impedance. This is why, in order to avoid confusion and facilitate the comparison with other texts, in this section I (still staying with the SI units used through my series) will use the fraction $1/\varepsilon_0 c$, instead of its equivalent Z_0, for the free-space wave impedance and write the coefficients in a form that makes the transfer to the Gaussian units trivial: it is sufficient to replace all $(qq'/4\pi\varepsilon_0)_{\text{SI}}$ with $(qq')_{\text{Gaussian}}$. In the (rare) cases when I spell out the charge values, I will use a different font: $q \equiv \mathscr{Z}e$, $q' \equiv \mathscr{Z}'e$.

[32] A more careful analysis shows that this approximation is actually quite reasonable up to much higher frequencies of the order of γ^2/τ.

[33] Please note the font-marked difference between this variable ($\mathscr{q}$) and the particle's electric charge (q).

The most important feature of the result (10.77)–(10.78) is the frequency-independent ('white') spectrum of the radiation, very typical for any rapid leaps, which may be approximated as delta-functions of time[34]. (Note, however, that Eq. (10.77) implies a fixed value of $\mathscr{q}$, so that the statistics of this parameter, to be discussed in a minute, may 'color' the radiation.) Note also the 'doughnut-shaped' angular distribution of the radiation, typical for non-relativistic systems, with the symmetry axis directed along the momentum transfer vector $\mathscr{q}$. In particular, this means that in typical cases when $\mathscr{q} \ll p$, i.e. the vector $\mathscr{q}$ is nearly normal to the vector $\mathbf{p}_{\text{ini}}$ (see, e.g. the example shown in figure 10.13), the bremsstrahlung produces a significant radiation flow in the direction back to the particle source—the fact significant for the operation of x-ray tubes.

Now integrating the result (10.77) over all wave propagation angles, just as we did for the instant radiation power in section 8.2, we get the following spectral density of the particle energy loss,

$$-\frac{d\mathscr{E}}{d\omega} = \oint_{4\pi} I(\omega)\, d\Omega = \frac{2}{3\pi c}\frac{q^2}{4\pi\varepsilon_0}\frac{\mathscr{q}^2}{m^2c^2}. \tag{10.79}$$

The main new feature of bremsstrahlung (as of most scattering problems[35]), is the necessity to take into account the randomness of the impact parameter b (figure 10.13). For elastic ($\beta_{\text{ini}} = \beta_{\text{fin}} \equiv \beta$) Coulomb collisions we can use the so-called Rutherford formula for the differential cross-section of scattering[36]

$$\frac{d\sigma}{d\Omega'} = \left(\frac{qq'}{4\pi\varepsilon_0}\right)^2 \left(\frac{1}{2pc\beta}\right)^2 \frac{1}{\sin^4(\theta'/2)}. \tag{10.80}$$

Here $d\sigma = 2\pi b db$ is the elementary area of the sample cross-section (as visible from the direction of incident particles) corresponding to particle scattering into an elementary body angle[37]

[34] This is the foundation, in particular, of the so-called *high-harmonic generation* (HHG), discovered in 1977, which takes place at the irradiation of materials by intensive laser beams. The high electric field of the beam strips valent electrons from initially neutral atoms and accelerates them, just to slam them back into the remaining ions as the field's polarity changes in time. The electrons change their momentum sharply during their recombination with the ions, resulting in a bremsstrahlung-like radiation. The spectrum of radiation from each such event obeys Eq. (10.77), but since the ionization/acceleration/recombination cycles repeat periodically with the frequency ω_0 of the laser field, the final spectrum consists of many equidistant lines, with frequencies $n\omega_0$. The bremsstrahlung mechanism does not provide for a classical cutoff $\omega_{\max} = n_{\max}\omega_0$ of the spectrum, but such a limitation is imposed by quantum mechanics: $\hbar\omega_{\max} \sim \mathscr{E}_{\text{p}}$, where the so-called *ponderomotive energy* $\mathscr{E}_{\text{p}} = (eE_0/\omega_0)^2/4m_e$ is the average kinetic energy given to a free electron by the periodic electric field of the laser, with amplitude E_0. In practice, $n_{\max}$ may be as high as ~100, enabling alternative compact sources of x-ray radiation. For a detailed quantitative theory of this effect, see, for example, [5].

[35] See, e.g. *Part CM* section 3.5 and *Part QM* section 3.3.

[36] See, e.g. *Part CM* Eq. (3.73) with $\alpha = qq'/4\pi\varepsilon_0$. In the form used in Eq. (10.80), the Rutherford formula is also valid for small-angle scattering of relativistic particles, the criterion being $|\Delta\boldsymbol{\beta}| \ll 2/\gamma$.

[37] The angle θ' and differential $d\Omega'$, describing the direction of scattered *particles* (see figure 10.13) should not be confused with the parameters θ and $d\Omega$ describing the direction of the *radiation* emitted at the scattering event.

$$d\Omega' = 2\pi \sin\theta' |d\theta'|. \tag{10.81}$$

Differentiating the geometric relation, which is evident from figure 10.13b,

$$\mathscr{q} = 2p \sin\frac{\theta'}{2}, \tag{10.82}$$

we may represent Eq. (10.80) in a more convenient form

$$\frac{d\sigma}{d\mathscr{q}} = 8\pi\left(\frac{qq'}{4\pi\varepsilon_0}\right)^2 \frac{1}{u^2\mathscr{q}^3}. \tag{10.83}$$

Now combining Eqs. (10.79) and (10.83) we obtain

$$-\frac{d\mathscr{E}}{d\omega}\frac{d\sigma}{d\mathscr{q}} = \frac{16}{3}\frac{q^2}{4\pi\varepsilon_0}\left(\frac{qq'}{4\pi\varepsilon_0 mc^2}\right)^2 \frac{1}{c\beta^2}\frac{1}{\mathscr{q}}. \tag{10.84}$$

This product is called the *differential radiation cross-section.* When integrated over all values of $\mathscr{q}$ (which is equivalent to averaging over all values of the impact parameter), it gives a convenient measure of the radiation intensity. Indeed, after the multiplication by the volume density n of independent scattering centers, the integral yields the particle's energy loss per unit bandwidth of radiation per unit path length, $-d^2\mathscr{E}/d\omega dx$. A minor problem here is that the integral of $1/\mathscr{q}$ formally diverges at both infinite and zero values of $\mathscr{q}$. However, these divergences are very weak (logarithmic), and the integral converges for virtually any reason unaccounted for by our simple analysis. The standard, although slightly approximate, way to account for these effects is to write

$$-\frac{d^2\mathscr{E}}{d\omega dx} \approx \frac{16}{3} n \frac{q^2}{4\pi\varepsilon_0}\left(\frac{qq'}{4\pi\varepsilon_0 mc^2}\right)^2 \frac{1}{c\beta^2} \ln\frac{\mathscr{q}_{max}}{\mathscr{q}_{min}}, \tag{10.85}$$

and then plug, instead of $\mathscr{q}_{max}$ and $\mathscr{q}_{min}$, the scales of the most important effects limiting the range of the momentum transfer magnitude. In classical-mechanics analysis, according to Eq. (10.82), $\mathscr{q}_{max} = 2p \equiv 2mu$. To estimate $\mathscr{q}_{min}$, let us note that the very small momentum transfer takes place when the impact parameter b is very large and hence the effective scattering time $\tau \sim b/v$ is very long. Recalling the condition of the low-frequency approximation, we may associate $\mathscr{q}_{min}$ with $\tau \sim 1/\omega$ and hence with $b \sim u\tau \sim v/\omega$. Since for the small scattering angles $\mathscr{q}$ may be estimated as the impulse $F\tau \sim (qq'/4\pi\varepsilon_0 b^2)\tau$ of the Coulomb force, we obtain the estimate $\mathscr{q}_{min} \sim (qq'/4\pi\varepsilon_0)\omega/u^2$, and Eq. (10.85) should be used with

$$\ln\frac{\mathscr{q}_{max}}{\mathscr{q}_{min}} = \ln\left(\frac{2mu^3}{\omega}\Big/\frac{qq'}{4\pi\varepsilon_0}\right). \tag{10.86}$$

This is the *Bohr's formula* for what is called the *classical bremsstrahlung.* We see that the low momentum cutoff indeed makes the spectrum slightly colored, with more energy going to lower frequencies. There is even a formal divergence at $\omega \to 0$;

however, this divergence is integrable, so it does not present a problem for finding the total energy radiative losses $(-d\mathscr{E}/dx)$ as an integral of Eq. (10.86) over all radiated frequencies ω. A larger problem for this procedure is the upper integration limit, $\omega \to \infty$, at which the integral diverges. This means that our approximate description, which considers the collision as an elastic process, becomes invalid and needs to be amended by taking into account the difference between the initial and final kinetic energies of the particle due to radiation of the energy quantum $\hbar\omega$ of the emitted photon, so that

$$\frac{p_{\text{ini}}^2}{2m} - \frac{p_{\text{fin}}^2}{2m} = \hbar\omega, \qquad \text{i.e.}\ \frac{p_{\text{ini}}^2}{2m} = \mathscr{E}, \quad \frac{p_{\text{fin}}^2}{2m} = \mathscr{E} - \hbar\omega. \tag{10.87}$$

As a result, taking into account that the minimum and maximum values of $\mathscr{q}$ correspond to, respectively, the parallel and antiparallel alignments of the vectors $\mathbf{p}_{\text{ini}}$ and $\mathbf{p}_{\text{fin}}$, we obtain

$$\begin{aligned}\ln \frac{\mathscr{q}_{\max}}{\mathscr{q}_{\min}} &= \ln \frac{p_{\text{ini}} + p_{\text{fin}}}{p_{\text{ini}} - p_{\text{fin}}} \equiv \ln \frac{(p_{\text{ini}} + p_{\text{fin}})^2/2m}{\left(p_{\text{ini}}^2 - p_{\text{fin}}^2\right)/2m} \\ &= \ln \frac{[\mathscr{E}^{1/2} + (\mathscr{E} - \hbar\omega)^{1/2}]^2}{\hbar\omega},\end{aligned} \tag{10.88}$$

Plugged into Eq. (10.85), this expression yields the so-called *Bethe–Heitler formula* for *quantum bremsstrahlung*[38]. Note that at this approach, $\mathscr{q}_{\max}$ is close to that of the classical approximation, but $\mathscr{q}_{\min} \sim \hbar\omega/u$, so that

$$\frac{\mathscr{q}_{\min}\big|_{\text{classical}}}{\mathscr{q}_{\min}\big|_{\text{classical}}} \sim \frac{\alpha \mathscr{Z}\mathscr{Z}'}{\beta}, \tag{10.89}$$

where $\mathscr{Z}$ and $\mathscr{Z}'$ are the particles' charges in units of e, and α is the *fine structure* ('Sommerfeld') *constant*,

$$\alpha \equiv \frac{e^2}{4\pi\varepsilon_0\hbar c}\bigg|_{\text{SI}} = \frac{e^2}{\hbar c}\bigg|_{\text{Gaussian}} \approx \frac{1}{137} \ll 1, \tag{10.90}$$

which is one of the basic notions of quantum mechanics[39]. Due to the smallness of the constant, the ratio (10.89) is smaller than 1 for most cases of practical interest, and since the integral of (10.84) over $\mathscr{q}$ is limited by the largest of all possible cutoffs $\mathscr{q}_{\min}$, it is the Bethe–Heitler formula which should be used.

Now nothing prevents us from calculating the total radiative losses of energy per unit length:

---

[38] The modifications of this formula necessary for the relativistic case description are surprisingly minor—see, e.g. chapter 15 of [6]. For even more detail, the standard reference monograph on bremsstrahlung is [7].

[39] See, e.g. *Part QM* sections 6.3, 9.3, 9.5, and 9.7.

$$-\frac{d\mathscr{E}}{dx}=\int_0^\infty\left(-\frac{d^2\mathscr{E}}{d\omega dz}\right)d\omega$$
$$=\frac{16}{3}n\frac{q^2}{4\pi\varepsilon_0 c}\left(\frac{qq'}{4\pi\varepsilon_0 mc^2}\right)^2\frac{1}{\beta^2}2\int_0^{\omega_{\max}}\ln\frac{\mathscr{E}^{1/2}-(\mathscr{E}-\hbar\omega)^{1/2}}{(\hbar\omega)^{1/2}}d\omega, \tag{10.91}$$

where $\hbar\omega_{\max}=\mathscr{E}$ is the maximum energy of the radiation quantum. By introducing the dimensionless integration variable $\xi\equiv\hbar\omega/\mathscr{E}=2\hbar\omega/(mu^2/2)$, this integral is reduced to a table one[40], and we obtain

$$-\frac{d\mathscr{E}}{dx}=\frac{16}{3}n\frac{q^2}{4\pi\varepsilon_0 c}\left(\frac{qq'}{4\pi\varepsilon_0 mc^2}\right)^2\frac{1}{\beta^2}\frac{u^2}{\hbar}=\frac{16}{3}n\left(\frac{q'^2}{4\pi\varepsilon_0\hbar c}\right)\left(\frac{q^2}{4\pi\varepsilon_0}\right)^2\frac{1}{mc^2}. \tag{10.92}$$

Following my usual style, at this point I would give you an estimate of the losses for a typical case; however, let me first discuss a parallel energy loss mechanism, the so-called *Coulomb losses*, due to the transfer of mechanical impulse from the scattered particle to the scattering centers. (This energy eventually goes into an increase of the thermal energy of the scattering medium, rather than to the electromagnetic radiation.)

Using Eqs. (9.139) for the electric field of a linearly moving charge q, we can readily find the momentum it transfers to the counterpart charge q':[41]

$$\Delta p'=\left|(\Delta p')_y\right|=\left|\int_{-\infty}^{+\infty}(\dot{p}')_y\,dt\right|=\left|\int_{-\infty}^{+\infty}q'E_y\,dt\right|$$
$$=\frac{qq'}{4\pi\varepsilon_0}\int_{-\infty}^{+\infty}\frac{\gamma b}{(b^2+\gamma^2u^2t^2)^{3/2}}dt=\frac{qq'}{4\pi\varepsilon_0}\frac{2}{bu}. \tag{10.93}$$

Hence, the kinetic energy acquired by the scattering particle (and hence to the loss of the energy $\mathscr{E}$ of the incident particle) is

$$-\Delta\mathscr{E}=\frac{(\Delta p')^2}{2m'}=\left(\frac{qq'}{4\pi\varepsilon_0}\right)^2\frac{2}{m'u^2b^2}. \tag{10.94}$$

Such elementary energy losses have to be summed up over all collisions, with random values of the impact parameter b. At the scattering center density n, the number of collisions per small path length dz per small range db is $dN=n\,2\pi b\,db\,dx$, so that

$$-\frac{d\mathscr{E}}{dx}=\int(-\Delta\mathscr{E})dN=n\left(\frac{qq'}{4\pi\varepsilon_0}\right)^2\frac{2}{m'u^2}2\pi\int_{b_{\min}}^{b_{\max}}\frac{db}{b}$$
$$=4\pi n\left(\frac{qq'}{4\pi\varepsilon_0}\right)^2\frac{\ln B}{m'u^2},\quad\text{where } B\equiv\frac{b_{\max}}{b_{\min}}. \tag{10.95}$$

[40] See, e.g., Eq. (A.41).

[41] According to Eq. (9.139), $E_z=0$, and the net impulse of the longitudinal force $q'E_x$ is also zero.

Here the logarithmic integral over b was treated similarly to that over $\mathscr{q}$ in the bremsstrahlung theory. This approach is adequate, because the ratio $b_{\max}/b_{\min}$ is much larger than 1. Indeed, $b_{\min}$ may be estimated from $(\Delta p')_{\max} \sim p = \gamma mu$. For this value, Eq. (10.93) with $q' \sim q$ gives $b_{\min} \sim r_c$ (see Eq. (8.41) and its discussion), which is, for elementary particles, of the order of 10^{-15} m. On the other hand, for the most important case when the charges q' belong to electrons (which, according to Eq. (10.94), are the most efficient Coulomb energy absorbers due to their extremely low mass m'), $b_{\max}$ may be estimated from condition $\tau = b/\gamma u \sim 1/\omega_{\max}$, where $\omega_{\max} \sim 10^{16}$ s^{-1} is the characteristic frequency of electron transitions in atoms. (Below this frequency, our classical analysis of the scatterer's motion is invalid.) From here, we have the estimate $b_{\max} \sim \gamma u/\omega_{\max}$, so that

$$B \equiv \frac{b_{\max}}{b_{\min}} \sim \frac{\gamma u}{r_c \omega_0}, \tag{10.96}$$

for $\gamma \sim 1$ and $u \sim c \approx 3 \times 10^8$ m s^{-1} giving $b_{\max} \sim 3 \times 10^{-8}$ m, so that $B \sim 10^9$ (give or take a couple orders of magnitude—this does not change the estimate $\ln B \approx 20$ too much)[42].

Now we can compare the Coulomb losses (10.95) with those due to the bremsstrahlung, given by Eq. (10.92):

$$\frac{-d\mathscr{E}|_{\text{radiation}}}{-d\mathscr{E}|_{\text{Coulomb}}} \sim \alpha \mathscr{Z}\mathscr{Z}' \frac{m'}{m} \beta^2 \frac{1}{\ln B}. \tag{10.97}$$

Since $\alpha \sim 10^{-2} \ll 1$, for non-relativistic particles ($\beta \ll 1$) the bremsstrahlung losses of energy are much lower (this is why we did not need to rush with their estimate), and only for ultra-relativistic particles may the relation be opposite.

According to Eq. (10.95), for electron–electron scattering ($q = q' = -e$, $m' = m_e$),[43] at the value $n = 6 \times 10^{26}$ m^{-3} typical for air at ambient conditions, the characteristic length of energy loss,

$$l_c \equiv \frac{\mathscr{E}}{(-d\mathscr{E}/dx)}, \tag{10.98}$$

for electrons with kinetic energy $\mathscr{E} = 6$ keV is close to 2×10^{-4} m $= 0.2$ mm. (This is why we need a high vacuum in particle accelerators and electron microscope columns!) Since $l_c \propto \mathscr{E}^2$, more energetic particles penetrate deeper into matter until the bremsstrahlung steps in and limits this trend at very high energies.

---

[42] A quantum analysis (carried out by H Bethe in 1940) replaces, in Eq. (10.95), $\ln B$ with $\ln(2\gamma^2 mu^2/\hbar\langle\omega\rangle) - \beta^2$, where $\langle\omega\rangle$ is the average frequency of the atomic quantum transitions' weight by their oscillator strength. This refinement does not change the estimate given below. Note that both the classical and quantum formulas describe a fast increase (as $1/\beta$) of the energy loss rate $(-d\mathscr{E}/dx)$ at $\gamma \to 1$, and its slow increase (as $\ln\gamma$) at $\gamma \to \infty$, so that the losses have a minimum at $(\gamma - 1) \sim 1$.

[43] Actually, the above analysis has neglected the change of momentum of the incident particle. This is legitimate at $m' \ll m$, but for $m = m'$ the change approximately doubles the energy losses. Still, this does not change the order of magnitude of the estimate.

10.5 Density effects and Cherenkov radiation

For condensed matter, the Coulomb loss estimate made in the last section is not quite suitable, because it is based on the upper cutoff $b_{\max} \sim \gamma u/\omega_{\max}$. For the example given above, the incoming electron velocity u is close to 5×10^7 m s^{-1}, and for the typical value $\omega_{\max} \sim 10^{16}$ s^{-1} ($\hbar\omega_{\max} \sim 10$ eV), this cutoff $b_{\max} \sim 5 \times 10^{-9}$ m = 5 nm. Even for air at ambient conditions, this cutoff is larger than the average distance (~2 nm) between the molecules, so that at the high end of the impact parameter range, at $b \sim b_{\max}$, the Coulomb loss events in adjacent molecules are not quite independent, and the theory needs corrections. For condensed matter, with much higher particle density n, most collisions satisfy the condition

$$nb^3 \gg 1, \tag{10.99}$$

and the treatment of Coulomb collisions as independent events is inadequate. However, the condition (10.99) enables the opposite approach: treating the medium as a continuum. In the time domain formulation, used in the previous sections of this chapter, this would be a very complex problem, because it would require an explicit description of the medium dynamics. Here the frequency-domain approach, based on the Fourier transform in both time and space, helps a lot, provided that functions $\varepsilon(\omega)$ and $\mu(\omega)$ are considered known—either calculated or taken from experiment. Let us have a good look at this approach, because it gives some interesting (and practically important) results.

In chapter 6, we have used the macroscopic Maxwell equations to derive Eqs. (6.118), which describe the time evolution of potentials in a linear medium with frequency-independent ε and μ. Looking for all functions participating in Eqs. (6.118) in the form of plane-wave expansion[44]

$$f(\mathbf{r}, t) = \int d^3k \int d\omega f_{\mathbf{k},\omega} e^{i(\mathbf{k}\cdot\mathbf{r}-\omega t)}, \tag{10.100}$$

and requiring all coefficients at similar exponents to be balanced, we obtain their Fourier image[45]:

$$(k^2 - \phi^2\varepsilon\mu)\, \varphi_{\mathbf{k},\omega} = \frac{\rho_{\mathbf{k},\omega}}{\varepsilon}, \quad (k^2 - \omega^2\varepsilon\mu)\, \mathbf{A}_{\mathbf{k},\omega} = \mu\, \mathbf{j}_{\mathbf{k},\omega}. \tag{10.101}$$

As was discussed in chapter 7, in such a Fourier form, the macroscopic Maxwell theory remains valid even for the dispersive (but linear!) media, so that Eq. (10.101) may be generalized as

[44] All integrals here and below are in infinite limits, unless specified otherwise.

[45] As was discussed in section 7.2, the Ohmic conductivity of the medium (generally, also a function of frequency) may be readily incorporated into the dielectric permittivity: $\varepsilon(\omega) \rightarrow \varepsilon_{\text{ef}}(\omega) + i\sigma(\omega)/\omega$. In this section, I will assume that such incorporation, which is especially natural for high frequencies, has been performed, so that the current density $\mathbf{j}(\mathbf{r}, \mathrm{t})$ describes only stand-alone currents—for example, the current (10.105) of the incident particle.

$$[k^2 - \omega^2\varepsilon(\omega)\mu(\omega)]\,\varphi_{\mathbf{k},\omega} = \frac{\rho_{\mathbf{k},\omega}}{\varepsilon(\omega)}, \quad [k^2 - \omega^2\varepsilon(\omega)\mu(\omega)]\,\mathbf{A}_{\mathbf{k},\omega} = \mu(\omega)\;\mathbf{j}_{\mathbf{k},\omega}, \quad (10.102)$$

The evident advantage of these equations is that their formal solution is trivial:

$$\varphi_{\mathbf{k},\omega} = \frac{\rho_{\mathbf{k},\omega}}{\varepsilon(\omega)[k^2 - \omega^2\varepsilon(\omega)\mu(\omega)]}, \qquad \mathbf{A}_{\mathbf{k},\omega} = \frac{\mu(\omega)\;\mathbf{j}_{\mathbf{k},\omega}}{[k^2 - \omega^2\varepsilon(\omega)\mu(\omega)]}, \qquad (10.103)$$

so that the 'only' remaining things to do are to, first, calculate the Fourier transforms of the functions $\rho(\mathbf{r}, t)$ and $\mathbf{j}(\mathbf{r}, t)$, describing stand-alone charges and currents, using the transform reciprocal to Eq. (10.100) with one factor $1/2\pi$ per each scalar dimension,

$$f_{\mathbf{k},\omega} = \frac{1}{(2\pi)^4}\int d^3r \int dt f(\mathbf{r}, t)e^{-i(\mathbf{k}\cdot\mathbf{r}-\omega t)}, \qquad (10.104)$$

and then carry out the integration (10.100) of Eqs. (10.103).

For our current problem of a single charge q, uniformly moving in the medium with velocity $\mathbf{u}$,

$$\rho(\mathbf{r}, t) = q\delta(\mathbf{r} - \mathbf{u}t), \qquad \mathbf{j}(\mathbf{r}, t) = q\mathbf{u}\delta(\mathbf{r} - \mathbf{u}t), \qquad (10.105)$$

the first task is easy:

$$\begin{aligned}\rho_{\mathbf{k},\omega} &= \frac{q}{(2\pi)^4}\int d^3r \int dt \;\; q\delta(\mathbf{r} - \mathbf{u}t)e^{-i(\mathbf{k}\cdot\mathbf{r}-\omega t)} \\ &= \frac{q}{(2\pi)^4}\int e^{i(\omega t - \mathbf{k}\cdot\mathbf{u}t)}dt = \frac{q}{(2\pi)^3}\delta(\omega - \mathbf{k}\cdot\mathbf{u}).\end{aligned} \qquad (10.106)$$

Since the expressions (10.105) for $\rho(\mathbf{r}, t)$ and $\mathbf{j}(\mathbf{r}, t)$ differ only by a constant factor $\mathbf{u}$, it is clear that the absolutely similar calculation for the current gives

$$\mathbf{j}_{\mathbf{k},\omega} = \frac{q\mathbf{u}}{(2\pi)^3}\delta(\omega - \mathbf{k}\cdot\mathbf{u}). \qquad (10.107)$$

Let us summarize what we have got by now, plugging Eqs. (10.106) and (10.107) into Eqs. (10.103):

$$\begin{aligned}\varphi_{\mathbf{k},\omega} &= \frac{1}{(2\pi)^3}\frac{q\delta(\omega - \mathbf{k}\cdot\mathbf{u})}{\varepsilon(\omega)[k^2 - \omega^2\varepsilon(\omega)\mu(\omega)]}, \\ \mathbf{A}_{\mathbf{k},\omega} &= \frac{1}{(2\pi)^3}\frac{\mu(\omega)q\mathbf{u}\delta(\omega - \mathbf{k}\cdot\mathbf{u})}{[k^2 - \omega^2\varepsilon(\omega)\mu(\omega)]} = \varepsilon(\omega)\mu(\omega)\mathbf{u}\varphi_{\mathbf{k},\omega}.\end{aligned} \qquad (10.108)$$

Now, at the last calculation step, namely the integration (10.100), we are starting to pay a heavy price for the easiness of the first steps. This is why we need to consider well what exactly we need from it. First of all, for the calculation of power losses, the electric field is more convenient to use than the potentials, so let us calculate the Fourier images of $\mathbf{E}$ and $\mathbf{B}$. Plugging the expansion (10.100) into the basic relations (6.7), and again requiring the balance of exponent's coefficients, we obtain

$$\mathbf{E}_{\mathbf{k},\omega} = -\,i\mathbf{k}\varphi_{k,\omega} + i\omega\mathbf{A}_{\mathbf{k},\omega} = i[\omega\varepsilon(\omega)\mu(\omega)\mathbf{u} - \mathbf{k}]\ \varphi_{\mathbf{k},\omega},$$
$$\mathbf{B}_{\mathbf{k},\omega} = i\mathbf{k}\times\mathbf{A}_{\mathbf{k},\omega} = i\varepsilon(\omega)\mu(\omega)\mathbf{k}\times\mathbf{u}\varphi_{\mathbf{k},\omega}, \tag{10.109}$$

so that Eqs. (10.100) and (10.108) yield

$$\begin{aligned}\mathbf{E}(\mathbf{r},t) &= \int d^3k\int d\omega\ \mathbf{E}_{\mathbf{k},\omega}e^{i(\mathbf{k}\cdot\mathbf{r}-\omega t)}\\ &= \frac{iq}{(2\pi)^3}\int d^3k\int d\omega\ \frac{[\omega\varepsilon(\omega)\mu(\omega)\mathbf{u}-\mathbf{k}]\ \delta(\omega-\mathbf{k}\cdot\mathbf{u})}{\varepsilon(\omega)[k^2-\omega^2\varepsilon(\omega)\mu(\omega)]}e^{i(\mathbf{k}\cdot\mathbf{r}-\omega t)}.\end{aligned} \tag{10.110}$$

This formula may be rewritten as the temporal Fourier integral (10.51), with a space-dependent amplitude

$$\begin{aligned}\mathbf{E}_\omega(\mathbf{r}) &= \int \mathbf{E}_{\mathbf{k},\omega}e^{i\mathbf{k}\cdot\mathbf{r}}d^3k\\ &= \frac{iq}{(2\pi)^3}\int\frac{[\omega\varepsilon(\omega)\mu(\omega)\mathbf{u}-\mathbf{k}]\ \delta(\omega-\mathbf{k}\cdot\mathbf{u})}{\varepsilon(\omega)[k^2-\omega^2\varepsilon(\omega)\mu(\omega)]}e^{i\mathbf{k}\cdot\mathbf{r}}d^3k\ .\end{aligned} \tag{10.111}$$

Let us calculate the Cartesian components of this partial Fourier image $\mathbf{E}_\omega$, at a point separated by distance b from the particle's trajectory. Selecting the coordinates and time origin as shown in figure 9.11a, we have $\mathbf{r} = \{0, b, 0\}$, so that only E_x and E_y are different from zero. In particular, according to Eq. (10.111),

$$\begin{aligned}(E_x)_\omega = &\frac{iq}{(2\pi)^3\varepsilon(\omega)}\int dk_x\int dk_y\int dk_z\frac{\omega\varepsilon(\omega)\mu(\omega)u-k_x}{k^2-\omega^2\varepsilon(\omega)\mu(\omega)}\delta(\omega-k_xu)\\ &\times\exp\left\{ik_yb\right\}.\end{aligned} \tag{10.112}$$

The delta-function kills one integral (over k_x) of the three, and we obtain

$$\begin{aligned}(E_x)_\omega = &\frac{iq}{(2\pi)^3\varepsilon(\omega)u}\left[\omega\varepsilon(\omega)\mu(\omega)u-\frac{\omega}{u}\right]\\ &\times\int\exp\left\{ik_yb\right\}dk_y\int\frac{dk_z}{\omega^2/u^2+k_y^2+k_z^2-\omega^2\varepsilon(\omega)\mu(\omega)}.\end{aligned} \tag{10.113}$$

The internal integral (over k_z) may be readily reduced to the table integral $\int d\xi/(1+\xi^2)$ in infinite limits, equal to π.[46] The result may be represented as

$$(E_x)_\omega = -\frac{i\pi\ q\kappa^2}{(2\pi)^3\omega\varepsilon(\omega)}\int\frac{\exp\left\{ik_yb\right\}}{\left(k_y^2+\kappa^2\right)^{1/2}}dk_y, \tag{10.114}$$

where the parameter κ (generally, a complex function of frequency) is defined as[47]

[46] See, e.g. Eq. (A.32*a*).

[47] Of course, the frequency-dependent parameter $\kappa(\omega)$ should not be confused with the dc low-frequency dielectric constant $\kappa \equiv \varepsilon(0)/\varepsilon_0$, which was discussed in chapter 3.

$$\kappa^2(\omega) \equiv \omega^2\left(\frac{1}{u^2} - \varepsilon(\omega)\mu(\omega)\right). \tag{10.115}$$

The last integral may be expressed via the modified Bessel function of the second kind[48]:

$$(E_x)_\omega = -\frac{iqu\kappa^2}{(2\pi)^2\omega\varepsilon(\omega)}K_0(\kappa b). \tag{10.116}$$

A very similar calculation yields

$$\left(E_y\right)_\omega = \frac{q\kappa}{(2\pi)^2\varepsilon(\omega)}K_1(\kappa b). \tag{10.117}$$

Now, instead of rushing to make the final integration (10.51) over ω to calculate $\mathbf{E}(t)$, let us realize that what we need is actually only the total energy loss through the whole time of particle's passage over an elementary distance dx. According to Eq. (4.38), the energy loss per unit volume is

$$-\frac{d\mathscr{E}}{dV} = \int \mathbf{j}\cdot\mathbf{E}\,dt, \tag{10.118}$$

where $\mathbf{j}$ is the current of the bound charges in the medium, and should not be confused with the stand-alone particle's current (10.105). This integral may be readily expressed via the partial Fourier image $\mathbf{E}_\omega$ and the similarly defined image $\mathbf{j}_\omega$, just as was done at the derivation of Eq. (10.54):

$$\begin{aligned}-\frac{d\mathscr{E}}{dV} &= \int dt\int d\omega e^{-i\omega t}\int d\omega' e^{-i\omega' t}\mathbf{j}_\omega\cdot\mathbf{E}_{\omega'}\\ &= 2\pi\int d\omega\int d\omega'\,\mathbf{j}_\omega\cdot\mathbf{E}_{\omega'}\delta(\omega+\omega') = 2\pi\int \mathbf{j}_\omega\cdot\mathbf{E}_{-\omega}\,d\omega.\end{aligned} \tag{10.119}$$

In our approach, the effective Ohmic conductivity $\sigma_{ef}(\omega)$ is incorporated into the complex permittivity $\varepsilon(\omega)$ just as was discussed in section 7.2, so that we may use Eq. (7.46) to write

$$\mathbf{j}_\omega = \sigma_{ef}(\omega)\mathbf{E}_\omega = -i\omega\varepsilon(\omega)\mathbf{E}_\omega. \tag{10.120}$$

As a result, Eq. (10.119) yields

$$-\frac{d\mathscr{E}}{dV} = -2\pi i\int \varepsilon(\omega)\mathbf{E}_\omega\cdot\mathbf{E}_{-\omega}\omega\,d\omega = 4\pi\mathrm{Im}\int_0^\infty \varepsilon(\omega)\,|E_\omega|^2\,\omega\,d\omega. \tag{10.121}$$

(The last step was possible due to the property $\varepsilon(-\omega) = \varepsilon^*(\omega)$, which was discussed in section 7.2.)

48 As a reminder, the main properties of these functions are listed in section 2.7—see, in particular, figure 2.22, and Eqs. (2.157) and (2.158).

Finally, just as in the last section, we have to average the energy loss rate over random values of the impact parameter b:

$$-\frac{d\mathscr{E}}{dx} = \int\left(-\frac{d\mathscr{E}}{dV}\right)d^2b \approx 2\pi\int_{b_{\min}}^{\infty}\left(-\frac{d\mathscr{E}}{dV}\right)b\, db$$
$$= 8\pi^2\int_{b_{\min}}^{\infty} b\, db\int_0^{\infty}\left(|E_x|_\omega^2 + |E_y|_\omega^2\right)\mathrm{Im}\varepsilon(\omega)\omega\, d\omega. \tag{10.122}$$

Note that due to the divergence of the functions $K_0(\xi)$ and $K_1(\xi)$ at $\xi \to 0$ we have to cut the resulting integral over b at some $b_{\min}$ where our theory loses legitimacy. (On that limit, we are not doing much better than in the past section). Plugging in the calculated expressions (10.116) and (10.117) for the field components, swapping the integrals over ω and b, and using the recurrence relations (2.142), which are valid for any Bessel functions, we finally obtain:

$$-\frac{d\mathscr{E}}{dx} = \frac{2}{\pi}q^2\mathrm{Im}\int_0^{\infty}(\kappa^* b_{\min})K_1(\kappa^* b_{\min})K_0(\kappa^* b_{\min})\frac{d\omega}{\omega\varepsilon(\omega)}. \tag{10.123}$$

This general result is valid for an arbitrary linear medium, with arbitrary dispersion relations $\varepsilon(\omega)$ and $\mu(\omega)$. (The last function participates in Eq. (10.123) only via Eq. (10.115), which defines the parameter κ.) To obtain more concrete results, some particular model of the medium should be used. Let us explore the Lorentz oscillator model, which was discussed in section 7.2, in its form (7.33) suitable for the transition to a quantum-mechanical description of atoms:

$$\varepsilon(\omega) = \varepsilon_0 + \frac{nq'^2}{m}\sum_j \frac{f_j}{\left(\omega_j^2 - \omega^2\right) - 2i\omega\delta_j}, \quad \text{with} \sum_j f_j = 1; \quad \mu(\omega) = \mu_0. \tag{10.124}$$

If the damping of the effective atomic oscillators is low, $\delta_j \ll \omega_j$, as it typically is, and the particle's speed u is much lower than the typical wave's phase velocity v (and hence c!), then for most frequencies Eq. (10.115) gives

$$\kappa^2(\omega) \equiv \omega^2\left(\frac{1}{u^2} - \frac{1}{v^2(\omega)}\right) \approx \frac{\omega^2}{u^2}, \tag{10.125}$$

i.e. $\kappa \approx \kappa^* \approx \omega/u$ is real. In this case, Eq. (10.123) may be shown to reduce to Eq. (10.95) with

$$b_{\max} = \frac{1.123u}{\langle\omega\rangle}. \tag{10.126}$$

The good news here is that both approaches (the microscopic analysis of section 10.4 and the macroscopic analysis of this section) give essentially the same result. The same fact may be also perceived as bad news: the treatment of the medium as a continuum does not give any new results here. The situation somewhat changes at relativistic velocities, at which such treatment provides noticeable corrections (called *density effects*), in particular reducing the energy loss estimates.

Let me, however, skip these details and focus on a much more important effect described by our formulas. Consider the dependence of the electric field components on the impact parameter b, i.e. on the closest distance between the particle's trajectory and the field observation point. If $\kappa^2 > 0$, then κ is real, and we can use, in Eqs. (10.116) and (10.117), the asymptotic formula (2.158),

$$K_n(\xi) \rightarrow \left(\frac{\pi}{2\xi}\right)^{1/2} e^{-\xi}, \qquad \text{at } \xi \rightarrow \infty, \tag{10.127}$$

to conclude that the complex amplitudes E_ω of both components E_x and E_y of the electric field decrease with b exponentially. However, let us consider what happens at frequencies where $\kappa^2(\omega) < 0$, i.e.

$$\varepsilon(\omega)\mu(\omega) \equiv \frac{1}{v^2(\omega)} < \frac{1}{u^2} < \frac{1}{c^2} \equiv \varepsilon_0\mu_0. \tag{10.128}$$

(This condition means that the particle's velocity is larger than the phase velocity of waves, at this particular frequency.) In these intervals the parameter $\kappa(\omega)$ is purely imaginary[49], so that the functions $\exp\{\kappa b\}$ in the asymptotics (10.127) of Eqs. (10.116) and (10.117) become just phase factors and the field components fall very slowly:

$$|E_x(\omega)| \propto |E_y(\omega)| \propto \frac{1}{b^{1/2}}. \tag{10.129}$$

This means that the Poynting vector drops as $1/b$, so that its flux through a surface of a round cylinder of radius b, with the axis on the particle trajectory (i.e. power flow from the trajectory), does not depend on b. This is wave emission—the famous *Cherenkov radiation*[50].

The direction $\mathbf{n}$ of its propagation may be readily found by taking into account that at large distances from the particle's trajectory the emitted wave has to be locally planar and transverse ($\mathbf{n}\perp\mathbf{E}$), so that the *Cherenkov angle* θ between the vector $\mathbf{n}$ and the particle's velocity $\mathbf{u}$ may be simply found from the ratio of the electric field components—see figure 10.14a:

$$\tan\theta = -\frac{E_x}{E_y}. \tag{10.130}$$

---

[49] Strictly speaking, the inequality $\kappa^2(\omega) < 0$ does not make sense for a medium with a complex product $\varepsilon(\omega)\mu(\omega)$, and hence complex $\kappa^2(\omega)$. However, in a typical medium where particles can propagate over substantial distances, the imaginary part of the product $\varepsilon(\omega)\mu(\omega)$ does not vanish only in very limited frequency intervals, much narrower than the intervals which we are now discussing—please have one more look at figure 7.5.

[50] This radiation was observed experimentally by P Cherenkov (in older Western texts, 'Čerenkov') in 1934, with the observations explained by I Frank and E Tamm in 1937. Note, however, that the effect was predicted theoretically as early as in 1889 by the same O Heaviside, whose name was mentioned in this course so many times—and whose genius I believe is still underappreciated.

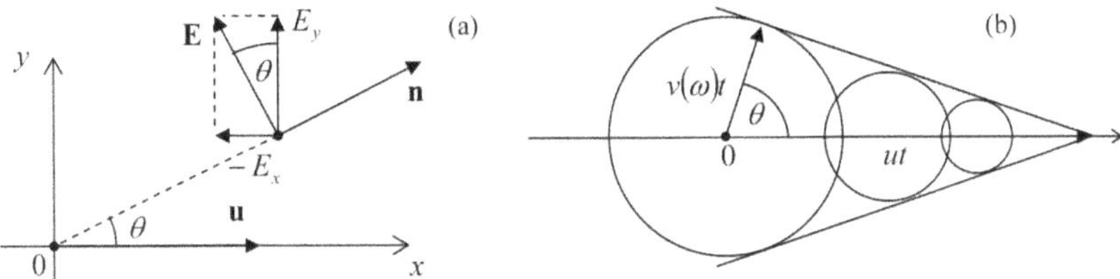

Figure 10.14. The Cherenkov radiation's propagation angle θ and (b) its interpretation.

This ratio may be calculated by plugging the asymptotic formula (10.127) into Eqs. (10.116) and (10.117) and calculating their ratio:

$$\tan\theta = -\frac{E_x}{E_y} = \frac{i\kappa u}{\omega} = [\varepsilon(\omega)\mu(\omega)u^2 - 1]^{1/2} = \left(\frac{u^2}{v^2(\omega)} - 1\right)^{1/2}, \qquad (10.131a)$$

so that

$$\cos\theta = \frac{v(\omega)}{u} < 1. \qquad (10.131b)$$

Remarkably, this direction does not depend on the emission time t_{ret}, so that the radiation of frequency ω, at each instant, forms a hollow cone led by the particle. This simple result allows an evident interpretation (figure 10.14b): the cone is just the set of all observation points that may be reached by 'signals' propagating with the speed $v(\omega) < u$ from all previous points of the particle's trajectory.

This phenomenon is closely related to the so-called *Mach cone* in fluid dynamics, except that in the Cherenkov radiation there is a separate cone for each frequency (of the range in which $v(\omega) < u$): the smaller the $\varepsilon(\omega)\mu(\omega)$ product (i.e. the larger is wave velocity $v(\omega) = 1/[\varepsilon(\omega)\mu(\omega)]^{1/2}$) and the broader the cone, the earlier the corresponding 'shock wave' arrives at an observer. Please note that the Cherenkov radiation is a unique radiative phenomenon: it takes place even if a particle moves without acceleration, and (in agreement with our analysis in section 10.2), is impossible in the free space, where $v = c$ is always larger than u.

The intensity of the Cherenkov radiation intensity may be also readily found by plugging the asymptotic expression (10.127), with imaginary κ, into Eq. (10.123). The result is

$$-\frac{d\mathscr{E}}{dx} \approx \left(\frac{\mathscr{Z}e}{4\pi}\right)^2 \int_{v(\omega)<u} \omega \left(1 - \frac{v^2(\omega)}{u^2}\right) d\omega. \qquad (10.132)$$

For non-relativistic particles ($u \ll c$), the Cherenkov radiation condition $u > v(\omega)$ may be fulfilled only in relatively narrow frequency intervals where the product $\varepsilon(\omega)$ $\mu(\omega)$ is very large (usually, due to optical resonance peaks of the electric permittivity —see figure 7.5 and its discussion). In this case the emitted light consists of a few nearly monochromatic components. In contrast, if the condition $u > v(\omega)$, i.e. $u^2/\varepsilon(\omega)$ $\mu(\omega) > 1$ is fulfilled in a broad frequency range (as it is for ultra-relativistic particles in condensed media), the radiated power is clearly dominated by higher frequencies

Figure 10.15. The Cherenkov radiation glow from the Advanced Test Reactor of the Idaho National Laboratory in Arco, ID. (Adapted from http://en.wikipedia.org/wiki/Cherenkov_radiation under the Creative Commons' CC-BY-SA-2.0 license.)

of the range—hence the famous bluish color of the Cherenkov radiation glow from water nuclear reactors—see figure 10.15.

The Cherenkov radiation is broadly used for the detection of radiation in high energy experiments for particle identification and speed measurement (since it is easy to pass the particles through layers of various density and hence of various dielectric constant values)—for example, in the so-called Ring Imaging Cherenkov (RICH) detectors that have been designed for the DELPHI experiment[51] at the Large Electron–Positron Collider (LEP) in CERN.

A little bit counter-intuitively, the formalism described in this section is also very useful for the description of an apparently rather different effect—the so-called *transition radiation* that takes place when a charged particle crosses a border between two media[52]. The effect may be understood as the result of the time dependence of the electric dipole formed by the moving charge q and its mirror image q' in the counterpart medium—see figure 10.16. In the non-relativistic limit, this effect allows a straightforward description combining the electrostatics picture of section 3.4 (see figure 3.9 and its discussion) and Eq. (8.27), corrected for the media polarization effects. However, if the particle's velocity u is comparable with

[51] See, e.g. http://delphiwww.cern.ch/offline/physics/delphi-detector.html. For a broader view at radiation detectors (including the Cherenkov ones), the reader may be referred, for example, to the classical text [8] and a newer treatment [9].

[52] The effect was predicted theoretically in 1946 by V Ginzburg and I Frank, and only later observed experimentally.

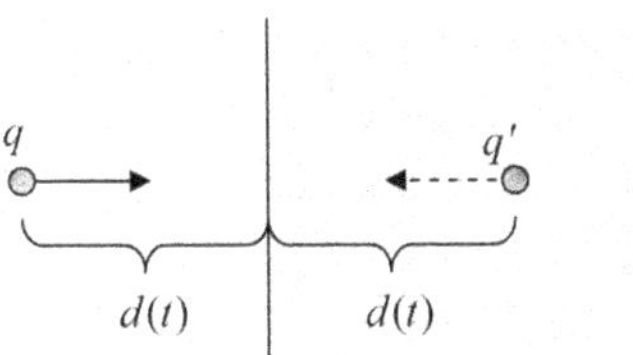

Figure 10.16. The physics of transition radiation.

the phase velocity of waves in either medium, the adequate theory of the transition radiation becomes very close to that of the Cherenkov radiation.

In comparison with the Cherenkov radiation, the transition radiation is rather weak and its practical use (mostly for the measurement of the relativistic factor γ, to which the radiation intensity is proportional) requires multi-layered stacks[53]. In these systems, the radiation emitted at sequential borders may be coherent, and the system's physics becomes close to that of the undulators discussed in section 10.4.

10.6 Radiation's back-action

An attentive reader could notice that so far our treatment of charged particle dynamics has never been fully self-consistent. Indeed, in section 9.6 we have analyzed the particle's motion in various external fields, ignoring the fields radiated by the particle itself, while in section 8.2 and earlier in this chapter these fields have been calculated (admittedly, just for a few simple cases) but, again, their back-action on the emitting particle has been ignored. Only in few cases have we taken the back effects of the radiation implicitly, via energy conservation. However, even in these cases, the near-field components of the fields, such as the first term in Eq. (10.19), which affect the moving particle most, have been ignored.

At the same time, it is clear that in sharp contrast to electrostatics, the interaction of a moving point charge with its own field cannot be always ignored. As the simplest example, if an electron is made to fly through a resonant cavity, thus inducing oscillations in it, and then is forced (say, by an appropriate static field) to return to it before the oscillations have decayed, its motion will certainly be affected by the oscillating fields, just as if they had been induced by another source. There is no conceptual problem with applying the Maxwell theory to such 'field-particle rendezvous' effects; moreover, it is the basis of the engineering design of such vacuum electron devices as klystrons, magnetrons, and undulators.

A problem arises only when no clear 'rendezvous' point is enforced by boundary conditions, so that the most important self-field effects are at $R \equiv |\mathbf{r} - \mathbf{r}'| \rightarrow 0$, the most evident example being the radiation of a particle in free space, described earlier in this chapter. We already know that such radiation takes away a part of the charge's kinetic

[53] See, e.g. section 5.3 in [9].

energy, i.e. has to cause its deceleration. One should wonder, however, whether such self-action effects might be described in a more direct, non-perturbative way.

As a first attempt, let us try a phenomenological approach based on the already derived formulas for radiation power $\mathscr{P}$. For the sake of simplicity, let us consider a non-relativistic point charge q in free space, so that $\mathscr{P}$ is described by Eq. (8.27), with the electric dipole moment's derivative over time equal to $q\mathbf{u}$:

$$\mathscr{P} = \frac{Z_0 q^2}{6\pi c^2}\dot{u}^2 \equiv \frac{2}{3c^3}\frac{q^2}{4\pi\varepsilon_0}\dot{u}^2. \tag{10.133}$$

The most naïve approach would be to write the equation of the particle's motion in the form

$$m\dot{\mathbf{u}} = \mathbf{F}_{\text{ext}} + \mathbf{F}_{\text{self}}, \tag{10.134}$$

and try to calculate the radiation back-action force $\mathbf{F}_{\text{self}}$ by requiring its instant power, $-\mathbf{F}_{\text{self}} \cdot \mathbf{u}$, to be equal to $\mathscr{P}$. However, this approach (say, for a 1D motion) would give a very unnatural result,

$$F_{\text{self}} \propto \frac{\dot{u}^2}{u}, \tag{10.135}$$

that might diverge at some points of the particle's trajectory. This failure is clearly due to the retardation effect: as the reader may recall, Eq. (10.133) results from the analysis of radiation fields in the far zone, i.e. at *large* distances R from the particle (e.g. from the second term in Eq. (10.19)) when the non-radiative first term (which is much larger at *small* distances, $R \to 0$) is ignored.

Before exploring the effects of this term, let us, however, make one more try at Eq. (10.133), considering its *average* effect on some periodic motion of the particle. (A possible argument for this step is that at the periodic motion the retardation effects may perhaps be averaged out—just at the transfer from Eq. (8.27) to Eq. (8.28).) To calculate the average, let us write

$$\overline{\dot{u}^2} \equiv \frac{1}{\mathcal{T}}\int_0^{\mathcal{T}} \dot{\mathbf{u}} \cdot \dot{\mathbf{u}}\, dt, \tag{10.136}$$

and integrate this identity over the motion period $\mathcal{T}$ by parts:

$$\begin{aligned}\overline{\mathscr{P}} &= \frac{2}{3c^3}\frac{q^2}{4\pi\varepsilon_0}\overline{(\dot{\mathbf{u}})^2} = \frac{2}{3c^3}\frac{q^2}{4\pi\varepsilon_0}\frac{1}{\mathcal{T}}\left(\dot{\mathbf{u}} \cdot \mathbf{u}\,\Big|_0^{\mathcal{T}} - \int_0^{\mathcal{T}} \ddot{\mathbf{u}} \cdot \mathbf{u}\, dt\right) \\ &= -\frac{1}{\mathcal{T}}\int_0^{\mathcal{T}} \frac{2}{3c^3}\frac{q^2}{4\pi\varepsilon_0}\ddot{\mathbf{u}} \cdot \mathbf{u}\, dt\end{aligned}. \tag{10.137}$$

One the other hand, the back-action force would give

$$\overline{\mathscr{P}} = -\frac{1}{\mathscr{T}}\int_0^{\mathscr{T}} \mathbf{F}_{\text{self}} \cdot \mathbf{u}\ dt. \tag{10.138}$$

These two averages coincide if[54]

$$\mathbf{F}_{\text{self}} = \frac{2}{3c^3}\frac{q^2}{4\pi\varepsilon_0}\ddot{\mathbf{u}}. \tag{10.139}$$

This is the so-called *Abraham–Lorentz force* of self-action. Before going after a more serious derivation of this formula, let us estimate its scale, representing Eq. (10.139) as

$$\mathbf{F}_{\text{self}} = m\tau\ddot{\mathbf{u}}, \quad \text{with } \tau \equiv \frac{2}{3mc^3}\frac{q^2}{4\pi\varepsilon_0}, \tag{10.140}$$

where the constant τ evidently has the dimension of time. Recalling the definition (8.41) of the classical radius r_c of the particle, Eq. (10.140) for τ may be rewritten as

$$\tau = \frac{2}{3}\frac{r_c}{c}. \tag{10.141}$$

For the electron, τ is of the order of 10^{-23} s, so that the right-hand side of Eq. (10.140) is very small. This means that in most cases the Abrahams–Lorentz force is either negligible or leads to the same results as the perturbative treatments of energy loss we have used earlier in this chapter.

However, Eq. (10.140) brings some unpleasant surprises. For example, let us consider a 1D oscillator with the own frequency ω_0. For it, Eq. (10.134), with the back-action force given by Eq. (10.140), takes the form

$$m\ddot{x} + m\omega_0^2 x = m\tau\dddot{x}. \tag{10.142}$$

Looking for the solution to this linear differential equation in the usual exponential form, $x(t) \propto \exp\{\lambda t\}$, we obtain the following characteristic equation,

$$\lambda^2 + \omega_0^2 = \tau\lambda^3. \tag{10.143}$$

It may look like that for any 'reasonable' value of $\omega_0 \ll 1/\tau \sim 10^{23}\ \text{s}^{-1}$, the right-hand side of this nonlinear algebraic equation may be treated as a perturbation. Indeed, looking for its solutions in the natural form $\lambda_\pm = \pm i\omega_0 + \lambda'$, with $|\lambda'| \ll \omega_0$, expanding both parts of Eq. (10.143) in the Taylor series in the small parameter λ' and keeping only the terms linear in λ', we obtain

[54] Just for the reader's reference, this formula may be readily generalized to the relativistic case:

$$F^\alpha_{\text{self}} = \frac{2}{3mc^3}\frac{q^2}{4\pi\varepsilon_0}\left[\frac{d^2p^\alpha}{d\tau^2} + \frac{p^\alpha}{(mc)^2}\left(\frac{dp_\beta}{d\tau}\frac{dp^\beta}{d\tau}\right)\right],$$

the so-called *Abraham–Lorentz–Dirac force*.

$$\lambda' \approx -\frac{\omega_0^2\tau}{2}. \qquad (10.144)$$

This means that the energy of free oscillations decreases in time as $\exp\{2\lambda' t\} = \exp\{-\omega_0{}^2\tau\ t\}$; this is exactly the radiative damping analyzed earlier. However, Eq. (10.143) is deceptive; it has the third root corresponding to unphysical, exponentially growing (so-called *run-away*) solutions. It is easiest to see this for a free particle, with $\omega_0 = 0$. Then Eq. (10.143) becomes the very simple

$$\lambda^2 = \tau\lambda^3, \qquad (10.145)$$

and it is easy to find all its three roots explicitly: $\lambda_1 = \lambda_2 = 0$ and $\lambda_3 = 1/\tau$. While the first two roots correspond the values $\lambda_\pm$ found earlier, the last one describes an exponential (and extremely rapid!) acceleration.

In order to remove this artifact, let us try to develop a self-consistent approach to the back-action effects, taking into account the near-field terms of particle fields. For that, we need to somehow overcome the divergence of Eqs (10.10) and (10.19) at $R \to 0$. The most reasonable way to do this is to spread the particle's charge over a ball of radius a, with a spherically symmetric (but not necessarily constant) density $\rho(r)$, and in the end of the calculations trace the limit $a \to 0$.[55] Again sticking to the non-relativistic case (so that the magnetic component of the Lorentz force is not important), we should calculate

$$\mathbf{F}_{\text{self}} = \int_V \rho(\mathbf{r})\mathbf{E}(\mathbf{r},\ t)d^3r, \qquad (10.146)$$

where the electric field is that of the charge itself, with the field of any elementary charge $dq = \rho(r)d^3r$ described by Eq. (10.19).

In order to make analytical calculations doable, we need to make assumption $a \ll r_c$, treat the ratio $R/r_c \sim a/r_c$ as a small parameter, and expand the resulting the right-hand side of Eq. (10.146) into the Taylor series in small R. This procedure yields

$$\mathbf{F}_{\text{self}} = -\frac{2}{3}\frac{1}{4\pi\varepsilon_0}\sum_{n=0}^{\infty}\frac{(-1)^n}{c^{n+2}n!}\frac{d^{n+1}\mathbf{u}}{dt^{n+1}}\int_V d^3r\int_V d^3r'\rho(r)R^{n-1}\rho(r'). \qquad (10.147)$$

The distance R cancels only in the term with $n = 1$,

$$\mathbf{F}_1 = \frac{2}{3c^3}\frac{\ddot{\mathbf{u}}}{4\pi\varepsilon_0}\int_V d^3r\int_V d^3r'\rho(r)\rho(r') \equiv \frac{2}{3c^3}\frac{q^2}{4\pi\varepsilon_0}\ddot{\mathbf{u}}, \qquad (10.148)$$

showing that we have recovered (now in an *apparently* legitimate fashion) Eq. (10.139) for the Abrahams–Lorentz force. One could argue that in the limit $a \to 0$ the terms higher in $R \sim a$ (with $n > 1$) could be ignored. However, we have to

[55] Note: this operation cannot be interpreted as describing a quantum spread due to the finite extent of a point particle's wavefunction. In quantum mechanics, parts of wavefunction of the same charged particle do *not* interact with each other!

notice that the main contribution into series (10.147) is *not* described by Eq. (10.148) for $n = 1$, but is given by the larger term with $n = 0$:

$$\mathbf{F}_0 = -\frac{2}{3}\frac{1}{4\pi\varepsilon_0}\frac{\dot{\mathbf{u}}}{c^2}\int_V d^3r\int_V d^3r'\frac{\rho(r)\rho(r')}{R} \equiv -\frac{4}{3}\frac{\dot{\mathbf{u}}}{c^2}\frac{1}{4\pi\varepsilon_0}\frac{1}{2}\int_V d^3r\int_V d^3r'\frac{\rho(r)\rho(r')}{R} \equiv -\frac{4}{3c^2}\dot{\mathbf{u}}U, \quad (10.149)$$

where U is the electrostatic energy (1.59) of the static charge self-interaction. This term may be interpreted as the inertial 'force'[56] $(-m_{\text{ef}}\mathbf{a})$ with the following effective *electromagnetic mass*:

$$m_{\text{ef}} = \frac{4}{3}\frac{U}{c^2}. \quad (10.150)$$

This is the famous (or rather infamous) *4/3 problem* that does not allow one to interpret the electron's mass as that of its electric field. An (admittedly, rather formal) resolution of this paradox is possible only in quantum electrodynamics with its renormalization techniques—beyond the framework of this course. Note, however, that these issues are only important for motions with frequencies of the order of $1/\tau \sim 10^{23}\ \text{s}^{-1}$, i.e. at energies $\mathscr{E} \sim \hbar/\tau \sim 10^8$ eV, while other quantum electrodynamics effects may be observed at much lower frequencies, starting from $\sim 10^{10}\ \text{s}^{-1}$. Hence the 4/3 problem is by no means the only motivation for the transfer from classical to quantum electrodynamics.

However, the reader should not think that his or her time spent on this course has been lost: quantum electrodynamics incorporates virtually all classical electrodynamics results and the basic transition to it is surprisingly straightforward[57]. So, I look forward to welcoming the readers to the next, quantum-mechanics part of this series.

10.7 Problems

Problem 10.1. Derive Eqs. (10.10) from Eqs. (10.1) by a direct (but careful!) integration.

Problem 10.2. Derive the radiation-related parts of Eqs. (10.19) and (10.20) from the Liénard–Wiechert potentials (10.10) by direct differentiation.

Problem 10.3. A point charge q that was in a stationary position on a circle of radius R is carried over, along the circle, to the opposite position on the same diameter (see the figure below) as fast as only physically possible, and then is kept steady at this new position. Calculate and sketch the time dependence of its electric field $\mathbf{E}$ at the center of the circle.

[56] See, e.g. *Part CM* section 4.6.

[57] See, e.g. *Part QM* chapter 9.

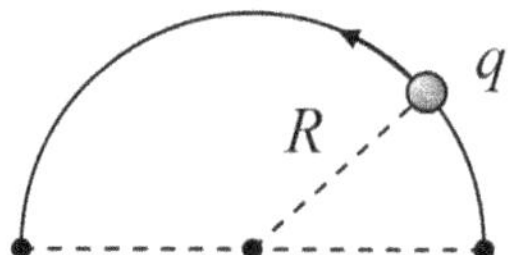

Problem 10.4. Express the total power of electromagnetic radiation by a relativistic particle with electric charge q and rest mass m, moving with velocity $\mathbf{u}$, via the external Lorentz force $\mathbf{F}$ exerted on the particle.

Problem 10.5. A relativistic particle with the rest mass m and electric charge q, initially at rest, is accelerated by a constant force $\mathbf{F}$ until it reaches a certain velocity u, and then moves by inertia. Calculate the total energy radiated during the acceleration.

Problem 10.6. Calculate

(i) the instantaneous power, and
(ii) the power spectrum
of the radiation emitted into a unit solid angle by a relativistic particle with charge q, performing 1D harmonic oscillations with frequency ω_0 and displacement amplitude a.

Problem 10.7. Analyze the polarization and the spectral contents of the synchrotron radiation propagating in the direction perpendicular to the particle's rotation plane. How do the results change if not one, but $N > 1$ similar particles move around the circle, at equal angular distances?

Problem 10.8. Calculate and analyze the time dependence of the energy of a charged relativistic particle performing synchrotron motion in a constant and uniform magnetic field $\mathbf{B}$, and hence emitting the synchrotron radiation. Qualitatively, what is the particle's trajectory?

Hint: You may assume that the energy loss is relatively slow ($-d\mathscr{E}/dt \ll \omega_c\mathscr{E}$), but should spell out the condition of validity of this assumption.

Problem 10.9. Analyze the polarization of the synchrotron radiation propagating within the particle's rotation plane.

Problem 10.10.* The basic quantum theory of radiation shows that the electric dipole radiation by a particle is allowed only if the change of its angular momentum's magnitude L at the transition is of the order of the Planck's constant $\hbar$.

(i) Estimate the change of L of an ultra-relativistic particle due to its emission of a typical single photon of the synchrotron radiation.
(ii) Does the quantum mechanics forbid such a radiation? If not, why?

Problem 10.11. A relativistic particle moves along axis z, with velocity u_z, through an undulator—a system of permanent magnets providing (in the simplest model) a perpendicular magnetic field, whose distribution near the axis z is sinusoidal[58]:

$$\mathbf{B} = \mathbf{n}_y B_0 \cos k_0 z.$$

Assuming that the field is so weak that it causes only very small deviations of the particle's trajectory from the straight line, calculate the angular distribution of the resulting radiation. What condition does this assumption impose on the system's parameters?

Problem 10.12. Discuss possible effects of the interference of the undulator radiation from different periods of its static field distribution. In particular, calculate the angular positions of the power density maxima.

Problem 10.13. An electron, launched directly toward a plane surface of a perfect conductor, is instantly absorbed by it at the collision. Calculate the angular distribution and the frequency spectrum of the electromagnetic waves radiated at this collision, if the initial kinetic energy T of the particle is much larger than the conductor's workfunction ψ.[59] Is your result valid near the conductor's surface?

Problem 10.14. A relativistic particle, with the rest mass m and electric charge q, flies ballistically with velocity u, by an immobile point charge q', with an impact parameter b so large that the deviations of its trajectory from the straight line are negligible. Calculate the total energy loss due to the electromagnetic radiation during the passage. Formulate the conditions of validity of your result.

References

[1] Koch E *et al* (eds) 1983–1991 *Handbook on Synchrotron Radiation* vols 1–5 (Amsterdam: North-Holland)

[2] Hofmann A 2007 *The Physics of Synchrotron Radiation* (Cambridge: Cambridge University Press)

[3] Luchini P and Motz H 1990 *Undulators and Free-electron Lasers* (Oxford: Oxford University Press)

[4] Salin E *et al* 2000 *The Physics of Free Electron Lasers* (Berlin: Springer)

[5] Lewenstein M *et al* 1994 *Phys. Rev.* A **49** 2117

[6] Jackson J 1999 *Classical Electrodynamics* 3rd edn (New York: Wiley)

[7] Heitler W 1954 *The Quantum Theory of Radiation* 3rd edn (Oxford: Oxford University Press)

[8] Knoll G 2010 *Radiation Detection and Measurement* 4th edn (New York: Wiley)

[9] Kleinknecht K 1999 *Detectors for Particle Radiation* (Cambridge: Cambridge University Press)

[58] As the Maxwell equation for $\nabla \times \mathbf{H}$ shows, such a field distribution cannot be created in any non-vanishing volume of free space. However, it may be created on a line—e.g. on a particle's straight trajectory.

[59] See section 2.9, in particular figure 2.27a.

IOP Publishing

Classical Electrodynamics
Lecture notes
Konstantin K Likharev

Appendix A

Selected mathematical formulas

This appendix lists selected mathematical formulas that are used in this lecture course series, but not always remembered by students (and some instructors:-).

A.1 Constants

- Euclidean circle's *length-to-diameter ratio*:

$$\pi = 3.141\,592\,653\ldots; \qquad \pi^{1/2} \approx 1.77. \tag{A.1}$$

- *Natural logarithm base*:

$$e \equiv \lim_{n\to\infty}\left(1 + \frac{1}{n}\right)^n = 2.718\,281\,828\ldots; \tag{A.2a}$$

from that value, the logarithm base conversion factors are as follows ($\xi > 0$):

$$\frac{\ln \xi}{\log_{10}\xi} = \ln 10 \approx 2.303, \qquad \frac{\log_{10}\xi}{\ln \xi} = \frac{1}{\ln 10} \approx 0.434. \tag{A.2b}$$

- The *Euler* (or 'Euler–Mascheroni') *constant*:

$$\gamma \equiv \lim_{n\to\infty}\left(1 + \frac{1}{2} + \frac{1}{3} + \ldots \frac{1}{n} - \ln n\right) = 0.577\,156\,649\,0\ldots; \quad e^{\gamma} \approx 1.781. \tag{A.3}$$

A.2 Combinatorics, sums, and series

(i) *Combinatorics*

- The number of different *permutations*, i.e. *ordered* sequences of k elements selected from a set of n distinct elements ($n \geqslant k$), is

$$^{n}P_k \equiv n \cdot (n-1) \cdots (n-k+1) = \frac{n!}{(n-k)!}; \tag{A.4a}$$

in particular, the number of different permutations of *all* elements of the set ($n = k$) is

$$^kP_k = k \cdot (k-1) \cdots 2 \cdot 1 = k!. \tag{A.4b}$$

- The number of different *combinations*, i.e. *unordered* sequences of k elements from a set of $n \geqslant k$ distinct elements, is equal to the binomial coefficient

$$^nC_k \equiv \binom{n}{k} \equiv \frac{^nP_k}{^kP_k} = \frac{n!}{k!(n-k)!}. \tag{A.5}$$

In an alternative, very popular 'ball/box language', ^nC_k is the number of different ways to put in a box, in an arbitrary order, k balls selected from n distinct balls.

- A generalization of the binomial coefficient notion is the multinomial coefficient,

$$^nC_{k_1,k_2,\ldots k_l} \equiv \frac{n!}{k_1!k_2!\ldots k_l!}, \qquad \text{with } n = \sum_{j=1}^{l} k_j, \tag{A.6}$$

which, in the standard mathematical language, is a number of different permutations in a multiset of l distinct element types from an n-element set which contains k_j ($j = 1, 2,\ldots l$) elements of each type. In the 'ball/box language', the coefficient (A.6) is the number of different ways to distribute n distinct balls between l distinct boxes, each time keeping the number (k_j) of balls in the jth box fixed, but ignoring their order inside the box. The binomial coefficient ^nC_k (A.5), is a particular case of the multinomial coefficient (A.6) for $l = 2$ - counting the explicit box for the first one, and the remaining space for the second box, so that if $k_1 \equiv k$, then $k_2 = n - k$.

- One more important combinatorial quantity is the number $M_n^{(k)}$ of ways to place n *indistinguishable* balls into k distinct boxes. It may be readily calculated from Eq. (A.5) as the number of different ways to select $(k-1)$ partitions between the boxes in an imagined linear row of $(k - 1 + n)$ 'objects' (balls in the boxes *and* partitions between them):

$$M_n^{(k)} = {}^{n-1+k}C_{k-1} \equiv \frac{(k-1+n)!}{(k-1)!n!}. \tag{A.7}$$

(ii) *Sums and series*

- *Arithmetic progression*:

$$r + 2r + \cdots + nr \equiv \sum_{k=1}^{n} kr = \frac{n(r + nr)}{2}; \tag{A.8a}$$

in particular, at $r = 1$ it is reduced to the sum of n first natural numbers:

$$1 + 2 + \cdots + n \equiv \sum_{k=1}^{n} k = \frac{n(n+1)}{2}. \tag{A.8b}$$

- Sums of squares and cubes of n first natural numbers:

$$1^2 + 2^2 + \cdots + n^2 \equiv \sum_{k=1}^{n} k^2 = \frac{n(n+1)(2n+1)}{6}; \tag{A.9a}$$

$$1^3 + 2^3 + \cdots + n^3 \equiv \sum_{k=1}^{n} k^3 = \frac{n^2(n+1)^2}{4}. \tag{A.9b}$$

- The *Riemann zeta function*:

$$\zeta(s) \equiv 1 + \frac{1}{2^s} + \frac{1}{3^s} + \cdots \equiv \sum_{k=1}^{\infty} \frac{1}{k^s}; \tag{A.10a}$$

the particular values frequently met in applications are

$$\begin{aligned} &\zeta\left(\frac{3}{2}\right) \approx 2.612, \quad \zeta(2) = \frac{\pi^2}{6}, \quad \zeta\left(\frac{5}{2}\right) \approx 1.341, \\ &\zeta(3) \approx 1.202, \quad \zeta(4) = \frac{\pi^4}{90}, \quad \zeta(5) \approx 1.037. \end{aligned} \tag{A.10b}$$

- Finite geometric progression (for real $\lambda \neq 1$):

$$1 + \lambda + \lambda^2 + \cdots + \lambda^{n-1} \equiv \sum_{k=0}^{n-1} \lambda^k = \frac{1 - \lambda^n}{1 - \lambda}; \tag{A.11a}$$

in particular, if $\lambda^2 < 1$, the progression has a finite limit at $n \to \infty$ (called the *geometric series*):

$$\lim_{n\to\infty} \sum_{k=0}^{n-1} \lambda^k = \sum_{k=0}^{\infty} \lambda^k = \frac{1}{1 - \lambda}. \tag{A.11b}$$

- *Binomial sum* (or the 'binomial theorem'):

$$(1 + a)^n = \sum_{k=0}^{n} {}^nC_k a^k, \tag{A.12}$$

where ${}^nC_k$ are the binomial coefficients defined by Eq. (A.5).

- The *Stirling formula*:

$$\lim_{n\to\infty} \ln(n!) = n(\ln n - 1) + \frac{1}{2}\ln(2\pi n) + \frac{1}{12n} - \frac{1}{360n^3} + \ldots; \qquad \text{(A.13)}$$

  for most applications in physics, the first term[1] is sufficient.
- The *Taylor* (or 'Taylor–Maclaurin') *series*: for any infinitely differentiable function $f(\xi)$:

$$\begin{aligned}\lim_{\tilde{\xi}\to 0} f(\xi + \tilde{\xi}) &= f(\xi) + \frac{df}{d\xi}(\xi)\,\tilde{\xi} + \frac{1}{2!}\frac{d^2 f}{d\xi^2}(\xi)\,\tilde{\xi}^2 + \cdots \\ &= \sum_{k=0}^{\infty}\frac{1}{k!}\frac{d^k f}{d\xi^k}(\xi)\,\tilde{\xi}^k;\end{aligned} \qquad \text{(A.14}a\text{)}$$

  note that for many functions this series converges only within a limited, sometimes small range of deviations $\tilde{\xi}$. For a function of several arguments, $f(\xi_1,\xi_2,\ldots,\xi_N)$, the first terms of the Taylor series are

$$\begin{aligned}\lim_{\tilde{\xi}_k\to 0} f(\xi_1 + \tilde{\xi}_1,\ \xi_2 + \tilde{\xi}_2,\ \cdots) = f(\xi_1,\ \xi_2,\ \cdots) \\ + \sum_{k=1}^{N}\frac{\partial f}{\partial \xi_k}(\xi_1,\ \xi_2,\ \cdots)\,\tilde{\xi}_k \\ + \frac{1}{2!}\sum_{k,k'=1}^{N}\frac{\partial^2 f}{\partial_k \xi\ \partial \xi_{k'}}\tilde{\xi}_k\tilde{\xi}_{k'} + \cdots\end{aligned} \qquad \text{(A.14}b\text{)}$$

- The *Euler–Maclaurin formula*, valid for any infinitely differentiable function $f(\xi)$:

$$\begin{aligned}\sum_{n=0}^{n-1} f(k) = \int_0^n f(\xi)dx - \frac{1}{2}[f(n) - f(0)] + \frac{1}{6}\cdot\frac{1}{2!}\left[\frac{df}{d\xi}(n) - \frac{df}{d\xi}(0)\right] \\ + \frac{1}{30}\cdot\frac{1}{4!}\left[\frac{d^3 f}{d\xi^3}(n) - \frac{d^3 f}{d\xi^3}(0)\right] \\ - \frac{1}{42}\cdot\frac{1}{6!}\left[\frac{d^5 f}{d\xi^5}(n) - \frac{d^5 f}{d\xi^5}(0)\right] + \cdots;\end{aligned} \qquad \text{(A.15}a\text{)}$$

  the coefficients participating in this formula are the so-called *Bernoulli numbers*[2]:

$$\begin{aligned}&B_1 = \frac{1}{2}, \quad B_2 = \frac{1}{6}, \quad B_3 = 0, \quad B_4 = \frac{1}{30}, \quad B_5 = 0, \\ &B_6 = \frac{1}{42}, \quad B_7 = 0, \quad B_8 = \frac{1}{30}, \quad \cdots\end{aligned} \qquad \text{(A.15}b\text{)}$$

[1] Actually, this leading term was derived by A de Moivre in 1733, before J Stirling's work.

[2] Note that definitions of B_k (or rather their signs and indices) vary even in the most popular handbooks.

A.3 Basic trigonometric functions

- Trigonometric functions of the sum and the difference of two arguments[3]:

$$\cos(a \pm b) = \cos a \cos b \mp \sin a \sin b, \tag{A.16a}$$

$$\sin(a \pm b) = \sin a \cos b \pm \cos a \sin b. \tag{A.16b}$$

- Sums of two functions of arbitrary arguments:

$$\cos a + \cos b = 2\cos\frac{a+b}{2}\cos\frac{b-a}{2}, \tag{A.17a}$$

$$\cos a - \cos b = 2\sin\frac{a+b}{2}\sin\frac{b-a}{2}, \tag{A.17b}$$

$$\sin a \pm \sin b = 2\sin\frac{a \pm b}{2}\cos\frac{\pm b - a}{2}. \tag{A.17c}$$

- Trigonometric function products:

$$2\cos a \cos b = \cos(a+b) + \cos(a-b), \tag{A.18a}$$

$$2\sin a \cos b = \sin(a+b) + \sin(a-b), \tag{A.18b}$$

$$2\sin a \sin b = \cos(a-b) - \cos(a+b); \tag{A.18c}$$

For the particular case of equal arguments, $b = a$, these three formulas yield the following expressions for the squares of trigonometric functions, and their product:

$$\cos^2 a = \frac{1}{2}(1 + \cos 2a), \qquad \sin a \cos a = \frac{1}{2}\sin 2a, \\ \sin^2 a = \frac{1}{2}(1 - \cos 2a). \tag{A.18d}$$

- Cubes of trigonometric functions:

$$\cos^3 a = \frac{3}{4}\cos a + \frac{1}{4}\cos 3a, \qquad \sin^3 a = \frac{3}{4}\sin a - \frac{1}{4}\sin 3a. \tag{A.19}$$

[3] I am confident that the reader is quite capable of deriving the relations (A.16) by representing the exponent in the elementary relation $e^{i(a \pm b)} = e^{ia}e^{\pm ib}$ as a sum of its real and imaginary parts, Eqs. (A.18) directly from Eqs. (A.16), and Eqs. (A.17) from Eqs. (A.18) by variable replacement; however, I am still providing these formulas to save his or her time. (Quite a few formulas below are included because of the same reason.)

- Trigonometric functions of a complex argument:

$$\begin{aligned}\sin(a+ib) &= \sin a \cosh b + i \cos a \sinh b,\\ \cos(a+ib) &= \cos a \cosh b - i \sin a \sinh b.\end{aligned} \tag{A.20}$$

- Sums of trigonometric functions of n equidistant arguments:

$$\sum_{k=1}^{n}\left\{\begin{matrix}\sin\\ \cos\end{matrix}\right\}k\xi = \left\{\begin{matrix}\sin\\ \cos\end{matrix}\right\}\left(\frac{n+1}{2}\xi\right)\sin\left(\frac{n}{2}\xi\right)\Big/\sin\left(\frac{\xi}{2}\right). \tag{A.21}$$

A.4 General differentiation

- Full differential of a product of two functions:

$$d(fg) = (df)g + f(dg). \tag{A.22}$$

- Full differential of a function of several independent arguments, $f(\xi_1, \xi_2, \ldots, \xi_n)$:

$$df = \sum_{k=1}^{n}\frac{\partial f}{\partial \xi_k}d\xi_k. \tag{A.23}$$

- Curvature of the Cartesian plot of a 1D function $f(\xi)$:

$$\kappa \equiv \frac{1}{R} = \frac{|\,d^2f/d\xi^2\,|}{[1+(df/d\xi)^2]^{3/2}}. \tag{A.24}$$

A.5 General integration

- Integration *by parts* - immediately follows from Eq. (A.22):

$$\int_{g(A)}^{g(B)} f\ dg = fg\Big|_A^B - \int_{f(A)}^{f(B)} g\ df. \tag{A.25}$$

- Numerical (approximate) integration of 1D functions: the simplest *trapezoidal rule*,

$$\begin{aligned}\int_a^b f(\xi)d\xi &\approx h\left[f\left(a+\frac{h}{2}\right)+f\left(a+\frac{3h}{2}\right)+\cdots+f\left(b-\frac{h}{2}\right)\right]\\ &= h\sum_{n=1}^{N} f\left(a-\frac{h}{2}+nh\right), \qquad h \equiv \frac{b-a}{N}.\end{aligned} \tag{A.26}$$

has relatively low accuracy (error of the order of $(h^3/12)d^2f/d\xi^2$ per step), so that the following *Simpson formula*,

$$\int_a^b f(\xi)d\xi \approx \frac{h}{3}[f(a) + 4f(a+h) + 2f(a+2h) + \cdots + 4f(b-h) + f(b)],$$
$$h \equiv \frac{b-a}{2N}, \tag{A.27}$$

whose error per step scales as $(h^5/180)d^4f/d\xi^4$, is used much more frequently[4].

A.6 A few 1D integrals[5]

(i) *Indefinite integrals*:

- Integrals with $(1+\xi^2)^{1/2}$:

$$\int (1+\xi^2)^{1/2}d\xi = \frac{\xi}{2}(1+\xi^2)^{1/2} + \frac{1}{2}\ln|\xi + (1+\xi^2)^{1/2}|, \tag{A.28}$$

$$\int \frac{d\xi}{(1+\xi^2)^{1/2}} = \ln|\xi + (1+\xi^2)^{1/2}|, \tag{A.29a}$$

$$\int \frac{d\xi}{(1+\xi^2)^{3/2}} = \frac{\xi}{(1+\xi^2)^{1/2}}. \tag{A.29b}$$

- Miscellaneous indefinite integrals:

$$\int \frac{d\xi}{\xi(\xi^2 + 2a\xi - 1)^{1/2}} = \cos^{-1}\frac{a\xi - 1}{|\xi|(a^2+1)^{1/2}}, \tag{A.30a}$$

$$\int \frac{(\sin\xi - \xi\cos\xi)^2}{\xi^5}d\xi = \frac{2\xi\sin 2\xi + \cos 2\xi - 2\xi^2 - 1}{8\xi^4}, \tag{A.30b}$$

$$\int \frac{d\xi}{a + b\cos\xi} = \frac{2}{(a^2-b^2)^{1/2}}\tan^{-1}\left[\frac{(a-b)}{(a^2-b^2)^{1/2}}\tan\frac{\xi}{2}\right], \quad \text{for } a^2 > b^2. \tag{A.30c}$$

$$\int \frac{d\xi}{1+\xi^2} = \tan^{-1}\xi. \tag{A.30d}$$

---

[4] Higher-order formulas (e.g. the *Bode rule*), and other guidance including ready-for-use codes for computer calculations may be found, for example, in the popular reference texts by W H Press *et al* [1]. In addition, some advanced codes are used as subroutines in the software packages listed in the same section. In some cases, the Euler–Maclaurin formula (A.15) may also be useful for numerical integration.

[5] A powerful (and free) interactive online tool for working out indefinite 1D integrals is available at http://integrals.wolfram.com/index.jsp.

(ii) *Semi-definite integrals*:

- Integrals with $1/(e^{\xi} \pm 1)$:

$$\int_a^{\infty} \frac{d\xi}{e^{\xi}+1} = \ln\left(1+e^{-a}\right), \tag{A.31a}$$

$$\int_{a>0}^{\infty} \frac{d\xi}{e^{\xi}-1} = \ln\frac{1}{1-e^{-a}}. \tag{A.31b}$$

(iii) *Definite integrals*:

- Integrals with $1/(1+\xi^2)$:[6]

$$\int_0^{\infty} \frac{d\xi}{1+\xi^2} = \frac{\pi}{2}, \tag{A.32a}$$

$$\int_0^{\infty} \frac{d\xi}{(1+\xi^2)^{3/2}} = 1; \tag{A.32b}$$

more generally,

$$\int_0^{\infty} \frac{d\xi}{(1+\xi^2)^n} = \frac{\pi}{2}\frac{(2n-3)!!}{(2n-2)!!} \equiv \frac{\pi}{2}\frac{1\cdot 3\cdot 5\ldots(2n-3)}{2\cdot 4\cdot 6\ldots(2n-2)}, \quad \text{for } n = 2,\ 3,\ \ldots \tag{A.32c}$$

- Integrals with $(1-\xi^{2n})^{1/2}$:

$$\int_0^1 \frac{d\xi}{(1-\xi^{2n})^{1/2}} = \frac{\pi^{1/2}}{2n}\Gamma\left(\frac{1}{2n}\right)\Big/\Gamma\left(\frac{n+1}{2n}\right), \tag{A.33a}$$

$$\int_0^1 (1-\xi^{2n})^{1/2} d\xi = \frac{\pi^{1/2}}{4n}\Gamma\left(\frac{1}{2n}\right)\Big/\Gamma\left(\frac{3n+1}{2n}\right), \tag{A.33b}$$

where $\Gamma(s)$ is the *gamma-function,* which is most often defined (for Re $s > 0$) by the following integral:

$$\int_0^{\infty} \xi^{s-1}e^{-\xi}\, d\xi = \Gamma(s). \tag{A.34a}$$

[6] Eq. (A.32a) follows immediately from Eq. (A.30d), and Eq. (A.32b) from Eq. (A.29b)—a couple more examples of the (intentional) redundancy in this list.

The key property of this function is the recurrence relation, valid for any $s \neq 0, -1, -2, \ldots$:

$$\Gamma(s+1) = s\Gamma(s). \tag{A.34b}$$

Since, according to Eq. (A.34*a*), $\Gamma(1) = 1$, Eq. (A.34*b*) for non-negative integers takes the form

$$\Gamma(n+1) = n!, \quad \text{for } n = 0,\ 1,\ 2,\ \cdots \tag{A.34c}$$

(where $0! \equiv 1$). Because of this, for integer $s = n + 1 \geqslant 1$, Eq. (A.34*a*) is reduced to

$$\int_0^\infty \xi^n e^{-\xi} d\xi = n!. \tag{A.34d}$$

Other frequently met values of the gamma-function are those for positive semi-integer arguments:

$$\begin{aligned} &\Gamma\left(\frac{1}{2}\right) = \pi^{1/2}, \quad \Gamma\left(\frac{3}{2}\right) = \frac{1}{2}\pi^{1/2}, \quad \Gamma\left(\frac{5}{2}\right) = \frac{1}{2}\cdot\frac{3}{2}\pi^{1/2}, \\ &\Gamma\left(\frac{7}{2}\right) = \frac{1}{2}\cdot\frac{3}{2}\cdot\frac{5}{2}\pi^{1/2}, \quad \ldots. \end{aligned} \tag{A.34e}$$

- Integrals with $1/(e^\xi \pm 1)$:

$$\int_0^\infty \frac{\xi^{s-1}d\xi}{e^\xi + 1} = (1 - 2^{1-s})\,\Gamma(s)\zeta(s), \quad \text{for } s > 0, \tag{A.35a}$$

$$\int_0^\infty \frac{\xi^{s-1}d\xi}{e^\xi - 1} = \Gamma(s)\zeta(s), \quad \text{for } s > 1, \tag{A.35b}$$

where $\zeta(s)$ is the Riemann zeta-function—see Eq. (A.10). Particular cases: for $s = 2n$,

$$\int_0^\infty \frac{\xi^{2n-1}d\xi}{e^\xi + 1} = \frac{2^{2n-1} - 1}{2n}\pi^{2n}B_{2n}, \tag{A.35c}$$

$$\int_0^\infty \frac{\xi^{2n-1}d\xi}{e^\xi - 1} = \frac{(2\pi)^{2n}}{4n}B_{2n}. \tag{A.35d}$$

where B_n are the Bernoulli numbers—see Eq. (A.15). For the particular case $s = 1$ (when Eq. (A.35a) yields uncertainty),

$$\int_0^\infty \frac{d\xi}{e^\xi + 1} = \ln 2. \tag{A.35e}$$

- Integrals with $\exp\{-\xi^2\}$:

$$\int_0^\infty \xi^s e^{-\xi^2} d\xi = \frac{1}{2}\Gamma\left(\frac{s+1}{2}\right), \quad \text{for } s > -1; \tag{A.36a}$$

for applications the most important particular values of s are 0 and 2:

$$\int_0^\infty e^{-\xi^2} d\xi = \frac{1}{2}\Gamma\left(\frac{1}{2}\right) = \frac{\pi^{1/2}}{2}, \tag{A.36b}$$

$$\int_0^\infty \xi^2 e^{-\xi^2} d\xi = \frac{1}{2}\Gamma\left(\frac{3}{2}\right) = \frac{\pi^{1/2}}{4}, \tag{A.36c}$$

though we will also run into the cases $s = 4$ and $s = 6$:

$$\begin{aligned} \int_0^\infty \xi^4 e^{-\xi^2} d\xi &= \frac{1}{2}\Gamma\left(\frac{5}{2}\right) = \frac{3\pi^{1/2}}{8}, \\ \int_0^\infty \xi^6 e^{-\xi^2} d\xi &= \frac{1}{2}\Gamma\left(\frac{7}{2}\right) = \frac{15\pi^{1/2}}{16}; \end{aligned} \tag{A.36d}$$

for odd integer values $s = 2n + 1$ (with $n = 0, 1, 2,\ldots$), Eq. (A.36*a*) takes a simpler form:

$$\int_0^\infty \xi^{2n+1} e^{-\xi^2} d\xi = \frac{1}{2}\Gamma(n+1) = \frac{n!}{2}. \tag{A.36e}$$

- Integrals with cosine and sine functions:

$$\int_0^\infty \cos(\xi^2)\, d\xi = \int_0^\infty \sin(\xi^2)\, d\xi = \left(\frac{\pi}{8}\right)^{1/2}. \tag{A.37}$$

$$\int_0^\infty \frac{\cos\xi}{a^2 + \xi^2} d\xi = \frac{\pi}{2a} e^{-a}. \tag{A.38}$$

$$\int_0^\infty \left(\frac{\sin\xi}{\xi}\right)^2 d\xi = \frac{\pi}{2}. \tag{A.39}$$

- Integrals with logarithms:

$$\int_0^1 \ln\frac{a + (1-\xi^2)^{1/2}}{a - (1-\xi^2)^{1/2}} d\xi = \pi[a - (a^2 - 1)^{1/2}], \quad \text{for } a \geqslant 1. \tag{A.40}$$

$$\int_0^1 \ln\frac{1 + (1-\xi)^{1/2}}{\xi^{1/2}} d\xi = 1. \tag{A.41}$$

- Integral representations of the Bessel functions of integer order:

$$J_n(\alpha) = \frac{1}{2\pi}\int_{-\pi}^{+\pi} e^{i(\alpha \sin\xi - n\xi)}d\xi,$$
$$\text{so that } e^{i\alpha\sin\xi} = \sum_{k=-\infty}^{\infty} J_k(\alpha)e^{ik\xi}; \qquad \text{(A.42}a\text{)}$$

$$I_n(\alpha) = \frac{1}{\pi}\int_0^{\pi} e^{\alpha\cos\xi}\cos n\xi \;\; d\xi. \qquad \text{(A.42}b\text{)}$$

A.7 3D vector products

(i) *Definitions*:

- *Scalar* ('dot-') *product*:

$$\mathbf{a}\cdot\mathbf{b} = \sum_{j=1}^{3} a_j b_j, \qquad \text{(A.43)}$$

where a_j and b_j are vector components in any orthogonal coordinate system. In particular, the vector squared (the same as the norm squared):

$$a^2 \equiv \mathbf{a}\cdot\mathbf{a} = \sum_{j=1}^{3} a_j^2 \equiv \|\mathbf{a}\|^2. \qquad \text{(A.44)}$$

- *Vector* ('cross-') *product*:

$$\mathbf{a}\times\mathbf{b} \equiv \mathbf{n}_1(a_2b_3 - a_3b_2) + \mathbf{n}_2(a_3b_1 - a_1b_3) + \mathbf{n}_3(a_1b_2 - a_2b_1)$$
$$= \begin{vmatrix} \mathbf{n}_1 & \mathbf{n}_2 & \mathbf{n}_3 \\ a_1 & a_2 & a_3 \\ b_1 & b_2 & b_3 \end{vmatrix}, \qquad \text{(A.45)}$$

where $\{\mathbf{n}_j\}$ is the set of mutually perpendicular unit vectors[7] along the corresponding coordinate system axes[8]. In particular, Eq. (A.45) yields

$$\mathbf{a}\times\mathbf{a} = 0. \qquad \text{(A.46)}$$

(ii) *Corollaries* (readily verified by Cartesian components):

- Double vector product (the so-called *bac minus cab* rule):

$$\mathbf{a}\times(\mathbf{b}\times\mathbf{c}) = \mathbf{b}(\mathbf{a}\cdot\mathbf{c}) - \mathbf{c}(\mathbf{a}\cdot\mathbf{b}). \qquad \text{(A.47)}$$

[7] Other popular notations for this vector set are $\{\mathbf{e}_j\}$ and $\{\hat{\mathbf{r}}_j\}$.

[8] It is easy to use Eq. (A.45) to check that the direction of the product vector corresponds to the well-known 'right-hand rule' and to the even more convenient *corkscrew rule*: if we rotate a corkscrew's handle from the first operand toward the second one, its axis moves in the direction of the product.

- Mixed scalar–vector product (the *operand rotation rule*):

$$\mathbf{a} \cdot (\mathbf{b} \times \mathbf{c}) = \mathbf{b} \cdot (\mathbf{c} \times \mathbf{a}) = \mathbf{c} \cdot (\mathbf{a} \times \mathbf{b}). \tag{A.48}$$

- Scalar product of vector products:

$$(\mathbf{a} \times \mathbf{b}) \cdot (\mathbf{c} \times \mathbf{d}) = (\mathbf{a} \cdot \mathbf{c})(\mathbf{b} \cdot \mathbf{d}) - (\mathbf{a} \cdot \mathbf{d})(\mathbf{b} \cdot \mathbf{c}); \tag{A.49a}$$

  in the particular case of two similar operands (say, $\mathbf{a} = \mathbf{c}$ and $\mathbf{b} = \mathbf{d}$), the last formula is reduced to

$$(\mathbf{a} \times \mathbf{b})^2 = (ab)^2 - (\mathbf{a} \cdot \mathbf{b})^2. \tag{A.49b}$$

A.8 Differentiation in 3D Cartesian coordinates

- Definition of the *del* (or 'nabla') vector-operator ∇:[9]

$$\nabla \equiv \sum_{j=1}^{3} \mathbf{n}_j \frac{\partial}{\partial r_j}, \tag{A.50}$$

  where r_j is a set of linear and orthogonal (*Cartesian*) coordinates along directions $\mathbf{n}_j$. In accordance with this definition, the operator ∇ acting on a *scalar* function of coordinates, $f(\mathbf{r})$,[10] gives its gradient, i.e. a new *vector*:

$$\nabla f \equiv \sum_{j=1}^{3} \mathbf{n}_j \frac{\partial f}{\partial r_j} \equiv \mathbf{grad}\, f. \tag{A.51}$$

- The *scalar product* of del by a *vector* function of coordinates (a *vector field*),

$$\mathbf{f}(\mathbf{r}) \equiv \sum_{j=1}^{3} \mathbf{n}_j f_j(\mathbf{r}), \tag{A.52}$$

  compiled formally following Eq. (A.43), is a *scalar* function—the *divergence* of the initial function:

$$\nabla \cdot \mathbf{f} \equiv \sum_{j=1}^{3} \frac{\partial f_j}{\partial r_j} \equiv \mathbf{div}\, \mathbf{f}, \tag{A.53}$$

[9] One can run into the following notation: $\nabla \equiv \partial/\partial \mathbf{r}$, which is convenient is some cases, but may be misleading in quite a few others, so it will be not used in these notes.

[10] In this, and four next sections, all scalar and vector functions are assumed to be differentiable.

while the *vector product* of ∇ and **f**, formed in a formal accordance with Eq. (A.45), is a new vector - the *curl* (in European tradition, called rotor and denoted **rot**) of **f**:

$$\nabla \times \mathbf{f} \equiv \begin{vmatrix} \mathbf{n}_1 & \mathbf{n}_2 & \mathbf{n}_3 \\ \frac{\partial}{\partial r_1} & \frac{\partial}{\partial r_2} & \frac{\partial}{\partial r_3} \\ f_1 & f_2 & f_3 \end{vmatrix} = \mathbf{n}_1\left(\frac{\partial f_3}{\partial r_2} - \frac{\partial f_2}{\partial r_3}\right) + \mathbf{n}_2\left(\frac{\partial f_1}{\partial r_3} - \frac{\partial f_3}{\partial r_1}\right) + \mathbf{n}_3\left(\frac{\partial f_2}{\partial r_1} - \frac{\partial f_1}{\partial r_2}\right) \equiv \mathbf{curl}\,\mathbf{f}. \quad \text{(A.54)}$$

- One more frequently met 'product' is **(f·∇)g**, where **f** and **g** are two arbitrary vector functions of **r**. This product should be also understood in the sense implied by Eq. (A.43), i.e. as a vector whose *j*th Cartesian component is

$$[(\mathbf{f} \cdot \nabla)\,\mathbf{g}]_j = \sum_{j'=1}^{3} f_{j'} \frac{\partial g_j}{\partial r_{j'}}. \quad \text{(A.55)}$$

A.9 The Laplace operator $\nabla^2 \equiv \nabla \cdot \nabla$

- Expression in Cartesian coordinates—in the formal accordance with Eq. (A.44):

$$\nabla^2 = \sum_{j=1}^{3} \frac{\partial^2}{\partial r_j^2}. \quad \text{(A.56)}$$

- According to its definition, the Laplace operator acting on a *scalar* function of coordinates gives a new scalar function:

$$\nabla^2 f \equiv \nabla \cdot (\nabla f) = \mathbf{div}(\mathbf{grad}\, f) = \sum_{j=1}^{3} \frac{\partial^2 f}{\partial r_j^2}. \quad \text{(A.57)}$$

- On the other hand, acting on a *vector* function (A.52), the operator ∇^2 returns another *vector*:

$$\nabla^2 \mathbf{f} = \sum_{j=1}^{3} \mathbf{n}_j \nabla^2 f_j. \quad \text{(A.58)}$$

Note that Eqs. (A.56)–(A.58) are only valid in Cartesian (i.e. orthogonal and linear) coordinates, but generally not in other (even orthogonal) coordinates—see, e.g. Eqs. (A.61), (A.64), (A.67) and (A.70) below.

A.10 Operators ∇ and ∇^2 in the most important systems of orthogonal coordinates[11]

(i) *Cylindrical*[12] *coordinates* $\{\rho, \varphi, z\}$ (see figure below) may be defined by their relations with the Cartesian coordinates:

$$r_1 = \rho \cos\varphi, \quad r_2 = \rho \sin\varphi, \quad r_3 = z. \tag{A.59}$$

- Gradient of a scalar function:

$$\nabla f = \mathbf{n}_\rho \frac{\partial f}{\partial \rho} + \mathbf{n}_\varphi \frac{1}{\rho}\frac{\partial f}{\partial \varphi} + \mathbf{n}_z \frac{\partial f}{\partial z}. \tag{A.60}$$

- The Laplace operator of a scalar function:

$$\nabla^2 f = \frac{1}{\rho}\frac{\partial}{\partial \rho}\left(\rho \frac{\partial f}{\partial \rho}\right) + \frac{1}{\rho^2}\frac{\partial^2 f}{\partial \varphi^2} + \frac{\partial^2 f}{\partial z^2}, \tag{A.61}$$

- Divergence of a vector function of coordinates ($\mathbf{f} = \mathbf{n}_\rho f_\rho + \mathbf{n}_\varphi f_\varphi + \mathbf{n}_z f_z$):

$$\nabla \cdot \mathbf{f} = \frac{1}{\rho}\frac{\partial(\rho f_\rho)}{\partial \rho} + \frac{1}{\rho}\frac{\partial f_\varphi}{\partial \varphi} + \frac{\partial f_z}{\partial z}. \tag{A.62}$$

- Curl of a vector function:

$$\nabla \times \mathbf{f} = \mathbf{n}_\rho\left(\frac{1}{\rho}\frac{\partial f_z}{\partial \varphi} - \frac{\partial f_\varphi}{\partial z}\right) + \mathbf{n}_\varphi\left(\frac{\partial f_\rho}{\partial z} - \frac{\partial f_z}{\partial \rho}\right) + \mathbf{n}_z \frac{1}{\rho}\left(\frac{\partial(\rho f_\varphi)}{\partial \rho} - \frac{\partial f_\rho}{\partial \varphi}\right). \tag{A.63}$$

- The Laplace operator of a vector function:

$$\nabla^2 \mathbf{f} = \mathbf{n}_\rho\left(\nabla^2 f_\rho - \frac{1}{\rho^2} f_\rho - \frac{2}{\rho^2}\frac{\partial f_\varphi}{\partial \varphi}\right) + \mathbf{n}_\varphi\left(\nabla^2 f_\varphi - \frac{1}{\rho^2} f_\varphi + \frac{2}{\rho^2}\frac{\partial f_\rho}{\partial \varphi}\right) + \mathbf{n}_z \nabla^2 f_z. \tag{A.64}$$

[11] Some other orthogonal curvilinear coordinate systems are discussed in *Part EM*, section 2.3.

[12] In the 2D geometry with fixed coordinate z, these coordinates are called *polar*.

(ii) *Spherical coordinates* $\{r, \theta, \varphi\}$ (see figure below) may be defined as:

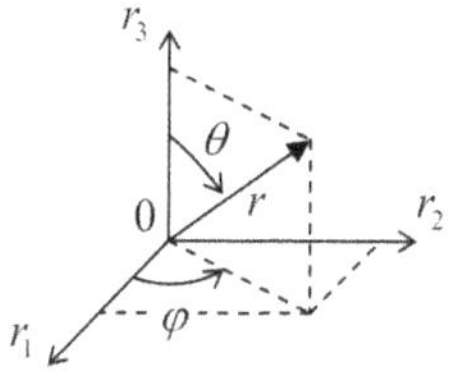

$$r_1 = r\sin\theta\cos\varphi,\quad r_2 = r\sin\theta\sin\varphi,\quad r_3 = r\cos\theta. \tag{A.65}$$

- Gradient of a scalar function:

$$\nabla f = \mathbf{n}_r\frac{\partial f}{\partial r} + \mathbf{n}_\theta\frac{1}{r}\frac{\partial f}{\partial \theta} + \mathbf{n}_\varphi\frac{1}{r\sin\theta}\frac{\partial f}{\partial \varphi}. \tag{A.66}$$

- The Laplace operator of a scalar function:

$$\nabla^2 f = \frac{1}{r^2}\frac{\partial}{\partial r}\left(r^2\frac{\partial f}{\partial r}\right) + \frac{1}{r^2\sin\theta}\frac{\partial}{\partial\theta}\left(\sin\theta\frac{\partial f}{\partial\theta}\right) + \frac{1}{(r\sin\theta)^2}\frac{\partial^2 f}{\partial\varphi^2}. \tag{A.67}$$

- Divergence of a vector function $\mathbf{f} = \mathbf{n}_r f_r + \mathbf{n}_\theta f_\theta + \mathbf{n}_\varphi f_\varphi$:

$$\nabla\cdot\mathbf{f} = \frac{1}{r^2}\frac{\partial(r^2 f_r)}{\partial r} + \frac{1}{r\sin\theta}\frac{\partial(f_\theta\sin\theta)}{\partial\theta} + \frac{1}{r\sin\theta}\frac{\partial f_\varphi}{\partial\varphi}. \tag{A.68}$$

- Curl of a similar vector function:

$$\begin{aligned}\nabla\times\mathbf{f} = \mathbf{n}_r\frac{1}{r\sin\theta}\left(\frac{\partial(f_\varphi\sin\theta)}{\partial\theta} - \frac{\partial f_\theta}{\partial\varphi}\right) + \mathbf{n}_\theta\frac{1}{r}\left(\frac{1}{\sin\theta}\frac{\partial f_r}{\partial\varphi} - \frac{\partial(rf_\varphi)}{\partial r}\right)\\ + \mathbf{n}_\varphi\frac{1}{r}\left(\frac{\partial(rf_\theta)}{\partial r} - \frac{\partial f_r}{\partial\theta}\right).\end{aligned} \tag{A.69}$$

- The Laplace operator of a vector function:

$$\begin{aligned}\nabla^2\mathbf{f} = \mathbf{n}_r\left(\nabla^2 f_r - \frac{2}{r^2}f_r - \frac{2}{r^2\sin\theta}\frac{\partial}{\partial\theta}(f_\theta\sin\theta) - \frac{2}{r^2\sin\theta}\frac{\partial f_\varphi}{\partial\varphi}\right)\\ + \mathbf{n}_\theta\left(\nabla^2 f_\theta - \frac{1}{r^2\sin^2\theta}f_\theta + \frac{2}{r^2}\frac{\partial f_r}{\partial\theta} - \frac{2\cos\theta}{r^2\sin^2\theta}\frac{\partial f_\varphi}{\partial\varphi}\right)\\ + \mathbf{n}_\varphi\left(\nabla^2 f_\varphi - \frac{1}{r^2\sin^2\theta}f_\varphi + \frac{2}{r^2\sin\theta}\frac{\partial f_r}{\partial\varphi} + \frac{2\cos\theta}{r^2\sin^2\theta}\frac{\partial f_\theta}{\partial\varphi}\right).\end{aligned} \tag{A.70}$$

A.11 Products involving ∇

(i) *Useful zeros*:

- For any scalar function $f(\mathbf{r})$,

$$\nabla \times (\nabla f) \equiv \mathbf{curl}(\mathbf{grad} f) = 0. \tag{A.71}$$

- For any vector function $\mathbf{f}(\mathbf{r})$,

$$\nabla \cdot (\nabla \times \mathbf{f}) \equiv \mathbf{div}(\mathbf{curl} f) = 0. \tag{A.72}$$

(ii) The *Laplace operator* expressed via the curl of a curl:

$$\nabla^2 \mathbf{f} = \nabla(\nabla \cdot \mathbf{f}) - \nabla \times (\nabla \times \mathbf{f}). \tag{A.73}$$

(iii) Spatial differentiation of a product of a *scalar* function by a *vector* function:

- The scalar 3D generalization of Eq. (A.22) is

$$\nabla \cdot (f\,\mathbf{g}) = (\nabla f) \cdot \mathbf{g} + f(\nabla \cdot \mathbf{g}). \tag{A.74a}$$

- Its vector generalization is similar:

$$\nabla \times (f\,\mathbf{g}) = (\nabla f) \times \mathbf{g} + f(\nabla \times \mathbf{g}). \tag{A.74b}$$

(iv) Spatial differentiation of products of *two vector* functions:

$$\nabla \times (\mathbf{f} \times \mathbf{g}) = \mathbf{f}(\nabla \cdot \mathbf{g}) - (\mathbf{f} \cdot \nabla)\mathbf{g} - (\nabla \cdot \mathbf{f})\mathbf{g} + (\mathbf{g} \cdot \nabla)\mathbf{f}, \tag{A.75}$$

$$\nabla(\mathbf{f} \cdot \mathbf{g}) = (\mathbf{f} \cdot \nabla)\mathbf{g} + (\mathbf{g} \cdot \nabla)\mathbf{f} + \mathbf{f} \times (\nabla \times \mathbf{g}) + \mathbf{g} \times (\nabla \times \mathbf{f}), \tag{A.76}$$

$$\nabla \cdot (\mathbf{f} \times \mathbf{g}) = \mathbf{g} \cdot (\nabla \times \mathbf{f}) - \mathbf{f} \cdot (\nabla \times \mathbf{g}). \tag{A.77}$$

A.12 Integro-differential relations

(i) For an *arbitrary surface* S limited by closed contour C:

- The *Stokes theorem*, valid for any differentiable vector field $\mathbf{f}(\mathbf{r})$:

$$\int_S (\nabla \times \mathbf{f}) \cdot d^2\mathbf{r} \equiv \int_S (\nabla \times \mathbf{f})_n d^2 r = \oint_C \mathbf{f} \cdot d\mathbf{r} \equiv \oint_C f_\tau dr, \tag{A.78}$$

where $d^2\mathbf{r} \equiv \mathbf{n} d^2 r$ is the elementary area vector (normal to the surface), and $d\mathbf{r}$ is the elementary contour length vector (tangential to the contour line).

(ii) For an *arbitrary volume* V limited by closed surface S:

- *Divergence* (or 'Gauss') *theorem*, valid for any differentiable vector field $\mathbf{f}(\mathbf{r})$:

$$\int_V (\nabla \cdot \mathbf{f})\, d^3r = \oint_S \mathbf{f} \cdot d^2\mathbf{r} \equiv \oint_S f_n d^2r. \tag{A.79}$$

- *Green's theorem*, valid for two differentiable scalar functions $f(\mathbf{r})$ and $g(\mathbf{r})$:

$$\int_V (f\, \nabla^2 g - g\nabla^2 f)\, d^3r = \oint_S (f\, \nabla g - g\nabla f)_n d^2r. \tag{A.80}$$

- An identity valid for any two scalar functions f and g, and a vector field $\mathbf{j}$ with $\nabla\cdot\mathbf{j} = 0$ (all differentiable):

$$\int_V [f(\mathbf{j} \cdot \nabla g) + g(\mathbf{j} \cdot \nabla f)]\; d^3r = \oint_S fgj_n d^2r. \tag{A.81}$$

A.13 The Kronecker delta and Levi-Civita permutation symbols

- The *Kronecker delta symbol* (defined for integer indices):

$$\delta_{jj'} \equiv \begin{cases} 1, \text{ if } j' = j, \\ 0, \text{ otherwise.} \end{cases} \tag{A.82}$$

- The *Levi-Civita permutation symbol* (most frequently used for 3 integer indices, each taking one of values 1, 2, or 3):

$$\varepsilon_{jj'j''} \equiv \begin{cases} +1, & \text{if the indices follow in the 'correct' ('even')} \\ & \text{order: } 1 \to 2 \to 3 \to 1 \to 2 \ldots, \\ -1, & \text{if the indices follow in the 'incorrect' ('odd')} \\ & \text{order: } 1 \to 3 \to 2 \to 1 \to 3 \ldots, \\ 0, & \text{if any two indices coincide.} \end{cases} \tag{A.83}$$

- Relation between the Levi-Civita and the Kronecker delta products:

$$\varepsilon_{jj'j''}\varepsilon_{kk'k''} = \sum_{l,l',l''=1}^{3} \begin{vmatrix} \delta_{jl} & \delta_{jl'} & \delta_{jl''} \\ \delta_{j'l} & \delta_{j'l'} & \delta_{j'l''} \\ \delta_{j''l} & \delta_{j''l'} & \delta_{j''l''} \end{vmatrix}; \tag{A.84a}$$

summation of this relation, written for 3 different values of $j = k$, over these values yields the so-called *contracted epsilon identity*:

$$\sum_{j=1}^{3} \varepsilon_{jj'j''}\varepsilon_{jk'k''} = \delta_{j'k'}\delta_{j''k''} - \delta_{j'k''}\delta_{j''k'}. \tag{A.84b}$$

A.14 Dirac's delta-function, sign function, and theta-function

- Definition of 1D *delta-function* (for real $a < b$):

$$\int_a^b f(\xi)\delta(\xi)d\xi = \begin{cases} f(0), & \text{if } a < 0 < b, \\ 0, & \text{otherwise,} \end{cases} \tag{A.85}$$

where $f(\xi)$ is any function continuous near $\xi = 0$. In particular (if $f(\xi) = 1$ near $\xi = 0$), the definition yields

$$\int_a^b \delta(\xi)d\xi = \begin{cases} 1, & \text{if } a < 0 < b, \\ 0, & \text{otherwise.} \end{cases} \tag{A.86}$$

- Relation to the *theta-function* $\theta(\xi)$ and *sign function* $\text{sgn}(\xi)$

$$\delta(\xi) = \frac{d}{d\xi}\theta(\zeta) = \frac{1}{2}\frac{d}{d\xi}\,\text{sgn}(\xi), \tag{A.87a}$$

where

$$\begin{aligned} \theta(\xi) &\equiv \frac{\text{sgn}(\xi) + 1}{2} = \begin{cases} 0, & \text{if } \xi < 0, \\ 1, & \text{if } \xi > 1, \end{cases} \\ \text{sgn}(\xi) &\equiv \frac{\xi}{|\xi|} = \begin{cases} -1, & \text{if } \xi < 0, \\ +1, & \text{if } \xi > 1. \end{cases} \end{aligned} \tag{A.87b}$$

- An important integral[13]:

$$\int_{-\infty}^{+\infty} e^{is\xi}ds = 2\pi\delta(\xi). \tag{A.88}$$

- 3D generalization of the delta-function of the radius-vector (the 2D generalization is similar):

$$\int_V f(\mathbf{r})\delta(\mathbf{r})d^3r = \begin{cases} f(0), \text{ if } 0 \in V, \\ 0, \quad \text{otherwise;} \end{cases} \tag{A.89}$$

it may be represented as a product of 1D delta-functions of Cartesian coordinates:

$$\delta(\mathbf{r}) = \delta(r_1)\delta(r_2)\delta(r_3). \tag{A.90}$$

[13] The coefficient in this equation may be readily recalled by considering its left-hand part as the Fourier-integral presentation of function $f(s) \equiv 1$, and applying Eq. (A.85) to the reciprocal Fourier transform

$$f(s) \equiv 1 = \frac{1}{2\pi}\int_{-\infty}^{+\infty} e^{-is\xi}[2\pi\delta(\xi)]d\xi.$$

A.15 The Cauchy theorem and integral

Let a complex function $f(z)$ be analytic within a part of the complex plane z, that is limited by a closed contour C and includes point z'. Then

$$\oint_C f(z)dz = 0, \tag{A.91}$$

$$\oint_C f(z)\frac{dz}{z - z'} = 2\pi i f(z') \tag{A.92}$$

The first of these relations is usually called the *Cauchy integral theorem* (or the 'Cauchy–Goursat theorem'), and the second one—the *Cauchy integral* (or the 'Cauchy integral formula').

A.16 Literature

(i) Properties of some *special functions* are briefly discussed at the relevant points of the lecture notes; in the alphabetical order:

- Airy functions: *Part QM* section 2.4;
- Bessel functions: *Part EM* section 2.7;
- Fresnel integrals: *Part EM* section 8.6;
- Hermite polynomials: *Part QM* section 2.9;
- Laguerre polynomials (both simple and associated): *Part QM* section 3.7;
- Legendre polynomials, associated Legendre functions: *Part EM* section 2.8, and *Part QM* section 3.6;
- Spherical harmonics: *Part QM* section 3.6;
- Spherical Bessel functions: *Part QM* sections 3.6 and 3.8.

(ii) For *more formulas*, and their discussions, I can recommend the following handbooks (in the alphabetical order of the authors)[14]:

- *Handbook of Mathematical Formulas* [2][15];
- *Tables of Integrals, Series, and Products* [3];
- *Mathematical Handbook for Scientists and Engineers* [4];
- *Integrals and Series* volumes 1 and 2 [5].

A popular textbook *Mathematical Methods for Physicists* [6] may be also used as a formula manual. Many formulas are also available from the symbolic calculation parts of commercially available software packages listed in section (iv) below.

(iii) Probably the most popular collection of *numerical calculation codes* are the twin manuals by W Press *et al* [1]:

[14] On a personal note, perhaps 90% of all formula needs throughout my research career were satisfied by a tiny, wonderfully compiled old book: H. Dwight, *Tables of Integrals and Other Mathematical Data*, 4th ed., Macmillan, 1961, whose copies, rather amazingly, are still available on the Web.

[15] An updated version of this collection is now available online at http://dlmf.nist.gov/.

- *Numerical Recipes in Fortran* 77;
- *Numerical Recipes* [in C++ - KKL].

My lecture notes include very brief introductions into numerical methods of differential equation solution:

- ordinary differential equations: *Part CM* section 3.9, and
- partial differential equations: *Part CM* section 8.5, and *Part EM* section 2.8, which include references to literature for further reading.

(iv) The following are the most popular *software packages* for numerical and symbolic calculations, all with plotting capabilities (in alphabetical order):

- *Maple* (http://www.maplesoft.com/);
- *MathCAD* (http://www.ptc.com/products/mathcad/);
- *Mathematica* (http://www.wolfram.com/products/mathematica/index.html);
- *MATLAB* (http://www.mathworks.com/products/matlab/).

References

[1] Press W *et al* 1992 *Numerical Recipes in Fortran 77* 2nd edn (Cambridge: Cambridge University Press); Press W *et al* 2007 *Numerical Recipes* 3rd edn (Cambridge: Cambridge University Press)

[2] Abramowitz M and Stegun I (eds) 1965 *Handbook of Mathematical Formulas* (New York: Dover)

[3] Gradshteyn I and Ryzhik I 1980 *Tables of Integrals, Series, and Products* 5th edn (New York: Academic)

[4] Korn G and Korn T 2000 *Mathematical Handbook for Scientists and Engineers* 2nd edn (New York: Academic)

[5] Prudnikov A *et al* 1986 *Integrals and Series* vol 1 (Boca Raton, FL: CRC Press); Prudnikov A *et al* 1986 *Integrals and Series* vol 2 (Boca Raton, FL: CRC Press)

[6] Arfken G *et al* 2012 *Mathematical Methods for Physicists* 7th edn (New York: Academic)

IOP Publishing

Classical Electrodynamics
Lecture notes
Konstantin K Likharev

Appendix B

Selected physical constants[1]

Table B.1.

Symbol	Quantity	SI value and unit	Gaussian value and unit	Relative rms uncertainty
c	speed of light in free space	$2.997\ 924\ 58 \times 10^{8}$ m s^{-1}	$2.997\ 924\ 58 \times 10^{10}$ cm s^{-1}	0 (defined value)
G	gravitation constant	$6.674\ 1 \times 10^{-11}$ m^{3} kg^{-1} s^{-2}	$6.674\ 1 \times 10^{-8}$ cm ^{3} g^{-1} s^{-2}	$\sim 5 \times 10^{-5}$
$\hbar$	Planck constant	$1.054\ 5718\ 0 \times 10^{-34}$ J s	$1.054\ 571\ 80 \times 10^{-27}$ erg s	$\sim 1 \times 10^{-8}$
e	elementary electric charge	$1.602\ 176\ 2 \times 10^{-19}$ C	$4.803\ 203 \times 10^{-10}$ statcoulomb	$\sim 6 \times 10^{-9}$
m_e	electron's rest mass	$0.910\ 938\ 35 \times 10^{-30}$ kg	$0.910\ 938\ 35 \times 10^{-27}$ G	$\sim 1 \times 10^{-8}$
m_p	proton's rest mass	$1.672\ 621\ 90 \times 10^{-27}$ kg	$1.672\ 621\ 90 \times 10^{-24}$ G	$\sim 1 \times 10^{-8}$
μ_0	magnetic constant	$4\pi \times 10^{-7}$ N A^{-2}	–	0 (defined value)
ε_0	electric constant	$8.854\ 187\ 817 \times 10^{-12}$ F m^{-1}	–	0 (defined value)
k_B	Boltzmann constant	$1.380\ 648 \times 10^{-23}$ J K^{-1}	$1.380\ 650\ 9 \times 10^{-16}$ erg K^{-1}	$\sim 2 \times 10^{-6}$

Comments:

1. The fixed value of c was defined by an international convention in 1983, in order to extend the official definition of a second (as 'the duration of 9 192 631 770 periods of the radiation corresponding to the transition

[1] The listed numerical values of the constants are from the most recent (2014) International CODATA recommendation (see, e.g. http://physics.nist.gov/cuu/Constants/index.html), besides a newer result for k_B—see [1].

between the two hyperfine levels of the ground state of the cesium-133 atom') to that of a meter. The values are back-compatible with the legacy definitions of the meter (initially, as 1/40 000 000th of the Earth's meridian length) and the second (for a long time, as $1/(24 \times 60 \times 60) = 1/86\,400$th of the Earth's rotation period), within the experimental errors of those measures.

2. ε_0 and μ_0 are not really the fundamental constants; in the SI system of units one of them (say, μ_0) is selected arbitrarily[2], while the other one is defined via the relation $\varepsilon_0\mu_0 = 1/c^2$.
3. The Boltzmann constant k_B is also not quite fundamental, because its only role is to comply with the independent definition of the kelvin (K), as the temperature unit in which the triple point of water is exactly 273.16 K. If temperature is expressed in energy units k_BT (as is done, for example, in *Part SM* of this lecture note series), this constant disappears altogether.
4. The dimensionless *fine structure* ('Sommerfeld's) *constant* α is numerically the same in any system of units:

$$\alpha \equiv \left\{ \begin{matrix} e^2/4\pi\varepsilon_0\hbar c & \text{in SI units} \\ e^2/\hbar c & \text{in Gaussian units} \end{matrix} \right\} \approx 7.297\,352\,566 \times 10^{-3} \approx \frac{1}{137.035\,999\,14},$$

and is known with a much smaller rms uncertainty (currently, $\sim 3 \times 10^{-10}$) than those of the component constants.

References

[1] Gaiser C *et al* 2017 *Metrologia* **54** 280

[2] Newell D 2014 *Phys. Today* **67** 35–41

[2] Note that the selected value of μ_0 may be changed (a bit) in a few years—see, e.g. [2].

Bibliography

This section presents a partial list of the textbooks and monographs used in the work on the EAP series[1,2].

Part CM: Classical Mechanics

Fetter A L and Walecka J D 2003 *Theoretical Mechanics of Particles and Continua* (New York: Dover)

Goldstein H, Poole C and Safko J 2002 *Classical Mechanics* 3rd edn (Reading, MA: Addison Wesley)

Granger R A 1995 *Fluid Mechanics* (New York: Dover)

José J V and Saletan E J 1998 *Classical Dynamics* (Cambridge: Cambridge University Press)

Landau L D and Lifshitz E M 1976 *Mechanics* 3rd edn (Oxford: Butterworth-Heinemann)

Landau L D and Lifshitz E M 1986 *Theory of Elasticity* (Oxford: Butterworth-Heinemann)

Landau L D and Lifshitz E M 1987 *Fluid Mechanics* 2nd edn (Oxford: Butterworth-Heinemann)

Schuster H G 1995 *Deterministic Chaos* 3rd edn (New York: Wiley)

Sommerfeld A 1964 *Mechanics* (New York: Academic)

Sommerfeld A 1964 *Mechanics of Deformable Bodies* (New York: Academic)

Part EM: Classical Electrodynamics

Batygin V V and Toptygin I N 1978 *Problems in Electrodynamics* 2nd edn (New York: Academic)

Griffiths D J 2007 *Introduction to Electrodynamics* 3rd edn (Englewood Cliffs, NJ: Prentice-Hall)

Jackson J D 1999 *Classical Electrodynamics* 3rd edn (New York: Wiley)

Landau L D and Lifshitz E M 1984 *Electrodynamics of Continuous Media* 2nd edn (Auckland: Reed)

Landau L D and Lifshitz E M 1975 *The Classical Theory of Fields* 4th edn (Oxford: Pergamon)

Panofsky W K H and Phillips M 1990 *Classical Electricity and Magnetism* 2nd edn (New York: Dover)

---

[1] The list does not include the sources (mostly, recent original publications) cited in the ends of the chapters, and the mathematics textbooks and handbooks listed in appendix A, section A.16.

[2] Recently some high-quality teaching materials on advanced physics have become available online, including: R Fitzpatrick's text on classical electromagnetism (farside.ph.utexas.edu/teaching/jk1/Electromagnetism.pdf); B Simons' 'lecture shrunks' on advanced quantum mechanics (www.tcm.phy.cam.ac.uk/~bds10/aqp.html); and D Tong's lecture notes on several advanced topics (www.damtp.cam.ac.uk/user/tong/teaching.html).

doi:10.1088/978-0-7503-1404-6ch13

Stratton J A 2007 *Electromagnetic Theory* (New York: Wiley)
Tamm I E 1979 *Fundamentals of the Theory of Electricity* (Paris: Mir)
Zangwill A 2013 *Modern Electrodynamics* (Cambridge: Cambridge University Press)

Part QM: Quantum Mechanics

Abers E S 2004 *Quantum Mechanics* (London: Pearson)
Auletta G, Fortunato M and Parisi G 2009 *Quantum Mechanics* (Cambridge: Cambridge University Press)
Capri A Z 2002 *Nonrelativistic Quantum Mechanics* 3rd edn (Singapore: World Scientific)
Cohen-Tannoudji C, Diu B and Laloë F 2005 *Quantum Mechanics* (New York: Wiley)
Constantinescu F, Magyari E and Spiers J A 1971 *Problems in Quantum Mechanics* (Amsterdam: Elsevier)
Galitski V *et al* 2013 *Exploring Quantum Mechanics* (Oxford: Oxford University Press)
Gottfried K and Yan T-M 2004 *Quantum Mechanics: Fundamentals* 2nd edn (Berlin: Springer)
Griffith D 2005 *Quantum Mechanics* 2nd edn (Englewood Cliffs, NJ: Prentice Hall)
Landau L D and Lifshitz E M 1977 *Quantum Mechanics (Nonrelativistic Theory)* 3rd edn (Oxford: Pergamon)
Messiah A 1999 *Quantum Mechanics* (New York: Dover)
Merzbacher E 1998 *Quantum Mechanics* 3rd edn (New York: Wiley)
Miller D A B 2008 *Quantum Mechanics for Scientists and Engineers* (Cambridge: Cambridge University Press)
Sakurai J J 1994 *Modern Quantum Mechanics* (Reading, MA: Addison-Wesley)
Schiff L I 1968 *Quantum Mechanics* 3rd edn (New York: McGraw-Hill)
Shankar R 1980 *Principles of Quantum Mechanics* 2nd edn (Berlin: Springer)
Schwabl F 2002 *Quantum Mechanics* 3rd edn (Berlin: Springer)

Part SM: Statistical Mechanics

Feynman R P 1998 *Statistical Mechanics* 2nd edn (Boulder, CO: Westview)
Huang K 1987 *Statistical Mechanics* 2nd edn (New York: Wiley)
Kubo R 1965 *Statistical Mechanics* (Amsterdam: Elsevier)
Landau L D and Lifshitz E M 1980 *Statistical Physics, Part 1* 3rd edn (Oxford: Pergamon)
Lifshitz E M and Pitaevskii L P 1981 *Physical Kinetics* (Oxford: Pergamon)
Pathria R K and Beale P D 2011 *Statistical Mechanics* 3rd edn (Amsterdam: Elsevier)
Pierce J R 1980 *An Introduction to Information Theory* 2nd edn (New York: Dover)
Plishke M and Bergersen B 2006 *Equilibrium Statistical Physics* 3rd edn (Singapore: World Scientific)
Schwabl F 2000 *Statistical Mechanics* (Berlin: Springer)
Yeomans J M 1992 *Statistical Mechanics of Phase Transitions* (Oxford: Oxford University Press)

Multidisciplinary/specialty

Ashcroft W N and Mermin N D 1976 *Solid State Physics* (Philadelphia, PA: Saunders)
Blum K 1981 *Density Matrix and Applications* (New York: Plenum)
Breuer H-P and Petruccione E 2002 *The Theory of Open Quantum Systems* (Oxford: Oxford University Press)
Cahn S B and Nadgorny B E 1994 *A Guide to Physics Problems, Part 1* (New York: Plenum)

Cahn S B, Mahan G D and Nadgorny B E 1997 *A Guide to Physics Problems, Part 2* (New York: Plenum)

Cronin J A, Greenberg D F and Telegdi V L 1967 *University of Chicago Graduate Problems in Physics* (Reading, MA: Addison Wesley)

Hook J R and Hall H E 1991 *Solid State Physics* 2nd edn (New York: Wiley)

Joos G 1986 *Theoretical Physics* (New York: Dover)

Kompaneyets A S 2012 *Theoretical Physics* 2nd edn (New York: Dover)

Lax M 1968 *Fluctuations and Coherent Phenomena* (London: Gordon and Breach)

Lifshitz E M and Pitaevskii L P 1980 *Statistical Physics, Part 2* (Oxford: Pergamon)

Newbury N *et al* 1991 *Princeton Problems in Physics with Solutions* (Princeton, NJ: Princeton University Press)

Pauling L 1988 *General Chemistry* 3rd edn (New York: Dover)

Tinkham M 1996 *Introduction to Superconductivity* 2nd edn (New York: McGraw-Hill)

Walecka J D 2008 *Introduction to Modern Physics* (Singapore: World Scientific)

Ziman J M 1979 *Principles of the Theory of Solids* 2nd edn (Cambridge: Cambridge University Press)

www.ingramcontent.com/pod-product-compliance
Lightning Source LLC
Chambersburg PA
CBHW082120210326
41599CB00031B/5824

* 9 7 8 0 7 5 0 3 1 9 2 1 8 *